Tantrasaṅgraha of Nīlakaṇṭha Somayājī

For other titles published in this series, go to
http://www.springer.com/series/4142

Sources and Studies in the History of Mathematics and Physical Sciences

Tantrasaṅgraha of Nīlakaṇṭha Somayājī

K. Ramasubramanian

and

M.S. Sriram

Springer

HINDUSTAN BOOK AGENCY

Authors:
K. Ramasubramanian
Cell for Indian Science and Technology in Sanskrit
Department of Humanities and Social Sciences
IIT Bombay
Powai, Mumbai 400076
India
kramas@iitb.ac.in

M.S. Sriram
Department of Theoretical Physics
University of Madras
Guindy Campus
Chennai 600025
India
sriram.physics@gmail.com

Sources Managing Editor:
Jed Z. Buchwald
California Institute of Technology
Division of the Humanities and Social Sciences
MC 101–40
Pasadena, CA 91125
USA

Associate Editors:
J.L. Berggren
Simon Fraser University
Department of Mathematics
University Drive 8888
V5A 1S6 Burnaby, BC
Canada

Jesper Lützen
University of Copenhagen
Institute of Mathematics
Universitetsparken 5
2100 Koebenhaven
Denmark

ISBN 978-0-85729-035-9 e-ISBN 978-0-85729-036-6
DOI 10.1007/978-0-85729-036-6
Springer London Dordrecht Heidelberg New York

British Library Cataloguing in Publication Data
A catalogue record for this book is available from the British Library

Library of Congress Control Number: 2010932514

Mathematics Classification Number: 01A32, 85-03

Printed on acid-free paper

Springer is part of Springer Science+Business Media (www.springer.com)

समर्पणम्

नानायुक्त्यात्तशोभं न मितमनधिकं तन्त्रजालाभिनीतं
ग्रासच्छायादिकार्ये बुधभृगुविषये युक्तमार्गं दिशन्तम्।
प्राणैषीत् ग्रन्थरत्नं गणितविदमलो तन्त्रसङ्क्षिप्तिकाख्यं
गार्ग्यः श्रीनीलकण्ठः कृतबहुलकृतिर्भाग्यमेतद्बुधानाम्॥

व्याख्यालेख्यविशेषवृत्तिभरितं वो धीजुषां सम्मुदे
प्राकाश्यन्ननु तन्त्रसङ्ग्रहममुं प्रीत्या नयामोऽधुना।
श्रीमन्माधवकेरलीयगणितज्योतिर्विदां संहतेः
सन्तुष्ट्यै च समार्पयाम विबुधाः हृष्टान्तरङ्गा वयम्॥

The scholarly world is indeed fortunate that Nīlakaṇṭha—an astronomer-mathematician hailing from the Garga lineage and blessed with a clear intellect—composed among several other works a treatise called Tantrasaṅgraha, which is considered to be a gem among works [in astronomy], resplendent with a variety of yuktis, neither terse nor too elaborate, and which gives far more accurate procedures for solving the problems involving eclipses, shadow measurements, [the application of the equation of centre] in the case of Mercury and Venus, and so on.

We are now extremely happy to bring out this treatise, Tantrasaṅgraha, along with translation, annotation and detailed mathematical exposition for the intellectual delight of the scholarly community. We are also immensely pleased to dedicate this work of ours to Mādhava and the galaxy of other astronomers that this great lineage has produced.

Foreword

In the history of mathematics and astronomy in India, the Kerala school which flourished during the fourteenth–seventeenth centuries CE, has a unique position. Mādhava of Saṅgamagrāma, Parameśvara, Nīlakaṇṭha Somayājī, Jyeṣṭhadeva and Śaṅkara Vāriyar were among the luminaries of this school, which made original contributions in mathematics, formulating the infinite series for the trigonometric functions and π, that antedated similar achievements of European mathematicians by a couple of centuries. The origin of calculus also can be traced to this school.

In astronomy too, the Kerala school had significant achievements. The versatile astronomer Nīlakaṇṭha Somayājī (1444-1545) produced several works on astronomy, of which the *Tantrasaṅgraha* (about 430 verses in *anuṣṭubh* metre in eight sections or *prakaraṇas*) is a comprehensive treatise. He introduced in this elegant work a major revision of the traditional Indian planetary model, a detailed geometrical picture of which is discussed in his two small but lucid works—*Siddhāntadarpaṇa* (31 verses) and *Golasāra* (56 verses). According to Nīlakaṇṭha's geometrical picture of planetary motion, the five planets (Mercury, Venus, Mars, Jupiter and Saturn) move in eccentric orbits around the mean Sun, which in turn orbits around the Earth. Such a formulation was put forward by the European (Danish) astronomer Tycho Brahe, nearly a century later. *Tantrasaṅgraha* is also known for its other innovations introduced by Nīlakaṇṭha.

In March 2000, the Department of Theoretical Physics, University of Madras, organized a conference in collaboration with the Indian Institute of Advanced Studies, Shimla, to celebrate the 500th anniversary of *Tantrasaṅgraha*. Though the importance of this text was known to historians of Indian astronomy for quite some time, and several research papers have been published on the original ideas presented in this work, there was a great need for an accurate English translation of this seminal treatise, with detailed notes in modern notation. Profs K. Ramasubramanian and M. S. Sriram, who have the linguistic and subject expertise, have fulfilled this need admirably. As may be noted from the current volume, every attempt has been made by the authors to make the work as self-contained as possible by giving detailed explanations as well as several explanatory appendices besides a glossary and bibliography. Historians of astronomy, both Indian and foreign, are most grateful indeed to them for their devoted efforts in bringing out this publication.

The authors are already well known for their studies and publications in the area of Indian mathematics and astronomy. Together with another savant, M. D. Srinivas of the Centre for Policy Studies, Chennai, they were involved in preparing a detailed explanatory notes for *Gaṇita-yuktibhāṣā* of Jyeṣṭhadeva, which was published in two volumes by the Hindustan Book Agency, New Delhi with a critical edition of the text and English translation by K. V. Sarma, an eminent scholar who published several works on Indian astronomy. A reprint of this work was also brought out recently by Springer, the noted international publishers, to make it available for the international readership. In fact, Jyeṣṭhadeva, who was a junior contemporary of Nīlakaṇṭha, at the commencement of his work states that his aim in composing the work is to explain the calculational procedures given in *Tantrasaṇgraha*. It is but fitting, therefore, that Profs Ramasubramanian and Sriram, who were involved with the production of explanatory notes of *Gaṇita-yuktibhāṣā*, are the authors of the present volume on *Tantrasaṅgraha*.

The Hindustan Book Agency has been rendering yeoman service to scholars interested in the history of mathematics, by bringing out several volumes in its series 'Culture and History of Mathematics'. I am happy that in collaboration with Springer it is publishing the present work on *Tantrasaṅgraha*, which, I am sure, will be of great value to historians of science in general and of astronomy in particular. It is my fond hope that several other timeless works of this type will emerge from the pens of these erudite authors in future.

Bangalore
March 2010

B. V. Subbarayappa
Former President, International Union of History and Philosophy of Science

Preface

Tantrasaṅgraha composed in 1500 CE by the Kerala astronomer Nīlakaṇṭha Somayājī, has long been recognized as an important Indian text in astronomy. It is a comprehensive text which discusses all aspects of mathematical astronomy such as the computation of the longitudes and latitudes of planets, various diurnal problems, the determination of time, eclipses, the visibility of planets etc. There are two critical editions of the Sanskrit text, by S. K. Pillai in 1958 and K. V. Sarma in 1977, which between them include the commentaries *Laghu-vivṛti* in prose for the entire text, and *Yukti-dīpikā* in verses for the first four chapters, both of which are composed by Śaṅkara Vāriyar. The need has long been felt for an English translation of the work, with detailed explanatory notes in modern notation, so that the work is accessible to a larger audience. It is with this objective that we began a project on *Tantrasaṅgraha*, funded by the Indian National Science Academy (INSA), in 2000.

Meanwhile, along with M. D. Srinivas (Centre for Policy Studies, Chennai), we were involved in preparing detailed explanatory notes for *Gaṇita-yukti-bhāṣa* (GYB) of Jyeṣṭhadeva, edited and translated by K. V. Sarma, and published in 2008 by the Hindustan Book Agency and reprinted in 2009 by Springer. Though the work on GYB caused delay in the publication of the present work, it was very rewarding as GYB gives detailed explanations of most of the algorithms in *Tantrasaṅgraha*, and provides valuable insights on many topics covered in that work.

Scholars in the area of the history of astronomy in general, and Indian astronomy in particular, form the natural readership for this work. However, keeping the larger readership—anyone wanting to know the methods of Indian astronomy—in mind, we have attempted to make it as self-contained as possible, so that any motivated person with a sound background in mathematics at the final school (+2, as it is termed in India) level and interested in spherical astronomy will find it useful. We have also included a glossary of frequently occurring Sanskrit terms and several appendices that should serve to clarify many concepts relevant to the topics in the main text.

The modification of the traditional Indian planetary model by Nīlakaṇṭha in *Tantrasaṅgraha* is what attracted us to the work initially. But this topic is dealt with all too briefly in it, as it is a *Tantra* text devoted mainly to computational algorithms. However, Nīlakaṇṭha has discussed his model extensively, along with his geometrical picture of planetary motion, in other works. In collaboration with M. D. Srinivas, we had made an incisive study of this model and published a paper on it in the Indian journal *Current Science* way back in 1994. Since then we have had occasions to study this in more detail. Appendix F, on the traditional Indian planetary model and its revision by Nīlakaṇṭha Somayājī, of which M. D. Srinivas is a co-author, reflects our current understanding on this subject.

We are deeply indebted to Prof. M. D. Srinivas—our collaborator on the different aspects of studies on Indian astronomy and mathematics that we have been doing for almost two decades now—for meticulously going through the entire manuscript and offering several valuable suggestions. We would also like to acknowledge the suggestions given by the two anonymous referees for improving the manuscript. We are grateful to (the late) Prof. K. V. Sarma with whom we have had an extensive collaboration, especially during the preparation of *Yuktibhāṣā*, and who has been a source of great inspiration for us. We would like to thank Profs C. S. Seshadri and R. Sridharan of the Chennai Mathematical Institute for their continued support and encouragement. Our special thanks go to Prof. B. V. Subbarayappa, Bangalore, the doyen of the history of science in India, for having readily agreed to write the Foreword to this work.

Our heart-felt thanks are due also to Profs S. Balachandra Rao of Bangalore, and V. Srinivasan of Hyderabad (currently with the University of Madras) for their constant and vociferous support to us over all the years in all our work on Indian astronomy. We also thank Profs P. M. Mathews, G. Bhamati, M. Seetharaman, S. S. Vasan, K. Raghunathan, A. S. Vytheeswaran, R. Radhakrishnan, and Dr Sekhar Raghavan associated with the Department of Theoretical Physics, University of Madras, for their kind and active interest in our work over the years.

We would like to acknowledge the keen interest expressed by Swami Atmapriyananda, Belur Math, in promoting studies in Indian astronomy and mathematics. We are indeed grateful to Profs David Mumford of Brown University and Manjul Bhargava of Princeton University for their kind encouragement and enthusiastic support. It is a pleasure to thank Profs S. M. R. Ansari, A. K. Bag, Jitendra Bajaj, V. Balakrishnan, A. V. Balasubramanian, Rajendra Bhatia, S. G. Dani, Sinniruddha Dash, Amartya Datta, P. C. Deshmukh, P. P. Divakaran, Raghavendra Gadagkar, George Joseph, Rajesh Kocchar, S. Madhavan, Madhukar Mallayya, N. Mukunda, Roddam Narasimha, M. G. Narasimhan, Jayant Narlikar, C. K. Raju, Sundar Sarukkai, B. S. Shylaja, Navjyoti Singh, S. P. Suresh, T. Trivikraman, Mayank Vahia, Padmaja Venugopal and K. Vijayalakshmi, as well as Profs Mohammad Bagheri, Subhash Kak, Agathe Keller, Francois Patte, Kim Plofker, T. R. N. Rao, S. R. Sarma and Michio Yano for their kind interest in our work in Indian astronomy and mathematics in general, and this work in particular.

One of the authors (Ramasubramanian) would like to acknowledge the unstinting support and encouragement received from the former Director of IIT Bombay, Prof. Ashok Misra, and other IIT Bombay fraternity members, particularly Profs. S. D. Agashe, Rangan Banerjee, Jayadeva Bhat, S. M. Bhave, Amitabha

Gupta, Devang Khakhar, Malhar Kulkarni, Ravi Kulkarni, H. Narayanan, Prabhu Ramachandran, Krithi Ramamritham, H. S. Shankar, G. Sivakumar, A. K. Suresh, Sivaramakrishnan and Jugal Verma. Special thanks are also due to K. Mahesh, U. K. V. Sarma, R. Venketeswara Pai, Dinesh Mohan Joshi and Vanishri Bhat for their devoted assistance at different stages of the preparation of the manuscript. In fact, we are particularly indebted to Vanishri for spending countless hours in setting a uniform style for all the figures that appear in the book and U. K. V. Sarma for carefully proof reading the entire text.

This work is the outcome of a project sanctioned by INSA, New Delhi, during October 2000–March 2004. We would also like to place on record our gratitude to the Sir Dorabji Tata Trust and the National Academy of Sciences, India, for their financial assistance by way of projects, which was extremely useful in offering fellowship to the project staff as well as in the production of the manuscript in a camera-ready form. We are deeply indebted to INSA for the financial support as well as for readily granting the permission to publish the work. Our special thanks go to Jainendra Jain and Devendra Jain of the Hindustan Book Agency, New Delhi, for graciously coming forward to publish this work in collaboration with Springer, London.

Finally, the authors are grateful to the copy-editor(s) for going through the manuscript meticulously, and making valuable comments and suggestions.

विकृति-आश्वयुजशुक्लदशमी, कल्यब्द ५११२

K. Ramasubramanian

Cell for Indian Science and Technology in Sanskrit
IIT Bombay

October 17, 2010

M. S. Sriram

Department of Theoretical Physics
University of Madras

Contents

List of Figures

List of Tables

Introduction

Tantrasaṅgraha and its importance

Tantrasaṅgraha, a comprehensive treatise on astronomy, was composed by the renowned Kerala astronomer Nīlakaṇṭha Somayājī (1444–1545 CE) of Tṛkkaṇṭiyūr. It ranks along with *Āryabhaṭīya* of Āryabhaṭa (499 CE) and *Siddhāntaśiromaṇi* of Bhāskarācārya (1150 CE) as one of the major works which significantly influenced further work on astronomy in India. In *Tantrasaṅgraha*, Nīlakaṇṭha introduced a major revision of the traditional Indian planetary model. He arrived at a unified theory of planetary latitudes and a better formulation of the equation of centre for the interior planets (Mercury and Venus) than was available, either in the earlier Indian works or in the Greco-European or Islamic traditions of astronomy, till the work of Kepler.[1] Besides this, the work also presents many important innovations in mathematical techniques related to accurate sine tables, use of series for sine and cosine functions, and a systematic treatment of spherical astronomical problems. The relations of spherical trigonometry stated here are exact, and are applied with care to diurnal problems, eclipses etc. The explanations of the procedures of *Tantrasaṅgraha* are to be found in the commentaries *Laghu-vivṛti* and *Yukti-dīpikā* by Śaṅkara Vāriyar, as well as the seminal Malayalam work *Yuktibhāṣā* of Jyeṣṭhadeva.

The present work and its context

There have been two critical editions of *Tantrasaṅgraha*, the first by Surnad Kunjan Pillai in 1958 and the second by K. V. Sarma in 1977. While the former includes the

[1] In his other works *Āryabhaṭīya-bhāṣya*, *Golasāra*, *Siddhāntadarpaṇa* and *Grahasphuṭānayane vikṣepavāsana*, Nīlakaṇṭha also discusses the geometrical model implied by his theory according to which the planets go around the Sun, which itself orbits around the Earth. See Appendix F for more details.

commentary *Laghu-vivṛti* in the form of prose for the whole text, the latter includes the elaborate commentary *Yukti-dīpikā* (for the first four chapters) in the form of verses.[2] Both these commentaries are by Śaṅkara Vāriyar. There is very little difference in the text between the two editions. While the main text, *Tantrasaṅgraha*, as edited by K. V. Sarma, is based on 12 manuscripts, the commentary, *Yukti-dīpikā*, is based on only four manuscripts.[3] The textual verses, as well as the references to the citations from *Laghu-vivṛti* and *Yukti-dīpikā*, that are reproduced in the present work are based on the above two editions of *Tantrasaṅgraha*.

We have gone through the entire *Laghu-vivṛti* commentary in the process of preparing the translation and explanatory notes. Some important portions of *Yukti-dīpikā* having a direct bearing on the contents of the main text have also been cited in our explanations. For the most part, *Laghu-vivṛti* gives a plain and direct description of the verses of the text in simple prose without excursions into related topics. Nevertheless, it does offer very valuable insights on several occasions and clarifies the contents of many verses, which would have been unclear otherwise. However, the commentary *Yukti-dīpikā* is of a different nature. Here Śaṅkara Vāriyar transcends the confines of immediate utility and discusses several related issues that would greatly enhance one's understanding of the subject. Many verses in *Yukti-dīpikā* reveal several aspects of the Indian thinking on astronomy and mathematics. Besides these two commentaries, we have also consulted the astronomy part of Jyeṣṭhadeva's *Yuktibhāṣā* which has proved to be extremely useful in understanding the contents of *Tantrasaṅgraha*. In fact, according to Jyeṣṭhadeva—as stated by him at the very commencement of the work—the main purpose of *Yuktibhāṣā*[4] is to elucidate the procedures enunciated in *Tantrasaṅgraha*. We have made extensive use of this work while preparing the explanatory notes on certain topics such as the planetary model, spherical astronomical problems, visibility corrections, the eclipses and so on.

There is an earlier translation of *Tantrasaṅgraha* by V. S. Narasimhan, which was published in the *Indian Journal of History of Science* as a supplement in three parts during 1998.[5] Narasimhan has also presented some explanatory notes to his translation. However, the author does not seem to have carefully studied the commentaries of Śaṅkara Vāriyar in preparing the translation. He also did not have the benefit of consulting an edited version of the astronomy part of *Yuktibhāṣā*. Often, his translation and explanations do not really bring out the exact content of the verses of *Tantrasaṅgraha*. This has been one of the motivating factors for undertaking the present work.

---

[2] See {TS 1958} and {TS 1977}.

[3] {TS 1977}, p. xlii.

[4] {GYB 2008}, p. 1; p. 313.

[5] {TS 1999}, pp. S1–S146.

The Kerala school of astronomy and mathematics

Kerala has had a long, continuous and vigorous tradition of astronomy and mathematics from very early times. *Āryabhaṭīya* (c. 499 CE) of Āryabhaṭa, which sets the tone for all further work on mathematical astronomy in India, appears to have become popular in Kerala soon after its composition. The astronomical parameters in *Āryabhaṭīya* were revised by a group of astronomers who had gathered at the celebrated centre of learning at Tirunāvāy in northern Kerala during 683–684 CE. The revised system called *Parahita-gaṇita* was enunciated by Haridatta in his *Grahacāra-nibandhana.*

Laghubhāskarīya and *Mahābhāskarīya* of Bhāskara I (c. 630), which expounded the Āryabhaṭan school in detail, were also popular in Kerala. Govindasvāmin (c. 800) wrote an elaborate commentary on *Mahābhāskarīya* and his student Śaṅkaranārāyaṇa (c. 850) wrote one on *Laghubhāskarīya.* Udayadivākara (11th century), who also probably hailed from Kerala, wrote a detailed commentary, *Sundarī* on *Laghubhāskarīya,* which contains the method for solving quadratic indeterminate equations (*varga-prakṛti*). This method is ascribed to Jayadeva (10–11th century) and is the same as the famous *Cakravāla* algorithm, expounded in detail by Bhāskara II in his *Bījagaṇita* (c. 1150).

The Kerala tradition enters a new phase with Mādhava of Saṅgamagrāma (c. 1340–1425). His known works like *Veṇvāroha, Sphuṭacandrāpti* and *Agaṇitagrahacāra* may not be major works conceptually, but all the later astronomers from Kerala invariably attribute to him the path-breaking results on infinite series for the inverse tangent, sine and cosine functions, plus many innovations in astronomical calculations.

Parameśvara of Vaṭaśśeri (c. 1360–1455), a student of Mādhava, was a prolific writer, who authored about 30 works. Emphasizing the need for revising the planetary parameters through observations, he thoroughly revised the *Parahita* system and introduced the *Dṛggaṇita* system. Nīlakaṇṭha in his *Jyotirmīmāṃsā* and *Āryabhaṭīya-bhāṣya* mentions that Parameśvara observed eclipses for 55 years continuously and revised these parameters so that the observations and calculations tally with each other. Apart from *Dṛggaṇita,* the other important works of Parameśvara are *Goladīpikā, Bhaṭadīpikā* (a commentary on *Āryabhaṭīya*), *Siddhāntadīpikā* (a super-commentary on Govindasvāmin's *Mahābhāskarīya-bhāṣya*), *Grahaṇamaṇḍana* on eclipses and *Mahābhāskarīya-bhāṣya,* his own commentary on *Mahābhāskarīya* of Bhāskara I. Parameśvara also happens to be one of the few astronomers to discuss in detail the geometrical model of planetary motion implied by the calculational procedure in Indian astronomy in his *Siddhāntadīpikā, Bhaṭadīpikā* and *Goladīpikā.*

Nīlakaṇṭha Somayājī (c. 1444–1545), of whom a brief biographical sketch is provided in the next section, was a disciple of Dāmodara, who was the son and disciple of Parameśvara. Suffice it to say here that the innovations in the planetary model and spherical astronomical calculations made by Nīlakaṇṭha in his *Tantrasaṅgraha* and other works were considered as major advances by the later astronomers of Kerala.

Jyeṣṭhadeva (c. 1500–1610), of Parakroḍa or Paronnoṭṭu family, was also a pupil of Dāmodara, and seems later to have received instructions from Nīlakaṇṭha Somayājī. In his *Yuktibhāṣā* (c. 1530), we have an elaborate and systematic exposition of the rationale of the algorithms employed in Indian mathematics in Part I, and those employed in Indian astronomy in Part II. Though it claims to explain the contents of *Tantrasaṅgraha* and to provide the rationale for the calculational procedures in it, *Yuktibhāṣā* is indeed an independent work (especially Part I). This treatise occupies a unique place in Indian mathematics and astronomy for two reasons: (i) it is exclusively devoted to proofs and demonstrations, including those for the infinite series for inverse tangent, sine and cosine functions, and (ii) it is written in the local language of Kerala, Malayalam. Perhaps it is one of the reasons for the title of the book, *Yuktibhāṣā*.[6] Though there are innumerable commentaries on *Siddhāntaśiromaṇi*, *Līlāvatī* etc. in the regional languages of India, we are yet to find a major work of this nature among them.

Śaṅkara Vāriyar of Tirukkuṭaveli (c. 1500–1560) was a disciple of Nīlakaṇṭha Somayājī and was also deeply influenced by Jyeṣṭhadeva. He is the author of two commentaries on *Tantrasaṅgraha*, namely *Laghu-vivṛti*, in prose, and a far more elaborate one, *Yukti-dīpikā*, which is composed entirely in verses. He has also authored a commentary, *Kriyākramakarī*, on *Līlāvatī* of Bhāskara II. There are similarities in the treatment of various topics in *Yukti-dīpikā* and *Yuktibhāṣā*. It has also been noted that there are several verses in common between *Yukti-dīpikā* and *Kriyākramakarī*.

Citrabhānu (c. 1475–1550), the author of *Karaṇāmṛta*, was also a disciple of Nīlakaṇṭha, whereas Acyuta Piṣāraṭi (c. 1550-1621) of Tṛkkaṇṭiyūr was a student of Jyeṣṭhadeva. Acyuta has written a *karaṇa* work, *Karaṇasāra* and a more detailed work on planetary theory, *Sphuṭanirṇayatantra*. He also discusses the 'reduction to the ecliptic' in detail in his *Rāśigolasphuṭanīti*. *Karaṇapaddhati* of Putumana Somayājī (c. 1660–1740) is an important later work in the *karaṇa* form. *Sadratnamāla* of Rājā Śaṅkara Varman (c. 1800–38) is a compendium of the Kerala school of mathematics and astronomy. The Kerala tradition continued even into modern times, with some works incorporating a few results of modern positional astronomy.

Nīlakaṇṭha and his works

Nīlakaṇṭha, generally referred to with the title *Somayājī* or *Somasutvan*, hailed from Tṛkkaṇṭiyūr (Sanskritized into *Śrīkuṇḍapura* or *Śrīkuṇḍagrāma*) near Tirūr in south Malabar, a famous seat of learning in Kerala during the middle ages. He is also called *Gārgya-kerala*, as he belonged to the *Garga-gotra* and hailed from Kerala. The name of his Illam—as the house of a Nambuthiri Brahmin is called—

[6] The word *bhāṣā*, when derived using *karmavyutpatti*, refers to the language that is spoken *bhāṣyate iti bhāṣā*. It is also possible that the title stems from the derivation—*yuktayaḥ atra bhāṣyanta iti* (the rationales are being elucidated here)— *Yuktibhāṣā*.

was Kelallūr. Hence he was known locally as *Kelallūr Comātiri*.[7] From several references in his writings it is known that he was intimately connected with and was patronized by Kauṣītaki Āḍhya Netranārāyaṇa, known locally as Azhvānceri Tamparkkāl, the religious head of the Nambuthiri Brahmins of Kerala. Nīlakaṇṭha informs us in his writings that he studied *Vedānta* under one Ravi and *Jyotiṣa* under Dāmodara, son of Parameśvara. He refers to Parameśvara as his *Paramaguru* and is indebted to him for many results and insights.

The phrases '*he viṣṇo nihitaṃ kṛtsnaṃ*' and '*lakṣmīśanihitadhyānaiḥ*' occurring in the first and the last verses of *Tantrasaṅgraha*, though each literally conveying the meanings appropriate to the context, also give the *Kali-ahargaṇa* (the number of days elapsed since the beginning of the *Kaliyuga*) corresponding to the dates of the commencement and completion of the work. This has been pointed out by the commentator Śaṅkara Vāriyar in his commentary *Laghu-vivṛti*. The numbers represented by the two phrases in *Kaṭapayādi* notation are 1680548 and 1680553. These correspond in the Gregorian calendar to March 22, 1500 and March 27, 1500 CE respectively. Nīlakaṇṭha states in the commentary on *Siddhānta-darpaṇa* that he was born on *Kali* day 1660181 which corresponds to June 17, 1444 CE. That he lived to a ripe old age, even to become a centenarian, is attested by a reference to him in *Praśnasāra*, a Malayalam work on astrology.

The erudition of Nīlakaṇṭha in several branches of Indian philosophy and learning such as *Vedānta, Mīmāṃsā, Dharmaśāstras, Purāṇas* etc. is quite evident from the frequent references to them in his works, particularly *Āryabhaṭīya-bhāṣya* and *Jyotirmīmāṃsā*. This is in addition to the citations from *Jyotiṣa* works beginning from *Vedāṅga-jyotiṣa* down to the treatises of his own times.

Besides *Tantrasaṅgraha*, Nīlakaṇṭha composed many other works. *Āryabhaṭīya-bhāṣya*, composed by him late in his life, is perhaps the most elaborate commentary on *Āryabhaṭīya*, and is yet to be translated and studied in detail. He himself calls it a *Mahābhāṣya*,[8] which is amply justified considering the wealth of information and explanations in it. In this work, he summarizes the prevalent knowledge of mathematics and astronomy, in India in general and Kerala in particular, and supplements it with his own insights. Apart from the detailed explanations of mathematical results and procedures presented in the text, this work also discusses the geometrical model of planetary motion, eclipses and even some 'physical' concepts.

Golasāra is a short work in 56 verses containing many details not covered in *Tantrasaṅgraha*. The importance of *Siddhānta-darpaṇa* lies in the fact that the author presents therein the astronomical constants as verified through his own observations and investigations, and which can be taken as the final figures accepted by him for his own times. It is noteworthy that Nīlakaṇṭha himself wrote a commentary on it. There is also a small but important tract written by Nīlakaṇṭha, entitled *Grhasphuṭānayane vikṣepavāsanā*, where he presents a succinct but definitive account of his cosmological model of planetary motion.

---

[7] *Comātiri* is the Malayalam version of the Sanskrit word *Somayājī*.

[8] {ABB 1930}, p. 180.

Computational methods for determining the exact time on the basis of the shadow cast by the Sun or the Moon were important in Indian astronomy, so much so that there are many tracts exclusively devoted to this aspect. The *Candracchāyā-gaṇita* of Nīlakaṇṭha belongs to this genre of tracts and sets out the procedures for the computation both of the *kramacchāyā*, the shadow from the time, and the *viparītacchāyā*, the time from the shadow.

Grahaṇanirṇaya is a work on lunar and solar eclipses, which is quoted by Nīlakaṇṭha himself and later authors, the manuscripts of which are yet to be discovered. *Sundararāja-praśnottara* is another work whose manuscripts are yet to come to light. Sundararājā was a contemporary of Nīlakaṇṭha hailing from Tamilnadu, and author of a detailed commentary on the *Vākyakaraṇa* or *Vākya-pañcādhyāyī*, which is a manual on the basis of which almanacs were computed in Tamilnadu. Sundararājā sought clarifications on many topics in astronomy from Nīlakaṇṭha, the answers to which formed the work *Sundararāja-praśnottara*. Sundararājā has explicitly stated this in his commentary on *Vākyakaraṇa*. So this work is different in nature from texts and commentaries, and will be a valuable addition to the corpus of literature on Indian astronomy if brought to light.

Jyotirmīmāṃsā of Nīlakaṇṭha has a unique place in the history of Indian astronomy, as it is a major work which focuses exclusively on epistemological issues concerning the science of astronomy.[9] It rebuts the claim of many scholars that Indian astronomy did not have a scientific methodology worth the name. It strongly emphasizes the role of observation and experimentation in revising astronomical parameters and testing any theory. Confronting the credulous view that the numbers of planetary revolutions given by Āryabhaṭa are immutable since they form part of 'divine instruction', Nīlakaṇṭha points out that by the expression 'divine instruction' is not meant any direct instruction by gods, but only the chastening of the intellect through divine grace, as a result of which the author could organize his thoughts logically. This is what he has to say regarding the authority of established texts:

पञ्चसिद्धन्तास्तावत् क्वचित्काले प्रमाणमेव इत्यवगन्तव्यम्। अपि च यः सिद्धान्तः दर्शनाविसंवादी भवति सोऽन्वेषणीयः। दर्शनसंवादश्च तदानीन्तनैः परीक्षकैः ग्रहणादौ विज्ञातव्यः ।

One has to accept that [each of] the five *siddhāntas* had been authoritative at one time [though they might not be so now]. Therefore one has to look for a system which tallies with observation. The said tallying has to be verified by contemporary experimenters at the time of eclipses etc.

It had long been recognized that eclipses are very sensitive to the parameters associated with the motion of the Sun and the Moon, and also the latitude and longitude of a place. It is precisely for this reason that Nīlakaṇṭha stresses the need for utilizing eclipses to revise the parameters so that future eclipses could be computed and predicted with accuracy.

---

[9] See {JM 1977}, p. 6. See also M. D. Srinivas, 'Indian approach to science: The case of *Jyotiḥśāstra*', *PPST Bulletin*, 19–20, Madras 1990.

Nīlakaṇṭha quotes the following passage from a *Mīmāṃsā* text which expresses his ideas regarding the maintenance and furtherance of astronomical tradition, and the role of observations and logical inference in it.

गणितोन्नीतस्य चन्द्रादेः देशविशेषान्वयस्य प्रत्यक्षेण संवादः, ततो निश्चितान्वयस्य परस्य गणितलिङ्गोपदेशः, ततस्तस्याप्तोपदेशावगतान्वयस्य अनुमानं संवादः, परस्मै चोपदेशः इति सम्प्रदायाविच्छेदात् प्रामाण्यम्।

The correlation of the computed Moon etc., with actual observation at a particular place, the revision of computation on the basis of such correlation, logical inference therefrom being transmitted as tradition, it being again correlated [with observation and again revised] and transmitted further down to others—this is how tradition is continued without interruption, and hence its [continued] authoritativeness.

Summary of *Tantrasaṅgraha*

Tantrasaṅgraha, the magnum opus of Nīlakaṇṭha, is composed in eight chapters or *prakaraṇas* consisting of 432 verses.[10] The division of the chapters is along the same lines as in any other typical text in Indian astronomy such as *Sūryasiddhānta* or *Siddhāntaśiromaṇi.* The development of the subject is not only systematic, often beginning with the basics, but also quite comprehensive. All the procedures needed for the computation of quantities of physical interest, such as the longitudes and latitudes of planets, various diurnal problems, the determination of time, eclipses, the visibility of planets etc., are thoroughly discussed. However, explanations are not provided, save on some occasions, as the work belongs to the *Tantra* class of texts[11] which are intended to be mainly algorithmic in nature. Explanations are to be found in the two commentaries on the text, namely *Laghu-vivṛti* and *Yukti-dīpikā*, and also in *Yuktibhāṣā* composed by Jyeṣṭhadeva.

It is in *Tantrasaṅgraha* that Nīlakaṇṭha explicitly formulates his revision of the traditional planetary model. Some of the procedures adopted by the earlier texts for calculating quantities of astronomical interest like the latitude of a place, *lagna*etc. are improvised or made exact. Brevity, clarity, exactness and comprehensiveness are hallmarks of this work. In what follows, we provide a brief chapter-wise summary of the text.

[10] Most of the verses are in *anuṣṭubh* metre, which has 8 syllables per quarter (*pāda*).

[11] Based on the epoch chosen for calculations (beginning of *a kalpa, a yuga* or an appropriate recent date), the Indian astronomical texts are broadly classified into three types: *Siddhānta, Tantra* and *Karaṇa*. While *Siddhāntas* provide theoretical explanations besides presenting algorithms, *Tantras* are mainly algorithmic with explicit formulae but do not explain the procedures. *Karaṇas* often dispense with even the formulae, substituting them with abbreviated procedures and tables to be used in computations.

Madhyamādhikāra (Computation of mean positions)

The work begins with an invocation to Lord Viṣṇu: *he viṣṇo nihitaṃ* ... This is also the *Kali* chronogram of the date of commencement of the work, which turns out to be 1680548, and corresponds to *Mīna* 26, 4600 *gatakali* (elapsed *Kali* years) according to the Indian calendar, which corresponds to March 22, 1500 CE according to the Gregorian calendar. Then, various time units, like the *sāvana-dina* (mean civil day), the *nākṣatra-dina* (sidereal day), lunar month, solar month, *yuga* etc., are defined. Smaller units of time, such as the *tithi* (the time period during which the elongation of the Moon from the Sun increases by 12 degrees), the *nāḍī* (one-sixtieth of a day), the *prāṇa* (21600 *prāṇas* constitute a day) etc. are also defined.

The *adhimāsa* (intercalary month) and its nature are then explained. The *kṣayamāsa* and its incorporation in the calendar are also discussed. The number of revolutions of the Sun, Moon and the five planets (Mercury, Venus, Mars, Jupiter and Saturn) in a large period called a *Mahāyuga*, consisting of 4320000 years, are given. Also the number of revolutions made by the apsides of these planets and their nodes in a *Mahāyuga* are listed. Here it is noteworthy that while specifying the number of revolutions of the inner planets, the word *svaparyayāḥ* is used, signifying the fact that the revolution number given refers to their own revolutions and not of their *śīghrocca* as specified in the earlier texts. The significance of this departure has profound implications with respect to the computation of the longitudes of the inner planets, which is explained in the next section.

After stating the *yugasāvanadinas*, the number of days in a *Mahāyuga*, the procedure for finding the *Ahargaṇa*, the number of days elapsed since a given epoch, is explained. *Ahargaṇa* is the antecedent of the modern Julian day. The mean longitude of a planet at sunrise on any given day can be calculated given the *Ahargaṇa* and the revolution number of the planet. This is actually valid for the mean sunrise at Laṅkā, a fictitious place on the Earth's equator, whose longitude is the same as that of Ujjayinī. The correction to the mean longitudes due to the difference in longitude (terrestrial) between the given place and Laṅkā is the *Deśāntara-saṃskāra*.

In some of the earlier Indian texts like *Āryabhaṭīya*, the mean longitudes of all the planets were taken to be zero at the beginning of the *Kaliyuga*, which is of course a rough approximation. Noticing this fact, most of the later texts give corrections which are called *Dhruvas*.[12] Nīlakaṇṭha also specifies the values of the *Dhruvas* for the planets and their 'apsides' (*mandoccas*) at the beginning of the *Kaliyuga*. The latter are essential for the calculation of the true longitudes or the *sphuṭa-graha*.

[12] *Dhruvas* are the initial positions of the planets at the beginning of an epoch. Hence, their values will vary with a change in the epoch. Even for the same epoch, *Dhruvas* differ from text to text.

Sphuṭādhikāra (Computation of true positions)

The second chapter of *Tantrasaṅgraha* commences with a discussion on the construction of the sine table. In the Indian texts the quadrant of a circle is normally divided into 24 equal parts and sines of multiples of $3^\circ 45'$ are explicitly stated.

In constructing the sine table, Nīlakaṇṭha follows the method of Āryabhaṭa, involving the second differences of the sine values, and this essentially amounts to making use of the differential equation

$$\frac{d^2}{dx^2}\sin x = -\sin x.$$

However, Nīlakaṇṭha's choice of the first sine value being more accurate, his sine table is far more accurate than the ones provided in earlier texts. Generally, the sines of intermediate angles are determined using the first-order interpolation known as the *trairāśika*. But in *Tantrasaṅgraha* we find several methods discussed for finding more accurate values of sines using the series expansion for $\sin\theta$. The inverse problem of finding the arc from the sine is also discussed.

An epicycle model is used to obtain the *manda* correction to the planetary longitudes. This is essentially the 'equation of centre' in modern parlance and takes into account the eccentricity of the planetary orbit. The *manda*-corrected mean position is called the *manda-sphuṭa-graha* or simply the *manda-sphuṭa*. The actual distance of the *manda-sphuṭa* from the centre of the concentric is called the *manda-karṇa* (*manda* hypotenuse). Interestingly, the epicycle radius is assumed to be proportional to the *manda-karṇa*. Then the *manda-karṇa* can be determined by an iterative process. A formula for the *aviśiṣṭa-manda-karṇa* which is found without iterations appears for the first time in the text and is ascribed to Mādhava. While the *manda* correction gives the true (geocentric) position in the case of the Sun and the Moon, in the case of actual planets (referred to as *tārā-grahas*) it gives their true heliocentric position. The term *tārā-graha* includes both the interior planets (Mercury and Venus) and the exterior ones (Mars, Jupiter and Saturn).

Besides the *manda* correction, one more correction, namely the *śīghra*, has to be applied in the case of the *tārā-grahas* to get the true longitudes or the *sphuṭa-grahas*. By applying this correction, we essentially convert the true heliocentric longitude of the planet to the geocentric longitude. In the earlier Indian astronomical texts (as well as in Ptolemy's *Almagest*), the *manda* correction for Mercury and Venus was wrongly applied to the mean Sun, which has no physical significance whatsoever. It is Nīlakaṇṭha who sets this procedure right, for the first time in the history of astronomy, by applying the *manda* correction to the actual mean heliocentric longitude of Mercury/Venus. In his other works like *Āryabhaṭīya-bhāṣya*, *Siddhānta-darpaṇa*, *Grahasphuṭānayane vikṣepavāsanā* etc., Nīlakaṇṭha discusses the implications of this procedure for calculating planetary positions, and describes a geometrical model in which all the five planets (Mercury, Venus, Mars, Jupiter and Saturn) move in eccentric orbits around the mean Sun, which in turn orbits around the Earth. Being a *tantra* text, *Tantrasaṅgraha* does not give details of the geometrical model of planetary motion.

It may further be mentioned here that as both the *manda* and the *śīghra* corrections involve the inverse sine function, the expression for the geocentric longitude of the planet also involves it too. Hence, if one wants to find the instantaneous velocity of the planet called *tātkālikagati*, one will have to find the time derivative of this function. It is indeed remarkable that the exact formula for the derivative of the inverse sine function is given in *Tantrasaṅgraha*. If M is the *manda-kendra* (mean anomaly which is the difference between the mean planet and the apogee or aphelion), then the content of the relevant verse can be expressed in mathematical form as

$$\frac{d}{dt}\left(\sin^{-1}\left(\frac{r}{R}\sin M\right)\right) = \frac{r\cos M\,\frac{dM}{dt}}{\sqrt{R^2 - r^2\sin^2 M}}.$$

This verse appears in the context of finding the true rate of motion of the planet (instantaneous velocity) from its average rate of motion (mean velocity).

The time interval between the transits of the mean Sun and the true Sun across the meridian is called the 'equation of time'. The application of the equation of time is exactly formulated for all the planets in the second chapter of *Tantrasaṅgraha* and is the same as in modern astronomy. There is one more correction to the sunrise at a given place, due to the latitude of a place. This is the ascensional difference, which is also described correctly in this chapter. These corrections are also included in this chapter to ensure that one has a complete procedure for obtaining the true longitudes of the planets at the true local sunrise.

Tripraśnādhikāra (Time, place and direction)

The chapter *Tripraśnādhikāra*, dealing with the three problems of time, direction and place from the *chāyā* (the shadow of a gnomon), has always received great attention. The importance given to this topic by Nīlakaṇṭha can be gauged just from the extent of this chapter. In *Tantrasaṅgraha*, it contains 117 verses, which is more than a quarter of the text containing merely 432 verses. Here Nīlakaṇṭha deals with the diurnal problems (mostly related to the motion of the Sun and the shadow cast by it) in an exhaustive manner. It is here that his mastery over spherical trigonometry comes to the fore.

The gnomon, invariably taken to be 12 units in length (normally referred to as *aṅgulas* in astronomical texts), plays a central role in the diurnal problems. As the first problem, the cardinal directions are determined using the shadow of the gnomon, incorporating the correction due to the variation of Sun's declination during the day. The procedure is the same as the one given by Bhāskara II in his *Siddhāntaśiromaṇi*.

The second important problem is the determination of the latitude of a place. The equinoctial midday shadow is $12\tan\phi$, where ϕ is the latitude of the observer's location. Thus, in principle, ϕ can be determined by measuring the equinoctial shadow. However, one needs to take into account the correction due to the finite size of the Sun and its parallax. Nīlakaṇṭha has given the exact formula for the correction that needs to be applied in order to take the above two effects into account in the determination of the latitude of the place through shadow measurements.

When the zenith distance of the Sun is z, the length of the shadow is simply $12\tan z$. There is an elaborate discussion on the determination of the shadow at any instant after sunrise and the inverse problem of the determination of the time from the shadow. Here again, the corrections to z due to the finite size of the Sun and its parallax are discussed.

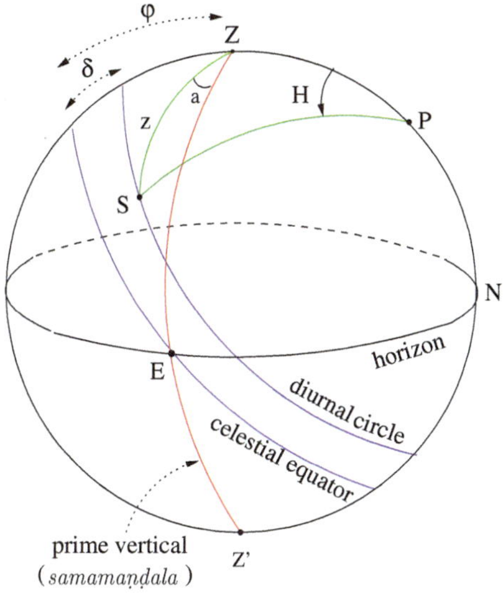

Fig. 1 The five quantities involved in the 'Ten Problems'.

Next we find a section dealing with the *Daśapraśna* or the 'Ten Problems', which in short may be explained as follows. Consider the five quantities, the zenith distance of the Sun (z), its hour angle (H), its declination (δ), its amplitude (a) and the latitude of the place (ϕ) in Fig. 1. There are ten different ways to choose any two out of the five. The subject matter of the *daśapraśna* is to determine any two of them, given the other three. Perhaps it is for the first time that a problem of this type is posed and systematically solved. The expressions for the two unknowns in terms of the three known quantities are exact spherical trigonometrical results. The text *Yuktibhāṣā* of Jyeṣṭhadeva gives a systematic derivation of all these results.

The calculation of the *lagna* is another important problem discussed in this chapter. The *lagna* (ascendant) is the longitude of the point of the ecliptic intersecting the eastern horizon at any given time. The procedure for its calculation as delineated in earlier astronomical texts involves certain approximations. It is remarkable that an exact procedure for finding the *lagna* is discussed here, by introducing the concepts of (i) the *kālalagna*, which is the time elapsed after the rising of the vernal equinox, and (ii) the *dṛkkṣepa*, which is the zenith distance of the point of the ecliptic 90° away from the *lagna*.

All the diurnal problems involving the Sun depend upon the tropical longitude of the Sun, referred to as the *sāyana*,[13] whereas the true longitude considered in the earlier chapter refers to the *nirayaṇa* longitude.[14] The difference between them is due to the precession of the equinoxes, termed the *ayanacalana* (motion of the equinoxes). The author appears to believe in the oscillation of the equinox Γ, whereas according to modern astronomy Γ moves continuously westwards at the rate of approximately $50.2'$ per year.

Candragrahaṇa (Lunar eclipse)

A lunar eclipse occurs when the longitude of the Moon is 180° away from the Sun, and the Moon happens to be close to one of its nodes. As usual, the exact instant of conjunction is determined by an iterative process, starting from the longitudes of the Sun and the Moon and their rates of motion at sunrise. It is interesting to note that several corrections for improving the accuracy of the results over the earlier treatments of the problem are suggested by Nīlakaṇṭha in this chapter.

The mean distance of the Moon from the Earth is given as 34380 *yojanas*.[15] The mean Earth–Sun distance is taken to be 459620 *yojanas*, based on the assumption that the linear velocities of the Sun and the Moon are the same.[16] The actual distances of the Sun and the Moon at any time are to be found by taking the eccentricities of the orbits into account. A further correction is specified which amounts to taking into account the 'evection' term, in the case of the Moon. It may be recalled that the evection term for the Moon makes its first appearance in India in *Laghumānasa* of Mañjulācārya. There is a similar correction suggested for the Sun. The resultant distance which is called the *dvitīya-sphuṭa-yojana-karṇa* is used in all the relevant calculations.

The angular diameters of the Moon, the Sun and the Earth's shadow, and the length of the Earth's shadow are all calculated as in the earlier texts. The linear diameters of the Sun and the Moon are given as 4410 and 315 *yojanas*. Then the criteria for partial and total eclipses are clearly enunciated. The first and second half-durations of the eclipse are computed iteratively. It is interesting to note that the instant of maximum obscuration is taken to be different from the instant of opposition. An expression for this time difference is also given, perhaps for the first time.

A feature that is noteworthy throughout the text is Nīlakaṇṭha's concern for detail. When the *sparśa* or the first contact is very close to the sunrise, it may not

---

[13] The longitude determined with the vernal equinox Γ, as the zero-point of the ecliptic.

[14] The longitude determined with respect to a fixed reference star—usually the beginning point of *Aśvinī*—as the zero-point of the ecliptic.

[15] A *yojana* is a unit of distance employed in Indian astronomy.

[16] It may be mentioned here that the linear velocities of all the planets, (not only the Sun and the Moon) are taken to be the same in Indian astronomy as a first approximation. This fact, though implicit in the procedures given in *Tantrasaṅgraha*, is explicitly mentioned at the very beginning of *Yuktibhāṣā*.

be visible. The exact criterion for visibility is given, taking the Moon's parallax into account. There is a similar discussion when the *mokṣa* or the last contact is close to sunset.

Generally, in deriving the expression for the *bimbāntara* or the separation between the centres of the Earth's shadow and the Moon's disc, a 'planar' approximation is made by replacing the arc by the chord. But here Nīlakaṇṭha gives the exact expression for separation between the discs which does not involve this approximation. Towards the end of the chapter, he discusses the concept of *valana*, which is essentially the angle between the line joining the centres of the Moon and the shadow, and the vertical direction. This has two components, the *ākṣavalana* (inclination due to latitude) and the *āyanavalana* (inclination due to obliquity of the ecliptic). The expressions given in the text for these two are only approximate. They are actually employed in depicting the geometrical representation of the motion of the shadow as well as the evolution of the eclipse.

Ravigrahaṇa (Solar eclipse)

A solar eclipse is far more sensitive to the Moon's latitude, and the apparent longitudes of the Sun and the Moon, than a lunar eclipse. The parallaxes of the Sun and the Moon play an important role in the computation of a solar eclipse and their effect is treated in great detail in this chapter. The terms *lambana* and *nati* refer to the parallaxes in longitude and latitude (that is, the projections along and perpendicular to the ecliptic) respectively. *Lambana* introduces a correction to the *parvānta* (instant of conjunction), the time at which the longitudes of the Sun and the Moon are equal, which is given by

$$4 \cos z_v \sin(\lambda_v - \lambda_s)$$

in *nāḍikās*. Here λ_v is the longitude of the *vitribhalagna* (*lagna* + 90°, also referred to as nonagesimal), z_v its zenith distance and λ_s the Sun's longitude (which is the same as that of the Moon).

The above expression for parallax correction is under the assumption that the horizontal parallax is equal to one-fifteenth of the daily motion of the object. Though the formula given above is found in the earlier texts, it is noteworthy that the expression for $\cos z_v$ given here is exact. The effect of parallaxes in longitude and latitude is to modify the half-durations of the eclipse, as the first contact, the last contact and the instant of conjunction are all affected by them. Hence an iterative process is used for the computation of the half-durations.

Further, a more accurate method for determining the instant of conjunction is discussed here, using the notion of the *dṛkkarṇa*. This refers to the actual physical distance of the celestial object from the observer on the surface of the Earth. Consider for instance the Moon, which is at a distance $D_m = CM$ from the centre of the Earth (see Fig. 2). D_m is the *dvitīya-sphuṭa-yojana-karṇa* mentioned earlier, which is calculated from the mean distance, after taking the 'eccentricity' (*manda-*

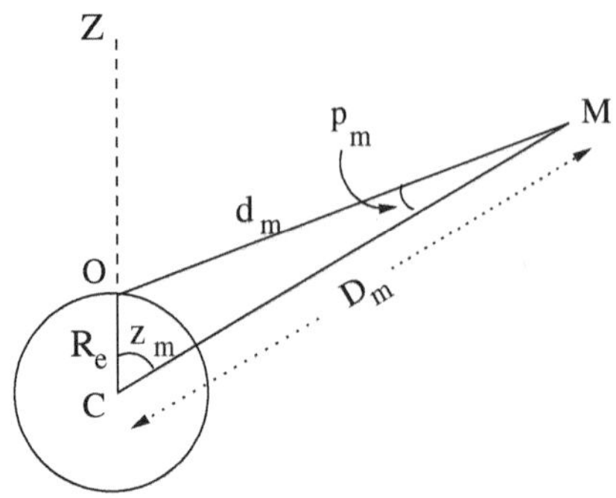

Fig. 2 The actual distance of the celestial object from the observer.

saṃskāra) and 'evection' (*dvitīya-saṃskāra*) corrections into account. Then the *dṛkkarṇa*, d_m, is given by

$$d_m = OM = \sqrt{(D_m - R_e \cos z_m)^2 + (R_e \sin z_m)^2},$$

where R_e is the Earth's radius, and z_m is the geocentric zenith distance of the Moon. Then the expression for Moon's parallax will be

$$p_m \approx \sin p_m = \frac{R_e}{d_m} \sin z_m,$$

where z_m is calculated exactly by taking the Moon's latitude into account. A similar expression is used in order to take into account the effect of solar parallax also.

Thus the *parvānta* (the instant of conjunction of the Sun and the Moon) is to be calculated by taking all these corrections due to parallax and *dṛkkarṇa* into account. Further, the actual distances of the Sun and Moon from the observer are needed to find their angular diameters and the latitude of the Moon. Then the criteria for the visibility of a partial eclipse and a total eclipse are discussed along the usual lines. Mention is also made of an 'annular eclipse', when the angular diameter of the Sun happens to be more than that of the Moon.

Vyatīpāta

We have a *vyatīpāta* when the magnitudes of the declinations of the Sun and the Moon are equal and their gradients are in opposite directions. This concept seems to be peculiar to Indian astronomy. Issues such as the occurrence/non-occurrence of *vyatīpāta*, its duration etc. are discussed in detail in this chapter.

The exact expression for the declination of the Moon in terms of its longitude and latitude is given in the text, perhaps for the first time in Indian astronomy. Further, Nīlakaṇṭha gives an alternative expression for the Moon's declination in terms of the instantaneous inclination of its orbit with the equator. This expression, though not exact, is reasonably accurate when the inclination of the orbit with the ecliptic is small, which is actually the case (mean value $\approx 5^\circ$). The expression for the instantaneous inclination (with the equator) itself is exact, and non-trivial. This is useful in the formulation of the criteria for the occurrence of *vyatīpāta*. Further, an iterative process for determining the time interval between a desired instant and the middle of the *vyatīpāta* is also discussed.

Dṛkkarma (Reduction to observation)

This chapter is devoted to *Dṛkkarma*, or the determination of the visibility of planets. For this, the *lagna* corresponding to the time when the planet is rising or setting is to be determined. If the planet has a latitude, one needs to determine the correction $\Delta\lambda$ to the longitude λ of the planet in order to find the *lagna*. The standard expression for this in terms of the *ākṣavalana* and *āyanavalana* is first given. Then an exact expression for $\Delta\lambda$ is given, which is similar to the expression for the *caraprāṇa* (ascensional difference), with the latitude of the planet replacing the terrestrial latitude of the place, and the zenith distance of the *dṛkkṣepa* replacing the Sun's declination. The criterion for the visibility of the planet is considered next. The difference between the *kālalagna* of the Sun and the planet has to be greater than a value specified for each planet for it to be visible. This criterion, specified in the text, seems to be purely empirical.

Sṛṅgonnati (Elevation of the lunar horn)

The 'evection', which is the second correction (*dvitīya-saṃskāra*) for the Moon—the first being the *manda* or eccentricity correction—is discussed in this chapter. The true physical distance of the Moon, called the *dvitīya-sphuṭa-yojana-karṇa* is calculated taking this correction into account. This is used in the calculation of the angular diameter of the Moon's disc and other quantities.

The distance between the centres of the solar and lunar discs and the angular separation between them is calculated exactly. This is for an observer on the surface of the Earth and includes the effect of parallax. Lengthy as the procedure is, it reveals Nīlakaṇṭha's impressive geometrical insights, especially as it amounts to calculating the distance between two points in space, as in three-dimensional coordinate geometry. The angle of separation determines the lunar phase.

Sṛṅgonnati[17] is the elevation of the lunar horn, or the angle between the line of cusps and the horizontal plane. The expression for *sṛṅgonnati* given here appears to be valid only when the Moon is on the horizon. Further, a graphical representation of the Moon's disc, line of cusps etc. also provided in this chapter. The time interval between sunset and moonrise is to be determined through an iterative procedure.

---

[17] *Sṛṅga* is horn and *unnati* is elevation.

Nīlakaṇṭha takes up the issue of planetary distances in a couple of verses at the very end of the last chapter. Here he seems to suggest that the *kakṣyā-vyāsārdha*, or the mean distances in *yojanas*, obtained from the principle that all planets cover equal distances in equal times, should be understood as the mean *śīghrocca*–planet distance, and not as the mean Earth–planet distances as comprehended in earlier texts.

Chapter 1
मध्यमप्रकरणम्
Mean longitudes of planets

१.१ मङ्गलाचरणम्
1.1 Invocation

हे विष्णो निहितं कृत्स्नं जगत् त्वय्येव कारणे ।
ज्योतिषां ज्योतिषे तस्मै नमो नारायणाय ते ॥ १ ॥

he viṣṇo nihitaṃ kṛtsnaṃ jagat tvayyeva kāraṇe |
jyotiṣāṃ jyotiṣe tasmai namo nārāyaṇāya te || 1 ||

O *Viṣṇu*! the entire universe is embodied in thee, who art the very cause of it. My salutations to thee *Nārāyaṇa*, who art the source of radiance of all the radiating objects.

It is a time-honoured practice in Indian tradition to commence any worthwhile undertaking with a *maṅgalācaraṇam*. Literally the word *maṅgalācaraṇam* means 'doing something good' or 'doing something for the sake of good'. In this context, it means both.

Here the author Nīlakaṇṭha, adhering to this traditional practice, commences the composition of the text *Tantrasaṅgraha* with an invocation to Lord *Viṣṇu* seeking His divine blessings for the successful completion of the work. Thus, the very act can be conceived to be good (having a prayerful attitude) and it is for the sake of good (viz., completion of the work) also. The first quarter of the verse is the chronogram for the *Kalyahargaṇa* (the count of days from the beginning of the *Kaliyuga*) of the date of commencement of the work, which is 1680548, which corresponds to March 22, 1500 CE.

The date of composition of *Tantrasaṅgraha*

In *Laghu-vivṛti* it is noted that the date of commencement of *Tantrasaṅgraha* is indicated in the first quarter of the invocatory verse.

आचार्येण इमं श्लोकम् आदितो ब्रुवता प्रथमपादेन प्रबन्धारम्भकल्यहर्गणश्च अक्षरसङ्ख्यया उपदिष्टः ।

The *Ācārya*,[1] by composing this verse (*maṅgalācaraṇam*) in the very beginning, has also indicated the *Kalyahargaṇa* corresponding to the date of commencement of the work, in its first quarter (*pāda*), through *akṣarasaṅkhyā*.[2]

As mentioned earlier, in the *Kaṭapayādi* system of numeration, the first *pāda* (quarter) of the invocatory verse—*he viṣṇo nihitaṃ kṛtsnam*—refers to the number 1680548. Assuming that the beginning of the *Kaliyuga* is on February 17/18, 3102 BCE, this number corresponds to March 22, 1500 CE according to the Gregorian calendar and *Mīna* 26, 4600 *gatakali* (elapsed *Kali* years) according to the Indian calendrical system. Similarly, the *pāda—lakṣmīśanihitadhyānaiḥ*—occuring in the last verse of the work, is identified by Śaṅkara Vāriyar as giving the date of completion of the work. This quarter in the *Kaṭapayādi* system is equivalent to the number 1680553, which correponds to *Meṣa* 1, 4601 *gatakali*. Thus, from these indication given by the author we understand that the work was completed in just five days. The year of composition of the work can be thus fixed as 1500 CE.

Now we move on to briefly explain the purpose of *maṅgalācaraṇam* as enunciated by Śaṅkara Vāriyar in his *Laghu-vivṛti* and *Yukti-dīpikā*.

The purpose of the *maṅgalācaraṇam*

The purpose of the *maṅgalācaraṇam* is an important topic of discussion in itself and is debated at great length in some of the texts on Indian logic.[3] Here, it has been succinctly stated by Śaṅkara Vāriyar in his *Laghu-vivṛti* as follows:[4]

मङ्गलाचारयुक्तानां विनिपातो न विद्यते ।

Those who follow the practice of starting a work with invocation do not suffer a fall.

Expanding the same idea along similar lines, Śaṅkara Vāriyar in his *Yukti-dīpikā* observes:

प्रह्वीभावो नमस्कारशब्दार्थः सर्वविन्मतः ।
सर्वानुग्राहके विश्वोत्कृष्टे युक्तः स तादृशः ॥
अतस्तादृङ्नमस्कारात् सिद्ध्येत् सर्वं यथेप्सितम् ॥ [5]

It is accepted by all the scholars that the *namaskāra*[6] is an expression of one's humility. Such an expression is quite appropriate towards Him, the one who is the most superior

---

[1] Commentators usually refer to the author of the text, as *Ācārya*. Here for instance, Śaṅkara Vāriyar refers to Nīlakaṇṭha as *Ācārya*.

[2] The word *akṣarasaṅkhyā* refers to the *Kaṭapayādi* system of numeration. An exposition of this system with a few illustrations is presented in Appendix A.

[3] See, for instance, *Tarkasaṅgraha-dīpikā* by Annaṃbhaṭṭa or *Muktāvalī* of Viśvanātha Pañcānana.

[4] The source of this quotation cited by Śaṅkara Vāriyar in his *Laghu-vivṛti* is not known.

[5] {TS 1977}, p. 2.

[6] It could be either the use of the word *namaḥ* or the physical act of bending down forward.

and who blesses all the beings in the world. Thus by paying devout homage to Him, all undertakings get accomplished as desired.

The commentator conveys the idea that prayer involves recognition of one's limitations. This very act brings in a certain strength and fortitude. Thus, by the act of prayer, a person not only acknowledges the limitations but also gains a certain inner strength to face a variety of obstacles that he would come across in accomplishing the task that he has undertaken.

Viṣṇu as the cause

Another interesting discussion that is found in the commentary of this verse is regarding the description of the *Viṣṇu* as the *kāraṇa* (the cause of the universe). For any effect produced there are at least two types of causes that can be conceived of: (i) the material cause (*upādāna-kāraṇa*) and (ii) the efficient cause (*nimitta-kāraṇa*). Here Nīlakaṇṭha describes *Viṣṇu* as the *kāraṇa* (cause) of the entire universe. The concept of *Īśvara* being both the efficient and the material cause of the universe is central to Śaṅkara's philosophy, *Advaita-vedānta*. Nīlakaṇṭha, being a master of several disciplines of knowledge,[7] including philosophy, purposefully uses the word *kāraṇa* without any qualifier to convey the fact that *Viṣṇu* is both the material and the efficient cause of the universe. This concept is explained with simile in *Yukti-dīpikā* as follows:

मृत्तिकायां यथा भाति कृत्स्नं कुम्भघटादिकम् ।
नारायणे तथा भाति जगदेतच्चराचरम् ॥
जगत्कारणता तस्य निर्विवादा स्थिता ततः ॥

As the pot and other things [made out of clay] shine (or owe their existence) due to clay [the material cause], so too the entire universe shines in *Nārayaṇa* [as he is both the material and efficient cause]. Therefore the fact that he is the cause of the universe remains unquestioned.

It is further observed that the adjective *jyotiṣāṃ jyotiḥ* used to describe *Nārāyaṇa*—in the third quarter of the verse—has its basis in *Śrīkṛṣṇa*'s saying in *Bhagavadgītā*: (XV.12)

ज्योतिष्ट्वं ज्योतिषां तस्य ज्ञेया भगवदुक्तितः ।
'यदादित्यगतं तेजो जगद्भासयतेऽखिलम् ।
यच्चन्द्रमसि यच्चाग्नौ तत्तेजो विद्धि मामकम्' ॥ [8]

That He is the source of brilliance of all the brilliant objects has to be understood through the statement of *Bhagavān* himself. [He states:] May you understand that the *tejas* (brilliance)

---

[7] This fact is brought out by Śaṅkara Vāriyar explicitly in his invocatory verses to *Laghu-vivṛti*. Śaṅkara Vāriyar observes:

नारायणं जगदनुग्रहजागरूकं श्रीनीलकण्ठमपि सर्वविदं प्रणम्य ।

[8] {TS 1977}, pp. 1–2.

of the Sun which illumines the world, the *tejas* in the Moon and in the fire etc. all belong to me.

१.२ सावननाक्षत्रदिनमानम्

1.2 Measurement of civil and sidereal day

रवेः प्रत्यग्भ्रमं प्राहुः सावनाख्यं दिनं नृणाम् ।
आर्क्षमृक्षभ्रमं तद्वत् ज्योतिषां प्रेरको मरुत् ॥ २ ॥

raveḥ pratyagbhramaṃ prāhuḥ sāvanākhyaṃ dinaṃ nṛṇām |
ārkṣamṛkṣabhramaṃ tadvat jyotiṣāṃ prerako marut || 2 ||

It is said that the *sāvanadina*, the civil day of humans, is the time taken by the Sun for [completing] one westward revolution. Likewise the *ārkṣa*[*dina*], the sidereal day, is the time taken by the stars for one [complete westward] revolution. The impeller of the celestial objects is the *marut* [wind called *pravaha*].

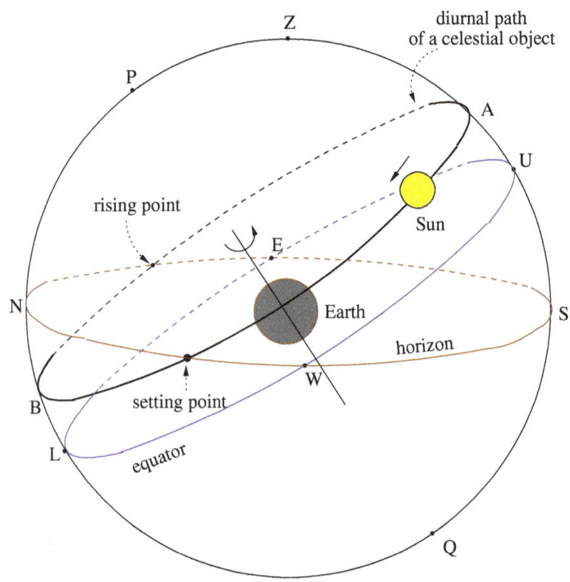

Fig. 1.1 The diurnal motion of a celestial object.

As the Earth rotates around its axis from west to east, the apparent motion of the celestial objects is from east to west (see Fig. 1.1). This apparent westward motion of the objects as seen by the terrestrial observer is described as *pratyagbhrama*. The word *pratyak* means west, and *bhrama* is motion, and hence *pratyagbhrama* is westward motion. In modern spherical astronomy, this apparent westward motion is termed the 'diurnal motion'.

The time taken by the Sun to complete one revolution westwards is defined to be the *sāvana-dina/sāvana-vāsara* (civil day). *Sūryasiddhānta* defines the civil day as follows:

उदयादुदयं भानोः भूमिसावनवासरः ।[9]

[Time interval between] one sunrise and the next sunrise is a terrestrial civil day.

Civil day and sidereal day

The time taken by the stars to complete one revolution around the Earth westwards is defined to be a *sidereal day*. This corresponds to the time interval between successive meridian transits of a particular star, which is the same as the time taken by the Earth to complete one rotation around its own axis. It is precisely this rotation of Earth which makes the stars appear to have a diurnal motion. Apart from the diurnal motion, which is westward, the stars do not have any eastward motion of their own.[10] However, this is not true of the Sun, Moon and other planets. They have both diurnal (which is westward) motion and eastward motion relative to the stellar background.

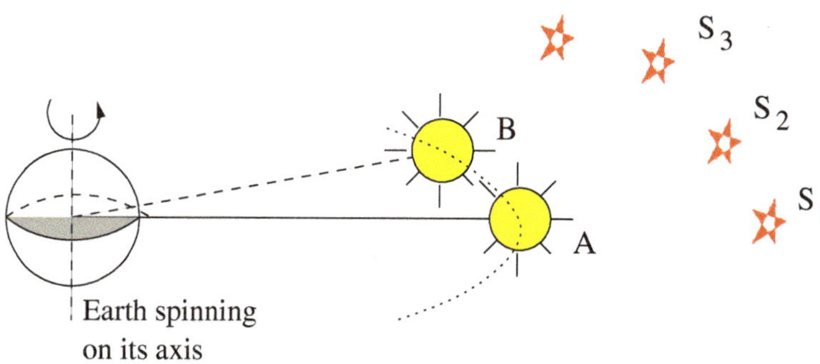

Fig. 1.2 The eastward motion of the Sun in the background of stars.

Suppose the Sun is near a star on a particular day. After one sidereal day (≈23h, 56m and 4s), the star would have completed one revolution around the Earth. But the Sun would not have completed one full revolution around the Earth because of its eastward motion of nearly 1° per day in the stellar background. This situation is schematically depicted in Fig. 1.2.

Here, A and B represent the positions of the Sun and S_1, S_2, S_3 are different stars in the stellar background. On a particular day, let us assume that the Sun at the point

---

[9] {SSI 1995} (I.36), p. 24.

[10] As the relative positions of the stars seem fixed—which in fact have 'proper' motion, according to modern physics—they are assumed to be stationary objects in the sky. In fact, it is only because of the 'fixed' background provided by these stars that the motion of other celestial objects such as Sun, Moon, planets etc. can be studied as motion with respect to them.

A and the star S_1 are in the same direction at a particular instant. Let us further assume that at that instant both of them are in meridian transit (i.e. cutting across the observer's meridian). After one sidereal day, the star S_1 would appear on the meridian again. But the Sun would not appear on the meridian because it has moved to the east to the point *B* by nearly 1°. So the Earth has to rotate by this amount before the Sun appears on the meridian again. It takes approximately 4 minutes ($\frac{24}{360}$h) for this to happen.[11] Thus, the sidereal day plus this time interval is a civil day, which is the same as the time interval between two successive meridian transits of the Sun.[12]

The cause of diurnal motion

In the last quarter of verse 2, it is stated that the celestial objects are impelled by a flow of *marut* (wind). This is termed the *pravaha-marut* and the diurnal motion of the celestial sphere is ascribed to it in *Yukti-dīpikā*.

प्रत्यहं भ्रमणं प्रत्यङ्मुखं दृष्टं द्युचारिणाम् ।
परतोऽथ स्वतस्तेषां प्राङ्मुखं चानुमीयते ॥
भगोलं भ्रामयेन्नित्यं प्रत्यक् प्रवहमारुतः ।
तद्वशादेव तत्स्थानि ग्रहर्क्षाणि मुहुर्मुहुः ॥
अतः प्रत्यङ्मुखं तेषां भ्रमणं परतो मतम् ।[13]

The daily westward motion of the celestial objects observed is due to an external agency, whereas their eastward motion is inferred to be of their own accord.

The *pravaha-vāyu* continuously rotates the *bhagola* again and again westwards, and because of this the stars and planets situated in it [keep rising and setting]. Hence it is mentioned that their westward motion is due to an external agency.

The use of the word *muhurmuhuḥ* (again and again) indicates that *pravaha-marut* generates a continuous and uniform westward motion of the celestial objects. The very word *pravaha* is suggestive of this meaning and much more. It consists of two parts, a prefix and a verb.

प्रवह = प्र (prefix) + वह (verb)

The verb *vaha* means flow and the prefix *pra* means special and/or great. The speciality of the wind is that it flows perennially without cessation. It is also great, because it is able to carry all the innumerable celestial objects along with it. The earlier Indian texts also attribute the motion of the celestial objects to the *pravaha* wind and this concept finds mention even in Āryabhaṭa's *Āryabhaṭīya*.

---

[11] This is obtained by the rule of three, as it takes nearly 24 hours for one rotation of the Earth (corresponding to 360 degrees).

[12] In this verse, only the mean civil day is defined. This is because the time interval between two successive meridian transits of the Sun is not constant, but varies over the year. A more detailed discussion of this topic can be found in Sections 11 and 12 of Chapter 2.

[13] {TS 1977}, p. 4.

Āryabhaṭa on diurnal motion

It is well known that Āryabhaṭa was the first Indian astronomer to have suggested that the apparent diurnal motion of the celestial objects is actually due to the rotation of the Earth. It is described thus in *Āryabhaṭīya*:

अनुलोमगतिर्नौस्थः पश्यत्यचलं विलोमगं यद्वत् ।
अचलानि भानि तद्वत् समपश्चिमगानि लङ्कायाम् ॥ [14]

Just as a person in a boat moving in the forward direction observes the stationary objects (trees etc. on the bank) to be moving in the opposite direction, so also the stationary stars seem to move directly westward for an observer in Laṇka.[15]

This verse clearly describes the rotation of the Earth. Āryabhaṭa has presented an apt example to describe relative motion and seems to explain the observed phenomenon of diurnal motion of celestial objects as being actually due to the rotation of the Earth around its axis.

The very next verse of *Āryabhaṭīya* gives the standard description:

उदयास्तमयनिमित्तं नित्यं प्रवहेण वायुना क्षिप्तः ।
लङ्कासमपश्चिमगः भपञ्जरः सग्रहो भ्रमति ॥ [16]

For the sake of the rising and setting [of all the celestial objects] the stellar sphere, along with the planets being blown by the *Pravaha* wind, moves westward at a uniform rate.

Inferring the eastward motion of planets

While explaining the role played by the *pravaha-marut*, *Yukti-dīpikā* describes how the eastward motion of the celestial objects in the background of stars can be inferred from the conjunction of these objects with different stars whose relative positions are fixed in the sky:

ज्योतिश्चक्रे यदा तेन नक्षत्रेण युतो ग्रहः ॥
कालान्तरे पुनस्तस्मात् संयुक्तोऽन्येन दृश्यते ।
नक्षत्राणि स्थिराण्येव स्वप्रदेशेषु तु स्वतः ॥
न कदाचित् क्वचिद्यान्ति स्वदेशात् गत्यभावतः ।
प्रदेशान्तरसंयोगो नोपपन्नोप्यगच्छतः ॥
ततः प्रतीच्यनक्षत्रात् प्राच्यनक्षत्रयोगतः ।
प्रत्यहं प्राङ्मुखी भुक्तिः ग्रहाणामनुमीयते ॥ [17]

The planet which is found to be in conjunction with a particular star at one time is found to be in conjunction with some other star at a later time. The stars are fixed in their own places. Because they do not have any motion, they never leave their location.

[14] {AB 1976}, (*Golapāda*, 9), p. 119.

[15] A 'fictitious' place on the equator.

[16] {AB 1976}, (*Golapāda*, 10), p. 119.

[17] {TS 1977}, p. 4.

Moreover, conjunction of an object [with another object] at a different location is not possible if it does not move. Since the planets are found to be in conjunction with a star in the east, after their conjunction with a star in the west, it can be inferred that they have a daily motion in the eastward direction.

Continuing with the commentary on the eastward motion of the Sun, the Moon and the planets, *Yukti-dīpikā* makes a very interesting observation, which amounts to a clear recognition of the fact that the speed of the planet changes continuously.

भिन्ना प्रतिक्षणं चासौ भगोलावयवैर्मिता ।
भगोलमध्यवर्तिन्यां भूमौ द्रष्टुरवस्थितेः ॥[18]

The eastward motion of the planets in the units of parts (arcs) of the *bhagola* (celestial sphere) differs from instant to instant, because the location of the observer is on the Earth, which is situated at the centre of the *bhagola*.

१.३ परिच्छिन्नकालमानम्

1.3 Measurement of smaller units of time

भ्रमणं पूर्यते तस्य नाडीषष्ट्या मुहुर्मुहुः ।
विनाडिकापि षष्ट्यंशो नाड्या गुर्वक्षरं ततः ॥ ३ ॥
प्राणो गुर्वक्षराणां स्यात् दशकं चक्रपर्यये ।
खखषड्घनतुल्यास्ते वायुः समजवो यतः ॥ ४ ॥

bhramaṇaṃ pūryate tasya nāḍīṣaṣṭyā muhurmuhuḥ |
vināḍikāpi ṣaṣṭyaṃśo nāḍyā gurvakṣaraṃ tataḥ || 3 ||
prāṇo gurvakṣarāṇāṃ syāt daśakaṃ cakraparyaye |
khakhaṣaḍghanatulyāste vāyuḥ samajavo yataḥ || 4 ||

Its revolution (the revolution of the stellar sphere) gets completed in 60 *nāḍīs* again and again. A *vināḍi* is one-sixtieth part of a *nāḍī*. One-sixtieth of a *vināḍī* is a *gurvakṣara*. Ten *gurvakṣaras* constitute one *prāṇa*. In one complete revolution [of the *vāyu* (wind)], there are 21600 *prāṇās* [and this remains fixed] since the *vāyu* moves with uniform speed.

Prāṇa as a basic unit of time

Here the *prāṇa* has been introduced as a fundamental unit of time. Though the word *prāṇa* is generally used to refer to life, in the context of astronomy it has to be understood as a time unit which is 4 sidereal seconds (see Table 1.1). The relation between the use of the word *prāṇa* to refer to life/life energy and its definition as above, as a time unit, can be understood from the fact that the average time taken by a healthy person for one inhalation and exhalation is nearly 4 seconds.

---

[18] {TS 1977}, p. 4.

The period of revolution of the stellar sphere is defined to be 60 *nāḍīs*. This implies that the duration of a sidereal day is taken to be 60 *nāḍikās*.[19] Units of time smaller than a *nāḍi*/*nāḍikā* defined in the above verse are shown in Table 1.1.

Name of unit	Its measure (sidereal)	Modern equivalent (sidereal)
ghaṭikā/*nāḍikā*	$\frac{1}{60}$ day	24 minutes
vināḍikā	$\frac{1}{60}$ *nāḍikā*	24 seconds
gurvakṣara	$\frac{1}{60}$ *vināḍikā*	0.4 seconds
prāṇa	10 *gurvakṣara*	4 seconds

Table 1.1 Units of time from day to *prāṇa*.

Uniform diurnal motion

In *Laghu-vivṛti* it is explained that the word *muhurmuhuḥ* – which literally means 'again and again' – has been specifically employed here to indicate the fact that the stellar sphere always moves with uniform speed (60 *ghaṭikās* for all cycles):

मुहुर्मुहुः इत्यनेन सर्वेषां तद्भगणानां मिथः कालसाम्यं दर्शितम्।

By the use of the word *muhurmuhuḥ* it is shown that the period of revolution of it (the stellar sphere) remains the same [at all times].

Units of time smaller than the *prāṇa*

Units of time smaller than the *gurvakṣara* have also been used in Indian astronomy texts. For instance, Vaṭeśvara in his *Vaṭeśvara-siddhānta* observes:

कमलदलनतुल्यः काल उक्तस्त्रुटिस्तत्
शतमिहलवसंज्ञः तच्छतं स्यान्निमेषः।
सदलजलधिभिस्तैः गुर्विहैवाक्षरं तत्
कृतपरिमितकाष्ठा तच्छरार्धेन चासुः ॥[20]

The time [taken by a sharp needle] to pierce [a petal of] a lotus is called a *truṭi*; one hundred times that is called a *lava*; one hundred times that is a *nimeṣa* (i.e., the time required for blinking the eyes); four and half times that is a long syllable or *gurvakṣara* (i.e., the time required by a healthy person to pronounce a long syllable); four times that is a *kāṣṭha* and one half of five times that is an *asu* (*prāṇa*).

The units of time that are orders of magnitude smaller than a *prāṇa*—starting with a *kāṣṭha* up to a *truṭi*—described in the above verse are given in Table 1.2.

[19] This is a sidereal day which is slightly less than the civil day, which is equal to the time interval between two mean sunrises. The difference is about 4 minutes and is due to the eastward motion of the Sun by about 1° per day.

[20] {VS 1986}, (1.1.7), p. 2.

Name of unit	Its measure	Modern equivalent (sidereal) in seconds
kāṣṭha	$\frac{2}{5}$ *prāṇa*	1.6
gurvakṣara	$\frac{1}{4}$ *kāṣṭha*	0.4
nimeṣa	$\frac{2}{9}$ *gurvakṣara*	0.889
lava	$\frac{1}{100}$ *nimeṣa*	8.889×10^{-4}
truṭi	$\frac{1}{100}$ *lava*	8.889×10^{-6}

Table 1.2 Units of time much smaller than a *prāṇa* described in *Vaṭeśvara-siddhānta.*

As regards the units of time, the following significant observation is made in *Yukti-dīpikā*:

कालो यतः परिच्छेद्यः ज्योतिश्चक्रपरिभ्रमात् ।[21]

Since the time gets delimited by the motion of the planets in the celestial sphere.

This in turn is based on Āryabhaṭa's understanding of the nature of time.

कालोऽयमनाद्यन्तो ग्रहभैरनुमीयते क्षेत्रे।[22]

This time, which is without beginning and end, is measured with the help of the [motion of] planets and the asterisms on the celestial sphere.

It seems that in the Indian astronomers' conception 'time' is limitless or unbounded and the smaller units of time, like the day, week, fortnight, month etc., are all limitations in time, created by us—for our own convenience—based purely upon the motion of the celestial objects.

१.४ चान्द्रमानम्
1.4 Lunar reckoning of time

पूर्वपक्षः शशाङ्कस्य विप्रकर्षो रवेः स्मृतः ।
सन्निकर्षोऽपरः पक्षः सितवृद्धिक्षयौ ययोः ॥ ५ ॥
मासस्ताभ्यां मतश्चान्द्रः त्रिंशत्तिथ्यात्मकः स च ।

pūrvāpakṣaḥ śaśāṅkasya viprakarṣo raveḥ smṛtaḥ |
sannikarṣo'paraḥ pakṣaḥ sitavṛddhikṣayau yayoḥ || 5 ||
māsastābhyāṃ mataścāndraḥ triṃśattithyātmakaḥ sa ca |

The [period of] separation of the Moon from the Sun is termed the *pūrva-pakṣa* (the first fortnight) and its approach [towards the Sun] is the *apara-pakṣa* (the other fortnight). These two fortnights, in which the brightness of the Moon (the phase that looks bright) increases and decreases, constitute a lunar month. It is made up of 30 *tithis*.

[21] {TS 1977}, p. 2.

[22] {AB 1976}, (*Kālakriyāpāda*, 11), p. 98.

Verses 5–8 in *Tantrasaṅgraha* are devoted to the description of the Indian calendar. To begin with, a *pakṣa* (fortnight) and a *cāndramāsa*[23] (a lunar month) are defined. Both these units are primarily based on the motion of the Moon relative to the Sun.

The Indian calendrical system is based on the motion of the Sun and the Moon. In other words, it is luni-solar in nature, i.e. based on the positions of both the Sun and the Moon against the background of different *rāśis* (zodiacal signs) and *nakṣatras* (asterisms) along the ecliptic.

The luni-solar nature of the Indian calendrical system is evident from the fact that some of the social and religious functions/festivals are celebrated according to *tithis, nakṣatras* etc. (which are essentially based upon the motion of the Moon relative to the Sun) while others that depend on *saṅkrānti, ayana* etc., are based on the motion of the Sun alone. For instance, festivals like *Rāma-navamī, Gaṇeśa-caturthī* etc., are based on Moon's position relative to the Sun, whereas others like *Makara-saṅkrānti, Viṣu* etc., are based on the Sun's position against the background of stars.

Lunar month and *tithi*

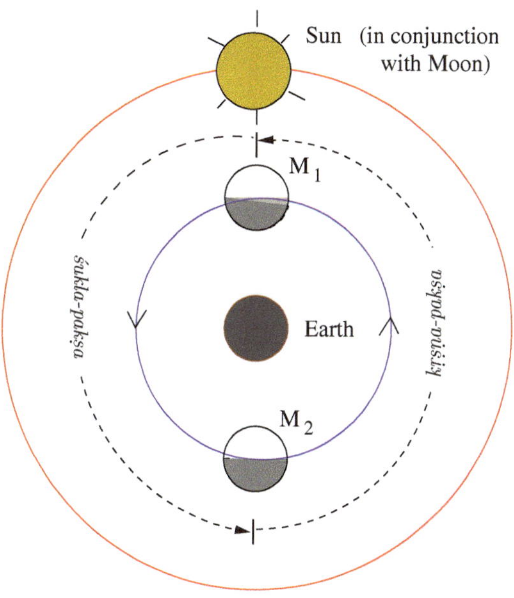

Fig. 1.3 Lunar month consisting of bright and dark fortnights.

Both the Sun and the Moon are in continuous motion as seen by an observer on the Earth. The angular distance covered by them against the background of stars

[23] A *cāndramāsa* consists of two fortnights.

each day is roughly 1° and 13° respectively. Since the Moon moves much faster than the Sun, the angular separation between them keeps increasing with time. A *tithi* is the time unit during which the angular separation between the Sun and the Moon increases precisely by 12°. A lunar month consists of 30 *tithis* and two fortnights, the *śukla* (bright) and *kṛṣṇa* (dark) as indicated in Fig. 1.3.

Suppose at a particular instant the Sun is in conjunction with the Moon (M_1 in Fig. 1.3). This instant is taken to be the ending moment of the *amāvāsyā*, or the new Moon day. As the Moon's angular velocity is much greater than that of the Sun, the angular separation between them keeps increasing. When it becomes exactly 12°, that corresponds to the ending moment of the first *tithi*, namely the *pratipad*. Similarly, when the angular separation becomes exactly 24°, it corresponds to the ending moment of the second *tithi*, namely the *dvitīyā*, and so on. The names of the different *tithis* constituting a lunar month are listed in Table 1.3.

Śukla-pakṣa		*Kṛṣṇa-pakṣa*	
Name of *tithi*	Angular separation bet. Moon and Sun	Name of *tithi*	Angular separation bet. Moon and Sun
Prathamā	0° – 12°	*Prathamā*	180° – 192°
Dvitīyā	12° – 24°	*Dvitīyā*	192° – 204°
Tṛtīyā	24° – 36°	*Tṛtīyā*	204° – 216°
Caturthī	36° – 48°	*Caturthī*	216° – 228°
Pañcamī	48° – 60°	*Pañcamī*	228° – 240°
Ṣaṣṭhī	60° – 72°	*Ṣaṣṭhī*	240° – 252°
Saptamī	72° – 84°	*Saptamī*	252° – 264°
Aṣṭamī	84° – 96°	*Aṣṭamī*	264° – 276°
Navamī	96° – 108°	*Navamī*	276° – 288°
Daśamī	108° – 120°	*Daśamī*	288° – 300°
Ekādaśī	120° – 132°	*Ekādaśī*	300° – 312°
Dvādaśī	132° – 144°	*Dvādaśī*	312° – 324°
Trayodaśī	144° – 156°	*Trayodaśī*	324° – 336°
Caturdaśī	156° – 168°	*Caturdaśī*	336° – 348°
Pūrṇimā	168° – 180°	*Amāvāsyā*	348° – 360°

Table 1.3 The names of the 30 *tithis* and their angular ranges.

Pūrṇimā and *Amāvāsyā*

When the angular separation becomes exactly 180°, the Moon (M_2 in Fig. 1.3) will be in opposition to the Sun and it corresponds to the ending moment of the *pūrva-pakṣa*, or the first half of the lunar month. The first fortnight consists of 15 *tithis*. It is also referred to as the *śukla-pakṣa* (white fortnight) or the *sita-pakṣa* (bright fortnight)[24] in view of the fact that the brightness of the Moon keeps increasing during this period. The fifteenth *tithi* of the bright fortnight is the full Moon day,

[24] The terms *śukla* and *sita* mean white or bright.

called the *pūrṇimā*. The etymological derivation of the word *pūrṇimā*, along with a couple of slight variations of it, are given below:

- पूर्णिमा ≡ पूर्णिः (=पूरणं) मिमीते इति ।
- पूर्णमासी ≡ पूर्णो मासः (चान्द्रः) यत्र । (पूर्ण + मास + ङीष्)
- पौर्णमासी ≡ पूर्णो मासः अस्यां (तिथौ) । (पूर्ण + मास + अण् + ङीष्)

The other half of the lunar month is called the *apara-pakṣa*. It is also known as the *kṛṣṇa-pakṣa* or *asita-pakṣa*[25] (dark fortnight), as the phase of the Moon keeps decreasing during this period. When the angular separation between the Sun and the Moon becomes exactly 360° or 0°, the Moon looks completely dark and once again it is in conjunction with the Sun. During the dark fortnight, the angular separation between the Sun and the Moon keeps increasing from 180° to 360°. Like the bright fortnight, the dark fortnight also consists of 15 *tithis*. The fifteenth *tithi* of the dark fortnight is the new Moon day, called the *amāvāsyā*. It is also known by the name *amāvāsī* and the derivation of these terms is as follows:

- अमावास्या ≡ अमा (=सह) चन्द्रार्कौ वसतः यत्र ।
  (अमा + वस् + ण्यत् (आधारे) + टाप्)
- अमावासी ≡ अमा (=सह) चन्द्रार्कौ वसतः यत्र ।
  (अमा + वस् + ण्यत् (आधारे) + ङीष्)

Commencement of a lunar month/year

The two fortnights, bright and dark, together consisting of 30 *tithis*, form a *cāndra-māsa* (lunar month). A normal lunar year has twelve lunar months. The names of the twelve lunar months are: *Caitra, Vaiśākha, Jyeṣṭha, Āṣādha, Śrāvaṇa, Bhādrapada, Āśvayuja, Kārtika, Mārgaśira, Puṣya, Māgha* and *Phālguna*. During most *Caitra* months the Moon will be close to the star *Citrā* (Spica), on the full Moon day of the month. Similarly, during most *Vaiśākha* months, the moon will be near to the star *Viśākhā* on the full moon day in that month. This is the reason for the nomenclature.

Regarding the commencement of a lunar month there are two different systems, namely the *Amānta* system and the *Pūrṇimānta* system. In the *Amānta* system, a lunar month commences with the ending moment of the new Moon day or equivalently the beginning of the bright fortnight, whereas in the *Pūrṇimānta* system a lunar month commences with the ending moment of the full Moon day or equivalently the beginning of the dark fortnight. In both systems, the names of the lunar months being the same, the bright fortnights will share the same name, while the dark fortnights will have different names; i.e. the *caitra-śukla-pakṣa* of the *Amānta* system will be the same as the *caitra-śukla-pakṣa* of the *Pūrṇimānta* system, though the *caitra-kṛṣṇa-pakṣa* of the *Amānta* system will be the *vaiśākha-kṛṣṇa-pakṣa* of the *Pūrṇimānta* system.

---

[25] The terms *kṛṣṇa* and *asita* mean dark or black.

The *Amānta* system is more popular in south India, whereas the *Pūrṇimānta* system is so in the north. For instance, places like Tamil Nadu, Andhra Pradesh, Karnataka etc. follow the *Amānta* system and places in the North like Uttar Pradesh, Bihar, Rajasthan etc. follow the *Pūrṇimānta* system. As a result, the commencement of the lunar year also differs by about 15 days. The commencement of the lunar year, *yugādi* (as it is popularly called in the south), is celebrated a fortnight earlier in the north.

१.५ सौरमानम्

1.5 Solar reckoning of time

सौराब्दो भास्करस्यैव ज्योतिश्चक्रपरिभ्रमः ॥ ६ ॥
मासस्तु राशिभोगः स्यात् अयने चापि तद्गती ।

saurābdo bhāskarasyaiva jyotiścakraparibhramaḥ || 6 ||
māsastu rāśibhogaḥ syāt ayane cāpi tadgatī |

The [time required for one] complete revolution of the Sun around the ecliptic is a solar year. The period for which it (the Sun) dwells in a *rāśi* is a solar month. The two *ayanās* are nothing but its motion [towards the north and south].

In the above verse, the word *jyotiścakra* refers to the apparent path traced by the Sun in the celestial sphere, as seen from the Earth. This is the same as the 'ecliptic' in modern spherical astronomy. The time taken by the Sun to go around the ecliptic once, thereby covering 360° (*cakra*), is defined as a *saurābda*, a solar year. What is referred to here is the *sidereal* year,[26] which corresponds to the time interval between two successive transits of the Sun across the same star along the ecliptic.

The ecliptic is actually inclined to the celestial equator, as shown in Fig. 1.4. The angle of inclination, denoted by ε, is currently around $23\frac{1}{2}^{\circ}$ and is called as the obliquity of the ecliptic. However, in the Indian tradition, most of the texts on astronomy including *Tantrasaṅgraha* take the angle of inclination to be 24°.

Rāśi division of the ecliptic and solar month

The ecliptic is divided into twelve equal parts, each corresponding to 30°, called *rāśis*. The *rāśis Meṣa* (Aries), *Vṛṣabha* (Taurus), *Mithuna* (Gemini) etc. as indicated in Fig. 1.4*a* are known as *sāyana-rāśis* whereas the ones depicted in Fig. 1.4*b* are known as *nirayaṇa-rāśis*. In Indian astronomy, the beginning point of the '*Meṣa-rāśi*' known as '*Meṣādi*' (first point of Aries) is a fixed point on the ecliptic, which is 180° away from the location of the star 'Spica'. This point is different from the vernal equinox[27] (the beginning point of the *sāyana-meṣa*) because the

---

[26] This is different from the *tropical* year, which is marked by the successive transits of the Sun across the vernal equinox.

[27] It can be shown by computing backwards that the vernal equinox and *Meṣādi* were coincident around fifteen centuries ago.

latter drifts continuously westwards along the ecliptic at the rate of nearly $50''$ per year. In other words, the '*Meṣādi*' moves continuously eastwards with respect to the vernal equinox as indicated in Fig. 1.4*b*. This phenomenon of the westward motion of the equinox is known as the 'precession of equinoxes' in modern astronomy. A closely related but different model of motion of the equinoxes is described by the name *ayanacalana* (motion of equinoxes) in the works of Indian astronomy. In this, the equinox executes an oscillatory motion, moving both eastwards and westwards from *Meṣādi* to a maximum extent of 24°. This phenomenon is called the 'trepidation of the equinoxes'.

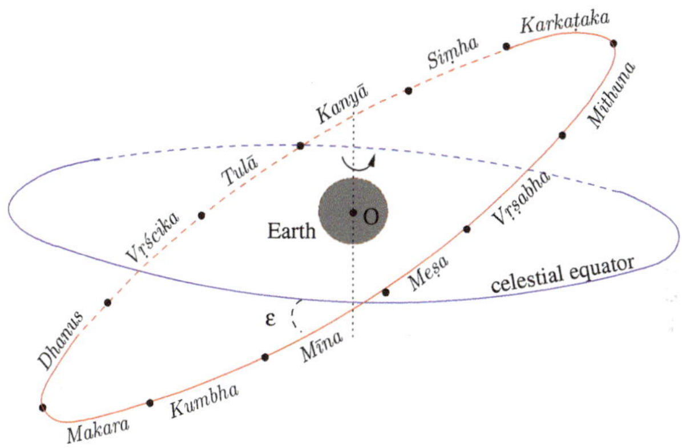

Fig. 1.4*a* The *rāśi* division of the ecliptic, with markings of *sāyana-rāśi*.

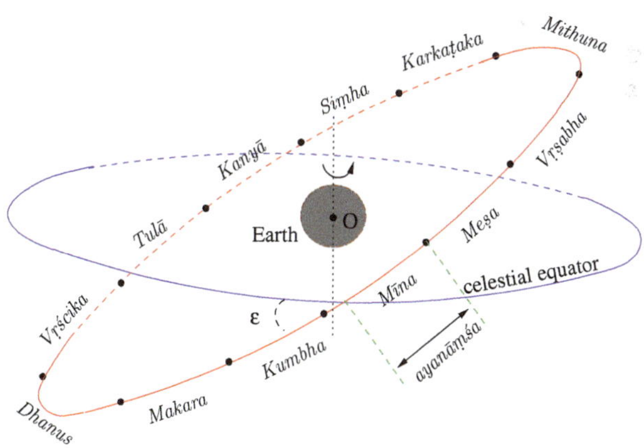

Fig. 1.4*b* The *rāśi* division of the ecliptic, with markings of *nirayaṇa-rāśi*.

At the beginning of the year 2008, '*Meṣādi*' is situated nearly 23° 58′ from the vernal equinox. A schematic sketch of this is shown in Fig. 1.4*b*. The *rāśis Meṣa*,

Vṛṣabha etc. marked here are called *nirayaṇa-rāśis*, in contrast to the markings in Fig. 1.4*a*.

The time taken by the Sun to travel across one *rāśi*, which is a 30° segment on the ecliptic, is defined to be a *sauramāsa* or solar month. The names of the solar months are the same as those of the lunar months. The solar *caitramāsa* is the solar month during which the Sun is in *Mīna-rāśi* (Pisces sign). Similarly, the Sun is in *Meṣa-rāśi* during the solar *vaiśākhamāsa* and so on.

Uttarāyaṇa and *Dakṣiṇāyana*

The ecliptic intersects the celestial equator at two points S_1 and S_3 (see Fig. 1.5). At these points, known as the equinoctial points, the Sun is on the equator. The point S_1—at which the Sun is moving northwards—is called the spring equinox or the vernal equinox and this occurs around March 21st each year.

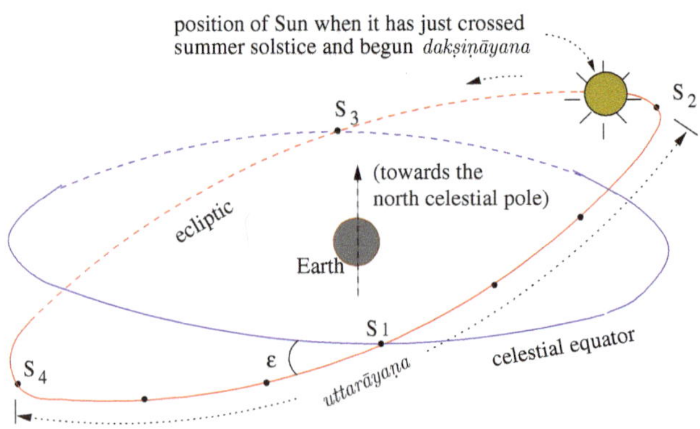

Fig. 1.5 *Uttarāyaṇa* and *Dakṣiṇāyana*.

It may be observed from the figure that before reaching the vernal equinox the Sun lies below or south of the equator. After crossing the vernal equinox it lies to the north of the celestial equator, till it reaches the position S_3. The point S_3 is known as the autumnal equinox and it occurs around September 23rd. At the autumnal equinox, the Sun transits from the northern to the southern hemisphere.

When the Sun is at S_2, it is the summer solstice, which occurs around June 21st. At the summer solstice, the Sun is at the maximum distance from the equator towards the north, and the duration of daytime will be maximum for all observers having a northern latitude. It will be minimum for all observers having a southern latitude. When the Sun is at S_4, it is the winter solstice, which occurs around December 21st. At the winter solstice, the Sun is at the maximum distance towards the south of the equator. On this day, the duration of daytime is minimum for all observers having a northern latitude.

Further, it may be observed from Fig. 1.5 that between the winter solstice (S_4) and the summer solstice (S_2), the Sun steadily moves northwards and this time interval is known as the *uttarāyaṇa/saumyāyana*. The choice of this terminology stems from the fact that *uttara/saumya* mean *north* and *ayana* means *motion*. Similarly, between the summer solstice (S_2) and the winter solstice (S_4), the Sun steadily moves southwards and it is known as *dakṣiṇāyana/yāmyāyana*. These two *ayanās* together constitute a *saurābda* or a solar year.

The commentary *Laghu-vivṛti* defines the *sauramāsa* and the two *ayanās* as follows:

सौराब्दस्य द्वादशांशभूतः यः सौरमासः, सः तस्यैव भास्करस्य ज्योतिश्चक्रद्वादशांशभूतस्य राशेर्भोगकालः स्यात्। तस्यैव उदग्दक्षिणादिगभिमुखगतिद्वयं सौम्ययाम्यरूपमयनद्वितयमपि ॥

The solar month which is one-twelfth of a solar year is equal to the time spent by the Sun in a *rāśi*, which in turn is one-twelfth of the *jyotiścakra* (ecliptic). The motion of the same (Sun) along the north and the south directions is termed *saumyāyana* and *yāmyāyana*.[28]

१.६ अधिमासलक्षणम्

1.6 Definition of an intercalary month

त्रयोदशस्य चैत्रादिद्वादशानामियं भिदा ॥ ७ ॥
मेषाद्येकैकराशिस्फुटगतिदिनकृत्सङ्क्रमैकैकगर्भाः
चान्द्राश्चैत्रादिमासाः इह न यदुदरे सङ्क्रमः सोऽधिमासः ।
संसर्पः स्यात् स चांहस्पतिरुपरि यदि ग्रस्तसङ्क्रातियुग्मः
तौ चाब्दर्त्वङ्गभूतौ सह सुचिरभवौ सोऽधिमासोऽत्र पश्चात् ॥ ८ ॥

trayodaśasya caitrādidvādaśānāmiyaṃ bhidā || 7 ||
meṣādyekaikarāśisphuṭagatidinakṛtsaṅkramaikaikagarbhāḥ
cāndrāścaitrādimāsāḥ iha na yadudare saṅkramaḥ so'dhimāsaḥ |
saṃsarpaḥ syāt sa cāṃhaspatirupari yadi grastasaṅkrātiyugmaḥ
tau cābdartvaṅgabhūtau saha sucirabhavau so'dhimāso'tra paścāt ||8||

The following is the difference between the twelve lunar months, *Caitra* etc., and the thirteenth month.

Lunar months which include only one *saṅkrama* (transit) of the true Sun into any of *Meṣādi rāśis* are the usual *Caitra* etc. Those lunar months which do not include a *saṅkrama* are *adhimāsas*. The same *adhimāsa* is referred as a *saṃsarpa* if it is followed by an *aṃhaspati*, a lunar month which has two *saṅkramas* in it. These two lunar months, a *saṃsarpa* and an *aṃhaspati*, which always occur together in pairs and keep recurring, are considered to be part of the year and seasons. Here, that (month without a *saṅkrama*) which occurs later (after an *aṃhaspati*) is the [actual] *adhimāsa*.

In India, three types of calendrical system are followed: solar, lunar and luni-solar. While the solar calendar is based only on solar months and the lunar on lunar

---

[28] The term *ayana* essentially means motion; the words *saumya* and *uttara* refer to the north direction. Hence the term *uttarāyaṇa* refers to the northerly motion of the Sun.

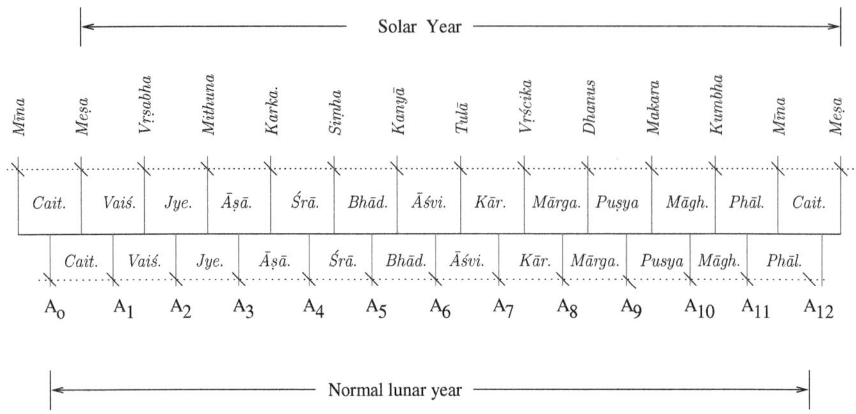

Fig. 1.6 A normal lunar year consisting of 12 lunar months.

months, the luni-solar is based on both. It is well known that a solar year has nearly 365.25 days. If twelve lunar months constitute a lunar year, it would have nearly 354 days. Hence, there is a shortage of nearly 11.25 days in such a lunar year. If the lunar and solar calendars are to be mutually linked, it is necessary to introduce an additional thirteenth month in some lunar years to align the lunar calendar with the solar one. The additional thirteenth month occurring in some lunar years is known as the 'intercalary month', or *adhikamāsa*. In the Indian calendrical system, there is a definite, well-defined procedure for introducing the *adhikamāsa*. This is described in the first half of the above verse.

In Fig. 1.6, the markings $A_0, A_1, A_2, A_3, \ldots$ below the horizontal line represent the occurrence of the new Moons. The vertical lines above the horizontal line marked with *Mīna, Meṣa*, etc., represent the *saṅkramas* or solar transits. A lunar month by definition is the time interval between two successive new Moons or full moons. Here we consider the *Amānta* system. Normally each lunar month will include one *saṅkrama*, i.e. transit of true Sun from one *rāśi* to another. Under this circumstance, both the solar and lunar year will consist of 12 months each. This situation is schematically sketched in Fig. 1.6.

However, when the rate of motion of the Sun is slower than average,[29] it may so happen that in between two sucessive new moons or full moons there is no *saṅkrama/saṅkrānti* (solar transit). Such a lunar month is called an *adhimāsa* (an intercalary month). Approximately, an *adhimāsas* occur once in three years. Somewhat more precisely, they occur 7 times in a span of 19 years. If an *adhimāsa* occurs, then that particular lunar year will have 13 lunar months.

Here it must be noted that in the Indian calendrical system, all the units of time, namely solar year, lunar year, solar month, lunar month, *adhimāsa*, day and *tithi*, are determined based on the positions of the true Sun and the Moon. This is in

[29] This will occur when Sun is near its apogee.

contrast to the modern calendar, where the day and the year are based on the position of the mean Sun.

Lunar year with an *adhimāsa*

In Fig. 1.6, A_0 refers to the *amāvāsyā* just preceding the *Meṣa saṅkrānti* that marks the beginning of a solar year. Similarly, A_{12} marks the *amāvāsyā* just before the next *Meṣa saṅkrānti*. By definition, a lunar year is the time period between these two *amāvāsyās*. It is evident from the above definition that a lunar year always commences before the solar year.

Usually there will be one *saṅkrānti* between two *amāvāsyās*. But as mentioned earlier, because of the non-uniform motion of the Sun and the Moon, during the course of a lunar year it may so happen that no *saṅkrānti* occurs between two *amāvāsyās*. In other words, there are two *amāvāsyās* occuring within a solar month. Such a situation is depicted in Fig. 1.7, wherein *Śrāvaṇa* happens to be the solar month in which two *amāvāsyās* occur.

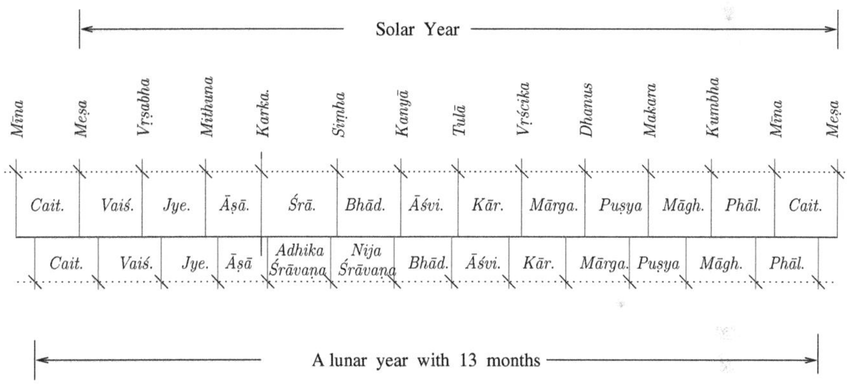

Fig. 1.7 A lunar year including an *adhikamāsa*.

The lunar year shown in Fig. 1.7 has 13 lunar months instead of the normal 12. The extra month is called an *adhikamāsa* (*adhika* = excess). Conventionally, the name of the *adhikamāsa* is the same as the name of the solar month with two *amāvāsyās*. The 'true' (= *nija*) lunar month with the same name follows this *adhikamāsa*. In the figure depicted, since it is the solar *Śrāvaṇa* which has two *amāvāsyās* we have marked the lunar month following the *Āṣāḍha-māsa* as *Adhika-śrāvaṇa* and the month following that as *Nija-śrāvaṇa*.

Further, it may be mentioned here that generally an *adhikamāsa* is considered inauspicious and no festivals are observed during that period. With this in mind, sometimes the adjectives *mala* (inauspicious) and *śuddha* (auspicious) are used instead of *adhika* and *nija*. As far as the pattern of occurrence is concerned, generally one observes that an *adhikamāsa* occurs after 33 months and the cycle repeats almost exactly once in 19 years. *Māgha-māsa* cannot be an *adhikamāsa* because the angular velocity of the Sun is quite large during this period (December–January)—since currently the Sun approaches its perihelion around 3rd January.

Lunar year with a *saṃsarpa* and an *aṃhaspati*

Very rarely, one also comes across a lunar year in which two *saṅkramās* take place within a lunar month. Such a lunar month is referred to as an *aṃhaspati* (see Fig. 1.8). It has been observed that if an *aṃhaspati* occurs then it is invariably preceeded and succeeded by an *adhimāsa*. Of these two *adhimāsas*, the one which occurs earlier is called a *saṃsarpa*. This *saṃsarpa–aṃhaspati* pair is taken to be part of the lunar year. In otherwords, they form part of the twelve *caitrādi* lunar months constituting a lunar year. The other lunar month without a *saṅkrānti* which occurs after an *aṃhaspati*, is considered to be an actual *adhimāsa*, a thirteenth lunar month which does not form part of the lunar calendar year. One such instance is shown in Fig. 1.8.

Here, the lunar month following *Bhādrapada* is without a *saṅkrānti*. Later, we have a lunar month with two *saṅkrāntis* (*Makara* and *Kumbha*), immediately followed by another lunar month without a *saṅkrānti*. In this case, the earlier lunar month without a *saṅkrānti* is the *saṃsarpa*, corresponding to *Āśvayuja*, and the later one with two *saṅkrāntis* is the *aṃhaspati*, corresponding to *Māgha*. Both are treated as other lunar months in that year, whereas the lunar month following *Māgha* is *Adhika-phālguna*.

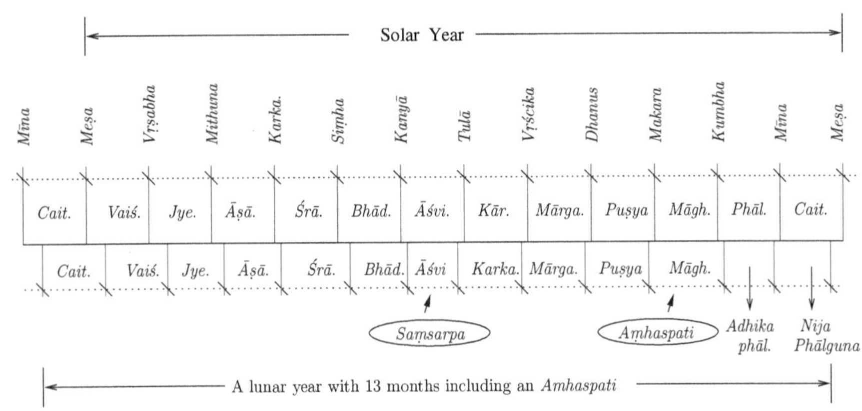

Fig. 1.8 A lunar year including a *saṃsarpa* and an *aṃhaspati*.

The reason for considering a *saṃsarpa* and an *aṃhaspati*—the former not having a *saṅkrānti* and the latter having two *saṅkrāntis*—to be an integral part of the lunar year, that is, treating them as any other lunar month, is explained in *Laghu-vivṛti* as follows:

यदा पुनः कबलीकृतसङ्क्रमद्वितयः चान्द्रमासः उपरि विद्यते तदा तत्पुरोगतोऽसौ सङ्क्रमरहितश्चान्द्रः न त्रयोदशेति व्यपदिश्यते चैत्रादिलक्षणसद्भावात्।[30]

यद्येवं पूर्वस्यासङ्क्रमस्य कथं चैत्रादित्वं तल्लक्षणाभावात् इति चेत्, उपरितनद्विसङ्क्रमायत्तत्वादिति ब्रूमः। अत एव हि तयोः सहभावित्वं अन्योन्याश्रयत्वं च। पूर्वस्य असङ्क्रमगर्भत्वदोषस्य तदूर्ध्वगतद्विसङ्क्रमत्वेन परिहृतत्वात्, पश्चात्तनासङ्क्रमे पुनर्नैवं तदुपरि कस्यचिदपि द्विसङ्क्रमस्य अभावात्।

अतः सङ्क्रमरहितयोः पश्चात्तन एव अधिमासः न पूर्वः संसर्पत्वात्; तयोः मध्यगतः पुनः द्विसङ्क्रमः अंहस्पतिरित्युच्यते; पौरस्त्यस्यासङ्क्रमत्वदोषस्य तदूर्ध्वगतानां च सङ्क्रमान्तरगर्भत्वस्य तत्पर्यन्तं प्रवृत्तत्वात् तेनैव परिहृतत्वात्; अत एव हि अस्य अंहस्पतित्वमपि। तत्र संसर्पांहस्पती द्वावपि अब्दर्त्वोरङ्गभूतौ सहभाविनौ सुचिरकालभाविनौ च । अधिमासः पुनः न कदाचिदपि तदङ्गभूत इति।

If there is a lunar month with two *saṅkramas* ahead, then the earlier lunar month without a *saṅkrama* is not considered to be a 13th lunar month as it is included into the *caitrādi* [usual 12 lunar months].

If you ask how it is that the lunar month without a *saṅkrama* is included in the *caitrādi* lunar months and not considered as an *adhimāsa*, we say it is because of the lunar month with two *saṅkramas* occurring later. In fact, because of this, there is co-occurrence and dependence on each other of these two months. The 'error' due to the earlier *asaṅkrama* (month with absence of transition) is nullified by the subsequent *dvisaṅkrama* (month with two transitions). Moreover, there is no *dvisaṅkrama* occurring after the later *asaṅkrama*.

Therefore, of the two lunar months in which *saṅkrama* does not take place, the later one is the *adhimāsa* and not the former; because the former one is *saṃsarpa*. The one which includes a *dvisaṅkrama* and which lies in between the two *asaṅkramas* is referred to as an *aṃhaspati* because it has nullified the error due to the *asaṅkrama* and the following inclusion of extra *saṅkramas* that had accumulated so far. Thus we find the quality of an *aṃhaspati*[31] in this. These two lunar months, the *saṃsarpa* and the *aṃhaspati*, which form part of the [lunar] year and the seasons, always occur in pairs and will keep recurring over time. But the *adhimāsa* never forms a part of that (lunar year and seasons).

१.७ अधिमासस्वरूपम्

1.7 Nature of the intercalary month

अर्केन्द्वोः स्फुटतः सिद्धाः त्रयो मासा मलिम्लुचाः ।
इति च ब्रह्मसिद्धान्ते मलमासास्त्रयः स्मृताः ॥ ९ ॥
द्वाभ्यां द्वाभ्यां वसन्तादिः मध्वादिभ्यामृतुः स्मृतः ।
मध्वादिभिस्तपस्यान्तैः वर्षं द्वादशभिः स्मृतम् ॥ १० ॥
त्रयोदशभिरप्येकं वर्षं स्यादधिमासके ।
स्वोत्तरेणाधिमासस्य सम्बन्धो मुनिभिः स्मृतः ॥ ११ ॥

30 The reading in the text happens to be चैत्रादिलक्षणाभावात्, which seems to be inappropiate.

31 The term अंहः refers to the error/sin; and the term पति refers to head/leader; hence the term अंहस्पति means the destroyer of errors.

भानुना लङ्घितो मासः ह्यनर्हः सर्वकर्मसु ।
षष्टिभिर्दिवसैर्मासः कथितो बादरायणैः ॥ १२ ॥
इति केषुचिदब्देषु सन्ति मासास्त्रयोदश ।
श्रूयते चर्तुयागादिष्वयमेव त्रयोदश ॥ १३ ॥

arkendvoḥ sphuṭataḥ siddhāḥ trayo māsā malimlucāḥ |
iti ca brahmasiddhānte malamāsāstrayaḥ smṛtāḥ || 9 ||
dvābhyāṃ dvābhyāṃ vasantādiḥ madhvādibhyāmṛtuḥ smṛtaḥ |
madhvādibhistapasyāntaiḥ varṣaṃ dvādaśabhiḥ smṛtam || 10 ||
trayodaśabhirapyekaṃ varṣaṃ syādadhimāsake |
svottareṇādhimāsasya sambandho munibhiḥ smṛtaḥ || 11 ||
bhānunā laṅghito māsaḥ hyanarhaḥ sarvakarmasu |
ṣaṣṭibhirdivasairmāsaḥ kathito bādarāyaṇaiḥ || 12 ||
iti keṣucidabdeṣu santi māsāstrayodaśa |
śrūyate cartuyāgādiṣvayameva trayodaśa || 13 ||

The three months (*saṃsarpa, aṃhaspati* and *adhimāsa*), which are obtained from the true motions of the Sun and the Moon, are impure (inauspicious); Therefore in *Brahmasiddhānta*, the three months are considered to be *malamāsas* (impure months).

The *madhvādimāsas*[32] in pairs are said to constitute the *vasantādi ṛtus*.[33] The *madhvādi* 12 months ending with *tapasya* is [generally] said to constitute a [lunar] year. If there is an *adhimāsa*, then even 13 months constitute a year. The *munis* (wise ones) have associated the *adhimāsa* with the later one.

The [lunar] month which has been by-passed by the Sun—the month in which a *saṅkrama* does not take place—is not suitable for any [auspicious] activities. The followers of *Bādarāyaṇa* have stated that the month consists of 60 days (*tithis*).

Thus we find that there are 13 months in certain years. It is this month, the *adhimāsa*, which is referred to as the 13th month in the context of seasons and sacrifices [in *śrutis*].

In verse 11, it is stated that 'the *munis* have associated the *adhimāsa* with the later one.' This statement is with reference to a lunar year in which both *saṃsarpa* and *aṃhaspati* occur. According to another view ascribed to Bādarāyaṇa, the *adhika* (extra) and *nija* (true) months together constitute a lunar month consisting of 60 *tithis*.

Reference to *adhimāsa* and *aṃhaspati* in the *Vedas*

There are several passages in the *Vedas* referring to the names of the months, seasons, etc.[34] Some of these passages present a list consisting of 13 names, while others present 14, whereas certain others present 24 names. While using the term *madhvādibhyām* in verse 10, the author presumably has the following passage occurring in *Taittirīya-saṃhitā* in his mind, which provides a list of names of the 12 regular lunar months and 2 special months.

---

[32] The names of the list of months commencing with *Madhu* is provided in the following section.

[33] The six *ṛtus* (seasons) are: *vasanta, grīṣma, varṣa, śarat, hemanta,* and *śiśira.*

[34] See for instance, *Ṛg-veda* 1.25.8; *Yajur-veda, Taittirīya-Brāhmaṇa* 3.10.1; *Itareya-Brāhmaṇa* 1.1.

मधुश्च माधवश्च शुक्रश्च शुचिश्च नभश्च नभस्यश्चेषश्चोर्जश्च सहश्च सहस्यश्च तपश्च तपस्यश्चोपयामगृहीतोस्यंसर्पोस्यंहस्पत्यायत्वा ॥

[*Taittirīya-saṃhitā* 1.4.18]

The twelve lunar months that form a part of the normal lunar year, the *caitradi*, are named as *madhu, mādhava, śukra, śuci, nabha, nabhasya, iṣa, ūrja, saha, sahasya, tapa* and *tapasya*. The two names *saṃsarpa* and *aṃhaspati* at the end of the list refer to the special lunar months discussed above. The former corresponds to the *asaṅkrama* (no transit) occurring just before the *aṃhaspati* and the latter is *aṃhaspati* itself, which is a *dvisaṅkrama* (two transits).

Commenting upon a different viewpoint held regarding the *adhimāsa* by those belonging to the school of Bādarāyaṇa, *Laghu-vivṛti* notes:

बादरायणाः पुनः अधिमासेन सह तदुत्तरं चान्द्रमासं तिथिषष्ट्यात्मकं प्रतिपन्नाः। तत्रापि तन्मासपूर्वार्धं सर्वकर्मस्वयोग्यत्वात् परित्यज्यत एव। अत एव हि केषुचित् पुस्तकेषु 'पूर्वार्धं तु परित्यज्य उत्तरार्धं प्रशस्यते' इति पठ्यते।

The Bādarāyaṇas combining an *adhimāsa* with the succeeding lunar month consider a month consisting of 60 *tithis*. Even then, the first half of that month is left out as it is not suitable for any [auspicious] activities. It is therefore stated in certain texts: 'The later half is considered leaving out the earlier one.'

१.८ दिव्यदिनादिः

1.8 Days of the God etc.

दिव्यं दिनं तु सौराब्दः पितॄणां मास ऐन्दवः ।
सर्वेषां वत्सरोऽह्नां स्यात् षष्ट्युत्तरशतत्रयम् ॥ १४ ॥

divyaṃ dinaṃ tu saurābdaḥ pitṝṇāṃ māsa aindavaḥ |
sarveṣāṃ vatsaro'hnāṃ syāt ṣaṣṭyuttaraśatatrayam || 14 ||

A solar year [of humans] is [said to constitute] a day of the Gods, and a lunar month of the *pitṛs*. An year consists of 360 days[35] [in their own measure] for all of them.

As mentioned earlier, a solar year is made of two *ayanas*. Thus six months, corresponding to the northern motion for us, are equal to a day for the *devas*; and six months, corresponding to the southern motion for us, forms the period of night for them. In the same way, it is stated that a lunar month consisting of two fortnights is one complete day for the *pitṛs*. Obviously, the bright and dark fortnights would form the day and night for the *pitṛs*.

---

[35] Here the day refers to solar day and (*sauradina*) and not the civil day (*sāvanadina*). The duration of a *sauradina* is equal to the time taken by the Sun to move by one degree along the ecliptic, and on an average this would be greater than that of a civil day.

१.९ ग्रहयुगभगणाः

1.9 Number of revolutions of planets in a *Mahāyuga*

दिव्याब्दानां सहस्राणि द्वादशैकं चतुर्युगम् ।
सूर्यस्य पर्ययास्तस्मादयुतघ्नरदार्णवाः ॥ १५ ॥
खाश्विदेवेषुसप्ताद्रिशराश्चेन्दोः, कुजस्य तु ।
वेदाङ्गाहिरसाङ्काश्विकराः, ज्ञस्य स्वपर्ययाः ॥ १६ ॥
नागवेदनभस्सप्तरामाङ्कस्वरभूमयः ।
व्योमाष्टरूपवेदाङ्गपावकाश्च बृहस्पतेः ॥ १७ ॥
अष्टाङ्गदस्रनेत्राश्विखाद्रयो भृगुपर्ययाः ।
भास्कराङ्गरसेन्द्राश्च शनेः, शश्युच्चपातयोः ॥ १८ ॥
नेत्रार्काष्टाहिवेदाश्च खखरामरदाश्विनः ।

divyābdānāṃ sahasrāṇi dvādaśaikaṃ caturyugam |
sūryasya paryayāstasmādayutaghnaradārṇavāḥ || 15 ||
khāśvideveṣusaptādriśarāścendoḥ, kujasya tu |
vedāṅgāhirasāṅkāśvikarāḥ, jñasya svaparyayāḥ || 16 ||
nāgavedanabhassaptarāmāṅkasvarabhūmayaḥ |
vyomāṣṭarūpavedāṅgapāvakāśca bṛhaspateḥ || 17 ||
aṣṭāṅgadasranetrāśvikhādrayo bhṛguparyayāḥ |
bhāskarāṅgarasendrāśca śaneḥ, śaśyuccapātayoḥ || 18 ||
netrārkāṣṭāhivedāśca khakharāmaradāśvinaḥ |

Twelve thousand years of the Gods correspond to one *caturyuga*. Therefore the number of revolutions of the Sun [in a *caturyuga*] is ten thousand multiplied by 432. The number of revolutions of the Moon is 57753320; and that of Mars is 2296864; the number of Mercury's own revolutions is 17937048; that of Jupiter is 364180. The number of revolutions of Venus is 7022268; that of Saturn is 146612; those of the apogee and the node of the Moon are 488122 and 232300 respectively.

It is mentioned that 12000 years of the Gods corresponds to one *caturyuga* (group of four *yugas*), which is also referred to as a *Mahāyuga*. Since one solar year is taken to be a day of the Gods, and a year is assumed to consist of 360 days, the number of revolutions made by the Sun in a *Mahāyuga* is equal to $12000 \times 360 = 4320000$. As the period of revolution of the Sun is the same as the sidereal year, the number of sidereal years in a *Mahāyuga* is also equal to 4320000.

The *yugas* that constitute a *Caturyuga/Mahāyuga*

The term *caturyuga* refers to a group of four *yugas*. The four *yugas* which constitute a *caturyuga* are *Kṛtayuga, Tretāyuga, Dvāparayuga* and *Kaliyuga*. The details regarding the periods of these *yugas* are not presented in the text or in the commentaries *Laghu-vivṛti* or *Yukti-dīpikā*. But nevertheless it is an interesting division of time, and hence we present the details—as given in *Sūryasiddhānta* and other important works of astronomy—in the form of Table 1.4.

No.	Name of *Yuga*	Duration (years)
1	*Kṛtayuga*	1728000
2	*Tretāyuga*	1296000
3	*Dvāparayuga*	864000
4	*Kaliyuga*	432000

Table 1.4 The four *yugas* constituting a *Mahāyuga* and their durations as given in *Sūryasiddhānta*.

It may be noted that the periods of these four *yugas* are in the ratio 4:3:2:1 respectively. Because of this pattern the total number of years in a *Mahāyuga* happens to be just 10 times the number of years in the *Kaliyuga*. However, some scholars are of the view that Āryabhaṭa might have employed a different scheme for the division of the four *yugas*, in which all of them are taken to be of equal periods. This view is based on the use of the term '*pāda*', which means quarter, in the expression '*kalpāderyugapādāḥ ga ca*' (... [and] three-fourths of the *yuga* [have elapsed] since the beginning of the *Kalpa* [till the beginning of the current *Kaliyuga*]).[36]

Revolutions completed by the planets in a *Mahāyuga*

In verses 16–18a, Nīlakaṇṭha gives the number of revolutions completed by the planets in a *Mahāyuga*. In doing so, he has adopted the *Bhūta-saṅkhyā*[37] system of numeration. Table 1.5 presents the number of revolutions completed by the planets in a *Mahāyuga*, along with their Sanskrit equivalents as given in the text.

Planet	Number of revolutions in *bhūtasaṅkhyā-paddhati*	in numerals
Sun	अयुतघ्नरदार्णवाः	4320000
Moon	खाश्विदेवेषुसप्ताद्रिशराः	57753320
Mercury	नागवेदनभस्सप्तरामाङ्कस्वरभूमयः	17937048
Venus	अष्टाङ्गदस्रनेत्राश्विखाद्रयः	7022268
Mars	वेदाङ्गाहिरसाङ्काश्विकराः	2296864
Jupiter	व्योमाष्टरूपवेदाङ्गपावकाः	364180
Saturn	भास्कराङ्गरसेन्द्राः	146612
Moon's apogee	नेत्रार्काष्टाहिवेदाः	488122
Moon's node	खखरामरदाश्विनः	232300

Table 1.5 The number of revolutions completed by the planets in a *Mahāyuga*.

36 काहो मनवो ढ, मनुयुगाः श्ख, गतास्ते च, मनुयुगाः छ्ना च।
कल्पादेर्युगपादा ग च, गुरुदिवसाच्च भारतात् पूर्वम्॥– {AB 1976} (*Gītikāpāda*, 5), p. 9

37 For details regarding the *Bhūta-saṅkhyā* system, the reader is referred to Appendix A.

Nīlakaṇṭha's modification of the traditional model for Mercury and Venus

While presenting the number of revolutions completed by the planets in a *Mahāyuga*, Nīlakaṇṭha makes a clear departure from the traditional planetary model in the case of Mercury and Venus. This is clearly indicated by the use of the adjective *sva* (one's own) with the word *paryaya* (revolution). This special usage is commented upon in *Laghu-vivṛti* thus:

ज्ञस्य स्वपर्ययाः पुनः नागवेदनभस्सप्तरामाङ्कस्वरभूमयः। अत्र स्वशब्देन पर्यायाणां भास्कराचार्याद्यभिमतं स्वशीघ्रोच्चसम्बन्धित्वं बुधस्य निरस्तम्।...भृगुपर्ययास्तु अष्टाङ्गदस्रनेत्राश्विखाद्रयः।

And the number of Mercury's own revolutions is 17937048. Here by the use of the word *sva* (its own), the association of this number of revolutions with the *śighrocca* of Mercury as done by Bhāskarācārya and others has been discarded. ... And the number of revolutions of Venus is 7022268.

The same idea is highlighted in *Yukti-dīpikā* as follows:

बुधभार्गवयोर्वृत्तं शैघ्र्यनीचोच्चवृत्तताम् ।
याति, तस्मिन् स्वगत्यैव चरतो बुधशुक्रयोः ॥
युगोत्थाः पर्ययास्त्वेते तत् स्वशब्देन दर्शितम् ।[38]

The circles of Mercury and Venus (the actual orbits in which they move) become the *śaighrya-nīcocca-vṛtta* [in the *śighra-saṃskāra*]. The fact that the numbers of revolutions given here are the revolutions of the Mercury and Venus, moving in these circles with their own velocities, is indicated by the word *sva*.

In fact, this novel view regarding the motion of interior planets—according to which what were traditionally identified as the *śighroccas* are now identified with the mean planets themselves—became the starting point of a major revision of the traditional planetary model. This will be explained in Chapter 2 as well as in Appendix F.

१.१० युगसावनदिनादिः

1.10 Number of civil days in a *Mahāyuga* etc.

खखाक्षात्यष्टिगोसप्तस्वरेषुशशिनो युगे ॥ १९ ॥
सावना दिवसाः, चार्क्षा मार्ताण्डभगणाधिकाः ।
अधिमासाः खनेत्राग्निरामनन्देषुभूमयः ॥ २० ॥
अयुतघ्नाब्धिवस्वेकशरा मासा रवेः स्मृताः ।
खव्योमेन्दुयमाष्टाभ्रतत्त्वतुल्यास्तिथिक्षयाः ॥ २१ ॥
खखषण्णवगोनन्दनेत्रशून्यरसेन्दवः ।
तिथयः, चान्द्रमासाः स्युः सूर्येन्दुभगणान्तरम् ॥ २२ ॥

khakhākṣātyaṣṭigosaptasvareṣuśaśino yuge || 19 ||
sāvanā divaṣāh, cārkṣā mārtāṇḍabhagaṇādhikāḥ |

---

[38] {TS 1977}, pp. 9–10.

adhimāsāḥ khanetrāgnirāmanandeṣubhūmayaḥ || *20* ||
ayutaghnābdhivasvekaśarā māsā raveḥ smṛtāḥ |
khavyomenduyamāṣṭābhratattvatulyāstithikṣayāḥ || *21* ||
khakhaṣaṇṇavagonandanetraśūnyarasendavaḥ |
tithayaḥ, cāndramāsāḥ syuḥ sūryendubhagaṇāntaram || *22* ||

The number of civil days in a *Mahāyuga* is 1577917500; and the number of sidereal days (*ārkṣāḥ*) is [equal to] this number increased by the number of revolutions of the Sun. The number of *adhimāsas* is 1593320.

The number of solar months is stated to be the product of 5148 and *ayuta* (10000). The number of *kṣayatithis* (unreckoned *tithis*) is 25082100. The number of [actual] lunar *tithis* is 1602999600; the number of lunar months will be equal to the difference in the number of revolutions of the Sun and the Moon.

The number of civil days in a *Mahāyuga* is the number of sunrises that take place in it. Similarly the number of sidereal days is equal to the number of star-rises that take place in a *Mahāyuga*. Since the stars do not have any eastward motion, and the Sun completes one full revolution eastwards once in a sidereal year, the number of sidereal days in a sidereal solar year will be greater than the number of civil days by one unit. Hence, in a *Mahāyuga*, the total number of sidereal days will be exceeding the total number of civil days by exactly the number of solar years or the revolutions of the Sun. That is,

$$\begin{aligned}\text{Sidereal days} &= \text{Civil days} + \text{No. of Sun's revolutions}\\ &= 1577917500 + 4320000\\ &= 1582237500.\end{aligned}$$

On the other hand, if we know the number of sidereal days in a *Mahāyuga* then the number civil days (*sāvana-dinas*) may be obtained by subtracting the number of revolutions of the Sun from the former:

$$\text{Civil days} = \text{Sidereal days} - \text{No. of Sun's revolutions}. \qquad (1.1)$$

As there are 12 solar months in a solar year, the number of solar months in a *Mahāyuga* is 51480000. The number of lunar months is the number of conjuctions of the Sun and the Moon. Hence, it is equal to the difference in the number of revolutions of the Sun and the Moon, which is 53433320. As there are 30 *tithis* in a lunar month, the number of lunar *tithis* is 30 times this number, which is 160299960. As an intercalary month or *adhimāsa* is introduced to match the solar and lunar calendars, the number of *adhimāsas* is the difference between the number of lunar and solar months and is 1593320.

As there are 30 *tithis* in a lunar month whose duration is less than 30 civil days, the average duration of a *tithi* is less than a civil day. Because of this, a certain number of *tithis*, known as '*kṣayatithis*', or '*avamadinas*', are to be dropped from the calendar in order to have a concurrence between the number of civil days and the number of reckoned *tithis* in a *yuga*. In fact, the number of *kṣayatithis* in a *Mahāyuga* is 25082100, which is exactly the difference in number between the *tithis* and the civil days (see Table 1.6).

Number of risings and settings of planets in a *Mahāyuga*

By definition, the number of sunrises in a *Mahāyuga* is the same as the number of civil days. Hence from (1.1) we have:

$$\text{No. of sunrises} = \text{Sidereal days} - \text{No. of Sun's revolutions.} \qquad (1.2)$$

A similar relation must be valid for other planets also. In *Yukti-dīpikā* it is observed:

रव्यादेरुदयाः स्वस्वभगणोनार्क्षपर्यायाः ।[39]

The numbers of risings of the Sun and other planets in a *Mahāyuga* are equal to their own revolutions subtracted from the number of revolutions of the stars (sidereal days).

$$\text{No. of planet-rises} = \text{Sidereal days} - \text{No. of planet's revolutions.} \qquad (1.3)$$

For the sake of convenience we present the numbers for various relevant time units in a *Mahāyuga* stated above in the form of a table (see Table 1.6).

No. (in a *Mahāyuga*)	Rationale behind the number
No. of solar months	$= 12 \times$ No. of solar years in a *Mahāyuga*
	$= 12 \times 4320000$
	$= 51840000$.
No. of lunar months	= Difference in the No. of revolutions made by the Sun and the Moon in a *Mahāyuga*
	$= 57753320 - 4320000$
	$= 53433320$.
No. of *adhimāsas*	= Difference in the No. of lunar months and solar months in a *Mahāyuga*
	$= 53433320 - 51840000$
	$= 1593320$.
No. of *avamadinas* or *kṣayatithis*	= Difference in the No. of lunar days and civil days in a *Mahāyuga*
	$= (53433320 \times 30) - 1577917500$
	$= 25082100$.
No. of normal *tithis*	$= 30 \times$ No. of lunar months in a *Mahāyuga*
	$= 30 \times 53433320$
	$= 1602999600$.

Table 1.6 Number of solar months, lunar months, *adhimāsas*, *avama-dinas* etc. in a *Mahāyuga*.

In *Yukti-dīpikā*, the above set of verses (19b–22) of *Tantrasaṅgraha* are commented upon elaborately. This commentary runs to more than 100 verses and touches upon several related issues. In the following we make a brief mention of some of them:

---

[39] {TS 1977}, p. 10.

1. *Kālapramāṇādhāraḥ* (The basis for reckoning time): Having mentioned the cause for the eastward motion of the planets, the following interesting remark is made regarding the basis for different units employed for reckoning time.

   ग्रहाणां प्राङ्मुखी भुक्तिः स्वव्यापारकृतोदिता ।
   दृक्संवादविशुद्धासौ गृह्यते गोलवित्तमैः ॥
   यत्कृतो निखिलः कालः वर्षमासदिनात्मकः ।
   ग्रहभुक्तिं विना यस्मात् न कालो जातु जायते ॥[40]

   The eastward motion of the planets is stated to be due to their own *vyāpara* (action) [and not due to the *pravaha-vāyu*]. This (eastward motion) is noted down [in terms of revolution numbers] by experts in spherics only after careful observation and verification. The different units of time like the year, month, day etc. are all dependent on the [motion of planets] because without the motion of planets the concept of time does not arise.

2. *Bhagaṇa-parīkṣaṇam* (Verification of the revolution numbers): There is a lengthy, detailed and thorough discussion on this topic. Starting with observation and inference, the different methods employed in measuring and verifying these, such as conjuctions with celestial objects, are described at great length.
3. *Bhagaṇa-nānātvopapattiḥ* (Reconciliation of differences in parameters between different texts): Having described the procedure for finding/verifying the revolution numbers etc., Śaṅkara Vāriyar proceeds to reconcile the discrepancies one may observe, when comparing different texts. He attributes the differences to variation in the accuracy of measurement. Further, he emphasizes that the purpose of a text is only to acquaint the reader with the procedures and not to give him a false impression about the ultimate accuracy of the parameter values mentioned therein. It is precisely for this reason that the parameter values are specified in a separate section (*saṅkhyābhāga*) in texts such as *Āryabhaṭīya*.

   उच्चनीचपरिध्यादिः अर्थापत्त्या नियम्यते ।
   अतीन्द्रियाश्च भगणाः नियम्यन्तेऽनुमानतः ॥
   तेषां शास्त्रेषु नानात्वं परीक्षातारतम्यतः ।
   सुग्रहत्वाय तत्सिद्धं सर्वं यत्रोपदिश्यते ॥
   अतः सर्वेषु शास्त्रेषु न्यायमार्गप्रदर्शिषु ।
   सङ्ख्याभागं पृथक्कृत्य बभणुर्गोलचिन्तकाः ॥[41]

   The dimensions of the epicycles etc. are fixed by the process of *arthāpatti* (presumptive reasoning). The number of revolutions of planets which are not directly accessible to the senses are fixed through the process of inference.

   The difference in the number of revolutions from text to text is due to differences in measurement. In order to facilitate understanding [of future generations], whatever is obtained is stated as such.

   Therefore (since the parameters have to be updated from time to time), in all the texts which purport to explain the rationale of the procedures, the experts in spherics have

---

[40] {TS 1977}, p. 11.

[41] {TS 1977}, pp. 17–18.

stated the parameter values in a separate section called *saṅkhyā-bhāga*, distinguishing it from the rest.

Besides these issues which are of immediate relevance in understanding the procedures described, certain other topics such as the theory of the rotation of the Earth,[42] the distance moved by the planets in their orbits, etc., are also discussed. According to the general principle stated in most Indian texts, *all the planets move equal distances in equal intervals of time*, or, equivalently, *the linear velocity of all the planets is the same*. This is reiterated in *Yukti-dīpikā*.

१.११ अहर्गणानयनम्

1.11 Finding the number of days elapsed since an epoch

द्वादशघ्नान् कलेरब्दान् मासैश्चैत्रादिभिर्गतैः ।
संयुक्तान् पृथगाहत्याप्यधिमासैस्ततो हृतैः ॥ २३ ॥
सौरमासैर्युगोक्तैस्तैः अधिमासैर्युतान् गतैः ।
मासांश्च त्रिंशता हत्वा तिथीर्युक्त्वा गताः पृथक् ॥ २४ ॥
तिथिक्षयैर्निहत्यातो युगोक्ततिथिभिर्हृतान् ।
अवमाञ्छोधयेच्छेषः सावनो द्युगणः कलेः ॥ २५ ॥
सप्तभिः क्षपिते शेषात् शुक्रादिः स्याद्दिनाधिपः ।

dvādaśaghnān kalerabdān māsaiścaitrādibhirgataiḥ |
saṃyuktān pṛthagāhatyāpyadhimāsaistato hṛtaiḥ || 23 ||
sauramāsairyugoktaistaiḥ adhimāsairyutān gataiḥ |
māsāṃśca triṃśatā hatvā tithīryuktvā gatāḥ pṛthak || 24 ||
tithikṣayairnihatyāto yugoktatithibhirhṛtān |
avamāñchodhayecchesaḥ sāvano dyugaṇaḥ kaleḥ || 25 ||
saptabhiḥ kṣapite śeṣāt śukrādiḥ syāddinādhipaḥ |

The number of years elapsed since the beginning of the *Kaliyuga* multiplied by twelve and added to the *caitrādi* lunar months elapsed in the present year [is stored separately and] is multiplied by *adhimāsas* and divided by the number of solar months in a *Mahāyuga*. [This gives the *adhimāsas* elapsed.] This is added to the number of solar months elapsed, multiplied by 30 and the resultant is added to the number of *tithis* elapsed [in the current lunar month] and stored separately (A').

This (A') is multiplied by the number of *kṣayatithis*, and divided by the total number of *tithis* in a *Mahāyuga*, and this quantity, referred to as the *avama*, has to be subtracted

---

[42] After giving the number of sidereal days in a *Mahāyuga*, Śaṅkara Vāriyar states that this number is exactly the same as the number of eastward revolutions made by the Earth in a *Mahāyuga* (*Yukti-dīpikā*, Chapter 1, verse 73):

आर्क्षाः खखेषुसप्ताग्निनेत्राब्ध्यष्टेषुभूमयः । ऋक्षपर्ययतुल्यं च प्राङ्मुखं भ्रमणं भुवः ॥

The number of sidereal days is equal to 1582237500. The number of eastward revolutions of the Earth is same as the revolutions made by the stars [westwards].

This idea of the rotation of the Earth (eastwards) has been given by Śaṅkara Vāriyar as perhaps a tribute to Āryabhaṭa, who proposed the idea for the first time in the Indian tradition.

[from A' itself]. The result is the number of civil days [*Ahargaṇa*] elapsed since the beginning of the *Kaliyuga*. From the remainder obtained by dividing [the *Ahargaṇa*] by 7, the Lord of the day, beginning with *śukra* is to be found.

The term *Ahargaṇa*[43] literally means a count of days. It is a positive integer which gives the number of civil days that have elapsed since a given epoch,[44] till the date for which the *Ahargaṇa* is calculated.[45] In *Tantrasaṅgraha*, the epoch has been chosen to be the beginning of the *Kaliyuga*. Hence, the *Ahargaṇa* computed by the procedure given in the text gives the number of civil days that have elapsed since the beginning of the *Kaliyuga*, which is taken to be the sunrise of February 18, 3102 BCE as per the Julian calendar.

In the following, we shall explain the procedure for finding the *Ahargaṇa*, as given in the above verses, and this will be followed by a few illustrative examples.

Procedure for finding *Ahargaṇa*

Let p represent the number of years that have elapsed since the beginning of the *Kaliyuga* and q represent the number of lunar months that have elapsed since the beginning of the present lunar year. Then the quantity

$$m = 12p + q, \tag{1.4}$$

represents the number of solar months that have elapsed since the beginning of the *Kaliyuga*, which is also the number of lunar months excluding the number of *adhimāsas*. The number of *adhimāsas* that have elapsed since the beginning of the *Kaliyuga* till the desired date is found from the number of *adhimāsas* in a *Mahāyuga*, and employing the rule of three:

$$\begin{aligned} 51840000 &: 1593320 \\ m &: ? \end{aligned} \tag{1.5a}$$

If a be the number of *adhimāsas* elapsed, then

$$a = \frac{m \times 1593320}{51840000}. \tag{1.5b}$$

The number of lunar months l that have elapsed since the beginning of the *Kaliyuga* is given by

$$l = m + \lfloor a \rfloor, \tag{1.6}$$

---

43 The word *ahaḥ* means a day and the term *gaṇa* refers to a group.

44 The choice of the epoch can be the beginning of the *kalpa*, the beginning of the *Kaliyuga* or any desired date on which the planetary positions are known.

45 The computation of the *Ahargaṇa* plays a crucial role in determining the mean positions of the planets, as will be seen in Section 1.12.

where $[a]$ is the integral part of a. To find the number of *tithis* that have elapsed since the beginning of the *Kaliyuga* till the desired date, we need to simply multiply l by 30 and add to this the number of *tithis* that have elapsed in the current lunar month. If s be the number of *tithis* that have elapsed in the present month, then the total number of *tithis* A' that have elapsed till the required date since the beginning of the *Kaliyuga* is given by

$$A' = l \times 30 + s. \tag{1.7}$$

Once again, by the rule of three, the number of *kṣayatithis* elapsed since the beginning of the *Kaliyuga* is found:

$$\begin{aligned} 1602999600 &: 25082100 \\ A' &: ? \end{aligned} \tag{1.8a}$$

If k is the number of *kṣayatithis* that have elapsed since the beginning of the *Kaliyuga*, then

$$k = \frac{A' \times 25082100}{160299600}. \tag{1.8b}$$

Now the *Ahargaṇa* A, the number of civil days elapsed since the beginning of the *Kaliyuga* is given by

$$A = A' - [k], \tag{1.9}$$

where $[k]$ is the integral part of k. There is a possibility of round off errors which may occur at different stages in the computation of the *Ahargaṇa*. We discuss these before moving on to some illustrative examples for finding the *Ahargaṇa*.

Resolution of likely errors in the calculation of the *Ahargaṇa*

In the procedure for finding *Ahargaṇa*, or the number of civil days elapsed, we round off the fractional part and use the integers in further calculation, at least in two places, namely the computation of the *adhimāsas* and *kṣayatithis*. In doing so, it is quite likely that this rounding off may lead to errors. The following discussion would be useful in removing the errors.

1. **Error in the computation of *adhimāsas***: The number of *adhimāsas* obtained using (1.5*b*), has a fractional part that is indicative of the proximity of the *adhimāsa* to the date for which *Ahargaṇa* is calculated. The closer the value of the fraction to unity, the closer the *adhimāsa* will be to the date for which the *Ahargaṇa* is being computed. As per the calculational procedure, we have to use only the number of *adhimāsas* that have completely elapsed and hence we round off the number and choose the closest integer. This could introduce an error at times, which can be easily dealt with as shown in the examples discussed below.
2. **Error in the computation of *kṣayatithis***: The average duration of a *tithi* is less than that of a civil day. In the computation of the *Ahargaṇa* we are trying to find the number of civil days elapsed from the measure of *tithis* elapsed. From the

computed value of the number of *tithis* elaspsed, the number of *kṣayatithis* (k) has to be computed in order to obtain the *Ahargaṇa*. The quantity k obtained using (1.8*b*) essentially represents the excess of the *tithis* that has to be subtracted from the total number of *tithis*, A', elapsed since the beginning of the *Kaliyuga*. As we are interested only in the integral number of *tithis* that have to be subtracted (to get A from A'), we round off k to the nearest integer. However, in doing so, when the fraction is close to unity, an error is likely to occur. This can be at most one day, and can be easily rectified by comparing the weekday that is obtained from the computation, and the actual weekday for which the *Ahargaṇa* is being computed. The idea behind this is explained below.

3. **Fixing the error**: In *Tantrasaṅgraha* the beginning of the *Kaliyuga* is taken as mean sunrise on February 18, 3102 BCE, a Friday. This fact, is implicit in verse 26. Suppose that the value of A obtained, when divided by 7, leaves a remainder of 0, 1, 2, ..., 6. Then it means that the day on which A has been computed must be Friday, Saturday, ..., Thursday. If the actual weekday differs from the computed one, then it implies that there has been an error in rounding off k. The new k is obtained by adding ± 1 to the old k. This is demonstrated in Example 3 below.
4. **Popular eras and conversion factors**: Different kinds of eras (*saṃvats*) have been popular in different parts of India.[46] Most of the Indian *pañcāṅgas* would mention the *Śaka* and *Vikrama* besides the most popular *Kali saṃvat*. The relationship between the three is given by

$$\textit{Śaka}\,0 = \textit{Vikrama}\,135 = \textit{Kali}\,3179.$$

Regarding the convention adopted in the *pañcāṅgas*, it may be mentioned that, whether it is *Śaka*, *Vikrama* or *Kali*, the value given always corresponds to the number of years elapsed since the commencement of epoch, and not the number of the year currently in progress.

Example 1:

Find the *Kalyahargaṇa* corresponding to *Phālguṇa-kṛṣṇa-trayodaśī*, *Śaka* 1922 (March 22, 2001 CE).

Number of *Kali* years elapsed, p	$= 1922 + 3179$
	$= 5101$
Number of lunar months elapsed in the present year, q	$= 11$
No. of lunar months elapsed (excluding *adhimāsas*), m	$= (5101 \times 12) + 11$
	$= 61223$
No. of *adhimāsas* corresponding to m lunar months, a	$= \dfrac{61223 \times 1593320}{51840000}$

[46] For instance, the *Kollam* era has been popular in Kerala, whereas the *Śaka and Vikrama* eras have been popular in northern India.

	$= 1881.70969$
Since we are interested in the integral part, we take a	$= 1881$
No. of lunar months elapsed (including *adhimāsas*), l	$= 61223 + 1881$
	$= 63104$
No. of *tithis* elapsed in the present lunar month, s	$= 15+12 = 27$
No. of *tithis* elapsed (including *kṣayatithis*), A'	$= (63104 \times 30) + 27$
	$= 1893147$
Number of *kṣayatithis*, k (corresponding to A')	$= \dfrac{1893147 \times 25082100}{1602999600}$
	$= 29622.03008$
We round off the above fraction and take k to be	$= 29622$
Kalyahargaṇa is given by, A $(= A' - k)$	$= 1893147 - 29622$
	$= 1863525$
	$= (266217 \times 7) + 6$

The remainder 6 implies that the day has to be a Thursday. March 22, 2001 happens to have been a Thursday, and hence the computed value of the *Ahargaṇa* is correct. Thus the number of civil days elapsed since the beginning of the *Kaliyuga* till *Phālguṇa-kṛṣṇa-trayodaśī, Śaka* 1922 is 1863525.

Example 2:

Find the *Kalyahargaṇa* corresponding to *Nija-āṣāḍha-kṛṣṇa-navamī, Śaka* 1891 (August 6, 1969 CE).

Number of *Kali* years elapsed, p	$= 1891 + 3179$
	$= 5070$
Number of lunar months elapsed in the present year, q	$= 3$
No. of lunar months elapsed (excluding *adhimāsas*), m	$= (5070 \times 12) + 3$
	$= 60843$
No. of *adhimāsas* corresponding to m lunar months a	$= \dfrac{60843 \times 1593320}{51840000}$
	$= 1870.03$
Since we are interested in the integral part, we take a	$= 1870$
No. of lunar months elapsed (including *adhimāsas*), l	$= 60843 + 1870$
	$= 62713$
No. of *tithis* elapsed in the present lunar month s	$= 15 + 8 = 23$
No. of *tithis* elapsed (including *kṣayatithis*), A'	$= (62713 \times 30) + 23$
	$= 1881413$

Number of *kṣayatithis* k (corresponding to A') $= \dfrac{1881413 \times 250821000}{1602999600}$

$= 29438.42844$

We round off the above fraction and take k to be $= 29438$

Kalyahargaṇa is given by A $(= A' - k)$ $= 1881413 - 29438$

$= 1851975$

$= (264567 \times 7) + 6$

The remainder 6 implies that the day has to be a Thursday. But August 6, 1969 happened to be a Wednesday. Hence the computed value of the *Ahargaṇa* is incorrect by one day. The error is due to the error in the computation of k. Hence we round off k to the next integer and take its value to be 29439, and so the actual *Ahargaṇa* is given by $1881413 - 29439 = 1851974$. Thus the number of civil days elapsed since the beginning of the *Kaliyuga* till *Nija-āṣāḍha-kṛṣṇa-navamī, Śaka* 1891 is 1851974.

Note:

1. In this example, a was found to be 1882.0170. A very small value of the decimal part indicates that an *adhimāsa* has just occurred. In fact, in the previous lunar month there was no *saṅkrānti* and it was an *adhimāsa* referred to as *Adhika-āṣāḍham.*
2. In Example 1, rounding off the value of k obtained to the closest integer gave the correct value of the *Ahargaṇa.* But in this example when it was rounded off to the closest integer there was an error in the *Ahargaṇa* by one day and it was fixed by comparing the result obtained with the day of the week. The source of the error can be explained as follows.
3. By using the rule of three for finding the *adhimāsas* and *kṣayatithis,* it is implicitly assumed that they occur periodically. Since they do not occur with exact periodicity, and it is fixed depending upon the occurrence or absence of true *saṅkrānti* in the true lunar month, care has to be taken when the value of a is close to an integer. If there is an error in the choice of a, the *Ahargaṇa* would differ from the actual value by nearly 30 days, and if there is an error in the choice of k then we will miss the *Ahargaṇa* by one day. These errors can be easily fixed from the knowledge of the occurrence or otherwise of the *adhimāsa* near the desired date and the day of the week respectively.

In the last example we find out the *Kalyahargaṇa* corresponding to August 18, 1947 CE. This forms an interesting example, for there was an *adhimāsa* in the year 1947 CE.

Example 3:

Find the *Kalyahargaṇa* corresponding to *Nija-śrāvaṇa-śukla-dvitīyā, Śaka* 1869 (August 18, 1947 CE).

Number of *Kali* years elapsed, p	$= 1869 + 3179$
	$= 5048$
Number of lunar months elapsed in the present year, q	$= 4$
No. of lunar months elapsed (excluding *adhimāsas*), m	$= (5048 \times 12) + 4$
	$= 60580$
No. of *adhimāsas* corresponding to m lunar months, a	$= \dfrac{60580 \times 1593320}{51840000}$
	$= 1861.9468$
Since the decimal part is close to 1, we take a to be	$= 1862$
No. of lunar months elapsed (including *adhimāsas*), l	$= 60580 + 1862$
	$= 62442$
No. of *tithis* elapsed in the present lunar month, s	$= 1$
No. of *tithis* elapsed (including *kṣayatithis*), A'	$= (62442 \times 30) + 1$
	$= 1873261$
Number of *kṣayatithis*, k (corresponding to A')	$= \dfrac{1873261 \times 25082100}{1602999600}$
	$= 29310.87427$
We round off the above fraction and take k to be	$= 29311$
Kalyahargaṇa is given by $A\ (= A' - k)$	$= 1873261 - 29311$
	$= 1843950$
	$= (263421 \times 7) + 3$

The remainder 3 implies that the day has to be a Monday. August 18, 1947 was a Monday. Thus the number of civil days elapsed since the beginning of the *Kaliyuga* till *Nija-śrāvaṇa-śukla-dvitīyā, Śaka* 1869 is 1843950.

Note: In this example, a was found to be 1861.9468. A very high value of the decimal part indicates that an *adhimāsa* was in the vicinity. In fact, an *adhimāsa* had just then occurred and that is why a has to be taken to be 1862, and the present month is referred to as *Nija-śrāvaṇa*, preceded by an *Adhika-śrāvaṇa*.

१.१२ अहर्गणात् मध्यमानयनम्

1.12 Finding the mean positions from *Ahargaṇa*

द्युगणात् भगणाभ्यस्तात् भूदिनैर्भगणा गताः ॥ २६ ॥
द्वादशघ्नाश्च तैरेव शेषादाप्ताश्च राशयः ।

मुहुश्च त्रिंशता षष्ट्या निघ्नाद् भागादयश्च तैः ॥ २७ ॥
कल्यादिध्रुवयुक्तं तत् मध्यं स्यादुदयोद्भवम् ।

dyugaṇāt bhagaṇābhyastāt bhūdinairbhagaṇā gatāḥ || 26 ||
dvādaśaghnāśca taireva śeṣādāptāśca rāśayaḥ |
muhuśca triṃśatā ṣaṣṭyā nighnād bhāgādayaśca taiḥ || 27 ||
kalyādidhruvayuktaṃ tat madhyaṃ syādudayodbhavam |

The *Ahargaṇa* multiplied by the revolutions and divided by the total number of civil days [in a *Mahāyuga*] gives the revolutions that have elapsed. Multiplying the remainder by twelve [and dividing by the same divisor], the *rāśis* are obtained. The remainder again multiplied by 30 and 60 [and dividing by the same divisor] gives the degrees etc. elapsed. The result, added to the mean longitude of the planet at the beginning of the *Kaliyuga*, gives the mean planet at sunrise.

It is straightforward to obtain the mean longitudes of the planets from the *Ahargaṇa*. In Indian astronomy, the longitudes are generally expressed in *rāśis*, *aṃśas*, *liptās* and *viliptās*. (i.e., signs, degrees, minutes and seconds). In what follows we first explain the procedure described in the above verses, before taking up a numeric example.

Let A be the *Ahargaṇa* and N the number of revolutions completed by the planet in a *Mahāyuga*. Then, the number of revolutions including the fractional part covered by the planet since the epoch, till the mean sunrise, is given by

$$n = \frac{A \times N}{1577917500} = I_1 + f_1, \tag{1.10}$$

where I_1 represents the integral part of n, and f_1 the fractional part. The integral part gives the number of revolutions completed by the planet since the beginning of *Kaliyuga*. It is from the fractional part f_1 that the *rāśis* etc. elapsed are obtained. For this, the fractional part f_1 is first multiplied by 12 and this gives the number of *rāśis* elapsed in the present revolution. Let

$$f_1 \times 12 = I_2 + f_2.$$

Here the integral part I_2 gives the number of *rāśis* that the planet has traversed. The fractional part f_2 multiplied by 30 gives the number of degrees elapsed in the present *rāśi*. Let

$$f_2 \times 30 = I_3 + f_3,$$

where the integral part I_3 gives the number of degrees covered by the planet in the present *rāśi*. The integral part of the fractional part f_3 multiplied by 60 gives the number of minutes covered in the present degree, and so on. Hence the mean longitude of the planet is given by

$$\theta_0 = I_2 \text{ signs} + I_3 \text{ degrees} + I_4 \text{ minutes}, \tag{1.11}$$

where I_4 represents the integral part of $f_3 \times 60$.

Illustrative example

Now we illustrate the above procedure with an example. Suppose we want to find the mean longitude of the Moon at mean sunrise on January 14, 2002. The *Ahargaṇa* A corresponding to this date is found to be 1863823. The number of revolutions N completed by the Moon in a *Mahāyuga* is given to be 57753320. Therefore the number of revolutions completed by the Moon is

$$n = \frac{1863823 \times 57753320}{1577917500} = 68217.7402447. \tag{1.12}$$

Here 68217 represents the complete number of revolutions made by the Moon since the beginning of the *Kaliyuga*. From the fractional part 0.7402447 we get the number of *rāśis* etc. covered.

$$0.7402447 \times 12 = 8 + 0.8829364.$$

This shows that 8 *rāśis* have been covered and the Moon is in the 9th one, namely *Dhanus*. To get the degrees etc. we multiply the fractional part by 30. Thus we have

$$0.8829364 \times 30 = 26 + 0.488092.$$

This means that the mean Moon has crossed 26° in *Dhanū-rāśi* (Sagittarius sign). The fractional part of the above expression further multiplied by 60 gives 29.28552. The fractional part of this can further be multiplied by 60 to get the seconds etc. Thus the mean longitude of the Moon on January 14, 2002 is found to be

$$\theta_0 = 8 \text{ signs} + 26 \text{ degrees} + 29 \text{ minutes} = 8^r 26^\circ 29'. \tag{1.13}$$

Note: The mean longitude of the Moon obtained above corresponds to the longitude of the Moon at mean sunrise for an observer situated on the meridian passing through Ujjayinī. For other observers, the *Deśāntara* correction has to be applied, which is explained in the following verses.

१.१३ देशान्तरसंस्कारः
1.13 Correction due to difference in longitude

लङ्कामेरुगरेखायां उज्जयिन्यादितस्ततः ॥ २८ ॥
पूर्वापरदिशोः कार्यं कर्म देशान्तरोद्भवम् ।

laṅkāmerugarekhāyāṃ ujjayinyāditastataḥ || 28 ||
pūrvāparadiśoḥ kāryaṃ karma deśāntarodbhavam |

The *deśāntara-karma* has to be done for those places which lie to the east or west of Ujjayanī, which itself is situated in the meridian passing through the Laṅkā and Meru.

In Indian astronomical texts, the meridian passing through Ujjayinī (23°11′N, 75°46′E, of the Greenwich meridian) is taken to be the reference meridian. For the sake of convenience, we refer to it hereafter as the standard meridian. The mean positions of the planets obtained from the *Ahargaṇa* correspond to the positions of the planets at the mean sunrise for an observer along the standard meridian. For observers lying to the east of the standard meridian, the sunrise would be earlier; and for the observers lying west it will be later. Hence the mean longitudes of the planets obtained using the *Ahargaṇa* will not be the mean longitudes at the mean sunrise at the observer's location. To obtain the mean longitudes at sunrise at the observer's location, a correction due to the difference in terrestrial longitude has to be made and this correction is called the *Deśāntara-saṃskāra.*

In this context, there is an elaborate discussion in *Yukti-dīpikā* on some related and interesting issues such as: how the Earth is situated in the space, how it is supported etc.

Does the Earth remain unsupported in the sky ?

In the following we shall present some of the verses of *Yukti-dīpikā*, with a translation.

ननु तिष्ठेत् कथं भूमिः खमध्येऽसौ निराश्रया ।
गुरुत्वान्नभसस्तस्याः पतनं न भवेत् कथम् ॥
पतन्ति यानि वस्तूनि यत्र तत्र पतत्वसौ ।
अधः पतन्ति दृश्यन्ते सर्वा अपि विहायसः ॥
समर्थयामहे तस्मात् अधस्तात् पतनं भुवः ।
अत्रोच्यतेऽखिलं वस्तु पततीह यथा भुवि॥
तद्वन्नान्यत्र पतनं भूगोलस्य तु कल्प्यताम् ।
विश्वाश्रयोऽयं भूगोलः यतो विश्वम्भरैव भूः ॥
पतनं सर्ववस्तूनां भूपृष्ठावधि दृश्यते ।
दिवः पतत्यास्तु भुवः प्रतिष्ठा न क्वचित् भवेत् ॥
भूगोले सर्ववस्तूनि पतन्त्येव समन्ततः ।
न पुनः क्वापि भूगोलः सर्वतोऽधो गतो यतः ॥
तत्रापि घनभूमध्यं स्वशक्त्या धारयेदधः ।
भूपृष्ठावयवान् सर्वान् नियम्य व्यस्तदिग्गतात् ॥
भूपृष्ठावयवाः सर्वे गोलार्धान्तरितत्वतः ।
समन्तात् प्रतिबध्नन्ति मिथः पिपतिषां च ताम् ॥
खमध्यात् पतनाभावः भूगोलस्य ततो ध्रुवः ।
गोलाकारत्वमप्यस्य समन्तात् समगौरवात् ॥
न तिष्ठति क्वचित् किञ्चित् आश्रयेण विना यदि ।
तादृशोप्याश्रयोस्त्यस्य नभ एव निरन्तरम् ॥

आश्रयाश्रयिभावश्च मिथो भूतेषु दृश्यते ।
मूर्तामूर्तविभागोऽपि तत्राकिञ्चित्करो भवेत् ॥[47]

How is it that the Earth stands in space without any support? Being heavier than the sky, how is it that it does not fall? All those objects which fall from the sky, wherever they may reach (on the Earth), they are always found to fall below. Therefore we justify that the Earth should also fall below.

Here it is said: Do not presume that the Earth will fall somewhere as the objects fall on the Earth. This is because it (the Earth) is the *āśraya* for the entire world and hence it is also referred as the *viśvambharā*. All the objects are observed to fall till they reach the surface of the Earth. For the falling Earth, there will be no *pratiṣṭhā*, support from where it would be stopped from falling further down.

All the objects keep falling on the Earth from all sides. The Earth does not fall because it is in the lowest position vis-à-vis all objects. Even then [it has to be understood that] the centre of the Earth, by its own force, will keep all the objects situated on the surface of it in different directions bound downwards (to it).

Different parts on the surface of the Earth are at a distance equal to the radius of the sphere [from the centre of the Earth]. These parts, [which are] spread all over, try to bind themselves mutually as they tend to fall. Therefore certainly the Earth does not fall from the centre of space. The circular shape of the Earth is also because of the equal distribution of weight in all directions.

If [you say that] no object can stand without any support, even in that case space itself serves as permanent support for the Earth. It is found among objects that they can be mutually supportive [the supporter can become the supported and vice versa]. It does not matter that one has a manifest form (*mūrta*, referring to the Earth) and the other has an unmanifest form (*amūrta*, referring to space).

१.१४ देशान्तरकालः

1.14 Duration corresponding to difference in longitude

खखदेवा भुवो वृत्तं त्रिज्याप्तं लम्बकाहतम् ॥ २९ ॥
स्वदेशजं, ततः षष्ट्या हृतं चक्रांशकाहतम् ।
खखदेवहृतं भागाद्यन्तरं त्वक्षभागयोः ॥ ३० ॥
स्वदेशसमयाम्योदग्रेखायां देशयोर्ययोः ।
तदन्तरालदेशोत्थयोजनैः सम्मिते स्वके ॥ ३१ ॥
भूवृत्ते नाडिकैका स्यात् कालो देशान्तरोद्भवः ।
निमीलनान्तरं यद्वा स्वदेशसमरेखयोः ॥ ३२ ॥
देशान्तरभवः कालः इन्दोरुन्मीलनादपि ।
प्रागेव दृश्यते प्रत्यक्, पश्चात् प्राच्यां ग्रहः सदा ॥ ३३ ॥
देशान्तरघटीक्षुण्णा मध्या भुक्तिर्द्युचारिणाम् ।
षष्ट्या भक्तमृणं प्राच्यां रेखायाः पश्चिमे धनम् ॥३४॥

khakhadevā bhuvo vṛttaṃ trijyāptaṃ lambakāhatam || 29 ||
svadeśajaṃ, tataḥ ṣaṣṭyā hṛtaṃ cakrāṃśakāhatam |
khakhadevahṛtaṃ bhāgādyantaraṃ tvakṣabhāgayoḥ || 30 ||
svadeśasamayāmyodagrekhāyāṃ deśayoryayoḥ |

---

[47] {TS 1977} pp. 68–69.

tadantarāladeśotthayojanaiḥ sammite svake || 31 ||
bhūvṛtte nāḍikaikā syāt kālo deśāntarodbhavaḥ |
nimīlanāntaraṃ yadvā svadeśasamarekhayoḥ || 32 ||
deśāntarabhavaḥ kālaḥ indorunmīlanādapi |
prāgeva dṛśyate pratyak, paścāt prācyāṃ grahaḥ sadā || 33 ||
deśāntaraghaṭīkṣuṇṇā madhyā bhuktirdyucāriṇām |
ṣaṣṭyā bhaktamṛṇaṃ prācyāṃ rekhāyāḥ paścime dhanam ||34||

The [measure of the] circumference (*vṛtta*) of the Earth [at the equator] 3300, when divided by *trijyā* and multiplied by the Rsine of colatitude (*lambaka*) is [the circumference of the latitudinal circle] at one's own place. The same divided by 60, multiplied by 360 and again divided by 3300, gives the distance of separation between two places in the latitudinal circle [corresponding to one *ghaṭikā*] in degrees and so on (*bhāgādi*).

The *deśāntara-kāla* will be one *nāḍikā* between the places on the two meridians (*yāmyodagrekhā*), the one passing through the observer and the other separated from it by a distance in *yojanas*, measured along the latitudinal circle, corresponding to the above circular measure.

Or, the difference between the time of obscuration of the Moon in [the meridian passing through] one's own place, and the one passing through the meridian (*samarekhā*) [at another location], is the *deśāntara-kāla* (time difference due to the difference in the longitude). This *deśāntara* can also be determined from the release of the Moon [at the two longitudes]. The planet is always seen earlier in the west and later in the east. Hence, the *deśāntara-kāla*, multiplied by the mean daily motion of the planet and divided by 60, must be subtracted in the east and added in the west.

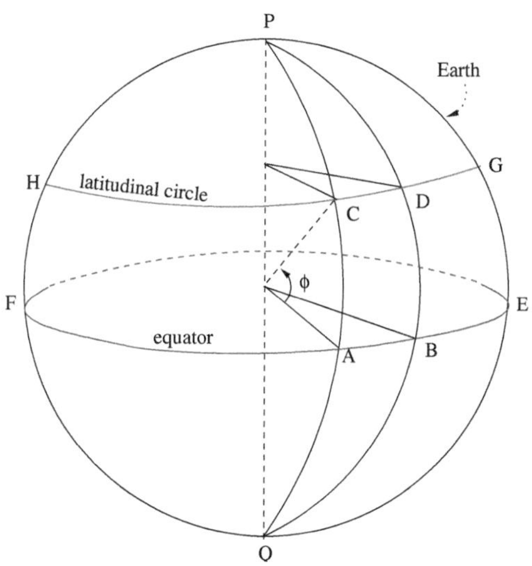

Fig. 1.9 *Deśāntara-saṃskāra* to find the longitude of a planet at sunrise at one's own place.

In Indian astronomy the linear distances are measured in *yojanas*. In Fig. 1.9, *PCAQ* is the prime meridian through Ujjayinī. *PDBQ* is the meridian through the observer. *ABEF* represents the terrestrial equator whose circumference C_e is given to be 3300 *yojanas*. *CDGH* is a latitudinal circle corresponding to a latitude ϕ. The radius of this circle (a small circle) is the radius of the sphere multiplied by $\cos\phi$. The circumference of this latitudinal circle C_0 or *svadeśabhūmiparidhi* in *yojanas* is stated to be:

$$C_0 = \frac{3300 \times R\cos\phi}{R} = 3300\cos\phi. \tag{1.14}$$

As mentioned earlier (Section 1.3), the time taken by the stellar sphere (or the Earth) to rotate through 360° is 60 *ghaṭikās* and this duration corresponds to one full rotation of the latitudinal circle C_0. Hence the distance along the latitudinal circle that corresponds to one *ghaṭikā* is $\frac{C_0}{60}$. In other words, the distance of separation whose measure in *yojanas* is equal to $\frac{C_0}{60}$ corresponds to a difference of one *ghaṭikā* in local time.

Further, the text also mentions that the distance in *bhāgādis* between two places in the latitudinal circle corresponding to a separation of one *ghaṭikā* can be expressed as

$$\frac{C_0 \times 360}{60 \times 3300} = 6\cos\phi. \tag{1.15}$$

Let t_0 be the time of an event, such as the obscuration of the Moon, for an observer in the standard meridian. Let δt be the difference in the sunrise times between the observer on the standard meridian and an observer elsewhere. Then the local time t at which the observer will observe the event is given by

$$\begin{aligned} t &= t_0 - \delta t \qquad \text{(if the observer is to the west)} \\ t &= t_0 + \delta t \qquad \text{(if the observer is to the east).} \end{aligned} \tag{1.16}$$

Let d be the distance of separation between the given place and the standard meridian along the latitudinal circle. Then

$$\delta t = \frac{d}{3300\cos\phi} \times 60 \qquad \text{in } ghaṭikās. \tag{1.17}$$

It is suggested that δt can be determined from the difference in times corresponding to the beginning or end of a lunar eclipse at these places (with respect to their local sunrise times). This particular physical phenomenon is chosen probably because the beginning of obscuration and the release of the Moon are sharply defined events.

Let $\Delta\theta$ be the daily motion of the planet, that is, the angle covered by it in 60 *ghaṭikās*. Then the angle covered by it in a time δt is given by

$$\delta\theta = \frac{\delta t \times \Delta\theta}{60}. \tag{1.18}$$

Here δt is called the *deśāntarakāla* and the term *deśāntara-saṃskāra* refers to the application of $\delta\theta$ to the mean longitude of the planet obtained from the *Ahargaṇa*, to get the mean longitude at sunrise at the observer's location.

The above correction has to be applied positively to the mean longitude of the planet obtained from the *Ahargaṇa*, if the meridian passing through the observer lies to the west of the standard meridian; and negatively if it lies to the east of it. If θ_0 is the mean longitude of the planet obtained from the *Ahargaṇa*, then the longitude at sunrise at the observer's location is given by

$$\theta = \theta_0 + \delta\theta \qquad \text{(the observer is to the west)}$$
$$\theta = \theta_0 - \delta\theta \qquad \text{(the observer is to the east).} \qquad (1.19)$$

As the sunrise takes place earlier for observers to the east of the prime meridian, and later for the observers to the west, the corrections have the signs as indicated above.

१.१५ कल्यादिध्रुवाः

1.15 Initial positions at the beginning of the *Kaliyuga*

षड्वेदेष्वब्धिवेदास्तु विलिप्तादि ध्रुवो विधोः ।
प्राणात्यष्ट्यङ्कनेत्राग्नितुल्यं चन्द्रोच्चमध्यमम् ॥ ३५ ॥
सप्तसागरशैलेन्दुभवा लिप्तादयोऽसृजः ।
षट्त्रिंशल्लिप्तिका शोध्या विदो जीवे तु योजयेत् ॥ ३६ ॥
पङ्क्त्यर्कतुल्यलिप्तादि सिते राशिः षडंशकाः ।
विश्वतुल्याः कलाश्च स्वं नखात्यष्टिभवाः शनेः ॥ ३७ ॥
पाते तु मण्डलाच्छुद्धे नखाकृतिरसा अपि ।

ṣaḍvedeṣvabdhivedāstu viliptādi dhruvo vidhoḥ |
prāṇātyaṣṭyaṅkanetrāgnitulyaṃ candroccamadhyamam || 35 ||
saptasāgaraśailendubhavā liptādayo'sṛjaḥ |
ṣaṭtriṃśalliptikā śodhyā vido jīve tu yojayet || 36 ||
paṅktyarkatulyaliptādi site rāśiḥ ṣaḍaṃśakāḥ |
viśvatulyāḥ kalāśca svaṃ nakhātyaṣṭibhavāḥ śaneḥ || 37 ||
pāte tu maṇḍalācchuddhe nakhākṛtirasā api |

The correction to the initial position of the Moon [at the beginning of the *Kaliyuga*] is $4^\circ 45' 46''$; of the Moon's apogee it is $3^r 29^\circ 17' 5''$; of Mars in minutes etc. it is $11^r 17^\circ 47'$; For Mercury 36 seconds have to be subtracted. In the case of Jupiter $12^\circ 10'$ has to be added. In the case of Venus add $1^r 6^\circ 13'$; and for Saturn $11^r 17^\circ 20'$ [has to be added]; in the case of the node of the Moon, $6^r 22^\circ 20'$ has to be added to the longitude obtained by subtracting the mean longitude from the *maṇḍala* (360°).

The term *dhruva* refers to the epochal position of the planets, i.e. the mean longitudes of the planets at the beginning of a given epoch. The epoch could be the beginning of the *Kaliyuga*, or any other date chosen by the astronomer. In modern parlance, the *dhruva* is the same as the initial value. The mean longitude of a planet (*madhyama-graha*) is obtained by adding the *dhruva* to the product of the daily motion of the planet and the time elapsed (the *Ahargaṇa*) since the epoch. From this, the true position (*sphuṭa-graha*) can be calculated by applying *saṃskāras* (corrections).

In specifying the *dhruvas* of the planets, one observes that there are slight variations from text to text. Some of the important Indian astronomical texts such as *Āryabhaṭīya*, *Sūryasiddhānta* etc. have assumed that the five planets, namely Mercury, Venus, Mars, Jupiter and Saturn, and the Sun and the Moon were in conjunction with the beginning point of *Meṣa-rāśi* at the commencement of the present *Kaliyuga*. In other words, their mean longitudes at the beginning of the *Kaliyuga* are taken to be $0^r 0^\circ 0'$. This is an assumption.[48] For instance, the following verses in *Sūryasiddhānta* specify the *dhruvas* of the planets at the beginning of the *Kaliyuga*:

अस्मिन् कृतयुगस्यान्ते सर्वे मध्यगता ग्रहाः ।
विनेन्दुपातमन्दोच्चान् मेषादौ तुल्यतामिताः ॥
मकरादौ शशाङ्कोच्चं तत्पातस्तु तुलादिगः ।
निरंशत्वं गताश्चान्ये नोक्तास्ते मन्दचारिणः ॥ [49]

But for the apogees and nodes, the mean positions of all the planets at the end of the *Kṛtayuga* were at the beginning of *Meṣa rāśi* ($0^r 0^\circ 0'$). The apogee of the Moon was at the beginning of *Makara* (Capricorn) *rāśi* (270°) and its node was at the beginning of *Tulā* (Libra) *rāśi* (180°). The positions of the nodes and apogees of the other planets are not mentioned [separately], since their rate of motion is very slow.

Here one may wonder why the verse gives the epochal positions at the end of the *Kṛtayuga* and not at the beginning of the *Kaliyuga*. The positions at the beginning of the *Kaliyuga* would be the same as the positions at the end of the *Kṛtayuga*, as the planets make an integral number of revolutions in the intervening period. This is because the number of revolutions made by the planets in a *Mahāyuga* is even and the combined duration of *Tretāyuga* and *Dvāparayuga* is exactly half that of a *Mahāyuga* (see Section 1.9.1).

Dhruvas given in *Tantrasaṅgraha*

In contrast to the *dhruvas* specified in *Sūryasiddhānta*, non-zero epochal positions (at the beginning of the *Kaliyuga*) are specified by Nīlakaṇṭha in his *Tantrasaṅgraha*. The values given by him are listed in Table 1.7.

Need for changing the *dhruva*

The reason for choosing the epochal positions to be different from those in *Sūryasiddhānta* and other texts is given in *Yukti-dīpikā*. It is pointed out that the rate of motion of the planets might have changed over time, and hence, the epochal

[48] It is possible that the astronomers would have arrived at it by doing back-computation. That is, computing their position by moving backwards in time, based on their present position and rate of motion observed by them.

[49] {SSI 1995} (I. 57–8), p. 37.

Name of planet		*Dhruvas*
in Sanskrit	in English	(in *rāśis* etc.)
विधुः	Moon	$+4^{\circ}\,45'\,46''$
विधूच्चम्	Moon's apogee	$+3^{r}\,29^{\circ}\,17'\,5''$
असृक्	Mars	$+11^{r}\,17^{\circ}\,47'\,0''$
वित्	Mercury	$-36'\,0''$
जीवः	Jupiter	$+12^{\circ}\,10'\,0''$
सितः	Venus	$+1^{r}\,6^{\circ}\,13'\,0''$
शनिः	Saturn	$+11^{r}\,17^{\circ}\,20'\,0''$
विधुपातः	Moon's node	$+6^{r}\,22^{\circ}\,20'\,0''$

Table 1.7 *Dhruvas* of the planets at the beginning of the *Kaliyuga* as given in *Tantrasaṅgraha*.

positions at the beginning of the *Kaliyuga* based on the current observed rates (determined from back-computations) should be taken to be the correct ones.

कल्यादौ न निरंशत्वं भगणादेर्द्युचारिणाम् ।
गतिभेदात्तु दृक्सिद्धाः तत्रैषां स्युर्ध्रुवास्ततः ॥[50]

The mean positions of the planets at the beginning of *Kaliyuga* is not $0^{r}\,0^{\circ}\,0'$. Since there is a difference in the rate of motion, whatever is determined on the basis of [current] observations are to be considered as the *dhruvas*.

१.१६ युगान्तरपरिकल्पनम्
1.16 Introducing an alternative *yuga*

कल्यादिध्रुवका ह्येते युगभोगसमन्विताः ॥ ३८ ॥
तत्तद्युगे ध्रुवा ज्ञेयाः षडश्वेष्वब्दकं युगम् ।
भगणात् खखभूताश्वैर्युगभोगस्त्ववाप्यते ॥
अष्टघ्नयुगभोगाः स्वं अतः कल्यादिजे ध्रुवे ॥ ३९ ॥

kalyādidhruvakā hyete yugabhogasamanvitāḥ || 38 ||
tattadyuge dhruvā jñeyāḥ ṣaḍaśveṣvabdakaṃ yugam |
bhagaṇāt khakhabhūtāśvairyugabhogastvavāpyate ||
aṣṭaghnayugabhogāḥ svaṃ ataḥ kalyādije dhruve || 39 ||

These are the initial positions of the planets at the beginning of the *Kaliyuga*. These, added to the angle covered by the planets in [all the] *yugas* [elapsed before the commencement of the present *yuga*], give the initial positions of the planets at the beginning of the present *yuga*. The *yuga* [defined here] consists of 576 years. The revolutions [completed by the planets in a *Mahāyuga*] divided by 7500 gives the angle covered by the planets in a *yuga* [of 576 years]. This angle multiplied by 8 has to be added to the initial positions at the

[50] {TS 1977}, p. 73. The prose order for the second half of the verse is as follows – ततः (= तस्मात् कालात्) गतिभेदात् तु, तत्र (= कल्यादौ) एषां (= ग्रहाणाम्) दृक्सिद्धाः (= इदानीं यन्त्रैः परीक्ष्य साधिताः एव) ध्रुवाः स्युः ।

beginning of the *Kaliyuga* [to obtain the *dhruvas*], at the beginning of end of the eight *yugas* [of 576 years] after the beginning of the *Kaliyuga*.

Here a shorter *yuga* of 576 years which is $\frac{1}{7500}th$ of a *Mahāyuga* is defined for computational convenience. The beginning of the ninth such *yuga* (or the end of 8 such *yugas* or 4608 years) after the *Kaliyuga*'s beginning is in 1507 CE, which is close to the date of composition of *Tantrasaṅgraha*. That is why the method of obtaining the *dhruvas* 4608 years after the beginning of the *Kaliyuga* is spelt out.

१.१७ कल्यादौ मन्दोच्चाः

1.17 The *Mandoccas* at the beginning of the *Kaliyuga*

स्वररवयः खाकृतयः द्विनगभुवोऽशीतिरभ्रजिनाः ।
भौमान्मन्दोच्चांशाः वसुतुरगा भास्करस्यापि ॥ ४० ॥

svararavayaḥ khākṛtayaḥ dvinagabhuvo'śītirabhrajināḥ |
bhaumānmandoccāṃśāḥ vasuturagā bhāskarasyāpi || 40 ||

The *mandoccas* of the planets beginning with Mars are 127, 220, 172, 80 and 240 degrees respectively. And for the Sun [the *mandocca*] is 78 degrees.

The term *mandocca* refers to the direction of that point on the planetary orbit where the planet has the least angular velocity. In modern parlance, it refers to the direction of aphelion in the case of the five planets and the apogee in the case of the Sun and the Moon.

Like the planets, the *mandoccas* of the planets are also in continuous motion. But, since their rate of motion is very small (hardly a few minutes over hundreds of years), they can be taken to be fixed for practical purposes. The longitudes of the *mandoccas* at the beginning of the *Kaliyuga*, referred to as *mandoccāṃśāḥ*[51] in the above verse, are listed in Table 1.8.

Name of planet		Its *Mandocca*
in Sanskrit	in English	(in degrees)
कुजः	Mars	127
बुधः	Mercury	220
गुरुः	Jupiter	172
शुक्रः	Venus	80
शनिः	Saturn	240
सूर्यः	Sun	78

Table 1.8 Longitudes of the *mandoccas* of the planets at the beginning of the *Kaliyuga*.

In the revised planetary model of Nīlakaṇṭha, discussed in the next chapter, the Sun is the *śīghrocca* of all the planets, including the interior planets Mercury and

[51] The term *aṃśāḥ* in this context refers to degrees.

Venus, which had separate *śīghroccas* in the traditional planetary model. This is emphasized in *Laghu-vivṛti* in the present context:

ननु कुजादीनां शीघ्रोच्चस्यापि विद्यमानत्वात् तदपि वक्तव्यमेवेति चेत् न, रविमध्यमस्यैव शीघ्रोच्चत्वात् तस्य प्रदर्शितत्वाच्च ।

The planets Mars etc. also have *śighroccas*. If so, is it not true that their longitudes should also be mentioned? No, [they need not be mentioned] because the mean Sun is the *śighrocca* for all these planets and its longitude (the procedure to obtain it) has already been shown.

The commentary *Yukti-dīpikā* on this chapter ends with the following verse, where Śaṅkara Vāriyar acknowledges that the Malayalam work *Yuktibhāṣā* of Jyeṣṭhadeva[52] happens to be the basis for the commentary *Yukti-dīpikā*.

इत्येष परक्रोडावासद्विजवरसमीरितो योऽर्थः ।
स तु तन्त्रसङ्ग्रहस्य प्रथमेऽध्याये मया कथितः ॥[53]

The principles expounded by the reputed *dvija* living in the *Parakroḍa* have been explained by me thus in the first chapter of *Tantrasaṅgraha*.

[52] Jyeṣṭhadeva has been referred to as *parakroḍāvāsa*—one who lives in Parakroḍa.

[53] {TS 1977}, p. 77.

Chapter 2
स्फुटप्रकरणम्
True longitudes of planets

२.१ केन्द्रलक्षणं पदलक्षणं च
2.1 Definition of the anomaly and the quadrant

स्वोच्चोनो विहगः केन्द्रं तत्र राशित्रयं पदम् ।
ओजे पदे गतैष्याभ्यां बाहुकोटी समेऽन्यथा ॥ १ ॥

svoccono vihagaḥ kendraṃ tatra rāśitrayaṃ padam|
oje pade gataiṣyābhyāṃ bāhukoṭī same'nyathā || 1 ||

The *ucca* subtracted from the planet is the *kendra* (anomaly). Three *rāśis* constitute a *pada* (quadrant). In the odd quadrants, the *bāhu* and *koṭi* [are to be found] from the angle covered and to be covered [respectively]. In the even quadrants it is otherwise.

The procedure for obtaining the *madhyama-graha* i.e. the mean longitude of a planet from the *Ahargaṇa*, was explained in the previous chapter. Two corrections, namely *manda-saṃskāra* and *śīghra-saṃskāra*, have to be applied to the *madhyama-graha* to obtain the *sphuṭa-graha* or the true longitude of the planet. In these two *saṃskāras*, to be described later in this chapter, two angles, namely the *manda-kendra* (*manda* anomaly or mean anomaly) and the *śīghra-kendra* (*śīghra*-anomaly or anomaly of conjunction or solar anomaly) play important roles. In the above verse, the *kendras* and their sines and cosines (known as *bāhus* and *koṭis*) pertaining to both the *saṃskāras* are dealt with. For this, two quantities, namely the *ucca* and the *kendra*, are introduced.

Ucca and *kendra*

The *ucca* and *kendra* essentially refer to the apsis and anomaly respectively. These two terms are generally used with the adjectives *manda* and *śīghra* and appear in the two processes of correction, namely *manda-saṃskāra* and *śīghra-saṃskāra*.

The *manda-saṃskāra*[1] is a procedure to obtain the correction for the eccentricity of the planetary orbit. The terms *ucca* and *kendra* used in this context refer to the direction of the *mandocca* (apogee/aphelion of the planet) and the *manda-kendra* respectively.

Similarly, *ucca* and *kendra* used in the context of *śīghra-saṃskāra*—the process by which the geocentric longitudes of the planets are obtained from their heliocentric longitudes[2]—refer to the directions of the *śīghrocca* and *śīghra-kendra* respectively. If θ_0 refers to the longitude of the mean planet, and θ_m that of its *mandocca*, then the *manda-kendra*, θ_{mk}, is defined as

$$\theta_{mk} = \theta_0 - \theta_m. \tag{2.1}$$

If θ_{ms} is the longitude of the *manda-sphuṭa-graha*, that is, the mean longitude of the planet corrected by *manda-saṃskāra*, and θ_s that of the *śīghrocca*, then the *śīghra-kendra*, θ_{sk}, is defined as

$$\theta_{sk} = \theta_{ms} - \theta_s. \tag{2.2}$$

In the second quarter of the above verse, it is mentioned that three *rāśis* constitute a *pada*. Since *rāśi* is a 30° division on the ecliptic, by definition the term *pada* refers to a quadrant. In Fig. 2.1*a*, *APB* represents a *pada*. Before explaining the second half of the verse, it would be useful to introduce the concepts of *bāhu* and *koṭi*, which are frequently employed in this and the following chapters.

Bāhu and *Koṭi*

In Indian astronomical texts, the terms *bāhu*[3] and *koṭi* are used in association with either *cāpa* or *jyā*. The terms *cāpa* and *jyā* literally mean bow and string respectively. In this context, they refer to the arc of a circle and the chord associated with it. Sometimes instead of the term *cāpa*, *dhanus* is also used to refer to the arc of a circle.

In Fig. 2.1*a*, the arc *PAL* represents the *cāpa* and *PQL* is the *jyā* associated with the *cāpa* (arc). Though literally the term *jyā* refers to the chord *PL*, in most situations *PQ*, which is half of *PL*, is referred to as the *jyā* (Rsine) of the arc *PA*. Since *PQ* is only half of *PL*, it must actually be referred to as the *jyārdha*. However, since only *PQ* is involved in planetary computations (as will be clear later), the term *jyā* itself is used to refer to the semi-chord *PQ*, for the sake of brevity in the use of terminology. Hence the terms *bāhucāpa* and *bāhujyā* or Rsine refer to the arc *AP* and the semi-chord *PQ* in the figure, respectively. The terms *koṭicāpa* and *koṭijyā*

---

[1] The significance of this is explained in detail in Appendix F. The equivalent of this correction in modern astronomy is the equation of centre.

[2] For details refer to Sections 2.26–28 and Appendix F.

[3] The literal meaning of *bāhu* is hand. Similarly, *koṭi* means side. In this context, the term *koṭi* refers to the side which is perpendicular to *bāhu*.

or Rcosine refer to the arc *PB* and the segment *OQ* (perpendicular to the chord *PL*), respectively.

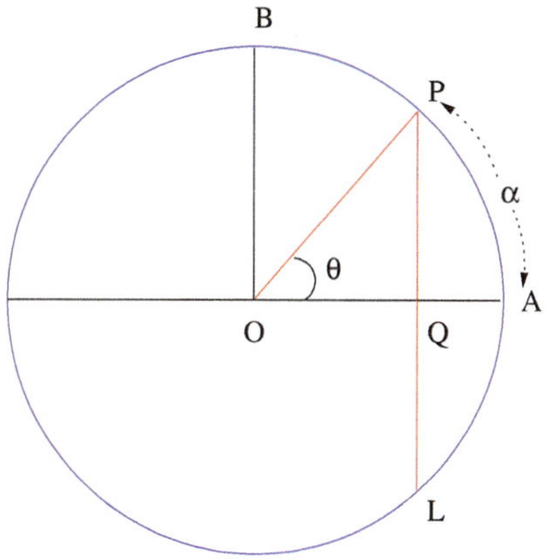

Fig. 2.1*a* *Bāhu* and *cāpa*.

Relation between the *jyās* and the sine and cosine function

Let R be the radius of the circle shown in Fig. 2.1*a*. Now, the quantities which are designated by the terms *bāhucāpa*, *bāhujyā*, *koṭicāpa* and *koṭijyā* are listed below:

bāhucāpa = $R\theta$ = the length of the arc *AP* corresponding to the angle θ.
bāhujyā = $R\sin\theta = R\times$ the *sine* of the angle θ.
koṭicāpa = $R(90-\theta)$ = the length of the arc corresponding to the angle $(90-\theta)$.
koṭijyā = $R\cos\theta = R\times$ the *cosine* of the angle θ.

In the following, we give the relationship between *sine* of an angle, θ, and the *jyā* of the corresponding arc, $\alpha = R\theta$, normally expressed in minutes. In Fig. 2.1*a*, let the length of the arc AP be α. Then we have the following relation between the *jyās* and the modern *sine* and *cosine* functions:

$$\begin{aligned} b\bar{a}hujy\bar{a}\;\alpha &= R\sin\theta \\ ko\d{t}ijy\bar{a}\;\alpha &= R\cos\theta. \end{aligned} \tag{2.3}$$

Normally the circumference of the circle is taken to be 21600 units (the number of minutes in 360°), so that an angle of $1'$ corresponds to an arc length of 1 unit. Hence the radius $R = \frac{21600}{2\pi} \approx 3437.7468$, which is approximately 3438 minutes. In Indian astronomical and mathematical texts, the radius of the circle R is referred to as the *trijyā*. This is because R is the *jyā* corresponding to the arc whose length is equal to three *rāśis* (5400′). In other words, *tri-rāśi-jyā* is shortened to *trijyā*.

Finding the *bāhu* and *koṭijyās* in different quadrants

The sine or cosine of an angle greater than 90° can always be determined in terms of an angle less than 90°. This is the essence of the second half of the verse wherein it is stated that:

- if the *kendra* is in the odd quadrant, i.e. its value lies in the range $0^\circ - 90^\circ$ or $180^\circ - 270^\circ$, then the *bāhu* and *koṭi* are to be determined from the angles already covered and to be covered in that quadrant, respectively.
- if the *kendra* is in the even quadrant, i.e. its value lies in the range $90^\circ - 180^\circ$ or $270^\circ - 360^\circ$, then the *bāhu* and *koṭi* are to be determined from the angles to be covered and already covered in that quadrant, respectively.

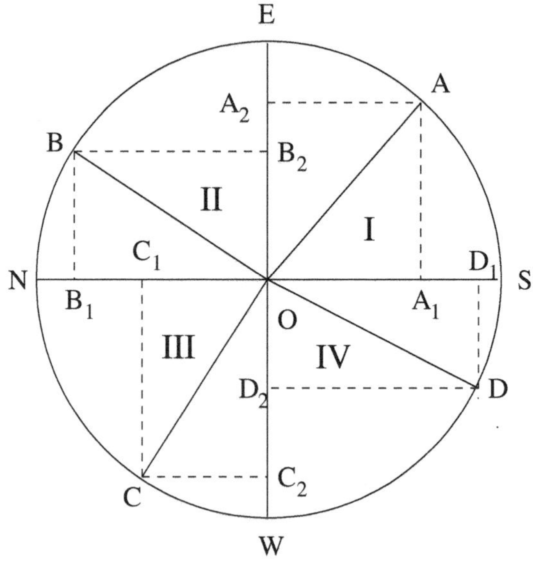

Fig. 2.1*b* *Bāhu* and *koṭi* when the *kendra* is in different quadrants.

We explain this concept further with the help of Fig. 2.1*b*. In the following we use K to denote the *kendra*. Then,

1. If K is in the first quadrant, i.e. $K = A\hat{O}A_1$, $R\sin A\hat{O}A_1 = AA_1$, $R\cos A\hat{O}A_1 = R\sin A\hat{O}A_2 = AA_2$.
2. If K is in the third quadrant, i.e. $K = C\hat{O}A_1$, $|R\sin C\hat{O}A_1| = R\sin C\hat{O}C_1 = CC_1$ and $|R\cos C\hat{O}A_1| = R\sin C\hat{O}C_2 = CC_2$.

Hence, in the above cases, the *bāhu* and *koṭi* are determined from the angles covered and to be covered respectively in the odd quadrant.

3. If K is in the second quadrant, i.e. $K = B\hat{O}A_1$, $|R\sin B\hat{O}A_1| = R\sin B\hat{O}B_1 = BB_1$, and $|R\cos B\hat{O}A_1| = R\sin B\hat{O}B_2 = BB_2$.
4. If K is in the fourth quadrant, $|R\sin D\hat{O}A_1| = R\sin D\hat{O}D_1 = DD_1$ and $|R\cos D\hat{O}A_1| = R\sin D\hat{O}D_2 = DD_2$.

Thus the *bāhu* and the *koṭi* are determined from the angles to be covered and covered respectively in the even quadrants. Here, only the procedure to find the magnitudes of the Rsines and Rcosines is given. Their signs (whether they have to be applied positively or negatively) will be stated separately in each context in which they are being employed.

The concepts of the *manda-kendra* and *śīghra-kendra* are explained in *Laghu-vivṛti* as follows:

तत्र प्रथमाध्यायोक्तप्रकारेण त्रैराशिकानीता भगणादिका ये ग्रहमध्यमाः तेभ्यो भगणानपास्य शिष्टेभ्यो राश्यादिभ्यो भागात्मकमुपदिष्टं स्वं स्वं मन्दोच्चं विशोध्य यच्छिष्यते, तदिह मन्दकेन्द्रमित्यभिधीयते। यदा पुनः मन्दफलेन स्फुटीकृतात् कुजादीनां मध्यमात् शीघ्रोच्चभूतं रविमध्यमं विशोध्यते, तदा तत्र अवशिष्टं शीघ्रकेन्द्रं भवति।

From the mean positions of the planets (*madhyama-grahas*), obtained using the rule of three described in Chapter 1, which includes an integral number of revolutions, *rāśis*, etc. [the fractional part], subtract the integral number of revolutions. From the remaining *rāśis* etc. [which represents the mean longitude of the planet] when its own *mandocca* is subtracted, the remainder obtained is said to be the *manda-kendra*. When the mean Sun, which is the *śīghrocca*, is subtracted from the *manda* corrected longitudes of Mars etc., the remainder obtained is the *śīghra-kendra*.

Note: Here it is specifically mentioned that the *śīghrocca* is the mean Sun for all the five planets while defining *śīghra-kendra*. The significance of this is explained later in sections 2.26–28 and also in Appendix F, during the discussion of *śīghra-saṃskāra* for the inner planets.

The complementarity between the sine and the cosine functions is also succinctly put forth in the commentary *Laghu-vivṛti*:

बाहुधनुर्विहीनं राशित्रयं कोटिधनुः। तद्विहीनं राशित्रयं बाहुधनुः ॥

The arc of the *bāhu* subtracted from 90° is the arc of the *koṭi*. That (arc of the *koṭi*) subtracted from 90° is the arc of the *bāhu*.

२.२ ज्याग्रहणं चापीकरणञ्च

2.2 Computation of the Rsines and the arcs

लिप्ताभ्यस्तत्त्वनेत्राप्ताः गता ज्याः शेषतः पुनः ।
गतगम्यान्तरघ्नाद्य हृतास्तत्त्वयमैः क्षिपेत् ॥ २ ॥
दोःकोटिज्ये नयेदेवं ज्याभ्यश्चापं विपर्ययात् ।

liptābhyastattvanetrāptāḥ gatā jyāḥ śeṣataḥ punaḥ |
gatagamyāntaraghnācca hṛtāstattvayamaiḥ kṣipet || 2 ||
doḥkoṭijye nayedevaṃ jyābhyaścāpaṃ viparyayāt |

By dividing the minutes [of arc] by 225, the number of *jyās* that have elapsed is obtained. Multiply the remainder by the difference between the (tabular) Rsine values of the elapsed and the next, divide by 225 and add the result to the elapsed *jyā*, to obtain the *bāhu* and *koṭi*. From the *jyās* the arcs can be obtained by the reverse process.

As already explained, in Indian astronomical and mathematical works the circumference of a circle is taken to be $360^\circ = 21600'$. Therefore the length of the arc corresponding to each quadrant will be $5400'$. This length is divided into 24 equal segments, each segment corresponding to $225'$. In Fig. 2.2, the points P_i ($i = 1,2,\ldots,24$) represent the end points of the 24 segments represented by the arcs $P_{i-1}P_i$. The set of *jyās*, $J_i = P_iN_i$, ($i = 1,2,\ldots,24$) corresponding to the 24 *cāpas* P_0P_i, are explicitly stated in many texts such as *Āryabhaṭīya* and *Sūryasiddhānta*.[4] A method for obtaining more accurate values of these tabulated *jyās* will be presented in the next verse.

Let S_i represent the length of the 24 segments P_0P_i, in minutes of arc and J_i, the *jyā* corresponding to it. That is,

$$\begin{aligned} S_i &= P_0P_i = i \times 225, \\ \text{and} \quad J_i &= P_iN_i. \qquad (i = 1,2,\ldots,24) \end{aligned} \tag{2.4}$$

The above verse gives an interpolation formula to find out the *jyā* corresponding to any length of arc between 0 and $5400'$ from the set of 24 *jyās* listed in Table 2.1 (page 64). Suppose S is the length of an arc in minutes that lies between S_i and S_{i+1}. That is,

$$S = S_i + r, \qquad O \leq r < 225, \tag{2.5}$$

where $S_i = i \times 225$. Since the *jyā* corresponding to the nearest arc lengths S_i and S_{i+1} on either side of S are known, the *jyā* corresponding to S is obtained by the rule of three. It is given by

$$jy\bar{a}\, S = J_i + \frac{r \times (J_{i+1} - J_i)}{225}. \tag{2.6}$$

[4] The following verses in *Sūryasiddhānta* (II. 17–22) give the values of the 24 *jyās*:

तत्त्वाश्विनोङ्काब्धिकृता रूपभूमिधरर्त्तवः । खाङ्काष्टौ पञ्चशून्येशा बाणरूपगुणेन्दवः ॥
शून्यलोचनपञ्चैकाश्छिद्ररूपमुनीन्दवः । वियच्चन्द्रातिधृतयो गुणरन्ध्राम्बराश्विनः ॥
मुनिषड्यमनेत्राणि चन्द्राग्निकृतदस्रकाः । पञ्चाष्टविषयाक्षीणि कुञ्जराश्विनगाश्विनः ॥
रन्ध्रपञ्चाष्टकयमाः वस्वद्र्यङ्कयमास्तथा। कृताष्टशून्यज्वलनाः नागाद्रिशशिवह्नयः ॥
षट्पञ्चलोचनगुणाः चन्द्रनेत्राग्निवह्नयः । यमाद्रिवह्निज्वलनाः रन्ध्रशून्यार्णवाग्नयः ॥
रूपाग्निसागरगुणाः वस्वग्निकृतवह्नयः । प्रोज्झ्योत्क्रमेण व्यासार्धात्
उत्क्रमज्यार्धपिण्डकाः ॥

The 24 *jyā* values in the above verses have been given using the *Bhūtasaṅkhyā* system of representing numbers (see Appendix A).

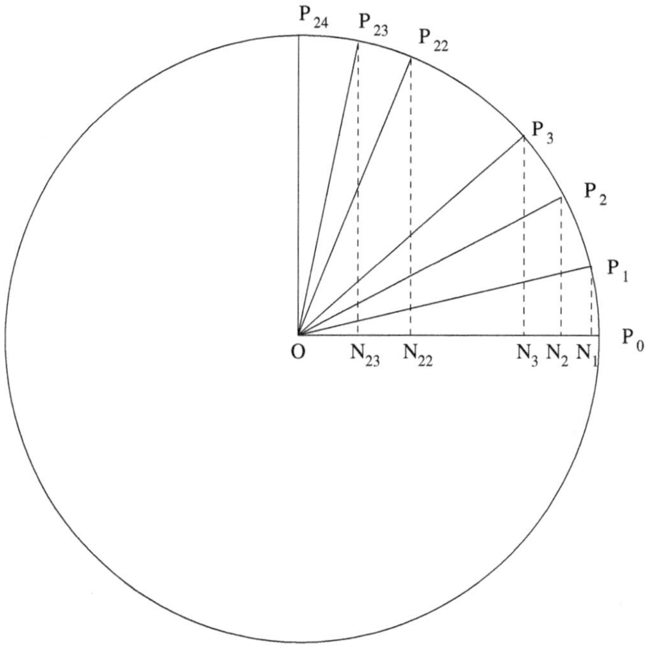

Fig. 2.2 Determination of the *jyā* corresponding to the arc lengths which are multiples of 225′.

Illustrative example

Suppose the arc length $S = 1947$. Find the *jyā* corresponding to it.

The given arc length $S = 1947$ lies between S_8 and S_9, as $S_8 = 8 \times 225 = 1800$ and $S_9 = 2025$. Hence, S can be written as $S = S_8 + 147$. The *jyā* corresponding to arc length S is given by:

$$jy\bar{a}\, S = J_8 + \frac{147 \times (J_9 - J_8)}{225}.$$

For instance, we may use the values of Mādhava, quoted in *Laghu-vivṛti*, $J_8 = 1718'52''24'''$ and $J_9 = 1909'54''35'''$. Then,

$$\begin{aligned} jy\bar{a}\, 1947 &= 1718'52''24''' + \frac{147 \times (1909'54''35''' - 1718'52''34''')}{225} \\ &= 1843'41''02'''. \end{aligned}$$

This is the value of *jyā* (1947) obtained by the first-order interpolation, while the actual value is $1844'34''09'''$.

२.३ पठितज्यानयनम्
2.3 Computation of the tabular Rsines

विलिप्तादशकोना ज्या राश्यष्टांशधनुःकलाः ॥ ३ ॥
आद्यज्यार्धात् ततो भक्ते सार्धदेवाश्विभिस्ततः ।
त्यक्ते द्वितीयखण्डज्या द्वितीया ज्या च तद्युतिः ॥ ४ ॥
ततस्तेनैव हारेण लब्धं शोध्यं द्वितीयतः ।
खण्डात् तृतीयखण्डज्या द्वितीयस्तद्युतो गुणः ॥ ५ ॥
तृतीयः स्यात् ततश्चैवं चतुर्थाद्याः क्रमाद् गुणाः ।

viliptādaśakonā jyā rāśyaṣṭāṃśadhanuḥkalāḥ || 3 ||
ādyajyārdhāt tato bhakte sārdhadevāśvibhistataḥ |
tyakte dvitīyakhaṇḍajyā dvitīyā jyā ca tadyutiḥ || 4 ||
tatastenaiva hāreṇa labdhaṃ śodhyaṃ dvitīyataḥ |
khaṇḍāt tṛtīyakhaṇḍajyā dvitīyastadyuto guṇaḥ || 5 ||
tṛtīyaḥ syāt tataścaivaṃ caturthādyāḥ kramād guṇāḥ |

The *jyā* of one-eighth of the arc, corresponding to a *rāśi* (expressed) in minutes, is $10''$ short of that (length of the arc in minutes). The quantity obtained by dividing the first *jyārdha* by $233\frac{1}{2}$, and subtracting it from the same, is the *dvitīyakhaṇḍajyā*. This added to it (the first *jyā*) is the second *jyā*. The result obtained by dividing that (the second *jyā*) by the same divisor ($233\frac{1}{2}$) is to be subtracted from the second *khaṇḍajyā*. This is the *tṛtīyakhaṇḍajyā*. This added to the second is the third *guṇa*.[5] From that, the fourth *guṇa* etc. have to be obtained in order.

As mentioned earlier, some texts like *Āryabhaṭīya* and *Sūryasiddhānta* give the table of 24 *jyās* from which the *jyā* of any length of arc can be found, as illustrated through an example in the previous section. In the verses above, a procedure for finding more accurate values of the 24 *jyās* is described.[6] For this, two new terms, namely the *khaṇḍajyā* (Rsine difference) and the *piṇḍajyā* (whole Rsine) are introduced.

With reference to Fig. 2.2, they are defined as follows:

$$\begin{aligned} \textit{piṇḍajyā} &= P_iN_i = J_i & i &= 1,2,\ldots,24, \\ \textit{khaṇḍajyā} &= P_{i+1}N_{i+1} - P_iN_i = \Delta_i & i &= 1,2,\ldots,23. \end{aligned} \tag{2.7}$$

The term *piṇḍajyā* essentially refers to the whole or the tabulated *jyā*. They are 24 in number, represented by $J_1, J_2, \ldots, J_{24}$ and are expressed in minutes of arc. The last *piṇḍajyā*, namely $P_{24}N_{24} = P_{24}O$, is referred to as *trijyā*, and its length is equal to the radius of the circle. The difference between the successive *piṇḍajyās* are referred to as the *khaṇḍajyās*. In these verses the first *piṇḍajyā* and the procedure for generating the successive *piṇḍajyās* from that are given.

---

[5] The term *guṇa* has various meanings. In this verse and in verse 5a, it could be assigned the meaning rope, in which case it is the same as the word *jyā*. But in verses 6, 8 etc. of this chapter it is used to mean a multiplier (i.e. numerator).

[6] In fact, the procedure is the same as in *Āryabaṭīya*, but for the values of the first *jyā* (which is taken to be $224'50''$ instead of $225'$) and the divisor (which is taken to be $233\frac{1}{2}$ instead of $225'$).

The length of the first *piṇḍajyā* is stated to be one-eighth of a *rāśi* expressed in minutes minus 10 seconds; thus P_1N_1 (in Fig. 2.2) is equal to $224'\,50''$. This is also equal to the first *khaṇḍajyā*. Thus we have

$$jy\bar{a}\, P_0P_1 = P_1N_1 = J_1 = 224'\,50'' = \Delta_1. \tag{2.8}$$

This can be understood as follows. In Fig. 2.2,

$$P_0\hat{O}P_1 = \frac{90}{24} = 3.75^\circ = 225' = 0.65949846\,\text{radian}. \tag{2.9}$$

The first *piṇḍajyā* is often taken to be $225'$ in some earlier Indian texts like *Āryabhaṭīya* and *Sūryasiddhānta* based on the approximation,

$$R\sin\alpha \approx R\alpha = 225'. \tag{2.10}$$

In contrast to the above approximation, which of course is reasonably good for small α, the above set of verses present the value of the first *piṇḍajyā* based on a better approximation,

$$\sin\alpha \approx \alpha - \frac{\alpha^3}{3!}. \tag{2.11}$$

In fact, it is later stated explicitly in the text (see verse 17 of this chapter) that this is the approximation that has been employed in arriving at the value of $224'\,50''$ for the first *piṇḍajyā*. Thus,

$$P_1N_1 = R\sin\alpha \approx \frac{21600}{2\pi}\left(\alpha - \frac{\alpha^3}{6}\right) = 224.8389' \approx 224'50''. \tag{2.12}$$

In the following, we give the procedure outlined in the text for obtaining the successive *khaṇḍajyās* and *piṇḍajyās*, along with the rationale behind it. The second *khaṇḍajyā* Δ_2 is defined as

$$\begin{aligned} \Delta_2 &= J_2 - J_1 \\ &= R(\sin 2\alpha - \sin\alpha), \end{aligned} \tag{2.13}$$

where $P\hat{O}P_2 = 2\alpha$. Now, $\sin 2\alpha = 2\sin\alpha\cos\alpha$. Hence,

$$\Delta_2 = R\sin\alpha(2\cos\alpha - 1). \tag{2.14}$$

Rewriting the above expression we have,

$$\Delta_2 = R\sin\alpha[1 - 2(1-\cos\alpha)]. \tag{2.15}$$

For $\alpha = 225'$, we have

$$2(1-\cos\alpha) \approx 0.004282153. \tag{2.16}$$

This is approximated in the text by

$$\frac{1}{233\frac{1}{2}} \approx 0.004282655. \tag{2.17}$$

Hence

$$\begin{aligned}\Delta_2 &= R\sin\alpha\left(1-\frac{1}{233\frac{1}{2}}\right),\\ \text{or}\qquad \Delta_2 &= J_1 - \frac{J_1}{233\frac{1}{2}}\\ &= \Delta_1 - \frac{J_1}{233\frac{1}{2}}\\ &\approx 224'50'' - 57.77''\\ &\approx 223'52''.\end{aligned} \tag{2.18}$$

The second *piṇḍajyā* is given by

$$\begin{aligned}J_2 &= J_1 + \Delta_2\\ &= 224'50'' + 223'52''\\ &= 448'42''.\end{aligned} \tag{2.19}$$

The third *khaṇḍajyā* Δ_3 is defined as

$$\Delta_3 = J_3 - J_2 = R(\sin 3\alpha - \sin 2\alpha). \tag{2.20}$$

Rewriting the above expression we get

$$\begin{aligned}\Delta_3 &= R\left[\sin(2\alpha+\alpha) - \sin 2\alpha\right]\\ &= R\left[(\sin 2\alpha\cos\alpha + \cos 2\alpha\sin\alpha) - \sin 2\alpha\right]\\ &= R\left[(\sin 2\alpha\cos\alpha + (2\cos^2\alpha - 1)\sin\alpha) - \sin 2\alpha\right]\\ &= R\left[2\sin 2\alpha\cos\alpha - \sin\alpha - \sin 2\alpha\right]\\ &= R\left[\sin 2\alpha - \sin\alpha - 2\sin 2\alpha(1-\cos\alpha)\right]\\ &= \Delta_2 - J_2\, 2(1-\cos\alpha).\end{aligned} \tag{2.21}$$

We have already noted that

$$2(1-\cos\alpha) \approx \frac{1}{233\frac{1}{2}}. \tag{2.22}$$

Hence the third *khaṇḍajyā* is given by

$$\Delta_3 = \Delta_2 - \frac{J_2}{233\frac{1}{2}}$$

$$\begin{aligned} &\approx 223'52'' - 1'55'' \\ &= 221'57''. \end{aligned} \tag{2.23}$$

Thus the third *piṇḍajyā* becomes

$$\begin{aligned} J_3 &= J_2 + \Delta_3 \\ &= 448'42'' + 221'57'' \\ &= 670'39'', \end{aligned} \tag{2.24}$$

and so on. In general, the ith *khaṇḍajyā* is given by

$$\Delta_i = \Delta_{i-1} - \frac{J_{i-1}}{233\frac{1}{2}}, \tag{2.25}$$

and the ith *piṇḍajyā* by

$$J_i = J_{i-1} + \Delta_i. \tag{2.26}$$

The iterative relation (2.25) follows from the easily verifiable relation for Δ_{i+1} given by

$$\begin{aligned} \Delta_{i+1} &= R\sin(i+1)\alpha - R\sin i\alpha \\ &= R\left[\sin i\alpha - \sin(i-1)\alpha - 2\sin i\alpha(1-\cos\alpha)\right], \\ &= \Delta_i - 2(1-\cos\alpha)R\sin i\alpha, \end{aligned} \tag{2.27}$$

and the above approximation (2.22) for $2(1-\cos\alpha)$. In fact, a recursion relation amounting to the above is stated a few verses later. The above iterative procedure is described in *Laghu-vivṛti* as follows:

ततो विलिप्तादशकेन विरहितात् तत्त्वनेत्रात् आद्यज्यार्धतुल्यात्, सार्धदेवाश्विभिः विभज्य यल्लिप्तादिफलं लभ्यते, तदाद्यद्वितीययोः खण्डज्ययोरन्तरं स्यात्। तदाद्यज्याखण्डतो विशोध्य शिष्टं द्वितीयखण्डज्या स्यात्। ततस्तद्युक्ता प्रथमखण्डज्या द्वितीयपिण्डज्या स्यात्। ततो द्वितीयपिण्डज्यातः पूर्वहारेणैव विभक्तं फलं द्वितीयतृतीययोः खण्डज्ययोरन्तरं स्यात्। ततः पुनः द्वितीयखण्डज्यातः विशोध्य शिष्टं तृतीयखण्डज्या स्यात्। तस्याः द्वितीयपिण्डज्यायाश्च योगः तृतीयपिण्डज्या स्यात्। अथ ततोप्युक्तप्रकारेण चतुर्थाद्याः गुणाः क्रमेण साध्याः।

Then, whatever is obtained in minutes etc. (*liptādi*) as the result when 225 diminished by 10 seconds, which is equal to the first Rsine, is divided by 233.5, will be the difference between the first and second *khaṇḍajyās*. This [result] when subtracted from the first *khaṇḍajyā* will be the second *khaṇḍajyā*. The first *khaṇḍajyā* added to this will then be the second *piṇḍajyā*. Then the result obtained by dividing the second *piṇḍajyā* by the above-mentioned divisor will be the difference between the second and third *khaṇḍajyās*. Again when this [result] is subtracted from the second *khaṇḍajyā*, [the quantity obtained] will be the third *khaṇḍajyā*. The sum of this and the second *piṇḍajyā* will be the third *piṇḍajyā*. From there on, the fourth Rsine etc. have to be obtained by the method stated above.

Laghu-vivṛti also prescribes more accurate values of the first Rsine (J_1), as well as the divisor.

अत्र विलिप्तारूपमेव ज्याचापान्तरमभिप्रेत्य विलिप्तादशकमुक्तम्। वस्तुतः पुनः अष्टत्रिंशत् तत्पराधिकं विलिप्तानवकमेवैतत्। अत एव भागहारोऽपि न तत्र सार्धदेवाश्वितुल्यः। अपि तु द्वात्रिंशद्विलिप्ताधिकदेवाश्वितुल्य इति।

Here with the intention of specifying the difference between the Rsine and the arc in terms of *viliptās* only, it was stated to be 10 *viliptās*. Actually it is 38 *tatparās* in excess of 9 *viliptās*. It is for this reason, the divisor is also not $233\frac{1}{2}$. But 233 (minutes) and 32 *viliptās*.

What is stated above is that the first Rsine (J_1) should be taken to be $225' - 0'9''38''' = 224'50''22'''$. Similarly the divisor should be taken to be $233 + \frac{32}{60}$ instead of $233\frac{1}{2}$. The values of tabular Rsines as calculated with these more accurate values of J_1 and the divisor are more or less the same as the tabular Rsines given by Mādhava, as we show below in Table 2.1.

२.४ प्रकारान्तरेण ज्यानयनम्

2.4 Another method for obtaining the Rsines

व्यासार्धं प्रथमं नीत्वा ततो वान्यान् गुणान् नयेत् ॥ ६ ॥
त्रीशघ्नचक्रलिप्ताभ्यः व्यासोऽर्थेष्वग्निभिर्हृतः ।
तद्दलाद्यज्ययोः कृत्योः भेदान्मूलमुपान्तिमा ॥ ७ ॥
अन्त्योपान्त्यान्तरं द्विघ्नं गुणो व्यासदलं हरः ।
आद्यज्यायास्तथापि स्यात् खण्डज्यान्तरमादितः ॥ ८ ॥
ताभ्यां तु गुणहाराभ्यां द्वितीयादेरपि क्रमात् ।
उत्तरोत्तरखण्डज्याभेदाः पिण्डगुणार्धतः ॥ ९ ॥
एवं सावयवा जीवाः सम्यङ्नीत्वा पठेत् क्रमात् ।

vyāsārdhaṃ prathamaṃ nītvā tato vānyān guṇān nayet || 6 ||
trīśaghnacakraliptābhyaḥ vyāso'rtheṣvagnibhirhṛtaḥ |
taddalādyajyayoḥ kṛtyoḥ bhedānmūlamupāntimā || 7 ||
antyopāntyāntaraṃ dvighnaṃ guṇo vyāsadalaṃ haraḥ |
ādyajyāyāstathāpi syāt khaṇḍajyāntaramāditaḥ || 8 ||
tābhyāṃ tu guṇahārābhyāṃ dvitīyāderapi kramāt |
uttarottarakhaṇḍajyābhedāḥ piṇḍaguṇārdhataḥ || 9 ||
evaṃ sāvayavā jī vāḥ samyaṅnītvā paṭhet kramāt |

Or else, the *guṇās* [the values of the *jyās*] may be obtained by first obtaining the *vyāsārdha* (radius). The number of seconds of arc in a circle multiplied by 113 and divided by 355 is the diameter.[7]

The square root of the difference between the squares of half of that (diameter) and the first *jyā* is the penultimate *jyā*. The difference between the last *jyā* and the penultimate

[7] Here, a clarifying note regarding the number 355 represented using the *Bhūtasaṅkhyā* system may be useful. In the string *artheṣvagni* employed to refer to this number, the word *artha* should not be taken to refer to *puruṣārtha*, in which case it would be referring to the number 4. On the other hand, it should be taken to be referring to 5 sense organs—through the derivation "arthyate anenetyartha.h" (through which things are sought after).

one multiplied by two is the *guṇa* (multiplier) and the radius is the *hāra* (divisor). From the *ādyajyā* [multiplying it with the multiplier and dividing by the divisor], the difference between the first two *khaṇḍajyās* is obtained. With the same multiplier and divisor, and multiplying the multiplier by the second *piṇḍajyā*, the third *piṇḍajyā* etc., the difference between the successive *khaṇḍajyās* are obtained. Having thus obtained the *jyās* with their parts (seconds etc.) they may be tabulated in a sequence.

Here a procedure for generating the *jyā* table (table of Rsines) by finding the differences of the successive *khaṇḍajyās* is described. As will be seen below, this procedure merely involves the knowledge of the first *jyā* (J_1) and *trijyā*. It may be recalled that the method described in the previous section (verses 4–6a) essentially made use of the following equations for generating the successive *piṇḍajyā* values given in Table 2.1.

$$J_{i+1} = J_i + \Delta_{i+1} \qquad (0 \leq i \leq 23) \tag{2.28}$$

$$\Delta_{i+1} = \Delta_i - \frac{J_i}{233\frac{1}{2}} \qquad (1 \leq i \leq 23), \tag{2.29}$$

where Δ_i and J_i $i = 1, 2, \ldots, 24$, refer to the *khaṇḍajyās* and *piṇḍajyās* respectively. Since $\Delta_1 = J_1$, is known, all the *jyās* can be generated using the above equations recursively. Equation (2.29) can be rewritten as

$$\Delta_i - \Delta_{i+1} = \frac{J_i}{233\frac{1}{2}}. \tag{2.30}$$

In the above verses (6–9) the recursion relation which is the basis of (2.30) is stated. Here the value of the last *jyā* (J_{24} = *trijyā*), which is the same as the radius of the circle, is first stated. Since J_1 is already known, with these two *jyās* (the first and the last), the value of the penultimate *jyā* (J_{23}) is found. Then the text defines a *guṇa* or multiplier and a *hāra* or divisor, using which a recursion relation is formulated; making use of this, all the tabular differences of the *khaṇḍajyās* and hence the values of the 24 *jyās* can be obtained. This method is quite instructive and may be described as follows. It has already been noted that the circumference of the circle is taken to be 21600. The diameter of this circle is stated to be:

$$D = \frac{21600 \times 113}{355}. \tag{2.31}$$

So, essentially, $\frac{355}{113} = 3.14159$ is taken to be the approximate value of π. Using (2.3), and the notation $\alpha = 225' = 3.75^\circ$, we have

$$\begin{aligned}
\sqrt{{J_{24}}^2 - {J_1}^2} &= R\sqrt{\sin^2 24\alpha - \sin^2\alpha} \\
&= \sqrt{(R\sin 90)^2 - (R\sin 3.75)^2} \\
&= R\sqrt{1 - \sin^2\alpha} \\
&= R\cos\alpha \\
&= R\sin(24\alpha - \alpha) \qquad (24\alpha = 90^\circ)
\end{aligned}$$

$$= R\sin 23\alpha$$
$$= J_{23}, \qquad (2.32)$$

where R is *trijyā*. Having obtained the penultimate *jyā* from the first and the last *jyās*, the multiplier and divisor are defined. *Laghu-vivṛti* puts them in very clear terms as follows:

तस्याः उपान्तिमज्यायाः अन्त्यज्यायाश्च व्यासार्धतुल्यायाः यदन्तरं, तद्विगुणितं गुणः; व्यासार्धतुल्यो हारः।

The difference between the penultimate *jyā* and the ultimate *jyā*, which is equal to the radius, multiplied by two is the multiplier. The radius is the divisor.

That is,

$$\begin{aligned} guṇa &= 2(R - R\sin 23\alpha), \\ hāra &= R. \end{aligned} \qquad (2.33)$$

Now the recursion relation to obtain the sine differences or the *khaṇḍajyās* can be written as follows:

$$\begin{aligned} \Delta_i - \Delta_{i+1} &= \frac{guṇa}{hāra} R\sin i\,\alpha \\ &= \frac{2(R - R\sin 23\alpha)}{R} R\sin i\,\alpha. \end{aligned} \qquad (2.34)$$

For instance, with $i = 1$, the above equation becomes

$$\begin{aligned} \Delta_1 - \Delta_2 &= \frac{2(R - R\sin 23\alpha)}{R} R\sin\alpha \\ &= R[2\sin\alpha - 2\sin 23\alpha\sin\alpha] \\ &= R[2\sin\alpha - (\cos 22\alpha - \cos 24\alpha)] \\ &= R[2\sin\alpha - \cos(24\alpha - 2\alpha) + 0] \\ &= R[2\sin\alpha - \sin 2\alpha]. \end{aligned} \qquad (2.35)$$

From the definition of *khaṇḍajyā*, we have

$$\begin{aligned} \Delta_1 - \Delta_2 &= (J_1 - J_0) - (J_2 - J_1) \\ &= 2J_1 - J_2. \end{aligned} \qquad (2.36)$$

Clearly (2.36) is the same as (2.35). In general,

$$\begin{aligned} \Delta_i - \Delta_{i+1} &= (J_i - J_{i-1}) - (J_{i+1} - J_i) \\ &= 2J_i - J_{i+1} - J_{i-1} \\ &= R\,[2\sin i\alpha - \sin(i+1)\alpha - \sin(i-1)\alpha]. \end{aligned} \qquad (2.37)$$

Using $\cos(90-\theta) = \sin\theta$, $\cos(90+\theta) = -\sin\theta$, we get

$$\begin{aligned}\Delta_i - \Delta_{i+1} &= R\,[2\sin i\alpha - \cos(24\alpha - (i+1)\alpha) + \cos(24\alpha + (i-1)\alpha)] \\ &= R\,[2\sin i\alpha - \cos(23-i)\alpha - \cos(23+i)\alpha] \\ &= R\,[2\sin i\alpha - 2\sin 23\alpha \sin i\alpha] \\ &= \frac{2(R - R\sin 23\alpha)}{R} R\sin i\alpha, \end{aligned} \qquad (2.38)$$

which is the recursion relation (2.34) for the *khaṇḍajyās* given in the text.

Commenting on the first line of the tenth verse, *evaṃ sāvayavā jīvāḥ samyaṅnītvā paṭhet kramāt*, Śaṅkara Vāriyar describes the accurate values of the 24 Rsines—which he attributes to Mādhava—in the following verses:

श्रेष्ठं नाम वरिष्ठानां हिमाद्रिर्वेदभावनः।
तपनो भानुसूक्तज्ञो मध्यमं विद्धि दोहनम्॥ १ ॥
धिगाज्यो नाशनं कष्टं छन्नभोगाशयाम्बिका।
मृगाहारो नरेशोऽयं वीरो रणजयोत्सुकः॥ २ ॥
मूलं विशुद्धं नालस्य गानेषु विरला नराः।
अशुद्धिगुप्ता चोरश्रीः शङ्कुकर्णो नगेश्वरः॥ ३ ॥
तनुजो गर्भजो मित्रं श्रीमानत्र सुखी सखे।
शशी रात्रौ हिमाहारो वेगज्ञः पथि सिन्धुरः॥ ४ ॥
छायालयो गजो नीलो निर्मलो नास्ति सत्कुले।
रात्रौ दर्पणमभ्राङ्गं नागस्तुङ्गनखो वली ॥ ५ ॥
धीरो युवा कथालोलः पूज्यो नारीजनैर्भगः।
कन्यागारे नागवल्ली देवो विश्वस्थली भृगुः॥ ६ ॥
तत्परादिकलान्तास्तु महाज्या माधवोदिताः।
स्वस्वपूर्वविशुद्धे तु शिष्टास्तत्खण्डमौर्विकाः॥ ७ ॥ इति ॥

Here the values of the 24 Rsines are given up to the thirds in the *Kaṭapayādi* notation. For instance, consider the first Rsine given by '*śreṣṭhaṃ nāma variṣṭhānāṃ*'. The three words here stand for 22, 50 and 224 respectively. Hence the value of the first Rsine is: $224'\,50''\,22'''$. The values of the other Rsines are deciphered in a similar manner. These have been arrived at by considering terms up to θ^{11} in the series expansion of $\sin\theta$ which was also derived by Mādhava:

$$\sin\theta = \theta - \frac{\theta^3}{3!} + \frac{\theta^5}{5!} - \frac{\theta^7}{7!} + \frac{\theta^9}{9!} - \frac{\theta^{11}}{11!} + \ldots.$$

Table of *jyās*

In Table 2.1, we reproduce the values of *jyās* corresponding to arc lengths which are multiples of $225'$, given in *Āryabhaṭīya*/*Sūryasiddhānta*, *Tantrasaṅgraha* and *Laghu-vivṛti* (considering more accurate values for the first *jyā* as well as the divisor). The values of *jyās* enunciated by Mādhava are also listed based on the verses '*śreṣṭhaṃ nāma variṣṭhānāṃ*...' cited in *Laghu-vivṛti*. In fact, the modern values presented in the last column show that the Mādhava's values are accurate up to the thirds. In *Yuktibhāṣā* it is noted that the *jyā* for any arc can be obtained without using the tabular values, by using the infinite series expansion for it.

Dhanu or *Cāpa*		Notation	Value of the *jyā* (in minutes, seconds and thirds)				
Symbol used	Length (min)	used	As in AR/SS	From TS	From LV	Given by Mādhava	Modern
S_1	225	J_1	225	224 50	224 50 21	224 50 22	224 50 21
S_2	450	J_2	449	448 42	448 42 58	448 42 58	448 42 57
S_3	675	J_3	671	670 39	670 40 16	670 40 16	670 40 16
S_4	900	J_4	890	889 44	887 45 17	889 45 15	889 45 15
S_5	1125	J_5	1105	1105 00	1105 01 41	1105 01 39	1105 01 38
S_6	1350	J_6	1315	1315 32	1315 34 11	1315 34 07	1315 34 07
S_7	1575	J_7	1520	1520 26	1520 28 41	1520 28 35	1520 28 35
S_8	1800	J_8	1719	1718 49	1718 52 32	1718 52 24	1718 52 24
S_9	2025	J_9	1910	1909 51	1909 54 46	1909 54 35	1909 54 35
S_{10}	2250	J_{10}	2093	2092 42	2092 46 19	2092 46 03	2092 46 03
S_{11}	2475	J_{11}	2267	2266 35	2266 40 10	2266 39 50	2266 39 50
S_{12}	2700	J_{12}	2431	2430 45	2430 51 40	2430 51 15	2430 51 14
S_{13}	2925	J_{13}	2585	2584 32	2585 38 37	2584 38 06	2584 38 05
S_{14}	3150	J_{14}	2728	2727 14	2727 21 31	2727 20 52	2727 20 52
S_{15}	3375	J_{15}	2859	2858 15	2858 23 42	2858 22 55	2858 22 55
S_{16}	3600	J_{16}	2978	2977 02	2977 11 30	2977 10 34	2977 10 33
S_{17}	3825	J_{17}	3084	3083 03	3083 14 23	3083 13 17	3083 13 16
S_{18}	4050	J_{18}	3177	3175 53	3176 05 07	3176 03 50	3176 03 49
S_{19}	4275	J_{19}	3256	3255 06	3255 19 50	3255 18 22	3255 18 21
S_{20}	4500	J_{20}	3321	3320 24	3320 38 11	3320 36 30	3320 36 30
S_{21}	4725	J_{21}	3372	3371 27	3371 43 24	3371 41 29	3371 41 29
S_{22}	4950	J_{22}	3409	3408 05	3408 22 20	3408 20 11	3408 20 10
S_{23}	5175	J_{23}	3431	3430 07	3430 25 35	3430 23 11	3430 23 10
S_{24}	5400	J_{24}	3438	3437 27	3437 47 29	3437 44 48	3437 44 48

Table 2.1 *Jyā* values corresponding to arc lengths which are multiples of $225'$. *Āryabhaṭīya*, *Sūryasiddhānta*, *Tantrasaṅgraha*, *Laghu-vivṛti* (with a more accurate first sine as well as the divisor) and Mādhava's values.

२.५ इष्टदोःकोटिज्यानयनम्

2.5 Obtaining the desired Rsines and Rcosines

इष्टदोःकोटिधनुषोः स्वसमीपसमीरिते ॥ १० ॥
ज्ये द्वे सावयवे न्यस्य कुर्यादूनाधिकं धनुः ।
द्विघ्नतल्लिप्तिकाप्तैकशरशैलशिखीन्दवः ॥ ११ ॥
न्यस्याच्छेदाय च मिथः तत्संस्कारविधित्सया ।
छित्वैकां प्राक् क्षिपेज्जह्यात् तद्धनुष्यधिकोनके ॥ १२ ॥
अन्यस्यामथ तां द्विघ्नां तथा स्यामिति संस्कृतिः ।

इति ते कृतसंस्कारे स्वगुणौ धनुषोस्तयोः ॥ १३ ॥
तत्राल्पीयःकृतिं[8] त्यक्त्वा पदं त्रिज्याकृतेः परः ।

iṣṭadoḥkoṭidhanuṣoḥ svasamīpasamīrite || 10 ||
jye dve sāvayave nyasya kuryādūnādhikaṃ dhanuḥ |
dvighnatalliptikāptaikaśaraśailaśikhīndavaḥ || 11 ||
nyasyācchedāya ca mithaḥ tatsaṃskāravidhitsayā |
chitvaikāṃ prāk kṣipejjahyāt taddhanuṣyadhikonake || 12 ||
anyasyāmatha tāṃ dvighnāṃ tathā syāmiti saṃskṛtiḥ |
iti te kṛtasaṃskāre svaguṇau dhanuṣostayoḥ || 13 ||
tatrālpīyaḥkṛtiṃ tyaktvā padaṃ trijyākṛteḥ paraḥ |

Having noted down the listed/tabulated values (*samīrita*) of the *dorjyās* (Rsines) and *koṭijyās* (Rcosines) corresponding to the two points which are on either side of the arc whose *dorjyā* and *koṭijyā* are desired, find the difference in the arc which may be in excess of or short of it. [The number] 13751 divided by twice this difference has to be stored [as divisor *D*] for dividing. This is done for mutual correction (i.e. for correcting the *dorjyā* in determining *koṭijyā* and vice versa). First divide one of them (the *dorjyā* or *koṭijyā* by *D*) and add or subtract this from the other (if the *dorjyā* is divided, apply it to the *koṭijyā* and if the *koṭijyā* is divided, apply it to the *dorjyā*) depending upon whether the difference is in excess or short. This result multiplied by two and operated as before (divided by *D* and applied to the *dorjyā* or *koṭijyā*) forms the process of correction. The correction thus carried out gives the exact value of the *dorjyā* or the *koṭijyā* of the desired arc. Of the two (*dorjyā* or *koṭijyā*) find the square of the *jyā* of the smaller one and subtract it from the square of the *trijyā*. The square root of the result gives the other (the *koṭijyā* or *dorjyā*).

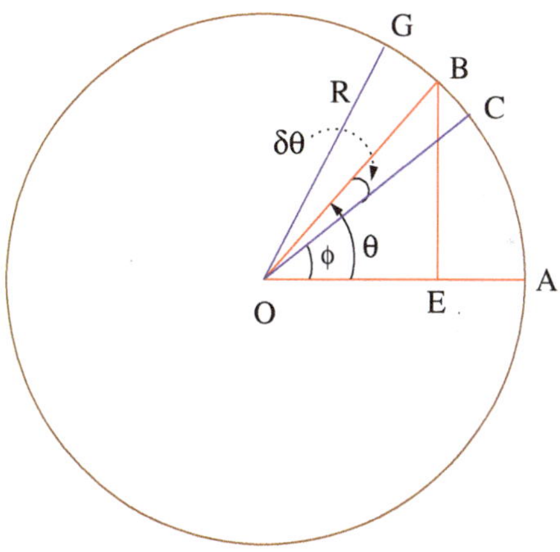

Fig. 2.3 Finding the *jyā* value corresponding to a desired arc.

In Fig. 2.3, *AB* is the arc whose *jyā* and *kotijyā* are desired to be found. The length of the arc $AB = R\theta$, where R is the *trijyā* and θ is the angle subtended by the arc at the centre O, expressed in radians. The *jyās* corresponding to the known arc lengths *AC* and *AG* are known from the *jyā* table (Table 2.1). The procedure for

[8] The reading in both the printed editions is: तत्राल्पीयः कृतिं। This however is grammatically incorrect. Hence we have provided the right compound form of the word above.

finding the *jyā* corresponding to the desired arc length AB from either of the two known *jyās* is described in the above verses.

It may be noted from the figure that the desired arc length $AB = R\theta$ is such that $i\alpha \leq R\theta \leq (i+1)\alpha$, for some integer $0 < i < 24$, where $\alpha = 225'$. Assume that point B is closer to C than G, i.e. $BC < BG$. Let $BC = R\delta\theta$. The problem is to find the *dorjyā* and *koṭijyā* corresponding to the arc AB, where $AB = i\alpha + R\delta\theta$.

The formulae for the two *jyās* involve an intermediate quantity (called the *hāraka*, or divisor (D) by the commentator), which is defined as:

$$D = \frac{13751}{2R\,\delta\theta}. \tag{2.39}$$

The number 13751 appearing in the numerator is essentially four times the radius R of the circle measured in minutes. In fact it is a good approximation too, as $2 \times 21600/\pi \approx 13750.98708$. Hence the above equation can be written as

$$D = \frac{4R}{2R\,\delta\theta} = \frac{2}{\delta\theta}. \tag{2.40}$$

While the *dorjyā* of an arc increases with the arc length, the *koṭijyā* decreases. Considering this, the text presents the following relations.

$$\begin{aligned}
dorjy\bar{a}(i\alpha + R\delta\theta) &= dorjy\bar{a}\; i\alpha + \frac{2}{D}\left(ko\d{t}ijy\bar{a}\; i\alpha - \frac{dorjy\bar{a}\; i\alpha}{D}\right) \\
&= dorjy\bar{a}\; i\alpha - \frac{(dorjy\bar{a}\; i\alpha)\delta\theta^2}{2} + (ko\d{t}ijy\bar{a}\; i\alpha)\,\delta\theta \\
&= dorjy\bar{a}\; i\alpha\left(1 - \frac{\delta\theta^2}{2}\right) + (ko\d{t}ijy\bar{a}\; i\alpha)\,\delta\theta, \qquad (2.41) \\
dorjy\bar{a}\,(i\alpha - R\delta\theta) &= dorjy\bar{a}\; i\alpha - \frac{2}{D}\left(ko\d{t}ijy\bar{a}\; i\alpha + \frac{dorjy\bar{a}\; i\alpha}{D}\right) \\
&= dorjy\bar{a}\; i\alpha - \frac{(dorjy\bar{a}\; i\alpha)\delta\theta^2}{2} - (ko\d{t}ijy\bar{a}\; i\alpha)\,\delta\theta \\
&= dorjy\bar{a}\; i\alpha\left(1 - \frac{R\delta\theta^2}{2}\right) - (ko\d{t}ijy\bar{a}\; i\alpha)\,\delta\theta, \qquad (2.42) \\
ko\d{t}ijy\bar{a}\,(i\alpha + R\delta\theta) &= ko\d{t}ijy\bar{a}\; i\alpha - \frac{2}{D}\left(dorjy\bar{a}\; i\alpha + \frac{ko\d{t}ijy\bar{a}\; i\alpha}{D}\right) \\
&= ko\d{t}ijy\bar{a}\; i\alpha - \frac{(ko\d{t}ijy\bar{a}\; i\alpha)\delta\theta^2}{2} - (dorjy\bar{a}\; i\alpha)\delta\theta \\
&= ko\d{t}ijy\bar{a}\; i\alpha\left(1 - \frac{R\delta\theta^2}{2}\right) - (dorjy\bar{a}\; i\alpha)\delta\theta, \qquad (2.43) \\
ko\d{t}ijy\bar{a}\,(i\alpha - R\delta\theta) &= ko\d{t}ijy\bar{a}\; i\alpha + \frac{2}{D}\left(dorjy\bar{a}\; i\alpha - \frac{ko\d{t}ijy\bar{a}\; i\alpha}{D}\right) \\
&= ko\d{t}ijy\bar{a}\; i\alpha - \frac{(ko\d{t}ijy\bar{a}\; i\alpha)\delta\theta^2}{2} + (dorjy\bar{a}\; i\alpha)\delta\theta
\end{aligned}$$

$$= \textit{koṭijyā}\, i\alpha \left(1 - \frac{R\delta\theta^2}{2}\right) + (\textit{dorjyā}\, i\alpha)\delta\theta. \qquad (2.44)$$

In *Laghu-vivṛti* the procedure for finding the *dorjyā* and *koṭijyā* of any arc length is explained clearly as follows:

ततस्तेन हारकेण भुजाज्यां कोटिज्यां वा एकां कर्तुमिष्टां प्रथमतः विभज्य लब्धं कलादिकं फलं अन्यस्यां, भुजायाः साध्यत्वे कोटिज्यायां तस्याः साध्यत्वे भुजाज्यायां च साध्येतरज्यायां तत्सम्बन्धिनो धनुषः ऊनाधिकत्ववशात् ऋणं धनं वा कुर्यात्। अथैवं कृतां तां द्विगुणितां कृत्वा पूर्वोक्तेनैव हारकेण विभज्य लब्धं यत् फलं तत्पुनरन्यस्यां साध्यज्यायामेव तं धनुषः ऊनाधिकवशादृणं धनं वा कुर्यात्। एवं कृता भुजाज्या कोटिज्या च परस्परलब्धफलसंस्कृते स्फुटे भवतः।

By that divisor divide the *dorjyā* or *koṭijyā*, whichever is desired to be found, and this may be added to or subtracted from the other one. That is, if the *dorjyā* is desired to be found, it may be applied to the *koṭijyā* and if the *koṭijyā* is to be found it may be applied to the *dorjyā*, the application being positive or negative depending upon whether the arc $R\delta\theta$ is added to or subtracted from $[i\alpha]$.

Then this quantity may be multiplied by two and divided by the same divisor. The result has to be applied to the desired *jyā* [i.e.,] if the *koṭijyā* is to be found it has to be applied to the *koṭijyā*, and if the *dorjyā* is to be found it has to be applied to the *dorjyā*, the application being positive or negative depending upon whether the arc $R\delta\theta$ is added to or subtracted from $[i\alpha]$. The *dorjyā* and *koṭijyā* thus applied to each other give the correct *jyā* of the desired arc.

If the arc length $i\alpha \pm R\delta\theta$ corresponds to an angle $\phi \pm \delta\theta$ (in radians), then equations (2.41) to (2.44) are equivalent to the following relations:

$$R\sin(\phi + \delta\theta) = R\sin\phi \left(1 - \frac{\delta\theta^2}{2}\right) + (R\cos\phi)\delta\theta, \qquad (2.45)$$

$$R\sin(\phi - \delta\theta) = R\sin\phi \left(1 - \frac{\delta\theta^2}{2}\right) - (R\cos\phi)\delta\theta, \qquad (2.46)$$

$$R\cos(\phi + \delta\theta) = R\cos\phi \left(1 - \frac{\delta\theta^2}{2}\right) - (R\sin\phi)\delta\theta, \qquad (2.47)$$

$$R\cos(\phi - \delta\theta) = R\cos\phi \left(1 - \frac{\delta\theta^2}{2}\right) + (R\sin\phi)\delta\theta. \qquad (2.48)$$

It is obvious that (2.45) to (2.48) are approximations of the standard trigonometric relations

$$R\sin(\phi + \delta\theta) = R(\sin\phi\cos\delta\theta + \cos\phi\sin\delta\theta), \qquad (2.49)$$

$$R\sin(\phi - \delta\theta) = R(\sin\phi\cos\delta\theta - \cos\phi\sin\delta\theta), \qquad (2.50)$$

$$R\cos(\phi + \delta\theta) = R(\cos\phi\cos\delta\theta - \sin\phi\sin\delta\theta), \qquad (2.51)$$

$$R\cos(\phi - \delta\theta) = R(\cos\phi\cos\delta\theta + \sin\phi\sin\delta\theta), \qquad (2.52)$$

when the approximations, $\cos\delta\theta = \left(1 - \frac{\delta\theta^2}{2}\right)$ and $\sin\delta\theta = \delta\theta$, for small $\delta\theta$ are used. These also happen to be the first two terms in the Taylor series expansion of

$\sin(\phi \pm \delta\theta)$ and $\cos(\phi \pm \delta\theta)$. Śaṅkara Vāriyar, however, has given an incorrect generalisation of these to higher orders in his *Laghu-vivṛti.*

If either the *dorjyā* or the *koṭijyā* of an arc is known, the other can be determined using the following relation. Let α be the length of the arc AB (in minutes) as shown in Fig. 2.4; then,

$$dorjy\bar{a}^2\alpha + ko\d{t}ijy\bar{a}^2\alpha = R^2, \tag{2.53}$$

which is the same as

$$\sin^2\theta + \cos^2\theta = 1. \tag{2.54}$$

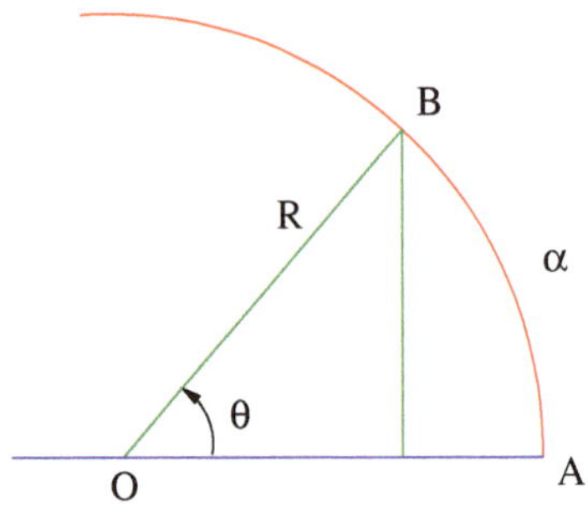

Fig. 2.4 Relation between the *dorjyā*, the *koṭijyā* and the *trijyā*.

२.६ इष्टज्यायाश्चापीकरणम्

2.6 Determining the length of the arc from the corresponding Rsine

ज्ययोरासन्नयोर्भेदभक्तस्तत्कोटियोगतः ॥ १४ ॥
छेदस्तेन हृता द्विघ्ना त्रिज्या तद्धनुरन्तरम् ॥

jyayorāsannayorbhedabhaktastatkoṭiyogataḥ || 14 ||
chedastena hṛtā dvighnā trijyā taddhanurantaram ||

The sum of the cosines divided by the difference of those two sines, which are close to each other, forms the *cheda* (divisor). Twice the *trijyā* divided by this is the difference between the corresponding arcs.

Consider Fig. 2.5*a*. P and Q are points along the circle whose distance from the point A are multiples of $\alpha = 225'$, that is $AP = i\alpha$, and $AQ = (i+1)\alpha$, where i is an integer. The *jyās* corresponding to the arcs AP and AQ are known from the table. The idea is to find the arc length (AB in minutes) corresponding to the given *jyā* (BN). Since the arc length AP is known, to determine AB we just need to find the length of the arc PB.

Let $A\hat{O}P = \theta_0$, $A\hat{O}B = \theta$ and $P\hat{O}B = \theta - \theta_0 = \delta\theta$. Then, according to the text the arc length PB is given by the following approximate formula:

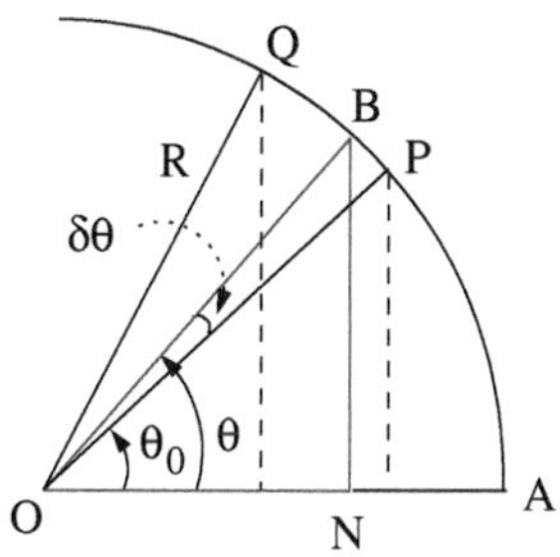

Fig. 2.5*a* Determination of the arc length corresponding to a given *jyā*.

$$PB = R\,\delta\theta \approx \frac{2R}{\left[\frac{(\cos\theta+\cos\theta_0)}{(\sin\theta-\sin\theta_0)}\right]}. \tag{2.55}$$

The rationale behind the above formula can be understood as follows. When $\delta\theta$ is small, $\sin\delta\theta \approx \delta\theta$ and $\cos\delta\theta \approx 1$. Hence, we have

$$\sin\theta = \sin(\theta_0+\delta\theta) \approx \sin\theta_0 + \cos\theta_0\;\delta\theta \tag{2.56a}$$
$$\sin\theta_0 = \sin(\theta-\delta\theta) \approx \sin\theta - \cos\theta\;\delta\theta. \tag{2.56b}$$

The above equations may be rewritten as

$$\sin\theta - \sin\theta_0 \approx \cos\theta_0\;\delta\theta \tag{2.56c}$$
$$\sin\theta - \sin\theta_0 \approx \cos\theta\;\delta\theta, \tag{2.56d}$$

from which we have

$$2(\sin\theta - \sin\theta_0) \approx (\cos\theta + \cos\theta_0)\;\delta\theta, \tag{2.57}$$

or,

$$\delta\theta \approx \frac{2(\sin\theta - \sin\theta_0)}{(\cos\theta + \cos\theta_0)}. \tag{2.58}$$

The above equation is the same as (2.55). We now proceed to explain another method—one that is most likely to have been employed by Indian astronomers—of arriving at the above expression for $\delta\theta$ with the help of a geometrical construction (see Fig. 2.5*b*). Here J is the midpoint of the arc PB and BN, JK and PM are perpendicular to OM. As the arc PB is small, it can be approximated by a straight line and K can be taken to be the midpoint of NM.

Then it can be easily seen from the figure that

$$BD = R(\sin\theta - \sin\theta_0)$$
$$\text{and} \qquad OK = \frac{R(\cos\theta + \cos\theta_0)}{2}. \tag{2.59}$$

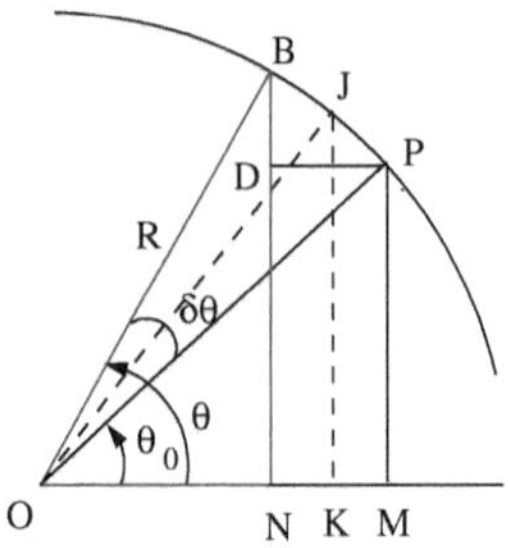

Fig. 2.5*b* Geometrical construction to determine the arc length corresponding to a given *jyā*.

Considering the similar triangles PBD and JOK, we have the relation

$$\frac{PB}{JO} = \frac{BD}{OK} \qquad \text{or} \qquad PB = JO \times \frac{BD}{OK}. \tag{2.60}$$

Using (2.59) in the above, we get

$$PB = \frac{2R}{\left[\frac{(\cos\theta + \cos\theta_0)}{(\sin\theta - \sin\theta_0)}\right]}, \tag{2.61}$$

which is the same as (2.55) given in the text.

The above verse is explained in *Laghu-vivṛti* as follows.

तत्र चापसन्धिपठितयोः निरन्तरयोर्जीवयोः यस्याः (जीवायाः) इष्टज्यां प्रत्यासन्नतरत्वं तस्याः इष्टज्यायाश्च यो भेदः तेन तयोः कोटिज्ययोः योगं विभजेत्। तत्र लब्धः छेदो नाम। ततस्तेन छेदेन द्विगुणितां त्रिज्यां विभजेत्। तत्र लब्धं इष्टज्या-तदासन्न-चापसन्धिज्ययोर्धनुषोः अन्तरम्; इष्टज्या-तदासन्न-चापसन्धिज्ययोरन्तरोत्थस्य ज्या-भागस्य धनुरित्यर्थः।

Of the two points whose *jyā* values are listed in the table, find the one which is closer to the desired *jyā* [whose arc value is to be found]. Then find the difference between these two *jyās* (*dorjyās*), and divide the sum of the *koṭijyās* by this difference. The result is called the *cheda*. Divide twice the *trijyā* by this *cheda*. The result obtained is the difference between the arcs lying between the desired *jyā*, and the *jyā* closest to it (as found from the table); that is, it gives the length of the arc corresponding to the difference in the *dorjyās*.

२.७ सूक्ष्मज्यानयनम्

2.7 Finding more accurate values of the desired Rsine

इति ज्याचापयोः कार्यं ग्रहणं माधवोदितम् ।
विधान्तरं च तेनोक्तं तयोः सूक्ष्मत्वमिच्छताम् ॥ १५ ॥
जीवे परस्परनिजेतरमौर्विकाभ्यां अभ्यस्यविस्तृतिदलेन विभज्यमाने।
अन्योन्ययोगविरहानुगुणे भवेतां यद्वा स्वलम्बकृतिभेदपदीकृते द्वे॥ १६ ॥

iti jyācāpayoḥ kāryaṃ grahaṇaṃ mādhavoditam|
vidhāntaraṃ ca tenoktaṃ tayoḥ sūkṣmatvamicchatām || 15 ||
jīve parasparanijetaramaurvikābhyāṃ abhyasyavistṛtidalena vibhajyamāne|
anyonyayogavirahānuguṇe bhavetāṃ yadvā svalambakṛtibhedapadīkṛte dve||16||

The [above] procedure for obtaining the *jyā* and *cāpa* has thus been explained by Mādhava. He has also given another method for those desirous of obtaining accurate values. Multiply each *jyā* (*dorjyā* of an arc length) by the other *jyā* (of another arc length) and divide them by the *trijyā*. Their sum or difference becomes (the *jyā*) of the sum or difference of the arcs. Or else, the square root of the difference of their own squares and that of the *lamba* [may be added and subtracted for getting the *jyā* of the sum or difference of the arcs].

The procedures for obtaining the Rsine and the arc described in the previous verses are attributed to Mādhava. Verse 16 essentially gives the $\sin(A+B)$ formula. This formula too is attributed to Mādhava and is explained in the commentary as follows:

···योगवियोगयोग्ये द्वे अपि अर्धज्ये परस्परस्य निजेतरज्याभ्यां स्वभुजाज्यां अन्यस्याः कोट्या अन्यभुजाज्यां स्वकोट्या च गुणयेत्। यदि स्वयं कोटिज्या, तर्हि तां अन्यस्य भुजाज्यया अन्यकोटिज्यां च स्वभुजया गुणयेत्। एवं कृतयोः द्वयोर्योगो वियोगो वा अभीष्टः कार्यः। ततो विस्तृतिदलेन विभजेत्। विभज्यमाने इति शानचा विभजनात् प्रागेव योगवियोगौ कर्तव्यौ इति दर्शयति। एवं कृतयोः योगो वियोगो वा स्फुटो भवति। अथवा द्वयोर्वर्गतः पृथगेकस्यैव तयोः साधारणस्य लम्बस्य वर्गमपनीय मूलीकृते उभे योगविरहानुगुणे भवेताम्। लम्बानयनं पुनः उभयोर्जीवयोः संवर्गतः त्रिज्यया हरणेन कर्तव्यम्।

...The *dorjyās* (Rsines) [of the arcs α and β] whose sum or difference is desired to be found have to be multiplied mutually with the other *jyā*. That is, the *dorjyā* of one (α) has to be multiplied by the *koṭijyā* of the other (β) and the *koṭijyā* of the one (α) has to be multiplied by the *dorjyā* of the other (β). The sum or difference of these two quantities has to be found as desired. Then it has to be divided by the *trijyā*. Here by using the suffix, *'śānac'* in the word *vibhajyamane* [the author] indicates that the addition or subtraction has to be done before division [by the *trijyā*]. This gives the correct value of the *dorjyā* of the sum or difference of the two arcs.

Alternatively, after subtracting the square of the *lamba/lambana* separately from the squares of the two *dorjyās* and taking the square root, the two quantities (thus obtained) become suitable for addition or subtraction. The *lamba* has to be obtained by multiplying the two *dorjyās* and dividing by the *trijyā*.

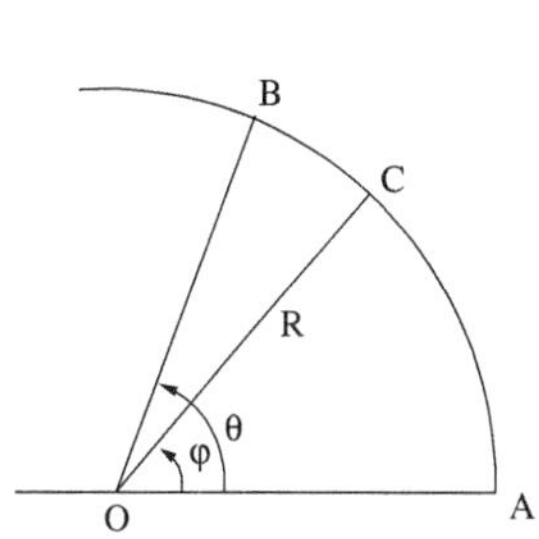

Fig. 2.6a Determination of the *jyā* corresponding to the sum or difference of two arcs.

Let α and β be the two arc lengths corresponding to the two angles θ and ϕ as shown in Fig. 2.6*a*. That is, $AB = \alpha$ and $AC = \beta$ respectively. Nīlakaṇṭha gives the following two formulae for finding the *jyā* of the sum or difference of these arc lengths.

$$dorjy\bar{a}\,(\alpha \pm \beta) = \frac{dorjy\bar{a}\,\alpha\; ko\d{t}ijy\bar{a}\,\beta \pm ko\d{t}ijy\bar{a}\,\alpha\; dorjy\bar{a}\,\beta}{trijy\bar{a}} \tag{2.62a}$$

$$dorjy\bar{a}\,(\alpha \pm \beta) = \sqrt{dorjy\bar{a}^2\alpha - lamba^2} \pm \sqrt{dorjy\bar{a}^2\beta - lamba^2}, \tag{2.62b}$$

where *lamba* in the above equation is defined by

$$lamba = \frac{dorjy\bar{a}\,\alpha\; dorjy\bar{a}\,\beta}{trijy\bar{a}}. \tag{2.63}$$

In terms of the angles θ and ϕ, *lamba* can be expressed as

$$lamba = \frac{R\sin\theta\; R\sin\phi}{R}. \tag{2.64}$$

The term *lamba* generally means a vertical line or a plumb-line. The expression for the *lamba* given above can be understood using a geometrical construction. For this consider two angles θ and ϕ such that $\theta > \phi$, as shown in Fig. 2.6*b*. Find $\sin\theta$ and $\sin\phi$. Draw lines XY and OZ perpendicular to each other as indicated in the figure. Now we consider a segment of length $R\sin\phi$ and place it inclined to OZ such that the segment BN makes an angle θ with BO.

Then draw a line NC such that $O\hat{C}N = \phi$. By construction, $B\hat{N}C = \theta - \phi$. Draw a perpendicular from B which meets the line NC at D. From the triangle NBD,

$$\sin(\theta - \phi) = \frac{BD}{R\sin\phi} \tag{2.65a}$$

Also in the triangle BCD,

$$\sin\phi = \frac{BD}{BC}. \tag{2.65b}$$

From (2.65*a*) and (2.65*b*)

$$BC = R\sin(\theta - \phi). \tag{2.65c}$$

Now, applying the sine rule to the triangle NBC, we get the following relation

$$\frac{NB}{\sin\phi} = \frac{BC}{\sin(\theta - \phi)} = \frac{NC}{\sin(180 - \theta)} \tag{2.66}$$

Since $NB = R\sin\phi$ (by construction) and $BC = R\sin(\theta - \phi)$ (see (2.65*c*)), from the above equation, the third side NC of the triangle must be equal to $R\sin(180 - \theta)$. That is $NC = R\sin(180 - \theta) = R\sin\theta$. Now it can be easily seen that NO in Fig. 2.6*b* represents the expression for the *lamba* given above.

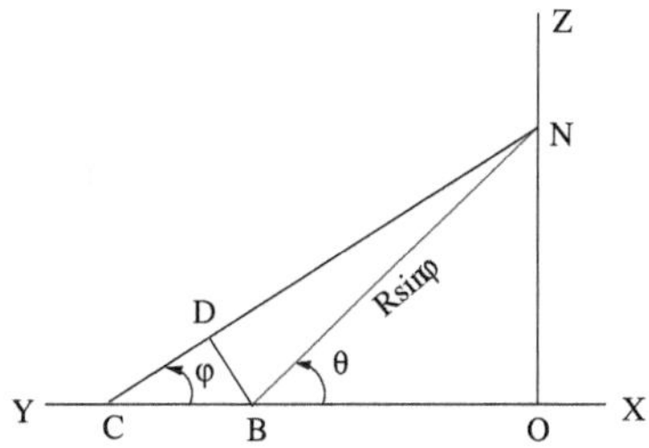

Fig. 2.6*b* Geometrical construction to understand the expression for the *lamba* given in Chapter 2, verse 16.

Using (2.3), and the above expression for the *lamba*, (2.62*a*) and (2.62*b*) reduce to the following equations respectively,

$$R\,\sin(\theta \pm \phi) = \frac{R\sin\theta\, R\cos\phi \pm R\cos\theta\, R\sin\phi}{R} \tag{2.67a}$$

$$R\,\sin(\theta \pm \phi) = R\sin\theta\cos\phi \pm R\cos\theta\sin\phi, \tag{2.67b}$$

which are the same as the standard formula used in planar trigonometry,

$$\sin(\theta \pm \phi) = \sin\theta\cos\phi \pm \cos\theta\sin\phi. \tag{2.68}$$

२.८ अल्पचापज्यानयनम्

2.8 Computation of the Rsine value of a small arc

शिष्टचापघनषष्ठभागतो विस्तरार्धकृतिभक्तवर्जितम् ।
शिष्टचापमिह शिञ्जिनी भवेत् स्पष्टता भवति चाल्पतावशात् ॥ १७ ॥

śiṣṭacāpaghanaṣaṣṭhabhāgato vistarārdhakṛtibhaktavarjitam |
śiṣṭacāpamiha śiñjinī bhavet spaṣṭatā bhavati cālpatāvaśāt || 17 ||

Divide one-sixth of the cube of the remaining arc by the square of the *trijyā*. This quantity when subtracted from the remaining arc becomes the *śiñjinī* (the *dorjyā* corresponding to the remaining arc). The value is accurate because of the smallness [of the arc].

In the above verse, Nīlakaṇṭha gives the approximation for the sine of an angle when it is small. If $\alpha = R\delta\theta$ is the length of a small arc along the circle, corresponding to an angle $\delta\theta$, then the above verse gives the following expression for its *dorjyā*:

$$\textit{dorjyā}\ \alpha = \alpha - \frac{\alpha^3}{6\ \textit{trijyā}^2}. \tag{2.69}$$

The above expression is equivalent to

$$R\sin\delta\theta = R\,\delta\theta - \frac{(R\,\delta\theta)^3}{6\,R^2}$$

$$\text{or} \qquad \sin\delta\theta = \delta\theta - \frac{(\delta\theta)^3}{6}. \tag{2.70}$$

Thus we find that $\sin\delta\theta$ is approximated by the first two terms in the series expansion for it. This gives fairly accurate results when $\delta\theta$ is small. That (2.70) is valid and yields accurate results only when the arc is small is clearly emphasised in *Laghu-vivṛti* as follows:

एवं कृतायास्तस्याः चापाल्पतावशादेव स्पष्टता भवति ।

The accuracy of this operation is due solely to the smallness of the arc.

२.९ इष्टज्यानयनम्

2.9 Computation of the desired Rsine

ऊनाधिकधनुर्ज्यां च नीत्वैवं पठितां न्यसेत् ।
ऊनाधिकधनुःकोटिजीवया तां समीपजाम् ॥ १८ ॥
निहत्य पठितां तस्याः कोट्या शिष्टगुणाञ्च तम् ।
तद्योगं वाथ विश्लेषं हरेद् व्यासदलेन तु ॥ १९ ॥
इष्टज्या भवति स्पष्टा तत्फलं स्यात् कलादिकम् ।
न्यायेनानेन कोट्याश्च मौर्व्याः कार्या सुसूक्ष्मता ॥ २० ॥

ūnādhikadhanurjyāṃ ca nītvaivaṃ paṭhitāṃ nyaset |
ūnādhikadhanuḥkoṭijīvayā tāṃ samīpajām || 18 ||
nihatya paṭhitāṃ tasyāḥ koṭyā śiṣṭaguṇañca tam |
tadyogaṃ vātha viśleṣaṃ hared vyāsadalena tu || 19 ||
iṣṭajyā bhavati spaṣṭā tatphalaṃ syāt kalādikam |
nyāyenānena koṭyāśca maurvyāḥ kāryā susūkṣmatā || 20 ||

Having also obtained the *dorjyā* of the arc which is in excess or deficit [from a multiple of 225 minutes], as described above (in the previous verse), keep it separately.

Multiply the nearest *dorjyā* [obtained from the tabulated Rsines] by the *koṭijyā* of the arc which is in excess or deficit. Also multiply the *koṭijyā* by the *dorjyā* of the arc which is in excess or deficit. The sum or difference of these two has to be divided by the radius (*trijyā*).

The desired *jyā* (*dorjyā*) can thus be found accurately. By the same procedure, the *koṭijyā* of any desired arc may be found accurately.

The above verses give the formulae for finding the *dorjyā* and *koṭijyā* of an arc of any desired length. To find this using the procedure given in Indian astronomical texts, the desired arc length is expressed as a sum of two arcs, say $\alpha + \delta\alpha$ where α is an integral multiple of 225 and $0 < \delta\alpha < 225$. The formulae given in the above verses are:

$$dorjy\bar{a}\,(\alpha \pm \delta\alpha) = \frac{dorjy\bar{a}\,\alpha\ ko\d{t}ijy\bar{a}\,\delta\alpha \pm ko\d{t}ijy\bar{a}\,\alpha\ dorjy\bar{a}\,\delta\alpha}{trijy\bar{a}} \tag{2.71}$$

$$ko\d{t}ijy\bar{a}\,(\alpha \pm \delta\alpha) = \frac{ko\d{t}ijy\bar{a}\,\alpha\ ko\d{t}ijy\bar{a}\,\delta\alpha \mp dorjy\bar{a}\,\alpha\ dorjy\bar{a}\,\delta\alpha}{trijy\bar{a}}, \tag{2.72}$$

which have already been commented upon. Since $\delta\alpha$ is always small (less than $225'$ or 3.75°), here it is suggested that the approximation (2.70) given in previous verse—which gives $\sin\delta\alpha$ correct to $O(\delta\alpha^3)$—may be used for determining the *dorjyā* $\delta\alpha$ in the above relation. Once the *dorjyā* is known, the corresponding *koṭijyā* may be found from the former using (2.53).

२.१० रविस्फुटः
2.10 True longitude of the Sun

त्र्यभ्यस्तबाहुकोटिभ्यां अशीत्याप्ते फले उभे ।
चापितं दोःफलं कार्यं स्वर्णं सूर्यस्य मध्यमे ॥ २१ ॥
केन्द्रोर्ध्वार्धे च पूर्वार्धे तत्कालार्कः स्फुटः स च ।
मध्यसावनसिद्धोऽतः कार्यः स्यादुदये पुनः ॥ २२ ॥

tryabhyastabāhukoṭibhyāṃ aśītyāpte phale ubhe |
cāpitaṃ doḥphalaṃ kāryaṃ svarṇaṃ sūryasya madhyame || 21 ||
kendrordhvārdhe ca pūrvārdhe tatkālārkaḥ sphuṭaḥ sa ca |
madhyasāvanasiddho'taḥ kāryaḥ syādudaye punaḥ || 22 ||

The *dorjyā* and *koṭijyā* [of the *manda-kendra* of the Sun] multiplied by 3 and divided by 80 form the *doḥphala* and *koṭiphala*. The arc corresponding to the *doḥphala* has to be applied to the longitude of the mean Sun positively or negatively depending upon whether the *manda-kendra* is within the six signs beginning with *Tulā* (Libra) or within the six signs beginning with *Meṣa* (Aries). The longitude thus obtained is the true longitude. Since this longitude corresponds to the true longitude at the mean sunrise, it has to be further corrected for the true sunrise.

These verses present an explicit expression for the *manda-phala* of the Sun. *Manda-phala* is a correction that needs to be applied to the mean longitude of the planet, called the *madhyama/madhyama-graha*, to obtain the *manda-sphuṭa-graha*. The significance of the *manda-phala*, whose equivalent in modern astronomy is known as the equation of centre, is explained in Appendix F.

If θ_0 be the mean longitude of the planet (here the Sun) at the mean sunrise, then the true longitude θ of the Sun at the mean sunrise is given by $\theta = \theta_0 \pm \Delta\theta$. The correction to the *madhyama* known as the *manda-phala*, $\Delta\theta$, (referred to as the arc of the *doḥphala* in verse 21) is given by

$$\textit{manda-phala} = \textit{cāpa}\left(\frac{3}{80}\textit{manda-kendrajyā}\right). \tag{2.73}$$

The term *manda-kendrajyā* in the above expression stands for the Rsine of the *manda-kendra* or mean anomaly which refers to the difference between the longitude of the mean planet and the *mandocca* (apogee). We denote it as $\theta_0 - \theta_m$, where θ_0 is the longitude of the mean planet and θ_m that of the *mandocca*. Now, the above equation translates to

$$\Delta\theta = \sin^{-1}\left(\frac{3}{80}|\sin(\theta_0 - \theta_m)|\right). \tag{2.74}$$

Here $\frac{3}{80}$ represents the ratio of the radii of the epicycle and the *manda-karṇa-vṛtta*, or '*manda*-hypotenuse circle', whose significance is explained in Appendix F. When the *manda-kendra* is within the six signs beginning with *Meṣa*, that is, $0 \leq (\theta_0 - \theta_m) \leq 180^\circ$, it is stated that the *manda* correction has to be applied negatively. On the other hand, when it is within the six signs beginning with *Tulā*, that is $180^\circ \leq (\theta_0 - \theta_m) \leq 360^\circ$, the correction is to be applied positively. Thus, the true longitude of the Sun is given by

$$\theta = \theta_0 - \sin^{-1}\left(\frac{3}{80}\sin(\theta_0 - \theta_m)\right). \tag{2.75}$$

२.११ चरप्राणाः

2.11 *Prāṇās* of the ascensional difference

संस्कृतायनभागादेः दोर्ज्या कार्या रवेस्ततः ।
चतुर्विंशतिभागज्याहतायास्त्रिज्यया हृतः ॥ २३ ॥
अपक्रमगुणोऽर्कस्य तात्कालिक इह स्फुटः ।
तत्त्रिज्याकृतिविश्लेषात् मूलं द्युज्याथ कोटिका ॥ २४ ॥
दोर्ज्यापक्रमकृत्योश्च भेदान्मूलमथापि वा ।
अन्त्यद्युज्याहता दोर्ज्या त्रिज्याभक्तेष्टकोटिका ॥ २५ ॥
त्रिज्याघ्नेष्टद्युजीवाप्ता चापितार्कभुजासवः ।
दोःप्राणलिप्तिकाभेदमविनष्टं तु पालयेत् ॥ २६ ॥
विषुवद्भाहता क्रान्तिः सूर्याप्ता क्षितिमौर्विका ।
त्रिज्याघ्नेष्टद्युजीवाप्ता चापिता स्युश्चरासवः ॥ २७ ॥

saṃskṛtāyanabhāgādeḥ dorjyā kāryā ravestataḥ |
caturviṃśatibhāgajyāhatāyāstrijyayā hṛtaḥ || 23 ||
apakramaguṇo'rkasya tātkālika iha sphuṭaḥ |
tattrijyākṛtiviśleṣāt mūlaṃ dyujyātha koṭikā || 24 ||
dorjyāpakramakṛtyośca bhedānmūlamathāpi vā |
antyadyujyāhatā dorjyā trijyābhakteṣṭakoṭikā || 25 ||
trijyāghneṣṭadyujīvāptā cāpitārkabhujāsavaḥ |
doḥprāṇaliptikābhedamavinaṣṭaṃ tu pālayet || 26 ||
viṣuvadbhāhatā krāntiḥ sūryāptā kṣitimaurvikā |
trijyāghneṣṭadyujīvāptā cāpitā syuścarāsavaḥ || 27 ||

The Rsine of the longitude of the Sun (*dorjyā*) corrected for the precession of the equinox (the *saṃskṛtāyana*) has to be determined. This, when multiplied by $R\sin 24^\circ$ and divided by the *trijyā*, gives the Rsine of the true declination of the Sun (the *apakramajyā*) at that instant of time. The square root of the difference of the squares of that and the *trijyā* is the *dyujyā*.

Then the *koṭikā* is obtained by finding the square root of the difference between the squares of the *dorjyā* and the *apakramajyā*. The *koṭikā* is also given by the product of the *antyadyujyā* ($R\cos 24$) and the *dorjyā* divided by the *trijyā*. This (the *koṭikā*) is multiplied by the *trijyā* and divided by the *dyujyā*. The arc of this is the right ascension of the Sun (the *arkabhujāsava*). The difference between the longitude and the right ascension in minutes is to be preserved such that it is not lost.

The equinoctial midday shadow (the *viṣuvadbhā*) multiplied by the Rsine of the declination (*krānti*) and divided by 12 is the *kṣitimaurvikā*. This is to be multiplied by the *trijyā* and divided by the desired *dyujyā*. The arc of that gives the ascensional difference in *prāṇas* (the *carāsava*).

While most of the quantities related to the diurnal motion of the Sun are discussed in the third chapter, some of those that are related to the determination of the true longitude of the Sun at true or actual sunrise for a given location are described here. Before explaining the above verses, it would be convenient to list the quantities defined here as follows:

Quantity	Its physical significance	Notation
apakramajyā	the Rsine of declination of the Sun	$R\sin\delta$
dyujyā	the radius of the diurnal circle of the Sun	$R\cos\delta$
arkabhujāsava	the right ascension of the Sun	α
carāsus	the ascensional difference	$\Delta\alpha$

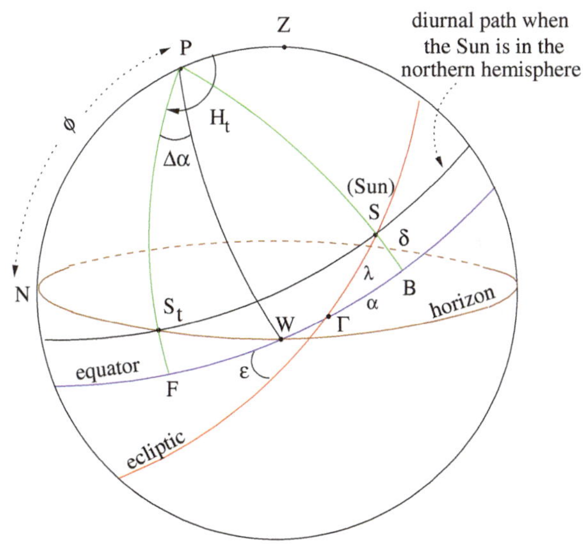

Fig. 2.7 Determination of the declination and right ascension of the Sun on any particular day.

In Indian astronomy texts, it is the *nirayaṇa* longitude or the longitude measured from a fixed star which is calculated. *Ayanāṃśa*, which is the amount of precession, has to be added to the *nirayaṇa* longitude to obtain the *sāyana* or tropical longitude λ. In Fig. 2.7, the celestial sphere is depicted for an observer at latitude ϕ, on a day when the Sun's declination is δ. Let λ and α be the Sun's (tropical) longitude and right ascension on that day.[9] The Rsine of the declination of the Sun, the *apakramajyā*, is given by

$$\begin{aligned}
apakramajy\bar{a} &= \frac{dorjy\bar{a} \times caturvi\d{m}\acute{s}atibh\bar{a}gajy\bar{a}}{trijy\bar{a}} \\
\text{or} \qquad R\sin\delta &= \frac{R\sin\lambda \times R\sin 24^\circ}{R}. \qquad (2.76)
\end{aligned}$$

This is the formula for declination,

$$\sin\delta = \sin\lambda \sin\varepsilon, \qquad (2.77)$$

which can be easily verified by considering the spherical triangle ΓSB in Fig. 2.7 and applying the sine formula. Here ε represents obliquity of the ecliptic whose value is taken to be 24° in most of the Indian astronomy texts. The *dyujyā* is the radius of the diurnal circle of the Sun, $R\cos\delta$, and it is given as

$$\begin{aligned}
dyujy\bar{a} &= \sqrt{trijy\bar{a}^2 - apakramajy\bar{a}^2} \\
\text{or} \qquad R\cos\delta &= \sqrt{R^2 - R^2\sin^2\delta}. \qquad (2.78)
\end{aligned}$$

Now a quantity, the *koṭikā*, is defined by the following two equivalent expressions:

$$\begin{aligned}
ko\d{t}ik\bar{a} &= \sqrt{R^2\sin^2\lambda - R^2\sin^2\delta} \\
ko\d{t}ik\bar{a} &= \frac{R\cos\varepsilon\, R\sin\lambda}{R}. \qquad (2.79)
\end{aligned}$$

The second of these follows from the first by substituting the expression for $R\sin\delta$ given in (2.77). The *arkabhujāsava* is the right ascension of the Sun and is the arc ΓB, which is given as:

$$arkabhuj\bar{a}sava = \alpha = c\bar{a}pa\left(\frac{ko\d{t}ik\bar{a} \times trijy\bar{a}}{dyujy\bar{a}}\right). \qquad (2.80)$$

Substituting the expressions for the *koṭikā* and the *dyujyā* in the above, we have

$$\alpha = R\sin^{-1}\left(\frac{R\cos\varepsilon\, R\sin\lambda}{R\cos\delta}\right) \qquad (2.81)$$

or

$$R\sin\alpha = \left(\frac{R\cos\varepsilon\, R\sin\lambda}{R\cos\delta}\right). \qquad (2.82)$$

This relation follows from the sine formula applied to the spherical triangle $P\Gamma S$, where the spherical angle $P\hat{\Gamma}S = 90 - \varepsilon$, the spherical angle $\Gamma\hat{P}S = \alpha$, arc $\Gamma S = \lambda$ and arc $PS = 90 - \delta$. Then

[9] The reader is referred to Appendix C on coordinate systems for details of these quantities.

$$\frac{\sin\lambda}{\sin\alpha} = \frac{\sin(90-\delta)}{\sin(90-\varepsilon)} = \frac{\cos\delta}{\cos\varepsilon}, \tag{2.83}$$

which is the same as the above.

As the axis of rotation of the Earth is perpendicular to the equator, the rotation angle measured along the equator is related to time and can be expressed in *prāṇas*. One *prāṇa* corresponds to one minute of arc along the equator. Since the right ascension is an arc measured along the equator, α is expressed in *prāṇas*.

The difference between the longitude of the Sun $\odot$ and its right ascension α figures in the equation of time described in the next set of verses (see also Appendix C). Hence $\alpha - \odot$, which is called the *prāṇaliptā* or *prāṇakalāntara*[10] is to be stored. This is the correction due to the obliquity of the ecliptic. This is explained in *Laghu-vivṛti* as follows:

...तत्र लब्धा सैव इष्टकोटिज्या। तां इष्टकोटिज्यां त्रिज्यया निहत्य इष्टद्युज्यया विभज्य लब्धं फलं चापीकुर्यात्। तच्च अर्कभुजासवो भवन्ति। तेषां अर्कभुजासूनां तत्कलानां च यदन्तरं तत् प्राणकलान्तरं नाम। तद्विनियोगमुत्तरत्र वक्ष्यामः। अत उक्तं अविनष्टं तु पालयेत् इति॥

What is obtained thus is the *iṣṭakoṭijyā*. That has to be multiplied by the *trijyā* and divided by the *iṣṭadyujyā*. The arc of the result obtained has to be found and that is known as the *arkabhujāsava*. The difference between the *arkabhujāsava* and the Sun's longitude measured in minutes is known as the *prāṇakalāntara*.[11] The utility of this will be stated later (verse 31). Hence it is stated that this has to be preserved such that it is not lost.

The great circle passing through *EPW* is known as the 6 o'clock circle, as the hour angle of any object lying on that circle corresponds to six hours. For an equatorial observer, whose latitude is zero, the horizon itself is the 6 o'clock circle and the Sun always rises on it. When the latitude of a place is not zero, the Sun does not rise on the 6 o'clock circle. In Fig. 2.7,

$$H_t = Z\hat{P}S_t = Z\hat{P}W + W\hat{P}S_t = 90^\circ + \Delta\alpha \tag{2.84a}$$

is the hour angle at sunset. It is greater than 90° when the Sun's declination is north and would be less than 90° when the declination is south. From the spherical triangle PZS_t, using the cosine formula it can be shown that

$$\begin{aligned} \cos H_t &= -\tan\phi\tan\delta \\ \text{or} \qquad \sin\Delta\alpha &= \tan\phi\tan\delta. \end{aligned} \tag{2.84b}$$

H_t expressed in minutes is the time interval in *prāṇās* between the meridian transit of the Sun and sunset. When $\delta = 0$, $H_t = 90^\circ = 5400$ *prāṇas* (6 hours). $\Delta\alpha$

---

[10] The terms *prāṇa* and *kalā* here refer to the right ascension and Sun's longitude expressed in minutes respectively. Hence the *prāṇakalāntara* is $\alpha - \odot$.

[11] It must be noted that Śaṅkara Vāriyar uses the term *prāṇakalāntara* instead of *prāṇaliptikā*. Nīlakaṇṭha himself has used the term *prāṇakalāntara* later in verse 31, where he discusses the application of the *prāṇakalāntara*.

is difference between H_t and 6 hours or 5400 *prāṇas*, and is termed the *carāsava* or the ascensional difference. It is clear that it is also the difference between sunrise and transit of the Sun across the 6 o'clock circle.

Expression for the *carāsus*

For giving the expression for the *carāsus* (ascensional difference), an intermediate quantity called the *kṣiti-maurvikā* (*kṣitijyā*, earth-sine) is defined as follows:

$$kṣitimaurvikā = \frac{viṣuvadbhā \times krānti}{12}. \tag{2.85}$$

The *viṣuvadbhā* refers to the equinoctial shadow of a stick of length 12 units. It will be shown in the next chapter that the equinoctial shadow for an observer at latitude ϕ is $12 \tan\phi$. The term *krānti* is the same as *apakramajyā* given earlier in (2.76). The expression for the *carāsus* is given by

$$carāsus = cāpa\left(\frac{kṣitimaurvikā \times trijyā}{dyujyā}\right). \tag{2.86}$$

Substituting for *kṣitimaurvikā* and *dyujyā* in the above expression we have

$$\Delta\alpha = (R\sin)^{-1}\left(\frac{R\tan\phi \; R\sin\delta}{R\cos\delta}\right), \tag{2.87}$$

which is the same as (2.84*b*). At the equator, where $\phi = 0$, $\Delta\alpha = 0$. Hence, the sunrise or sunset is exactly 6 hours before or after meridian transit. Since the *carāsus* ($\Delta\alpha$) is the interval between the sunrise at a given latitude and that at the equator, the knowledge of it is essential for finding the exact sunrise and sunset times at the observer's location. It is also needed for finding the longitude of planets at sunrise at any non-zero latitude.

२.१२ स्वदेशसूर्योदयकाले ग्रहाः

2.12 Longitude of the planets at sunrise at the observer's location

लिप्ताप्राणान्तरं भानोः दोःफलं च चरासवः ।
स्वर्णसाम्येन संयोज्या भिन्नेन तु वियोजयेत् ॥ २८ ॥
भानुमध्यमभुक्तिघ्नं चक्रलिप्ताहृतं फलम् ।
भानुमध्ये तु संस्कार्यं स्फुटभुक्त्याहतं स्फुटे ॥ २९ ॥
उदक्स्थेऽर्के चरप्राणाः शोध्याः स्वं याम्यगोलके ।
व्यस्तमस्ते तु संस्कार्या न मध्याह्नार्धरात्रयोः ॥ ३० ॥
युग्मौजपदयोः स्वर्णं रवौ प्राणकलान्तरम् ।
दोःफलं पूर्ववत् कार्यं रवेरेभिर्द्युचारिणाम् ॥ ३१ ॥

मध्यभुक्तिं स्फुटां वापि हत्वा चक्रकलाहृतम् ।
स्वर्णं कार्यं यथोक्तं तत् व्यस्तं वक्रगतौ स्फुटे ॥ ३२ ॥

liptāprāṇāntaram bhānoḥ doḥphalaṃ ca carāsavaḥ |
svarṇasāmyena saṃyojyā bhinnena tu viyojayet || 28 ||
bhānumadhyamabhuktighnaṃ cakraliptāhṛtaṃ phalam |
bhānumadhye tu saṃskāryaṃ sphuṭabhuktyāhataṃ sphuṭe|| 29 ||
udaksthe'rke caraprāṇāḥ śodhyāḥ svaṃ yāmyagolake |
vyastamaste tu saṃskāryā na madhyāhnārdharātrayoḥ || 30 ||
yugmaujapadayoḥ svarṇaṃ ravau prāṇakalāntaram |
doḥphalaṃ pūrvavat kāryaṃ raverebhirdyucāriṇām || 31 ||
madhyabhuktiṃ sphuṭāṃ vāpi hatvā cakrakalāhṛtam |
svarṇaṃ kāryaṃ yathoktaṃ tat vyastaṃ vakragatau sphuṭe|| 32 ||

The *prāṇakalāntara*, *doḥphala* (equation of centre) and *carāsus*, all in minutes, have each to be added or subtracted depending upon their signs. This quantity multiplied by the mean daily motion of the Sun and divided by 21600 has to be applied to the mean Sun, and the same has to be multiplied by the true daily motion of the Sun and applied to the true Sun [to get the longitudes of the mean and the true Sun respectively at the true sunrise at any given location].

When the Sun is to the north (has northern declination), then the *carāsus* have to be applied negatively and when it is to the south they have to be applied positively [this sign convention is to be adopted when the longitude is to be determined at sunrise]. The *carāsus* have to be applied in the reverse order at the sunset. They need not be applied [for determining the longitude] at midday or midnight.

The *prāṇakalāntara* has to be applied positively and negatively in the even and odd quadrants respectively. The *doḥphala* has to be applied as discussed earlier. With these quantities (namely *prāṇakalāntara*, *doḥphala* and *carāsus*), which are related to the Sun, the mean or true daily motions of the planets are to be multiplied and divided by 21600. These have to be applied positively or negatively as mentioned earlier [when the planet is in direct motion] and the application has to be done in the reverse order when the planet is in retrograde motion [to get the mean and true planets at true sunrise].

In the above verses, Nīlakaṇṭha gives the procedure for obtaining the mean or true longitudes of the planets at the true sunrise at the observer's location. The longitudes obtained from the *Ahargaṇa* give the mean and true positions of the planets at the mean sunrise, i.e. when the mean Sun is on the 6 o'clock circle, at the observer's location. To get the positions of the planets at the true sunrise, i.e. when the true Sun is on the observer's horizon, corrections have to be applied.

Of the two corrections that need to be applied, one is due to the fact that at sunrise the Sun is on the horizon and not on the 6 o'clock circle. The time difference between the sunrise and the instant when it is on the 6 o'clock circle (the *carāsus*) has been discussed earlier. Now, when the Sun has a northerly declination, sunrise is earlier than its transit across the 6 o'clock circle and *carāsavas* have to be applied negatively. Similarly, when the Sun has a southerly declination, sunrise is after its transit across the 6 o'clock circle and the *carāsus* have to be applied positively. The other two corrections are due to the fact that there is a time difference between the transits of the mean Sun and the true Sun across the meridian or the 6 o'clock circle. In fact, we shall see below that the expression for the sum of these two corrections given in the text is the same as the equation of time in modern astronomy (for more details refer to Appendix C).

Equation of time

The 'mean Sun' is a fictitious body which is moving along the equator uniformly with the average angular velocity of the true Sun. In other words, the right ascension of the mean Sun (denoted by R.A.M.S.) increases by 360° in the same time period as the longitude of the true Sun increases by 360°. As the R.A.M.S. increases uniformly, the time interval between the successive transits of the mean Sun across the meridian or the 6 o'clock circle is constant. This is the mean civil day. All the civil time measurements are with reference to the mean Sun. The time interval between the transits of the mean Sun and the true Sun across the meridian or the 6 o'clock circle is known as the equation of time and is given by

$$\begin{aligned} E &= H.A.M.S. - H.A.\odot \\ &= R.A.\odot - R.A.M.S. \\ &= \alpha - \alpha_{M.S}, \end{aligned} \tag{2.88}$$

where $\odot$ stands for the true Sun. It will also be used to refer to the longitude of the true Sun later. Since the dynamical mean Sun moves along the ecliptic uniformly with the average angular velocity of the true Sun—and both of them are assumed to meet each other at the equinox Γ—the longitude of the dynamical mean Sun or the mean longitude of the Sun (l) is the same as the R.A.M.S. Hence the equation of time will be $E = \alpha - l$. This can be rewritten as

$$E = (\alpha - \odot) + (\odot - l). \tag{2.89}$$

The first term in the equation of time is the *prāṇakalāntara* = $\alpha - \odot$. Now $\sin\alpha = \frac{\cos\varepsilon \sin\odot}{\cos\delta}$. As $\delta < \varepsilon$, $|\sin\alpha| < |\sin\odot|$. This implies that $\alpha < \odot$ when α and $\odot$ are in the odd quadrants and $\alpha > \odot$ when α and $\odot$ are in the even quadrants. Hence the *prāṇakalāntara* has to be applied positively and negatively in the even and odd quadrants respectively. The sign of the *doḥphala* $(\odot - l)$ has already been discussed earlier. It is negative in the first and second quadrants and positive in the third and fourth quadrants.

Application of corrections

The three corrections, namely the *prāṇakalāntara*, *doḥphala* and *carāsus*, have to be applied to the mean or true longitude of planets at mean sunrise at the equator (or the 6 o'clock circle) to obtain the mean or true longitude at true sunrise on the observer's horizon. The motion of a planet in one *prāṇa* is equal to its daily motion divided by 21600. The net correction would be the sum of the three quantities (taking appropriate signs into account) multiplied by the above ratio. When the planet is in retrograde motion, the longitude decreases with time. Hence, all the signs discussed above have to be reversed in such a situation.

Śaṅkara Vāriyar in his *Yukti-dīpikā* gives a graphic description of what is meant by *cara*, and how it is to be used in the determination of the duration of day and night at the observer's location (having non-zero latitude).

स्वाक्षे स्वक्षितिजे ज्ञेयावुदयास्तमयौ रवेः ।
उन्मण्डलक्षितिजयोः अन्तरालधनुश्चरम् ॥
सौम्ये पूर्वमुदेत्यर्कः पश्चात्तेनास्तमेति च ।
अतश्चरेण द्विघ्नेन दिनं तत्र तु वर्धते ॥

क्षीयते च निशा यस्मात् मिथोभिन्ने दिनक्षये ।
याम्ये पश्चादुदेत्यर्कः तेन प्रागस्तमेति च ॥
अतश्चरेण द्विघ्नेन तत्र तु क्षीयते दिनम् ।
वर्धते च निशा यस्मात् व्यस्तत्वं गोलयोर्मिथः ॥

चरप्राणगतिः स्वर्णं याम्योदग्गोलयोस्ततः ।
उदयेऽस्तमये व्यस्तं ग्रहे रव्युदयावधौ ॥
मध्यार्कप्रमितं तुल्यरूपमेव सदा दिनम् ।
तुल्यत्वात् तद्गतेर्भिन्नं स्फुटार्कप्रमितं दिनम् ॥ [12]

The rising and setting of the Sun has to be determined with respect to the horizon corresponding to the observer's own latitude. The length of the arc [of the diurnal circle] lying between the *unmaṇḍala* (6 o'clock circle) and the *kṣitija* (horizon) is referred to as the *cara*.

When the Sun has northern declination it rises earlier and sets later. Hence the duration of the day increases by twice the *cara*. Naturally the duration of the night decreases, and hence day and night have different durations. When the Sun has southern declination it rises later and sets earlier. Therefore the duration of the day decreases by twice the *cara* and that of the night increases. [While this is true for an observer in the northern hemisphere] the reverse happens in the southern hemisphere.

The *caraprāṇas* have to be applied negatively and positively when the Sun has northern and the southern declination respectively. This is true at sunrise and during sunset they have to be applied in the reverse order. Since the mean Sun moves with uniform velocity, the duration of the day will always be uniform when measured with respect to the mean Sun. But the duration will vary when measured with respect to the true Sun.

२.१३ दिनक्षपयोर्मानम्

2.13 Durations of the day and the night

अहोरात्रचतुर्भागे चरप्राणान् क्षिपेदुदक् ॥ ३३ ॥
याम्ये शोध्या दिनार्धं तत् रात्र्यर्धं व्यत्ययाद्भवेत् ।
दिनक्षपे द्विनिघ्ने ते चन्द्रादेः स्वैश्चरासुभिः ॥ ३४ ॥

ahorātracaturbhāge caraprāṇān kṣipedudak || 33 ||
yāmye śodhyā dinārdhaṃ tat rātryardhaṃ vyatyayādbhavet |
dinakṣape dvinighne te candrādeḥ svaiścarāsubhiḥ || 34 ||

In the north (when the declination of the Sun is towards north), the *caraprāṇā* has to be added to one-fourth of the *ahorātra* and in the south it has to be subtracted. This gives the

[12] {TS 1977}, p. 154.

half-duration of the day. The half-duration of the night is obtained by applying the *cara* in the reverse order. By multiplying these durations by two, the durations of the day and night are obtained respectively. For the Moon and others, the half-durations [of their own days and nights] have to be obtained from their own *caraprāṇas*.

While the time unit, namely a day, can be considered with respect to different planets, we first consider the Sun and the solar day. By definition, on an average, one-fourth of an *ahorātra* or mean solar day or civil day is 6 hours. To this, correction due to the *cara* has to be added or subtracted in order to find the 'actual' half-duration of the day, i.e. the time interval between sunrise and the meridian transit of the Sun. Recalling that one hour corresponds to 15°, the half-duration of the day (in hours) for an observer with latitude ϕ is given by

$$6+\frac{(R\sin^{-1})(R\tan\phi\tan\delta)\ \text{[in deg]}}{15}, \tag{2.90}$$

where the second term is positive or negative depending upon the sign of δ, i.e. depending on whether the Sun is in the northern or southern hemisphere. δ is obtained using the relation

$$R\sin\delta = R\sin\varepsilon\sin\lambda. \tag{2.91}$$

As pointed out later, it is noted in *Laghu-vivṛti* that λ at true sunrise should be used in the calculation to obtain the first half-duration of the day. Similarly λ at true sunset should be used to obtain the second half-duration of the day. This is explained in *Laghu-vivṛti* as follows:

अहोरात्रस्य षष्टिघटिकात्मकस्य यश्चतुर्भागः पञ्चदशघटिकारूपः तस्मिन् तात्कालिकात् सायनार्कादानीतान् चरप्राणान् नाडीकृत्य क्षिपेत् यद्युदग्गोलगतः सायनस्फुटार्कः। याम्यगोलगते पुनस्तस्मिन् ततोऽहोरात्रचतुर्भागाद् विशोधयेत्। एवं कृतोऽहोरात्रचतुर्भागः तस्मिन् दिने दिनार्धं भवति। रात्र्यर्धं पुनः ततो व्यत्ययाद् भवति।उदग्गोले चरप्राणविरहितः अहोरात्रचतुर्भागः रात्र्यर्धं, याम्यगोले तु तत्सहितः इति। एवं कृतं दिनार्धं रात्र्यर्धं च द्विगुणितं कृत्स्नं दिनमानं क्षपामानं च भवति।

The *caraprāṇas* obtained from the *sāyana* longitude λ of the Sun, when it is in the northern hemisphere ($0 < \lambda < 180$), converted into *nāḍīs*, have to be applied positively to one-fourth of the duration of the *ahorātra*, which is 15 *ghaṭikās*, the duration of the *ahorātra* itself being 60 *ghaṭikās*. If the Sun is in the southern hemisphere ($180 < \lambda < 360$), then the *caraprāṇas*, converted into *nāḍīs* have to be applied negatively to one-fourth of the duration of the *ahorātra*. Thus one-fourth of the *ahorātra* being corrected by the *caraprāṇa* gives the half-duration of the day. The half-duration of the night is obtained by carrying out the reverse process. The half-duration of the night, which was obtained by subtracting the *caraprāṇa* in the northern hemisphere, is to be obtained by its addition in the southern hemisphere. The half-durations of the day and night when multiplied by two give the durations of day and night.

To get the half-durations of the day and night more accurately, a better procedure is suggested.

अत्र पुनः औदयिकात् सायनार्कात् आनीतमेव चरं दिनपूर्वार्धे कार्यम्; आस्तमिकादानीतं च अपरार्धे ।रात्रिपूर्वार्धेऽपि आस्तमिकादानीतं; अपरार्धेऽपि औदयिकादानीत-

मिति। एवं कृतयोः दिनार्धयोः क्षपार्धयोश्च योगः दिनक्षपयोः स्फुटतरं मानमिति। अथ प्राणकलान्तरमपि दिनमानार्थं कर्तव्यम्। तेन दिनरात्र्योः उभयतः सायनार्कतः यत् प्राणकलान्तरद्वितयमानीतं तयोर्विवरमपि दिनक्षपामानयोः कर्तव्यमेव, येन दिनक्षपयोर्माने स्फुटतमे स्यातामिति।

The *cara* obtained from the *sāyana* Sun at sunrise (instead of mean sunrise at the equator) has to be applied in the forenoon and the one obtained from the *sāyana* Sun at sunset in the afternoon. Similarly, the *cara* obtained from the *sāyana* Sun at the sunset and sunrise have to be applied for obtaining the duration of the first and second half of the night respectively. The duration of the day and night obtained thus (rather than those obtained from the earlier method) would be more accurate.

The *prāṇakalāntara* correction should also be implemented in finding the duration of the day. The difference in the *prāṇakalāntaras* obtained from the *sāyana* Sun at sunrise and sunset has to be applied to obtain more accurate durations of day and night.

Duration of the day of the planets

The stars are considered to be fixed objects in the sky. The sidereal day is defined as the time interval between two successive rises of the star across the horizon and is equal to the time taken by the Earth to complete one revolution around its axis. A 'planet-day' is defined in a similar manner. The time interval between two successive sunrises is the 'sun-day' or a solar day. The time interval between two successive moonrises is the 'moon-day' or lunar day.[13]. Similarly the time interval between two successive rises of any particular planet is defined to be the duration of that 'planet-day'.

This concept of the day of planets may be understood with the help of Fig. 2.8. In Fig. 2.8*a*, we have depicted a situation where a star X, the Sun S and the Moon M are all in conjunction and are just about to rise above the horizon. After exactly one sidereal day ($\approx$ 23 h 56 m) the star X will be back on the horizon. However, the Sun and Moon, due to their orbital motion eastwards, will not be back on the horizon. They would have moved in their respective orbits through distances, given by their daily motions which are approximately 1° and 13° respectively. This situation is depicted in Fig. 2.8*b* where X, S' and M' represent the star, the Sun and the Moon respectively.

It may be noted here that the Moon is shown to be on the ecliptic. Though the orbit of the Moon is slightly inclined to the ecliptic, since its orbital inclination is very small (approximately 5°), the angular distance covered by the Moon in its orbit can be taken to be roughly the angular distance covered by it on the ecliptic. After one sidereal day the star X will be again on the horizon. Only when the earth rotates through an angle equal to the difference between the right ascensions of X and S' will the Sun be on the horizon. This is taken to be the arc XS' on the ecliptic itself. (This can only be approximate.) Similarly only when it rotates through an angle XM' will the Moon be on the horizon (in the same approximation). Hence the duration of a solar day is given by

[13] This definition of lunar day should not be confused with that of a *tithi* defined earlier.

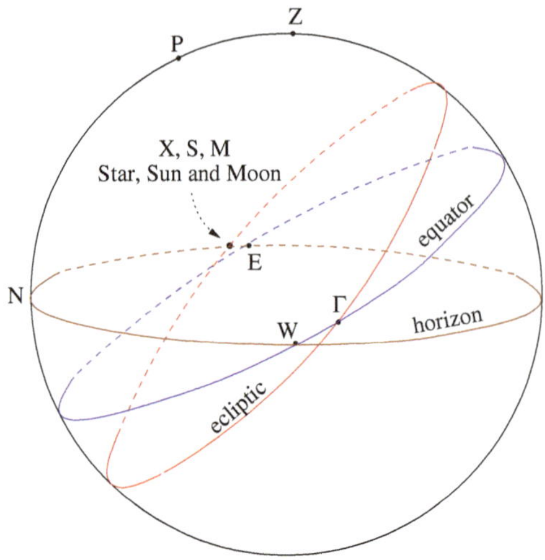

Fig. 2.8*a* The star X, Sun S and the Moon M at sunrise on a particular day.

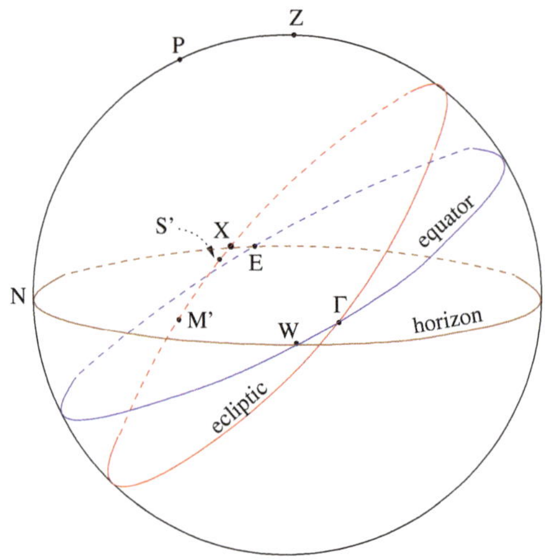

Fig. 2.8*b* The star X, Sun S' and the Moon M' exactly after one sidereal day.

$$\begin{aligned}\text{Solar day} &= \text{Sidereal day} + \text{Time taken by the earth to} \\ &\qquad\qquad \text{rotate through } XS' \\ &= 21600 + XS' (\text{in minutes of arc}) \\ &= 21600 + \text{Sun's daily motion (in } \textit{prāṇas}).\end{aligned}$$

In the above expression, the number 21600 represents the number of *prāṇas* (≈4 seconds) in a sidereal day, and XS' is expressed in minutes. Similarly the duration of the lunar day is given by

$$\begin{aligned}\text{Lunar day} &= \text{Sidereal day} + \text{Time taken by the earth to} \\ &\qquad\qquad\qquad\qquad \text{rotate through } XM' \\ &= 21600 + XM' \\ &= 21600 + \text{Moon's daily motion (in } pr\bar{a}\d{n}as). \end{aligned} \tag{2.92}$$

Similarly a 'planet day' can be defined for other planets also.

For finding half of the duration for which a planet is above the horizon, its own *caraprāṇa* has to be added to or subtracted from one-fourth of its own 'planet-day'. In *Laghu-vivṛti* there is a discussion on this:

चन्द्रादेः पुनः स्वैः स्वैः चरासुभिः उक्तवत् संस्कृतात् निजाहोरात्रचतुर्भागतः दिनार्धं रात्र्यर्धं च अवगन्तव्यम्। अहोरात्रश्च तस्य निजदिनस्फुटभोगप्राणाधिकचक्रकलातुल्य-प्रमाणः।नन्वेवं कृतस्यापि इन्दोः दिनक्षपयोर्मानस्य न स्फुटत्वम्। निजदोःफलकला-दिना अधिकोनत्वसंभवात्, सत्यम् ; अत एव हि तत्र स्वदिनान्तरितयोः स्फुटयोर्विधि-वत् कृतस्वचरप्राणकलान्तरयोः अन्तरेण चक्रकला सहितेन दिनासवः क्रियन्ते।

For the Moon and others (planets) their own durations of day and night have to be obtained from the quarter of their true *ahorātra* corrected for their own *carāsus*. The duration of their day is [nearly] equal to the sum of the their daily motion in *prāṇas* plus the number of minutes in 360 degrees. Even after applying this, the duration of the day or night of the Moon [and other planets] would not be correct as it may differ [from the actual value] by its own *doḥphala*. True; it is only to take this discrepancy into account that the [true] duration of a lunar day in minutes is obtained from the difference in the true positions of the Moon [and other planets], at intervals separated by the durations of their days corrected by *caraprāṇa* etc., added to the number of minutes in 360 degrees.

Ascensional difference in the case of the Moon and other planets:

It may be recalled that (2.67) gives the expression for finding the *caraprāṇa* in the case of the Sun. For the Moon and other planets the procedure to be adopted is stated in *Laghu-vivṛti* as follows:

स्वचरप्राणानयनमपि अमुनैव इष्टक्रान्त्या विक्षेपसंस्कृतया छायागणिते प्रदर्शितम्-
क्रान्तिज्या विषुवद्भाघ्ना क्षितिज्या द्वादशोद्धृता।
व्यासार्धघ्ना द्युजीवाप्ता चापितास्युस्चरासवः॥ [14]

For planets other than Sun the procedure for obtaining their own *caraprāṇa* from the declination corrected for the latitude of the planet has been shown by [the author] himself in [his] *Chāyāgaṇita*:

The sine of the declination multiplied by the *viṣuvadbhā* and divided by twelve is the *kṣitijyā*. This has to be multiplied by the *trijyā* and divided by the *dyujyā*. The arc of this is the *carāsava*.

In Fig. 2.9, S represents the position of the Sun on the observer's meridian on an equinoctial day. Since the motion of the sun takes place along the equator on

[14] {CCG 1976}, p. 16.

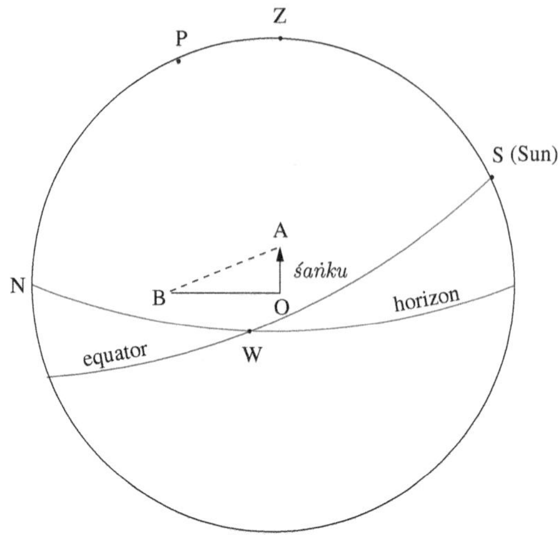

Fig. 2.9 Shadow of *śaṅku* on an equinoctial day.

an equinoctial day, the equator itself serves as the diurnal circle. The length of the shadow of a stick of length 12 units, when the Sun is on the observer's meridian on the equinoctial day, is termed the *viṣuvadbhā*. In the figure, *OA* represents the stick of length 12 units, referred to as a *śaṅku*. Since $ZS = \phi$, the latitude of the observer, $O\hat{A}B = \phi$. Hence,

$$viṣuvadbhā \;=\; 12\tan\phi. \tag{2.93}$$

From (2.85), *kṣitijyā* is given by $\frac{12\tan\phi \times R\sin\delta}{12}$. Also the ascensional difference *carāsava* of the planet is given by

$$carāsus = (R\sin)^{-1} R\tan\phi\tan\delta, \tag{2.94}$$

where δ is the declination of the planet. The declination δ of a planet with longitude λ and latitude β as depicted in Fig. 2.10 is given by

$$\begin{aligned}\sin\delta &= \cos\varepsilon\sin\beta + \sin\varepsilon\cos\beta\sin\lambda \\ &= \cos\varepsilon\sin\beta + \cos\beta\sin\delta_E,\end{aligned} \tag{2.95}$$

where δ_E is the declination of an object on the ecliptic with the same longitude as the planet. That is, $\sin\delta_E = \sin\varepsilon\sin\lambda$. Thus, in the case of planet having a latitude, a correction has to be applied to δ_E to obtain the actual declination δ. From the 'planet-day' and the *carāsava* of the planet, the time interval between the rising and setting of the planet which is the duration of the 'day' for the planet can be obtained.

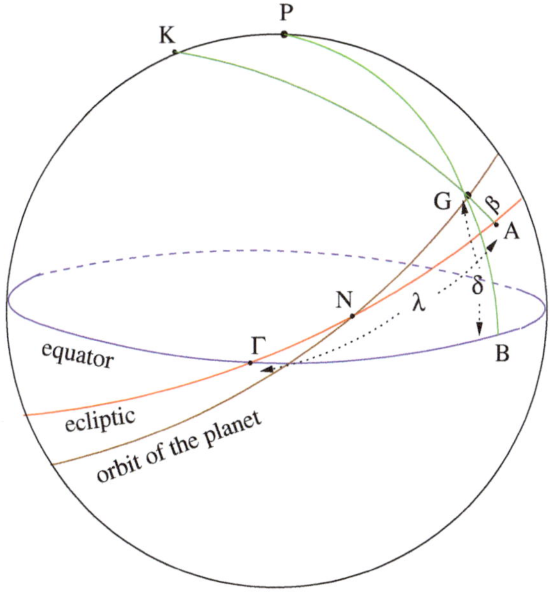

Fig. 2.10 Determination of *carapraṇa* for planets.

२.१४ चन्द्रस्फुटीकरणम्

2.14 Obtaining the true Moon

इन्दूच्चयोः स्वदेशोत्थरव्यानीतचरादिजम् ।
संस्कारं मध्यमे कृत्वा स्फुटीकार्यो निशाकरः ॥ ३५ ॥
दोःकोटिज्ये तु सप्तघ्ने अशीत्याप्ते फले उभे ।
चापितं दोःफलं कार्यं स्वमध्ये स्फुटसिद्धये ॥ ३६ ॥

indūccayoḥ svadeśottharavyānītacarādijam |
saṃskāraṃ madhyame kṛtvā sphuṭīkāryo niśākaraḥ || 35 ||
doḥkoṭijye tu saptaghne aśītyāpte phale ubhe |
cāpitaṃ doḥphalaṃ kāryaṃ svamadhye sphuṭasiddhaye || 36 ||

The mean position of the Moon and its apogee have to be corrected by the *carapraṇa* etc. obtained from the Sun, and then the *sphuṭa-karma* (procedure for the true longitude) has to be carried out.

The *dorjyā* and the *koṭijyā* multiplied by 7 and divided by 80 form the *doḥphala* and *koṭiphala*. The arc of the *doḥphala* has to be applied to the mean position to get the true position.

The mean positions of the planets obtained from the *Ahargaṇa* (count of days) correspond to their mean positions at the mean sunrise for an observer at Ujjayinī. To get their mean positions for other observers, corrections such as *deśāntara, cara* etc. have to be applied (see the previous section as well as Section 1.14). These are corrections to be carried out to get the mean position of the planet at the true sunrise at the observer's location. Verse 35 reemphasizes these corrections.

To get the true position of the planet at true sunrise, the equation of centre has to be applied to the mean planet at true sunrise. Verse 36 describes how this correction has to be implemented in the case of the Moon. The ratio of the mean radius of the epicycle and the radius of the deferent circle (the *trijyā*) is taken to be $\frac{7}{80}$ for the Moon. Hence according to the text the true longitude of the Moon, θ, is

$$\theta = \theta_0 - \sin^{-1}\left(\frac{7}{80}\sin(\theta_0 - \theta_m)\right),$$

where θ_0 is the mean longitude of the Moon and θ_m the longitude of the *mandocca*.

The procedure for obtaining the true longitude of the Moon is explained in the commentary as follows:

चन्द्रतदुच्चयोरपि इष्टद्युगणतः त्रैराशिकसिद्धे मध्यमे देशान्तरस्य रविदोःफलादीनां त्रयाणां च गतिं विधिवत् कृत्वा चन्द्रस्य मध्यतः तन्मन्दोच्चं विशोध्य शिष्टे तत्केन्द्रे भगणपूर्वोर्ध्वार्धगतत्वं च अवधार्य दोःकोट्योरुभयोरपि जीवे रविकेन्द्रोक्तवद्गृह्णीयात्।

तथा गृहीते दोःकोटिज्ये उभे अपि सप्तभिर्निहत्य अशीत्या विभज्य लब्धे दोःकोटिफले स्याताम्। तत्र कोटिफलस्य उपयोगं वक्ष्यामः। दोःफलं पुनश्चापीकृत्य तन्मध्यमे स्वकेन्द्रभगणपूर्वोर्ध्वार्धगतत्ववशात् ऋणं धनं वा कुर्यात्। एवं कृतश्चन्द्रमध्यमः स्वदेशस्फुटार्कोदयावधिकः स्फुटो भवति।

From the *deśāntara*, as well as the three corrections *manda-phala* etc. related to the Sun [obtaining the true sunrise time], the mean positions of the Moon and its apogee [at true sunrise time], are obtained from the *Ahargaṇa* by the rule of three. Then subtracting the apogee from the mean longitude, the *manda-kendra* of the Moon is determined. Depending upon the quadrant in which the *manda-kendra* lies, the *dorjyā* and *koṭijyā* have to be found following the procedure that was given for the Sun.

The *dorjyā* and *koṭijyā* obtained thus have to be multiplied by 7 and divided by 80 to get the *doḥphala* and *koṭiphala* respectively. The use of the *koṭiphala* will be stated later. The arc corresponding to the *doḥphala* is applied to the mean planet either positively or negatively depending upon the quadrant in which the *kendra* lies. These corrections applied to the mean Moon give its true position at the true sunrise at the observer's location.

२.१५ चरज्यादीनां चापीकरणम्

2.15 Finding the arc corresponding to *cara* etc.

ज्याचापान्तरमानीय शिष्टचापघनादिना ।
युक्त्वा ज्यायां धनुः कार्यं पठितज्याभिरेव वा ॥ ३७ ॥
त्रिखरूपाष्टभूनागरुद्रैः त्रिज्याकृतिः समा ।
एकादिघ्नया दशाप्ता या घनमूलं ततोऽपि यत् ॥ ३८ ॥
तन्मितज्यासु योज्याः स्युः एकद्व्याद्या विलिप्तिकाः ।
चरदोःफलजीवादेः एवमल्पधनुर्नयेत् ॥ ३९ ॥

jyācāpāntaramānīya śiṣṭacāpaghanādinā|
yuktvā jyāyāṃ dhanuḥ kāryaṃ paṭhitajyābhireva vā|| 37 ||
trikharūpāṣṭabhūnāgarudraiḥ trijyākṛtiḥ samā|

ekādighnyā daśāptā yā ghanamūlaṃ tato'pi yat || 38 ||
tanmitajyāsu yojyāḥ syuḥ ekadvyādyā viliptikāḥ|
caradoḥphalajīvādeḥ evamalpadhanurnayet || 39 ||

The arc corresponding to a *jyā* may be obtained either by finding the difference between the *jyā* and the arc as given in the verse [beginning] *śiṣṭacāpaghana* etc., and adding that (difference) to the *jyā*, or from the table of *jyās* listed earlier.

The square of the *trijyā* is 11818103 (in minutes). Multiply this by 1, 2 etc., divide by 10 and find the cube roots of these results. If the *jyā* (whose arc is to be found) has a measure equal to these (the above cube roots), then 1, 2, etc. seconds have to be added to them. Thus the arc of the R sine of small angles involved in the *caradoḥphala* may be obtained.

In Fig. 2.11, let PN represent the *jyā* whose corresponding arc length AP is to be determined. If R is the radius of the circle and $A\hat{O}P = \alpha$, then the length of the *jyā* corresponding to this angle is given by

$$jy\bar{a} = PN = l = R\sin\alpha. \tag{2.96}$$

When α is small we know that

$$\sin\alpha \approx \alpha - \frac{\alpha^3}{3!}.$$

$$\text{Hence,} \qquad R\sin\alpha \approx R\alpha - \frac{(R\alpha)^3}{6R^2}. \tag{2.97}$$

Or, the difference (D) between the *cāpa* (arc) and its *jyā* (Rsine) is given by

$$D \approx R\alpha - l = \frac{(R\alpha)^3}{6R^2}. \tag{2.98}$$

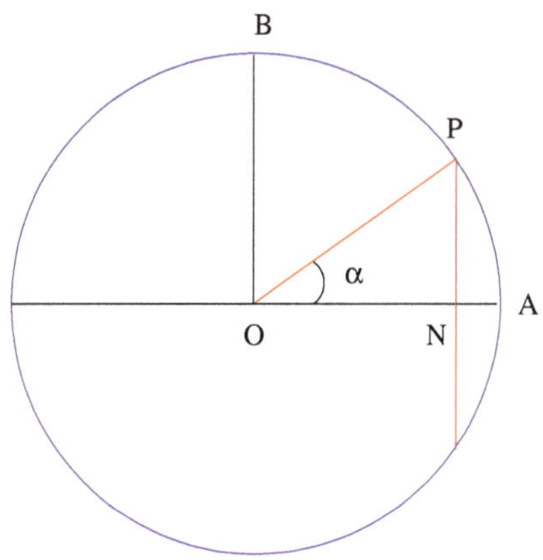

Fig. 2.11 Finding the arc length of a given *jyā* when it is very small.

An iterative procedure for obtaining the arc length corresponding to a given *jyā* is described in the above verses. This procedure is simple and also yields fairly accurate results for small angles. We may explain the procedure outlined here as follows.

As a first approximation, we take the arc length (which itself is very small) to be the *jyā* itself, i.e. $R\alpha \approx l$. Hence from (2.98) the difference between the arc length and its *jyā* becomes

$$D_1 = \frac{l^3}{6R^2}. \tag{2.99}$$

As a second approximation, we take the arc length to be $R\alpha = l + D_1$. Hence in the next approximation the difference (D_2) between the arc length and its *jyā* becomes

$$D_2 = \frac{(l + D_1)^3}{6R^2}. \tag{2.100}$$

As a third approximation, when we take $R\alpha = l + D_2$, we have

$$D_3 = \frac{(l + D_2)^3}{6R^2}. \tag{2.101}$$

In general,

$$D_i = \frac{(l + D_{i-1})^3}{6R^2}. \tag{2.102}$$

The above iteration process is continued till $D_i = D_{i-1}$, to a given level of accuracy. When this condition is satisfied, we have arrived at the required arc length corresponding to the given *jyā*, given by

$$R\alpha = l + D_i. \tag{2.103}$$

Aviśeṣakarma

The iterative procedure, known as *aviśeṣakarma*, to be employed is described in *Laghu-vivṛti* as follows:

चापीचिकीर्षितां ज्यां कार्त्स्न्येन घनीकृत्य षड्भिर्विभज्य लब्धं पुनः त्रिज्याकृत्या च विभजेत्। तत्र लब्धं लिप्तादिकं ज्याचापान्तरम्। अहार्यत्वे पुनः षष्ट्या निहत्य त्रिज्याकृत्या विभजेत्। तत्र लब्धं विलिप्तादिकं ज्याचापन्तरमिति।

नन्वत्र शिष्टचापघनेत्यादिना इष्टचापतः तज्ज्याचापान्तरं क्रियते। न पुनः इष्टज्यातः तच्चापान्तरम्। सत्यम् ; अत एव अत्र अविशेषकर्म क्रियते। तद्यथा - उक्तवदानीतम् इष्टज्याचापान्तरम् इष्टज्यायां प्रक्षिप्य पुनरपि तद्धनतः पूर्ववदानीतं ज्याचापान्तरं मुहुराद्यज्यायामेव प्रक्षिपेत् यावदविशेषः। अविशिष्टेन ज्याचापान्तरेण युक्ता इष्टज्या चापीकृता स्यादिति।

Find the cube of the given *jyā* and divide it by six. This may further be divided by the square of the *trijyā*. The result is the difference between the *jyā* and *cāpa* in minutes. If it is not divisible [if there is a fraction], then it has to be multiplied by 60 and then divided by the square of the *trijyā*. The result thus obtained will be the difference between the *jyā* and the *cāpa* in seconds.

Is it not true that, as per the procedure described in [the verse] *śiṣṭacāpaghana* ..., we find the difference between the *jyā* and *cāpa* from the given (known) *cāpa* and not from the given *jyā*? Yes, it is true. It is only because of this, that an iterative procedure (*aviśeṣakarma*) is followed here where the difference between the *jyā* and *cāpa* is to be found from the given *jyā*. It is as follows: The difference between the *jyā* and *cāpa* obtained as described earlier must be applied to the given *jyā* and from the cube of that the [next approximation to the] difference between the *jyā* and *cāpa* must be determined. This again has to be applied to the given *jyā*, and the process has to be repeated till the result becomes *aviśiṣṭa* (not different from the earlier). This difference added to the given *jyā* will be the required *cāpa*.

Finding the arc length corresponding to a given *jyā* from a look-up table

Apart from the iterative procedure described above, Nīlakaṇṭha also gives an ingenious way by which one can find out the arc length corresponding to a given *jyā*, when the *jyā* is small. Here the idea is to make use of a table of *jyās* and the differences $D_i's$, in order to obtain the required arc length and thereby avoid the iterative process. The procedure is as follows:

The difference between the *cāpa* and its *jyā* is given by

$$D = R\alpha - l \approx \frac{(R\alpha)^3}{6R^2} = \frac{(l)^3}{6R^2}. \tag{2.104}$$

In the above equation all the quantities are expressed in minutes. When the difference $D = 1''$, which is one-sixtieth of a minute, we obtain

$$\frac{(l)^3}{6R^2} = \frac{1}{60}. \tag{2.105}$$

This implies that when $D = 1''$ the corresponding *jyā* is given by

$$l_1 = \left(\frac{1.R^2}{10}\right)^{\frac{1}{3}}. \tag{2.106a}$$

Similarly when $D = 2''$, the corresponding *jyā* is given by

$$l_2 = \left(\frac{2.R^2}{10}\right)^{\frac{1}{3}}, \tag{2.106b}$$

and so on. In general, when $D = i''$, the corresponding *jyā* is given by

$$l_i = \left(\frac{i.R^2}{10}\right)^{\frac{1}{3}}. \tag{2.107}$$

Here, $l_i's$ correspond to the *jyās*, when the difference between the *jyā* and the *cāpa* (D) is i''. Hence, the lengths of the *cāpas*, A_is, corresponding to the *jyās*, l_i, are given by

$$A_i = l_i + i. \tag{2.108}$$

In Table 2.2, the *jyā* values are listed corresponding to the integral values of the difference between the *jyā* and the arc length, as given in *Laghu-vivṛti*. These are

Difference D = *cāpa*–*jyā* in seconds	Given value of *jyā* min	sec	Textual value of *cāpa* min	sec	Computed value of *cāpa* min	sec
1	105	43	105	44	105	43.56
2	133	11	133	13	133	12.42
3	152	26	152	29	152	29.04
4	167	46	167	50	167	49.80
5	180	43	180	48	180	47.34
6	192	02	192	08	192	07.02
7	202	08	202	15	202	14.82
8	211	20	211	28	211	27.12
9	219	47	219	56	219	55.14
10	227	38	227	48	227	46.80
11	234	58	235	09	235	07.98
12	241	52	242	04	242	03.18
13	248	24	248	37	248	35.88
14	254	36	254	50	254	48.90
15	260	31	260	46	260	44.58
16	266	10	266	26	266	24.78
17	271	36	271	53	271	51.12
18	276	48	277	06	277	04.86
19	281	50	282	09	282	07.20
20	286	40	286	60	286	59.10
21	291	22	291	43	291	41.46
22	295	55	296	17	296	14.94
23	300	18	300	41	300	40.26
24	304	36	304	60	304	58.02

Table 2.2 Look-up table from which the values of arc lengths of small *jyās* can be directly written down without performing any iteration, when the difference between the *jyā* and the *cāpa* is equal to integral number of seconds.

the l_is, $i = 1 \ldots 24$ in (2.107), which are listed in the second column. The third column gives the sum of columns 1 and 2. The fourth column gives the values of the arc length as computed by us using (2.108), which in turn involves the computation of the cube root of (2.107), for different values of i ($i = 1 \ldots 24$). In doing so, we have also used the exact value of the *trijyā* (in minutes), that is, $R = \frac{21600}{2\pi}$. Given the fact that some approximation in the *trijyā* value and the extraction of the cube root is involved in the computation of arc length, it is remarkable that the value given in the text differs at the most by $2''$ from the exactly computed value of the arc length. The idea behind listing these 24 *jyā* values is to avoid the iterative process outlined earlier, when the *jyā* value is small.

Finding the arc length from the look-up table

The procedure is explained in the commentary as follows:

अथवा एकद्व्यादिविलिप्तारूपं यज्ज्याचापान्तरं तद्विधायिनीः बह्वीः जीवाः पठित्वा तत्तुल्यासु अभीष्टज्यासु एकद्व्यादि विलिप्तारूपं ज्याचापान्तरं इष्टज्यायां प्रक्षिप्य तच्चापं कर्तव्यमिति। तत् कथमिति चेत् -

तत्र त्रिज्यायाः कृतिः त्रिखरूपाष्टभूनागरुद्रैस्तुल्यसङ्ख्या प्रसिद्धा। तत्र त्रिज्याकृतेः एकद्व्यादिनिहतायाः दशभिर्विभज्य लब्धात् फलात् घनमूलमानयेत्। तत्तत्तुल्यासु इष्टज्यासु क्रमादेकद्व्यादिविलिप्तिकाः ज्याचापान्तरत्वेन ग्राह्या इति। तथानीतज्याचापान्तरम् इष्टज्यायां प्रक्षिप्य चापीकरणं कार्यमिति। तद्यथा -

लवणं निन्द्यं कपिला गोपी चरराशयस्तवार्थितया ।
लघुनोद्दिष्टो राज्ञः प्रळयो धाम्नां त्रिनेत्र नरकपुरम् ॥
सवधूटीन्द्रो जलसूरद्रीहिमवान् गुरुस्त्रिशङ्कुवरः ।
वरदो वज्री तिलभूर्मेरुः कालेन तत्र नृपतिचरः ॥
तिलकं सान्द्रं धावतिसरित् न मे कुञ्जरो निवृत्तजरः ।
श्रेष्ठकळत्रममाशाधात्री धूपोऽग्नीनाम्बुतिलवनगः ॥
एवमार्यात्रयेणोक्ता मौर्विकाविकलादयः ।
चापीकरणमेताभिः सुकरं दोःफलेऽल्पके ॥
लवणादिषु जीवासु यया तुल्यं भुजाफलम् ।
तत्सङ्ख्या विकला क्षेप्याः तत्र चापप्रसिद्धये ॥ इति ॥

Or if the difference between the *jyā* and the arc length is equal to 1″, 2″, 3″ etc. then construct the table listing the *jyās* corresponding to these differences. If the *jyā* whose *cāpa* is to be determined happens to be (very close to) one of the values listed in the table, then add this difference between *jyā* and *cāpa* (1″, 2″, 3″ etc.) to the *jyā* to get the required *cāpa*. How should this be implemented?

It is well known that the square of the *trijyā* = 11818103. Multiply this by 1, 2, 3, etc., divide by 10, and take the cube roots of the resulting quantities [in minutes etc]. If the *jyā* whose *cāpa* is desired to be found happens to be one of the values [listed in the table], then it is to be understood that the corresponding difference between the *jyā* and *cāpa* is going to be only 1″, 2″, 3″, etc. The difference between the *jyā* and *cāpa*, obtained thus, may thus be added to the given *jyā* to get the desired *cāpa*. This may be done as follows.[15]

Thus the *jyās* in seconds and minutes are given in three *āryā* verses. For instance, the *lavaṇaṃ nindyam* and the *kapilā gopī* stand for 105′43″ and 133′11″, respectively. Finding the arc lengths from the *jyās*, when they are small, is quite simple making use of these values. If the *doḥphala* (whose arc length is to be calculated) is equal to one of the values listed, beginning with the *lavaṇa*, then the corresponding number of seconds have to be added to the *jyās* to get the corresponding *cāpa*.

In the commentary it is also stated that using the table and determining the arc lengths may not be as accurate as the result obtained by using the iterative procedure:

[15] The values of the *jyās* given in the succeeding verses *lavaṇaṃ* . . . , are listed in second column of Table 2.2.

यद्यपि सुसूक्ष्मचापीकरणोपायः पूर्वमेव प्रदर्शितः तथापि अल्पीयस्याः जीवायाः चापीकरणमेवं कर्तव्यम्, इति इहापि प्रदर्शितम्। अत उक्तम् - चरदोःफलजीवादेः एवमल्पधनुर्नयेत् - इति।

Though the procedure for obtaining more accurate values of the arc length has already been stated, for smaller *jyās* the arc lengths may be obtained by this method (from the look-up tables). That is why it is stated: The small arc length of the *cara-doḥphala* etc. should be obtained by this method.

The same idea is conveyed in *Yukti-dīpikā* in the following manner:

उक्तं चापघने षड्ढ्रिज्यावर्गे कलासमम् ॥
दशांशे तत्कृतेः चापज्यान्तरं विकलासमम् ।
एकादिघ्नात्ततस्त्रिज्यावर्गतो दशभिर्हृतात् ॥
घनमूलं तु यल्लब्धं तत्तुल्ये धनुषि स्थिते ।
एकद्व्याद्या विलिप्ताः स्युः चापज्याविवरोद्भवाः ॥
तदूनं चापमर्धज्या तद्युता ज्या च तद्धनुः ।
कार्योऽविशेषश्चापाप्तौ चापाल्पत्वे दृढं च तत् ॥ [16]

It has been stated implicitly (in verse 17 of the text) that the difference between the *jyā* and *cāpa* will be equal to $1'$ (one *kalā*), when the cube of the arc length is equal to six multiplied by square of the *trijyā*. The same will be equal to $1''$ (one *vikalā*) when the cube of the arc length is equal to one-tenth of the square of *trijyā*.

Now, the square of the *trijyā* divided by 10 is multiplied by 1,2,3, etc. Then the cube roots of the results are taken [and stored separately]. These correspond to the arc lengths, when the difference between the *jyā* and *cāpa* is equal to $1''$, $2''$, $3''$, etc., respectively. When differences are subtracted from the arc length we get the *jyā* and when they are added to the *jyā* we get the arc length. *Aviśeṣakarma* must be done in order to get accurate results for the *cāpa* from the *jyā* whose values are small.

In fact the accuracy of the tabulated results is of the order of 0.003%. For instance for a *cāpa* of $105'44''$, the listed *jyā* value is $105'43''$, whereas the exact Rsine value is $105'43.02''$. The percentage error is 0.0003%. This is not surprising considering the fact that for a small α the fractional error in retaining terms only up to α^3 in $\sin\alpha$ is $\frac{\alpha^5}{5!}$.

२.१६ मन्दशीघ्रकर्णानयनम्

2.16 Obtaining the *manda* and *śīghra* hypotenuses

आद्ये पदे चतुर्थे च व्यासार्धे कोटिजं फलम् ।
युक्त्वा त्यक्त्वान्ययोः तद्दोःफलवर्गैक्यजं पदम् ॥ ४० ॥
कर्णः स्यादविशेषोऽस्य कार्यो मन्दे चले न तु ।

ādye pade caturthe ca vyāsārdhe koṭijaṃ phalam|
yuktvā tyaktvānyayoḥ taddoḥphalavargaikyajaṃ padam || 40 ||
karṇaḥ syādaviśeṣo'sya kāryo mande cale na tu|

[16] {TS 1977}, p. 158.

Having added the *koṭiphala* to the radius (*vyāsārdha*) in the first and the fourth quadrants and having subtracted [the *koṭiphala*] from it (the radius) in the other two [quadrants] let the square root of the sum of the squares of this and the *doḥphala* be obtained. This is the *karṇa* and in the *manda* process this has to be further iterated upon, but not in the *śīghra* (*cala*).

The method given in the above verse for finding the *karṇa* can be explained with the help of an epicycle model represented in Fig. 2.12*a*. Here the mean planet P_0 is assumed to be moving on the deferent circle centred around O, and the true planet P is located on the epicycle such that PP_0 is parallel to OU (the direction of the *mandocca*). $O\Gamma$ represents the direction of *Aśvini nakṣatra* (*Meṣādi* or first point of Aries).

In Fig. 2.12*a* let R and r be the radii of the deferent circle and the epicycle respectively. OU represents the direction of the *mandocca* whose longitude is given by $\Gamma\hat{O}U = \theta_m$. The longitude of the mean planet P_0 is given by $\Gamma\hat{O}P_0 = \theta_0$. θ_{ms} represents the longitude of the *manda-sphuṭa-graha*. It is easily seen that

$$U\hat{O}P_0 = P\hat{P_0}N = \theta_0 - \theta_m, \tag{2.109}$$

where $(\theta_0 - \theta_m)$ is the *manda-kendra*. The *doḥphala* and the *koṭiphala* are given by

$$doḥphala = PN = |r\sin(\theta_0 - \theta_m)| \tag{2.110}$$

and

$$koṭiphala = P_0N = |r\cos(\theta_0 - \theta_m)|. \tag{2.111}$$

Now, the *manda-karṇa* K is the distance between the planet and the centre of the deferent circle. Clearly,

$$\begin{aligned} K &= OP \\ &= \left[(ON)^2 + (PN)^2\right]^{\frac{1}{2}} \\ &= \left[(R + r\cos(\theta_0 - \theta_m))^2 + (r\sin(\theta_0 - \theta_m))^2\right]^{\frac{1}{2}}. \end{aligned} \tag{2.112}$$

Here, $r\cos(\theta_0 - \theta_m) = \pm|r\cos(\theta_0 - \theta_m)|$ is positive in the first and fourth quadrants and negative in the second and third quadrants. That is why it is stated that the *koṭiphala* has to be added to the *trijyā* in the first and fourth quadrants and subtracted from it in the second and third quadrants.

It is also stated that the *karṇa* K has to be determined iteratively in the *manda-saṃskāra* to obtain the *aviśeṣa-karṇa* (iterated hypotenuse). This is because r in (2.112) is not a constant but is itself proportional to K. That is,

$$r = \frac{r_0}{R}K, \tag{2.113}$$

where r_0 is the radius of the epicycle whose value is specified in the text. The iterative procedure to determine K and r is discussed in the next section. In the *śīghra-saṃskāra*, r is fixed for each planet, and no iterative procedure is necessary to find K.

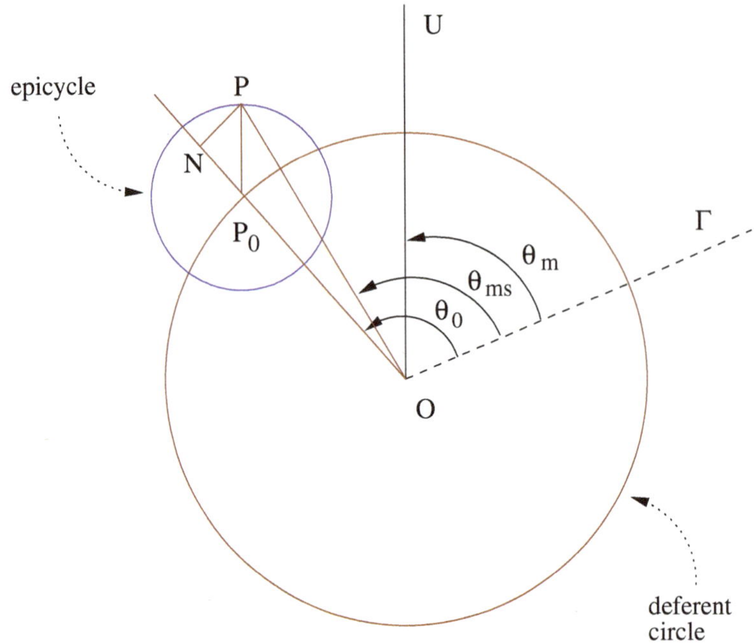

Fig. 2.12*a* Obtaining the *manda-karṇa* in the epicycle model.

In Fig. 2.12*a*, the longitude of the planet is given by $\Gamma\hat{O}P = \theta_{ms} = \theta$. Then $P\hat{O}P_0 = \theta_m - \theta$ is the difference between the mean and true planets. Now,

$$PN = OP\sin(P\hat{O}P_0) = K\sin(\theta_m - \theta). \tag{2.114}$$

PN is also given by

$$PN = PP_0\sin(P\hat{P}_0N) = r\sin(\theta_0 - \theta_m). \tag{2.115}$$

Equating the above two expressions for PN,

$$\begin{aligned} K\sin(\theta_m - \theta) &= r\sin(\theta_0 - \theta_m) \\ \text{or} \qquad \sin(\theta_m - \theta) &= \frac{r}{K}\sin(\theta_0 - \theta_m) \\ &= \frac{r_0}{R}\sin(\theta_0 - \theta_m). \end{aligned} \tag{2.116}$$

Thus the true planet θ can be obtained from the mean planet θ_0 from the above equation. It may be noted that (2.116) does not involve the *manda-karṇa* K.

While commenting on these verses, the eccentric and epicyclic models are described in *Yukti-dīpikā*. First, we give the verses explaining the eccentric model.

ग्रहोच्चसूत्रान्तरालं ग्रहवृत्तगतं भुजा ।
कोटिस्तत्केन्द्रतो दोर्ज्यामूलान्ता परिकल्प्यते ॥
केन्द्रान्तरं चान्त्यफलं स्यात् कक्ष्याग्रहवृत्तयोः ।
दोर्ज्यामूले तु कक्ष्यातो बहिरन्तर्गते क्रमात् ॥
कोटयन्त्यफलयोर्योगभेदाभ्यां स्फुटकोटिका ।
तयोर्वर्गयुतेर्मूलं कक्ष्याकेन्द्रग्रहान्तरम् ॥
कर्णः स एव विज्ञेयः प्रतिवृत्तकलामितः ।[17]

The distance of separation between the planet and the *uccasūtra* is the *dorjyā* measured with respect to the *grahavṛtta* (the circle in which the planet moves). The *koṭijyā* is equal to the distance of separation between the centre of the *grahavṛtta* and the foot of the *dorjyā* on the *uccasūtra*.

The distance of separation between the centres of the *grahavṛtta* and the *kakṣyāvṛtta* is the *antyaphala*. The *sphuṭakoṭikā* is obtained by adding or subtracting the *antyaphala* to or from the *koṭijyā* depending upon whether the foot of the *dorjyā* is outside or inside the *kakṣyāvṛtta*. The square root of the sum of the squares of the two [*dorjyā* and *sphuṭakoṭikā*] is the distance of separation between the centre of *kakṣyāvṛtta* and the planet. This has to be understood as the *karṇa* measured in terms of the *prativṛtta*.

In Fig. 2.12*b*, the circle centred around O' is called the *grahavṛtta*, or *prativṛtta* or *pratimaṇḍala* (the eccentric circle), and the one centred around O is the *kakṣyāvṛtta* (the deferent circle). OU represents the direction of the *mandocca*. These two circles, namely the *grahavṛtta* and the *kakṣyavṛtta*, have the same radius and their centres are displaced along the direction of the *mandocca* U. The dotted circle with its centre at the centre of the *kakṣyāvṛtta* is known as the *karṇamaṇḍala* or *karṇavrtta* (hypotenuse circle). The distance of separation between the centres of the *grahavṛtta* and the *kakṣyāvṛtta* is referred to as the *antyaphala*. If R is the radius of the *grahavṛtta* and $(\theta_0 - \theta_m)$ the *manda-kendra*, then the *dorjyā* and *koṭijyā* are given by

$$dorjy\bar{a} = PN = |R\sin(\theta_0 - \theta_m)| \tag{2.117}$$

and

$$ko\d{t}ijy\bar{a} = O'N = |R\cos(\theta_0 - \theta_m)|. \tag{2.118}$$

The *sphuṭakoṭikā* is defined by

$$\begin{aligned} sphu\d{t}ako\d{t}ik\bar{a} = ON &= ko\d{t}ijy\bar{a} \overset{+}{\sim} antyaphala \\ &= |R\cos(\theta_0 - \theta_m)| \overset{+}{\sim} r. \end{aligned} \tag{2.119}$$

It is stated that the '$\sim$' sign should be taken when both the edges of the *dorjyā* (points P and N in Fig. 2.12*b*) lie within the *kakṣyāvṛtta*, and '$+$' when at least one or both the edges of the *dorjyā* lie outside the *kakṣyāvṛtta*.

Actually, whether the '$+$' or the '$\sim$' sign has to be taken depends on whether P lies above or below the straight line perpendicular to OU passing through O', that is, when $(\theta_0 - \theta_m)$ is in the first/fourth quadrants or in the second/third quadrants respectively. If K represents the *karṇa* OP, then it is given by

[17] {TS 1977}, pp. 161–2.

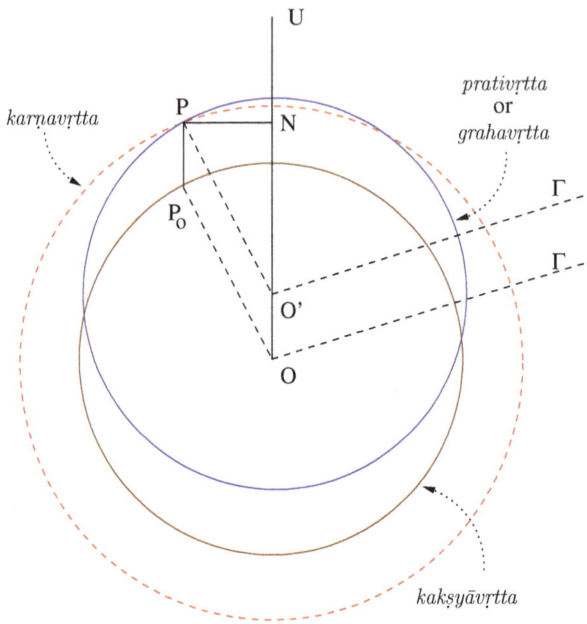

Fig. 2.12*b* Obtaining the *manda-karṇa* in the eccentric model.

$$
\begin{aligned}
K &= OP \\
&= (PN^2 + ON^2)^{\frac{1}{2}} \\
&= \left[dorjy\bar{a}^2 + sphu\d{t}ako\d{t}ik\bar{a}^2\right]^{\frac{1}{2}} \\
&= \left[(R\sin(\theta_0 - \theta_m))^2 + (|R\cos(\theta_0 - \theta_m)| \overset{+}{\sim} r)^2\right]^{\frac{1}{2}} \\
&= \left[(R\sin(\theta_0 - \theta_m))^2 + (R\cos(\theta_0 - \theta_m) + r)^2\right]^{\frac{1}{2}}. \qquad (2.120)
\end{aligned}
$$

K can be determined using the above formula, or by using equation (2.112), which are equivalent. This is explained in the following verses of *Yukti-dīpikā*:

कक्ष्यावृत्तस्य तन्नेमिस्थोच्चनीचस्य च द्वयोः ।
केन्द्रद्वयावभेदी यो मार्गस्तस्माद् ग्रहान्तरम् ॥
दोःफलं यत्तु तन्मूलान्तरं नीचोच्चकेन्द्रतः ।
कोटीफलं तद्युतोना त्रिज्या कक्ष्याख्यवृत्तितः ॥
क्रमाद् दोःफलमूले तु बहिरन्तर्गते सति ।
सा तु दोःफलमूलस्य कक्ष्याकेन्द्रस्य चान्तरम् ॥
तत्कृतौ दोःफलकृतिं युक्त्वा कर्णः पदीकृतः ।
एवं कर्णो द्विधा साध्यः स तु मान्दो विशिष्यते ॥ [18]

[18] {TS 1977}, p. 162.

The distance of separation between the planet, and the line passing through the centre of the *kakṣyāvṛtta* and the centre of the *uccanīcavṛtta* (epicycle) which moves on the circumference of the *kakṣyāvṛtta*, is the *doḥphala*. The distance of separation between the foot of the perpendicular [of the *dorjyā*] and the centre of the *uccanīcavṛtta* is the *koṭiphala*. Depending upon whether the foot of the *doḥphala* lies outside the *kakṣyāvṛtta* or inside, the *koṭiphala* has to be added to or subtracted from the *trijyā*. This gives the distance of separation between the centre of *kakṣyāvṛtta* and the foot of the *doḥphala*. The square root of the sum of the square of this (distance of separation) and the square of the *doḥphala* is the *karṇa*. In this way the *karṇa* can be obtained in two ways and it has to be iterated in the case of the *manda-saṃskāra*.

२.१७ अविशेषकर्णानयनम्

2.17 Obtaining the iterated hypotenuse

दोःकोटिफलनिघ्नाद्ये कर्णात् त्रिज्याहृते फले ॥ ४१ ॥
ताभ्यां कर्णः पुनस्साध्यः भूयः पूर्वफलाहतात् ।
तत्तत्कर्णात् त्रिभज्याप्तफलाभ्यामविशेषयेत् ॥ ४२ ॥

doḥkoṭiphalanighnādye karṇāt trijyāhṛte phale || 41 ||
tābhyāṃ karṇaḥ punassādhyaḥ bhūyaḥ pūrvaphalāhatāt|
tattatkarṇāt tribhajyāptaphalābhyāmaviśeṣayet || 42 ||

The *doḥphala* and the *koṭiphala* [initially obtained] are multiplied by the *karṇa* [obtained from them] and divided by *trijyā*. From these resulting *phalas*, the *karṇa* has to be obtained again. Further, the previous *phalas* must be multiplied by the corresponding *karṇas* and divided by the *trijyā*, and the process has to be repeated to get the *aviśeṣa-karṇa* (the hypotenuse which does not change on iteration).

It was shown earlier (2.112) that

$$K = \left[(R + r\cos(\theta_0 - \theta_m))^2 + (r\sin(\theta_0 - \theta_m))^2\right]^{\frac{1}{2}}. \tag{2.121}$$

Here the radius of the epicycle r itself is proportional to *karṇa* K (2.113) and therefore needs to be determined along with K iteratively.

Procedure for finding the iterated hypotenuse

We explain the procedure for finding the iterated hypotenuse or *aviśeṣa-karṇa* with the help of Fig. 2.12*a*. Let R, r be the radii of the deferent circle and the epicycle respectively. $U\hat{O}P_0$ is the *manda-kendra* $(\theta_0 - \theta_m)$. The quantities $r\sin(\theta_0 - \theta_m) = PN$ and $r\cos(\theta_0 - \theta_m) = P_0N$ are referred to as the *doḥphala* and *koṭiphala* respectively. Thus, in the first approximation, r is set equal to r_0 and the *doḥphala* and *koṭiphala* are taken to be $r_0\sin(\theta_0 - \theta_m)$ and $r_0\cos(\theta_0 - \theta_m)$ respectively. Let them be denoted d_1 and k_1. The *karṇa* OP which represents the distance of the planet from the centre of the *kakṣyāvṛtta* is given by

$$K_1 = \left[(R + k_1)^2 + d_1^2\right]^{\frac{1}{2}}. \tag{2.122}$$

Here K_1 is the first approximation to the *manda-karṇa*. Then, the *doḥphala* (d_2) and *koṭiphala* (k_2) are obtained as follows:

$$d_2 = \frac{K_1 \times d_1}{R} \qquad k_2 = \frac{K_1 \times k_1}{R}. \tag{2.123}$$

The second approximation to the *manda-karṇa*, K_2, is given by

$$K_2 = \left[(R + k_2)^2 + d_2^2\right]^{\frac{1}{2}}. \tag{2.124}$$

Then, the *doḥphala* (d_3) and *koṭiphala* (k_3) are obtained as follows:

$$d_3 = \frac{K_2 \times d_1}{R} \qquad k_3 = \frac{K_2 \times k_1}{R}. \tag{2.125}$$

The third approximation to the *manda-karṇa*, K_3, is obtained by

$$K_3 = \left[(R + k_3)^2 + d_3^2\right]^{\frac{1}{2}}. \tag{2.126}$$

The above process is carried out until $K_i \approx K_{i-1}$, to the desired accuracy. When this happens, K_i is referred to as the *aviśeṣa-karṇa*. This *aviśeṣa-karṇa* is to be used in *manda-saṃskāra* to obtain the *manda-phala*.

The rationale behind the iterative process used in obtaining the *aviśeṣa-karṇa* is explained in *Yukti-dīpikā* as follows:

मान्दं नीचोच्चवृत्तं तत्कर्णवृत्तकलामितम् ।
यतस्तत्कर्णवृद्धिक्षयानुसारीदमुच्यते ॥
मन्दकर्णे स्वलिप्ताभिः प्रमिते त्रिज्यया समे ।
पठितः परिधिर्मान्दः कर्णवृत्तकलामितः ॥
ऊनाधिके ततः कर्णे प्रतिवृत्तकलामिते ।
तेन कर्णेन दोःकोटिफले ताभ्यां तु तन्नयेत् ॥
अन्योन्याश्रयता चैषां अविशेषान्निरस्यते ।
तत्तत्कर्णाद्यदोःकोटिघाताभ्यां त्रिज्ययाहृते ॥
ताभ्यां त्रिभज्यया चापि प्राग्वत् कर्णं मुहुर्नयेत्।[19]

The *manda-nīcocca-vṛtta* (*manda* epicycle) is measured in terms of *karṇavṛtta* (hypotenuse circle) because it is said to increase or decrease in accordance with the *karṇavṛtta*. The tabulated value of the circumference of the *manda* circle is in the measure of the *karṇavṛtta*, when the *manda-karṇa* is taken to be the *trijyā*. When the *karṇa* increases and decreases and this value is measured in terms of *prativṛtta*, then the *doḥ* and *koṭiphala* have to be obtained from that *karṇa*. It is from them (*doḥ* and *koṭiphala*) that (the measure of *manda-nīcocca-vṛtta*) has to be obtained. This interdependence is eliminated by doing an iteration, the *aviśeṣakarma*. Multiplying the *doḥphala* and *koṭiphala* by *karṇa* and dividing it by the *trijyā* [the new *doḥphala* and *koṭiphala* are determined]. With the *trijyā* and these, once again the *karṇa* has to be obtained as explained earlier.

Now,

$$\sqrt{d_1^2 + k_1^2} = r_0. \tag{2.127}$$

[19] {TS 1977}, pp. 162–3.

From Fig. 2.12*a* and the equivalent of (2.125) it can be seen that for any *i*,

$$\sqrt{d_i^2 + k_i^2} = \frac{K_{i-1}}{R} \sqrt{d_1^2 + k_1^2} \tag{2.128}$$

$$= \frac{K_{i-1}}{R} r_0. \tag{2.129}$$

After a few iterations, the successive values of the radius and the *karṇa* start converging. That is,

$$\sqrt{d_{i-1}^2 + k_{i-1}^2} \approx \sqrt{d_i^2 + k_i^2} \to r$$

and

$$K_{i-1} \approx K_i \to K.$$

Hence

$$\frac{r}{K} = \frac{r_0}{R}. \tag{2.130}$$

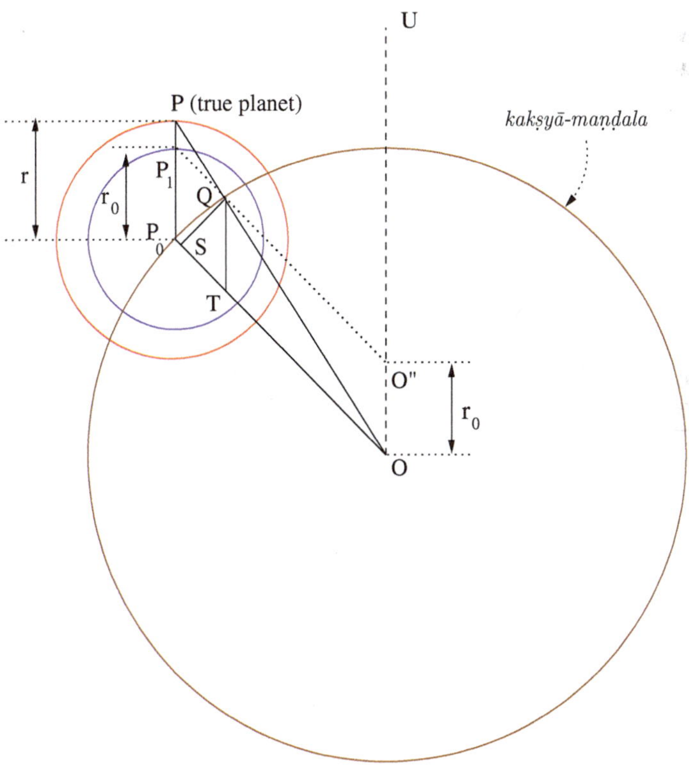

Fig. 2.12c Variation of the epicycle with the *karṇa* in the *manda* process and the *aviśiṣṭa-manda-karṇa*.

In Fig. 2.12*c*, P_0 is the mean planet moving in the *kakṣyāmaṇḍala* with O as the centre, and OU is the direction of the *mandocca*. Draw a circle of radius r_0 with P_0 as centre. Let P_1 be the point on this circle such that P_0P_1 is in the direction of the *mandocca* (parallel to OU). Let O'' be a point on the line OU, such that $OO'' = r_0$. Join P_1O'' and let that line meet the *kakṣyāmaṇḍala* at Q. Extend OQ and P_0P_1 so as to meet at P. The true planet is located at P. Then it can be shown that $OP = K$ and $P_0P = r$ are the actual *manda-karṇa* and the corresponding (true) radius of the epicycle as will result by the process of successive iteration.[20] Since P_1O'' is parallel to P_0O, the triangles OP_0P and $QO''O$ are similar and we have

$$\frac{r}{K} = \frac{P_0P}{OP} = \frac{O''O}{QO} = \frac{r_0}{R}. \tag{2.131}$$

The process of successive iteration to obtain K is essentially the following. In triangle OP_1P_0, with the angle $P_1\hat{P_0}O = 180° - (\theta_0 - \theta_m)$, the first approximation to the *karṇa* (*sakṛt-karṇa*) $K_1 = OP_1$ and the mean epicycle radius $r_0 = P_1P_0$ are related by

$$K_1 = \sqrt{R^2 + r_0^2 + 2r_0R\cos(\theta_0 - \theta_m)}. \tag{2.132}$$

In the RHS of (2.132), we replace r_0 by the next approximation to the radius of the epicycle

$$r_1 = \frac{r_0}{R}K_1, \tag{2.133}$$

and obtain the next approximation to the *karṇa*,

$$K_2 = \sqrt{R^2 + r_1^2 + 2r_1R\cos(\theta_0 - \theta_m)}, \tag{2.134}$$

and so on. This process is iterated till K_i and K_{i+1} become indistinguishable, and that will be the *aviśiṣṭa-karṇa* (iterated hypotenuse) K,[21] which is related to the corresponding epicycle radius r as in (2.133) by

$$r = \frac{r_0}{R}K. \tag{2.135}$$

२.१८ अविशेषकर्णानयने प्रकारान्तरम्
2.18 Another method of obtaining the iterated hypotenuse

विस्तृतिदलदोःफलकृतिवियुतिपदं कोटिफलविहीनयुतम् ।
केन्द्रे मृगकर्किगते स खलु विपर्ययकृतो भवेत् कर्णः ॥ ४३ ॥

---

[20] {MB 1960}, pp. 111–19.

[21] The term *viśeṣa* means 'distinction'. Hence, *aviśeṣa* is 'without distinction'. Therefore the term *aviśiṣṭa-karṇa* refers to that *karṇa* obtained after doing a series of iterations such that the successive values of the *karṇa* do not differ from each other.

तेन हृता त्रिज्याकृतिः अयत्नविहितोऽविशेषकर्णः स्यात् ।
इति वा कर्णः साध्यः मान्दे सकृदेव माधवप्रोक्तः ॥ ४४ ॥

vistṛtidaladoḥphalakṛtiviyutipadaṃ koṭiphalavihīnayutam|
kendre mṛgakarkigate sa khalu viparyayakṛto bhavet karṇaḥ || 43 ||
tena hṛtā trijyākṛtiḥ ayatnavihito'viśeṣakarṇaḥ syāt|
iti vā karṇaḥ sādhyaḥ mānde sakṛdeva mādhavaproktaḥ || 44 ||

The square of the *doḥphala* is subtracted from the square of the *trijyā* and its square root is taken. The *koṭiphala* is added to or subtracted from this depending upon whether the *kendra* (anomaly) is within 6 signs beginning from *Karki* (Cancer) or *Mṛga* (Capricorn). This gives the *viparyaya-karṇa*. The square of the *trijyā* divided by this *viparyaya-karṇa* is the *aviśeṣa-karṇa* (iterated hypotenuse) obtained without any effort [of iteration]. This is another way by which the [*aviśeṣa*]-*karṇa* in the *manda* process can be obtained as enunciated by Mādhava.

A method to determine the *manda-karṇa* without an iterative process is discussed here. This method is attributed to Mādhava of Saṅgamagrāma, the renowned mathematician and astronomer of the 14th century. A new quantity called the *viparyaya-karṇa* or *viparīta-karṇa* is introduced for this purpose. This *viparīta-karṇa* ('inverse' hypotenuse) is nothing but the radius of the *kakṣyāvṛtta* when the *manda-karṇa* is taken to be the *trijyā*, *R*.

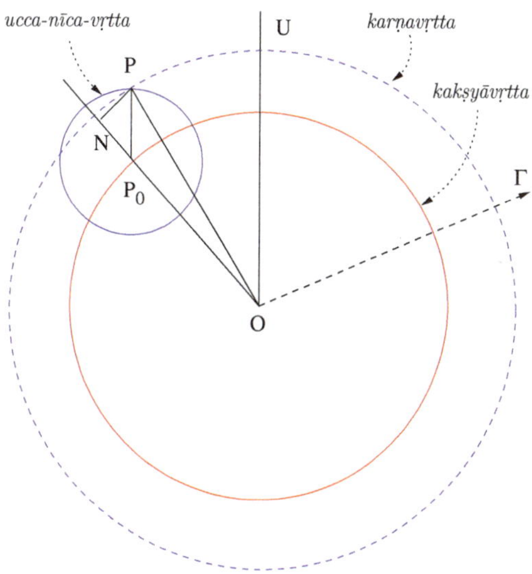

Fig. 2.13*a* Determination of the *viparīta-karṇa* when the *kendra* is in the first quadrant.

The rationale behind the formula given for *viparīta-karṇa* is outlined in the Malayalam text *Yuktibhāṣā*, and can be understood with the help of Figs. 2.13*a* and *b*. In these figures P_0 and P represent the mean and the true planet respectively. N denotes the foot of the perpendicular drawn from the true planet P to the line joining the centre of the circle and the mean planet. NP is equal to *doḥphala*. Let the radius of the *karṇavṛtta* OP be set equal to the *trijyā* R. Then the radius of the *uccanīcavṛtta* P_0P is r_0, as it is in the measurement of the *karṇavṛtta*. In this measurement, the radius of the *kakṣyāvṛtta* $OP_0 = R_v$, the *viparīta-karṇa*, and is given by

$$\begin{aligned} R_v &= ON \pm P_0N \\ &= \sqrt{R^2 - (r_0 \sin(\theta_0 - \theta_m))^2} \pm |r_0 \cos(\theta_0 - \theta_m)|. \end{aligned} \tag{2.136}$$

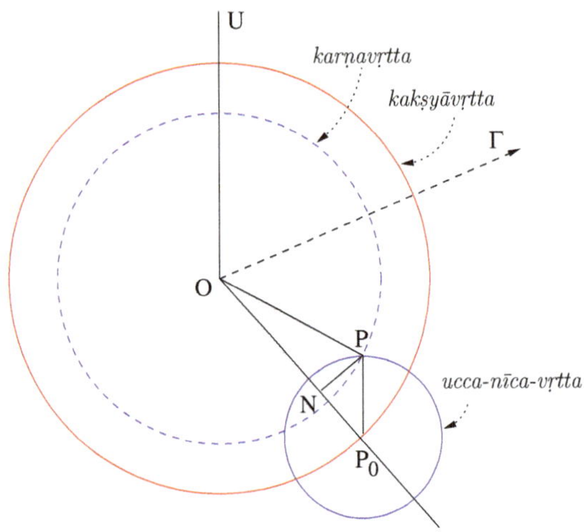

Fig. 2.13*b* Determination of the *viparīta-karṇa* when the *kendra* is in the third quadrant.

Here we should take the '−' sign when the *manda-kendra* is in the first and fourth quadrants $270 \leq (\theta_0 - \theta_m) < 90$ and the '+' sign when it is in the second and third quadrants $90 \leq (\theta_0 - \theta_m) < 270$. When the radius of the *kakṣyāvṛtta* is the *trijyā* R, the value of *manda-karṇa* is K, and when the radius of the *manda-karṇa* is R, the radius of the *kakṣyāvṛtta* is R_v. Hence

$$\frac{K}{R} = \frac{R}{R_v}$$

$$\text{or} \qquad K = \frac{R^2}{R_v}. \tag{2.137}$$

Thus the *aviśiṣṭa-manda-karṇa*, also referred to as the *aviśeṣa-karṇa*, is given by

$$\textit{aviśeṣa-karṇa} = \frac{\textit{trijyā}^{\,2}}{\textit{viparyaya-karṇa}}. \tag{2.138}$$

Since r_0 is a known quantity, for any given value of $(\theta_0 - \theta_m)$ R_v can be determined from (2.136). Once R_v is known, using (2.137) the *aviśiṣṭa-manda-karṇa*, K, can be found in one step without resorting to the tedious iterative process described in the previous section for its computation.

The formula for the *viparīta-karṇa* in (2.136) can also be understood from the geometrical construction in Fig. 2.12*c*. As the triangles OP_0P and OTQ are similar,

$$\frac{OT}{OQ} = \frac{OP_0}{OP}$$

$$\text{or} \qquad OT = \frac{R^2}{K}, \tag{2.139}$$

as $OQ = OP_0 = R$. Hence the *viparyaya-karṇa* $R_v = OT$. Also,

$$Q\hat{T}S = U\hat{O}P_0 = \theta_0 - \theta_m. \tag{2.140}$$

Hence, $QS = r_0 \sin(\theta_0 - \theta_m)$ and $ST = r_0 \cos(\theta_0 - \theta_m)$. Now

$$\begin{aligned} OT &= OS - ST \\ &= \sqrt{OQ^2 - SQ^2} - ST \\ &= \sqrt{R^2 - r_0^2 \sin^2(\theta_0 - \theta_m)} - r_0 \cos(\theta_0 - \theta_m), \end{aligned} \tag{2.141}$$

which is the same as (2.136).

२.१९ अविशेषकर्णेन अर्कस्फुटीकरणम्
2.19 Correcting the Sun using the iterated hypotenuse

त्रिज्याघ्नो दोर्गुणः कर्णभक्तः स्फुटभुजागुणः ।
तद्धनुः संस्कृतं स्वोच्चं नीचं वा युक्तितः स्फुटम् ॥ ४५ ॥

trijyāghno dorguṇaḥ karṇabhaktaḥ sphuṭabhujāguṇaḥ |
taddhanuḥ saṃskṛtaṃ svoccaṃ nīcaṃ vā yuktitaḥ sphuṭam || 45 ||

The true *dorjyā* is [equal to] the *dorjyā* multiplied by the *trijyā* and divided by the *karṇa*. The arc of this appropriately applied to the *ucca* or *nīca* gives the true position [of the planet].

This can be explained from Fig. 2.14*a*. Let $\phi = P\hat{O}U$ be the difference $(\theta - \theta_m)$ between the *manda-sphuṭa* and the *ucca*. Now

$$PN = P_0N_0,$$

$$\text{or} \qquad K \sin\phi = R \sin(\theta_0 - \theta_m). \tag{2.142a}$$

Hence

$$R \sin\phi = R \sin(\theta_0 - \theta_m) \frac{R}{K},$$

$$\text{or} \qquad \phi = (R \sin^{-1}) \left[R \sin(\theta_0 - \theta_m) \frac{R}{K} \right]. \tag{2.142b}$$

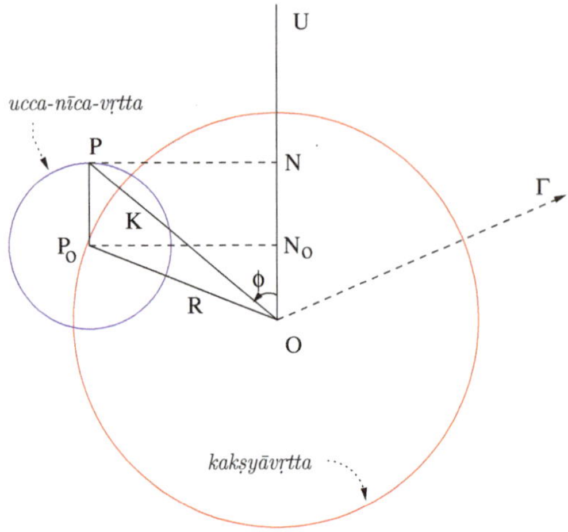

Fig. 2.14*a* The true position of the planet from the *ucca* and *nīca*.

Then the true planet ($\Gamma\hat{O}P$) is obtained as

$$\begin{aligned} \Gamma\hat{O}P &= \Gamma\hat{O}U + \phi \\ &= ucca + \phi. \end{aligned} \tag{2.143}$$

२.२० रविस्फुटात् तन्मध्यमानयनम्

2.20 Obtaining the mean Sun from the true Sun

अर्कस्फुटेनानयनं प्रकुर्यात् स्वमध्यमस्यात्र वितुङ्गभानोः ।
भुजागुणं कोटिगुणं च कृत्वा मृगादिकेन्द्रेऽन्त्यफलाख्यकोट्योः ॥ ४६ ॥
भेदः कुलीरादिगते तु योगः तद्वर्गयुक्तात् भुजवर्गतो यत् ।
पदं विपर्यासकृतः स कर्णः त्रिज्याकृतेस्तद्विहृतस्तु कर्णः ॥ ४७ ॥
तेनाहतामुच्चविहीनभानोः जीवां भजेद् व्यासदलेन लब्धम् ।
स्वोच्चे क्षिपेच्चापितमाद्यपादे चक्रार्धतः शुद्धमपि द्वितीये ॥ ४८ ॥
चक्रार्धयुक्तं तु तृतीयपादे संशोधितं मण्डलतश्चतुर्थे ।
एवं कृतं सूक्ष्मतरं हि मध्यं पूर्वं पदं यावदिहाधिकं स्यात् ॥ ४९ ॥
अन्त्यात् फलात् कोटिगुणं चतुर्थं त्वारभ्यते यद्यधिकात्र कोटिः ।
सर्वत्र विष्कम्भदलं श्रुतौ वा व्यासार्धके स्याद्विपरीतकर्णः ॥ ५० ॥

arkasphuṭenānayanaṃ prakuryāt svamadhyamasyātra vituṅgabhānoḥ|
bhujāguṇaṃ koṭiguṇaṃ ca kṛtvā mṛgādikendre'ntyaphalākhyakoṭyoḥ|| 46 ||
bhedaḥ kulīrādigate tu yogaḥ tadvargayuktāt bhujavargato yat|
padaṃ viparyāsakṛtaḥ sa karṇaḥ trijyākṛtestadvihṛtastu karṇaḥ || 47 ||
tenāhatāmuccavihīnabhānoḥ jīvāṃ bhajed vyāsadalena labdham|
svocce kṣipeccāpitamādyapāde cakrārdhataḥ śuddhamapi dvitīye|| 48 ||
cakrārdhayuktaṃ tu tṛtīyapāde saṃśodhitaṃ maṇḍalataścaturthe|

evaṃ kṛtaṃ sūkṣmataraṃ hi madhyaṃ pūrvaṃ padaṃ yāvadihādhikaṃ syāt|| 49 ||
antyāt phalāt koṭiguṇaṃ caturthaṃ tvārabhyate yadyadhikātra koṭiḥ|
sarvatra viṣkambhadalaṃ śrutau vā vyāsārdhake syādviparītakarṇaḥ|| 50 ||

The mean position of the Sun has to be obtained from the true position [as follows]. Having subtracted the longitude of the apogee from the true Sun, the *dorjyā* and *koṭijyā* are obtained. When the *manda-kendra* lies within the six signs beginning from *Mṛga*, the difference between the *antyaphala* and the *koṭijyā* has to be taken, and when it is within the six signs beginning from *Karka*, their sum has to be taken. The square root of the sum of the square of this and the square of the *dorjyā* is the *viparīta-karṇa*. The square of the *trijyā* divided by this *viparīta-karṇa* is the *karṇa*.

This (*karṇa*) is multiplied by the *dorjyā* obtained by subtracting the longitude of the apogee from the Sun, and divided by the *trijyā*. The arc of the result has to be applied positively to the longitude of the *mandocca* when the *manda-kendra* is in the first quadrant. 180 minus the arc, 180 (*cakrārdha*) plus the arc and 360 minus the arc have to be applied to the *mandocca* when the *manda-kendra* lies in the second, third and fourth quadrants respectively. The mean longitude obtained thus is accurate. In the first quadrant the *koṭijyā* is greater than the *antyaphala*. [Similarly] the fourth quadrant is said to commence when the *koṭiphala* becomes greater than the *antyaphala*. Always the *karṇa* bears the same relation to the *trijyā* as the *trijyā* to the *viparīta-karṇa* (inverse hypotenuse).

Normally the texts present the procedure for determining the true position of a planet from its mean position. The above set of verses present a procedure for solving the inverse problem, namely finding the mean Sun from its true position. We explain this procedure with the help of Fig. 2.14*b*. Here, the longitudes of the mean Sun, the true Sun and the *ucca* (apogee) are given by

$$\begin{aligned} \theta_0 &= \Gamma\hat{O}P_0 = P\hat{O'}P \\ \theta &= \Gamma\hat{O}P \\ \text{and} \qquad \theta_m &= \Gamma\hat{O}U = \Gamma\hat{O'}U, \end{aligned} \tag{2.144}$$

respectively. Further,

$$\begin{aligned} \theta - \theta_m &= N\hat{O}P \\ \theta_0 - \theta_m &= N\hat{O'}P = N\hat{O}P_0. \end{aligned} \tag{2.145}$$

Also, the *aviśiṣṭa-manda-karṇa* (iterated *manda* hypotenuse) $K = OP$ and the *vyāsārdha* $R = OP_0 = O'P$. The true epicycle radius $r = OO'$.

The word *antyaphala* used in the above verse has a special significance whose relation with the *manda-karṇa* may precisely be expressed as follows:

$$antyaphala = r_0 = \frac{r_0}{r}.r = \frac{R}{K}.r = \frac{R}{K}.OO'. \tag{2.146}$$

Now,

$$\begin{aligned} dorjy\bar{a} = R\sin N\hat{O}P &= \frac{R}{K}.K\sin N\hat{O}P \\ &= \frac{R}{K}.K\sin(\theta-\theta_m) = \frac{R}{K}.PN \end{aligned}$$

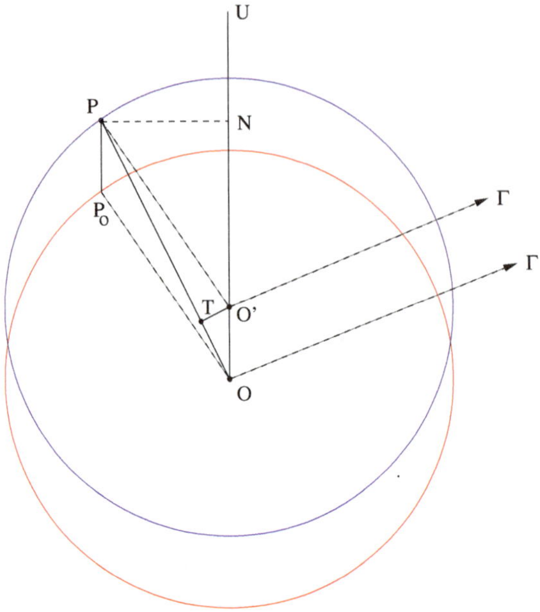

Fig. 2.14*b* Obtaining the *madhyama* (mean position) from the *sphuṭa* (true position).

$$\begin{aligned} koṭijyā = R\cos N\hat{O}P &= \frac{R}{K}.K\cos N\hat{O}P \\ &= \frac{R}{K}.K\cos(\theta - \theta_m) = \frac{R}{K}.ON. \end{aligned} \tag{2.147}$$

Hence the difference between the *koṭijyā* and the *antyaphala* is given by

$$\begin{aligned} koṭijyā - antyaphala &= R\cos(\theta - \theta_m) - r_0 \\ &= \frac{R}{K}(ON - OO') \\ &= \frac{R}{K}.O'N. \end{aligned} \tag{2.148}$$

Therefore,

$$\begin{aligned} \sqrt{(koṭijyā - antyaphala)^2 + (dorjyā)^2} &= \frac{R}{K}\sqrt{O'N^2 + PN^2} \\ &= \frac{R}{K}.O'P \\ &= \frac{R^2}{K}. \end{aligned} \tag{2.149}$$

The expression obtained above is the same as the *viparīta-karṇa* R_v appearing in (2.136). Now, using (2.147) and (2.148), this may be expressed as

$$\mathrm{R}_v = \sqrt{(R\cos(\theta-\theta_m)-r_0)^2 + R^2\sin^2(\theta-\theta_m)}. \tag{2.150}$$

Since the positions of the *ucca* and the true planet are known, R_v can be determined. Also, the *manda-karṇa* $K = \frac{R^2}{R_v}$ can be determined from $\theta - \theta_m$. Now

$$\begin{aligned} PN &= K\sin(\theta-\theta_m) \\ &= O'P\sin(N\hat{O'}P) \\ &= R\sin(\theta_0-\theta_m). \end{aligned} \tag{2.151}$$

Hence

$$\begin{aligned} madhyama - ucca &= \theta_0 - \theta_m \\ &= R\sin^{-1}\left[R\sin(\theta-\theta_m)\frac{K}{R}\right]. \end{aligned} \tag{2.152}$$

From this *madhyama* − *ucca* is obtained. When this is added to the *ucca*, the *madhyama* is obtained. When *sphuṭa* − *ucca* is positive, $O'N$ = *koṭijyā* − *antyaphala*. In the above, it is $R\sin$ (*madhyama* − *ucca*) which is found first in terms of $R\sin$(*sphuṭa* − *ucca*). The quadrant in which (*madhyama* − *ucca*) lies can be determined without any ambiguity from the geometry.

When it is in the second or third quadrants, $R\cos$(*sphuṭa* − *ucca*) is negative and $O'N$ = *koṭijyā*+ *antyaphala*. Of course, in all cases, the formula for R_v given above is valid. Now, when the true planet is to be found from the mean planet, it is not necessary to calculate the *manda-karṇa* K. However in the reverse case, when the mean planet is to be found from the true planet, it becomes necessary to first calculate K.

An elaborate explanation for the above verses is to be found in in *Yukti-dīpikā*.

कर्णवृत्ते स्फुटोच्चान्तरालज्या स्वकलामिता ।
तदुच्चसूत्रसंपातात् कोटिस्तत्केन्द्रगामिनी ॥
तदन्त्यफलयोर्योगो विश्लेषो वा यथोचितम् ।
प्रतिमण्डलकेन्द्रस्य दोर्ज्यामूलस्य चान्तरम् ॥
तद्दोर्वर्गयुतेर्मूलं प्रतिमण्डलकेन्द्रतः ।
ग्रहावधिव्यासदलं कर्णवृत्तकलामितम् ॥
व्यासार्धं प्रतिवृत्तस्य त्रिज्यैव स्वकलामिता ।
तदेव व्यस्तकर्णः स्यात् कर्णवृत्तकलामितम् ॥
त्रिज्यातुल्ये व्यस्तकर्णे कर्णो न्यूनाधिकस्ततः ।
ततस्त्रैराशिकेनात्र मन्दकर्णः स्फुटो भवेत् ॥
उच्चोनस्फुटतो दोर्ज्यां मन्दकर्णहतां हरेत् ।
त्रिज्यया तद्गुणां दोर्ज्यां व्यस्तकर्णेन वा हरेत् ॥
लब्धचापं धनर्णं स्यात् स्वोच्चे मध्यमसिद्धये ।

यादृशो नियमस्तत्र व्यस्तकर्णत्रिजीवयोः ॥
तादृशो नियमो वेद्यः त्रिजीवामन्दकर्णयोः ।
कर्णस्त्रैराशिकेनातः व्यस्तकर्णाद्विधीयते ॥
मध्यमात् स्फुटसंसिद्धिः दोःफलाद्यदि केवलात् ।
मध्यसिद्धिः स्फुटात् तस्मात् कर्णघ्नात् त्रिज्यया हृतात् ॥
अथ स्फुटोच्चान्तरदोर्गुणं श्रुतिहतं हरेत् ।
त्रिज्यया लब्धचापेन कृते स्वोच्चे स्वमध्यमः ॥
तदेव चापितं स्वोच्चे चक्रार्धं तेन वर्जितम् ।
चक्रार्धयुक्तं चक्राच्च त्यक्तं पदवशात् क्षिपेत् ॥ [22]

In the *karṇavṛtta* the *jyā* of the difference between the longitude of the true planet and its *mandocca* corresponds to the *dorjyā* in its own measure. The distance of separation between the point of intersection (N in the Fig. 2.14b) of the *jyā* with the *uccasūtra* (the apsis line) and the centre of the *karṇavṛtta* (O) corresponds to the *koṭijyā* (ON). The sum or difference of the *antyaphala* (OO') with this *koṭijyā*, as the case may be, gives the distance of separation between the centre of the *pratimaṇḍala* and the foot of the *dorjyā* (N). The square root of the sum of the squares of this ($O'N$) and the *dorjyā* (PN) gives the distance between the centre of *pratimaṇḍala* and the planet. This is the radius of the *pratimaṇḍala* in the measure of the *karṇavṛtta*. The radius of the *prativṛtta* with respect to its own measure is the *trijyā*. This (*trijyā*) will be the *vyasta-karṇa* (inverse-hypotenuse) in the measure of the *karṇavṛtta*. When the *vyasta-karṇa* is set equal to the *trijyā*, then the actual *karṇa* will be smaller or larger than that. Thus by the rule of three the true *manda-karṇa* is obtained.

The *dorjyā* obtained by subtracting the *mandocca* from the true Sun is multiplied by the *manda-karṇa* and divided by the *trijyā*. Or the *trijyā* multiplied by the *dorjyā* is divided by *vyasta-karṇa*. The arc of this is applied positively or negatively to the *mandocca* to get the mean Sun. It is to be understood that whatever is the relation between the *vyasta-karṇa* and the *trijyā*, the same relation is valid between the *trijyā* and the *manda-karṇa*. This is the reason why the *manda-karṇa* is obtained from the *vyasta-karṇa* by the rule of three.

As the true position of the planet is obtained from the mean position just by finding the *doḥphala*, the mean position is obtained from the true position by multiplying [the *dorjyā*] by the *manda-karṇa* and dividing by the *trijyā*. Then the *dorjyā* obtained by subtracting the *mandocca* from the true Sun is multiplied by the *manda-karṇa* and divided by the *trijyā*. The arc applied to the *mandocca* of the Sun will give the position of the mean Sun. Depending upon the quadrant, the same arc has to be applied to the *mandocca* after subtracting it from 180°, or adding 180° to it or subtracting it from 360°.

The procedure stated here is a slight variant of the one described earlier. Here, PN, ON and OO' are the *dorjyā*, the *koṭijyā* and the *antyaphala* respectively in the measure of the *karṇavṛtta* and are equal to $R\sin(\theta-\theta_m)$, $R\cos(\theta-\theta_m)$ and r_0 in the same measure. In this measure, the radius of the *pratimaṇḍala*, $O'P$, is the *vyasta-karṇa* or *viparita-karṇa*, R_v, given in (2.136). Then the *manda-karṇa*, K, in the measure of the *pratimaṇḍala* (when the radius is R, as usual) is determined from

$$\frac{K}{R}=\frac{R}{R_v}, \tag{2.153}$$

and *madhyama* − *ucca* is obtained as earlier.

---

[22] {TS 1977}, pp. 165–6.

२.२१ स्फुटान्मध्यमानयने प्रकारान्तरम्

2.21 Another method for getting the mean planet from the true planet

अर्केन्द्वोः स्फुटतो मृदूच्चरहितात् दोःकोटिजाते फले
नीत्वा कर्किमृगादितो विनिमयेनानीय कर्णं सकृत् ।
त्रिज्या दोःफलघाततः श्रुतिहृतं चापीकृतं तत् स्फुटे
केन्द्रे मेषतुलादिगे धनमृणं तन्मध्यसंसिद्धये ॥ ५१ ॥

arkendvoḥ sphuṭato mṛdūccarahitāt doḥkoṭijāte phale
nītvā karkimṛgādito vinimayenānīya karṇaṃ sakṛt |
trijyā doḥphalaghātataḥ śrutihṛtaṃ cāpīkṛtaṃ tat sphuṭe
kendre meṣatulādige dhanamṛṇaṃ tanmadhyasaṃsiddhaye|| 51 ||

Subtracting the longitude of their own *mandoccas* from the true positions of the Sun and the Moon, obtain their *doḥphala* and *koṭiphala*. Find the *sakṛt karṇa* (one-step hypotenuse) once by interchanging the sign [in the cosine term] depending upon whether the *kendra* is within the six signs beginning with *Karki* or *Mṛga*. Multiplying the *doḥphala* and *trijyā*, and dividing this product by the *karṇa* [here referred to as *śruti*], the arc of the result is applied to the true planet to obtain the mean planet. This arc has to be applied positively and negatively depending upon whether the *kendra* lies within the six signs beginning with *Meṣa* or *Tulā* respectively.

Now,

$$\begin{aligned} bāhuphala &= r_0 \sin(\theta - \theta_m) \\ koṭiphala &= r_0 \cos(\theta - \theta_m). \end{aligned} \tag{2.154}$$

Taking the one-step *karṇa* (*sakṛtkarṇa*) with the opposite sign in the *koṭiphala*, we have

$$karṇa = [(R - r_0 \cos(\theta - \theta_m))^2 + (r_0 \sin(\theta - \theta_m))^2]^{\frac{1}{2}}. \tag{2.155}$$

This is the same as the *viparīta-karṇa* R_v given by (2.150). In Fig. 2.14*b*, draw $O'T$ perpendicular to OP. Then in triangle $O'PT$,

$$\begin{aligned} O'T &= O'P \sin(O'\hat{P}T) \\ &= O'P \sin(P\hat{O}P_0) \\ &= R \sin(\theta_0 - \theta). \end{aligned} \tag{2.156}$$

$$\text{Also} \qquad O'T = r \sin(\theta - \theta_m). \tag{2.157}$$

Equating the above two expressions for $O'T$,

$$\begin{aligned} R \sin(\theta_0 - \theta) &= r \sin(\theta - \theta_0) \\ \text{or} \qquad R \sin(\theta_0 - \theta) &= r_0 \sin(\theta - \theta_0) \frac{R}{R_v}, \end{aligned} \tag{2.158}$$

where we have used (2.135) and (2.153). Hence,

$$\theta_0 - \theta = (R\sin)^{-1}\left[r_0 \sin(\theta - \theta_0)\frac{R}{R_v}\right]. \tag{2.159}$$

Thus the mean planet θ_0 can be obtained by adding the above difference to the true planet θ. $\theta_0 - \theta$ is positive when the *kendra* (anomaly) $\theta - \theta_m$ is within the six signs beginning with *Meṣa*, i.e. $0^\circ \leq \theta - \theta_m \leq 180^\circ$, and negative when the *kendra* is within the six signs beginning with *Tulā*, i.e. $180^\circ \leq \theta - \theta_m \leq 360^\circ$.

२.२२ मन्दकर्णानयने प्रकारान्तरम्

2.22 Another method for getting the *manda*-hypotenuse

मध्यतः स्फुटतश्चोच्चं उज्झित्वा तद्भुजे उभे ।
गृहीत्वाद्या तयोस्त्रिज्या हतान्याप्ता श्रुतिस्फुटा ॥ ५२ ॥

madhyataḥ sphuṭataścoccaṃ ujjhitvā tadbhuje ubhe |
gṛhītvādyā tayostrijyā hatānyāptā śrutisphuṭā|| 52 ||

Subtracting the *mandocca* from the mean and the true positions separately, obtain the two *dorjyās*. Of these, the former multiplied by the *trijyā* and divided by the latter gives the exact value of *śrutisphutā* (*aviśiṣṭa-manda-karṇa*).

In Fig. 2.14*b*, $R\sin(\theta_0 - \theta_m)$ and $R\sin(\theta - \theta_m)$ are the *dorjyās* corresponding to the *manda-kendras* of the mean and true planet respectively. It is noted from the figure that

$$\begin{aligned} PN &= K\sin(\theta - \theta_m) \\ &= R\sin(\theta_0 - \theta_m). \end{aligned} \tag{2.160}$$

Hence,

$$K = R \times \frac{R\sin(\theta_0 - \theta_m)}{R\sin(\theta - \theta_m)}$$

$$\text{or} \qquad \acute{s}ruti = \frac{trijy\bar{a} \times \bar{a}dy\bar{a}}{any\bar{a}}. \tag{2.161}$$

where *ādyā* and *anyā* refer to $R\sin(\theta_0 - \theta_m)$ and $R\sin(\theta - \theta_m)$ respectively, and the *aviśiṣṭa-manda-karṇa* is termed the *śrutisphuṭā* here.

२.२३ ग्रहतात्कालिकगतिः

2.23 Instantaneous velocity of a planet

चन्द्रबाहुफलवर्गशोधितत्रिज्यकाकृतिपदेन संहरेत् ।
तत्र कोटिफललिप्तिकाहतां केन्द्रभुक्तिरिह यद्य लभ्यते ॥ ५३ ॥

तद्विशोध्य मृगादिके गतेः क्षिप्यतामिह तु कर्कटादिके ।
तद्भवेत्स्फुटतरा गतिर्विधोरस्य तत्समयजा रवेरपि ॥ ५४ ॥

candrabāhuphalavargaśodhitatrijyakākṛtipadena saṃharet |
tatra koṭiphalaliptikāhatāṃ kendrabhuktirihayacca labhyate || 53 ||
tadviśodhya mṛgādike gateḥ kṣipyatāmiha tu karkaṭādike |
tadbhavetsphuṭatarā gatirvidhorasya tatsamayajā raverapi || 54 ||

Let the product of the *koṭiphala* (in minutes) and the daily motion of the *kendra* be divided by the square root of the square of the *bāhuphala* of the Moon subtracted from the square root of the *trijyā*. The quantity thus obtained has to be subtracted from the daily motion [of the Moon] if [the *kendra* lies within the six signs] beginning from *Makara* and is to be added to the daily motion if [the *kendra* lies within the six signs] beginning from *Karkaṭaka*. This will be a far more accurate (*sphuṭatarā*) value of the instantaneous velocity (*tatsamayajā gati*) of the Moon. For the Sun also [the instantaneous velocity can be obtained similarly].

The *bāhuphala* (or *doḥphala*) and *koṭiphala* are given by

$$\begin{aligned} bāhuphala &= r_0 \sin(\theta_0 - \theta_m) \\ \text{and} \qquad koṭiphala &= r_0 \cos(\theta_0 - \theta_m), \end{aligned} \tag{2.162}$$

where $\theta_0 - \theta_m$ is the *manda-kendra*; θ_0 and θ_m represent the longitude of the Moon and its *mandocca* respectively (see Fig. 2.12*a*). The term *kendrabhukti* refers to the daily motion of the *kendra* given by

$$kendrabhukti = \frac{\Delta(\theta_0 - \theta_m)}{\Delta t}, \tag{2.163}$$

where Δt refers to the time interval of one day and $\Delta(\theta_0 - \theta_m)$ represents the difference in the daily motion of the Moon and its *mandocca*. As the mean longitude and *mandocca* increase uniformly with time,

$$\frac{d}{dt}(\theta_0 - \theta_m) = \frac{\Delta}{\Delta t}(\theta_0 - \theta_m), \tag{2.164}$$

is a constant. It is stated here that a correction term has to be added to the above *kendrabhukti* to obtain a more accurate value of the rate of motion of the *kendra*. The correction factor is stated to be

$$\frac{koṭiphala \times kendrabhukti}{\sqrt{(trijyā^2 - bāhuphala^2)}} = -\frac{r_0 \cos(\theta_0 - \theta_m)\frac{\Delta(\theta_0 - \theta_m)}{\Delta t}}{\sqrt{R^2 - r_0^2 \sin^2(\theta_0 - \theta_m)}}. \tag{2.165}$$

Further, it is mentioned that the correction term is to be subtracted from the *kendrabhukti* when $\theta_0 - \theta_m$ is in the first and fourth quadrants (*Mṛgādi*) and it is to be added when it is in the second and third quadrants (*Karkādi*). This accounts for the negative sign in the RHS of the above equation (2.165).

Now the *manda-kendra* of the Moon's true longitude is given by

$$\theta - \theta_m = (\theta_0 - \Delta\theta) - \theta_m,$$

where the *manda* correction $\Delta\theta$ is given by

$$\Delta\theta = \sin^{-1}\left(\frac{r_0}{R}\sin(\theta_0 - \theta_m)\right), \tag{2.166}$$

as explained earlier. Hence,

$$\theta = \theta_0 - \sin^{-1}\left(\frac{r_0}{R}\sin(\theta_0 - \theta_m)\right). \tag{2.167}$$

Therefore,

$$\begin{aligned} \frac{d}{dt}\theta &= \frac{d\theta_0}{dt} - \frac{d}{dt}\sin^{-1}\left(\frac{r_0}{R}\sin(\theta_0 - \theta_m)\right) \\ &= \frac{d\theta_0}{dt} - \frac{r_0\cos(\theta_0 - \theta_m)\frac{d(\theta_0 - \theta_m)}{dt}}{\sqrt{R^2 - r_0^2\sin^2(\theta_0 - \theta_m)}}. \end{aligned} \tag{2.168a}$$

It may be mentioned here that in the case of all the planets, except the Moon, the rate of change of the *mandocca* is extremely small and can be neglected. That is, $\frac{d\theta_m}{dt} \approx 0$. Then the above equation reduces to

$$\frac{d}{dt}\theta = \frac{d\theta_0}{dt}\left(1 - \frac{r_0\cos(\theta_0 - \theta_m)}{\sqrt{R^2 - r_0^2\sin^2(\theta_0 - \theta_m)}}\right). \tag{2.168b}$$

Note:

1. It is remarkable that the author in this verse gives the correct form for the derivative of the inverse sine function. In his *Jyotirmīmāṃsa*, Nīlakaṇṭha mentions that this verse is due to his teacher *Dāmodara*.
2. The differentials of the sine and cosine functions were used in Indian astronomy at least from the time of Mañjulācārya in his *Laghu-mānasā*. Bhāskara II clearly makes use of them in his *Siddhāntaśiromaṇi*.
3. The significance of this verse lies in the fact that it is for the first time that the derivative of the arcsine function is being considered here in the context of discussing the *tātkālika-gati* or instantaneous rate of motion of the planet.

२.२४ नक्षत्रतिथ्यानयनम्

2.24 Finding *nakṣatra* and *tithi*

लिप्तीकृतो निशानाथः शतैर्भाज्योष्टभिः फलम् ।
अश्विन्यादीनि भानि स्युः षष्ट्या हत्वा गतागते ॥ ५५ ॥
गतगन्तव्यनाड्यः स्युः स्फुटभुक्त्योदयावधेः ।

अर्कहीनो निशानाथः लिप्तीकृत्य विभज्यते ॥ ५६ ॥
शून्याश्विपर्वतैर्लब्धाः तिथयो या गताः क्रमात् ।
भुक्त्यन्तरेण नाड्यः स्युः षष्ट्या हत्वा गतागते ॥ ५७ ॥
तिथ्यर्धहारलब्धानि करणानि बबादितः ।
विरूपाणि सिते पक्षे सरूपाण्यसिते विदुः ॥ ५८ ॥
विष्कम्भाद्या रवीन्द्वैक्यात् योगाश्चाष्टशतीहृताः ।
भुक्तियुक्त्या गतैष्याभ्यां षष्टिघ्नाभ्यां च नाडिकाः ॥ ५९ ॥

liptīkṛto niśānāthaḥ śatairbhājyoṣṭabhiḥ phalam |
aśvinyādīni bhāni syuḥ ṣaṣṭyā hatvā gatāgate || 55 ||
gatagantavyanāḍyaḥ syuḥ sphuṭabhuktyodayāvadheḥ |
arkahīno niśānāthaḥ liptīkṛtya vibhajyate || 56 ||
śūnyāśviparvatairlabdhāḥ tithayo yā gatāḥ kramāt |
bhuktyantareṇa nāḍyaḥ syuḥ ṣaṣṭyā hatvā gatāgate || 57 ||
tithyardhahāralabdhāni karaṇāni babāditaḥ |
virūpāṇi site pakṣe sarūpāṇyasite viduḥ || 58 ||
viṣkambhādyā ravīndvaikyāt yogāścāṣṭaśatīhṛtāḥ |
bhuktiyuktyā gataiṣyābhyāṃ ṣaṣṭighnābhyāṃ ca nāḍikāḥ || 59 ||

The longitude of the lord of the night (the Moon) in minutes is divided by 800. The quotient gives the number of *nakṣatras* that have elapsed beginning from the *Aśvini nakṣatra.* The remainder [which corresponds to the minutes covered by the Moon in the present *nakṣatra*] and the one which has to be covered multiplied by 60 and divided by the daily motion of the Moon [in minutes] at sunrise gives the *nāḍikās* that have elapsed and are yet to elapse in the present *nakṣatra.* The longitude of the Sun subtracted from that of the Moon, in minutes, is divided by $720'$. The quotient gives the number of *tithis* elapsed. The remainder and the quantity obtained by subtracting the remainder from $720'$, multiplied by 60 and divided by the difference in the daily motion of the Sun and the Moon, gives the number of *ghaṭikās* that have elapsed and are yet to elapse in the present *tithi.*

The same (difference in longitude between the Sun and the Moon) divided by half the divisor used in the *tithi* calculation gives the number of *karaṇas* elapsed, starting with *bava.* In the bright fortnight the *karaṇas* are without form and in the dark fortnight with form. The sum of the longitudes of the Sun and the Moon [in minutes] divided by 800 gives the *yogas*, starting with the *viṣkambha.* The remainder and the quantity obtained by subtracting the remainder from 800, multiplied by 60 and divided by the sum of the daily motion of the Sun and the Moon, gives the number of *ghaṭikās* that have elapsed and are yet to elapse in the present *yoga.*

The ecliptic is divided in to 27 equal parts called *nakṣatras* beginning with *Aśvinī* and ending with *Revatī.* Hence each *nakṣatra* corresponds to $\frac{21600}{27} =$ 800 minutes, along the ecliptic. The *nakṣatra* at any instant refers to the particular portion of the ecliptic in which the Moon is situated. Clearly, when the longitude of the Moon in minutes is divided by 800 the quotient gives the number of *nakṣatras* which have elapsed and the remainder corresponds to the minutes covered by the Moon in the present *nakṣatra.* When this is divided by the daily motion of the Moon in minutes at that time (taken to be the value at sunrise) and multiplied by 60, the result gives the *ghaṭikas* that have elapsed in the present *nakṣatra.* Similarly the *ghaṭikās* yet to elapse in the present *nakṣatra* can be calculated.

A *tithi* is the (variable) unit of time during which the difference between the longitudes of the Moon and the Sun increases by 12° or $720'$. Hence there are 30 *tithis* during a lunar month. Hence, when the difference in longitudes of Moon and the Sun in minutes is divided by 720′, the quotient gives the number of *tithis* elapsed in that month. The number of *ghaṭikās* (*nāḍikās*) which have elapsed and are yet to elapse in the present *tithi* are calculated in the manner indicated.

A *karaṇa* is half a *tithi* by definition and there are 60 *karaṇas* in a lunar month. The number of *karaṇas* that have elapsed can be calculated in the same manner as the number of *tithis*, except that the divisor is 360′ instead of 720′.

There are two types of *karaṇas*, namely *cala* (movable) and *sthira* (fixed). In this context these terms are used to mean repeating and non-repeating *karaṇas*. Of the 11 *karaṇas*, 7 are repeating and 4 are non-repeating. The 7 *cala-karaṇas* (moving *karaṇas*) have 8 cycles, thus forming 56 *karaṇas*. The 4 *sthira-karaṇas* (fixed *karaṇas*) occur just once each in a lunar month. The moving and fixed *karaṇas* together make up 60 *karaṇas* in a lunar month.

The names of the *karaṇas* and the pattern in which the *cala* and *sthirakaraṇas* occur are given in the following verses, quoted in *Laghu-vivṛti*:

बबबालवकौलवतैतिलगजवणिजाख्यविष्टिनामानि ।
सितपक्षस्यापरार्धात्[23] परिवर्तन्तेऽष्ट कृत्वोऽतः ॥
कृष्णचतुर्दश्यन्ते शकुनिः पर्वणि चतुष्पदः ।
प्रथमे तिथ्यर्धेऽन्त्ये नागः किंस्तुघ्नः प्रतिपदाद्यर्धे ॥

The *karaṇas* named *baba, bālava, kaulava, taitila, gaja, vaṇija* and *viṣṭi* repeat themselves eight times from the later half of the first *tithi*, *prathama*, of the bright fortnight. *Śakuni* occurs in the later half of the *caturdaśī* of the dark fortnight, *catuṣpada* and *nāga* in the first and second halves of [the following] *amavāsya* and *kiṃstughna* in the first half of the *prathamā* of the bright fortnight.

Śaṅkara Vāriyar also quotes the following verses which give the different names of both moving and fixed *karaṇas*. The moving *karaṇas* are: *siṃha, vyāghra, varāha, khara, ibha, paśu* and *viṣṭi*. The fixed *karaṇas* are: *pakṣī, catuṣpāt, nāga* and *kiṃstughna*.

तथा शुक्लप्रतिपदाद्यन्त्यार्धात् करणानि मुहुर्मुहुः ।
सिंहो व्याघ्रो वराहश्च खरेभपशुविष्टयः ॥
पक्षी चतुष्पान्नागश्च किंस्तुघ्नश्चान्ततः स्थिराः ।

The *yogas* involve the sum of the longitudes of the Sun and the Moon. There are 27 *yogās* in a 360° (21600’) cycle, each *yoga* corresponding to 800′. The number of *yogās* that have elapsed and the minutes or *ghatikās* that have elapsed and are yet to elapse in the present *yoga* are calculated in the same manner as in the case of the *tithis*, except that the sum of the longitudes of the Sun and Moon and the sum of their daily motion are involved here. In *Laghu-vivṛti*, the names of the *yogas* are listed in the following verses:

चन्द्रार्कस्फुटसंयोगे दृष्टा दस्रादिकारकाः ।
विष्कम्भः प्रीतिरायुष्मान् सौभाग्यः शोभनस्तथा ॥

---

[23] The incorrect reading सितपक्षत्यपरार्धात् in the printed edition ({TS 1958}, p. 40) has been modified as above.

अतिगण्डः सुकर्मा च धृतिः शूलं तथैव च ।
गण्डो वृद्धिर्ध्रुवश्चैव व्याघातो हर्षणस्तथा ॥
वज्रस्सिद्धिर्व्यतीपातो वरीयान् परिघः शिवः ।
सिद्धः साध्यः शुभः शुभ्रः ब्राह्मो माहेन्द्रवैधृतौ ॥

When the longitude of the Sun and the Moon are added the *dasrādikārakās* are seen. They are: *viṣkambha, prīti, ayuṣmān, saubhāgya, śobhana, atigaṇḍa, sukarmā, dhṛti, śūla, gaṇḍa, vṛddhi, dhruva, vyāghāta, harṣaṇa, vajra, siddhi, vyatīpāta, varīyān, parigha, śiva, siddha, sādhya, śubha, śubhra, brāhma, māhendra and vaidhṛti.*

२.२५ ग्रहसंस्कारप्रकारः

2.25 The scheme of correction for the planets

मान्दं शैघ्रं पुनर्मान्दं शैघ्रं चत्वार्यनुक्रमात् ।
कुजगुर्वर्कजानां हि कर्माण्युक्तानि सूरिभिः ॥ ६० ॥

māndaṃ śaighraṃ punarmāndaṃ śaighraṃ catvāryanukramāt|
kujagurvarkajānāṃ hi karmāṇyuktāni sūribhiḥ || 60 ||

The earlier *ācaryās* have stated that *manda, śīghra*, and again *manda* and *śīghra* are the four corrections which have to be applied in sequence to the planets Mars, Jupiter and Saturn [to obtain the true longitudes of the planets from their mean longitudes].

Though there are essentially only two corrections, namely *manda* and *śīghra* for the actual planets, that is Mercury, Venus, Mars, Jupiter and Saturn, the actual computation of their longitude involves a four-step procedure in most Indian texts. Nīlakaṇṭha, as we shall see below, prescribes this four-step process only in the case of the exterior planets, Mars, Jupiter and Saturn. The actual procedure prescribed in *Tantrasaṅgraha* is described in the next few verses.

२.२६ कुजगुरुमन्दस्फुटीकरणम्

2.26 The correction for Mars, Jupiter and Saturn

दोःकोटिज्याष्टमांशौ स्वखाब्ध्यंशोनौ शनेः फले ।
दोर्ज्या त्रिज्याप्तसप्तैक्यं गुणो मान्दे कुजेड्ययोः ॥ ६१ ॥
नवाभ्रयो द्व्यशीतिश्च हारौ दोःकोटिजीवयोः ।
पृथक्स्थे मध्यमे कार्यं दोःफलस्य धनुर्दलम् ॥ ६२ ॥
रविमध्यं विशोध्यास्मात् पृथक्स्थात् बाहुकोटिके ।
आनीय बाहुजीवायाः त्रिज्याप्तं गुरुमन्दयोः ॥ ६३ ॥
षोडशभ्यो नवभ्यश्च कुजस्यापि स्वदोर्गुणात् ।
त्रिज्याप्तं द्विगुणं शोध्यं त्रीषुभ्यः शिष्यते गुणः ॥ ६४ ॥
अशीतिरेव तेषां हि हारस्ताभ्यां फले उभे ।
आनीय पूर्ववत् कर्णं सकृत्कृत्वाथ दोःफलम् ॥ ६५ ॥

त्रिज्याघ्नं कर्णभक्तं यत् तद्धनुर्दलमेव च ।
मध्यमे कृतमान्दे तु संस्कृत्यातो विशोधयेत् ॥ ६६ ॥
मन्दोच्चं तत्फलं कृत्स्नं कुर्यात् केवलमध्यमे ।
तस्मात् पृथक्कृताच्छैघ्रं प्राग्वदानीय चापितम् ॥ ६७ ॥
कृतमान्दे तु कर्तव्यं सकलं स्यात् स्फुटः स च ।

doḥkoṭijyāṣṭamāṃśau khakhābdhyaṃśonau śaneḥ phale |
dorjyā trijyāptasaptaikyaṃ guṇo mānde kujeḍyayoḥ || 61 ||
navāgnayo dvyaśītiśca hārau doḥkoṭijīvayoḥ |
pṛthaksthe madhyame kāryaṃ doḥphalasya dhanurdalam || 62 ||
ravimadhyaṃ viśodhyāsmāt pṛthaksthāt bāhukoṭike |
ānīya bāhujīvāyāḥ trijyāptaṃ gurumandayoḥ || 63 ||
ṣoḍaśabhyo navabhyaśca kujasyāpi svadorguṇāt |
trijyāptaṃ dviguṇaṃ śodhyaṃ trīṣubhyaḥ śiṣyate guṇaḥ || 64 ||
aśītireva teṣāṃ hi hārastābhyāṃ phale ubhe |
ānīya pūrvavat karṇaṃ sakṛtkṛtvātha doḥphalam || 65 ||
trijyāghnaṃ karṇabhaktaṃ yat taddhanurdalameva ca |
madhyame kṛtamānde tu saṃskṛtyāto viśodhayet || 66 ||
mandoccaṃ tatphalaṃ kṛtsnaṃ kuryāt kevalamadhyame |
tasmāt pṛthakkṛtācchaighraṃ prāgvadānīya cāpitam || 67 ||
kṛtamānde tu kartavyaṃ sakalaṃ syāt sphuṭaḥ sa ca |

One-eighth of the *dorjyā* and *koṭijyā* (sine and cosine of the *manda-kendra*), diminished by one-fortieth of the same, form the *doḥphala* and *koṭiphala* in the case of Saturn. The *dorjyā* divided by the *trijyā* and added to 7, forms the *guṇa* (multiplier) for Mars and Jupiter. 39 and 82 are the *hāra* (divisor) for Mars and Jupiter respectively. Half of the arc of the *doḥphala* has to be applied to the mean longitude of the planet (P_0) to get the first corrected longitude (P_1).

Subtracting the longitude of the Sun (the *śīghrocca*) from this (P_1), the *dorjyā* and *koṭijyā* are obtained. Dividing the *dorjyā* by the *trijyā* and subtracting from 16 and 9, we get the multipliers for Jupiter and Mars respectively. The same (*dorjyā*) multiplied by 2 and subtracted from 53 forms the multiplier for Mars.

80 is the divisor for all of them (in the *śīghra-saṃskāra*). From them (the multiplier and divisor of all the three planets) after obtaining the *doḥphala* and *koṭiphala*, and the *sakṛtkarṇa* (once calculated hypotenuse), half of the *doḥphala* multiplied by the *trijyā* and divided by the *karṇa* is applied to the first corrected longitude (P_1). (The longitude thus obtained is, say, P_2.) From this (P_2), let the *mandocca* be subtracted and the full *manda-phala* be obtained; let that be applied to the original mean planet (P_0 to get say P_3). From that (P_3) let *śīghra-phala* be obtained as before, and let this be applied fully to the *manda*-corrected planet (P_3). The longitude obtained thus is the *sphuṭa* (the true longitude of the planet).

A detailed and comprehensive discussion of the planetary model, and the geometrical picture implied by it in the traditional scheme, as well as the modification introduced by Nīlakaṇṭha, can be found in Appendix F. Here and in the following sections we confine our explanation mainly to the computational scheme described in the verses of the text.

The computation of the *manda-sphuṭa* has already been described in the earlier verses in this chapter. Let θ_0, θ_m, θ_{ms} be the mean longitude and the longitudes of the *mandocca* and the *manda-sphuṭa* respectively. Also let R, r and K be radii of the deferent circle (*trijyā*), the epicycle and the *manda-karṇa-vṛtta* respectively. r is proportional to K and $\frac{r}{K} = \frac{r_0}{R}$ where, r_0 is the tabulated value of the radius of the epicycle. Then $\theta_{ms} - \theta_0$ is found from

$$
\begin{aligned}
K\sin(\theta_{ms} - \theta_0) &= -r\sin(\theta_0 - \theta_m) \\
\text{or} \qquad R\sin(\theta_{ms} - \theta_0) &= -\frac{r}{K}R\sin(\theta_0 - \theta_m) \\
&= -\frac{r_0}{R}R\sin(\theta_0 - \theta_m).
\end{aligned} \tag{2.169}
$$

$R\sin(\theta_0 - \theta_m)$ is the *dorjyā*, $r_0\sin(\theta_0 - \theta_m)$ is the *doḥphala* and $\theta_0 \sim \theta_{ms}$ is the 'arc' of the *dophala*. In the above verses $\frac{r_0}{R}$ for Saturn, Mars and Jupiter are specified to be

$$\frac{r_0}{R}\,(\text{Saturn}) = \frac{1}{8} - \frac{1}{320} = \frac{39}{320} \tag{2.170}$$

$$\frac{r_0}{R}\,(\text{Mars}) = \frac{7 + |\sin(\theta_0 - \theta_m)|}{39} \tag{2.171}$$

$$\text{and} \qquad \frac{r_0}{R}\,(\text{Jupiter}) = \frac{7 + |\sin(\theta_0 - \theta_m)|}{82}. \tag{2.172}$$

Note that r_0 is not constant for Mars and Jupiter, but varies with the *manda-kendra*, $\theta_0 - \theta_m$. When $\theta_{ms} - \theta_0$, found from the above equation, is added to θ_0, we obtain the *manda-sphuṭa-graha* (*manda*-corrected planet) θ_{ms}. The true geocentric longitude of the exterior planets is obtained from the *manda-sphuṭa* θ_{ms} as follows.

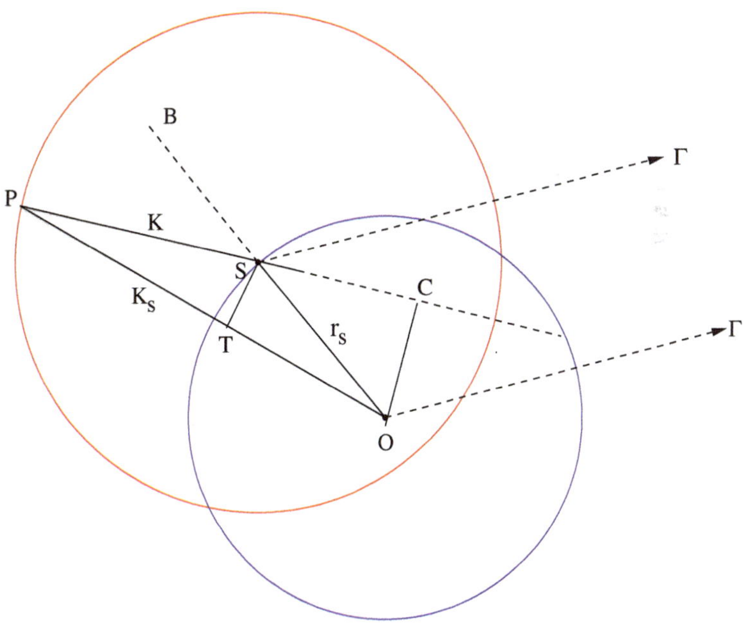

Fig. 2.15 Obtaining the *sphuṭa-graha* (geocentric longitude) from the *manda-sphuṭa-graha* (true heliocentric longitude) in the case of exterior planets .

In Fig. 2.15 the *śīghra-nīcocca-vṛtta* or *śīghra-vṛtta* or *śīghra*-circle is a circle with the *bhagolamadhya* (the centre of the Earth) as the centre at O. The radius of this circle is the *śīgrāntyaphala* r_s. The *śīghrocca* S, which is the mean Sun, is located on this circle. The planet P is located on the *manda-karṇa-vṛtta* of radius K with S as the centre, such that $\theta_{ms} = \Gamma\hat{S}P$ is the *manda-sphuṭa-graha.* Then the *śīghra-sphuṭa* (*śīghra*-corrected planet) is found in the same manner from the *manda-sphuṭa* as the *manda-sphuṭa* is found for the mean planet, the *madhyama-graha.*

Let θ_s be the longitude of the *śīghrocca*. That is, $\theta_s = \Gamma\hat{O}S$. Also from the figure,

$$\begin{aligned} \textit{śīghrocca}\ \theta_s &= \Gamma\hat{S}B \\ \textit{manda-sphuṭa}\ \theta_{ms} &= \Gamma\hat{S}P \\ \textit{śīghra-sphuṭa}\ \theta &= \Gamma\hat{O}P. \end{aligned} \tag{2.173}$$

Therefore

$$O\hat{S}C = P\hat{S}B = \theta_{ms} - \theta_s. \tag{2.174}$$

Further,

$$\begin{aligned} \textit{śīghrābhujaphala}\ OC &= r_s \sin(O\hat{S}C) \\ &= r_s \sin(\theta_{ms} - \theta_s) \\ \textit{śīghrakoṭiphala}\ SC &= r_s \cos(\theta_{ms} - \theta_s). \end{aligned} \tag{2.175}$$

Hence the *śīghra-karṇa* (*śīghra*-hypotenuse)

$$K_s = OP = \sqrt{(K + r_s \cos(\theta_{ms} - \theta_s))^2 + r_s^2 \sin^2(\theta_{ms} - \theta_s)}. \tag{2.176}$$

It can be easily seen that

$$O\hat{P}C = \theta_{ms} - \theta. \tag{2.177}$$

Also from the triangle POC,

$$OP \sin O\hat{P}C = OC. \tag{2.178}$$

Now using (2.175) to (2.177) in the above equation we have

$$\begin{aligned} K_s \sin(\theta_{ms} - \theta) &= r_s \sin(\theta_{ms} - \theta_s) \\ \text{or} \qquad R \sin(\theta_{ms} - \theta) &= \frac{R}{K_s} r_s \sin(\theta_{ms} - \theta_s). \end{aligned} \tag{2.179}$$

The arc corresponding to $\theta_{ms} - \theta$ is found from this. Subtracting $\theta_{ms} - \theta$ from the *manda-sphuṭa* θ_{ms}, we obtain the *śīghra-sphuṭa* θ. Here θ_{ms} is the true longitude of the planet with respect to S, which is taken to be the mean Sun. Hence θ_{ms} is essentially the true heliocentric longitude of the planet. So the true geocentric longitude θ is obtained from the true heliocentric longitude θ_{ms} using the above procedure. Now,

$$\acute{s}\bar{\imath}ghra\text{-}kendradorjy\bar{a} = R\sin(\theta_{ms} - \theta_s) \tag{2.180}$$

$$s\bar{\imath}ghrabhuj\bar{a}phala,\ r_s\sin(\theta_{ms} - \theta_s) = \frac{r_s}{R}R\sin(\theta_{ms} - \theta_s), \tag{2.181}$$

where *śīghra-kendradorjyā* is the Rsine of the *śīghra*-anomaly (anomaly of conjunction). In the *śīghra-saṃskāra*, the value of r_s is given in the text. Unlike in the calculation of the *manda-sphuṭa*, where the *manda-karṇa* K does not appear, here the *śīghra-karṇa* does appear in the computation of the *śīghra-sphuṭa*.

The values of $\frac{r_s}{R}$ for Mars, Jupiter and Saturn are given in the above verses as follows:

$$\frac{r_s}{R}\,(\text{Mars}) = \frac{53 - 2|\sin(\theta_{ms} - \theta_s)|}{80}, \tag{2.182}$$

$$\frac{r_s}{R}\,(\text{Jupiter}) = \frac{16 - |\sin(\theta_{ms} - \theta_s)|}{80}, \tag{2.183}$$

$$\frac{r_s}{R}\,(\text{Saturn}) = \frac{9 - |\sin(\theta_{ms} - \theta_s)|}{80}. \tag{2.184}$$

Planet	Range of ratio $\frac{r_s}{R}$	Average value (modern)
Mars	0.637–0.662	0.656
Jupiter	0.187–0.200	0.192
Saturn	0.100–0.115	0.105

Table 2.3 The range of variation in the ratio of the Earth–Sun to the planet–Sun distances for the exterior planets.

The range of variation of $\frac{r_s}{R}$ as obtained from the above equations along with the average value of the ratio of the Earth–Sun and planet–Sun distances as per modern astronomy are listed in Table 2.3. In Fig. 2.15,

$$\frac{\text{Earth–mean Sun distance}}{\text{planet–mean Sun distance}} = \frac{r_s}{K}, \tag{2.185}$$

where K varies depending upon the *manda-sphuṭa-graha* or the true heliocentric longitude. Taking the mean value of K to be R, the ratio would be $\frac{r_s}{R}$, which still depends upon $(\theta_{ms} - \theta_s)$. Even then, $\frac{r_s}{R}$ is always close to the average value of the ratio of the Earth–Sun and planet–Sun distances for each planet according to modern astronomy.

Āryabhaṭīya-bhāṣya and *Yuktibhāṣā* discuss the geometrical picture in detail. However they do not mention that $\frac{r_s}{R}$ is the ratio of the physical Earth–Sun to planet–Sun distances. There is an important later work of Nīlakaṇṭha, namely *Grahasphuṭānayane vikṣepavāsanā*, which indeed mentions this explicitly. This is discussed in detail in Appendix F.

The procedure for obtaining the *śīghra-sphuṭa* of these three planets, given in the above verses, is not a straightforward, two step process of (i) obtaining the *manda-*

sphuṭa first from the mean planet and then (ii) obtaining the *śīghra-sphuṭa* from the *manda-sphuṭa*. Instead, the following four-step procedure is prescribed:

1. Obtain the *manda-phala* from the mean planet θ_0. Apply half of this *manda-phala* to θ_0 to obtain the first corrected planet P_1.
2. Find the *śīghra-sphuṭa* taking P_1 as the *manda-sphuṭa*, using (2.179). Here it is understood that in the calculation of the *śīghra-karṇa*, the *manda-karṇa* is replaced by the *trijyā* R, so only the *śīghra-kendra* $(\theta_{ms} - \theta_s)$ and the value of r_s (which depends upon the *śīghra-kendra*) figure in the calculation of this *śīghra-sphuṭa*. This is the second corrected planet P_2.
3. Treating P_2 as the mean planet, the *manda-phala* is calculated with $P_2 - \theta_m$ as the anomaly. Apply the full *manda-phala* to θ_0. The resulting quantity is the third corrected planet P_3.
4. Treating P_3 as the *manda-sphuṭa*, the *śīghra-sphuṭa* P is calculated again using R instead of K in the calculation of the *śīghra-karṇa* K_s.

In fact, this four-step procedure to compute the true geocentric longitude is the standard one prescribed in many Indian texts. *Yuktibhāṣā* attempts to provide the rationale for this, though the arguments given there are not entirely clear. However the motivation for this procedure is clear enough and is as follows.

Now, the *manda* correction can be read off from a table, given the mean epicycle radius and the *manda-kendra*. But this is not so in the case of the *śīghra* correction, for the *śīghra-phala* depends not only on the *śīghra-kendra* but also on the *śīghra-karṇa*, which depends on the *manda-karṇa* (the distance SP in Fig. 2.15), which in turn is dependant on the *manda-kendra*. Hence, given the radius of the *śīghra-vṛtta*, the *śīghra-phala* cannot be read off from a table as a function of the *śīghra-kendra* alone, as it depends also on the *manda-karṇa* and hence on the *manda-kendra*. *Yuktibhāṣā* seems to argue that the four-step process is an attempt to stimulate, to some extent, the effect of the *manda-karṇa* in the *śīghra-phala*. Thus, in steps two and four above, the *śīghra-phala* is calculated using the *trijyā* instead of the *aviṣiṣṭa-manda-karṇa*.

२.२७ बुधस्फुटीकरणम्

2.27 The correction for Mercury

बुधमध्यात् स्वमन्दोच्चं त्यक्त्वा दोःकोटिजीवयोः ॥ ६८ ॥
षडंशाभ्यां फलाभ्यां तु कर्णः कार्योऽविशेषतः ।
दोःफलं केवलं स्वर्णं केन्द्रे जूकक्रियादिगे ॥ ६९ ॥
एवं कृतं हि तन्मध्यं स्फुटमध्यं बुधस्य तु ।
रविमध्यं ततः शोध्यं दोःकोटिज्ये ततो नयेत् ॥ ७० ॥
दोर्ज्यां द्विघ्ना त्रिभज्याप्ता शोध्यैकत्रिंशतो गुणः ।
मन्दकर्णहतः सोऽपि त्रिज्याप्तः स्यात् स्फुटो गुणः ॥ ७१ ॥
तद्धते बाहुकोटिज्ये खाहिभक्ते फले उभे ।

ताभ्यां कर्णं सकृन्नीत्वा त्रिज्याघ्नं दोःफलं हरेत् ॥ ७२ ॥
कर्णेनाप्तस्य यच्चापं कृत्स्नं तद्भानुमध्यमे ।
क्रमेण प्रक्षिपेज्जह्यात् केन्द्रे मेषतुलादिगे ॥ ७३ ॥
एवं शीघ्रफलेनैव संस्कृतं रविमध्यमम् ।
बुधः स्यात् स स्फुटः शुक्रोऽप्येवमेव स्फुटो भवेत् ॥ ७४ ॥

budhamadhyāt svamandoccaṃ tyaktvā doḥkoṭijīvayoḥ || 68 ||
ṣaḍaṃśābhyāṃ phalābhyāṃ tu karṇaḥ kāryo'viśeṣataḥ |
doḥphalaṃ kevalaṃ svarṇaṃ kendre jūkakriyādige || 69 ||
evaṃ kṛtaṃ hi tanmadhyaṃ sphuṭamadhyaṃ budhasya tu |
ravimadhyaṃ tataḥ śodhyaṃ doḥkoṭijye tato nayet || 70 ||
dorjyā dvighnā tribhajyāptā śodhyaikatriṃśato guṇaḥ |
mandakarṇahataḥ so'pi trijyāptaḥ syāt sphuṭo guṇaḥ || 71 ||
taddhate bāhukoṭijye khāhibhakte phale ubhe |
tābhyāṃ karṇaṃ sakṛnnītvā trijyāghnaṃ doḥphalaṃ haret|| 72 ||
karṇenāptasya yaccāpaṃ kṛtsnaṃ tadbhānumadhyame |
krameṇa prakṣipejjahyāt kendre meṣatulādige || 73 ||
evaṃ śīghraphalenaiva saṃskṛtaṃ ravimadhyamam |
budhaḥ syāt sa sphuṭaḥ śukro'pyevameva sphuṭo bhavet || 74 ||

From the *madhyamagraha* of Mercury, subtracting the *mandocca*, the *dorjyā* and *koṭijyā* are obtained. From one-sixth of these values, the *aviṣiṣṭa-manda-karṇa* is found iteratively. The *doḥphala* has to be added to or subtracted from the *madhyamagraha*, depending on whether the *manda-kendra* lies within 6 signs of *Meṣa* or *Tulā*. The value thus obtained is the *manda-sphuṭa-graha* of Mercury (say P_1).

Then subtracting the mean Sun (which is the *śīghrocca*) from this (P_1), obtain the *dorjyā* and *koṭijyā* (corresponding to the *śīghra-kendra*). The *dorjyā* multiplied by 2, divided by *trijyā* and subtracted from 31 forms the multiplier. This multiplier multiplied by the *aviṣiṣṭa-manda-karṇa* and divided by the *trijyā* forms the *sphuṭaguṇa* (true multiplier).

The *dorjyā* and *koṭijyā*, multiplied by the *sphuṭaguṇa* and divided by 80, form the *doḥphala* and *koṭiphala* respectively. From these two (the *doḥphala* and *koṭiphala*), obtain the *śīghra-karṇa* once (not iteratively) and divide the product of the *trijyā* and *doḥphala* by this *śīghra-karṇa*. The arc of this result is fully applied to the mean Sun. It is either added or subtracted depending upon whether the *śīghra-kendra* lies within 6 signs of *Meṣa* or *Tulā*. The mean Sun corrected by this *śīghra-phala* gives the true geocentric longitude of Mercury. The true geocentric longitude of Venus is obtained in a similar manner.

Unlike a four-step procedure employed for the exterior planets to obtain the *sphuṭa-graha* (true planet), in the case of interior planets only a two-step procedure is prescribed. First the *manda-sphuṭa-graha* (*manda*-corrected planet) is obtained from the *madhyama-graha* (mean planet) through *manda-saṃskāra* (*manda*-correction), that is, the equation of centre, and then the *sphuṭa-graha* is obtained through the *śīghra-saṃskāra* (*śīghra* correction).

The *manda-sphuṭagraha* of Mercury is obtained from the mean heliocentric planet following the same procedure as for the exterior planets. Here $\frac{r_0}{R}$ is specified as $\frac{1}{6}$, where r_0 is the mean radius of the epicycle. The *aviśiṣṭa-manda-karṇa* K is also calculated as described earlier. The procedure for obtaining the true geocentric longitude of Mercury from the *manda-sphuṭa-graha* as described in these verses can be understood from Fig. 2.16 (see also Appendix F).

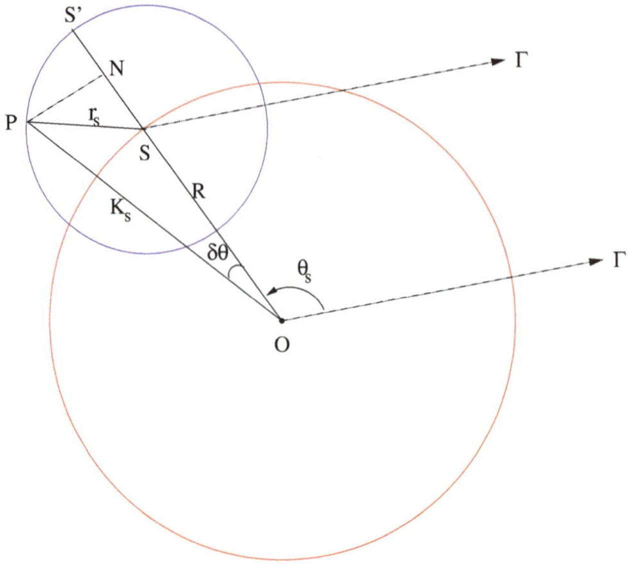

Fig. 2.16 Obtaining the *sphuṭa-graha* (the geocentric longitude) from the *manda-sphuṭa-graha* (the true heliocentric longitude) in the case of interior planets.

The mean Sun S is located on a circle of radius R with the centre of the Earth as the centre. Its longitude $\theta_s = \Gamma\hat{O}S$. Draw a circle of radius r_s around S. Mercury is located on this point such that its longitude is the *manda-sphuṭa-graha* $\theta_{ms} = \Gamma\hat{S}P$ with respect to S. Then $\theta = \Gamma\hat{O}P$ is the true geocentric longitude of Mercury called the *sphuṭa-graha* or simply the *sphuṭa*. Now,

$$\begin{aligned} \textit{śīghra-kendra} &= \theta_{ms} - \theta_s \\ &= \Gamma\hat{S}P - \Gamma\hat{O}S \\ &= \Gamma\hat{S}P - \Gamma\hat{S}S' \\ &= S'\hat{S}P. \end{aligned} \tag{2.186}$$

The radius of the epicycle r_s is given by

$$\frac{r_s}{R} = \frac{31 - 2|\sin\theta_{ms} - \theta_s|}{80} \times \frac{K}{R}, \tag{2.187}$$

and the *śīghra-karṇa* K_s is obtained from

$$\begin{aligned} K_s = OP &= \sqrt{ON^2 + PN^2} \\ &= \sqrt{(R + r_s\cos(\theta_{ms} - \theta_s))^2 + (r_s\sin(\theta_{ms} - \theta_s))^2}. \end{aligned} \tag{2.188}$$

The *śīghra* correction $S'OP = \delta\theta$ is found from the relation

$$\begin{aligned} OP\sin\delta\theta &= PN \\ &= r_s\sin(\theta_{ms}-\theta_s) \\ \text{or} \qquad K_s\sin\delta\theta &= r_s\sin(\theta_{ms}-\theta_s) \\ \text{or} \qquad R\sin\delta\theta &= r_s\sin(\theta_{ms}-\theta_s)\frac{R}{K_s}, \end{aligned} \tag{2.189}$$

where

$$\begin{aligned} dohphala &= r_s\sin(\theta_{ms}-\theta_s) \\ &= \frac{r_s}{R}R\sin(\theta_{ms}-\theta_s) \\ &= R\sin(\theta_{ms}-\theta_s)\times\left[(31-2|\sin(\theta_{ms}-\theta_s)|)\times\frac{K}{R}\right]\times\frac{1}{80} \\ &= dorjya\times sphuṭaguṇa\times\frac{1}{80}. \end{aligned} \tag{2.190}$$

Similarly,

$$\begin{aligned} koṭiphala &= r_s\cos(\theta_{ms}-\theta_s) \\ &= koṭijyā\times sphuṭaguṇa\times\frac{1}{80}. \end{aligned} \tag{2.191}$$

Adding the arc $\delta\theta$ obtained thus to the longitude of the mean Sun θ_s, we obtain the true geocentric longitude of Mercury, $\theta = \Gamma\hat{O}P = \theta_s + \delta\theta$.

In the earlier Indian texts, as was the case also in the Greco-European tradition up to Kepler, the equation of centre of the interior planet used to be applied wrongly to the mean Sun, which was taken as the mean planet in the case of interior planets. It is in *Tantrasaṅgraha* that the equation of centre is correctly applied to the mean heliocentric planet to obtain the true heliocentric planet, for the first time in the history of astronomy. We have already commented on this major modification that has been introduced for the interior planets in *Tantrasaṅgraha*, wherein the mean heliocentric planet is taken as the mean planet and the specified revolution number is noted as its own (*svaparyayāḥ*), and the mean Sun is taken as the *śīghrocca* for all the planets.

Now, ignoring the correction due to the eccentricity, the ratio of the Mercury–Sun to the Earth–Sun distance may be compared with the ratio $\frac{r_s}{R}$ given in (2.187):

$$\frac{\text{Mercury–Sun distance}}{\text{Earth–Sun distance}} = \frac{31-2|\sin(\theta_{ms}-\theta_s)|}{80}. \tag{2.192}$$

It may be noted that this ratio varies between $\frac{29}{80} = 0.362$ and $\frac{31}{80} = 0.387$, as compared with the average modern value of 0.387. The factor $\frac{K}{R}$ in $\frac{r_s}{R}$ in (2.187) takes into account the eccentricity of the planetary orbit.

Finally it may be mentioned that here, in calculating the true position of Mercury, only a two-step procedure is prescribed. The *śīghra-phala*, however, depends on the *manda-karṇa* and hence the *manda-kendra* also. Further, it is the iterated *manda-karṇa* that is involved in this calculation. A similar procedure is advocated for obtaining the true position of Venus.

२.२८ शुक्रस्फुटीकरणम्
2.28 The correction for Venus

मन्दकेन्द्रभुजा जीवा खजिनांशेन संयुता ।
मनवस्तस्य हारः स्यात् तद्भक्ते बाहुकोटिके ॥ ७५ ॥
स्यातां मन्दफले तस्य दोःफलं च स्वमध्यमे ।
कृत्वाऽविशेषकर्णं च क्रियतां शीघ्रकर्म च ॥ ७६ ॥
द्विघ्ना दोर्ज्या त्रिभज्याप्ता शोध्यास्यैकोनषष्टितः ।
गुणः सोऽपि स्फुटीकार्यः मन्दकर्णेन पूर्ववत् ॥ ७७ ॥
गुणः स मन्दकर्णघ्नः त्रिज्याप्तस्तस्य च स्फुटः ।
अशीत्याप्ते भुजाकोटी तद्घ्ने शीघ्रफले भृगोः ॥ ७८ ॥
दोःफलं त्रिज्यया हत्वा शीघ्रकर्णहृतं भृगोः ।
चापितं भास्वतो मध्ये संस्कुर्यात् स स्फुटः सितः ॥ ७९ ॥

mandakendrabhujā jīvā khajināṃśena saṃyutā |
manavastasya hāraḥ syāt tadbhakte bāhukoṭike || 75 ||
syātāṃ mandaphale tasya doḥphalaṃ ca svamadhyame |
kṛtvā'viśeṣakarṇaṃ ca kriyatāṃ śīghrakarma ca || 76 ||
dvighnā dorjyā tribhajyāptā śodhyāsyaikonaṣaṣṭitaḥ |
guṇaḥ so'pi sphuṭīkāryaḥ mandakarṇena pūrvavat || 77 ||
guṇaḥ sa mandakarṇaghnaḥ trijyāptastasya ca sphuṭaḥ |
aśītyāpte bhujākoṭī tadghne śīghraphale bhṛgoḥ || 78 ||
doḥphalaṃ trijyayā hatvā śīghrakarṇahṛtaṃ bhṛgoḥ |
cāpitaṃ bhāsvato madhye saṃskuryāt sa sphuṭaḥ sitaḥ || 79 ||

The 240th part of the Rsine of the *manda-kendra* added to 14 (forms the divisor). The *dorjyā* and the *koṭijyā* divided by this divisor form the *doḥphala* and *koṭiphala* in the *manda-saṃskāra*. After adding the arc of the *doḥphala* to the *madhyama-graha*, let the *aviṣiṣṭa-manda-karṇa* be found and *śīghra-saṃskāra* be carried out as set forth below.

The *dorjyā* (corresponding to the *śīghra-kendra*) multiplied by two, divided by the *trijyā*, and subtracted from 59, forms the multiplier. This multiplied by the *aviṣiṣṭa-manda-karṇa* and divided by *trijyā* forms the *sphuṭaguṇa*. The *dorjyā* and *koṭijyā* multiplied by the *sphuṭaguṇa* and divided by 80 are the *doḥphala* and *koṭiphala*. The arc of the *doḥphala* multiplied by the *trijyā* and divided by the *śīghra-karṇa* should be applied to the mean Sun. This gives the true longitude of the Venus.

The procedure for calculating the geocentric longitude of Venus is the same as for that of Mercury. The *manda-sphuṭagraha* is calculated taking the ratio of the epicycle to the deferent[24] to be

[24] It is interesting to note that the expression for the denominator given here, namely $14 + \frac{R|\sin(\theta_0 - \theta_m)|}{240}$, is such that the second term can be as large as the first one.

$$\frac{r_0}{R} = \frac{1}{14 + \frac{R|\sin(\theta_0 - \theta_m)|}{240}}. \tag{2.193}$$

The *śīghra-saṃskāra* is identical with that for Mercury, as shown in Fig. 2.16. In the same way as in (2.192), here we can set

$$\frac{\text{Venus–Sun distance}}{\text{Earth–Sun distance}} = \frac{r_s}{R} = \frac{59 - 2|\sin(\theta_{ms} - \theta_s)|}{80} \times \frac{K}{R}. \tag{2.194}$$

Ignoring the correction for eccentricity (taking $K = R$), we find that $\frac{r_s}{R}$ varies between $\frac{57}{80} = .712$ and $\frac{59}{80} = .737$, as compared with the average modern value of .723.

२.२९ ग्रहाणां दिनभुक्तिः
2.29 The daily motion of the planets

श्वस्तनेऽद्यतनाच्छुद्धे वक्रभोगोऽवशिष्यते ।
विपरीतविशेषोत्थचारभोगस्तयोः स्फुटः ॥ ८० ॥

śvastane'dyatanācchuddhe vakrabhogo'vaśiṣyate |
viparītaviśeṣotthacārabhogastayoḥ sphuṭaḥ || 80 ||

The longitude of the planet found for tomorrow is subtracted from the longitude of the planet today. The result [if positive] is the retrograde daily motion of the planet; if otherwise, the result gives the direct daily motion of the planet.

In this verse, essentially, the definition of direct/retrograde motion is given. By *bhoga* is meant daily motion, the angular distance travelled by the planet in one day as observed by an observer on the surface of the Earth.

Chapter 3
छायाप्रकरणम्
Gnomonic shadow

३.१ शङ्कुस्थापनम्
3.1 Positioning the gnomon

> शिलातलेऽपि वा भूमौ समायां मण्डलं लिखेत् ।
> तन्मध्ये स्थापयेच्छङ्कुं कल्पितं द्वादशाङ्गुलम् ॥ १ ॥
>
> *śilātale'pi vā bhūmau samāyāṃ maṇḍalaṃ likhet|*
> *tanmadhye sthāpayecchaṅkuṃ kalpitaṃ dvādaśāṅgulam ||1||*
>
> On the surface of a rock or a flat Earth surface, draw a circle, and place a gnomon (*śaṅku*) at the centre of it, whose length is taken to be twelve *aṅgulas*.

Preparation of a flat surface and a gnomon

The primary requirement for all measurements related to the shadow of a gnomon or *śaṅku*[1] is a flat surface. The following quote from *Laghu-vivṛti* explains how carefully the plane surface needs to be prepared for positioning the *śaṅku* on this surface. It also furnishes certain other details that are to be considered, before making the necessary markings on the surface, for various measurements—related to place, time and direction (*tripraśna*)—that will be discussed later.

> तत्र तावल्लम्बकादिना परीक्ष्य समीकृते शिलातले भूतले वा शङ्कुद्विगुणमानेन अन्यादृशेन वा व्यासार्धसूत्रेण तादृशं मण्डलमालिखेत्, यस्मिन् दिनमध्यभागसम्बन्धिनी रवेः छाया कृत्स्नाप्यन्तर्भावयितव्या स्यात्। तद्यथा -
>
> एकं व्यासार्धसूत्राग्रमवष्टभ्य क्वचित् परम्।
> परितो भ्रामयेत् भूमौ समवृत्तं यतो भवेत्॥
> केन्द्रतो विप्रकर्षस्य परितस्तुल्यताकृतम्॥ – इति ॥

[1] By convention, the length of the *śaṅku* is taken to be 12 *aṅgulas* (a unit of measurement).

एवमालिखितस्य मण्डलस्य केन्द्रावस्थितस्वकेन्द्रं निजद्वादशांशप्रमिताङ्गुलसम-विभक्तायां तदर्धपरिणाहं चाभीष्टशङ्कुं स्थापयेत्। शङ्कुलक्षणमुक्तं श्रीपतिना –

भ्रमविरचितवृत्तः तुल्यमूलाग्रभागः
द्विरददशनजन्मा सारदारूद्भवो वा।
सममृजुरवलम्बात् अव्रणः षड्कवृत्तः
समतल इह शस्तः शङ्कुरर्काङ्गुलोच्चः ॥ – इति ॥

On a surface of a stone or Earth that has been prepared to be flat (*samīkṛte*) by means of [a contrivance] such as a plumb-line, draw a circle whose radius is twice that of the height of the *śaṅku* or any other [appropriate] measure, so that the entire midday shadow [of the *śaṅku*] cast by the Sun falls within [the circumference of the circle]. This can be done as follows:

> Firmly fixing (*avaṣṭabhya*) one end of a rope whose length is the radius of the circle [desired to be drawn] at some point on the Earth [having a flat surface], rotate the other end so that a circle is obtained. [It must be verified that] the distance of separation (*viprakarṣa*) from the centre to the circumference is the same all around.

Place the desired gnomon (*abhīṣṭaśaṅku*)—whose length is equal to 12 *aṅgulas* in its own measure, with equally spaced divisions [marked along it], and is also equal to half the circumference of the circle—at the centre of the circle thus drawn. The characteristics of a *śaṅku* have been stated by Śrīpati thus:

> A stick which is 12 *aṅgulas* in length, prepared from elephant's tusk (*dviradadaśana*[2]) or else some good-quality wood (*sāra-dāru*), made perfectly circular by rotating [tools] so that it is uniform [in thickness] without any injury/dent (*avraṇa*),[3] with six circular markings and placed erect like a plumb-line on a flat surface is called a *śaṅku*.

Some important circles in the celestial sphere

In this context, *Yukti-dīpikā* presents a graphic description of how the celestial equator, prime vertical, ecliptic and other important great circles are situated with respect to each other in the celestial sphere for an equatorial observer[4] (see Fig. 3.1). We present the description here, as it will be quite useful for understanding the later verses in this chapter and elsewhere.

समोपरि निरक्षेषु घटिकामण्डलं भवेत् ।
समतिर्यग्गतं तस्मात् दक्षिणोत्तरमण्डलम् ॥

---

[2] The word द्विरद literally means that which has two teeth, and hence an elephant; दशन is generally used to refer to teeth, and in this context the tusk of an elephant.

[3] The word *avraṇa*, which literally means without injuries, is used as an adjective to indicate that there should not be any imperfections – which are likely to happen during the process of rendering it into cylindrical shape – in the *śaṅku* prepared.

[4] An observer whose latitude is zero.

निरक्षे घनभूमध्यपार्श्वस्थं क्षितिजं भवेत् ।
दक्षिणोत्तरसम्पातद्वये यस्य ध्रुवौ स्थिरौ ॥
ऊर्ध्वस्थात् स्वस्तिकाद् याम्ये सौम्ये चाधोगतादपि।
चतुर्विंशतिभागान्ते भवति क्रान्तिमण्डलम्॥
घटिकावृत्तमार्गेण भ्राम्यति प्रवहोऽन्वहम्।
पश्चान्मुखः समजवः ग्रहर्क्षाणि समीरयन्॥[5]

In places having zero latitude, the equator (*ghaṭikāmaṇḍala*) will be right above [passing through the zenith of the observer]. The one which is exactly perpendicular [to that] is the prime meridian (*dakṣiṇottara-maṇḍala*). The [horizontal plane] around the centre of the solid Earth would be the horizon for places with zero latitude. The intersection of the north–south circle and this (horizon) will always be the north and south celestial poles (*dhruvas*).

The ecliptic (*krāntimaṇḍala*) is situated 24° towards the south from the zenith which is right above the observer, and [similarly] to the north from the nadir which is right below the observer.[6] The wind called *pravaha* flows continuously along the equator, making the planets and stars move westwards at equal speed.

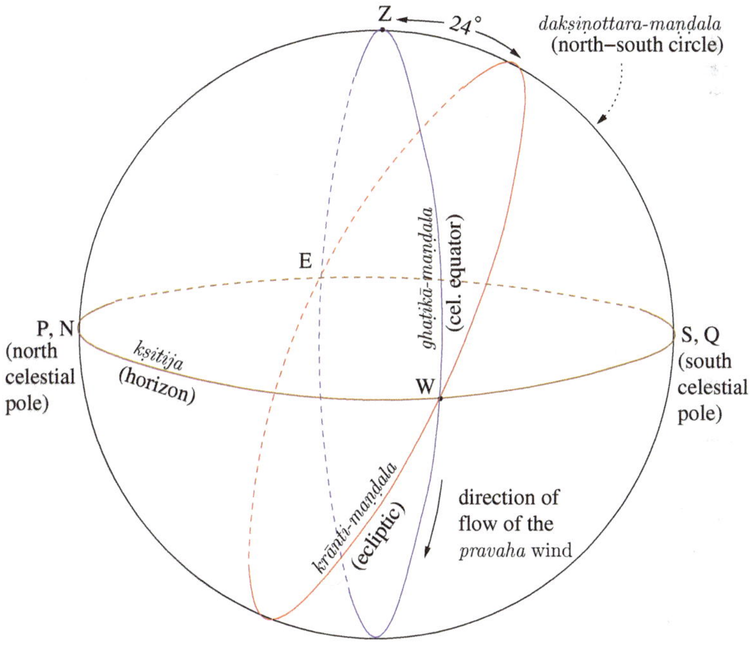

Fig. 3.1 Celestial sphere for an equatorial observer.

[5] {TS 1977}, pp. 187–8.

[6] This would be the case when the equinoxes are on the horizon coincident with the east and west points of the horizon.

३.२ पूर्वापरबिन्द्वानयनम्
3.2 Finding the east–west points

तच्छायाग्रं स्पृशेद्यत्र वृत्ते पूर्वापराह्णयोः ।
तत्र बिन्दू निधातव्यौ वृत्ते पूर्वापराभिधौ ॥ २ ॥

tacchāyāgraṃ spṛśedyatra vṛtte pūrvāparāhṇayoḥ|
tatra bindū nidhātavyau vṛtte pūrvāparābhidhau || 2 ||

The marks on the circle [drawn with the gnomon as the centre], known as the east and the west points, have to be made wherever the tip of the shadow of that (gnomon) grazes the circumference of the circle in the forenoon and the afternoon.

During the forenoon, the tip of the shadow will be entering into the circle from outside, and the point where it intersects the circumference is to be marked as the west point. Similarly, during the afternoon, as the tip moves out of the circle, it again intersects the circumference, and this point is to be noted as the east point. The line joining these two points will represent the exact east–west direction at that location, if it is assumed that the declination of the Sun remains constant during the day. But since the declination actually varies continuously, there is a need for a small correction, which is discussed in the following verse.

३.३ पूर्वापरबिन्दुशोधनम्
3.3 Correcting the east–west points

भेदात् पूर्वापरक्रान्त्योः छायाकर्णाङ्गुलाहतात् ।
लम्बकाप्तं पूर्वबिन्दोः नीत्वा कार्योऽत्र सोऽयनात् ॥ ३ ॥

bhedāt pūrvāparakrāntyoḥ chāyākarṇāṅgulāhatāt |
lambakāptaṃ pūrvabindoḥ nītvā kāryo'tra so'yanāt ||3||

The difference in the [Rsine of the] declinations determined in the forenoon and the afternoon[7] multiplied by the shadow-hypotenuse and divided by the Rcosine of the latitude of the place (*lambaka*) has to be applied to the east point and this [the sign of application, ±] depends upon the *ayana*.

Consider Fig. 3.2*a*. The points W' and E'' on the circle represent the points of intersection of the tip of the shadow with the circumference in the forenoon and afternoon respectively. If the declination of the Sun were to be constant during the course of the day, then $W'E''$ would be the west–east line. However, owing to the northward or southward motion of the Sun, the declination (δ) changes. Consequently, the tip of the eastern shadow point would have been shifted towards the south if the Sun has northward motion (δ increases) or north if the Sun has southward motion (δ decreases). So a correction Δ (see (ii) in Fig. 3.2*a*) has to be applied to E'' in order to obtain the actual east point E'. If the change in the declination from δ_1 to δ_2, then the magnitude of the correction Δ is stated to be

[7] At those instances when the tip of the shadow grazes the circumference.

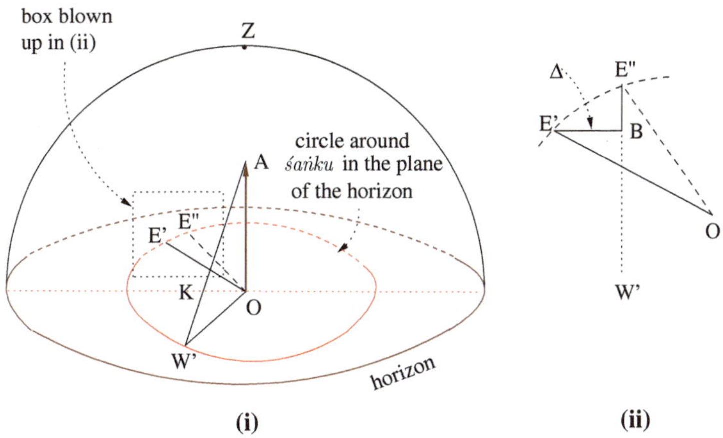

Fig. 3.2*a* Fixing the directions through shadow measurements.

$$\Delta = \frac{K(R\sin\delta_2 - R\sin\delta_1)}{R\cos\phi}, \tag{3.1}$$

where K is the hypotenuse of the shadow in *aṅgulas*[8] and $R\cos\phi$ is the *lambaka*, ϕ being the latitude of the place. The expression for Δ given here is the same as the one given earlier by Bhāskarācārya (c. 1150) in *Siddhāntaśiromaṇi*,[9] and may be understood as follows.

Consider the situation when the Sun has declination δ, zenith distance z and azimuth A (see Fig. 3.2*b*). OX is the gnomon, whose height is taken to be 12 *aṅgulas*. The length L of its *chāyā* (shadow) OY is given by

$$L = OY = XY\sin z = K\sin z, \tag{3.2}$$

where $K = XY$ is the *chāyā-karṇa* (shadow-hypotenuse). For future purposes we also note that

$$12 = K\cos z \qquad \text{or} \qquad K = \frac{12}{\cos z}. \tag{3.3}$$

Using (3.3) in (3.2) we have

$$L = 12\frac{\sin z}{\cos z}. \tag{3.4}$$

Chāyābhujā YQ is the perpendicular distance of the tip of the shadow from the east–west line and is given by

$$YQ = L\sin(A-90) = -L\cos A. \tag{3.5}$$

[8] The gnomon is taken to be 12 *aṅgulas*.

[9] {SSI 2000}, pp. 25–6.

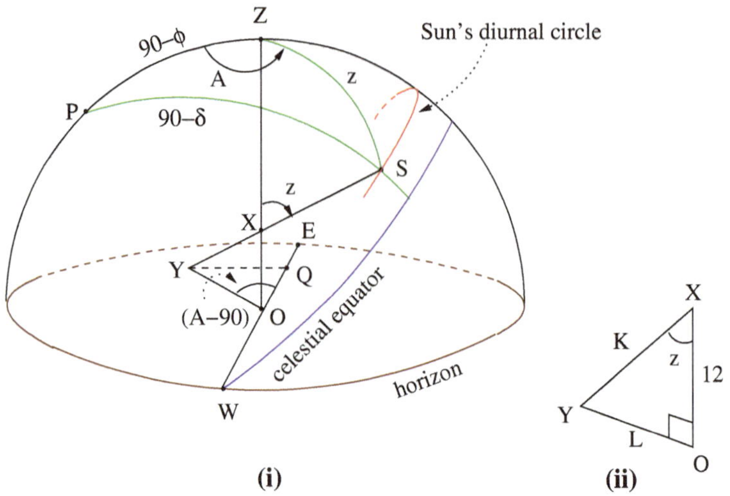

Fig. 3.2*b* Relation between the zenith distance of the Sun and the length of the shadow cast by a *śaṅku*.

In the spherical triangle *PZS*, $ZS = z$, $PS = 90^\circ - \delta$, $PZ = 90^\circ - \phi$ and $P\hat{Z}S = A$, where A is the azimuth (west). Now, using the cosine formula we have

$$\sin\delta = \cos z \sin\phi + \sin z \cos\phi \cos A. \tag{3.6}$$

Since the length of the shadow corresponding to the points W' and E'' are the same—both being exactly equal to the radius of the circle—the corresponding zenith distances of the Sun in the forenoon and the afternoon must be equal. However, the declination of the Sun changes from δ_1 to δ_2. If A_1 and A_2 are the azimuths corresponding to these positions of the Sun, then

$$\sin\delta_1 = \cos z \sin\phi + \sin z \cos\phi \cos A_1 \tag{3.7}$$
$$\sin\delta_2 = \cos z \sin\phi + \sin z \cos\phi \cos A_2. \tag{3.8}$$

Subtracting one from the other and doing some algebraic manipulations,

$$\begin{aligned} \sin\delta_2 - \sin\delta_1 &= \sin z \cos\phi\,(\cos A_2 - \cos A_1) \\ \frac{K\,(\sin\delta_2 - \sin\delta_1)}{\cos\phi} &= K \sin z\,(\cos A_2 - \cos A_1) \\ &= L\,(\cos A_2 - \cos A_1)\,. \end{aligned} \tag{3.9}$$

We have used (3.2) in arriving at the RHS of the above equation. Further, it may be noted that RHS is nothing but the difference in '*chayābhujās*' corresponding to δ_1 and δ_2. Hence, the LHS of the above equation represents the distance Δ by which the east point has to be displaced. Thus

$$\Delta = \frac{K(\sin\delta_2 - \sin\delta_1)}{\cos\phi}. \tag{3.10}$$

In other words, the true east point E' is the point on the circle which is at a distance Δ from the line $E''W'$ (see to (ii) in Fig. 3.2*a*). The true east–west line is $E'W'$. The point E' would be to the north (south) of E'', depending upon whether the Sun has northward (southward) motion.

Now we present the detailed discussion on the necessity for this correction and how it is to be implemented as given in *Yukti-dīpikā*:

तत्रायनवशात् सौम्ययाम्ययोर्गच्छतो रवेः ।
समपूर्वापरे स्यातां छायाग्रे नेष्टमण्डले ॥
प्रवेशनिर्यत्समयक्रान्तिभेदेन भेदतः ।
भिन्नकालरविक्रान्तिविवरोत्थे फले कृते ॥
शक्या कल्पयितुं तत्र समपूर्वापरस्थितिः ।
इष्टमण्डलनेमिस्थछायाग्राङ्गुलवर्गतः ॥
तच्छङ्क्वङ्गुलवर्गाढ्या छायाकर्णपदं भवेत् ।
छायाकर्णहतं कालद्वितयापक्रमान्तरम् ॥
लम्बकेन हरेल्लब्धं तद्वृत्ते क्रान्तिजं फलम् ।
तेनायनवशान्नेयं प्राच्यां छायाग्रमत्र तु ॥
अयनव्यत्ययात् पश्चात् छायाग्रं वाथमण्डले ।
समपूर्वापरं येन छायाग्रद्वितयं भवेत् ॥
छायाग्रद्वितयस्पृष्टं सूत्रं प्रागपरं ततः ।
समतिर्यग्गतं सूत्रं ततः स्यात् दक्षिणोत्तरम् ॥ [10]

Because of the northward and southward movement (*ayana*) [of the Sun], the tips of the shadow of the Sun on the desired circle may not be exactly along the east–west [direction].

Since the difference [from the exact east–west] is due to the difference in the declinations of the Sun at the times of entry and exit, once this difference (*vivarotthaphala*) is calculated, it would then be possible to determine the exact east–west direction.

The square of the tip of the shadow on the circumference of the desired circle [measured] in *aṅgulas*, added to the square of the *śaṅku* in *aṅgulas*, is stated to be the [square of] the *chāyākarṇa* (shadow-hypotenuse).

Let the *chāyākarṇa* (shadow-hypotenuse), multiplied by the difference in declinations calculated at the two different instants, be divided by the *lambaka*. The result obtained is due to [the change in] declination (*krāntijaphala*)[11] [to be applied] in that circle.

With that [result] depending upon the *ayana*, the tip of the shadow has to be shifted towards the east. If the *ayana* happens to be otherwise (*vyatyayāt*), then the tip of the shadow has to be shifted to the west. It is only then (*atha*) that the two tips of the shadows on the circle represent the exact east-west direction.

Then the line passing through the two tips of the shadows [obtained after the corrections] will represent the east–west direction. The line that is exactly perpendicular to this will then represent the north–south direction.

---

[10] {TS 1977}, pp. 188–9.

[11] The word क्रान्तिजफल should be considered as an example of मध्यमपदलोपिसमास, the *vigraha* of which should perhaps be done as follows – क्रान्तिभेदात् जायमानं फलम्; which means the result that is obtained because of the change in the declination.

३.४ स्वदेशे दिगानयनम्

3.4 Fixing the directions in one's own place

मध्यं कृत्वा तयोर्बिन्द्वोः तुल्ये वृत्ते समालिखेत् ।
तत्संश्लेषोत्थमत्स्येन ज्ञेये याम्योत्तरे दिशौ ॥ ४ ॥
तद्वृत्तमध्यमत्स्येन पूर्वापरदिशावपि ।
दिङ्मध्यमत्स्यसंसाध्याः चतस्रो विदिशोऽपि च ॥ ५ ॥
अधऊर्ध्वदिशौ ज्ञेये लम्बकेनैव नान्यथा ।

madhyaṃ kṛtvā tayorbindvoḥ tulye vṛtte samālikhet |
tatsaṃśleṣotthamatsyena jñeye yāmyottare diśau ||4||
tadvṛttamadhyamatsyena pūrvāparadiśāvapi |
diṅmadhyamatsyasaṃsādhyāḥ catasro vidiśo'pi ca ||5||
adhaūrdhvadiśau jñeye lambakenaiva nānyathā |

Draw two identical circles with these two points as centres. With the *matsya* (fish [figure]) that is formed by the intersection of these [two circles], the north and the south directions have to be determined.

[Again] with the *matsya* that is formed at the centre of that circle [at the centre of which *śaṅku* is placed], the east and the west directions [have to be determined]. And the four subordinate directions have to be determined by [drawing] *matsyas* in between [these] cardinal directions. The directions vertically above and below [i.e., zenith and nadir] can be determined only through the plumb-line (*lambaka*) and not by any other means.

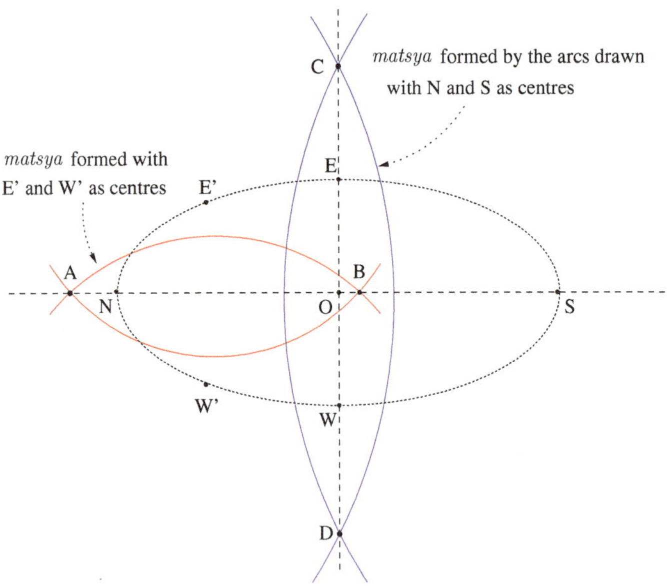

Fig. 3.3*a* Construction of *matsyas* for the determination of the cardinal directions.

In Fig. 3.3*a*, the *śaṅku* is placed at the centre of the circle marked O. The points where the tip of its shadow grazes the circumference of the circle in the forenoon and the afternoon are marked as W' and E'[12] respectively. With these two points as centres and radius greater than half of $E'W'$, arcs are to be drawn. The resulting figure appears like a fish and hence is referred to as a *matsya* (fish). A straight line is drawn through the points of intersection A and B of these two arcs. This line meets the circle at N and S and this represents the north–south direction.

With N and S as centres along north and south directions equidistant from O and radius greater than half the distance between them, two more arcs are again drawn. They intersect at points C and D, forming a *matsya*. With these two points a straight line is drawn that intersects the circle at E and W. This line represents the exact east–west direction. By constructing similar *matsyas* with N and E as centres, the north–east and south–west directions are determined. Similarly, by constructing a *matsya* with S and E as centres, the south–east and north–west directions get determined.

Thus the four cardinal and the four subordinate directions are determined by making shadow measurements and drawing *matsyas*. The direction vertically above the observer and the one below, denoted by the zenith and nadir, are to be determined only with the help of a plumb-line, referred to as a *lambaka*. A *lambaka* is a thread with a heavy object, made of wood or iron, having a fine tip—as indicated in Fig. 3.3*b*—tied to one of its ends.

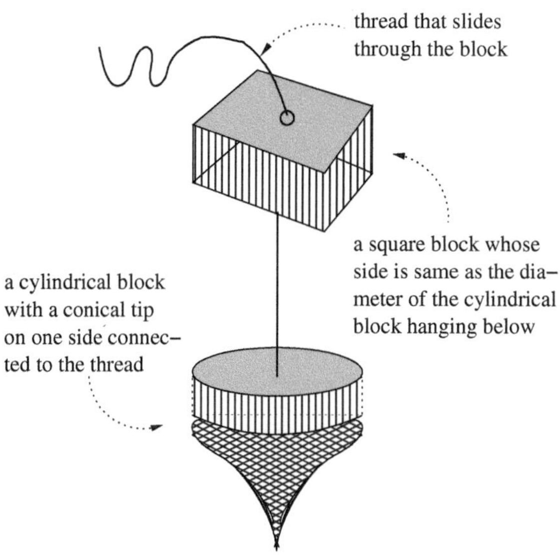

Fig. 3.3*b* Contrivance used as a plumb-line to determine the perpendicular to the horizon at the observer's location.

[12] E' in Fig. 3.3*a* represents the actual east point obtained after making the necessary correction prescribed in verse 3 of this chapter.

३.५ विषुवच्छाया
3.5 Equinoctial shadow

एकसूत्रगता छाया यस्मिन्नह्न्युदयास्तयोः ॥ ६ ॥
मध्याह्ने विषुवाख्यः स्यात् कालस्तस्मिन् दिने यतः ।
तस्मात् तद्दिनमध्याह्नछाया वैषुवती मता ॥ ७ ॥

ekasūtragatā chāyā yasminnahnyudayāstayoḥ ||6||
madhyāhne viṣuvākhyaḥ syāt kālastasmin dine yataḥ |
tasmāt taddinamadhyāhnachāyā vaiṣuvatī matā ||7||

Since the noon-time on a day in which the tips of the shadows at sunrise and sunset fall on a straight line (*ekasūtragatā*) is called *viṣuvat*, the shadow of the *śaṅku* at the noon-time on that day is called the *viṣuvacchāyā* (equinoctial shadow).

On an equinoctial day, the Sun is at one of the equinoxes and its declination is zero. On that day, the tip of the shadow traces a straight line parallel to the east–west line. The noon-time of the equinoctial day is called the *viṣuvat* and the noon-shadow on that day is called the *viṣuvacchāyā*.

It was shown in the previous section that the declination of the Sun at any time during the day is given by

$$\sin\delta = \cos z \sin\phi + \sin z \cos\phi \cos A. \tag{3.11}$$

On the equinoctial day, $\delta = 0$. Hence the above equation reduces to

$$\sin z \cos\phi \cos A = -\cos z \sin\phi. \tag{3.12}$$

Rewriting this we have

$$\begin{aligned} \tan z \cos A &= -\tan\phi \\ \text{or} \qquad 12\tan z \cos A &= -12\tan\phi. \end{aligned} \tag{3.13}$$

From (3.4) and (3.5) we have

$$ch\bar{a}y\bar{a}bhuj\bar{a} = -L\cos A = -12\tan z \cos A. \tag{3.14}$$

Hence from (3.13)

$$ch\bar{a}y\bar{a}bhuj\bar{a} = 12\tan\phi. \tag{3.15}$$

Since the RHS in the above equation is a *constant* for a given latitude, the *chāyābhujā* will be a constant on the equinoctial day and would be equal to *viśuvacchāyā*. This means that the tip of the shadow is at a constant distance from the east-west line. In other words, the tip of the shadow traces a straight line on the equinoctial day which is at a distance equal to $12\tan\phi$ from the east–west line passing through the base of the *śaṅku*.

Criterion for the noon-shadow to be declared equinoctial

The condition that must be satisfied for the noon-time shadow to be called the *viśuvacchāyā* is clearly spelt out in *Yukti-dīpikā*.

क्रान्त्योः भिन्नदिशोः साम्यं सायनास्तोदयार्कयोः ।
यस्मिन्नहनि तत्रार्को मध्याह्ने विषुवं व्रजेत् ॥
नान्यदास्तोदयक्रान्त्योः विदिशोः भेदसंभवात् ।
एकसूत्रगताच्छाया तत्र स्यादुदयास्तयोः ॥

The Sun would reach the equinox at noon on that day on which the declinations obtained at the sunrise and sunset from its *sāyana* longitude have opposite directions and are equal [in magnitude].

On no other day [can the Sun be at equinox at noon]. This is because the declinations having different directions will vary [in magnitude]. Only there [on that day] will the shadow at rising and setting lie on the same [straight] line.

Let δ_r and δ_s be the declinations of the Sun at sunrise and sunset. It is stated that the Sun is at the equinox at noon on that day on which these two declinations have opposite directions (one southern and the other northern) but equal magnitudes. In other words, the conditions to be satisfied for the noon-shadow to be declared equinoctial are

$$\delta_r = -\delta_s \qquad \text{and} \qquad |\delta_r| = |\delta_s|.$$

३.६ छायाशङ्कुकर्णानां सम्बन्धः

3.6 Relation between the gnomon, its shadow and the hypotenuse

तच्छङ्कुवर्गसंयोगमूलं कर्णोऽस्य वर्गतः ।
त्यक्त्वा शङ्कुकृतिं मूलं छायाशङ्कुर्विपर्ययात् ॥ ८ ॥
ज्ञेयो दोःकोटिकर्णेषु द्वाभ्यामन्योऽखिलेष्वपि ।

tacchaṅkuvargasaṃyogamūlaṃ karṇo'sya vargataḥ |
tyaktvā śaṅkukṛtiṃ mūlaṃ chāyāśaṅkurviparyayāt ||8||
jñeyo doḥkoṭikarṇeṣu dvābhyāmanyo'khileṣvapi |

The square root of the sum of the squares of that (shadow) and the *śaṅku* is the *karṇa* (hypotenuse). Subtracting the square of the *śaṅku* from the square of this (*karṇa*), and taking the square root the shadow is obtained. By [doing] the reverse process [subtracting the square of the shadow from the square of the *karṇa*], the *śaṅku* is obtained. It must be understood that among the *doḥ* (sine), *koṭi* (cosine) and *karṇa*, if any two are known then the other [third] can be determined in all the instances.

It can be easily seen from Fig. 3.2*b* or 3.4 that the triangle formed by the *śaṅku* (gnomon), the *chāyā* (shadow) and the *karṇa* (hypotenuse) is a right-angled triangle. In these figures OX represents the *śaṅku*, OY the *chāyā* and XY the *karṇa*:

$$\sqrt{(XY)^2 - (OY)^2} = OX$$

$$\sqrt{(XY)^2-(OX)^2} = OY$$
$$karṇa^2 = XY^2 = OY^2+OX^2 = chāyā^2 + śaṅku^2. \tag{3.16}$$

Thus, if any two quantities among *śaṅku*, *chāyā* and *karṇa* are known, the third can be determined using the above relation.

३.७ अक्षलम्बकौ
3.7 Rsine and Rcosine of the latitude

छायां तां त्रिज्यया हत्वा स्वकर्णेन हरेत् फलम् ॥ ९ ॥
अक्षजीवा तथा शङ्कुं कृत्वा लम्बकमानयेत् ॥

chāyāṃ tāṃ trijyayā hatvā svakarṇena haret phalam ||9||
akṣajīvā tathā śaṅkuṃ kṛtvā lambakamānayet ||

The [length of] that (equinoctial) shadow should be multiplied by *trijyā* and divided by its *karṇa*. The result is Rsine of latitude (*akṣajīvā*). By doing the same with *śaṅku* [instead of the shadow] the Rcosine of latitude (*lambaka*) may be obtained.

The procedure for obtaining the Rsine and Rcosine of the latitude of the observer is explained in the above verse. Consider Fig. 3.4. Here OX represents the *śaṅku*. On the equinoctial day, since the Sun is almost on the equator throughout the day, the zenith distance of the Sun as it crosses the prime meridian (at noon) will be equal to the latitude of the place.

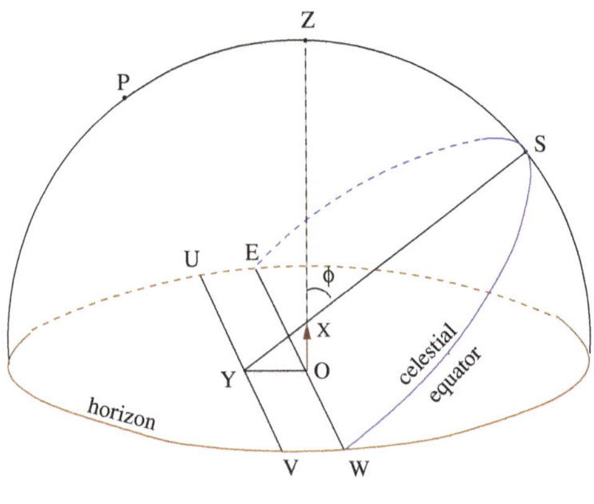

Fig. 3.4 Determination of the latitude from the equinoctial shadow of the *śaṅku*.

Considering the triangle OXY formed by the *śaṅku* OX, the *chāyā* OY and the *karṇa* XY, it can be easily seen that $O\hat{X}Y = \phi$. Hence,

$$\sin\phi = \frac{OY}{XY}, \qquad \cos\phi = \frac{OX}{XY}. \tag{3.17}$$

Now *akṣajyā* is $R\sin\phi$ and *lambaka* is $R\cos\phi$. Hence, multiplying the above equation by the *trijyā* we have

$$ak\d{s}ajy\bar{a} = \frac{trijy\bar{a} \times ch\bar{a}y\bar{a}}{kar\d{n}a}, \tag{3.18a}$$

$$lambaka = \frac{trijy\bar{a} \times \acute{s}a\dot{n}ku}{kar\d{n}a}. \tag{3.18b}$$

In *Yukti-dīpikā*, there is also a very interesting discussion on how the determination of latitude and the directions are mutually interdependent.

ननु लम्बेन सिद्धेन कर्तव्यं क्रान्तिजं फलम् ।
विषुवद्दिनमध्याह्नच्छायातोऽप्यक्षलम्बकौ ॥
तयोर्लम्बकाक्षयोः सिद्धिर्दिक्परिच्छेदपूर्विका ।
स पुनर्लम्बकाधीननिजक्रान्तिफलान्तरात् ॥
इति यत्तदुपायेऽस्मिंश्चक्रग्रस्ततया भवेत् ।
ततोऽयनान्ते मध्याह्ने दिनपूर्वापराह्णयोः ॥
तुल्यकालान्तरितयोः समपूर्वापरं भवेत् ।
छायाग्रद्वितयं क्रान्तिफलेनैवान्यदा सदा ॥
भानामस्तोदयाभ्यां वा समपूर्वापरे दिशौ ।

The *krāntijaphala*[13] is to be determined from the Rsine of colatitude (*lambaka*) already known. And the Rsines of latitude and colatitude are [to be found] from the noon-shadow on the equinoctial day. [However] for the determination of the latitude and co-latitude a clear demarcation (*pariccheda*) of the directions is a prerequisite. This [demarcation of directions] in turn depends upon the correction due to the difference in declinations, which [in turn] is dependent on the *lambaka*. Thus, the entire procedure [seems to be faulty as it] suffers from 'circularity' (*cakragrasta*).

Therefore [to circumvent this problem] at the solstices (*ayanānta*), obtain the two tips of the shadow (*chāyāgra-dvitayaṃ*) corresponding to two instants at forenoon and afternoon, equally separated from noon, which gives the exact east–west line [without making any correction for the change in declination]. Otherwise, at all other instances, the tips of the shadows [need to be determined] only after considering the correction due to difference in declinations. The exact east and west directions may also be determined from the rising and setting of the stars.

Here it is pointed out that the procedure for determining the latitude suffers from circularity. The determination of the latitude from the equinoctial noon–shadow is dependent on the knowledge of the exact east–west direction, which in turn requires the knowledge of the latitude.[14] However, at solstices the change in declination over

[13] This term refers to the magnitude of correction to be applied to the east/west point, due to the variation in the declination of the Sun between the forenoon and the afternoon, to obtain the correct east–west direction.

[14] From (3.10), it may be noted that ϕ appears in the denominator of the correction term to be used in the determination of the direction. On the other hand, *OY* in (3.17) is measured along the

a day is negligible. This is because the declination of the Sun at any instant is given by

$$\sin\delta = \sin\varepsilon \sin\lambda, \tag{3.19}$$

and the rate of change of declination is

$$\begin{aligned} \frac{d}{dt}\sin\delta &= \sin\varepsilon \frac{d}{dt}\sin\lambda \\ &= \sin\varepsilon \cos\lambda \frac{d\lambda}{dt}. \end{aligned} \tag{3.20}$$

Since $\cos\lambda = 0$ at solstices (as $\lambda = 90^\circ$ or 270°), $\frac{d}{dt}\sin\delta = 0$ at these points. Hence, to first order in Δt, the declination does not vary. Thus, when the Sun is close to the solstices, the exact east–west direction can be determined without considering the correction term given by (3.9), and thus the problem of circularity can be overcome.

३.८ स्फुटाक्षलम्बकौ

3.8 More accurate values of the Rsine and Rcosine of the latitude

अक्षज्यार्कगतिघ्नाप्ता खस्वरेष्वेकसायकैः [15] ॥ १० ॥
फलोनमक्षचापैः स्यात् अर्कबिम्बार्धसंयुतम्।
स्फुटं तज्ज्याक्षजीवापि तस्याः कोटिश्च लम्बकः ॥ ११ ॥

akṣjyārkagatighnāptā khasvareṣvekasāyakaiḥ ||10||
phalonamakṣacāpaiḥ syāt arkabimbārdhasaṃyutam|
sphuṭaṃ tajjyākṣajīvāpi tasyāḥ koṭiśca lambakaḥ ||11||

The *akṣajyā* is multiplied by the true daily motion of the Sun and divided by 51570. The result has to be subtracted from the latitude of the place (*akṣacāpa*) [and to this] the semi-diameter of the Sun has to be added. This is the true value [of the latitude]. The Rsine of this is the *akṣajīvā* and its complement is the *lambaka*.

The above verse gives the procedure for correcting the observed value of the latitude taking into account the effects of parallax and the finite diameter of the Sun. The effect of parallax is to increase the apparent zenith distance of the object, as may be seen from Fig. 3.5*a*.

Here, C represents the centre of the Earth, S the Sun and O the observer. R_e and d refer to the radius of the Earth and the distance of the Sun from the centre of the Earth respectively. Z represents the geocentric zenith of the observer. If z' and z are the apparent and the actual zenith distances of the Sun, then it is easily seen that

east–west direction. Thus we need to know ϕ for finding the exact east–west direction, and the east–west direction to know ϕ—hence the circularity.

[15] The word in both the printed editions is: खस्वराद्रोकसायकैः । The number represented by this code word (in *Bhūtasaṅkhyā* system) is 51770, whereas the number that fits into the present context is 51570, and we have indicated this correct reading here. In a similar context in Chapter 5, verse 10, we find the number (51570) occurring again. This indicates that our correction is justified.

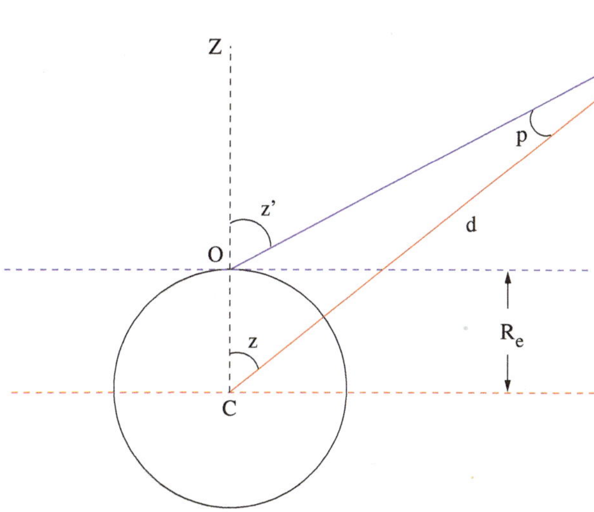

Fig. 3.5*a* The effect of parallax on the measurement of the latitude of the observer.

$$z = z' - p, \tag{3.21}$$

where $p = C\hat{S}O$ is the parallax of the Sun for an observer on the surface of the Earth. It is the angle subtended by the radius of the Earth at the centre of the Sun. In other words, it is the angle between the direction of the object as seen by the observer O and the direction of the object as seen from the Earth's centre (*bhūgola-madhya*, which is the standard reference point).

From the planar triangle COS we have

$$\sin p = \frac{R_e}{d} \sin z'. \tag{3.22}$$

Since $R_e \ll d$, p is small and the above equation can be written as

$$p = \frac{R_e}{d} \sin z'. \tag{3.23}$$

When $z' = 90°$, i.e. the celestial object is on the observer's horizon—which is the tangential plane passing through the observer, and not the centre of the Earth—it is easily seen that the correction due to parallax is maximum. It is called the horizontal parallax and is given by

$$P = \frac{R_e}{d}. \tag{3.24}$$

Using this in (3.23) we have

$$p = P \sin z'. \tag{3.25}$$

In Indian astronomical texts, the mean value of the horizontal parallax is taken to be one-fifteenth of the mean daily motion of the celestial object. This assumption is based on the fact that the mean value of the moon's horizontal parallax (which is found to be $\approx 52.5'$) in *Tantrasaṅgraha* is close to one-fifteenth of its mean daily motion, which is $790.6'$. As the linear velocities of all the planets are assumed to be the same in Indian texts, the mean parallax of any other object is also taken to be one-fifteenth of its daily motion.[16]

If D_{ms} represents the mean daily motion of the Sun, then the mean value of the parallax due to the Sun is given by

$$p_0 = \frac{D_{ms}}{15} \sin z'. \tag{3.26}$$

Multiplying and dividing the above equation by the *trijyā*, and substituting its approximate value 3438 (in minutes) in the denominator we have

$$p_0 = \frac{D_{ms}}{51570} R \sin z'. \tag{3.27}$$

As the distance of the Sun from the Earth keeps varying continuously, the value of the parallax also keeps varying. Hence, the true value of the parallax p at a particular instant can be obtained only by considering the actual distance of the Earth from the Sun at that instant. This is often achieved by multiplying the mean value of the parallax by the true daily motion and dividing by the mean daily motion:

$$p = p_0 \times \frac{\text{true daily motion } (D_{ts})}{\text{mean daily motion } (D_{ms})}. \tag{3.28}$$

Substituting for p_0 from (3.27), we have

$$p = \frac{D_{ts}}{51570} R \sin z'. \tag{3.29}$$

The above equation is the same as the expression given in the verse above for the correction to the observed latitude due to effect of parallax. When the Sun is on the prime meridian on an equinoctial day, then the zenith distance of the Sun is the same as the latitude of the place. Hence to obtain the correct latitude of the place, one has to subtract the parallax from the observed zenith distance.

The correction which arises owing to the finite size of the Sun is illustrated in Fig. 3.5*b*. Here *OA* is the *śaṅku*. *PSQ* represents the sectional view of the Sun, *S*

---

[16] A more detailed discussion on parallax and its application may be found in Chapters 4 and 5 on lunar and solar eclipses. According to *Tantrasaṅgraha*, the diameter of the Earth is 1050.4 *yojanas*. Hence its radius $R_e = \frac{1050.4}{2} = 525.2$ *yojanas*. Also the distance of the Moon is given to be 34380 *yojanas*. Therefore the horizontal parallax of the Moon is

$$\begin{aligned} \frac{R_e}{d} &= \frac{525.2}{34380} && \text{(in radians)} \\ &= \frac{525.2}{34380} \times \frac{180}{\pi} \times 60 \approx \frac{790.6}{15} && \text{(in minutes).} \end{aligned}$$

being its centre. If the Sun were a point source of light, then the tip of the shadow of the *śaṅku* would fall at S' and OS' would be the length of the shadow.

However, the rays that emerge from P and graze the tip of the *śaṅku* would fall at P'. All the rays that emerge from the section SP would be mapped to $S'P'$. Similarly the rays that emerge from SQ would be mapped to $S'Q'$. Since the rays emerging from the bottommost portion of the Sun Q, grazing the *śaṅku*, fall at Q', one would tend to think that OQ' should be the length of the shadow. Observationally, only OP' is the length of the shadow as it is only this region which does not receive any light from any part of the Sun, whereas the region $P'Q'$ would be partially illuminated. Hence, the apparent length of the shadow is determined by the rays emerging from the upper end of the Sun. If the Sun were a point object and located at S, then OS'

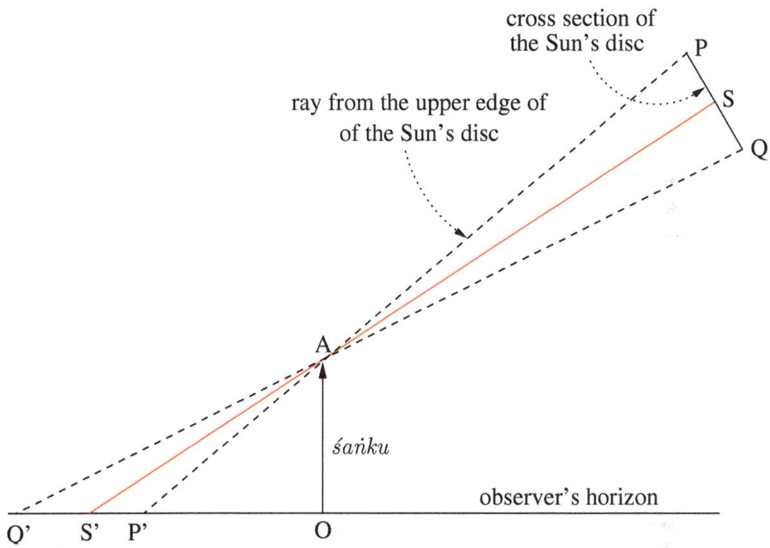

Fig. 3.5*b* Sectional view of the Sun and the shadow of the *śaṅku* generated by it.

would have been length of the shadow and $S'\hat{A}O$ would be the latitude of the place, after taking into account the correction due to parallax. Since the Sun is an extended object, and the length of the shadow observed is OP', it is easily seen that the angle by which the observed latitude must be corrected is $P'\hat{A}S'$. This angle is the same as $P\hat{A}S$, which is the semi-diameter of the Sun. Hence the correct latitude of a place is obtained by adding the semi-diameter of the Sun to the observed value.

In *Yukti-dīpikā* the formula for the correction due to parallax is explained. The physical reasoning offered is quite interesting and novel. The explanation for the correction due to the finite diameter of the Sun is also unusual, though it is convincing nevertheless.

Formula for the correction due to parallax

भूव्यासार्धं हि दृग्ज्यायां त्रिज्यायां लम्बयोजनम्।
कियत् तदिष्टदृग्ज्यायां स्वकक्ष्यायां भवेत् तदा ॥
इत्थं त्रैराशिकात् कल्प्यं दृग्ज्यालम्बनयोजनम् ।
मध्ययोजनकर्णेन त्रिज्यातुल्याः कला यदि ॥
इष्टयोजनलम्बेन कियत्यस्तत्कलास्तदा ।
स्फुटभुक्त्या हतास्ताश्च मध्यभुक्त्या विभाजिताः ॥
स्फुटलम्बनलिप्ताः स्युः इति त्रैराशिकत्रयात् ।
हारो गुणश्च त्रिज्यैका तयोराद्यद्वितीययोः ॥
मध्ययोजनकर्णश्च मध्यभुक्तिश्च हारकौ ।
अन्ययोर्गुणकस्त्वाद्ये तद्भूव्यासार्धयोजनम् ॥
गुणकारहते हारे हार एवात्र नो गुणः ।
मध्ययोजनकर्णघ्नमध्यभुक्तेस्ततो हतः ॥
नियतैरेव लम्बार्थं भूव्यासार्धस्य योजनैः ।
स एव हारकः खस्वरेष्वेकेषुमितो [17] मतः ॥
स्फुटभुक्तिर्गुणो याम्यां दृग्ज्यातः स्फुटलम्बनम् ।[18]

When the *dṛgjyā*[19] is [equal to] the *trijyā*, then the parallax in latitude (the *lamba/lambana*) in *yojanas* is equal to the semi-diameter of the Earth. Then for a desired *dṛgjyā* what will be the parallax in latitude in its own orbit? Thus the parallax in latitude in *yojanas* has to be obtained by using the rule of three.

If the *madhyayojanakarṇa* is equated to the minutes of the *trijyā* (3438), then what will be the desired parallax in latitude in minutes corresponding to the [above] parallax in *yojanas*. Those [minutes] are multiplied by the true daily motion and divided by the mean daily motion. Thus we get the true value of the parallax in latitude, measured in minutes, by [applying] the rule of three, three times [successively].

The *trijyā* alone is the divisor (*hāra*) and multiplier (*guṇa*), respectively, in the first and the second rule of three. The *madhyayojana-karṇa* and the mean daily motion are the divisors in the others [second and the third rules of three]. The radius of the Earth in *yojanas* is the multiplier in the first [rule of three].

If the divisor is multiplied by the multiplier, then only the divisor [remains] and not the multiplier. Therefore, the product of the *madhyayojana-karṇa* and the mean daily motion is divided by the fixed radius in *yojanas* of the Earth. That alone becomes the divisor [in obtaining the parallax in latitude], whose value is known to be equal to 51570. The true daily motion is the multiplier for them (i.e. the second and third rule of three). From these and the *dṛgjyā*, the true value of the parallax in latitude can be obtained.

The three rules of proportion given in the verses above may be written down as:

$$\begin{array}{llll} trijy\bar{a} & : R_e & :: \; i\d{s}\d{t}ad\d{r}gjy\bar{a} & : p_0 \text{ (in } yojanas\text{)?} \\ d & : R\,(=3438) & :: \; p_0 \text{ (in } yojanas\text{)} & : p_0 \text{ (in minutes) ?} \\ D_{ms} & : p_0 \text{ (in minutes)} & :: \; D_{ts} & : p \text{ (in minutes) ?} \end{array}$$

---

[17] This corresponds to the numeral 51570. The reading in both the printed editions is खस्वराद्र्येकेषुमितो, corresponding to 51770, which seems to be wrong. Refer also to footnote 13 above.

[18] {TS 1977}, pp. 191–2.

[19] The term *dṛgjyā* or *iṣṭadṛgjyā* refers to the Rsine of the apparent zenith distance ($R\sin z'$).

Combining these three we have

$$p = \left(R_e \times \frac{R\sin z'}{R}\right)\left(\frac{R}{d}\right)\left(\frac{D_{ts}}{D_{ms}}\right)$$
$$= D_{ts}\left[\frac{R_e}{d \times D_{ms}}\right] R\sin z'. \tag{3.30}$$

It is given that

$$\frac{d \times D_{ms}}{R_e} = 51570.^{20} \tag{3.31}$$

Hence the true parallax is given by

$$p = \frac{D_{ts}}{51570} \times R\sin z', \tag{3.32}$$

which is the same as (3.29).

Rationale behind the correction for the parallax

घनभूमध्यपार्श्वस्थं सर्वत्र क्षितिजं भवेत् ॥
उन्नतज्या ततः शङ्कुः तद्भुजा महती प्रभा ।
भूपृष्ठगामिनः शङ्कोः छाया स्याल्लम्बिता ततः ॥
भूव्यासार्धविहीनोऽसौ शङ्कुर्भूपृष्ठगोऽन्यतः[21] ।
छाया च वर्धते त्रिज्याकर्णे कोटिभुजत्वतः ॥
तल्लम्बनवशात् तस्याः यदाधिक्यं भवेदिह ।
दृक्सिद्धायास्त्यजेत् तत्तुच्छाया येन भगोलगा ॥
तत्त्रिज्यावर्गविश्लेषमूलं शङ्कुश्च कोटिका ।[22]

Everywhere [in all the computations], the plane which lies to either side of the centre of the solid Earth[23] is taken to be the horizon. The cosine of the zenith distance is the *śaṅku*. The Rsine of it is the large shadow (*mahatī prabhā*).

The shadow of the *śaṅku* located on the surface of the Earth will be an elongated one. Hence [the length of] this *śaṅku* which is reduced by the measure of the radius of the Earth [becomes] the same as the one on the Earth's surface. Otherwise, [were this correction not to be done, then] there would be an increase in the shadow [as calculated]. The hypotenuse taken to be the *trijyā* is obtained from the *koṭi* and *bhujā*. The increase in it (the shadow) that occurs here due to the *lambana* has to be subtracted from the observed value of the shadow. Thus the shadow corresponding to the *bhagola* [the celestial sphere with the centre of the Earth as its centre] is obtained. The square root of the square of it (shadow) subtracted from the square of the *trijyā* is the *śaṅku*, and it is the cosine of the latitude.

---

[20] This is because the horizontal parallax in minutes is $\frac{R_e}{d} \times R = \frac{1}{15} \times D_{ms}$, where $R = 3438$.

[21] The reading in both the printed editions is: शङ्कुर्भूपृष्ठतोऽन्यतः ।

[22] {TS 1977}, p. 192.

[23] That is, the one passing through the centre of the Earth and parallel to the observer's horizon, which is the tangential plane at the location of the observer.

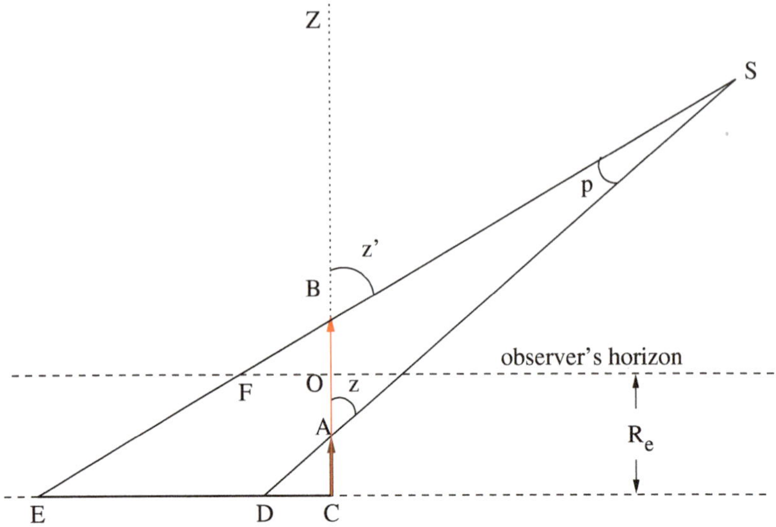

Fig. 3.6 The shadow of the two *śaṅkus*, one imagined to be from the centre to the surface of the Earth and the other at its centre.

We explain the content of the above verses with the help of Fig. 3.6. Here, O refers to the observer. OB is the actual *śaṅku* and OF will be the observed shadow. z' is the observed latitude and will be equal to ϕ, on the equinoctial day. CB and CA represent two hypothetical *śaṅkus* located at the centre of the Earth. Since all the measurements are made with respect to the centre of the Earth C as the standard reference point, the observed value of the latitude of the place must also be reduced to this reference point. The shadows cast by the two hypothetical *śaṅkus* are CE and CD respectively. The angle $C\hat{A}D$ on an equinoctial day gives the measure of the true latitude of the place. This angle is obtained by subtracting the parallax from the observed zenith distance. Thus the exact latitude of the place is $\phi = z = z' - p$.

Correction due to semi-diameter

बिम्बोर्ध्वनेम्याः प्रसृताः रश्मयः क्रशयन्ति[24] भाम् ॥
बिम्बव्यासार्धनिष्पन्नशङ्कुखण्डेन भास्वतः ।
वर्धयन्ति च दोःखण्डाः शङ्कुं भूपृष्ठवर्तिनम् ॥
बिम्बस्य घनमध्यान्तो भवेच्छङ्कुर्यतोऽपरः ।
फलयोरन्तरं शङ्कुच्छाययोस्तदृणं धनम् ॥
प्रत्यक्षसिद्धयोर्बिम्बघनमध्यगतौ यतः ॥[25]

[24] The reading in both the printed editions is: क्रशयन्ति। We feel that as this term does not convey any meaning; the correct reading should perhaps be कृशयन्ति।
[25] {TS 1977}, p. 192–3.

The rays that emerge from the upper part of the circumference [of the Sun] reduce the shadow (make it shorter). The rays emerging from the centre of the Sun increase the length of the *śaṅku* on the surface of the Earth by a measure determined by the semi-diameter of the Sun. This is because the *śaṅku* obtained by the rays emerging from the centre is different [from the actual *śaṅku*].

The difference in the *śaṅku*/shadow should be subtracted/added to obtain the correct value [of the *śaṅku*/shadow] since only those values corresponding to the centre of the disc (*bimbaghana-madhya*) [are to be considered].

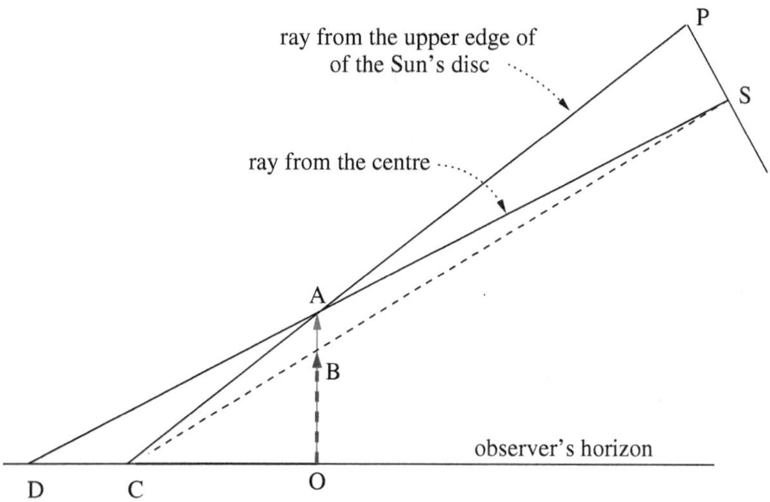

Fig. 3.7 The two *śaṅkus*, *OA* (actual) and *OB* (imaginary), drawn to explain the effect of an extended source of light on the shadow generated by the *śaṅku*.

If the Sun were a point object, then the length of the *śaṅku* would be *OB*, corresponding to the observed length of the shadow, *OC* (see Fig. 3.7). Since it is not so, the effect of the finite size of the Sun can be viewed as if the rays emerging from the centre of the Sun have increased the length of the *śaṅku* by a measure *AB*, which is determined by the semi-diameter of the Sun.

The value of the solar parallax given in *Tantrasaṅgraha* is far too large compared with the actual value. This is because Nīlakaṇṭha too has adopted the traditional viewpoint that the maximum value of parallax is one-fifteenth of the daily motion of the Sun. However, apart from this erroneous estimate, it is indeed remarkable that the problem has been correctly formulated and that the explanations given are quite sound and convincing. Also, the explanations provided by the above verses in *Yukti-dīpikā* are quite novel and give an idea of the methodology of the Kerala school of astronomers.

३.९ सममण्डलं उन्मण्डलं अग्रा च

3.9 The prime vertical, the celestial equator and the amplitude at rising

पूर्वापरायता रेखा प्रोच्यते सममण्डलम् ।
रेखा प्राच्यपरा साध्या विषुवद्भाग्रगा तथा ॥ १२ ॥
उन्मण्डलं च विषुवन्मण्डलं साभिधीयते ।
इष्टच्छायाग्रतद्रेखाविवरं त्वग्रसंज्ञितम् ॥ १३ ॥

pūrvāparāyatā rekhā procyate samamaṇḍalam |
rekhā prācyaparā sādhyā viṣuvadbhāgragā tathā ||12||
unmaṇḍalaṃ ca viṣuvanmaṇḍalaṃ sābhidhīyate |
iṣṭacchāyāgratadrekhāvivaraṃ tvagrasaṃjñitam ||13||

The line stretching from east to west is called the *samamaṇḍala*. Another line along the east–west direction has to be drawn, which will be same as the path traced by the tip of the shadow on the equinoctial day. This line is called the *unmaṇḍala* or *viṣuvanmaṇḍala*. The [perpendicular] distance of separation between the desired tip of the shadow [at any instant] and this line is called the *agrā*.

The term *samamaṇḍala* refers to the great circle passing through the zenith and the east and the west points on the horizon. In modern spherical astronomy it is referred to as the prime vertical. Here, *samamaṇḍala* refers to the east–west line. The terms *unmaṇḍala* and *viṣuvanmaṇḍala* are generally used to refer to the 6 o'clock circle and the celestial equator respectively. However, in the above verses, Nīlakaṇṭha has employed them synonymously to refer to the path traced by the tip of the shadow on an equinoctial day, which is a line parallel to the east–west line. We explain this with the help of Fig. 3.8.

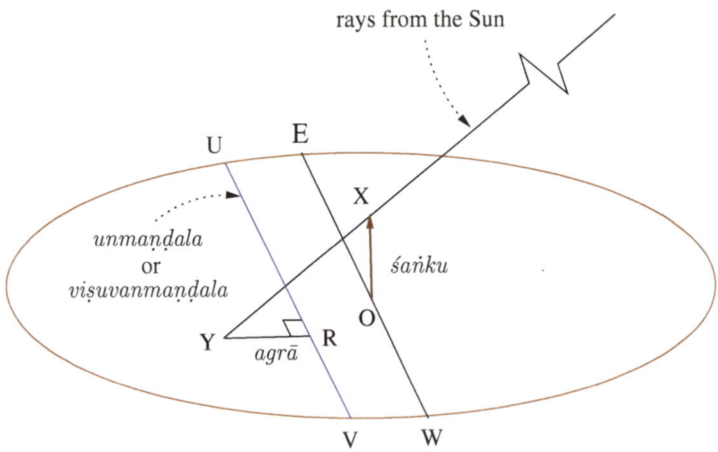

Fig. 3.8 Schematic representation of the *śaṅku*, the *unmaṇḍala* and the *agrā*.

In this figure, the line UV which runs parallel to the east–west line is referred as the *unmaṇḍala* or *viṣuvanmaṇḍala.* This line is the locus traced by the shadow of the tip of the *śaṅku* on the equinoctial day. The distance of separation—along the north–south direction—of the tip of the shadow from the *unmaṇḍala* is defined as the *agrā* YR in the figure. These are explained in *Yukti-dīpikā* in the following verses. It is also mentioned that the *agrā* and other quantities defined here will be used in the *chāyā-bhramaṇa-karma* (the process of movement of shadows), which will be dealt with later.

द्रष्टुः पूर्वापरारेखा प्रोच्यते सममण्डलम्।
तस्यास्तु विषुवद्भाग्रे रेखा पूर्वापरापरा ॥
उन्मण्डलं च विषुवन्मण्डलं साभिधीयते ।
इष्टच्छायाग्रविषुवद्रेखाभेदोऽग्रसंज्ञितः ।
एतत्त्रयोपयोगस्तु छायाभ्रमणकर्मणि ॥ [26]

The east–west line of the observer is stated to be the *samamaṇḍala.* Another east–west line, which is [drawn] at a distance of separation equal to the equinoctial shadow (*viṣuvadbhā*), is called the *unmaṇḍala* or the *viṣuvanmaṇḍala.* The distance of separation between the tip of the shadow and the *viṣuvanmaṇḍala* is known as the *agrā.* These three quantities will be used in the study of the motion of the [tip of] the shadow.

३.१० राशीनां लङ्कोदयप्राणाः स्वदेशोदयप्राणाश्च

3.10 The rising time of *rāśis* at Laṅkā and one's own place

राश्यन्तापक्रमैः कोटिः प्राणाः प्राग्वच्चरासवः ।
प्राणान् लङ्कोदयान् प्राहुः स्वोदयाश्चरसंस्कृताः ॥ १४ ॥
चरमाद्यन्त्ययोः शोध्यं पदयोर्योज्यमन्ययोः ।
एवं कृतास्तु विश्लिष्टा राशीनामुदयासवः ॥ १५ ॥
ओजयोस्तु क्रमेणैव युग्मयोरुत्क्रमेण च ।

rāśyantāpakramaiḥ koṭiḥ prāṇāḥ prāgvaccarāsavaḥ |
prāṇān laṅkodayān prāhuḥ svodayāścarasaṃskṛtāḥ ||14||
caramādyantyayoḥ śodhyaṃ padayoryojyamanyayoḥ |
evaṃ kṛtāstu viśliṣṭā rāśīnāmudayāsavaḥ ||15||
ojayostu krameṇaiva yugmayorutkrameṇa ca |

From the [Rsines of the] declinations (*apakrama*[27]) at the end of the *rāśis,* [their] Rcosines, the corresponding right ascensions (*prāṇas*) and the ascensional differences [are obtained] as earlier. These *prāṇas* [computed from the difference in the right ascensions] are said to be the durations of risings of the *rāśis* at Laṅkā. These [durations] corrected for the ascensional difference (*cara*) give the durations of risings at one's own place.

In the first and the last quadrants [i.e. for the first three and last three *rāśis*], the ascensional difference has to be subtracted. In the other quadrants, it has to be added. The durations thus obtained after corrections are the actual durations of risings of the *rāśis* [at one's own

---

[26] {TS 1977}, p. 193.

[27] The word *apakrama/krānti* generally refers to declination. It is also frequently used to refer to the Rsine of declination as in Chapter 2, verse 3.

place] and [they] are taken in order, in the odd quadrants and in the reverse order in the even quadrants.

The problem of finding the time required for the rising of different *raśis* for an observer on the equator amounts essentially to finding the right ascensions (R.A.s) corresponding to the end points of the first three *rāśis*. Then by making use of symmetry, the rising time of other *rāśis* can be easily determined. In Chapter 2, verses 23–27, the expression for right ascension (α) has already been given. An alternative expression of the same is given in the commentary *Laghu-vivṛti* as follows:

यदुक्तं प्राक्, 'विषुवद्भाहता क्रान्तिः' इत्यादि, ततः सायनस्फुटार्कं प्रथमद्वितीयतृतीय-राश्यन्तर्गतं परिकल्प्य तद्दोर्ज्यां चतुर्विंशतिभागज्यया निहत्य त्रिज्यया विभज्य लब्धं इष्टापक्रमं पृथग्वर्गीकृत्य त्रिज्यावर्गतो विशोध्य तत्कोटिमपि चानयेत्। अथ दोर्ज्येष्टापक्रमयोः वर्गान्तरमूलं इष्टकोटिज्या च स्यात्। ततः तां कोटिज्यां त्रिज्यया निहत्य इष्टद्युज्यया विभजेत्। तत्र लब्धस्य यच्चापं तदेव मेषादीनां राशीनां क्रमेण लङ्कोदयप्राणाः स्युः।

Obtaining the *sāyaṇa* longitudes of the Sun corresponding to the end points of the first, second and the third *rāśis*, as explained in the verses beginning with '*viṣuvadbhāhatā krāntiḥ*', multiply their Rsines by the *jyā* corresponding to 24° and divide by the *trijyā* to obtain the *iṣṭāpakrama* ($R\sin\delta$). The *koṭi* of this [known as the *iṣṭadyujyā* $= R\cos\delta$] should be obtained by subtracting its square from the square of the *trijyā* [and taking the square root of the difference]. The *iṣṭakoṭijyā* is the square root of the difference between the squares of *dorjyā* ($R\sin\lambda$) and *iṣṭāpakrama*. This *iṣṭakoṭijyā* should be multiplied by the *trijyā* and divided by the *iṣṭadyujyā*. The arc corresponding to the result obtained will be the rising time of the *Meṣādi rāśis* (Aries and other signs), in order, for an observer at Laṅkā.

This essentially gives the right ascension α (R.A.) as:

$$R\sin\alpha = \frac{\sqrt{R^2\sin^2\lambda - R^2\sin^2\delta}}{R\cos\delta}\,R. \tag{3.33}$$

Derivation of the alternative expression for right ascension

In Fig. 3.9, S represents the Sun, O the centre of the celestial sphere and P the celestial north pole. ASB refers to the ecliptic and $AN'C$ the equator. Let λ (the arc AS) be the longitude of the Sun and δ (SN') its declination. PSN' is the vertical passing through the Sun. The angle $S\hat{A}N' = \varepsilon$ is the obliquity of the ecliptic, and $S\hat{O}N' = \delta$ the declination of the Sun. The arc AN' is the right ascension, α. Draw SN perpendicular to ON'. Let NM and $N'M'$ be perpendiculars to OA. Then, SM is also perpendicular to OA. Let R be the radius of the celestial sphere. Now, considering the triangle SON, we have

$$SN = OS\,\sin\delta = R\sin\delta,$$
$$\text{and} \qquad ON = OS\cos\delta = R\cos\delta. \tag{3.34}$$

From the triangle SOM,

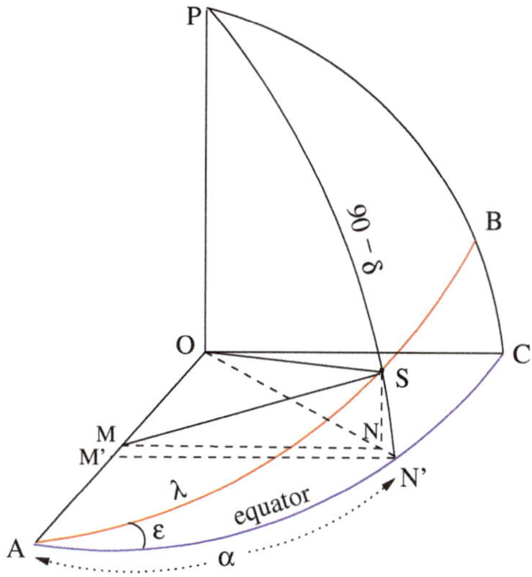

Fig. 3.9 Obtaining the right ascension α in terms of the declination δ and the longitude λ.

$$SM = OS \sin\lambda = R\sin\lambda. \tag{3.35}$$

In the triangle SMN,

$$MN = \sqrt{SM^2 - SN^2}. \tag{3.36a}$$

Using (3.34) and (3.35) in the above,

$$MN = R\sqrt{\sin^2\lambda - \sin^2\delta}. \tag{3.36b}$$

We are actually interested in arriving at an expression for the right ascension of the Sun which is the arc corresponding to $M'N'$, which is AN'. Since the triangles $OM'N'$ and OMN are similar, we have

$$\frac{M'N'}{ON'} = \frac{MN}{ON}$$

$$\text{or} \quad M'N' = MN \times \frac{ON'}{ON}$$

$$= MN \times \frac{R}{R\cos\delta}. \tag{3.37a}$$

In fact, $M'N' = R\sin AN'$, where AN' is the right ascension (α) of the Sun. Using this and (3.37*a*), the above relation reduces to

$$R\sin\alpha = R\sqrt{\sin^2\lambda - \sin^2\delta} \times \frac{R}{R\cos\delta}. \tag{3.37b}$$

Thus we have arrived at an expression for right ascension of the Sun in terms of its longitude and declination as given above in *Laghu-vivṛti*. Usually, the longitude of the Sun is found from the *ahargaṇa*. Once the longitude is known, its declination can be found using the relation

$$\sin\delta = \sin\varepsilon \sin\lambda, \tag{3.38}$$

where ε refers to the obliquity of the ecliptic. Substituting (3.38) in (3.37*b*) we have

$$\sin\alpha = \frac{\cos\varepsilon \sin\lambda}{\cos\delta}. \tag{3.39}$$

Duration of risings of *rāśis* at the equator

Let δ_1, δ_2 and δ_3 be the declinations of the Sun, when $\lambda = 30°$, $60°$ and $90°$ respectively. The corresponding right ascensions α_1, α_2 and α_3 can be obtained using (3.39). This expression, which gives the right ascension in angular measure, may be conveniently expressed in other measures too. The relation between the different measures is given by

$$90° \;=\; 6 \text{ hours} \;=\; 15\ \textit{ghaṭikās} \;=\; 5400\ \textit{prāṇas}. \tag{3.40}$$

Since the right ascensions are known at the end of each of the first three *rāśis*, the durations of rising for the first three *rāśis*, T_1, T_2 and T_3, for an observer on the equator, are obtained using the relations

$$T_1 = \alpha_1; \qquad T_2 = \alpha_2 - \alpha_1; \qquad T_3 = \alpha_3 - \alpha_2. \tag{3.41}$$

The symmetry of the problem clearly suggests that it would be enough to compute the rising time of the first three *rāśis* to know the rising time of other *rāśis* too. Thus the rising times of other sets of three *rāśis* have necessarily to be equal to those of the first set in either the direct order or in reversed order. If T_4, T_5 and T_6 represent the rising times of the second set of three *rāśis*, then they are given by,

$$T_4 = T_3; \qquad T_5 = T_2; \qquad T_6 = T_1. \tag{3.42}$$

The relation between the rising times of the second set of 6 *rāśis* and those of the first set is given by

$$T_{6+i} = T_i; \qquad (i = 1, 2, \ldots, 6). \tag{3.43}$$

Duration of risings of *rāśis* at one's own place

The rising times of *rāśis* at one's own place (t_i), having non-zero latitude, differ from the rising times at the equator (T_i). The former can be obtained from the latter using the relation

$$t_i = T_i - \Delta\beta_i \qquad (i = 1, 2, 3), \tag{3.44}$$

where $\Delta\beta_i$ are the 'ascensional differences' (differences between *caraprāṇas*). This can be understood with the help of Fig. 3.10*a*. Here Z, K and P represent the zenith, the pole of the ecliptic and the north celestial pole respectively. S_1, S_2 and S_3 are the positions of the Sun on the ecliptic, when its longitude is equal to 30, 60 and 90 degrees (the ends of the first three *raśis*). S_1', S_2' and S_3' are the points of intersection of the horizon and the diurnal circles of the Sun.

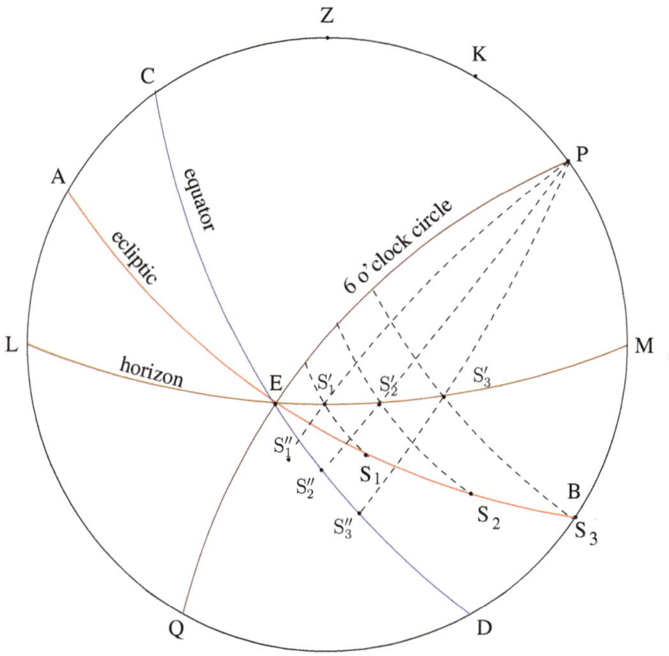

Fig. 3.10*a* Duration of rising of the first three *rāśis* for observers with non-zero latitude.

When the Sun is at the end points of the first, second and the third *rāśis* respectively, the great circles passing through PS_i' meet the celestial equator at S_i''. Then $\Delta\alpha_1 = ES_1''$, $\Delta\alpha_2 = ES_2''$ and $\Delta\alpha_3 = ES_3''$ are the *caraprāṇas* corresponding to the first three *rāśis*. $\Delta\beta_i$ are their differences.

$$\Delta\beta_1 = \Delta\alpha_1 \tag{3.45}$$
$$\Delta\beta_2 = \Delta\alpha_2 - \Delta\alpha_1 \tag{3.46}$$
$$\Delta\beta_3 = \Delta\alpha_3 - \Delta\alpha_2. \tag{3.47}$$

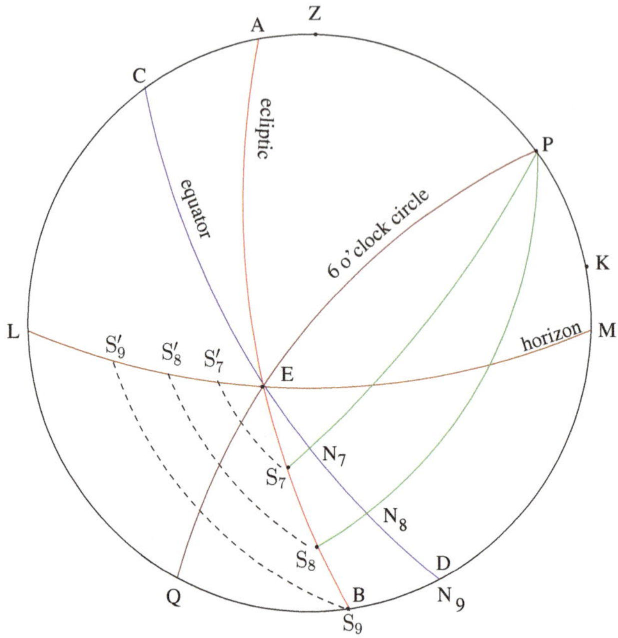

Fig. 3.10*b* Duration of rising of the three *rāśis* in the second set.

Rationale behind the difference in the duration of risings of *rāśis* at the equator and other locations

Yukti-dīpikā presents a detailed discussion as to why the duration of the risings of the *rāśis* for observers situated on the equator and elsewhere differ. We reproduce excerpts from it along with a free translation.

Definition of *tiryagvṛtta*

याम्यायनान्त[28]क्षितिजसम्पाताक्रान्तनेमिकम्।
घटीवृत्ततिरश्चीनं तद्ध्रुवद्वयसङ्गतम् ॥
तिर्यग्वृत्तं तु तत् तस्य क्षितिजेन तु संयुतौ। [29]
सौम्यायनाद्यभाद्यंशः युगपत् तद्द्वयं स्पृशेत्॥ [30]

The circle which runs perpendicular to the equator (*ghaṭīvṛtta*) and passes through its poles, and which also passes through the point of intersection of the midpoint of *yāmyāyana* and the horizon, is the *tiryagvṛtta* (this is actually the 6 o'clock circle). The

[28] अन्तशब्दः अत्र मध्यवाचीति भाति।(The word '*anta*' here seems to have been employed in the sense of '*madhya*'.)

[29] विद्यमानमिति योजनीयम् ।

[30] {TS 1977}, p. 194.

beginning point of the first *rāśi* in the *saumyāyana* would also coincide with the point of intersection of that (6 o'clock circle) and the horizon at the same instant.

Laṅkodayāsava and *svodayāsava*

प्रवहप्रेरणात् राशेः नित्यः[31] प्रत्यङ्मुखं यतः ।
राशेरन्त्योंशको नैव युगपत् तद्द्वयं स्पृशेत् ॥
तिर्यग्वृत्तस्य संस्पर्शात् प्राग्यतः क्षितिजं स्पृशेत् ।
क्षितिजस्य नतत्वेन तिर्यग्वृत्तादुदङ्मुखम् ॥
तिर्यग्वृत्ते तु भाद्यन्ते स्पर्शकालोत्थमन्तरम् ।
लङ्कोदयासवस्ते स्युः क्षितिजे स्वोदयासवः ॥
तिर्यग्वृत्तस्थसंस्पर्शात् प्रागेव क्षितिजस्पृशः ।
भान्तस्य तद्द्वयस्पर्शान्तरं स्वचरखण्डतः ॥
अतोऽयनाद्यराशेस्तु येऽमी लङ्कोदयासवः ।
हीनाः स्वचरखण्डेन ते स्वदेशोदयासवः ॥[32]

Since the *rāśis* have continuous westward motion due to the *pravaha* wind, the end point of the [first] *rāśi* will not intersect the two circles [namely, the horizon and the 6 o'clock circle][33] at the same time.

[Further,] since the horizon is [more] inclined towards the north than the *tiryagvṛtta*, [the end point of the *rāśi*] will intersect the horizon before it intersects the *tiryagvṛtta*. The time interval between [two successive] end points [of the *rāśis*] intersecting the 6 o'clock circle is the duration of rising at Laṅkā, called the *Laṅkodayāsava*, and the time interval corresponding to the intersections of the end points [of the *rāśis*] with the horizon is the duration of rising at one's own place, called the *Svodayāsava*.

The difference between the two time intervals is due to the *cara-khaṇḍa* (ascensional difference). Hence the *carakhaṇḍa* of the first *rāśi* subtracted from its duration of rising at *Laṅka* gives its duration of rising at the observer's location.

Application of *cara*

सौम्ययाम्यायनगतराशीनां व्यत्ययात् स्थितेः ।
चरखण्डैः स्वकैः शुद्धिक्षेपौ कार्यौ विपर्ययात् ॥
तिर्यग्वृत्तक्षितिजयोः भान्तस्पर्शान्तरं चरम् ।
यदा तज्ज्या द्युवृत्तस्था घटीवृत्तमिता भवेत् ॥
द्युज्या तदा चरज्या तु द्युवृत्तप्रमिता भवेत् ।[34]

[31] नित्यम् इति पाठेन भवितव्यम् इति भाति । भ्रमणम् इति तु अध्याहार्यम् ।

[32] {TS 1977}, p. 194.

[33] For an equatorial observer, the horizon and the 6 o'clock circle coincide with each other as the north celestial pole P coincides with the north point on the horizon. However, for a non-equatorial observer, P lies above the horizon and the two circles are inclined at an angle equal to the latitude of the observer ϕ.

[34] {TS 1977}, pp. 194–5.

Since the *rāśis* in the northern and the southern hemispheres are positioned differently (viz., have declinations with opposite signs) the *carakhaṇḍas* have to be applied inversely, that is, they have to be subtracted in the north and added in the south.

The time interval between the end point [of the *rāśi*] intersecting the horizon and the 6 o'clock circle is the ascensional difference, the *cara*. The Rsine of it, measured in terms of the dimensions of the diurnal circle, when converted in terms of the dimensions of the equator [by dividing it by Rcosine of the declination] becomes the *carajyā*.

Rationale behind the application of the *cara*

सौम्यायने तु सौम्याग्रं राशीनां क्षितिजे स्वके ॥
निरक्षोदयतः प्रागेवोदेति स्वचरासुभिः ।
उन्मण्डलक्षितिजयोः तत्र तावन्मितेऽन्तरे ॥
प्रवहप्रेरणात् प्राच्यां उद्गच्छत्क्रान्तिमण्डलम् ।
आधस्त्यवृत्तसम्पातं प्रागेव कुरुते ततः ॥
पश्चादेवोर्ध्ववृत्तेन चरप्राणा यदन्तरम् ।
याम्यायने तु याम्याग्रं राशीनां स्वैश्चरासुभिः ॥
निरक्षोदयतः पश्चात् उदेति क्षितिजे स्वके ।
अतो लङ्कोदयासुभ्यः त्यज्यन्ते स्वचरासवः ॥
सौम्येऽयनेऽन्यथा याम्ये स्वोदयप्राणसिद्धये ।
तत आद्यन्तपदयोः निरक्षोदयमानतः ॥
शोधनं स्वचरासूनां पदयोर्मध्ययोर्युतिः ।[35]

In the *saumyāyana* [when the Sun has a northerly motion], the northern tips of the *rāśis* rise earlier at the observer's horizon than at the horizon for an equatorial observer. This difference in the time of rising measured between the observer's horizon and the 6 o'clock circle is given by the ascensional difference.

By the impulse provided by the *pravaha* wind, the ecliptic which is set into motion, first comes into contact with the circle which is below [that is, the horizon]. Thereafter it comes into contact with the circle which is above [that is, the equatorial horizon], and the time interval between them is the ascensional difference.

In the *yāmyāyana* [when the Sun has southerly motion], the southern tips of the *rāśis* rise later at the observer's horizon than at the horizon for an equatorial observer, with the difference in times being equal to the ascensional difference.

Hence, for obtaining the duration of rising [of the *rāśis*] at one's own place, the ascensional difference is subtracted from the duration of rising at Laṅkā for *rāśis* in the *saumyāyana* and otherwise (is added) for those in the *yāmyāyana*. Thus it is seen that, [for the *rāśis*] in the first and the last quadrants, the ascensional difference has to be subtracted from the duration of rising at the equator, whereas in the second and the third it has to be added.

Let α_i be the right ascension at the end of the i-th *rāśi*. Then the time of rising of the i-th *rāśi* at the equator would be

$$T_i = (\alpha_i - \alpha_{i-1}) \qquad (i = 1, 2, \ldots, 12). \tag{3.48}$$

---

[35] {TS 1977}, p. 196.

The end point of the i-th *rāśi* intersects the local horizon $\Delta\alpha_i$ *prāṇas* earlier than the instant at which it intersects the 6 o'clock circle (the equatorial horizon), where

$$\Delta\alpha_i = \sin^{-1}(\tan\phi\tan\delta_i) \qquad (i = 1, 2, \ldots, 12), \tag{3.49}$$

δ_i being the declination corresponding to the end point of the ith *rāśi*. Since $\Delta\alpha_i$ is positive for $i = 1, \ldots, 6$, and negative when $i = 7, \ldots, 12$, it may be noted that the sign is incorporated in the above expression for $\Delta\alpha_i$.

Thus, if t_i is the duration of the rising of the i-th *rāśi* at the desired location, then it is given by

$$t_i = T_i - \Delta\beta_i, \tag{3.50}$$

where $\Delta\beta_i$ are given by (3.45) to (3.47)

Taking the sign also into account, δ_i keeps increasing in the first and the fourth quadrants, whereas it keeps decreasing in the second and third quadrants. Hence $\Delta\beta_i$ is positive when $i = 1, 2, 3, 10, 11, 12$, and negative when $i = 4, 5, \ldots, 9$. Hence the ascensional difference [magnitude of $\Delta\beta_i$] has to be subtracted from the duration of rising at the equator for the first and fourth quadrants, whereas it has to be added in the second and third.

३.११ इष्टशङ्कुः छाया च

3.11 Big gnomon and the gnomonic shadow at a desired time

प्राक्कपाले गतान् प्राणान् गम्यान् मध्यन्दिनात् परम् ॥ १६ ॥
विन्यस्यार्कचरप्राणाः शोध्या भानावुदग्गते।
योज्या दक्षिणगे तेभ्यो जीवा ग्राह्या यथोदितम् ॥ १७ ॥
व्यस्तं कृत्वा चरज्यां च द्युज्याघ्नां त्रिज्यया हरेत्।
लम्बकघ्नात् फलात् त्रिज्याहृतः शङ्कुर्विवस्वतः ॥ १८ ॥
तत्त्रिज्याकृतिविश्लेषात् मूलं छाया महत्यपि।

prākkapale gatān prāṇān gamyān madhyandināt param ||16||
vinyasyārkacaraprāṇāḥ śodhyā bhānāvudaggate|
yojyā dakṣiṇage tebhyo jīvā grāhyā yathoditam || 17 ||
vyastaṃ kṛtvā carajyāṃ ca dyujyāghnāṃ trijyayā haret|
lambakaghnāt phalāt trijyāhṛtaḥ śaṅkurvivasvataḥ || 18 ||
tattrijyākṛtiviśleṣāt mūlaṃ chāyā mahatyapi|

In the eastern part of the hemisphere [i.e. in the forenoon] the *prāṇas* that have elapsed [since sunrise], and in the afternoon [i.e. when the Sun has crossed the meridian] the *prāṇas* that are yet to elapse [till sunset], are found, and the results are stored [separately]. [From them] the ascensional differences (*caraprāṇā*) are subtracted when the Sun is to the north [of the ecliptic] and added when it is to the south [of the ecliptic]. The Rsine of the result has to be obtained as described [earlier].

To this the Rsine of the *cara* is applied in the reverse order and [the sum] is multiplied by the *dyujyā* and divided by the *trijyā*. The result multiplied by the *lambaka* and divided by the *trijyā* is the [*mahā*]*śaṅku* of the Sun. The square root of the difference between the squares of the *trijyā* and this (the *mahāśaṅku*) gives the *mahācchāyā*.

In the above verses, the procedure for the determination of the zenith distance of the Sun at a desired time is given. The terms *mahāśaṅku* and *mahācchāyā* used in the verses refer to the Rcosine and Rsine of the zenith distances of the Sun respectively. The prefix *mahā* is to distinguish them from the gnomon (*śaṅku*) and its shadow (*chāyā*). If z is the zenith distance of the Sun and R the *trijyā*, then

$$mah\bar{a}\acute{s}a\dot{n}ku = R\cos z \qquad \text{and} \qquad mah\bar{a}cch\bar{a}y\bar{a} = R\sin z.$$

These are also simply referred to at times as the *śaṅku* and the *chāyā*. We will see that the expression for the cosine of the zenith distance implied here is the same as the one obtained using modern spherical trigonometry, in terms of the latitude of the place (ϕ), the declination of the Sun (δ) and its hour angle (H).

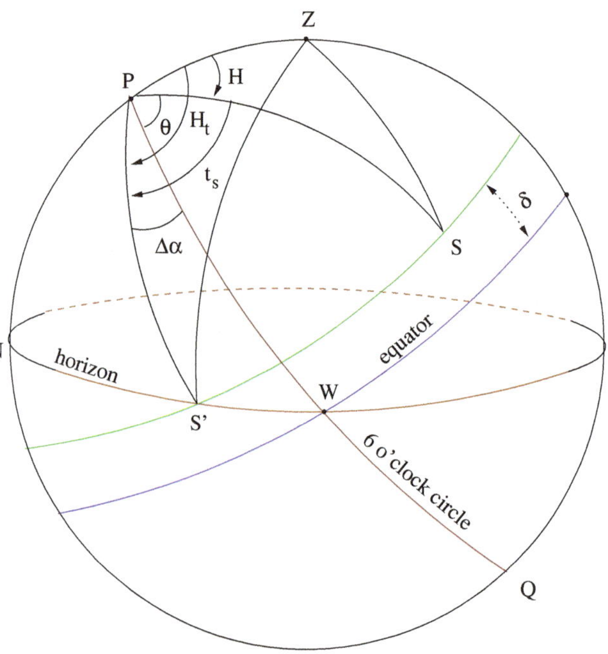

Fig. 3.11 The diurnal path of the Sun with a northern declination δ.

In Fig. 3.11, P and Z are the celestial north pole and the zenith of the observer. S and S' refer to the positions of the Sun at some instant in the afternoon and at the sunset. $\theta = S\hat{P}W$ is the angle covered by the Sun in reaching the 6 o'clock circle from S. Considering the spherical triangle PZS and applying the cosine formula

$$\cos z = \sin\phi \sin\delta + \cos\phi \cos\delta \cos H, \tag{3.51}$$

as $PZ = 90^\circ - \phi$ and $PS = 90^\circ - \delta$. It can be easily seen from the figure that $H = 90 - \theta$, since $Z\hat{P}W = 90^\circ$. Hence, the above equation becomes

$$\cos z = \sin\phi\sin\delta + \cos\phi\cos\delta\sin\theta. \tag{3.52}$$

In the following we show that the expression given by Nīlakaṇṭha is exactly the same as the above equation.

For this, we initially need to find out θ as per the procedure in the text. If d represents the half-duration of the day, $\Delta\alpha$ the ascensional difference, t_r the time elapsed since sunrise, and t_s the time that is yet to elapse till sunset, then

$$\theta = t_r - \Delta\alpha \qquad \text{(forenoon)} \tag{3.53}$$
$$\theta = t_s - \Delta\alpha \qquad \text{(afternoon)}. \tag{3.54}$$

$\Delta\alpha$ is positive when the Sun has a northern declination and negative otherwise. Hence the magnitude of $\Delta\alpha$ has to be subtracted from t_r or t_s for northern declinations, and added for southern declinations.

Having determined θ, it is stated that the zenith distance can be found using the relation

$$R\cos z = (R\sin\theta + R\sin\Delta\alpha)\left(\frac{R\cos\phi}{R}\right)\left(\frac{R\cos\delta}{R}\right), \tag{3.55}$$

which reduces to

$$\cos z = (\sin\theta + \sin\Delta\alpha)\cos\phi\cos\delta. \tag{3.56}$$

Considering the spherical triangle, PZS', where $PZ = 90^\circ - \phi$, $ZS' = 90^\circ$ and $Z\hat{P}S' = H_t$ (the hour angle of the Sun at the sunset), it follows from the cosine formula that

$$\cos H_t = -\tan\phi\tan\delta. \tag{3.57}$$

Using the fact that $H_t = 90 + \Delta\alpha$, the above equation reduces to

$$\sin\Delta\alpha = \tan\phi\tan\delta. \tag{3.58}$$

Substituting for $\sin\Delta\alpha$ in (3.56), we see that it reduces to (3.52), showing that the expression given by Nīlakaṇṭha for *mahāśaṅku* is nothing but the standard spherical astronomical result for the Rcosine of the zenith distance of the Sun at any time during the day.

The *mahācchāyā* is defined as the square root of the difference between the squares of the *trijyā* and the *mahāśaṅku*. Since the *mahāśaṅku* is $R\cos z$, it is obvious that the *mahācchāyā* is $R\sin z$. The value of the zenith distance obtained using (3.55) is for an observer situated at the centre of the Earth. Here it is also assumed that the Sun is a point object. To obtain the apparent value of the zenith distance (z') for an observer on the surface of the Earth from the theoretical value (z), corrections have to be implemented which take into account the effect of solar parallax and finite diameter of the Sun. It is precisely these corrections that are prescribed in the following verses.

३.१२ क्रमच्छायानयने संस्कारः

3.12 Correction to be implemented in the direct process of finding the shadow

छायायास्त्र्यगनागाप्तं[36] लिप्ताव्यासार्धतस्त्यजेत् ॥ १९ ॥
शिष्टेन शङ्कुमाहत्य त्रिज्याप्तं त्यज्यतामिह।
छायायाश्छायया ऽहत्य त्रिज्याप्तं शेषतोऽपि च[37] ॥ २० ॥
क्षिपेच्छङ्कौ सुसूक्ष्मोऽयं शङ्कुश्च महती प्रभा।
छायां द्वादशभिर्हत्वा शङ्कुभक्तेष्टशङ्कुभा ॥ २१ ॥

chāyāyāstryaganāgāptaṃ liptāvyāsārdhatastyajet ||19||
śiṣṭena śaṅkumāhatya trijyāptaṃ tyajyatāmiha|
chāyāyāśchāyayā'hatya trijyāptaṃ śeṣato'pi ca||20||
kṣipecchaṅkau susūkṣmo'yaṃ śaṅkuśca mahatī prabhā|
chāyāṃ dvādaśabhirhatvā śaṅkubhakteṣṭaśaṅkubhā||21||

The result obtained by dividing the shadow by 873 is to be subtracted from the semi-diameter [of the Sun] expressed in minutes. Multiplying the *śaṅku* by the remainder (say $\Delta\theta$) and dividing by the *trijyā*, the result should be subtracted from the [*mahā*]*chāyā*. The same ($\Delta\theta$) multiplied by [*mahā*]*chāyā* and divided by the *trijyā* should be applied positively to the [*mahā*]*śaṅku*. These are far more accurate [values] of [*mahā*]-*śaṅku* and *mahācchāya*. By multiplying the [*mahā*]*chāyā* by 12 and dividing by [*mahā*]*śaṅku*, the shadow of the *iṣṭaśaṅku* (the usual 12 *aṅgula*) is obtained.

What are referred to repeatedly in the above verses as the *chāyā* and the *śaṅku* are actually the *mahācchāyā* (great 'shadow') and the *mahāśaṅku* (great 'gnomon')—to be precisely defined a little later in this section. Finding the *prabhā*,[38] which in the present context means shadow $= R\sin z$, essentially means finding the zenith distance of the Sun. The effect due to the solar parallax and the finite diameter of the Sun cannot be neglected in an accurate measurement of the zenith distance. The procedure to obtain the theoretical value of the zenith distance (with reference to the centre of the Earth) was described in the previous section. A correction has to be applied to obtain the observed value from the theoretical value. If z' and z are the observed and the theoretical zenith distances of the Sun, then the relation between the two given in the text may be expressed as:

---

[36] The reading in the published text edited by K. V. Sarma is त्र्यङ्गनागाप्तम्। The number represented by this word is 863. Sarma has also given त्र्यगनागाप्तम् as another reading in the footnote, which corresponds to the number 873. The footnote reading seems to be the correct one as it agrees closely with the actual computed value of 872. This is confirmed from the commentary *Laghu-vivṛti*, where it is stated: एवमानीतायाः छायायाः त्र्यगनागैर्विभज्य लब्धम् फलं ··· ।

[37] The phrase *śeṣato'pi ca* can perhaps be integrated with the rest of the verse as follows: लिप्ताव्यासार्धतः त्यक्त्वा यत् फलं लब्धं तत् स्थलद्वये विन्यसेत्। एकत्र महाशङ्कुना निहत्य त्रिज्यया विभज्य लब्धम् फलं महाच्छायातो विशोधयेत्। शेषतोऽपि च, अर्थात् अन्यत्र विन्यस्तात् फलात्...

[38] Though the word *prabhā* actually means rays, it has been used in verse 21 as a synonym for *chāyā*, shadow. Occasionally, in Sanskrit literature, one finds '*kāraṇavācakaśabda*' (here, *prabhā*), being used in the place of *kāryavācakaśabda* (here *chāyā*).

$$R\cos z' = R\cos z + \Delta\theta\left(\frac{R\sin z}{R}\right), \tag{3.59}$$

$$R\sin z' = R\sin z - \Delta\theta\left(\frac{R\cos z}{R}\right). \tag{3.60}$$

The second terms in the RHSs of the above equations are the correction factors. The quantity $\Delta\theta$ occurring here is given by

$$\Delta\theta = d_s - p, \tag{3.61}$$

where d_s and p are the semi-diameter of the Sun's disc (expressed in minutes) and its parallax respectively. The parallax of the Sun is given as

$$p = \frac{R\sin z}{873}. \tag{3.62}$$

It is easily seen that (3.59) and (3.60) are nothing but the expressions for $\cos(z - \Delta\theta)$ and $\sin(z - \Delta\theta)$ when $\Delta\theta$ is small. Hence these equations imply that

$$z' = z - \Delta\theta = z - d_s + p. \tag{3.63}$$

The rationale behind the above expression has been discussed earlier (see section 3.8) in the context of explaining the procedure for obtaining more accurate values of the latitude. The same argument applies here. Owing to the finite size of the Sun, the apparent zenith distance gets reduced by the semi-diameter of the Sun. Also, there is an increase due to parallax. It has already been shown that

$$p = P\frac{R\sin z}{R}, \tag{3.64}$$

where P is the horizontal parallax. Further, it has been mentioned that

$$P = \frac{\text{Daily motion of Sun}}{15}. \tag{3.65}$$

The mean daily motion, according to *Tantrasaṅgraha*, is found[39] to be $59'8''$. This divided by 15 gives the horizontal parallax, which amounts to nearly $3'57''$. Substituting this value of P in (3.64), and taking $R = 3438$, we get

$$p = \frac{R\sin z}{872}. \tag{3.66}$$

[39] It is computed as follows. The number of civil days in a *Mahāyuga* is given by 1577917500. This when divided by the total number of sidereal years (= 4320000) gives the number of civil days per sidereal year, which turns out to be 365.25868056. Since the Sun covers an angle of 360° in this period, the mean angle covered per day is given by

$$\text{Mean daily motion} = \frac{360}{365.25868056} \approx 0.98560285947 \approx 59'8''.$$

It may be noted that the above expression for parallax differs from the expression given in the text (3.62) by only one unit in the denominator.

Some of the concepts related to the *mahāśaṅku* and the *mahācchāyā* are elaborated in *Yukti-dīpikā*. We reproduce a few important ones among them in the following:

Definition of *mahāśaṅku* and *mahācchāyā*

द्रष्टृमध्यं ग्रहस्पृष्टं यद्वृत्तं परिकल्प्यते।
तत्र ग्रहक्षितिजयोः अन्तरं शङ्कुरिष्यते।
तत्कोटिश्च महाच्छाया नभोमध्यग्रहान्तरम् ॥ [40]

Conceiving of a circle centred around the observer (*draṣṭṛ-madhyaṃ*) and passing through the planet (*grahaspṛṣṭaṃ*), [the Rsine of the angle of] the separation between the planet on that [circle] and the horizon is the [*mahā*]*śaṅku*. The Rcosine of it is the *mahācchāyā*, which is the [Rsine of the angle of] separation between the planet and the zenith.

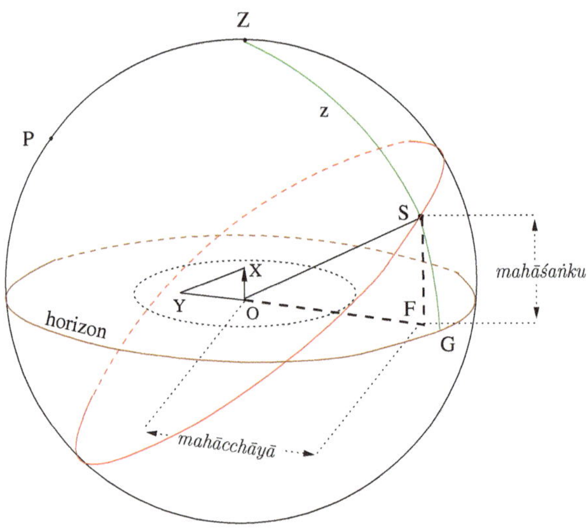

Fig. 3.12 Concept of the *mahāśaṅku* and the *mahācchāyā*.

We explain the concept of *mahāśaṅku* and *mahācchāyā* with the help of Fig. 3.12. Here, S is the Sun and F is the foot of perpendicular drawn from the Sun to the horizon. The angle $O\hat{X}Y = F\hat{S}O = z$ is the zenith distance of the Sun and $O\hat{Y}X = F\hat{O}S = a = 90 - z$ is the altitude of the Sun. The triangles OXY and FSO are similar. In the triangle OXY, OX represents the usual *śaṅku* of 12 units in height and OY is its shadow, referred to as the *chāyā*. Since OXY and FSO are similar, SF $(= R\cos z)$ and $FO (= R\sin z)$ are referred to as the *mahāśaṅku* and the *mahācchāyā* respectively.

[40] {TS 1977}, p. 198.

Correction due to the finite size of the object

स्वबिम्बघनमध्यान्तात् नतज्या स्यात् खमध्यतः ।
महाच्छाया तदासन्नबिम्बनेम्यवधिस्ततः ॥
बिम्बव्यासार्धधनुषः तत्खण्डज्यातदन्तरम् ।
कृत्स्नज्याकोटिसंवर्गात् खण्डज्या त्रिज्ययोद्धृता ॥
छायाशङ्कोश्च दोःकोटिरूपत्वं सम्मतं मिथः ।
अतोऽत्र बिम्बव्यासार्धात् छायाशङ्कुहतात् पृथक् ॥
शङ्कुतच्छाययोः स्वर्णं त्रिज्ययाप्तं फलद्वयम् ।
एवं दृग्विषये शङ्कुच्छाययोरिष्यते विधिः ॥ [41]

The [Rsine of the angle of] separation between the zenith and the centre of the object that is observed (the *svabimbaghanamadhya*) is the *nata-jyā* and is [the same as] the *mahācchāyā*. [However], the [apparent value of the] *mahācchāyā* would correspond to the edge of the disc [which is closer to the zenith]. Therefore, the difference between the two is equal to the *khaṇḍajyā* corresponding to the arc subtended by the semi-diameter of the object.

The *khaṇḍajyā* is multiplied by the sine and cosine (of the zenith distances) and divided by the *trijyā*. [The two results are stored separately]. The *chāyā* and *śaṅku* can be thought of as the sine and cosine of each other. Therefore, the product of the semi-diameter and the *chāyā* and that of the semi-diameter and the *śaṅku*, found separately and divided by the *trijyā*, are applied to the *śaṅku* and the *chāyā* positively and negatively respectively to get the desired result. This is the procedure to obtain the observed values (the *dṛgviṣaye*) of the *chāyā* and *śaṅku*.

If d_s be the semi-diameter of the Sun (expressed in minutes), then the procedure to obtain the observed value of the zenith distance (z') from the theoretical value (z) as described in the above verses is given by

$$R\cos z' = R\cos z + d_s\left(\frac{R\sin z}{R}\right), \tag{3.67}$$

$$R\sin z' = R\sin z - d_s\left(\frac{R\cos z}{R}\right). \tag{3.68}$$

३.१३ महाशङ्कोः गतगन्तव्यप्राणाः

3.13 Time elapsed or to be elapsed from the *mahāśaṅku*

शङ्कुच्छाये त्रिजीवाघ्ने महत्यौ कर्णसंहृते ।
लम्बकाक्षज्ययोः स्वर्णमन्योन्योत्थफलं यथा ॥ २२ ॥
तथा नृच्छाययोः कार्यं विपरीतप्रभाविधौ ।
व्यासार्धघ्नात् ततः शङ्कोः लम्बकाप्तं त्रिजीवया ॥ २३ ॥
हत्वा द्युज्याविभक्ते तत् चरज्या स्वर्णमेव च ।

[41] {TS 1977}, p. 199.

याम्योदग्गोलयोस्तस्य चापे व्यस्तं चरासवः ॥ २४ ॥
संस्कार्या गतगम्यास्ते पूर्वापरकपालयोः ।

śankucchāye trijīvāghne mahatyau karṇasaṃhṛte|
lambakākṣajyayoḥ svarṇamanyonyotthaphalaṃ yathā||22||
tathā nṛcchāyayoḥ kāryaṃ viparītaprabhāvidhau|
vyāsārdhaghnāt tataḥ śaṅkoḥ lambakāptaṃ trijīvayā||23||
hatvā dyujyāvibhakte tat carajyā svarṇameva ca|
yāmyodaggolayostasya cāpe vyastaṃ carāsavaḥ||24||
saṃskāryā gatagamyāste pūrvāparakapālayoḥ|

The *śaṅku* and *chāyā* multiplied by the *trijyā* and divided by the *karṇa* become the greater ones [the *mahāśaṅku* and *mahācchāyā* respectively]. As the addition and subtraction was done in the case of the *lambaka* and *akṣa*, with the quantities obtained from each other, so too the results have to be applied in finding the *śaṅku* (referred to by the word *nṛ* above) and its shadow in the reverse process (the *viparīta-prabhā*).

The *śaṅku* is multiplied by the radius (the *trijyā*) and divided by the *lambaka*. This is further multiplied by the *trijyā* and divided by the *dyujyā*. To this quantity, the *carajyā* is applied positively or negatively depending upon whether the Sun is in the southern or the northern hemisphere. To the arc of the result, the arc of the ascensional difference has to be applied in the reverse order. This gives the time that has elapsed or is yet to elapse in the eastern or in the western half of the hemisphere.

These verses give the procedure for finding the time that has elapsed since sunrise, or that is yet to elapse till sunset, from a knowledge of the zenith distance of the Sun and its ascensional difference. Essentially it is the reverse process of what is described in the earlier verses 17 and 18 and is termed the *viparīta-prabhā-vidhi*.[42] For that the *mahāśaṅku* and *mahācchāyā* are found first.

It was shown earlier (see (3.2) and (3.3)) that

$$\textit{śaṅku} = 12 = K\cos z \qquad (3.69a)$$

and

$$\textit{chāyā} = L = K\sin z, \qquad (3.69b)$$

where K is the *karṇa*. Hence,

$$\textit{mahāśaṅku} = R\cos z = \frac{\textit{saṅku}}{\textit{karṇa}}\,\textit{trijyā} \qquad (3.70a)$$

and

$$\textit{mahācchāyā} = R\sin z = \frac{\textit{chāyā}}{\textit{karṇa}}\,\textit{trijyā}. \qquad (3.70b)$$

The procedure for obtaining the observed value of the zenith distance from the theoretical value was described in the previous section. Here the reverse process is described. This reverse process of obtaining the theoretical value from the observed value is then carried out.

If z' and z are the observed and the theoretical zenith distances of the Sun, then the prescription given above may be expressed using the modern notation as

$$R\cos z = R\cos z' - \Delta\theta\left(\frac{R\sin z}{R}\right), \qquad (3.71)$$

---

[42] The term refers to the reverse process of determining the time from the observed shadow.

$$R\sin z = R\sin z' + \Delta\theta\left(\frac{R\cos z}{R}\right). \tag{3.72}$$

These relations are essentially the same as (3.67) and (3.68) except for the reversal of the signs of the correction terms. Both the pairs of relations amount to

$$z = z' + \Delta\theta = z' + d_s - p. \tag{3.73}$$

This correction is essentially the same as the one employed in the determination of latitude. In fact, this is remarked in *Laghu-vivṛti*:

विषुवच्छायया अक्षलम्बकानयनं इष्टच्छायया महाच्छायाशङ्कोरानयनं चैकरूपमेवेति।

The methods to find the *akṣa* ($R\sin\phi$) and *lambaka* ($R\cos\phi$) from the equinoctial shadow, and the *mahācchāyā* ($R\sin z$) and *mahā-śaṅku* ($R\cos z$) from the shadow at the desired (arbitrary) time are identical.

Now an intermediate quantity x is defined as

$$\begin{aligned} x &= \frac{\textit{śaṅku} \times \textit{trijyā}}{\textit{lambaka}} \times \frac{\textit{trijyā}}{\textit{dyujyā}}, \\ &= \frac{R\cos z \times R}{R\cos\phi} \times \frac{R}{R\cos\delta}. \end{aligned} \tag{3.74}$$

Then another quantity y $(= R\sin\theta)$ is defined as

$$y = x \pm \textit{carajyā}, \tag{3.75}$$

or
$$R\sin\theta = \frac{R\cos z}{\cos\phi\cos\delta} \pm |R\sin\Delta\alpha|, \tag{3.76}$$

where the sign '+' has to be used when the Sun has southern declination and '−' when it has northern declination.

We have already pointed out in Section 3.11 that the above expression for $R\sin\theta$ follows from the cosine formula in spherical trigonometry.

$$R\sin\theta = \frac{R\cos z}{\cos\phi\cos\delta} - R\sin\Delta\alpha, \tag{3.77}$$

where $\sin\Delta\alpha = \tan\phi\tan\delta$ is the *carajyā*. To θ, the arc of the ascensional difference, $\Delta\alpha$, has to be applied in the reverse order.[43] That is, negatively when the Sun has southern declination and positively when it has northern declination. This gives the time that has elapsed since sunrise (t_r) or that is yet to elapse till sunset (t_s) depending upon whether the Sun is in the eastern or the western part of the horizon. Thus we have

$$t_{r,s} = \theta + \Delta\alpha. \tag{3.78}$$

Substituting for θ from (3.77) we have,

[43] Here reverse refers to the reverse order of application of the sine of the ascensional difference as given in the previous equation.

$$t_{r,s} = (R\sin)^{-1}\left(\frac{R\cos z}{\cos\phi\cos\delta} - R\sin\Delta\alpha\right) + \Delta\alpha. \tag{3.79}$$

It can be seen from Fig. 3.11 that the time that is yet to elapse till sunset, when the Sun is in the northern hemisphere, is given by the sum of the arc lengths θ and $\Delta\alpha$. Using the expression for θ obtained from (3.52), we get

$$t_{r,s} = (\sin)^{-1}\left(\frac{\cos z}{\cos\phi\cos\delta} - \sin\Delta\alpha\right) + \Delta\alpha. \tag{3.80}$$

Thus we see that the expression given in the text for determining the time during any part of the day is exact.

३.१४ शङ्कुच्छायातः स्फुटसूर्यावगमनम्

3.14 Determination of the true Sun from the shadow of the gnomon

क्रान्त्यक्षचापयोगाच्च भेदाद्वा याम्यसौम्ययोः ॥ २५ ॥
जीवा मध्यन्दिनच्छाया ततो वार्कस्फुटं नयेत् ।
मध्यार्कनतभागेभ्यः स्वाक्षभागान् विशोधयेत् ॥ २६ ॥
शङ्कोरुदग्गता भा चेत् याम्यक्रान्तिर्हि शिष्यते ।
स्वाक्षभागान्नताश्चोना नतांस्तर्हि विशोधयेत् ॥ २७ ॥
उदक्क्रान्तिस्तदा शिष्टा नत्यक्षयुतिरन्यदा ।
तज्ज्या त्रिज्याहता भक्ता क्रान्त्या परमया रवेः ॥ २८ ॥
दोर्ज्या तच्चापमेव स्यात् सौम्ये गोलेऽयनेऽपि च।
रविस्तत्रायने भिन्ने राशिषट्कं तदूनितम् ॥ २९ ॥
याम्ये गोलेऽयने चापि राशिषट्कयुतं रविः।
तदूनं मण्डलं भानुः याम्यस्थे चोत्तरायणे ॥ ३० ॥

krāntyakṣacāpayogācca bhedādvā yāmyasaumyayoḥ||25||
jīvā madhyandinacchāyā tato vārkasphuṭaṃ nayet|
madhyārkanatabhāgebhyaḥ svākṣabhāgān viśodhayet||26||
śaṅkorudaggatā bhā cet yāmyakrāntirhi śiṣyate|
svākṣabhāgānnatāśconā natāṃstarhi viśodhayet||27||
udakkrāntistadā śiṣṭā natyakṣayutiranyadā|
tajjyā trijyāhatā bhaktā krāntyā paramayā raveḥ||28||
dorjyā taccāpameva syāt saumye gole'yane'pi ca|
ravistatrāyane bhinne raśiṣaṭkaṃ tadūnitam||29||
yāmye gole'yane cāpi raśiṣaṭkayutaṃ raviḥ|
tadūnaṃ maṇḍalaṃ bhānuḥ yāmyasthe cottarāyaṇe||30||

Depending upon whether the Sun is in the southern or the northern hemisphere, the sum of, or the difference between, the latitude [of the place] and the declination [of the Sun] is found. The Rsine of this angle is a measure of the length of the midday shadow. The position of the true Sun can be obtained from this also.

Subtract the latitude of the place from the zenith distance of the Sun at the noon. If the shadow [at noon] lies to the north of the *śaṅku*, then the remainder gives the southern declination of the Sun. If the zenith distance is smaller than the latitude of the place [and the shadow at the noon lies to the north of the *śaṅku*], then the zenith distance has to be subtracted from the latitude. The remainder gives the northern declination of the Sun. Or else the sum of the latitude and the zenith distance gives the northern declination.

The Rsine of it multiplied by the *trijyā* and divided by the Rsine of the maximum declination gives the Rsine of the longitude of the Sun. The arc of this gives the longitude of the Sun if both the *ayana* and the *gola* are north [Sun has northerly motion and also lies in the northern hemisphere]. If the *ayana* is different the arc subtracted from six *rāśis* [gives the longitude of the Sun]. When both the *gola* and the *ayana* are south, then six *rāśis* added to the arc gives the longitude of the Sun. If the Sun is in the southern hemisphere and the *ayana* is north, then the arc subtracted from a full circle (360 degrees) is [the longitude of] the Sun.

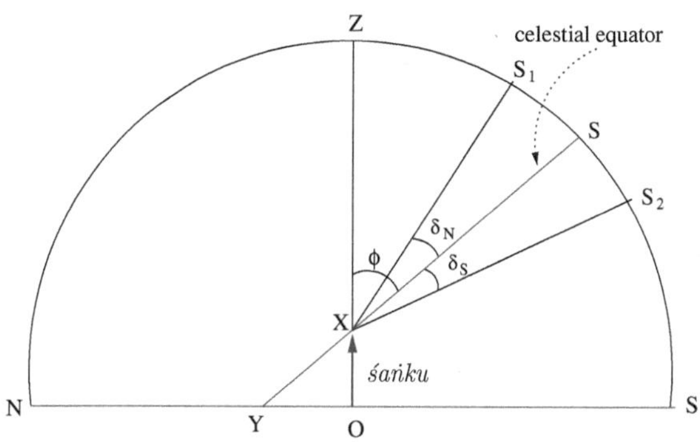

Fig. 3.13 The zenith distance of the Sun during meridian transit.

In the above verses it is explained how the true *sāyana* longitude of the Sun is obtained by a simple method which essentially involves measuring the midday shadow of the *śaṅku*. This method makes use of the relation

$$\sin\delta = \sin\varepsilon \sin\lambda, \tag{3.81}$$

where ε is the obliquity of the ecliptic, λ the *sāyana* (tropical) longitude of the Sun and δ its declination. ε is a fixed quantity and is taken to be 24°. Hence the true longitude can be obtained using the above relation if the declination of the Sun is determined.

For determining the declination of the Sun, we use the relationship between the latitude of the place, the midday zenith distance (z) of the Sun and its declination. This relationship may be explained with the help of Fig. 3.13. In this figure Z represents the zenith of the observer, OX the *śaṅku* and the line YXS the equator. S, S_1 and S_2 represent the positions of the Sun at the local noon as it passes across the prime meridian on different days. The angles subtended by the arc lengths SS_1 and SS_2 at X, δ_N and δ_S in the figure, represent the northern and southern declination respectively. If ϕ represents the latitude of the observer ZS, then it can be easily seen from the figure that the following relations are satisfied.

When the Sun is on the equator ($\delta = 0$),

$$z = \phi. \tag{3.82}$$

When the Sun is in the southern hemisphere (δ south),

$$z = \phi + \delta_s. \tag{3.83}$$

When the Sun is in the northern hemisphere (δ north),

$$z = \phi - \delta_N \qquad (\phi > \delta), \tag{3.84}$$

$$z = \delta_N - \phi \qquad (\phi < \delta). \tag{3.85}$$

When $\phi < \delta$, then the Sun would be to the north of the zenith at the meridian transit, which is not shown in the figure above.

३.१५ अयनचलनम्

3.15 Motion of equinoxes

करणागतसूर्यस्य छायानीतस्य चान्तरम् ।
आयनं चलनं ज्ञेयं तात्कालिकमिदं स्फुटम् ॥ ३१ ॥
छायार्कादधिकेऽन्यस्मिन् शोध्यं योज्यं विपर्यये ।
उदग्विषुवदादित्वसिद्धये करणागते ॥ ३२ ॥
मेषादिके ग्रहे कार्यं अंशादिकमिदं खलु ।
वृद्धिः क्षयश्च दिव्याब्दैः पञ्चभिः स्यात् धनर्णयोः ॥ ३३ ॥
दशांशोनाब्दतुल्या स्यात् गतिस्तस्य कलात्मिका ।
सप्तविंशतिभागान्तं चलनं चापनक्रयोः ॥ ३४ ॥
सिद्धान्तेषूदितं तस्य छाययापि विनिर्णयः ।

karaṇāgatasūryasya chāyānītasya cāntaram |
āyanaṃ calanaṃ jñeyaṃ tātkālikamidaṃ sphuṭam || 31 ||
chāyārkādadhike'nyasmin śodhyaṃ yojyaṃ viparyaye |
udagviṣuvadāditvasiddhaye karaṇāgate || 32 ||
meṣādike grahe kāryaṃ aṃśādikamidaṃ khalu |
vṛddhiḥ kṣayaśca divyābdaiḥ pañcabhiḥ syāt dhanarṇayoḥ||33||
daśāṃśonābdatulyā syāt gatistasya kalātmikā |
saptaviṃśatibhāgāntaṃ calanaṃ cāpanakrayoḥ || 34 ||
siddhānteṣūditaṃ tasya chāyayāpi vinirṇayaḥ |

The difference between true longitudes of the Sun determined by the procedures given in the text and the one determined from the shadow [as described in the previous verse] is equal to the actual motion of the *ayana* at that instant of time.

If the other longitude [determined through the textual procedure] is greater than the longitude determined from the shadow, the difference has to be subtracted, otherwise added.

When the planet is in *Meṣādi*, it is this (*ayanāṃśa*) which needs to be applied in minutes etc. The increase and decrease will be there in five divine years (*divyābdas*) in both the positive and negative directions. The motion of it will be one-tenth reduced from the number of years[44] in *kalās*. It has been mentioned in *siddhāntas* that the motion of the *Dhanus* and *Makara* (Sagittarius and Capricorn) is up to 27°. This can be verified using the shadow techniques.

In Fig. 3.14, the celestial equator, the ecliptic and their poles (P and K respectively) are shown. Here V is the vernal equinox and A is the autumnal equinox. M is the *Meṣādi*, which is the beginning point of the *rāśi* division. This point is fixed with respect to the stars. The equinoxes V and A are in motion with respect to the stars, and hence with reference to M also.

According to *Tantrasaṅgraha*, this motion can be westward or eastward. This is known as the 'trepidation of the equinoxes', where the equinoxes execute an oscillatory motion with respect to *Meṣādi* (the first point of Aries) M. If S represents the Sun, the longitude measured with M as the reference point is the '*nirayaṇa* longitude', $l_t = MS$. This is what is calculated by following the procedure given in the texts. The longitude with V as the reference point is the '*sāyana* (tropical) longitude,' $l_s = VS$. In the figure below, we have shown V to the west of M, and hence $l_s = l_t + \Delta$, where $\Delta = VM$ is the motion of the equinoxes. According to the text, it is possible that V would be east of M at some time. Then $l_s = l_t - \Delta$, where Δ is the eastward motion of the equinoxes. It is stated that the motion of the equinoxes is $54''$ per year. It will move westwards to the maximum extent of 27°. This will take place in $27 \times \frac{3600}{54} = 1800$ years, which is 5 divine years as a divine year is made up of 360 solar years. Then it will move eastwards by 27° in 1800 years. Hence the period of oscillations or trepidation of the equinoxes is 3600 years.

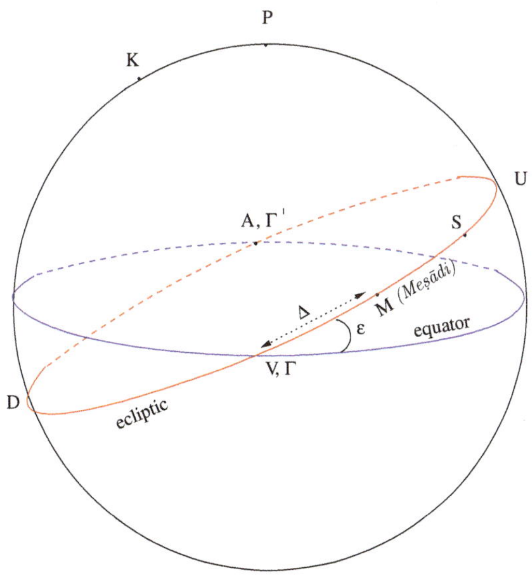

Fig. 3.14 The equinoctial and the solsticial points (*ayanāntas*).

[44] The number of years is taken to be 60 (a 60-year cycle). Hence the motion of the *ayana* in this period will be 54 minutes (*kalās*).

According to modern astronomy, V has a continuous westward motion (or 'retrograde motion') with respect to M at the rate of nearly $50.2''$ per year. This is referred to as the 'precession of the equinoxes'. Just by observations over even thousands of years it would not be possible to conclude whether the equinoxes precess westwards continuously or oscillate. Hence it is not surprising that 'trepidation of the equinoxes' is advocated here. The verses in *Yukti-dīpikā* that clearly define the *ayana*, its motion and how it can be inferred by observations are presented below.

Definition of *ayana*

अयनं नाम घटिकाक्रान्तिवृत्तान्तरं परम् ।

The [point of] maximum [angle of] separation between the equator and the ecliptic is the *ayana*.

In the above definition, the term *ayana* refers to the two solstices on the ecliptic. These are the points where the Sun has the maximum north and south declinations during the course of a year. In Fig. 3.14 they are marked as U and D. The points V and A represent the vernal and the autumnal equinoxes. In Indian astronomy, the two halves of the ecliptic, namely the paths DVU and UAD as indicated in the figure, are called the *saumyāyana* and the *yāmyāyana* respectively. The terms *saumya* and *yāmya* mean the north and the south directions. Since the Sun moves towards the north pole P as it moves along DVU, and towards the south pole Q as it moves along UAD, these two paths are called the *saumyāyana* and the *yāmyāyana*.

Thus the point U refers to the end of the *saumyāyana* and the beginning of the *yāmyāyana*. Similarly the point D refers to the end of *yāmyāyana* and the beginning of the *saumyāyana*. These are the points on the ecliptic, where the Sun changes its direction of motion.

Oscillatory motion of *ayana*

यस्मिन् प्रदेशे घटिकाक्रान्तिमण्डलयोर्यदा ॥
विप्रकर्षः परो दृष्टः स तु कालान्तरे ततः ।
पुरस्ताच्च परस्ताच्च चलति प्राक्प्रदेशतः ॥
चापान्ते मिथुनान्ते च विप्रकर्षः परोऽत्र यः ।
स चिरेण तयोः सप्तविंशत्यंशान्तरे ततः ॥
प्रदेशान्तरसम्बन्धः न युक्तः चलनं विना ।
अतोऽयनस्य चलनं वृत्तयोरनुमीयते ॥ [45]

The point of maximum separation [between the equator and the ecliptic], observed along a particular direction, shifts its positions either forward or backward. The maximum separation which occurs at the end of the *Dhanus* and *Mithuna rāśis* [and observed along a particular direction, after a long period of time] is at 27° from the earlier position. The association of the *ayana* with a different direction is not possible without motion. Hence the motion of the *ayana* on the circles is inferred.

[45] {TS 1977}, p. 205.

३.१६ नत्यपक्रमाभ्याम् अक्षः

3.16 Latitude of the place from the zenith distance and declination

क्रान्त्यर्कनतिभेदोऽक्षो याम्ये गोले युतिः पुनः ॥ ३५ ॥
छायायामपि सौम्येऽर्केऽप्यन्यथा स्यात् तदन्तरम्।

krāntyarkanatibhedo'kṣo yāmye gole yutiḥ punaḥ||35||
chāyāyāmapi saumye'rke'pyanyathā syāt tadantaram|

The latitude is the difference between the midday zenith distance and declination when the Sun is in the southern hemisphere. It is the sum [of the two] when the shadow and the Sun are in the northern hemisphere. Otherwise [when the Sun is in the northern hemisphere and the shadow is towards the south], it is the difference between them.

From (3.83) to (3.85), we get $\phi = z - \delta_S$ (Sun in south), $\phi = z + \delta_N$ (Sun in north and $\phi > \delta_N$), and $\phi = \delta_N - z$ (Sun in north $\phi < \delta_N$) where z is the midday zenith distance.

३.१७ शङ्कुच्छायातः दिगवगमनम्

3.17 Determination of the directions from the shadow of the gnomon

सायनार्कभुजाजीवा परमक्रान्तिताडिता ॥ ३६ ॥
लम्बकाप्ताग्रजीवा स्यात् छायाकर्णहता हृता ।
त्रिज्ययाग्राङ्गुलं याम्ये विषुवद्भायुतं भुजा ॥ ३७ ॥
सौम्याथ सौम्यगोलेऽपि न्यूनमग्राङ्गुलं यदि ।
शोधयेद्विषुवद्भायाः सौम्यो बाहुस्तदापि च ॥ ३८ ॥
विषुवद्भां त्यजेत् तस्मात् रवावुत्तरगेऽधिकात् ।
याम्य एव तदा बाहुः तच्छायाकृतिभेदतः ॥ ३९ ॥
मूलं कोटिः श्रुतिः छाया त्रिभिस्त्र्यश्रं भवेदिदम् ।
भ्रामयित्वाथ तत् त्र्यश्रं यावच्छायानुगा श्रुतिः ॥ ४० ॥
कोटया पूर्वापरे ज्ञेये, बाहुना दक्षिणोत्तरे ।

sāyanārkabhujājīvā paramakrāntitāḍitā || 36 ||
lambakāptāgrajīvā syāt chāyākarṇahatā hṛtā |
trijyayāgrāṅgulaṃ yāmye viṣuvadbhāyutaṃ bhujā || 37 ||
saumyātha saumyagole'pi nyūnamagrāṅgulaṃ yadi |
śodhayedviṣuvadbhāyāḥ saumyo bāhustadāpi ca || 38 ||
viṣuvadbhāṃ tyajet tasmāt ravāvuttarage'dhikāt |
yāmya eva tadā bāhuḥ tacchāyākṛtibhedataḥ || 39 ||
mūlaṃ koṭiḥ śrutiḥ chāyā tribhistryaśraṃ bhavedidam |
bhrāmayitvātha tat tryaśraṃ yāvacchāyānugā śrutiḥ || 40 ||
koṭyā pūrvāpare jñeye, bāhunā dakṣiṇottare |

The Rsine of the *sāyana* longitude of the Sun multiplied by the maximum declination and divided by the cosine of the latitude of the place is the *agrajīvā*. [This] multiplied by the hypotenuse of the shadow and divided by the *trijyā* is the *agrajīvā* in *aṅgulas*. The result added to the equinoctial shadow gives the *bhujā* [Rsine of the shadow] when the Sun is in the southern hemisphere.

And [when the Sun is] in the northern hemisphere, (a) if the *agrajīvā* in *aṅgulas* is less than the *viṣuvadbhā*, it has to be subtracted from the *viṣuvadbhā* to get the *bhujā* corresponding to the shadow of the *śaṅku* which lies to the north [of the east–west line]; (b) if the northern declination is sufficiently large, the *viṣuvadbhā* has to be subtracted from it (the *agrajīvā* in *aṅgulas*). Then the *bhujā* corresponding to the shadow of the *śaṅku* will be to the south [of the east–west line].

The square root of the difference between the *chāyā* and the *bāhu* is the [*chāyā*]*koṭi*. The three form a [right] triangle. The triangle is rotated [around the *śaṅku*] such that the hypotenuse (*chāyā*) is in the direction of the *chāyā*. Then the *koṭi* is understood to be along the east–west line, and the *bāhu* along the north–south [line].

The term *agrajīvā* (*arkāgrā*) refers to the perpendicular distance of the Sun from the east/west line in the plane of the horizon at the time of the rising/setting of the Sun. In Fig. 3.15(b), S_tB represents the *arkāgrā*. The expression for the *arkāgrā* ($R\cos A_t$) can be obtained in terms of other quantities as follows.

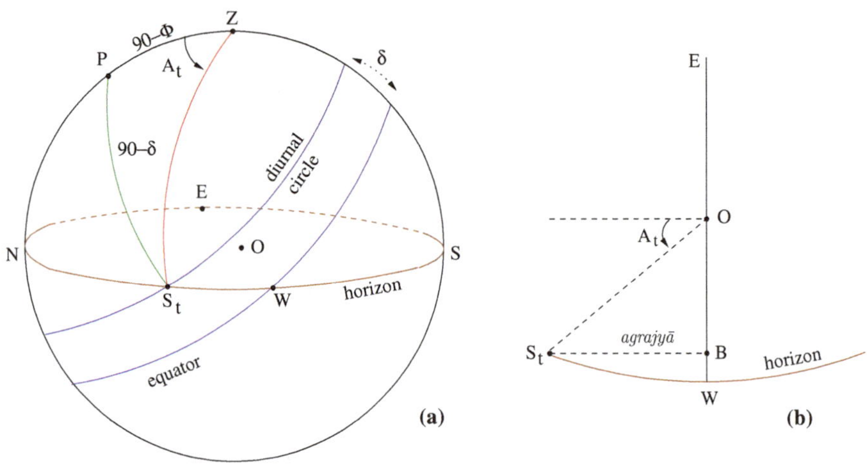

Fig. 3.15 *Arkāgrā* at sunset when the Sun has northerly declination: (a) as seen on the celestial sphere, (b) as seen in the plane of the horizon.

Consider the spherical triangle PZS_t. Here

$$PS_t = 90 - \delta; \qquad PZ = 90 - \phi; \qquad ZS_t = 90; \qquad P\hat{Z}S_t = A_t. \tag{3.86}$$

Using the cosine formula we have

$$\cos(90-\delta) = \cos(90-\phi)\cos 90 + \sin(90-\phi)\sin 90 \cos A_t. \tag{3.87}$$

Rewriting the above equation we get an expression for the *arkajīvā*,

$$R\cos A_t = \frac{R\sin\delta}{\cos\phi}. \tag{3.88}$$

From the above equation it is seen that the *arkāgrā* is known, once the declination of the Sun is known. The latitude of the place is already known by the measurement of shadow on the equinoctial day. Expressing the declination in terms of the true longitude of the Sun (3.88) reduces to

$$arkāgrā = |R\cos A_t| = \frac{R\sin\varepsilon\sin\lambda}{\cos\phi}. \tag{3.89}$$

Thus we see that the *arkāgrā* is known, if the *sāyana* longitude of the Sun is known at the time of rising or setting. The above expression for the *arkāgrā* is in terms of minutes of arc. It may be expressed in *aṅgulas* by multiplying it by the hypotenuse of the shadow K, which is measured in *aṅgulas*, and dividing it by the *trijyā*:

$$\begin{aligned} agrajyā = |K\cos A_t| &= K\left|\frac{\sin\delta}{\cos\phi}\right| \\ &= \frac{K}{R}\left|\frac{R\sin\varepsilon\sin\lambda}{\cos\phi}\right|. \end{aligned} \tag{3.90}$$

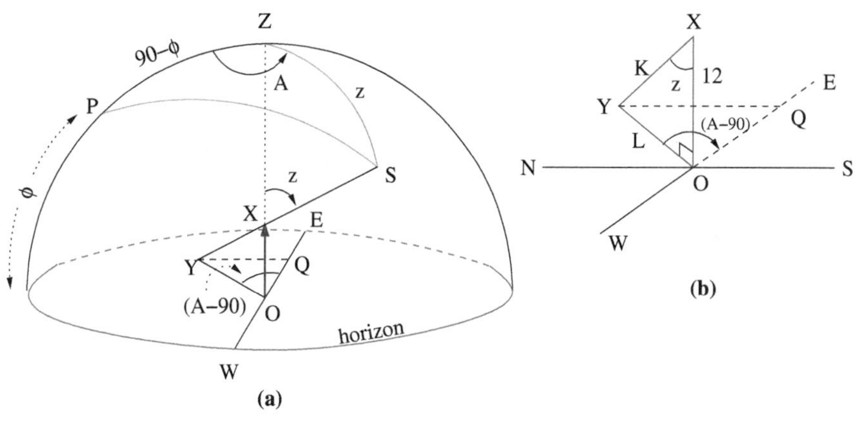

Fig. 3.16 Determination of the *chāyābhujā* from the *śaṅkucchāyā*.

Let the zenith distance and azimuth of the Sun be z and A respectively, as indicated in Fig. 3.16. Then considering the spherical triangle PZS,

$$\begin{aligned} \cos(90-\delta) &= \cos(90-\phi)\cos z + \sin(90-\phi)\sin z\cos A \\ \text{or} \qquad \sin\delta &= \sin\phi\cos z + \cos\phi\sin z\cos A. \end{aligned} \tag{3.91}$$

Multiplying this by the *karṇa*, K, and dividing by $\cos\phi$,

$$K\frac{\sin\delta}{\cos\phi} = K\tan\phi\cos z + K\sin z\cos A. \tag{3.92}$$

From (3.90), it is clear that the magnitude of the LHS of the above equation is nothing but the *agrajīvā* or the *agrajyā*.

Consider Fig. 3.16(b), where the *śaṅku*, *chāyā* etc. are more clearly depicted. Here, YQ is the distance between the tip of the shadow and the east–west line and is known as the *chāyābhujā*. In other words, *chāyābhujā* is the projection of the shadow along the north–south direction. In this figure,

$$chāyā = OY = K\sin z, \tag{3.93}$$

$$\text{hence,} \qquad chāyābhujā = YQ = -K\sin z\cos A. \tag{3.94}$$

Also for the situation depicted in the Fig. 3.16(b), the azimuth $A > 90°$, and hence $-\cos A$ is positive. Also, the *śaṅku* $= 12 = K\cos z$; and the equinoctial midday shadow, the *viṣuvadbhā*, $= 12\tan\phi$. Hence (3.92) can be rewritten as

$$-K\sin z\cos A = 12\,\tan\phi - K\frac{\sin\delta}{\cos\phi}\,. \tag{3.95}$$

It was already shown that the *arkāgrā* (in *aṅgulas*) is $|\frac{K\sin\delta}{\cos\phi}|$. It may further be noted that the second term in the RHS of the above equation is positive when δ is negative (i.e. the Sun is in the southern hemisphere) and it is negative when δ is positive. Hence the *chāyābhujā* is the sum of the *arkāgrāṅgula* and the *visuvadbhā*, when the Sun is in the southern hemisphere. When δ is positive (i.e. the Sun is in the northern hemisphere) and the *arkāgrāṅgula* is less than the *viṣuvadbhā*, the *chāyābhujā* is obtained by subtracting the former from the latter. In both these cases, the shadow is to the north of the east–west line, as indicated in Fig. 3.16.

Again, when the declination of the Sun is to the north, and the *arkāgrāṅgula* is greater than the *viṣuvadbhā*, then in this case $A < 90°$, and $K\sin z\cos A$ is positive. Then

$$chāyābhujā = arkāgrāṅgula - viṣuvadbhā. \tag{3.96}$$

In this case, the shadow is to the south of the east–west line. The *koṭi* of the *chāyā* is defined by *koṭi* $= OQ = K\sin z\sin A$.

The *chāyābhujā*, *koṭi* and *chāyā* form a right-angled triangle with the *chāyā* as the hypotenuse. From the physical shadow *chāyā*, the *chāyābhujā* can be determined from (3.95) as $K = \sqrt{12^2 + OY^2}$, δ and ϕ are known. Then, from the *chāyā* and *chāyābhujā* the *koṭi* can be found.

Construct a triangle with these as the sides, such that the intersection point of *chāyā* and the *koṭi* is at the base of the *śaṅku*. Rotate it such that the hypotenuse, OY, is actually along the physical shadow. Then the *koṭi*, OQ, is along the east–west line. Thus the east–west direction can be determined with the aid of the calculation described above. Similarly, the north–south direction would be along the *chāyābhujā*, YQ.

३.१८ छायाभ्रमणवृत्तपरिलेखः

3.18 Drawing the locus of the tip of shadow of the gnomon

छायाभ्रमणमप्येवं ज्ञेयमिष्टदिनोद्भवम् ॥ ४१ ॥
इष्टकालोद्भवां छायां बाहुं कोटिं च पूर्ववत्।
तत्तुल्याभिः शलाकाभिः तिसृभिः त्रिभुजं तथा ॥ ४२ ॥
कृत्वा पूर्वापरां कोटिं वृत्तमध्याद्यथादिशम्।
कृत्वा बाहुं च बाहोश्च छायायाश्चाग्रयोर्युतौ ॥ ४३ ॥
बिन्दुं कृत्वापराह्णेऽपि बिन्दुं तत्र प्रकल्पयेत्।
मध्यच्छायाशिरस्यन्यः तृतीयो बिन्दुरिष्यते ॥ ४४ ॥
लिखेद्वृत्तत्रयं तेन यथा मत्स्यद्वयं भवेत्।
तन्मत्स्यमध्यगे सूत्रे प्रसार्यैव तयोर्युतिः ॥ ४५ ॥
दृश्यते यत्र तन्मध्यं वृत्तं बिन्दुस्पृगालिखेत्।
छाया तन्नेमिगा तस्मिन् दिने स्यात् सर्वदापि च ॥ ४६ ॥

chāyābhramaṇamapyevaṃ jñeyamiṣṭadinodbhavam || 41 ||
iṣṭakālodbhavāṃ chāyāṃ bāhuṃ koṭiṃ ca pūrvavat |
tattulyābhiḥ śalākābhiḥ tisṛbhiḥ tribhujaṃ tathā || 42 ||
kṛtvā pūrvāparāṃ koṭiṃ vṛttamadhyādyathādiśam |
kṛtvā bāhuṃ ca bāhośca chāyāyāścāgrayoryutau || 43 ||
binduṃ kṛtvāparāhṇe'pi binduṃ tatra prakalpayet |
madhyacchāyāśirasyanyaḥ tṛtīyo binduriṣyate || 44 ||
likhedvṛttatrayaṃ tena yathā matsyadvayaṃ bhavet |
tanmatsyamadhyage sūtre prasāryaiva tayoryutiḥ || 45 ||
dṛśyate yatra tanmadhyaṃ vṛttaṃ binduspṛgālikhet |
chāyā tannemigā tasmin dine syāt sarvadāpi ca || 46 ||

The motion of the [tip of the] shadow on a desired day is to be determined as follows:

The *bāhu*, the *koṭi* and the *chāyā* are obtained at a desired instant as described earlier. With three sticks whose lengths are equal to the *bāhu*, the *koṭi* and the *chāyā* at some instant, a triangle is formed and it is placed such that the *koṭi* is along the east–west line with one tip of it at the centre of the circle. The *bāhu* also gets aligned in the appropriate direction (north–south direction). A point is marked at the intersection of the *bāhu* and the *chāyā*. A similar point is marked in the afternoon also. The tip of the midday shadow is taken to be the third point.

With these three points, three circles are drawn such that two fish figures are formed. The lines passing through the fish figures are extended and their point of intersection is found. With this point as the centre, draw a circle passing through the above three points. The [tip of the] shadow on that day will always be along the circle drawn.

In Fig. 3.17, *EW* and *NS* represent the east–west and the north–south lines on the horizon. The point *O* is the foot of the *śaṅku*. OAP_2 and OBP_1 are identical triangles[46] formed out of three sticks whose dimensions are the *bāhu*, the *koṭi* and the *chāyā* of the usual *śaṅku* of 12 *aṅgulas* at some instant. *OA* refers to the *koṭi*, AP_2 the *bāhu* of the shadow, and OP_2 the *chāyā*, which is the hypotenuse of the triangle whose sides are the *koṭi* and its *bāhu* at some instant during the day. If OP_2

[46] It is implicit that the variation in declination discussed in Section 3.3 is ignored.

is taken to be of unit length, then OA and AP_2 are the sine and the cosine of the azimuth of the Sun. Here, the length of the *chāyā* L is given by

$$chāyā = L = OP_2 = K \sin z = 12 \tan z. \tag{3.97}$$

The corresponding *bhujā* and *koṭi* are given by

$$bhujā = AP_2 = |L \cos A| = |K \sin z \cos A| \tag{3.98}$$
$$koṭi = OA = |L \sin A| = |K \sin z \sin A|. \tag{3.99}$$

Similar considerations apply for the sides of the triangle OBP_1 also.

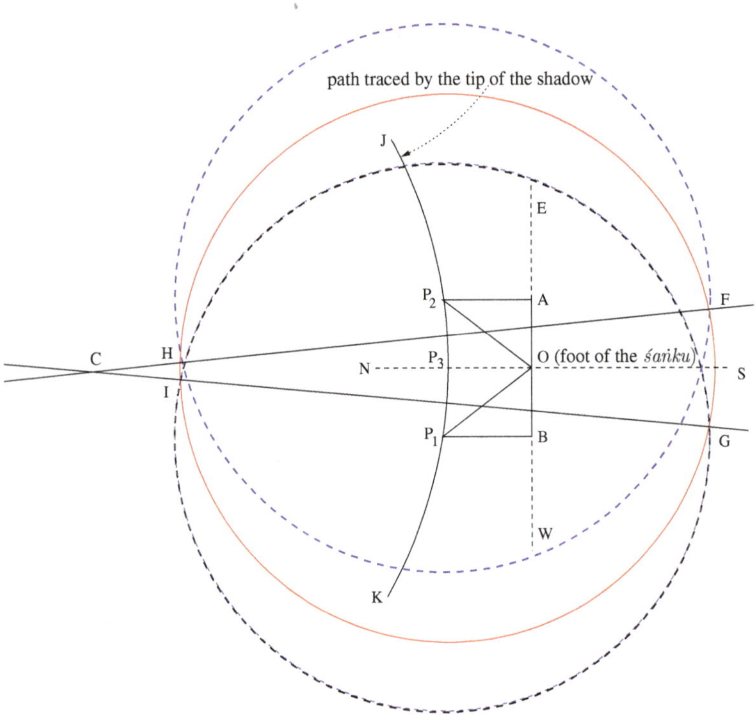

Fig. 3.17 The path traced by the tip of the shadow of the *śaṅku* on any day.

The point P_3 in Fig. 3.17 represents the tip of the midday shadow. With P_1, P_2 and P_3 as centres, three circles are drawn. The points of intersection of the circles with P_1 and P_3 as centres intersect at G and I and form a *matsya* (fish [figure]). Similarly, the points of intersection of the circles with P_2 and P_3 as centres intersect at F and H and form another *matsya*. We draw straight lines passing through the *matsyas* which intersect at point C (represented by solid lines FHC and GIC in the figure). These lines intersect at a point C. With C as the centre and with CP_3 as the radius we draw a circle. It can be shown that this circle will necessarily pass through points P_1

and P_2. According to the text, the path traced by the tip of the shadow of the *śaṅku* on that day is given by this circle represented by $JP_2P_3P_1K$. However, as we will show below, the path traced is a hyperbola and not a circle.

The circle described in the text

We illustrate the construction of the circle implicit in the verses by taking the declination of the Sun to be southerly, i.e. when δ is negative and the *chāyābhujā* (the distance between the tip of the shadow and the east–west line) is greater than the *viṣuvadbhā* (the equinoctial midday shadow).

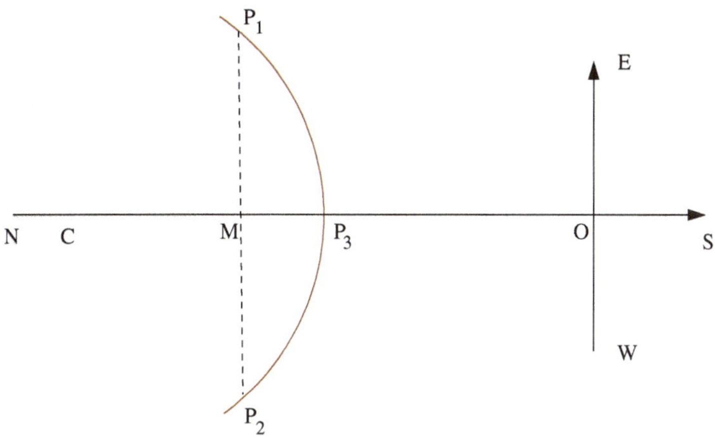

Fig. 3.18 The locus traced by the tip of the shadow.

As mentioned earlier (see Fig. 3.17), the point P_3 corresponds to the midday shadow and P_1, P_2 correspond to some common value of the zenith distance, z. Then

$$\begin{aligned}\text{midday shadow} = OP_3 &= 12\tan(\phi - \delta)\\ \textit{chāyābhujā} = OM = |x| &= 12\tan\phi - K\frac{\sin\delta}{\cos\phi}\\ &= 12\tan\phi - \frac{12\sin\delta}{\cos z\cos\phi}\\ &= \frac{12}{\cos\phi}\left(\sin\phi - \frac{\sin\delta}{\cos z}\right). \qquad (3.100)\end{aligned}$$

By construction, the lengths of the shadow (the *chāyā*) at P_1 and P_2 are equal and are given by

$$OP_1 = OP_2 = 12\tan z. \qquad (3.101)$$

Hence the magnitude of the cosine of the shadow is

$$
\begin{aligned}
chāyākoṭi &= MP_1 = MP_2 \\
\text{or} \qquad |y| &= \sqrt{chāyā^2 - bhujā^2} \\
&= \sqrt{12^2 \tan^2 z - OM^2}. \qquad (3.102)
\end{aligned}
$$

Let C be the centre of the circle which passes through P_1, P_2 and P_3. Let $CP_3 = R$. Now

$$
\begin{aligned}
CM &= OC - OM \\
&= CP_3 + OP_3 - OM \\
&= R + 12\tan(\phi - \delta) - |x|. \qquad (3.103)
\end{aligned}
$$

Then, considering the right-angled triangles CMP_1 and CMP_2, we have

$$
\begin{aligned}
CP_1^2 &= CP_2^2 \\
&= (R + 12\tan(\phi - \delta) - |x|)^2 + |y|^2. \qquad (3.104)
\end{aligned}
$$

Using the fact that $|x|^2 + |y|^2 = 12^2 \tan^2 z$,

$$
\begin{aligned}
CP_1^2 = CP_2^2 = R^2 &+ 2R(12\tan(\phi - \delta) - |x|) + 12^2 \tan^2(\phi - \delta) \\
&-2\,|x|\,12\tan(\phi - \delta) + 12^2 \tan^2 z. \qquad (3.105)
\end{aligned}
$$

As P_1, P_2, P_3 lie on a circle of radius R with C as centre,

$$
CP_1^2 = CP_2^2 = R^2. \qquad (3.106)
$$

Equating the two expressions for CP_1^2, from (3.105) and (3.106) we have

$$
R = \frac{12^2 \tan^2 z - 24\,|x|\,\tan(\phi - \delta)}{2(|x| - 12\tan(\phi - \delta))}, \qquad (3.107)
$$

where $|x|$ is given by (3.100) and R is the radius of the circle passing through the tip of the shadow at midday and the pair of shadow tips, corresponding to a given zenith distance.

The very fact that R is dependent on the zenith distance z implies that the tip of the shadow does not trace a circle over the day. Now we arrive at an expression which describes the locus traced by the tip of the shadow.

The actual curve traced by the tip of the shadow

In order to arrive at the locus of the shadow, we consider the north–south and east–west lines as the X and Y axes, and the base of the *śaṅku* as the origin as indicated in Fig. 3.19. Let the coordinates of the tip of the shadow of a 12-unit *śaṅku* at P be (x, y). Here these coordinates incorporate the sign also. For instance, in the figure

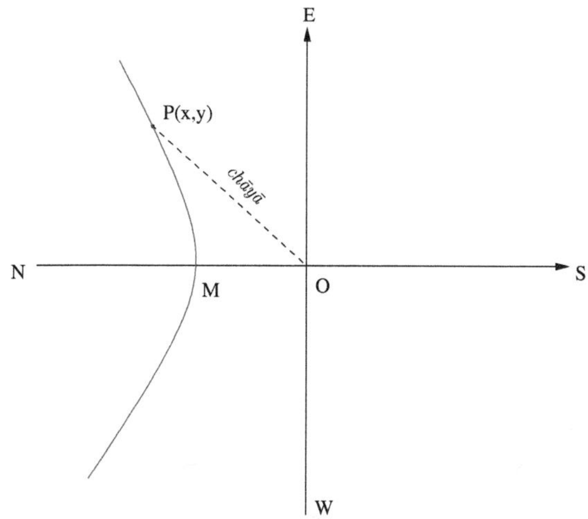

Fig. 3.19 To show that the curve traced by the tip of the shadow is a hyperbola.

x is negative whereas y is positive. $|x|$ and $|y|$ are the *chāyābhujā* and *chāyākoṭi*, whereas

$$OP = \sqrt{x^2 + y^2}, \tag{3.108}$$

is the *chāyā* (shadow). Now, the coordinates of the tip of the shadow, x and y, are given by

$$\begin{aligned} x &= K \sin z \cos A = 12 \tan z \cos A \\ \text{and} \qquad y &= K \sin z \sin A = 12 \tan z \sin A. \end{aligned} \tag{3.109}$$

Earlier we showed that

$$\begin{aligned} -K \sin z \cos A &= 12 \tan \phi - K \frac{\sin \delta}{\cos \phi} \\ &= 12 \tan \phi - \frac{12 \sin \delta}{\cos z \cos \phi}. \end{aligned} \tag{3.110}$$

Therefore,

$$\begin{aligned} -x &= 12 \tan \phi - \frac{12 \sin \delta}{\cos z \cos \phi} \\ \text{or} \qquad x + 12 \tan \phi &= \frac{12 \sin \delta}{\cos \phi} \sec z \\ \text{or} \qquad 12 \sec z &= (x + 12 \tan \phi) \frac{\cos \phi}{\sin \delta}. \end{aligned} \tag{3.111}$$

From (3.109), $\sqrt{x^2 + y^2} = 12 \tan z$. Also,

$$
\begin{aligned}
12^2 \sec^2 z &= 12^2 \tan^2 z + 12^2 \\
&= x^2 + y^2 + 12^2.
\end{aligned} \tag{3.112}
$$

Now from (3.111) and (3.112), we find

$$
(x + 12\tan\phi)^2 \frac{\cos^2\phi}{\sin^2\delta} = x^2 + y^2 + 12^2. \tag{3.113}
$$

After some straightforward manipulations, we find

$$
\left(x + \frac{12\sin\phi\cos\phi}{\cos^2\phi - \sin^2\delta}\right)^2 - \frac{y^2\sin^2\delta}{\cos^2\phi - \sin^2\delta} = 12^2 \frac{\sin^2\delta\cos^2\delta}{(\cos^2\phi - \sin^2\delta)^2}. \tag{3.114}
$$

This is the equation for a hyperbola, as $\cos^2\phi - \sin^2\delta > 0$ mostly except when $|\delta| > 90 - \phi$, which is possible only for latitudes $\phi > 66\frac{1}{2}^\circ$. Even for such high latitudes, this will be only for certain periods when the Sun becomes circumpolar, in which case the tip of the shadow traces an ellipse.

1. When $\delta = 0$, we have
$$
x = -12\tan\phi.
$$
This implies that the tip of the shadow traces a straight line parallel to the east–west line at a constant distance of $12\tan\phi$ (the *viṣuvadbhā*) from it, towards north for an observer in the northern hemisphere.[47] This is as expected.
2. The midday shadow is along the north–south line when $y = 0$. Then

$$
\begin{aligned}
x + \frac{12\sin\phi\cos\phi}{\cos^2\phi - \sin^2\delta} &= \frac{12\sin\delta\cos\delta}{\cos^2\phi - \sin^2\delta} \\
\text{or} \qquad x &= -12\left(\frac{\sin\phi\cos\phi - \sin\delta\cos\delta}{\cos^2\phi - \sin^2\delta}\right) \\
&= -12\tan(\phi - \delta).
\end{aligned} \tag{3.115}
$$

This is also as expected, since the zenith distance, z, at midday is $\phi - \delta$ and the length of the shadow is $12\tan z$.

Only one arm of the hyperbola would be relevant for a particular day as

$$
x + 12\tan\phi = 12\frac{\sin\delta}{\cos\phi}\sec z
$$

and the sign of RHS is determined by δ. We depict the relevant arm of the hyperbola when

- δ is negative (southern declination),
- when δ is positive (northern) with $\delta < \phi$,
- when δ is positive (northern) with $\delta > \phi$.

[47] The same will be towards the south for an observer in the southern hemisphere.

in Figs. (3.20)(a), (b) and (c) respectively.

It is important to note that Śaṅkara Vāriyar clearly states in *Yukti-dīpikā* that the path traced by the tip of the shadow is not actually a circle; and that in stating that it is a circle, Nīlakaṇṭha is merely following the tradition. In his own words:

छायाभ्रमणवृत्तं यदुक्तं प्रायिकमेव तत्।
वृत्तमार्गेण भाग्रस्य भ्रमणानुपपत्तितः।
पूर्वाचार्यानुरोधेन केवलं तदिहोदितम्॥ [48]

The statement made here that it is a circle is only approximate, since it has not been proved that the tip of the shadow of the *śaṅku* [throughout the course of the day] traces the path of a circle. It is stated here simply to maintain concordance with what has been stated by the earlier teachers.

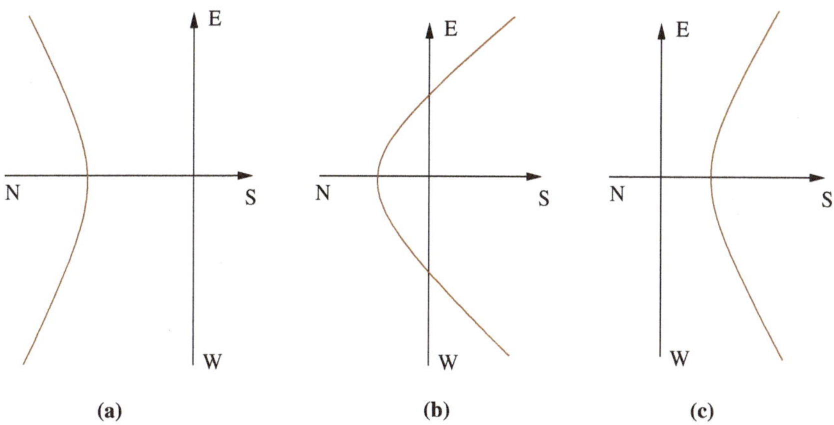

Fig. 3.20 The different hyperbolas obtained with the change in the declination of the Sun.

३.१९ प्रकारान्तरेण छायाभुजानयनम्

3.19 Another method for finding the Rsine of the shadow

अक्षज्याघ्नान्महाशङ्कोः शङ्क्वग्रं लम्बकाहृतम्।
सर्वदा दक्षिणं तद्धि योज्यमर्काग्रयापि तत् ॥ ४७ ॥
याम्ये गोले महाबाहुः सौम्ये चाग्रद्वयान्तरम्।
अधिकेऽत्रापि शङ्क्वग्रे याम्यः स्यादन्यथोत्तरः ॥ ४८ ॥
छायाकर्णहतः सोऽपि त्रिज्याभक्तोऽङ्गुलात्मकः।
विपरीतदिगप्येष पूर्वानीतसमोऽपि च ॥ ४९ ॥

[48] {TS 1977}, p. 214.

द्वादशघ्नोऽथवा बाहुः शङ्कुना महता हृतः ।
अङ्गुलात्मकमेवं वा छायाभ्यामथवा नयेत् ॥ ५० ॥

akṣajyāghnānmahāśaṅkoḥ śaṅkvagraṃ lambakāhṛtam |
sarvadā dakṣiṇaṃ taddhi yojyamarkāgrayāpi tat || 47 ||
yāmye gole mahābāhuḥ saumye cāgradvayāntaram |
adhike'trāpi śaṅkvagre yāmyaḥ syādanyathottaraḥ || 48 ||
chāyākarṇahataḥ so'pi trijyābhakto'ṅgulātmakaḥ |
viparītadigapyeṣa pūrvānītasamo'pi ca || 49 ||
dvādaśaghno'thavā bāhuḥ śaṅkunā mahatā hṛtaḥ |
aṅgulātmakamevaṃ vā chāyābhyāmathavā nayet || 50 ||

The *akṣajyā* multiplied by the *mahāśaṅku* and divided by the *lambaka* is the *śaṅkvagrā* and always lies to the south. [If the Sun is] in the southern hemisphere, the *arkāgrā* has to be added to that (*śaṅkvagrā*) to get the *mahābāhu* [and] in the northern the difference between the two *agrās* [gives the *mahābāhu*]. Here again, if the *śaṅkvagrā* is greater than the *arkāgrā*, then [the *mahābāhu*] will be to the south, otherwise to the north.

This multiplied by the *chāyākarṇa* and divided by the *trijyā* gives the *mahābāhu* in *aṅgulas*. Though this (the *mahābāhu*) obtained is in the opposite direction (the *viparītadigapi*), it is the same [in magnitude] as the one that was obtained earlier. Otherwise, this may be obtained by multiplying the *mahābāhu* by twelve and dividing by the *mahāśaṅku*. Thus, this can also be obtained from the shadows measured in *aṅgulas*.

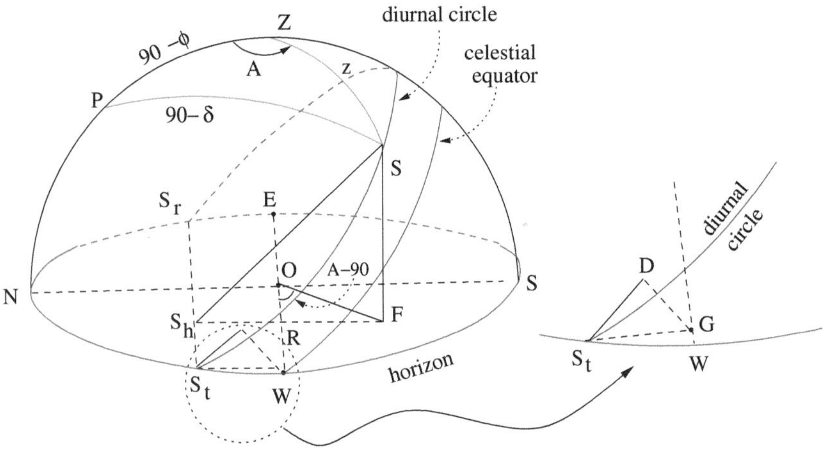

Fig. 3.21*a* *Śaṅkvagrā* when the declination is north (+ve) and $A > 90°$.

In Fig. 3.21*a*, the Sun rises at S_r, moves along the diurnal circle and sets at S_t. If we assume that Sun's declination δ is constant through the day, S_rS_t would be parallel to the east–west line. From S_t, draw S_tG perpendicular to the east–west line meeting it at G. S_tG is the *arkāgrā* or *arkajīvā*.

Now the plane of the diurnal circle is inclined at an angle $90-\phi$ with the horizon. From G draw GD perpendicular to the plane of the diurnal circle meeting it at D. Join S_tD, which would be perpendicular to GD. Clearly, $D\hat{S_t}G = 90-\phi$ and $D\hat{G}S_t = \phi$. S_tDG is a latitudinal triangle (a right-angled triangle with the latitude as one of the angles). Now $GD = |R\sin\delta|$. Hence

$$arkāgrā = S_tG = \left|R\frac{\sin\delta}{\cos\phi}\right|. \tag{3.116}$$

Let S be the position of the Sun at some instant, when its zenith distance is z and the azimuth is A, as shown in the figure. Draw SF perpendicular to the horizon meeting it at F. Draw FS_h perpendicular to S_tS_r meeting it at S_h and the east–west line at R. SS_hF is also a latitudinal triangle with $S_h\hat{S}F = \phi$. Now

$$mahāśaṅku = SF = R\cos z \tag{3.117}$$

and

$$mahācchāyā = OF = R\sin z. \tag{3.118}$$

The distance between the base of the *śaṅku* F and the east–west line EW, denoted by RF, is known as the *mahābāhu* or the *chāyābāhu.*

$$\begin{aligned} mahābāhu = RF &= OF\sin(A-90) \\ &= -R\sin z\cos A \\ &= |R\sin z\cos A|. \end{aligned} \tag{3.119}$$

The perpendicular distance of the foot of the gnomon F from the line S_rS_t, denoted by S_hF, is known as the *śaṅkvagrā*. It is so named because it gives the distance of the foot of the *śaṅku* at any given time from the line passing through the rising and setting points of the Sun.

Now, in the right-angled triangle SS_hF, $S_h\hat{S}F = \phi$ and $S_hF = SF\frac{\sin\phi}{\cos\phi}$. Hence

$$śaṅkvagrā = mahāśaṅku \times \frac{akṣajyā}{lambaka} = R\cos z \times \frac{R\sin\phi}{R\cos\phi}, \tag{3.120}$$

as stated. The *śaṅkvargā* is always to the south of S_rS_t. Now

$$\begin{aligned} śaṅkvagrā &= S_hF \\ &= S_hR + RF \\ &= S_tG + RF \\ &= arkāgrā + mahābāhu \end{aligned}$$

or

$$mahābāhu = śaṅkvagrā - arkāgrā. \tag{3.121}$$

This is so when the declination is north and $A > 90°$. In Figs 3.21*b* and 3.21*c*, we depict the cases when declination δ is north (+ve) and $A < 90°$ and when the declination δ is south ($-$ve).

When $\delta > 0$ and $A < 90°$, we see that

$$\begin{aligned} arkāgrā &= S_hR \\ &= S_hF + RF \\ &= śaṅkvagrā + mahābāhu \end{aligned}$$

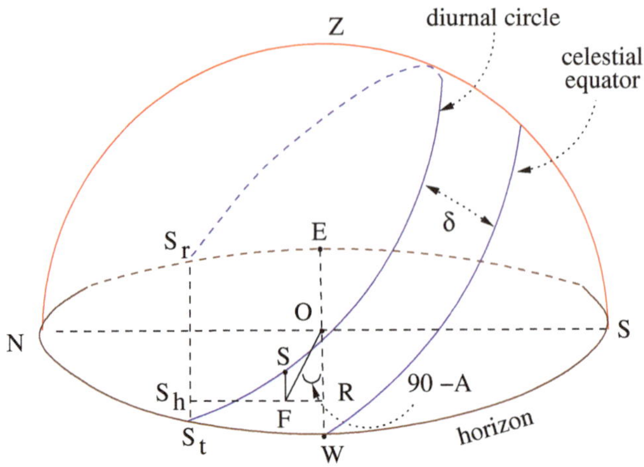

Fig. 3.21*b* *Śaṅkvagrā* when the declination is north (+ve) and $A < 90°$.

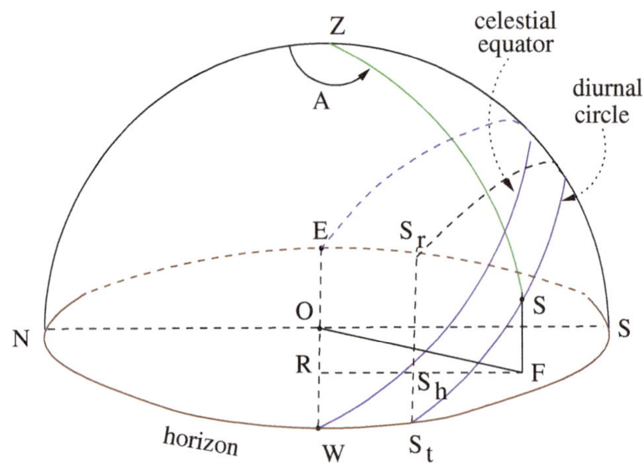

Fig. 3.21c *Śaṅkvagrā* when the declination is south (−ve).

$$\text{or} \qquad mah\bar{a}b\bar{a}hu = ark\bar{a}gr\bar{a} - \acute{s}a\dot{n}kvagr\bar{a}. \tag{3.122}$$

When the declination is south ($\delta < 0$), as shown in Fig. 3.21*c*,

$$mah\bar{a}b\bar{a}hu = RF = S_hF + S_hR \tag{3.123}$$

$$= \acute{s}a\dot{n}kvagr\bar{a} + ark\bar{a}gr\bar{a}. \tag{3.124}$$

All the cases can be combined in a single formula. Consider the spherical triangle *ZPS* in Fig. 3.21*a*. In this triangle,

$$\cos(90 - \delta) = \cos(90 - \phi)\cos z + \sin(90 - \phi)\sin z \cos A$$

$$\text{or} \qquad -R\sin z\cos A = R\cos z\frac{\sin\phi}{\cos\phi} - R\frac{\sin\delta}{\cos\phi}, \tag{3.125}$$

where the *arkāgrā* is $\left|\frac{R\sin\delta}{\cos\phi}\right|$, the *śaṅkvagrā* is $\left|\frac{R\cos z\sin\phi}{\cos\phi}\right|$ and the *mahābāhu* or *chāyā-bāhu* is $|R\sin z\cos A|$.

When the declination is north, it can be seen that the *chāyābāhu* RF is to the south when the *śaṅkvagrā* S_hF is greater than the *arkāgrā*, as in Fig. 3.21*a*, but it is to the north when the *śaṅkvagrā* is less than the *arkāgrā*, as in Fig. 3.21*b*. When the declination is south, the *śaṅkvagrā*, *arkāgrā* and *chāyābāhu* are all to the south as in Fig. 3.21*c*.

The *chāyābāhu* or *chāyābhujā* in *aṅgulas* is given by

$$\begin{aligned} chāyābāhu &= |K\sin z\cos A| \\ &= \left|\frac{12\sin z}{\cos z}\cos A\right|. \end{aligned} \tag{3.126}$$

Dividing the relation between the *mahābhāhu* (*chāyābāhu*), *arkāgrā* and *śaṅkvagrā* in (3.125) by $R\cos z$ and multiplying by 12, we have

$$\left|12\frac{\sin z}{\cos z}\cos A\right| = \left|12\tan\phi - K\frac{\sin\delta}{\cos\phi}\right| \tag{3.127}$$

$$\text{or} \quad chāyābāhu\,(aṅgulas) = |viṣuvadbhā \pm agrajīvā\,(aṅgulas)|. \tag{3.128}$$

This has been stated earlier.

३.२० सममण्डलशङ्कुः

3.20 Gnomon when the Sun is on the prime vertical

अक्षज्योना यदा क्रान्तिः सौम्या तां त्रिज्यया हताम्।
अक्षज्यया विभज्याप्तः शङ्कुः स्यात् सममण्डले ॥ ५१ ॥

akṣajyonā yadā krāntiḥ saumyā tāṃ trijyayā hatām|
akṣajyayā vibhajyāptaḥ śaṅkuḥ syāt samamaṇḍale || 51 ||

When the declination of the Sun is to the north and it is less than the latitude of the place, then the Rsine of declination multiplied by the *trijyā* and divided by Rsine of the latitude gives the *śaṅku* in *samamaṇḍala* [when the Sun is on the prime vertical].

The term *samamaṇḍala* refers to prime vertical, ZEZ' in Fig. 3.22. Let z_0 be the zenith distance of the Sun with declination δ, when it is on the prime vertical. Then the expression for the *śaṅku* is given to be

$$\begin{aligned} śaṅku &= \frac{krāntijyā \times trijyā}{akṣajyā} \\ \text{or} \qquad R\cos z_0 &= \frac{R\sin\delta \times R}{R\sin\phi}. \end{aligned} \tag{3.129}$$

In the spherical triangle PZS, $ZS = z_0$, $PZ = 90 - \phi$, $PS = 90 - \delta$ and $P\hat{Z}S = 90$.

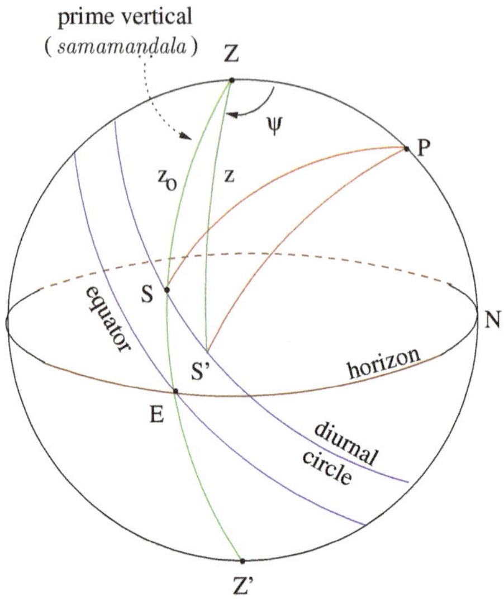

Fig. 3.22 Computation of the *samamaṇḍala-śaṅku*.

Applying the cosine formula, we have

$$\sin\delta = \cos z_0 \sin\phi,$$

which is the same as (3.129).

Significance of the condition that $\delta < \phi$ and it must be north

1. If the condition $\delta < \phi$ is not satisfied, then the Sun will never cross the prime vertical during its diurnal motion, and hence the expression for the *samamaṇḍala-śaṅku* (3.129) does not have any significance.
2. If the Sun does not have a northern declination, then it will not be above the horizon when it crosses the prime vertical.

The rule of three which is implicit in arriving at the formula given by (3.129) is explained in *Yukti-dīpikā* as follows.

उदगर्काग्रया तुल्ये शङ्क्वग्रे नित्यदक्षिणे ।
प्रतिक्षणं भिन्नरूपे यात्यर्कः सममण्डलम् ॥[49]

[49] {TS 1977}, p. 214. The prose order for this verse is: प्रतिक्षणं भिन्नरूपे नित्यदक्षिणे शङ्क्वग्रे उदगर्काग्रया तुल्ये (सति) अर्कः सममण्डलं याति।

When the *śaṅkvagrā*, which changes continuously, becomes equal to the *udagarkāgrā*, then the Sun reaches the prime vertical.

When the Sun is on the prime vertical, its azimuth is 90° and the *mahābāhu* is zero. Then the *śaṅkvagrā* is equal to the *arkāgrā*, as given by (3.125). That is

$$R\cos z_0 \frac{\sin\phi}{\cos\phi} = R\frac{\sin\delta}{\cos\phi}$$

$$\text{or} \qquad R\cos z_0 = R\frac{\sin\delta}{\sin\phi}. \tag{3.130}$$

तदर्थं न्यूनता क्रान्तेः सौम्यता कथ्यतेऽक्षतः[50] ।
क्रान्त्यक्षनियमो वेद्यः शङ्क्वग्रार्काग्रयोरपि[51] ॥

It is only for this to happen [that is, for the Sun to be on prime vertical] that it is stated that the declination has to be north and has to be less than the latitude of the place. [Further,] the condition on the declination and latitude is the same as that on the *śaṅkvagrā* and *arkāgrā*.

While the first half of the above verse is straightforward, the second half needs explanation. For this, let us consider the spherical triangle PZS' in Fig. 3.22. When the Sun is not on the prime vertical—as shown at S' in the figure—the angle $P\hat{Z}S' \neq 90$. Let us denote this angle by ψ and the zenith distance by z. That is, $P\hat{Z}S' = \psi$ and $ZS' = z$. Now using the cosine formula we have

$$\sin\delta = \cos z \sin\phi + \sin z \cos\phi \cos\psi$$

$$\text{or} \qquad \frac{\sin\delta}{\cos\phi} = \cos z \tan\phi + \sin z \cos\psi. \tag{3.131}$$

This can be written as

$$arkāgrā = śaṅkvagrā + X. \tag{3.132}$$

In the above expression, since $\sin z$ is always positive, the quantity X is positive only when $\psi < 90°$. In otherwords, the *śaṅkvagrā* is less than the *arkāgrā* only till the time when the Sun reaches the prime vertical from its rising.

सौम्यक्रान्तेर्यदाक्षेण तुल्यता जायते क्रमात् ।
द्रष्टुः समोपरि तदा यात्यर्कः सममण्डलम् ॥
उदक्क्रान्त्यल्पतायत्ता समशङ्क्वल्पता ततः ।
क्रान्तिः सौम्याक्षतुल्या चेत् समशङ्कुस्त्रिमौर्विका ॥
समोपरिष्टादर्कस्य दिनमध्ये प्रवृत्तितः ।
अक्षतो न्यूनसङ्ख्यायाः तस्याः शङ्कुं ततो नयेत् ॥

[50] The prose order for this half of the verse is: क्रान्तेः सौम्यता, अक्षतः न्यूनता (च) तदर्थं कथ्यते।

[51] The reading in the printed edition is: शङ्क्वर्काग्राग्रयोरपि।

शङ्कौ त्रिज्यासमे सौम्यक्रान्तिः अक्षसमा यदि ।
समशङ्कौ ततो न्यूने तत्क्रान्तिमपि चानयेत् ॥ [52]

As the northern declination [of the Sun] gradually increases and becomes equal to the latitude the Sun crosses the prime vertical right above the observer.

The measure of smallness of the *samamaṇḍala-śaṅku* depends upon the measure of the declination. When the declination is equal to the latitude, then the *samamaṇḍala-śaṅku* becomes equal to the *trijyā*. This is because the Sun crosses [the prime vertical] right above the observer at midday. From this [information], let the *samamaṇḍala-śaṅku* be ascertained [by the rule of three] when the declination is less than the latitude. When the *saumyakrānti* is equal to the *akṣa*, then the [*samamaṇḍala*] *śaṅku* is the *trijyā*. When the *samaśaṅku* is less than that (the *trijyā*), the corresponding *krānti* may be obtained.

The underlying mathematical equation, based on which the different cases have been described in the above verses, is given by the rule of three, and may be expressed as

$$R\cos z_0 = \frac{R\sin\delta}{R\sin\phi}R. \tag{3.133}$$

३.२१ समशङ्कुना अर्कस्फुटः

3.21 True longitude of the Sun from the *samaśaṅku*

अक्षज्याघ्नः परक्रान्त्या हृतः शङ्कुः स दोर्गुणः ।
तच्चापमेव भानुः स्यात् चक्रार्धं वा तदूनितम् ॥ ५२ ॥

akṣajyāghnaḥ parakrāntyā hṛtaḥ śaṅkuḥ sa dorguṇaḥ |
taccāpameva bhānuḥ syāt cakrārdhaṃ vā tadūnitam || 52 ||

The [*sama*]*śaṅku* [as defined in the previous verse] multiplied by Rsine of the latitude and divided by the Rsine of the maximum declination gives the *dorjyā*. The corresponding arc or that reduced from 180 degrees is the [longitude of] the Sun.

Here the term *dorjyā* refers to the Rsine of the longitude of the Sun. It is given by

$$\begin{aligned} dorjy\bar{a} &= \frac{\acute{s}a\dot{n}ku \times ak\d{s}ajy\bar{a}}{parakr\bar{a}nti} \\ \text{or} \qquad R\sin\lambda &= \frac{R\cos z_0 \times R\sin\phi}{R\sin\varepsilon}, \end{aligned} \tag{3.134}$$

where λ is the *sāyana* (tropical) longitude of the Sun, z_0 its zenith distance when it is on the prime vertical and ε the maximum declination, which is taken to be 24°. As $R\sin\delta = R\sin\lambda\sin\varepsilon$, the above relation is equivalent to $R\sin\delta = R\cos z_0\sin\phi$, which is the same as (3.130) derived earlier.

In (3.134) when $R\sin\lambda$ is known, the corresponding arc is not determined uniquely as $\sin(180^\circ - \lambda) = \sin\lambda$. This fact is expressed thus in *Yukti-dīpikā*.

---

[52] {TS 1977}, pp. 214–5.

दोर्ज्याचापसमो भानुः तत्काले प्रथमे पदे ।
द्वितीये तु पदे वेद्यं चक्रार्धं स्यात् तदूनितम् ॥ [53]

When the Sun is in the first quadrant, the longitude of it is equal to the arc of the *dorjyā*; if it is in the second quadrant, then the longitude is equal to 180 degrees minus the arc.

३.२२ समशङ्कोः अङ्गुलात्मकः कर्णः

3.22 Hypotenuse of the shadow from the *samaśaṅku* in *aṅgulas*

लम्बाक्षज्ये विषुवद्भार्कघ्ने क्रान्तिजीवया भक्ते ।
सममण्डलगे भानौ कर्णौ तावङ्गुलात्मकौ स्पष्टौ ॥ ५३ ॥

lambākṣajye viṣuvadbhārkaghne krāntijīvayā bhakte |
samamaṇḍalage bhānau karṇau tāvaṅgulātmakau spaṣṭau ||53||

The Rcosine and the Rsine of latitude multiplied separately by the *viṣuvadbhā* and twelve [respectively], when divided by the Rsine of the declination of the Sun give the true hypotenuse [of the *śaṅku*] in *aṅgulas* when the Sun is on the prime vertical.

In Fig. 3.23, *S* is the Sun on the prime vertical and *N* is the foot of perpendicular drawn from the Sun on to the horizon. *Z* is the zenith and *OX* the usual *dvādaśāṅgula-śaṅku* (the gnomon described earlier in this chapter, verses 1–3). Here *ZSE* represents the prime vertical.

The two expressions for the hypotenuse (*CX*) prescribed in the verse are:

$$karṇa = \frac{lambajyā \times viṣuvadbhā}{krāntijīvā}$$

$$\text{and} \qquad karṇa = \frac{akṣajyā \times arka}{krāntijīvā}. \tag{3.135}$$

The term *arka* literally means the Sun. In this context it refers however to the number 12. Therefore the above expressions for the *karṇa* reduce to

$$CX = \frac{R\cos\phi \times 12\tan\phi}{R\sin\delta}$$

$$\text{and} \qquad CX = \frac{R\sin\phi \times 12}{R\sin\delta}, \tag{3.136}$$

which are the same. Using the expression for the *samamaṇḍala-śaṅku* given by (3.130), the above equations reduce to

$$CX = \frac{12 \times R}{R\cos z_0}. \tag{3.137}$$

---

[53] {TS 1977}, p. 15.

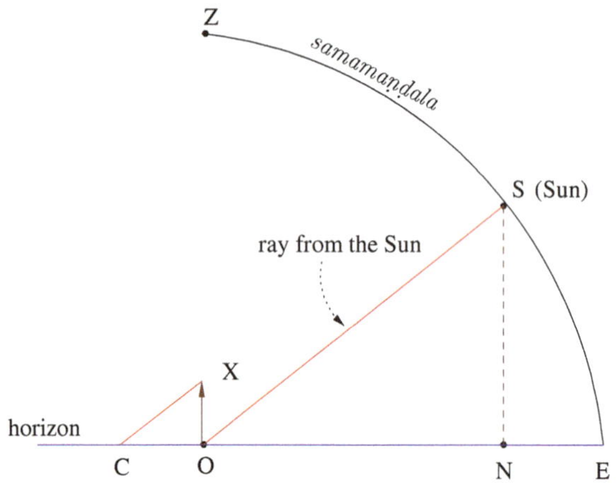

Fig. 3.23 Length of the hypotenuse in terms of the *samamaṇḍala-śaṅku*.

This is the *chāyākarṇa*, K, when the Sun is on the prime vertical. The rationale behind it can be understood with the help of Fig. 3.23. The two triangles COX and ONS are similar. Hence

$$\frac{CX}{OS} = \frac{OX}{NS}$$

$$\text{or} \qquad CX = \frac{OX \times OS}{NS}$$

$$= \frac{12 \times R}{R \cos z_0}. \tag{3.138}$$

३.२३ प्रकारान्तरेण कर्णानयनम्

3.23 Obtaining the hypotenuse of the shadow by a different method

मध्यच्छाया यदा मध्ये विषुवत्समरेखयोः ।
तन्मध्याह्नभवः कर्णः विषुवच्छायया हतः ॥ ५४ ॥
मध्याह्नाग्राङ्गुलैर्भक्तः कर्णः स्यात् सममण्डले ।

madhyacchāyā yadā madhye viṣuvatsamarekhayoḥ |
tanmadhyāhnabhavaḥ karṇaḥ viṣuvacchāyayā hataḥ || 54 ||
madhyāhnāgrāṅgulairbhaktaḥ karṇaḥ syāt samamaṇḍale |

When the [tip of the] midday shadow lies between the *viṣuvadrekhā* and the *samarekhā*, then the hypotenuse of the shadow is multiplied by the equinoctial shadow and divided by the *agrā* in *aṅgulas* corresponding to noon that day. This gives the *karṇa* in *aṅgulas* (when the Sun is) on the prime vertical.

The terms *viṣuvadrekhā* and *samarekhā* refer to the equinoctial line (UV) and the east–west line (EW) in Fig. 3.8 respectively. The condition that the midday shadow should lie between these two lines implies that $0 < \delta < \phi$, whose significance is explained in Section 3.18. As the zenith distance at noon is $(\phi - \delta)$, the *chāyākarṇa* K at midday is given by

$$K = \frac{12}{\cos(\phi - \delta)}. \tag{3.139}$$

It has already been noted that the equinoctial midday shadow, the *viṣuvacchāyā*, is $12\tan\phi$. Also, the *arkāgrā* or *agrajīvā* in *aṅgulas* on any day is

$$arkāgrā = K\frac{\sin\delta}{\cos\phi}. \tag{3.140}$$

Using (3.139) in the above equation, we have

$$arkāgrā = \frac{12}{\cos(\phi - \delta)}\frac{\sin\delta}{\cos\phi}. \tag{3.141}$$

The *karṇa* (in *aṅgulas*) when the Sun is on the prime vertical is stated to be

$$\begin{aligned} samamaṇḍala\text{-}karṇa &= chāyākarṇa \times \frac{viṣuvacchāyā}{agrajīvā} \\ &= \frac{12}{\cos(\phi-\delta)} \times \frac{12\tan\phi}{\left(\frac{12\sin\delta}{\cos(\phi-\delta)\cos\phi}\right)} \\ &= 12\frac{\sin\phi}{\sin\delta}. \end{aligned} \tag{3.142}$$

As the *samamaṇḍala-karṇa* (in *aṅgulas*) is $\frac{12}{\cos z_0}$, the above relation is equivalent to

$$\cos z_0 = \frac{\sin\delta}{\sin\phi}, \tag{3.143}$$

which was obtained earlier (3.130).

३.२४ समशङ्कुना गतैष्यप्राणाः

3.24 The duration elapsed and yet to elapse from the *samamaṇḍala-śaṅku*

समण्डलशङ्कुः लम्बघ्नः त्रिज्यया हृतः ॥ ५५ ॥
उन्मण्डलात्[54] द्युवृत्तज्या, त्रिज्याघ्ना द्युज्यया हृता ।
तच्चापं चरचापाढ्यं गतैष्यासव एव हि ॥ ५६ ॥

[54] The reading in both the printed editions is: उन्मण्डला ।

samamaṇḍalaśaṅkuḥ lambaghnaḥ trijyayā hṛtaḥ ||55||
unmaṇḍalāt dyuvṛttajyā, trijyāghnā dyujyayā hṛtā |
taccāpaṃ caracāpāḍhyaṃ gataiṣyāsava eva hi ||56||

The *samamaṇḍala-śaṅku* multiplied by the *lambaka* and divided by the *trijyā* is the *dyuvṛttajyā* from the *unmaṇḍala*. This multiplied by the *trijyā* and divided by the *dyujyā* gives the arc corresponding to it, and that added to the *cara* gives the *prāṇas* elapsed and yet to elapse.

In the above verse the procedure is given for determining the time elapsed since the sunrise till the Sun reaches the prime vertical. Consider Fig. 3.24. Here S is the Sun on the prime vertical and S_r is the sunrise point. SB is the part of the diurnal circle between the 6 o'clock circle and the prime vertical. The desired duration is obtained in two steps.

1. The duration corresponding to the diurnal motion of the Sun between the 6 o'clock circle and the prime vertical, the segment ES' corresponding to h, and
2. The duration corresponding to interval between the sunrise point and the 6 o'clock circle, which is the *cara* ($S'_r\hat{P}E$).

Let BS be the segment on the diurnal circle corresponding to ES'. The *dyuvṛttajyā* is the Rsine of h reduced to the diurnal circle, and is thus given by $R\cos\delta\sin h$.

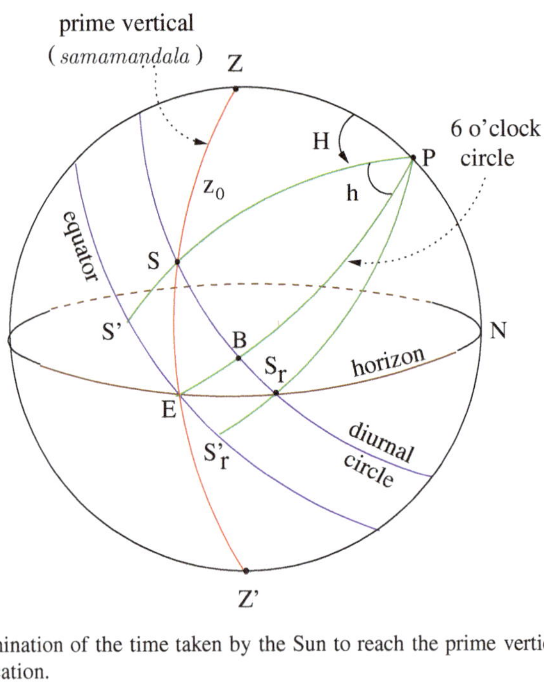

Fig. 3.24 Determination of the time taken by the Sun to reach the prime vertical from its rise at the observer's location.

The expression given for the *dyuvṛttajyā* is:

$$dyuvṛttajyā = \frac{samamaṇḍala\text{-}śaṅku \times lambaka}{trijyā}$$

$$\text{or} \qquad R\cos\delta\sin h = \frac{R\cos z_0 \times R\cos\phi}{R}. \qquad (3.144)$$

In Fig. 3.24, considering the spherical triangle *PSE*, we have $SE = 90 - z_0$, $PS = 90 - \delta$, $P\hat{E}S = 90 - \phi$ and $S\hat{P}E = h$. Applying the sine formula we get

$$\frac{\sin h}{\sin SE} = \frac{\sin P\hat{E}S}{\sin PS}$$

$$\text{or} \qquad \sin h = \frac{\cos z_0 \cos\phi}{\cos\delta}, \qquad (3.145)$$

which is the same as (3.144). The arc of this (h), converted into *prāṇas*, is added to the *cara* in *prāṇas* represented by the arc ES'_r, to obtain the time elapsed since sunrise till the Sun reaches the prime vertical.

३.२५ समशङ्कुना नतप्राणाः

3.25 Hour angle from the *samamaṇḍala-śaṅku*

लम्बघ्नः समशङ्कुः सः द्युज्याभक्तोऽथ तत्कृतिम् ।
त्यक्त्वा त्रिज्याकृतेर्मूलं चापितं हि नतासवः ॥ ५७ ॥

lambaghnaḥ samaśaṅkuḥ saḥ dyujyābhakto'tha tatkṛtim |
tyaktvā trijyākṛtermūlaṃ cāpitaṃ hi natāsavaḥ || 57 ||

The square of the product of the *samamaṇḍala-śaṅku* and the *lambaka* divided by the *dyujyā* is subtracted from the square of the *trijyā*. The arc of the square root of this is the *natāsavaḥ*.

The term *natāsavaḥ* refers to the hour angle of a celestial object, H, and the *nata-jyā* is $R\sin H$. In the above verse, Nīlakaṇṭha gives the expression for the Rsine of the hour angle (H) of the Sun when it is on the prime vertical:

$$nata\text{-}jyā = \left[trijyā^2 - \left(\frac{samamaṇḍala\text{-}śaṅku \times lambaka}{dyujyā} \right)^2 \right]^{\frac{1}{2}}$$

$$\text{or} \qquad R\sin H = \left[R^2 - \left(\frac{R\cos z_0 \times R\cos\phi}{R\cos\delta} \right)^2 \right]^{\frac{1}{2}}. \qquad (3.146)$$

We have already shown in (3.145) that

$$\sin h = \frac{\cos z_0 \cos\phi}{\cos\delta}. \qquad (3.147)$$

Since $H = 90^\circ - h$ (see Fig. 3.24), we have

$$\cos H = \frac{\cos z_0 \times \cos\phi}{\cos\delta}$$

$$\text{or} \qquad \sin H = \left[1 - \left(\frac{\cos z_0 \times \cos\phi}{\cos\delta}\right)^2\right]^{\frac{1}{2}}, \tag{3.148}$$

which is the same as (3.146).

३.२६ प्रकारान्तरेण नतप्राणाः

3.26 Hour angle by another method

> **सममण्डलगा छाया त्रिज्याघ्ना द्युज्यया हृता ।**
> **चापिता वा नतप्राणाः कोट्या वा सर्वदा तथा ॥ ५८ ॥**
>
> *samamaṇḍalagā chāyā trijyāghnā dyujyayā hṛtā |*
> *cāpitā vā natapraṇāḥ koṭyā vā sarvadā tathā || 58 ||*
>
> The arc of the product of the *samamaṇḍala-chāya* and the *trijyā* divided by the *dyujyā* is always the hour angle. This could also be obtained from the *koṭi*.

Here is another formula for the hour angle of the Sun when it is on the prime vertical. The term *samamaṇḍala-chāyā* refers to the *mahācchāyā* when the Sun is on the prime vertical. This is given by *ON* in Fig. 3.23. Since $ZS = Z\hat{O}S = O\hat{S}N = z_0$, the *chāyā* of the *samamaṇḍala-śaṅku*, *SN*, is given by $ON = OS \sin z_0 = R \sin z_0$. The expression for the *nata-jyā* given in the above verse is:

$$\textit{nata-jyā} = \frac{\textit{samamaṇḍala-chāyā} \times \textit{trijyā}}{\textit{dyujyā}}$$

$$\text{or} \qquad R \sin H = \frac{R \sin z_0 \times R}{R \cos\delta}. \tag{3.149}$$

We arrive at the same result using the spherical triangle *PZS* in Fig. 3.24 and applying the sine formula. We have

$$\frac{\sin ZPS}{\sin ZS} = \frac{\sin PZS}{\sin PS}. \tag{3.150}$$

Since $ZS = z_0$, $Z\hat{P}S = H$, $P\hat{Z}S = 90$ and $PS = 90 - \delta$ the above equation reduces to

$$\sin H = \frac{\sin z_0}{\cos\delta}, \tag{3.151}$$

which is the same as (3.149). In the fourth quarter of the above verse, it is stated that the hour angle can also be obtained from the *koṭi*. The term *koṭi* here refers to $R \cos z_0$. Hence, it is suggested that the *samamaṇḍala-chāyā* ($R \sin z_0$) can be obtained from the *samamaṇḍala-śaṅku* ($R \cos z_0$) using the relation

$$R \sin z_0 = \sqrt{R^2 - (R \cos z_0)^2}. \tag{3.152}$$

३.२७ समशङ्कोः क्षितिज्या

3.27 *Kṣitijyā* from the *samamaṇḍala-śaṅku*

अक्षज्याघ्नौ समौ शङ्कू त्रिज्यालम्बकभाजितौ ।
क्रान्त्यर्काग्रे तयोः कृत्योः भेदमूलं क्षितेर्गुणः ॥ ५९ ॥

akṣajyāghnau samau śaṅkū trijyālambakabhājitau |
krāntyarkāgre tayoḥ kṛtyoḥ bhedamūlaṃ kṣiterguṇaḥ || 59 ||

The product of the *akṣajyā* and *samamaṇḍala-śaṅku* [kept at two different places] divided by the *trijyā* and the *lambaka* are the *krānti* and the *arkāgrā* respectively. The square root of the difference of their squares is the *kṣitijyā*.

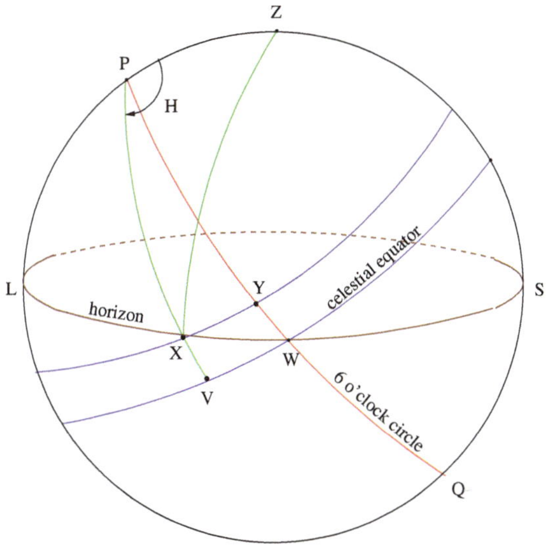

Fig. 3.25 Determination of the *kṣitijyā* from the *samamaṇḍala-śaṅku*.

In Fig. 3.25, X and Y represent the Sun when it is on the horizon, and on the 6 o'clock circle respectively. The Rsine of the arc length XY along the diurnal circle is the *kṣitijyā* or *kujyā* (earth-sine). In other words, the *kṣitijyā* is the Rsine of ascensional difference reduced to the diurnal circle. The arc $WV = W\hat{P}X$ is the *cara* or the ascensional difference. Hence

$$kṣitijyā = \frac{R\cos\delta \times R\sin WV}{R}. \qquad (3.153)$$

The expressions for the *krānti* and the *arkāgrā* as given in the above verse are:

$$krānti = \frac{akṣajyā \times samamaṇḍala\text{-}śaṅku}{trijyā}$$

$$\text{and} \qquad arkāgrā = \frac{akṣajyā \times samamaṇḍala\text{-}śaṅku}{lambaka}. \tag{3.154}$$

Or,

$$krānti = \frac{R\sin\phi \times R\cos z_0}{R}, \quad arkāgrā = \frac{R\sin\phi \times R\cos z_0}{R\cos\phi}. \tag{3.155}$$

Substituting for $R\cos z_0$ from (3.133), the above equations reduce to

$$krānti = R\sin\delta, \quad arkāgrā = \frac{R\sin\delta}{R\cos\phi}, \tag{3.156}$$

as expected. The *kṣitijyā* is said to be given by

$$kṣitijyā = \sqrt{arkāgrā^2 - krānti^2}. \tag{3.157}$$

Substituting for the *arkāgrā* and the *krānti* from (3.156), we find

$$kṣitijyā = \frac{R\sin\delta\, R\sin\phi}{R\cos\phi}. \tag{3.158}$$

Now, WV is the ascensional difference $\Delta\alpha$, which is given by

$$R\sin WV = R\sin\Delta\alpha = R\,\frac{\sin\phi\sin\delta}{\cos\phi\cos\delta}. \tag{3.159}$$

Hence the *kṣitijyā* given by (3.157) reduces to the standard form, $\cos\delta \times R\sin\Delta\alpha$.

३.२८ दशप्रश्नाः

3.28 The ten problems

इह शङ्कुनतक्रान्तिदिगग्राऽक्षेषु पञ्चसु ।
द्वयोर्द्वयोरानयनं दशधा स्यात् परैस्त्रिभिः ॥ ६० ॥
सशङ्कवो नतक्रान्तिदिगक्षाः सनतास्तथा ।
अपक्रमदिगग्राक्षा दिगक्षौ क्रान्तिसंयुतौ ॥ ६१ ॥
दिगक्षाविति नीयन्ते द्वन्द्वीभूयेतरैस्त्रिभिः ।

iha śaṅkunatakrāntidigagrā'kṣeṣu pañcasu |
dvayordvayorānayanaṃ daśadhā syāt paraistribhiḥ || 60 ||
saśaṅkavo natakrāntidigakṣāḥ sanatāstathā |
apakramadigagrākṣā digakṣau krāntisaṃyutau || 61 ||
digakṣāviti nīyante dvandvībhūyetaraistribhiḥ |

Out of the five quantities *śaṅku, nata, krānti, digagrā* and *akṣa*, any two of them can be determined from the other three and this happens in ten different ways. Pairs from the sequences (i) *śaṅku, nata, krānti, digagrā* and *akṣa*; (ii) *nata, krānti, digagrā* and *akṣa*; (iii) *krānti, digagrā* and *akṣa*; (iv) *digagrā* and *akṣa*; are [formed and] determined with the other three.

The modern equivalents of the five quantities listed in the above verses and the notation used to represent them are given in Table 3.1.

Sanskrit name	Modern equivalent Rsine of	Notation
śaṅku	zenith distance	$R\sin z$
nata	hour angle	$R\sin H$
krānti	declination	$R\sin\delta$
digagrā	amplitude	$R\sin a$
akṣa	latitude	$R\sin\phi$

Table 3.1 The five quantities associated with the problem of the *daśapraśna*.

Out of these five quantities, four (the exception being the latitude of the observer) keep continuously changing with the diurnal motion of the Sun. Further, it is noted that if any three of them are given the other two can be determined. The next 26 verses (up to verse 87 of this chapter) describe how this can be done in each of the ten different ways, which forms the subject matter of the *daśapraśnāḥ* (the ten problems).

Both the *krānti* and the *apakrama* refer to the Rsine of the declination of the Sun. Similarly, the amplitude, *digagrā* is referred to by other names such as the *āśāgrā* or the *arkāgrā*. The terms *āśā* and *dik* have the same meaning, namely direction. In this context, the term *āśāgrā* refers to the angle between the vertical circle passing through the Sun and the prime vertical passing through the zenith and the east–west points on the horizon. All these quantities are indicated in Fig. 3.26.

The order in which the ten pairs are selected, as given in verse 61 and the first half of verse 62, is shown in Table 3.2.

Set	Pairs formed from this set
$\{z,H,\delta,a,\phi\}$	(z,H), (z,δ), (z,a), (z,ϕ)
$\{H,\delta,a,\phi\}$	(H,δ), (H,a), (H,ϕ)
$\{\delta,a,\phi\}$	(δ,a), (δ,ϕ)
$\{a,\phi\}$	(a,ϕ)

Table 3.2 The ten pairs that can be formed out of the five quantities associated with the *daśapraśnāḥ*.

Verses 62–87 describe the explicit procedure for the solution of these 'ten problems'. In the explanatory notes for the same, we derive the stated procedures from modern spherical trigonometry. However, Nīlakaṇṭha would have used a different methodology to arrive at these results. In fact, the detailed demonstration of the solution of each of these problems is presented in Jyeṣṭhadeva's *Yuktibhāṣā*. In Appendix D we present the *Yuktibhāṣā* method of solving the ten problems by giving the full derivation for two of them.

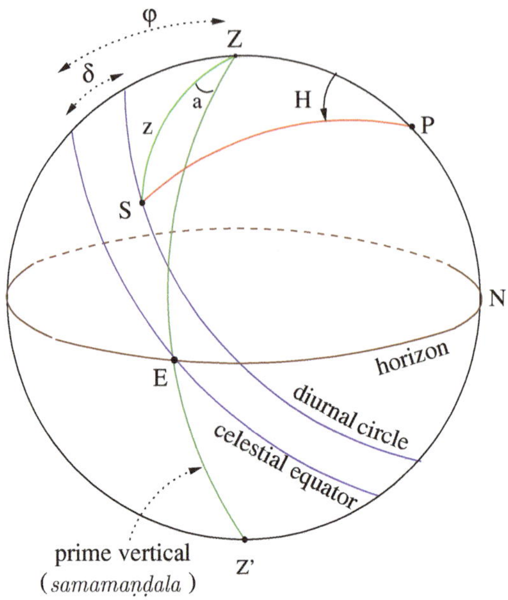

Fig. 3.26 Celestial sphere with markings of the five quantities, namely *śaṅku*, *nata*, *krānti*, *digagrā* and *akṣa*, associated with the *daśapraśnāḥ*.

३.२९ क्रान्ति-दिगग्राक्षैः शङ्कु-नत्यौ

3.29 Determination of the zenith distance and hour angle from the declination, amplitude and latitude (Problem 1)

आशाग्रा लम्बकाभ्यस्ता त्रिज्याभक्ता च कोटिका ॥ ६२ ॥
भुजाक्षज्या तयोर्वर्गयोगमूलं श्रुतिर्हरः ।
क्रान्त्यक्षवर्गौ तद्वर्गात् त्यक्त्वा कोटयौ तयोः पदे[55] ॥ ६३ ॥
कुर्यात् क्रान्त्यक्षयोर्घातं कोटयोर्घातं तथा परम् ।
सौम्ये गोले तयोर्योगात् भेदात् याम्ये तु घातयोः[56] ॥ ६४ ॥
आद्यघातेऽधिके सौम्ये योगभेदद्वयादपि ।
त्रिज्याघ्नात् हारवर्गाप्तः शङ्कुरिष्टदिगुद्भवः ॥ ६५ ॥

---

[55] The prose order: तयोः पदे कुर्यात्; (ते) कोटयौ (स्याताम्) ।

[56] The prose order: परम् (अनन्तरं) तयोः घातयोः योगात् सौम्ये गोले, याम्ये तु भेदात्, त्रिज्याघ्ना हारवर्गाप्तः इष्टदिगुद्भवः शङ्कुः (भवति)। (पुनः) सौम्ये गोले आद्यघातेऽधिके, (सति) योगभेदद्वयादपि (योगात् भेदाच्च त्रिज्यया निहत्य, हारकवर्गेण विभज्य, शङ्कू आनेतव्यौ)।

Here the commentator observes in *Yukti-dīpikā*: तत्र आद्यः पूर्वापरसूत्रतः सौम्यदिग्गतः द्वितीयो याम्यदिग्गतः।

छाया तत्कोटिराशाग्राकोटिघ्ना सा द्युजीवया [57] ।
भक्ता नतज्या क्रान्त्यक्षदिगग्राभिर्भवेदिति ॥ ६६ ॥
क्रान्त्यक्षघाते तत्कोटयोः घातात् याम्येऽधिके सति ।
नेष्टः शङ्कुर्भवेत् सौम्ये हाराच्चापक्रमेऽधिके ॥ ६७ ॥

āśāgrā lambakābhyastā trijyābhaktā ca koṭikā||62||
bhujākṣajyā, tayorvargayogamūlaṃ śrutirharaḥ|
krāntyakṣavargau tadvargāt tyaktvā koṭyau tayoḥ pade||63||
kuryāt, krāntyakṣayorghātaṃ koṭyorghātaṃ tathā param|
saumye gole tayoryogāt bhedāt yāmye tu ghātayoḥ||64||
ādyaghāte'dhike saumye yogabhedadvayādapi|
trijyāghnāt hāravargāptaḥ śaṅkuriṣṭadigudbhavaḥ||65||
chāyā tatkoṭirāśāgrākoṭighnā sā dyujīvayā|
bhaktā natajyā krāntyakṣadigagrābhirbhavediti||66||
krāntyakṣaghāte tatkoṭyoḥ ghātāt yāmye'dhike sati|
neṣṭaḥ śaṅkurbhavet saumye hārāccāpakrame'dhike||67||

The *āśāgrā* multiplied by the *lambaka* and divided by the *trijyā* is the *koṭi*. The *bhujā* is the *akṣajyā*. The square root of the sum of their squares is the hypotenuse and it is the *hara* [or *hāra*, the divisor, which will be used later].

Then find the square roots of the squares of the *krānti* and the *akṣa* subtracted from it. They form the *koṭis*. Similarly find the products of the *krānti* and the *akṣa* and also their *koṭis*.

The sum and the differences of the products are multiplied by the *trijyā* and divided by the square of the divisor [when the Sun is] in the northern and southern hemispheres respectively. This gives the *śaṅku* that is formed in the desired direction. If the first product is greater than the second one, in the northern hemisphere, then the *śaṅku* is obtained from both the sum and the difference.

Its (the *śaṅku*'s) *koṭi* (compliment) is the *chāyā* (the shadow). When that is multiplied by the *koṭi* (compliment) of the *āśāgrā* and divided by the *dyujyā*, the resultant is the *natajyā*. Thus the *śaṅku* and the *nata* can be obtained from the *krānti*, the *akṣa* and the *āśāgrā*.

In the southern hemisphere, when the product of the *krānti* and the *akṣa* is greater than the product of the *koṭis*, there is no *śaṅku* [i.e. no solution for z with $z < 90°$]. Similarly, in the nothern hemisphere, when the *apakrama* is greater than the divisor, there is no *śaṅku*.

Here, the problem is to obtain the zenith distance (*śaṅku*) and hour angle (*nata*) in terms of declination (*krānti*), latitude (*akṣa*) and amplitude (*āśāgrā*), that is, z and H are to be determined in terms of δ, ϕ and a. It is to be understood that the amplitude in Indian astronomy is always less than $90°$ and is measured towards either the north or the south from the prime vertical.

Formula for *śaṅku*

For convenience, we arrive at the required expression in three stages. In the process of arriving at the expression for the *śaṅku* ($R\cos z$) a number of intermediate quantities are defined, and these are taken up first.

[57] The prose order: तत्कोटिः छाया। सा (छाया) आशाग्राकोटिघ्ना द्युजीवया भक्ता नतज्या (स्यात्)।

Stage 1: Definition of *hāra*

Nīlakaṇṭha defines the divisor (*hara* or *hāra*) to be the hypotenuse of a triangle *ABC* shown in Fig. 3.27(a). The sides *AB* and *BC* are defined to be the *bhujā* and the *koṭi* respectively. The expressions for the *bhujā* and the *koṭi* are given by

$$bhuj\bar{a} = ak\d{s}ajy\bar{a} \qquad ko\d{t}i = \frac{\bar{a}\acute{s}\bar{a}gr\bar{a} \times lambaka}{trijy\bar{a}}$$

$$\text{or} \qquad AB = R\sin\phi \qquad BC = \frac{R\sin a \times R\cos\phi}{R}. \tag{3.160}$$

The divisor, denoted by K in the following, is the hypotenuse AC of this triangle and is given by

$$K = AC = \sqrt{AB^2 + BC^2}. \tag{3.161}$$

Hence the square of the divisor which will be used later is given by

$$K^2 = R^2(\sin^2\phi + \cos^2\phi\sin^2 a). \tag{3.162}$$

Note:

Often the *akṣajyā* is simply referred to as the *akṣa*, the *krāntijyā* as the *krānti* and so on in the above verses and the verses to follow, including the examples discussed below. That is, the Rsine of a coordinate is simply referred to by the coordinate itself.

Stage 2: Definition of the *koṭis* in terms of *hāra* and their products

The *koṭis* (k_1 and k_2) are defined by

$$k_1 = \sqrt{K^2 - (R\sin\delta)^2}, \tag{3.163}$$

$$k_2 = \sqrt{K^2 - (R\sin\phi)^2}. \tag{3.164}$$

Substituting for K^2 in the above expressions we have

$$k_1 = R\sqrt{\sin^2\phi + \cos^2\phi\sin^2 a - \sin^2\delta}, \tag{3.165}$$

$$\text{and} \qquad k_2 = R\sqrt{\sin^2\phi + \cos^2\phi\sin^2 a - \sin^2\phi}. \tag{3.166}$$

Hence $k_2 = R\cos\phi\sin a$ is the same as BC defined earlier. Further, the following two products (denoted by the symbols X and Y) are defined thus:

$$X = R(|\sin\delta|)R\sin\phi \tag{3.167}$$

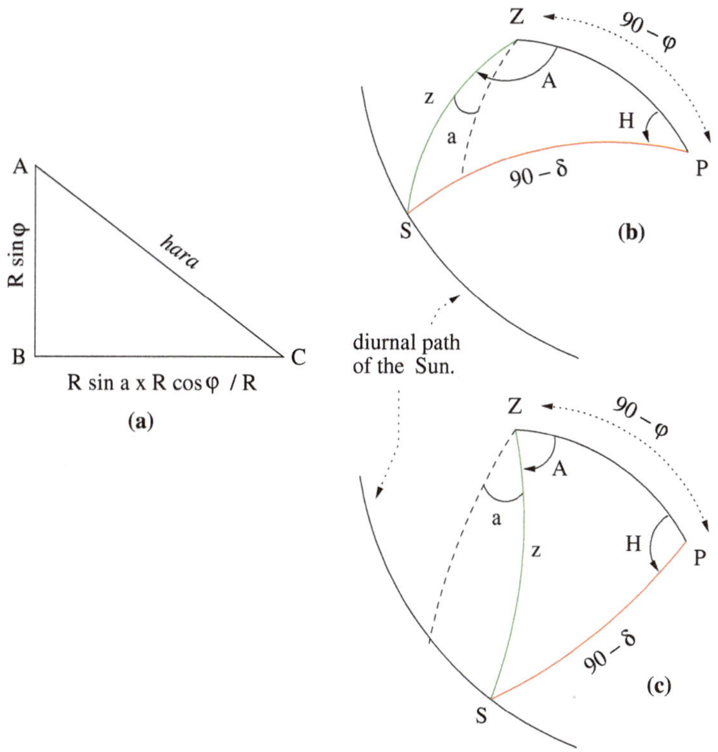

Fig. 3.27 Triangles used for arriving at the formula for the *śaṅku* ($R\cos z$) and the *nata* ($R\sin H$) in terms of the *krānti* (δ), *digagrā* (a) and *akṣa* (ϕ).

$$Y = k_1 k_2. \tag{3.168}$$

Substituting for k_1 and k_2, we find

$$Y = R^2 \cos\phi \sin a \sqrt{\sin^2\phi + \cos^2\phi \sin^2 a - \sin^2\delta}. \tag{3.169}$$

Stage 3: Expression for the *śaṅku*

The following results for the *śaṅku* are stated:

Case A: When the declination is north ($\delta > 0$)

$$\begin{aligned} R\cos z &= \frac{X \pm Y}{K^2} R \\ &= \frac{R^2(\sin\phi|\sin\delta| \pm \cos\phi \sin a \sqrt{\sin^2\phi + \cos^2\phi \sin^2 a - \sin^2\delta})}{R^2(\sin^2\phi + \cos\phi \sin^2 a)} R. \end{aligned} \tag{3.170}$$

This is valid when the first term in the numerator is greater than the magnitude of the second term ($X > Y$). When the first term is less than the magnitude of the second term ($X < Y$), only the +ve sign is to be considered.

Moreover, there is no solution for the *śaṅku* when

$$R\sin\delta > K = \sqrt{R^2\sin^2\phi + R^2\cos^2\phi\sin^2 a}, \tag{3.171}$$

since the term inside the square root in the discriminant becomes negative, and consequently the solution becomes complex.

Case B: When the declination is south ($\delta < 0$)

$$R\cos z = \frac{R^2(-\sin\phi|\sin\delta|) + \cos\phi\sin a\sqrt{\sin^2\phi + \cos^2\phi\sin^2 a - \sin^2\delta}}{R^2(\sin^2\phi + \cos^2\phi\sin^2 a)}R. \tag{3.172}$$

In this case, the magnitude of the first term is less than the second term. If it were to be otherwise, there would be no shadow.

Proof:

Though the expressions for *śaṅku* given by (3.170) and (3.172) seem to be involved, they can be derived in a straightforward manner by applying the cosine formula to the spherical triangle *PZS* shown in Fig. 3.27(b). Using the formula we have

$$\sin\delta = \cos z\sin\phi \pm \sin z\cos\phi\sin a, \tag{3.173}$$

when S is north or south of the prime vertical, corresponding to $A = 90 \mp a$. Now, making the substitution $\cos z = x$, we have

$$\sin\delta = x\sin\phi \pm \sqrt{1-x^2}\cos\phi\sin a. \tag{3.174}$$

Therefore,

$$\pm\sqrt{1-x^2}\cos\phi\sin a = \sin\delta - x\sin\phi. \tag{3.175}$$

Squaring the equation and rearranging the terms, we obtain the following quadratic equation in x:

$$(\sin^2\phi + \cos^2\phi\sin^2 a)\,x^2 - (2\sin\phi\sin\delta)\,x + (\sin^2\delta - \cos^2\phi\sin^2 a) = 0. \tag{3.176}$$

The roots of the above equation are:

$$x = \frac{\sin\phi\sin\delta \pm \sqrt{\sin^2\phi\sin^2\delta - (\sin^2\phi + \cos^2\phi\sin^2 a)(\sin^2\delta - \cos^2\phi\sin^2 a)}}{(\sin^2\phi + \cos^2\phi\sin^2 a)}. \tag{3.177}$$

Simplifying the expression within the square root sign in the above equation, we find

$$x = \cos z = \frac{(\sin\phi\sin\delta \pm \cos\phi\sin a\,\sqrt{\sin^2\phi + \cos^2\phi\sin^2 a - \sin^2\delta})}{(\sin^2\phi + \cos^2\phi\sin^2 a)}. \tag{3.178}$$

It can be seen that (3.178) is equivalent to (3.170) and (3.172). When the declination is north (δ positive), $|\sin\delta| = \sin\delta$. In that case, when $X > Y$, both the solutions for $\cos z$ in (3.178) are positive and $z < 90°$. These correspond to the situation with the Sun S to the north or south of the prime vertical with the same value of the *āśāgrā*, $R\sin a$, but with different values of the azimuth $A = 90° \pm a$. When $X < Y$, the second solution for $\cos z$ is negative or $z > 90°$, and there is no *śaṅku* as the Sun is below the horizon. Also, when $R\sin\delta > K$, the solutions for $\cos z$ are complex and there is no *śaṅku*.

When the declination is south (δ negative), $|\sin\delta| = -\sin\delta$, the first term in (3.178) becomes negative, and a physical solution for $\cos z$ (positive value) is possible only when the +ve sign is taken in the second term and $Y > X$.

Formula for the *nata*

The expression given for the *natajyā* ($R\sin H$) may be written as:

$$natajy\bar{a} = \frac{ch\bar{a}y\bar{a} \times \bar{a}\acute{s}\bar{a}gr\bar{a}ko\d{t}i}{dyujy\bar{a}}, \tag{3.179}$$

where the *chāyā* and the *āśāgrākoṭi* (the Rsine of amplitude) are defined to be compliments of the *śaṅku* and the *āśāgrā* respectively. That is,

$$\begin{aligned} ch\bar{a}y\bar{a} &= \sqrt{trijy\bar{a}^2 - \acute{s}a\dot{n}ku^2} \\ &= \sqrt{R^2 - (R\cos z)^2} = R\sin z, \end{aligned} \tag{3.180}$$

and
$$\begin{aligned} \bar{a}\acute{s}\bar{a}gr\bar{a}ko\d{t}i &= \sqrt{trijy\bar{a}^2 - \bar{a}\acute{s}\bar{a}gr\bar{a}^2} \\ &= \sqrt{R^2 - (R\sin a)^2} = R\cos a. \end{aligned} \tag{3.181}$$

Substituting for the *chāyā*, *āśāgrākoṭi* and *dyujyā* ($= R\cos\delta$), the expression for the *natajyā* becomes

$$R\sin H = \frac{R\sin z\, R\cos a}{R\cos\delta}. \tag{3.182}$$

Using the spherical triangle *PZS* and applying the sine formula, we have

$$\frac{\sin A}{\sin(90-\delta)} = \frac{\sin H}{\sin z}. \tag{3.183}$$

Since $A = (90 \pm a)$, the above equation reduces to

$$\sin H = \frac{\sin z \cos a}{\cos \delta}, \tag{3.184}$$

which is the same as (3.182).

उदाहरणम् (An example)

In *Laghu-vivṛti*, an example of the above procedure for finding z and H from δ, a and ϕ is discussed. First, the values of δ, a and ϕ are stated in the form of a verse.

क्रान्तिः शरद्व्यश्विमिता दिगग्रा याम्या भवेद् भूगुणवेददस्राः ।
महीधराम्भोधिरसैर्मितोऽक्षः तत्राशु शङ्कुं वद तन्नतं च ॥

The *krānti* is 225′, the *digagrā* which is to the south is 2431′ (and) the *akṣa* is 647′. Now tell me quickly what are the *śaṅku* and the *nata*.

In the above verse, the values given of the *krānti*, *digagrā* and *akṣa* given are those of the Rsines of the quantities. That is,

$$R\sin\delta = 225' \qquad R\sin a = 2431' \qquad R\sin\phi = 647'.$$

The procedure to be adopted in determining the *śaṅku* and the *nata* from the above values is described. The numerical values obtained in the intermediate stages of the calculation are also explicitly given. In the following we give the passage from *Laghu-vivṛti*:

तत्र तावत् महीधराम्भोधिरसप्रमितस्य अक्षस्य कोटिज्यारूपो लम्बकः त्रिज्यावृत्तगतः सप्तपर्वतगुणवह्निसङ्ख्यः। आशाग्रा च याम्या अध्यर्धराशिज्या भूगुणवेददस्रसङ्ख्या।[58] तामाशाग्रां लम्बकेन निहत्य त्रिज्यया विभज्य लब्धा तत्कोटिवृत्तगता कोटिज्या वस्वष्टवह्निदस्रसङ्ख्या। अक्षज्या च तद्भुजा ततस्तयोर्वर्गयोगमूलतुल्यस्तत्कर्णो वेदाश्वयुगदस्रसङ्ख्यः। अयमेव हारकत्वेन पश्चादुपादीयते। ततस्तस्यैकस्यैव हारकसंज्ञितस्य कर्णस्य वर्गतः पृथक् क्रान्तिवर्गं अक्षवर्गं च विशोध्य शिष्टमूलतुल्ये क्रान्त्यक्षकोट्यौ क्रमेण स्याताम्। तत्राक्षकोटिस्तावत् पूर्वानीताक्षकोटिज्यैव वस्वष्टवह्निदस्रतुल्या। क्रान्तिकोटिस्तु वेदाङ्गयुगदस्रतुल्या। अथ क्रान्त्यक्षयोर्घातो बाणाद्रिभूतशरमनुसङ्ख्यः। तत् कोट्योर्घातश्च दन्ताभ्रकृताष्टवसुभूतसङ्ख्य इतीयत्पर्यन्तं कर्म याम्यसौम्ययोरुभयोराशाग्रयोः समानमेव। अतःपरं याम्यायामाशाग्रायां तयोर्घातयोर्विश्लेषः। सौम्याशाग्रा पुनर्याम्यगोळे न सम्भवति। तयोर्याम्याशाग्राया तयोर्घातयोरन्तरं सप्तेषुवेदाष्टगुणाद्रिशरसङ्ख्यः। तस्मात् त्रिज्यया निहतात् हारकस्य पूर्वोदितस्य वर्गेणावाप्त इष्टशङ्कुर्गुणदस्रयमळाग्निसङ्ख्यः। तत् त्रिज्यावर्गविश्लेषमूलं तत्कोटिरूपेष्टच्छाया षण्णवरुद्रसङ्ख्या, तां पुनराशाग्राकोट्याध्यर्धराशिज्यातुल्यया निहत्य द्युज्यया विभजेत्। तत्र लब्धा नतज्या वसुवेदाष्टसङ्ख्या स्यात्। तस्या नतज्यायाश्चापं नतासव इति। अत्र च नतासून् दिनार्धाद्विशोध्य शिष्टासुभ्यो जीवां

[58] The reading found in the printed edition {TS 1958} is: 'आशाग्रा च याम्यादध्यर्धराशिज्या। भूगुणवेददस्रसङ्ख्या'। We have presented a corrected version above.

गृहीत्वा तत्र चरज्यां च याम्योदग्गोळयोः ऋणधने कृत्वा क्षितिजादुन्नतभागज्यामानीय द्युज्यालम्बकघातेन निहत्य त्रिज्यावर्गेण विभज्य लब्धोऽप्ययमेव शङ्कुः। तत्र छायाकोटिरूपाया द्युवृत्तगताया नतज्यायाश्च छायाबाहोश्च मिथस्तुल्यत्वाद् अयं शङ्कुर्वह्निकोणगत इति बोद्धव्यम्॥

In the measure of a circle whose radius is the *trijyā*, the value of the *lambaka*, which is in the form of the Rcosine corresponding to [the Rsine of] the *akṣa* whose measure is 647, is 3377. The *āśāgrā* is south. [And is equal to] the Rsine of one and a half *rāśis* (*adhyardha-rāśis*), whose numerical value is 2431. The *koṭijyā* that is obtained by multiplying that *āśāgrā* by the *lambaka* and dividing by the *trijyā*, along the *koṭivṛtta* is numerically equal to 2388. Since the *akṣajyā* is its *bhujā*, the *karṇa* given by the square root of the sum of the squares of those two is equal to 2474.

It is only this that will be considered as the divisor later. Then, having subtracted the squares of the *krānti* and the *akṣa* separately from the square of the *karṇa*, that was just described as the divisor [above], the square root of the remaining results will be equal to the *krānti-koṭi* and the *akṣa-koṭi*, respectively. There the *akṣa-koṭi* will be the same as the *akṣa-koṭijyā* 2388 obtained earlier. But the *krānti-koṭi* will be equal to 2464. Now the product of the *krānti* and the *akṣa* is 145575. The product of their *koṭis* is 5884032. Thus till now the process is the same for both the north and the south *āśāgrās*. From now on, if the *āśāgrā* is south, the difference of their [i.e. *krānti*, *akṣa* and *krānti-koṭi*, *akṣa-koṭi*] products [has to be taken]. The case of the *āśāgrā* being north does not arise in the southern hemisphere.

If the *āśāgrā* is south, the difference of those [two] products is 5738457. This [difference] multiplied by the *trijyā* and divided by the square of the divisor obtained earlier, is the desired *śaṅku*, whose value is 3223. The square root of the difference of the squares of that [*śaṅku*] and *trijyā* is the desired *chāyā* [shadow], which is the *koṭi* of that [*śaṅku*, and] whose value is 1196. This again has to be multiplied by the *āśāgrā-koṭi*, which is equal to 45°, and divided by the *dyujyā*. The *natajyā* thus obtained will be numerically equal to 848. The arc corresponding to that *natajyā* [in *prāṇas*] is called the *natāsus*. The same *śaṅku* is also obtained, by subtracting these *natāsus* from half the day, and taking the Rsine of that, and applying this *carajyā* either negatively or positively depending upon the southern or northern hemispheres [respectively], and taking the Rsine of the arc which is above the horizon, and multiplying [it] with the product of the *dyujyā* and the *lambaka* and dividing by the square of the *trijyā*. As the value of the *natajyā* in the measure of the *dyuvṛtta*, which is in the form of the Rcosine of the shadow, and the value of the Rsine of the shadow are same as each other, it is to be understood that this *śaṅku* is in the direction of the south–east.

The passage above explains the procedure involved in problem 1 with a numerical example. Here the Rsine values of the declination, the *āśāgrā* and the latitude are given to be

$$R\sin\delta = 225' \qquad R\sin a = 2431' \qquad R\sin\phi = 647'.$$

From these we have to find out the values of $R\cos z$ and $R\sin H$. This is done in several steps. First we obtain *lambaka* ($R\cos\phi$) from the given value of the *akṣajyā* ($R\sin\phi$).

$$\begin{aligned} lambaka &= \sqrt{(trijy\bar{a})^2 - (ak\d{s}ajy\bar{a})^2} \\ &= \sqrt{R^2 - (R\sin\phi)^2} \end{aligned}$$

$$= \sqrt{(3438)^2 - (647)^2}$$
$$\approx 3377.$$

Now a quantity, the *koṭi*, is defined in terms of the *lambaka* as follows.

$$\begin{aligned} koṭi &= \frac{\bar{a}\acute{s}\bar{a}gr\bar{a} \times lambaka}{trijy\bar{a}} \\ &= \frac{R\sin a \times R\cos\phi}{R} \\ &= \frac{2431 \times 3377}{3438} \\ &\approx 2388. \end{aligned}$$

Here one may conceive of a right-angled triangle (*ABC*) with one side *AB* (the *bhujā*) representing the given *akṣajyā* ($R\sin\phi$) and the other side *BC* representing the (*koṭi*) obtained above. Evidently, the square root of the sum of the squares of the two sides *AB* and *BC* gives the hypotenuse *AC* (the *karṇa*) denoted by K. That is

$$\begin{aligned} karṇa &= \sqrt{(bhuj\bar{a})^2 + (koṭi)^2} \\ &= \sqrt{(2388)^2 + (\,647)^2} \\ &\approx 2474. \end{aligned}$$

As the above value of *karṇa* will be used as the denominator in further calculations (see (3.170) or (3.172)), it is also called the divisor. Now the text prescribes that the squares of the $R\sin\phi$ (the *akṣa*) and $R\sin\delta$ (the *krānti*) be subtracted from the square of the divisor separately, and the square roots taken in order to get the *akṣakoṭi* and the *krāntikoṭi* respectively:

$$\begin{aligned} akṣakoṭi &= \sqrt{(karṇa)^2 - (akṣa)^2} \\ &= \sqrt{(2474)^2 - (647)^2} \\ &\approx 2388 \\ \text{and} \qquad kr\bar{a}ntikoṭi &= \sqrt{(karṇa)^2 - (kr\bar{a}nti)^2} \\ &= \sqrt{(2474)^2 - (225)^2} \\ &\approx 2464. \end{aligned}$$

Now the products of the *krānti* and the *akṣa* and their corresponding *koṭis* as given above are to be obtained for further calculations:

$$kr\bar{a}nti \times akṣa = 647 \times 225 = 145575$$
$$kr\bar{a}ntikoṭi \times akṣakoṭi = 2388 \times 2464 = 5884032.$$

As per the prescription given, if the amplitude (the *āśāgrā*) is towards the south (which happens to be the case in the present example), then the difference of the two products has to be obtained. That is $5884032 - 145575 = 5738457$. Now this value multiplied by the *trijyā* and divided by the square of the divisor gives the desired *śaṅku*.

$$\begin{aligned} \textit{śaṅku} \text{ (or) } R\cos z &= \frac{5738457 \times 3438}{(2474)^2} \\ &= \frac{19728815166}{6120676} \\ &\approx 3223. \end{aligned}$$

The square root of the difference of the squares of the *trijyā* and the *śaṅku* gives the shadow or *chāyā*.

$$\begin{aligned} \textit{chāyā} \text{ (or) } R\sin z &= \sqrt{(3438)^2 - (3223)^2} \\ &\approx 1196. \end{aligned}$$

In order to obtain the *nata-jyā* as defined in (3.179), the *chāyā* has to be multiplied by the *āśāgrākoṭi* ($R\cos a$), and divided by the *dyujyā* ($R\cos\delta$). The values of the latter two quantities are obtained from the given values of the *āśāgrā* and *krānti* simply by subtracting their squares from the square of the *trijyā* and taking the square root. That is,

$$\begin{aligned} \textit{āśāgrākoṭi} \text{ (or) } R\cos a &= \sqrt{(\textit{trijyā})^2 - (\textit{āśāgrā})^2} \\ &= \sqrt{(3438)^2 - (2431)^2} \\ &\approx 2431 \\ \text{and} \qquad \textit{dyujyā} \text{ (or) } R\cos\delta &= \sqrt{(\textit{trijyā})^2 - (\textit{krānti})^2} \\ &= \sqrt{(3438)^2 - (225)^2} \\ &\approx 3430. \end{aligned}$$

Therefore

$$\begin{aligned} \textit{nata-jyā} \text{ (or) } R\sin H &= \frac{\textit{chāyā} \times \textit{āśāgrākoṭi}}{\textit{dyujyā}} \\ &= \frac{R\sin z \times R\cos A}{R\cos\delta}. \\ &= \frac{1196 \times 2431}{3430} \\ &\approx 848. \end{aligned}$$

This completes the illustrative examples presented in the commentary *Laghu-vivṛti*. Now we proceed with Problem 2 given in the text.

३.३० नताशाग्राक्षैः शङ्कुपक्रमौ

3.30 Determination of the zenith distance and declination from the hour angle, amplitude and latitude (Problem 2)

नतलम्बकयोर्घातात् त्रिज्याप्तं तत् स्वदेशजम् ।
स्वदेशनतकोट्याप्तं नताक्षज्यावधात्तु यत् ॥ ६८ ॥
तदाशाग्रावधे कोट्योस्तयोर्घातं क्षिपेदथ[59] ।
शोधयेद्दक्षिणाग्रायां त्रिज्यया च ततो हरेत् ॥ ६९ ॥
लब्धात् स्वनतकोटिघ्नात् पृथक् त्रिज्याप्तवर्गितम् ।
युक्तं स्वनतवर्गेण तन्मूलेन हृतं फलम् [60] ॥ ७० ॥
पृथक्कृतात् भवेच्छङ्कुः छायातत्कोटिका भवेत् ।
छायाग्राकोटिसंवर्गात् द्युज्या लब्धा नतज्यया ॥ ७१ ॥
नतज्याद्युज्ययोस्तद्वत् छायाकोटित्रिजीवयोः ।
छायादिगग्राकोट्योश्च घात एको भवेत् ततः ॥ ७२ ॥
द्वयोरेकेन विहृतः तत्सम्बन्धीतरो भवेत् ।
द्युज्यात्रिजीवयोर्वर्गभेदमूलमपक्रमः ॥ ७३ ॥

natalambakayorghātāt trijyāptaṃ tat svadeśajam |
svadeśanatakoṭyāptaṃ natākṣajyāvadhāttu yat || 68 ||
tadāśāgrāvadhe koṭyostayorghātaṃ kṣipedatha |
śodhayeddakṣiṇāgrāyāṃ trijyayā ca tato haret || 69 ||
labdhāt svanatakoṭighnāt pṛthak trijyāptavargitam |
yuktaṃ svanatavargeṇa tanmūlena hṛtaṃ phalam || 70 ||
pṛthakkṛtāt bhavecchaṅkuḥ chāyātatkoṭikā bhavet |
chāyāgrākoṭisaṃvargāt dyujyā labdhā natajyayā || 71 ||
natajyādyujyayostadvat chāyākoṭitrijīvayoḥ |
chāyādigagrākoṭyośca ghāta eko bhavet tataḥ || 72 ||
dvayorekena vihṛtaḥ tatsambandhītaro bhavet |
dyujyātrijīvayorvargabhedamūlamapakramaḥ || 73 ||

The product of the *nata* and the *lambaka* divided by the *trijyā* is [the *nata*] at one's own place (*svadeśajam*). The product of the *nata* and the *akṣajyā* divided by the *koṭi* [of the *svadeśanata* is preserved]. The product of this and the *āśāgrā*, and the product of their *koṭis*, are found. Their sum or difference is taken, depending upon whether the *āśāgrā* is to the north or south, and this is divided by the *trijyā*.

The result is multiplied by the *koṭi* of the *svadeśanata* and kept separately [and is preserved at two places]. [At one place] it is divided by the *trijyā* and the square of it is added to the square of the *svadeśanata*. By the square root of the result, the product stored at the other place is divided. The result is the *śaṅku* and the *chāyā* is the *koṭi* of it. The result obtained by dividing the product of the *chāyā* and the *koṭi* of the *āśāgrā*, by the *natajyā* is the *dyujyā*.

The product of the *natajyā* and the *dyujyā*, and similarly that of the *chāyā-koṭi* and the *trijyā*, and in the same way that of the *chāyā* and the *āśāgrā*, are the same. Hence, of the two elements involved in the product, dividing by one gives the other related element. The *apakrama* is the square root of the difference between the *trijyā* and the *dyujyā*.

---

[59] Prose order: तस्य आशाग्रायाश्च वधे, तत्कोट्याः आशाग्राकोट्याश्च घातं क्षिपेत्। अथ अग्रायां दक्षिणायां शोधयेत्।

[60] Prose order: पृथक्कृतात् (= पृथक् स्थापितात्), तन्मूलेन हृतं फलं शङ्कुः भवेत्।

In these verses, the second problem, of finding the zenith distance (*śaṅku*) and the declination (*krānti*), in terms of the hour angle (*nata*), the latitude (*akṣa*) and the amplitude (*āśāgrā*), is considered. That is, z and δ are determined in terms of H, ϕ and a.

As in the earlier problem, before stating the formula for the *śaṅku* ($R\cos z$) a few intermediate quantities are defined here also. First, the *svadeśanata* and the *svadeśanata-koṭi*, which are compliments of each other, are defined as

$$\textit{svadeśanata} = \frac{\textit{natajyā} \times \textit{lambaka}}{\textit{trijyā}},$$
$$\text{or} \qquad R\sin h = \frac{R\sin H \times R\cos\phi}{R}. \tag{3.185}$$

Hence,

$$\textit{svadeśanata-koṭi} = R\cos h = \sqrt{R^2 - (R\sin h)^2}. \tag{3.186}$$

Two more quantities which depend on the *svadeśanata-koṭi*, which we denote by x and y, are defined thus

$$x = \frac{R\sin H \times R\sin\phi}{R\cos h}, \tag{3.187}$$

and

$$y = \frac{\sqrt{R^2 - x^2} \times R\cos a \pm x \times R\sin a}{R}, \tag{3.188}$$

'+' when the *āśāgrā* is north and '−' when it is south.

Suppose we choose *āśāgrā* to be north as shown in Fig. 3.28. Then substituting for x in the above equation and simplifying we have

$$y = \frac{R[\sin H \sin\phi \sin a + \cos H \cos a]}{\cos h}. \tag{3.189}$$

For convenience, we further define two quantities ρ and ξ as follows:

$$\rho = \left(\frac{y \times R\cos h}{R}\right)^2 \tag{3.190}$$
$$\xi = \sqrt{\rho + R^2 \sin^2 h}. \tag{3.191}$$

Now the *śaṅku* is given by

$$R\cos z = \frac{y \times R\cos h}{\xi}. \tag{3.192}$$

Substituting for y and ξ we have

$$R\cos z = \frac{R[\sin H \sin\phi \sin a + \cos H \cos a]}{\sqrt{[\sin\phi \sin a + \cot H \cos a]^2 \sin^2 H + \sin^2 H \cos^2\phi}}. \tag{3.193}$$

Simplifying further we have

$$R\cos z = \frac{R[\sin\phi\sin a + \cot H\cos a]}{\sqrt{[\sin\phi\sin a + \cot H\cos a]^2 + \cos^2\phi}} \qquad (\bar{a}\acute{s}\bar{a}gr\bar{a}\text{: north}). \qquad (3.194a)$$

When the *āśāgrā* is south,

$$y = R\frac{[\cos H\cos a - \sin H\sin\phi\sin a]}{\cos h}.$$

Following an identical procedure as above, we find

$$R\cos z = \frac{R[\cot H\cos a - \sin\phi\sin a]}{\sqrt{[\sin\phi\sin a - \cot H\cos a]^2 + \cos^2\phi}} \qquad (\bar{a}\acute{s}\bar{a}gr\bar{a}\text{: south}). \qquad (3.194b)$$

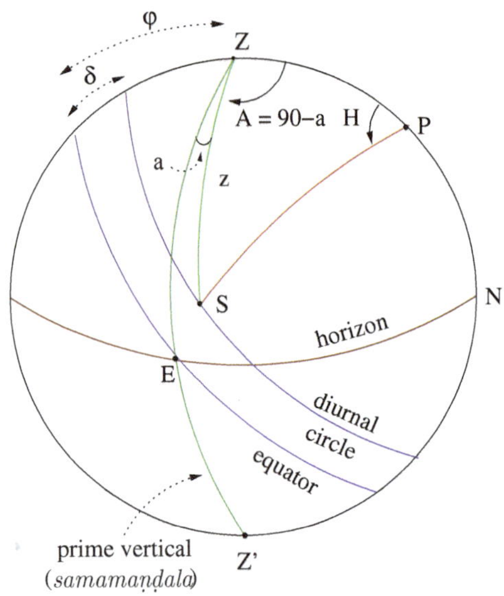

Fig. 3.28 Spherical triangle used for arriving at the formula for the *śaṅku* ($R\cos z$) and the *apakrama* ($R\sin\delta$), in terms of the *nata* (H), *āśāgrā* (a) and *akṣa* (ϕ).

Proof:

The expressions for the *śaṅku* given in (3.194) can be arrived at by applying the cotangent formula or the four-parts formula in spherical trigonometry to the spherical triangle PZS shown in Fig. 3.28. Taking H, $PZ = 90 - \phi$, A and z as the four parts,

$$\sin\phi\cos A = \cos\phi\cot z - \sin A\cot H. \tag{3.195}$$

In arriving at the above equation we have used the fact that $A = 90 - a$. Rewriting the above equation,

$$\cot z = \frac{1}{\cos\phi}[\sin\phi\cos A + \sin A\cot H]. \tag{3.196}$$

Or,

$$1 + \cot^2 z = \frac{1}{\cos^2\phi}(\cos^2\phi + [\sin\phi\cos A + \sin A\cot H]^2). \tag{3.197}$$

Since $\cos z = \frac{\cot z}{\sqrt{1+\cot^2 z}}$, we have

$$\cos z = \frac{[\sin\phi\cos A + \sin A\cot H]}{\sqrt{[\sin\phi\cos A + \sin A\cot H]^2 + \cos^2\phi}}. \tag{3.198}$$

When the *āśāgrā* is north, $A = 90 - a$ and $\cos A = \sin a$, $\sin A = \cos a$. Then (3.198) reduces to (3.194a). Similarly when the *āśāgrā* is south, $A = 90 + a$, and $\cos A = -\sin a$, $\sin A = \cos a$. Then (3.198) reduces to (3.194b).

Having obtained the formula for the *śaṅku*, the declination is determined using the relation

$$dyujy\bar{a} = \frac{ch\bar{a}y\bar{a} \times \bar{a}\acute{s}\bar{a}gr\bar{a}ko\d{t}i}{nataj y\bar{a}}, \tag{3.199}$$

where the *chāyā* and the *āśāgrākoṭi* are the same as defined in Problem 1 in (3.180) and (3.181). Substituting for the *chāyā*, *āśāgrākoṭi* and *natajyā*, the mathematical expression for the *dyujyā* is

$$R\cos\delta = \frac{R\sin z\, R\cos a}{R\sin H}. \tag{3.200}$$

The above expression can be easily obtained by using the spherical triangle *PZS* and applying the sine formula. It can be seen that the above equation is the same as (3.182). Further, it is stated that

$$natajy\bar{a} \times dyujy\bar{a} = ch\bar{a}y\bar{a}ko\d{t}i \times trijy\bar{a} = ch\bar{a}y\bar{a} \times \bar{a}\acute{s}\bar{a}gr\bar{a}ko\d{t}i. \tag{3.201}$$

That is,

$$R\sin H \times R\cos\delta = R\sin z\cos a \times R = R\sin z \times R\cos a. \tag{3.202}$$

In the above equation the middle term $R\sin z\cos a$ refers to the projection of the *chāyā* along the east–west line (*EW*) as shown in Fig. 3.29. Here the point *D* represents the foot of the perpendicular drawn from the Sun to the plane of the horizon. *SD* is the *śaṅku* ($R\cos z$) and the *chāyā*, its complement, is *OD* ($R\sin z$). In the planar right-angled triangle *COD*, $C\hat{O}D = a$, which is the spherical angle between the prime vertical and the vertical passing through the Sun. The projection of *chāyā* along the east–west line is the *chāyākoṭi* or *bhākoṭi* given by $R\sin z\cos a$.

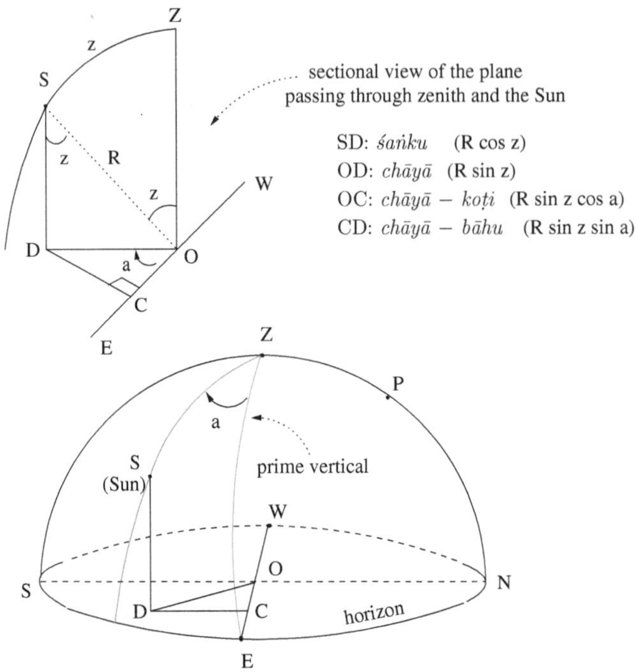

Fig. 3.29 Celestial sphere, and a sectional view of it to define the *chāyākoṭi* or *bhākoṭi*.

३.३१ नतापक्रमाक्षैः शङ्क्वाशाग्रे

3.31 Determination of the zenith distance and amplitude from the hour angle, declination and latitude (Problem 3)

> नतकोट्या हता द्युज्या विभक्ता त्रिभजीवया ।
> सौम्ययाम्यदिशोर्भूज्यायुतोना लम्बकाहता ॥ ७४ ॥
> त्रिज्याप्ता शङ्कुराशाग्राकोटिः द्युज्या च पूर्ववत् ।
>
> *natakoṭyā hatā dyujyā vibhaktātribhajīvayā |*
> *saumyayāmyadiśorbhūjyāyutonā lambakāhatā || 74 ||*
> *trijyāptā śaṅkurāśāgrākoṭiḥ, dyujyā ca pūrvavat |*
>
> The *koṭi* of the *natajyā* is multiplied by the *dyujyā* and divided by the *trijyā*. Depending upon whether the Sun is to the north or south, the *bhūjyā* is added to or subtracted from this and the result, multiplied by the *lambaka* and divided by the *trijyā*, becomes the *śaṅku*. The *koṭis* of the *āśāgrā* and the *dyujyā* [are to be obtained] as before.

Here, the zenith distance (*śaṅku*) and amplitude (*āśāgrā*) are obtained in terms of the hour angle (*nata*), the latitude (*akṣa*) and the declination (*krānti*). That is, z and a are obtained in terms of H, ϕ and δ.

In this problem, the expression for the *śaṅku* is not as complicated as in the earlier problems. Even then an intermediate quantity, denoted by the symbol x, is defined thus:

$$x = \frac{natajy\bar{a}ko\d{t}i \times dyujy\bar{a}}{trijy\bar{a}}$$

$$\text{or} \qquad x = \frac{R\cos H \times R\cos\delta}{R}. \tag{3.203}$$

It is then stated that the *bhūjyā* has to be added to or subtracted from this quantity. The result has to be multiplied by the *lambaka* and divided by the *trijyā* to get the expression for the *śaṅku*. That is,

$$\acute{s}a\dot{n}ku = \frac{(x \pm bh\bar{u}jy\bar{a})R\cos\phi}{R}. \tag{3.204}$$

Since the *bhūjyā* or *kṣitijyā* is given by $R\tan\phi|\sin\delta|$ (see Chapter 2, verse 27), the above expression reduces to

$$\begin{aligned} R\cos z &= \frac{R(\cos H\cos\delta \pm \tan\phi|\sin\delta|)R\cos\phi}{R} \\ &= R(\pm\sin\phi|\sin\delta| + \cos\phi\cos\delta\cos H). \end{aligned} \tag{3.205}$$

As the $+/-$ sign corresponds to a northerly/southerly declination, (3.205) is the same as

$$R\cos z = R(\sin\phi\sin\delta + \cos\phi\cos\delta\cos H). \tag{3.206}$$

Equation (3.206) follows by applying the cosine formula to the side ZS in the spherical triangle PZS in Fig. 3.27. The *āśāgrākoṭi* ($R\cos a$) is obtained using (3.200).

३.३२ नतापक्रमाशाग्राभिः शङ्क्वक्षौ

3.32 Determination of the zenith distance and latitude from the hour angle, declination and amplitude (Problem 4)

छायां नीत्वाथ तत्कोटिद्युज्यावर्गान्तरात् पदम् ॥ ७५ ॥
तच्छायाबाहुघातो यः शङ्कुक्रान्त्योर्वधोपि यः ।
क्रान्त्यग्रयोस्तु तुल्यदिशोस्तयोर्भेदोऽन्यथा युतिः ॥ ७६ ॥
उन्मण्डलक्षितिजयोः अन्तरेऽर्के च तद्युतिः ।
तद्धतां विभजेत् त्रिज्यां तच्छायाकोटिवर्गयोः ॥ ७७ ॥
अन्तरेण भवेदक्षः नताद्यैर्विदितैस्त्रिभिः ।

chāyāṃ nītvātha tatkoṭidyujyāvargāntarāt padam || 75 ||
tacchāyābāhughāto yaḥ śaṅkukrāntyorvadhopi yaḥ |
krāntyagrayostu tulyadiśostayorbhedo'nyathā yutiḥ || 76 ||
unmaṇḍalakṣitijayoḥ antare'rke ca tadyutiḥ |
taddhatāṃ vibhajet trijyāṃ tacchāyākoṭivargayoḥ || 77 ||
antareṇa bhavedakṣaḥ natādyairviditaistribhiḥ |

Having found the *chāyā*, the square root of the difference of the squares of the *chāyākoṭi* and the *dyujyā* is multiplied by the *chāyā-bāhu*. The sum/difference of the product of the *śaṅku* and the *krānti* and the (aforementioned) product is taken when the *krānti* and the *āśāgrā* have the different/same direction. If the Sun lies between the *unmaṇḍala* (6'0 clock circle) and the *kṣitija* (horizon), then it must always be added. The result is multiplied by the *trijyā* and divided by the difference of the squares of the *trijyā* and the *chāyākoṭi*. This gives the latitude in terms of the other three known quantities, the *nata* etc.

The fourth problem is devoted to the determination of the zenith distance (*śaṅku*) and the latitude (*akṣa*), in terms of the hour angle (*nata*), the amplitude (*āśāgrā*) and the declination (*krānti*). That is, z and ϕ are to be obtained in terms of H, a and δ. It has already been emphasized earlier (see (3.202)) that the *nata-jyā*, the *dyujyā* and the *chāyākoṭi* are related as follows:

$$\begin{aligned} & \textit{nata-jyā} \times \textit{dyujyā} = \textit{chāyākoṭi} \times \textit{trijyā} \\ \text{or} \quad & R \sin H \cos\delta = R \sin z \cos a. \end{aligned} \tag{3.207}$$

Hence the *chāyā* ($R\sin z$) and therefore the *śaṅku* are determined in terms H, a and δ. Now we only need to determine the latitude in terms of H, a and δ.

As in the earlier problems, a few intermediate quantities are defined. First an intermediate quantity (say u) is defined as follows:

$$\begin{aligned} u &= \sqrt{\textit{dyujyā}^2 - \textit{chāyākoṭi}^2} \\ &= \sqrt{(R\cos\delta)^2 - (R\sin H\cos\delta)^2} \\ &= R\cos H\cos\delta. \end{aligned} \tag{3.208}$$

This u has to be multiplied by the *chāyābāhu*, which is given by

$$\begin{aligned} \textit{chāyābāhu} &= \sqrt{\textit{chāyā}^2 - \textit{chāyākoṭi}^2} \\ &= \sqrt{(R\sin z)^2 - (R\sin H\cos\delta)^2} \\ &= \sqrt{(R\sin z)^2 - (R\sin z\cos a)^2} \qquad \text{[by (3.203)]} \\ &= R\sin z\sin a. \end{aligned} \tag{3.209}$$

It may be noted that the *chāyābāhu* is nothing but the projection of the shadow perpendicular to the east–west line (*CD* in Fig. 3.29). Further, another intermediate quantity, y, is defined as follows:

$$\begin{aligned} y &= u \times \textit{chāyābāhu} \pm \textit{śaṅkukrāntisaṃvarga} \\ &= R\cos H\cos\delta \times R\sin z\sin a \pm R\cos z\, R|\sin\delta|. \end{aligned} \tag{3.210}$$

Then the *akṣajyā* is given by the following expression:

$$\textit{akṣajyā} = \frac{y \times \textit{trijyā}}{(\textit{trijyā}^2 - \textit{chāyākoṭi}^2)}$$

$$\text{or} \qquad R\sin\phi = \frac{y \times R}{R^2 - (R\sin z\cos a)^2}. \tag{3.211}$$

Substituting for y, we have the following explicit expression for $\sin\phi$:

$$\sin\phi = \frac{\cos H\cos\delta\sin z\sin a \,\overset{+}{\sim}\, |\sin\delta|\cos z}{1-(\sin z\cos a)^2}. \tag{3.212}$$

Proof:

Here the above relation for $\sin\phi$ is derived by using standard spherical trigonometrical results. By applying the cosine formula for the spherical triangle PZS, as shown in Fig. 3.27, we have

$$\cos z = \sin\phi\sin\delta + \cos\phi\cos\delta\cos H. \tag{3.213}$$

This has to be solved for $\sin\phi$. Setting $x = \sin\phi$ (and $\cos\phi = \sqrt{1-x^2}$), we have

$$\begin{aligned} \cos z - x\sin\delta &= \sqrt{1-x^2}\cos\delta\cos H \\ &= \sqrt{1-x^2}\sqrt{\cos^2\delta - \sin^2 z\cos^2 a}, \end{aligned} \tag{3.214a}$$

where we have used the relation

$$\cos\delta\cos H = \sqrt{\cos^2\delta - \sin^2 z\cos^2 a}, \tag{3.214b}$$

which is a consequence of (3.207). Squaring the equation and rearranging terms, we obtain the following quadratic equation in x:

$$x^2(1-\sin^2 z\cos^2 a) - 2x\sin\delta\cos z - \sin^2 z\sin^2 a + \sin^2\delta = 0. \tag{3.215}$$

Solving the equation, we have

$$x = \frac{2\sin\delta\cos z \pm \sqrt{4\sin^2\delta\cos^2 z - 4(\sin^2\delta - \sin^2 z\sin^2 a)(1-\sin^2 z\cos^2 a)}}{2(1-\sin^2 z\cos^2 a)} \tag{3.216}$$

$$x = \frac{\sin\delta\cos z \pm \sqrt{\sin^2\delta\cos^2 z - \sin^2\delta + \mathscr{X} + \mathscr{Y} - \mathscr{Z}}}{1-\sin^2 z\cos^2 a}, \tag{3.217a}$$

where

$$\mathscr{X} = \sin^2\delta\sin^2 z\cos^2 a, \tag{3.217b}$$

$$\mathscr{Y} = \sin^2 z\sin^2 a, \tag{3.217c}$$

$$\text{and} \qquad \mathscr{Z} = \sin^4 z\sin^2 a\cos^2 a. \tag{3.217d}$$

The first pair of terms in the discriminant in (3.217a) reduces to $-\sin^2\delta\sin^2 z$. This with $\mathscr{X}$ reduces to $-\sin^2\delta\sin^2 z\sin^2 a$. It further reduces to $\sin^2 z\sin^2 a\cos^2\delta$, when combined with $\mathscr{Y}$. Combining this with $\mathscr{Z}$ and using (3.214b), the discriminant simply reduces to $\cos H\cos\delta\sin z\sin a$. Thus we have

$$x = \sin\phi = \frac{\sin\delta\cos z \pm \cos H\cos\delta\sin z\sin a}{1-\sin^2 z\cos^2 a}, \tag{3.218}$$

which is the same as (3.212).

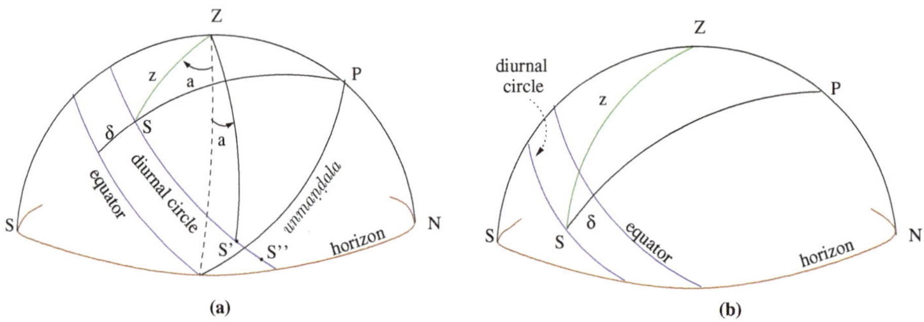

Fig. 3.30 The positions of the Sun towards the north and south of the prime vertical.

Now we discuss the signs as enunciated in the text. Consider the situation when the declination is northerly (δ positive), as shown in Fig. 3.30(a). Then the first term in the numerator is equal to $|\sin\delta|\cos z$ and is positive. For a northerly declination, the Sun can be south or north of the prime vertical. Then we have to take the '$-$' sign in (3.218), when the *āśāgrā* is north (Sun at S') and the '$+$' sign when the *āśāgrā* is south (Sun at S).

In the latter case, $H < 90°$ and $\cos H$ is positive, and we have to take the sum of the magnitudes of the two terms. In the former case, the second term is negative when $H < 90°$, so that x will be the difference of the magnitudes of the two terms. However, the second term is positive when $H > 90°$, and we have to add the magnitudes of the two terms. This corresponds to the situation when the Sun is between the *unmaṇḍala* and the horizon, as at S''.

When the declination is south (δ negative), the first term in x in (3.218) is negative. Also, $\cos H$ is positive as $H < 90°$. Then we have to take the '$+$' sign in front of the second term, as $x = \sin\phi$ has to be positive. In this, the *krānti* and the *āśāgrā* are both southerly and difference of the magnitudes of the two terms is to be taken.

३.३३ शङ्क्वाशाग्राक्षैः नतापक्रमौ

3.33 Determination of the hour angle and declination from the zenith distance, amplitude and latitude (Problem 5)

अक्षशङ्क्वोर्वधो यश्च यश्च भाबाहुलम्बयोः ॥ ७८ ॥
सौम्ययाम्यस्थिते भानौ तयोर्योगान्तरात् ततः ।
क्रान्तिस्त्रिज्याहृता प्राग्वत् नतज्यां च समानयेत् ॥ ७९ ॥

akṣaśaṅkvorvadho yaśca yaśca bhābāhulambayoḥ || 78 ||
saumyayāmyasthite bhānau tayoryogāntarāt tataḥ |
krāntistrijyāhṛtā prāgvat natajyāṃ ca samānayet || 79 ||

The sum or difference of the product of *akṣa* and *śaṅku* and the product of *chāyābāhu* and *lambaka* is taken depending upon whether the Sun is to the north or south. The result divided by the *trijyā* is the *krānti*. The *natajyā* may be obtained as before.

In this problem, the hour angle (*nata*) and declination (*krānti*) are obtained in terms of the zenith distance (*śaṅku*), amplitude (*āśāgrā*) and the latitude (*akṣa*) are given. That is, H and δ are obtained in terms of z, a and ϕ.

The expression for the *krānti* is given as

$$krānti = \frac{akṣa \times śaṅku \pm chāyābāhu \times lambaka}{trijyā} \tag{3.219}$$

$$R\sin\delta = \frac{R\sin\phi \times R\cos z \pm R\sin z \sin a \times R\cos\phi}{R}, \tag{3.220}$$

where '$+$' must be chosen when the Sun is to the north (of the prime vertical) and '$-$' otherwise. The above relation can be verified by applying the cosine formula to the side PS in the spherical triangle PZS shown in Fig. 3.27. We have

$$\sin\delta = \sin\phi\cos z + \cos\phi\sin z\cos A. \tag{3.221}$$

Now $A = 90^\circ \pm a$ and $\cos A = \mp\sin a$, depending on whether the Sun is to the south or north of the prime vertical. The expression for the *natajyā* has already been given in Problem 1, in terms of z, a and δ. Hence the *natajyā* can be obtained, as z and a are already known and δ has been determined.

३.३४ शङ्कपक्रमाक्षैः नताशाग्रे

3.34 Determination of the hour angle and amplitude from the zenith distance, declination and latitude (Problem 6)

त्रिज्यापक्रमघातो यः यश्च शङ्क्वक्षयोर्वधः ।
तयोर्योगान्तरं यत्तु गोलयोर्याम्यसौम्ययोः ॥ ८० ॥
भाबाहुर्लम्बकाप्तोऽस्मात् त्रिज्याघ्नाद्भाहृतेष्टदिक् ।

trijyāpakramaghāto yaḥ yaśca śaṅkvakṣayorvadhaḥ |
tayoryogāntaraṃ yattu golayoryāmyasaumyayoḥ ||80||
bhābāhurlambakāpto'smāt trijyāghnādbhāhṛteṣṭadik |

The sum or difference of the product of the *trijyā* and the *apakrama* and that of the *śaṅku* and the *akṣa* is found, depending upon whether the Sun is in the northern or the southern hemisphere. The result divided by the *lambaka* is the *bhābāhu*. This multiplied by the *trijyā* and divided by the *bhā* (*chāyā*) is the *iṣṭadik* (*āśāgrā*).

In the sixth problem the hour angle (*nata*) and amplitude (*āśāgrā*) are obtained in terms of the zenith distance (*śaṅku*), declination (*apakrama* or *krānti*) and the latitude (*akṣa*). That is, H and a are obtained in terms of z, δ and ϕ.

The expression for the *krānti* is given in two steps. Initially a term called the *bhābāhu* is defined thus:

$$\begin{aligned} bh\bar{a}b\bar{a}hu &= \frac{trijy\bar{a} \times kr\bar{a}nti \overset{+}{\sim} \acute{s}a\dot{n}ku \times ak\d{s}a}{lambaka} \\ &= \frac{R \times R|\sin\delta| \overset{+}{\sim} R\cos z \times R\sin\phi}{R\cos\phi}, \end{aligned} \tag{3.222}$$

where the sum/difference is considered when the declination is south/north. Then the *āśāgrā* is given by

$$\begin{aligned} \bar{a}\acute{s}\bar{a}gr\bar{a} &= \frac{bh\bar{a}b\bar{a}hu \times trijy\bar{a}}{ch\bar{a}y\bar{a}} \\ R\sin a &= \frac{[R \times R|\sin\delta| \overset{+}{\sim} R\cos z \times R\sin\phi]}{R\cos\phi \times R\sin z} \times R. \end{aligned} \tag{3.223}$$

Here '$+$' or '$\sim$' is to be taken when δ is negative or positive respectively.

Proof:

From (3.220),

$$\sin z \cos\phi \sin a = \pm\,(\cos z \sin\phi - \sin\delta), \tag{3.224}$$

when the *āśāgrā* is south/north. When the declination is south, $-\sin\delta = |\sin\delta|$, the *āśāgrā* is necessarily south and we have

$$\sin a = \frac{|\sin\delta| + \cos z \sin\phi}{\cos\phi \sin z}. \tag{3.225}$$

When the declination is north,

$$\sin\delta = |\sin\delta| \tag{3.226}$$

$$\text{and} \qquad \sin a = \frac{|\sin\delta| - \cos z \sin\phi}{\cos\phi \sin z}. \tag{3.227}$$

These coincide with the stated expression (3.223) for the *āśāgrā*.

३.३५ शङ्कुपक्रमाशाग्राभिः नताक्षौ

3.35 Determination of the hour angle and latitude from the zenith distance, declination and amplitude (Problem 7)

वर्गान्तरपदं यत् स्यात् छायाकोटिद्युजीवयोः ॥ ८१ ॥
तच्छायाबाहुयोगो यः शङ्कुक्रान्त्यैक्यवर्गतः ।
तेनाप्तं यत् फलं तस्मिन्नेव तत् स्वमृणं पृथक् ॥ ८२ ॥
तयोरल्पहता त्रिज्या महताप्ताक्षमौर्विका ।

vargāntarapadaṃ yat syāt chāyākoṭidyujīvayoḥ || 81 ||
tacchāyābāhuyogo yaḥ śaṅkukrāntyaikyavargataḥ |
tenāptaṃ yat phalaṃ tasminneva tat svamṛṇaṃ pṛthak ||82||
tayoralpahatā trijyā mahatāptākṣamaurvikā |

The square root of the difference of the squares of the *dyujyā* and the *chāyākoṭi* is found and it is added to the *chāyābāhu* [to get D]. The square of the sum of the *śaṅku* and the *krānti* is divided by this, (D) [to obtain say, C]. To the result (C), kept separately, D is added and subtracted. The smaller one multiplied by the *trijyā* and divided by the larger one gives the *akṣajyā*.

In Problem 7, the hour angle (*nata*) and latitude (*akṣa*) are obtained in terms of the zenith distance (*śaṅku*), the declination (*krānti*) and the amplitude (*āśāgrā*). That is, H and ϕ are obtained in terms of z, δ and a.

The expression for the hour angle has already been found in the earlier problems. Hence, in the above verses the expression for the *akṣa* alone is considered. As usual, it is convenient to define a few intermediate quantities. Initially a term, say x, is defined as:

$$\begin{aligned} x &= \sqrt{dyujy\bar{a}^2 - ch\bar{a}y\bar{a}ko\d{t}i^2} \\ &= \sqrt{(R\cos\delta)^2 - (R\sin H\cos\delta)^2} \\ &= R\cos H\cos\delta. \end{aligned} \tag{3.228}$$

The *chāyābāhu*, $R\sin z\sin a$, must be added to x. The result (D) is given by

$$D = R\cos H\cos\delta + R\sin z\sin a. \tag{3.229}$$

In the next step, another quantity (say C) is defined as

$$C = \frac{(R\cos z + R\sin\delta)^2}{D}. \tag{3.230}$$

Then the sum and difference of C and D are found. Of these two, obviously the difference $(C-D)$ will be smaller than the sum $(C+D)$. Then the *akṣa* is the ratio of the product of the smaller one and the *trijyā* to that of the larger one. That is,

$$ak\d{s}a = R\times\frac{(C-D)}{(C+D)}. \tag{3.231}$$

Substituting for C and D we have

$$R\sin\phi = R\times\frac{(R\cos z + R\sin\delta)^2-(R\cos H\cos\delta + R\sin z\sin a)^2}{(R\cos z + R\sin\delta)^2+(R\cos H\cos\delta + R\sin z\sin a)^2}. \tag{3.232}$$

Proof:

The above expression may be obtained by applying the cosine formula to the spherical triangle *PZS*, shown in Fig. (3.28), for the sides *ZS* and *PS*. We then have

$$\begin{aligned}\cos z &= \sin\phi\sin\delta+\cos\phi\cos\delta\cos H &(3.233)\\ \sin\delta &= \sin\phi\cos z+\cos\phi\sin z\cos A. &(3.234)\end{aligned}$$

Therefore

$$\begin{aligned}\cos z+\sin\delta &= \sin\phi(\cos z+\sin\delta)+\\ &\qquad \cos\phi(\cos H\cos\delta+\sin z\sin a).\\ \text{or}\quad (\cos z+\sin\delta)(1-\sin\phi) &= \cos\phi(\cos H\cos\delta+\sin z\sin a).\end{aligned} \tag{3.235}$$

As $D = R(\cos H\cos\delta+\sin z\sin a)$, squaring both sides and rewriting $\cos^2\phi$ as $(1-\sin^2\phi)$ in the RHS, we have

$$(\cos z+\sin\delta)^2(1-\sin\phi)^2 = (1-\sin^2\phi)\,\frac{D^2}{R^2}. \tag{3.236}$$

Therefore

$$\begin{aligned}\frac{(1-\sin^2\phi)}{(1-\sin\phi)^2} &= \frac{(\cos z+\sin\delta)^2}{D^2}R^2\\ \text{or}\quad \frac{(1-\sin^2\phi)-(1-\sin\phi)^2}{(1-\sin^2\phi)+(1-\sin\phi)^2} &= \frac{(\cos z+\sin\delta)^2-\frac{D^2}{R^2}}{(\cos z+\sin\delta)^2+\frac{D^2}{R^2}}.\end{aligned} \tag{3.237}$$

It may be easily verified that the LHS of the above equation reduces to $\sin\phi$, and hence

$$\sin\phi = \frac{(\cos z+\sin\delta)^2-(\cos H\cos\delta+\sin z\sin a)^2}{(\cos z+\sin\delta)^2+(\cos H\cos\delta+\sin z\sin a)^2}, \tag{3.238}$$

which is the same as (3.232).

३.३६ शङ्कुनताक्षैः अपक्रमाशाग्रे

3.36 Determination of the declination and amplitude from the zenith distance, hour angle and latitude (Problem 8)

त्रिज्याहताक्षशङ्कू स्वनतकोट्योद्धृतौ पृथक् ॥ ८३ ॥
ये तत्कोट्यौ च तत्त्रिज्यावर्गभेदपदीकृते[61] ।
मिथः कोटिघ्नयोर्योगात् याम्ये सौम्येऽन्तरात् तयोः ॥ ८४ ॥
त्रिज्यया विहृता द्युज्या क्रान्त्याशाग्रे तु पूर्ववत् ।
नतमण्डलदृश्यार्धमध्यतः सौम्ययाम्यता ॥ ८५ ॥

trijyāhatākṣaśaṅkū svanatakoṭyoddhṛtau pṛthak ||83||
ye tatkoṭyau ca tattrijyāvargabhedapadīkṛte |
mithaḥ koṭighnayoryogāt yāmye saumye'ntarāt tayoḥ ||84||
trijyayā vihṛtā dyujyā krāntyāśāgre tu pūrvavat |
natamaṇḍaladṛśyārdhamadhyataḥ saumyayāmyatā ||85||

The *akṣa* and *śaṅku* are multiplied by the *trijyā* and divided by the *svadeśa-natakoṭi*, and kept separately.

The square roots of the squares of these subtracted from the square of the *trijyā* become their *koṭis*. By finding the sum or difference of the cross products, depending upon whether [the position of the Sun is] to the south or north, and dividing by the *trijyā*, the *dyujyā* is obtained. The *krānti* and *āśāgrā* [are to be obtained] as stated earlier.

Here the north and south is with respect to the visible half of the 6 o'clock circle (*natamaṇḍala*).

In Problem 2, the *svadeśanatakoṭi* has been defined to be

$$svade\acute{s}anatako\d{t}i = R\cos h = R\sqrt{1-\sin^2 H\cos^2\phi}. \tag{3.239}$$

With this *svadeśanatakoṭi* as the divisor, two intermediate quantities x and y are defined as

$$x = \frac{R\times R\sin\phi}{R\cos h} \tag{3.240}$$

$$y = \frac{R\times R\cos z}{R\cos h}. \tag{3.241}$$

The *koṭis* of x and y, denoted by ρ and ξ, are given by

$$\begin{aligned}\rho &= \sqrt{R^2-x^2}\\ &= \sqrt{R^2-\frac{R^2\sin^2\phi}{\cos^2 h}}\\ &= \frac{R\cos\phi|\cos H|}{\cos h},\end{aligned} \tag{3.242}$$

$$\text{and} \qquad \xi = \sqrt{R^2-y^2}$$

---

[61] तयोः वर्गयोः, त्रिज्यावर्गयोश्च भेदं संपाद्य पदीकृते यल्लभ्येते ते तत्कोट्यौ।

$$= \sqrt{R^2 - \frac{R^2 \cos^2 z}{\cos^2 h}}$$

$$= \frac{R\sqrt{\sin^2 z - \sin^2 H \cos^2 \phi}}{\cos h}. \tag{3.243}$$

Now the *dyujyā* is stated to be

$$dyujy\bar{a} = \frac{x\xi \overset{+}{\sim} y\rho}{trijy\bar{a}}. \tag{3.244}$$

Substituting for ρ, ξ, x and y, we have

$$R\cos\delta = \frac{R\sin\phi\sqrt{\sin^2 z - \sin^2 H \cos^2 \phi} \overset{+}{\sim} \cos z R\cos\phi |\cos H|}{(1 - \sin^2 H \cos^2 \phi)}, \tag{3.245}$$

where it has been specifically mentioned that the sum or difference have to be taken depending upon whether we are dealing with the southern or the northern direction.

Usually, the terms southern and northern refer to declination of the Sun. But, here the terms southern and northern have to be understood in a different way, as is explicitly mentioned in the text. The directions north and south mentioned here are with reference to the *natamaṇḍala* or the 6 o'clock circle. The Sun is taken to be in the southern direction when it is above the 6 o'clock circle in the visible hemisphere. Then the positive sign must be chosen. The moment the Sun crosses the 6 o'clock circle, $H > 90$, and it is taken to be in the northern hemisphere and negative sign is prescribed. The $\overset{+}{\sim}$ prescription is understandable since we have $\cos H$ in the expression.

Proof:

From the cosine formula applied to the spherical triangle *PZS*, shown in Fig. 3.27, we have,

$$\cos z = \sin\phi \sin\delta + \cos\phi \cos\delta \cos H. \tag{3.246}$$

Putting $\cos\delta = x$, we get the following quadratic equation in x

$$x^2(\cos^2 H \cos^2 \phi + \sin^2 \phi) - 2x\cos z \cos\phi \cos H + (\cos^2 z - \sin^2 \phi) = 0. \tag{3.247}$$

Hence,

$$x = \frac{2\cos z \cos\phi \cos H \pm \sqrt{4\cos^2 z \cos^2 \phi \cos^2 H - 4\mathscr{A}\mathscr{C}}}{2(1 - \sin^2 H \cos^2 \phi)}, \tag{3.248a}$$

where

$$\mathscr{A} = (1 - \sin^2 H \cos^2 \phi) \qquad \text{and} \qquad \mathscr{C} = (\cos^2 z - \sin^2 \phi). \tag{3.248b}$$

On simplification, the discriminant reduces to

$$2\sin\phi\sqrt{\sin^2 z - \sin^2 H\cos^2\phi}. \tag{3.248c}$$

Therefore

$$\cos\delta = \frac{\cos z\cos\phi\cos H \pm \sin\phi\sqrt{\sin^2 z - \sin^2 H\cos^2\phi}}{(1-\sin^2 H\cos^2\phi)}, \tag{3.249}$$

which is the same as (3.245).

Note on the relative sign difference:

When the Sun is 'north' of the *natamaṇḍala*, in the sense of the text, $H > 90^\circ$ and the first term in the numerator of $\cos\delta$ is negative as $\cos H$ is negative. Hence the positive sign has to be taken in the second term, as $\cos\delta$ is positive. Hence the difference in magnitude between the two terms is to be taken when the Sun is 'north' of the *natamaṇḍala*. When $\cos H$ changes sign, then also the second term should be taken with the positive sign by continuity. But, as the first term is also positive now, the sum of the magnitude of the two terms is to be taken when the Sun is 'south' of the *natamaṇḍala*. These are in conformity with the prescription in the text.

३.३७ अपरैस्त्रिभिः क्रान्त्यक्षौ आशाग्राक्षौ च

3.37 Determination of the declination and latitude, and the amplitude and latitude, from the rest of the three (Problems 9, 10)

दिगग्रायास्तु तत्कोटिः[62] तच्छायाघाततो हृता ।
नतज्यया भवेद्द्युज्या तद्भुजा क्रान्तिरेव हि ॥ ८६ ॥
द्युज्यानतज्ययोर्घातात् अग्राकोटिः प्रभाहृता ।
अक्षः प्राग्वदिति प्रश्नदशकोत्तरमीरितम् ॥ ८७ ॥

digagrāyāstu tatkoṭiḥ tacchāyāghātato hṛtā |
natajyayā bhaveddyujyā tadbhujā krāntireva hi || 86 ||
dyujyānatajyayorghātāt agrākoṭiḥ prabhāhṛtā |
akṣaḥ prāgvaditi praśnadaśakottaramīritam || 87 ||

From the *āśāgrā*, its *koṭi* [is to be obtained]. The product of this and the *chāyā* when divided by the *natajyā* gives the *dyujyā*. Its *bhujā* is indeed the *krānti*.

---

[62] अत्र तत्कोटिः इत्यनन्तरम् आनेतव्या इति अध्याहर्यम्। अपि च, तत्कोटिः इत्यत्र, तच्छब्देन दिगग्रा गृह्यते। उत्तरत्र विद्यमानतच्छब्देन दिगग्राकोटिः ग्राह्या।

The product of the *dyujyā* and the *natajyā* divided by the *prabhā* [*chāyā*] becomes the *koṭi* of *āśāgrā*. The *akṣa* is found as stated earlier. Thus we have given the answers to the ten problems.

These two problems are concerned with finding δ and ϕ in terms of H, z and a, and a and ϕ in terms of H, z and δ respectively. According to the verses, δ can be found in terms of z, a and H, and a can be found in terms of z, H and δ through the relation

$$R\cos\delta = \frac{R\sin z\, R\cos a}{R\sin H}, \tag{3.250}$$

which can also be written as

$$R\cos a = \frac{R\sin H\, R\cos\delta}{R\sin z}. \tag{3.251}$$

Using the spherical triangle *PZS* (Fig. 3.27), and applying the sine formula, we have

$$\frac{\sin A}{\sin(90-\delta)} = \frac{\sin H}{\sin z}. \tag{3.252}$$

Since $A = (90 + a)$, the above equation reduces to

$$\frac{\sin H}{\sin z} = \frac{\cos a}{\cos\delta}, \tag{3.253}$$

which is nothing but the ratio of equations (3.250) and (3.251). ϕ is determined from z, δ, H and a through (3.238). As δ has been solved in terms of z, H and a, and a has been solved in terms of z, H and δ, ϕ can be found, in each of the problems, in terms of the three given quantities.

३.३८ इष्टदिक्छाया

3.38 Shadow along any direction

दिगग्रा विहृता यद्वा तत्कोटिघ्ना पलप्रभा [63] ।
तत्कोटिका तयोः कृत्योः योगमूलं स्वदृग्गुणः ॥ ८८ ॥
शङ्कुदृग्गुणयोः कृत्योः छायाकर्णो युतेः पदम् [64] ।
शङ्कुच्छाये त्रिजीवाघ्ने छायाकर्णहृते स्फुटे ॥ ८९ ॥
दृग्गुणाभिहतक्रान्तेः अक्षज्याप्तो ह्यपक्रमः ।
क्रान्तिदृग्गुणयोः कोटिः त्रिज्यावर्गान्तरात् पदम् ॥ ९० ॥

[63] अत्र तत्कोटिघ्ना, इत्यत्र तच्छब्देन दिगग्रा ग्राह्या। परन्तु उत्तरत्र तत्कोटिका इत्यत्र तच्छब्देन पलप्रभा ग्राह्या। अन्वयस्तु इत्थं - पलप्रभा, दिगग्रा विहृता तत्कोटिघ्ना (च), तत्कोटिका (पलप्रभा-कोटिका)। भवतीत्यध्याहर्यम्।

[64] Prose order: शङ्कुदृग्गुणयोः कृत्योः युतेः पदं छायाकर्णः।

मिथः कोटिहतत्रिज्याभक्तयोः क्रान्तिदृग्ज्ययोः ।
तयोर्योगान्तरं छायागोलयोः याम्यसौम्ययोः ॥ ९१ ॥

digagrā vihṛtā yadvā tatkoṭighnā palaprabhā|
tatkoṭikā tayoḥ kṛtyoḥ yogamūlaṃ svadṛgguṇaḥ||88||
śaṅkudṛgguṇayoḥ kṛtyoḥ chāyākarṇo yuteḥ padam|
śaṅkucchāye trijīvāghne chāyākarṇahṛte sphuṭe||89||
dṛgguṇābhihatakrānteḥ akṣajyāpto hyapakramaḥ|
krāntidṛgguṇayoḥ koṭiḥ trijyāvargāntarāt padam||90||
mithaḥ koṭihatatrijyābhaktayoḥ krāntidṛgjyayoḥ|
tayoryogāntaraṃ chāyāgolayoḥ yāmyasaumyayoḥ||91||

The product of the *palaprabhā* and the *koṭi* of the *āśāgrā* divided by the *āśāgrā* is the *koṭikā* of the *palaprabhā*. The square root of the sum of the squares of these is the *svadṛgguṇa*.

The square root of the sum of the squares of the *śaṅku* and the *svadṛgguṇa* is the *chāyākarṇa*. The *śaṅku* and the *chāyā* are multiplied by the *trijyā* and divided by the *chāyākarṇa* to get their *sphuṭa* values.

The *dṛgguṇa* multiplied by the *krānti* and divided by the *akṣajyā* gives the *apakrama*.[65] The *koṭis* of the *krānti* and the *sphuṭadṛgguṇa* are obtained [by subtracting their squares] from the square of the *trijyā* and taking the square root.

By cross-multiplying the *krānti* and *sphuṭadṛgguṇa* with the *koṭis* of each other, and finding their sum or difference, depending upon whether the Sun is in the southern or the northern hemisphere, the *chāyā* is obtained.

In the above verses an alternate expression for the *chāyā* ($R\sin z$) is given. This involves several steps and a number of intermediate terms are introduced. The term *palaprabhā* or *palabhā* refers to the shadow of the *śaṅku* on the equinoctial day at noon. It is given by

$$palabh\bar{a} = 12\tan\phi. \tag{3.254}$$

For convenience, we denote the *palabhā* by x and its *koṭikā* by y. The expression for the *palabhā-koṭikā* given by Nīlakaṇṭha is

$$y = \frac{palabh\bar{a} \times \bar{a}\acute{s}\bar{a}gr\bar{a}ko\d{t}i}{\bar{a}\acute{s}\bar{a}gr\bar{a}} = \frac{12\tan\phi \times R\cos a}{R\sin a}. \tag{3.255}$$

It may be noted that here both x and y are in the measure of *aṅgulas*. Two more quantities, namely the *svadṛgguṇa*, and *chāyākarṇa* are defined as follows:

$$\begin{aligned} svad\d{r}ggu\d{n}a &= \sqrt{x^2 + y^2} \\ &= 12\tan\phi\csc a \end{aligned} \tag{3.256}$$

$$\begin{aligned} ch\bar{a}y\bar{a}kar\d{n}a &= \sqrt{\acute{s}a\dot{n}ku^2 + svad\d{r}ggu\d{n}a^2} \\ K &= \sqrt{12^2 + (x^2 + y^2)} \\ &= 12\frac{\sqrt{\sin^2\phi + \cos^2\phi\sin^2 a}}{\cos\phi\sin a}. \end{aligned} \tag{3.257}$$

[65] Here the word *apakrama* does not refer to declination, but to a quantity related to that.

The term *chāyākarṇa* means the hypotenuse of the shadow. In the above expression for the *chāyākarṇa*, 12 being the height of the *śaṅku*, the *svadṛgguṇa* must represent the length of the shadow. We find that the length of the shadow is given as $12\tan\phi\csc a$. We know that the length of the shadow on any day at any given time is given by $12\tan z$, z being the zenith distance of the Sun at that instant. Now we try to get the condition under which

$$12\tan z = \frac{12\tan\phi}{\sin a}. \tag{3.258}$$

Rewriting the above equation we have

$$\cos z\sin\phi - \sin z\sin\phi\sin a = 0.$$

$$\text{or} \qquad \cos z\sin\phi + \sin z\sin\phi\cos(90+a) = 0. \tag{3.259}$$

Now, from the spherical triangle *PZS* shown in Fig. 3.27*b*, applying the cosine formula for the side *PS* we have

$$\begin{aligned}\sin\delta &= \cos z\sin\phi + \sin z\cos\phi\cos(90+a)\\ &= \cos z\sin\phi - \sin z\cos\phi\sin a.\end{aligned} \tag{3.260}$$

Thus from (3.256) and (3.258) we see that the expression for the *chāyā* given as $12\tan\phi\csc a$ is valid on the equinoctial day, when $\delta = 0$. Then the *sphuṭaśaṅku* and the *sphuṭacchāyā* or *sphuṭadṛgguṇa* are defined as follows.

$$\begin{aligned}\textit{sphuṭaśaṅku} &= \frac{\textit{śaṅku} \times \textit{trijyā}}{\textit{chāyākarṇa}}\\ &= \frac{12 \times R}{K}\end{aligned} \tag{3.261}$$

$$\begin{aligned}\textit{sphuṭadṛgguṇa} &= \frac{\textit{svadṛgguṇa} \times \textit{trijyā}}{\textit{chāyākarṇa}}\\ R\sin\theta &= \frac{12\tan\phi \times R}{\sin a \times K}.\end{aligned} \tag{3.262}$$

Of these two quantities, the latter and its *koṭi* are used in further calculations. Hence, for convenience, we have denoted it by $R\sin\theta$. Substituting for K, we have

$$R\sin\theta = \frac{R\sin\phi}{\sqrt{\sin^2\phi + \cos^2\phi\sin^2 a}}. \tag{3.263}$$

Another quantity related to the *sphuṭadṛgguṇa* is defined as

$$\begin{aligned}\textit{apakrama} &= \frac{\textit{sphuṭadṛgguṇa} \times \textit{krānti}}{\textit{akṣajyā}}\\ R\sin\psi &= \frac{R\sin\theta \times R\sin\delta}{R\sin\phi}\end{aligned}$$

$$= \frac{R\sin\delta}{\sqrt{\sin^2\phi + \cos^2\phi\sin^2 a}}. \tag{3.264}$$

Here we may note the following:

1. The term *apakrama* is usually used to refer to the declination or the Rsine of the declination of the celestial object. But in this context it refers to an entirely different quantity. In order to avoid misconception, the following observation is made in *Laghu-vivṛti*:

    अथेष्टक्रान्तिं दृग्ज्यया निहत्याक्षज्यया विभजेत् । तत्र लब्धः अपक्रमाधीनः छायाखण्डः ।

    The desired *krānti* ($R\sin\delta$) may be multiplied by the *dṛgjyā* (*sphuṭa-dṛgguṇa*) and divided by the *akṣajyā*. The result obtained is the *chāyākhaṇḍa*, which is dependent on the *apakrama*.

2. As we will be using the *chāyākhaṇḍa* and its *koṭi* later in the calculations, for convenience we denote this by $R\sin\psi$.

    The *koṭis* of the *chāyākhaṇḍa* (k_1) and the *sphuṭacchāyā* (k_2) are given by

$$k_1 = \sqrt{R^2 - (R\sin\psi)^2} = R\cos\psi \tag{3.265}$$

$$\text{and} \qquad k_2 = \sqrt{R^2 - (R\sin\theta)^2} = R\cos\theta. \tag{3.266}$$

Then it is stated that the *chāyā* ($R\sin z$) is given by

$$chāyā = \frac{k_1 \times sphuṭadṛgguṇa \pm k_2 \times chāyākhaṇḍa}{trijyā} \tag{3.267}$$

$$R\sin z = \frac{R\cos\psi \times R\sin\theta \pm R\cos\theta \times R\sin\psi}{R}. \tag{3.268}$$

Substituting the appropriate expressions for $R\sin\theta$ and $R\sin\psi$, and after some straightforward manipulations, we find

$$R\sin z = \frac{R(\sin\phi\sqrt{\sin^2\phi - \sin^2\delta + \cos^2\phi\sin^2 a} \pm \cos\phi\sin a\sin\delta)}{(\sin^2\phi + \cos^2\phi\sin^2 a)}. \tag{3.269}$$

Proof:

Considering the spherical triangle *PZS* shown in Fig. 3.27, and applying the cosine formula, we get

$$\sin\delta = \sin\phi\cos z + \cos\phi\sin z\cos A. \tag{3.270}$$

Making the substitution $\sin z = y$, rearranging the terms, and squaring both sides, we get

$$\begin{aligned}(1-y^2)\sin^2\phi &= (\sin\delta - y\cos\phi\cos A)^2 \\ &= \sin^2\delta + y^2\cos^2\phi\cos^2 A - 2y\sin\delta\cos\phi\cos A. \qquad (3.271)\end{aligned}$$

This leads to the quadratic equation

$$y^2(\sin^2\phi + \cos^2\phi\cos^2 A) - 2y\sin\delta\cos\phi\cos A + (\sin^2\delta - \sin^2\phi) = 0. \qquad (3.272)$$

Solving the above quadratic and noting that $\cos A = \pm\sin a$ we get

$$y = \sin z = \frac{(\sin\delta\cos\phi\cos A \pm \sin\phi\sqrt{\sin^2\phi - \sin^2\delta + \cos^2\phi\cos^2 A})}{(\sin^2\phi + \cos^2\phi\cos^2 A)}, \qquad (3.273)$$

which is the same as the expression given in the text (see (3.269)), as $\cos A = \pm\sin a$. In the above expression we need to take only the positive sign of the discriminant, as that is what corresponds to the physical situation. Otherwise, $\sin z$ would be negative when $\delta < \phi$, which is not possible when the Sun is above the horizon.

As per the prescription given in the text, in (3.269), '+' must be chosen when the Sun is in the southern hemisphere and '∼' when it is in the northern. When the Sun is in the southern hemisphere, $\delta < 0$ and $A > 90$. Hence the product of $\sin\delta$ and $\cos A = \cos(90+a)$—both individually being negative—is positive. When the Sun is in the northern hemisphere, $\delta > 0$. But A can be > 90, $= 90$ or < 90, depending upon whether the Sun has crossed the prime vertical, is on the prime vertical, or has yet to cross the prime vertical, when it is on the eastern part of the hemisphere. Hence the product of $\sin\delta$ and $\cos A$ can be both positive or negative. Hence it appears that the sum of two magnitudes should be taken when the declination and the *āśāgrā* are in the same direction (both north or both south), and the difference when the declination and the *āśāgrā* are in opposite directions.

३.३९ कोणशङ्कुच्छाया

3.39 Shadow when the amplitude is 45 degrees

भुजाक्षो लम्बवर्गार्धमूलं कोटिः श्रुतिस्तयोः[66] ।
हारः, क्रान्तिघ्नकोटयाश्च[67] दोःश्रुत्योः क्रान्तिहारयोः ॥ १२ ॥

[66] तयोः = भुजाकोटयोः = अक्षस्य लम्बवर्गार्धमूलस्य चेति यावत्। तयोः श्रुतिः हारः। कस्य हारः? दोःश्रुत्योः (सम्बन्धि) क्रान्तिघ्नकोटयाः, (एवं) क्रान्तिहारयोः (लब्ध) कोटिघ्नाक्षस्य च।
अत्र च दोः इति शब्देन अक्षज्या ग्राह्या। अपि च पृथक् पृथक् दोः (= अक्षज्यायाः), अक्षज्याश्रुतिसम्बन्धि क्रान्तिघ्नकोटयाश्च श्रुत्या (= पूर्वोक्तहारेण) हरणं इष्यते। एवं क्रान्तिहारयोर्लब्धा या कोटिः तस्याः तत्कोटिघ्नाक्षस्य च पूर्वोक्तहारेण हरणं इष्यते। यतो हि, वस्तुतः हारवर्गेण हरणं कार्यं इति गणितादवगम्यते। तच्च एवं पृथक् पृथक् हरणेनैव लभ्यते। एतच्च ग्रन्थे कण्ठतः न निर्दिष्टमिति भाति। युक्तिदीपिकायां, लघुविवृतौ च अयमंशः न स्पष्टीकृत इत्येतत् चिन्त्यम्। अतः अस्माभिः पूर्वोक्तरीत्या किञ्चित् क्लिष्टकल्पनया अर्थो वर्णितः। साधुत्वं सुधीभिश्चिन्त्यम्।
एवम् आप्तयोः याम्यदिशि ऐक्यम् उदग्दिशि भेदः। तत्र लब्धं फलं प्रभा।

[67] The reading in both the printed editions is क्रान्तिघ्नकोटयोश्च।

कोटिघ्नाक्षस्य चाप्तैक्यं याम्ये भेद उदक्प्रभा ।
अक्षकोट्यधिकायां तु क्रान्त्यां योगोऽप्युदक् प्रभा ॥ ९३ ॥
क्रान्त्यक्षयोश्च तत्कोट्योः वधात् भेदयुती नरः ।
तद्वद् द्विरुदगन्यत्राप्यभावः कोणयोर्द्वयोः ॥ ९४ ॥
अर्कघ्ने भाश्रुती शङ्कुभक्ते ते अङ्गुलात्मिके ।

bhujākṣo lambavargārdhamūlaṃ koṭiḥ śrutistayoḥ |
hāraḥ, krāntighnakoṭyāśca doḥśrutyoḥ krāntihārayoḥ || 92 ||
koṭighnākṣasya cāptaikyaṃ yāmye bheda udakprabhā |
akṣakoṭyadhikāyāṃ tu krāntyāṃ yogo'pyudak prabhā || 93 ||
krāntyakṣayośca tatkoṭyoḥ vadhāt bhedayutī naraḥ |
tadvad dvirudaganyatrāpyabhāvaḥ koṇayordvayoḥ || 94 ||
arkaghne bhāśrutī śaṅkubhakte te aṅgulātmike |

The *bhujā* is the *akṣa*. The *koṭi* is the square root of half the square of the *lambaka*. The hypotenuse formed by them is the divisor for [each factor in] the product of the *krānti*, and the *koṭi* which is obtained from the *doḥ* (sine) and the *śruti* (hypotenuse), and also for [each factor in] the product of the *akṣa* and the *koṭi*, obtained from the *krānti* and the divisor [individually]. The sum of the products gives the shadow in the south, and their difference that in the north.

If the *krānti* is greater than the *koṭi* of the *akṣa*, then the shadow is obtained even by taking the sum of the products [in the north].

The *nara* (or *śaṅku* = $R\cos z$) is the sum or the difference of the products of the *krānti* and *akṣa* and their *koṭis*. As in the earlier case, here again two *naras* are possible in the two *koṇas* of the northern hemisphere but not in the other (the southern). The shadow and hypotenuse are obtained in *aṅgulas* by multiplying by 12 and dividing by the *śaṅku*.

The term *koṇaśaṅku* refers to the Rsine of the zenith distance of the Sun when the *āśāgrā* is equal to 45°, that is, when the azimuth of the Sun is 45°. In the above verses, the expression for the *koṇaśaṅku* is given. To arrive at the expression for the *koṇaśaṅku*, two right-angled triangles as shown in Fig. 3.31 are considered. In the first triangle, the *bhujā* is defined to be the *akṣa* and the *koṭi* (denoted by k_1) is defined to be the square root of half the square of the *lambaka*. That is,

$$k_1 = \sqrt{\frac{\cos^2\phi}{2}} = \frac{\cos\phi}{\sqrt{2}}. \tag{3.274a}$$

Hence the hypotenuse (h), termed the *śruti*, is given by

$$h = R\sqrt{\sin^2\phi + \frac{\cos^2\phi}{2}}. \tag{3.274b}$$

Even for the other triangle, whose *bhujā* is stated to be the *krānti*, the hypotenuse is taken to be the same. Hence the *koṭi*, (marked as k_2) of the *krānti* in the second figure, is given by

$$k_2 = R\sqrt{\sin^2\phi + \frac{\cos^2\phi}{2} - \sin^2\delta}. \tag{3.275}$$

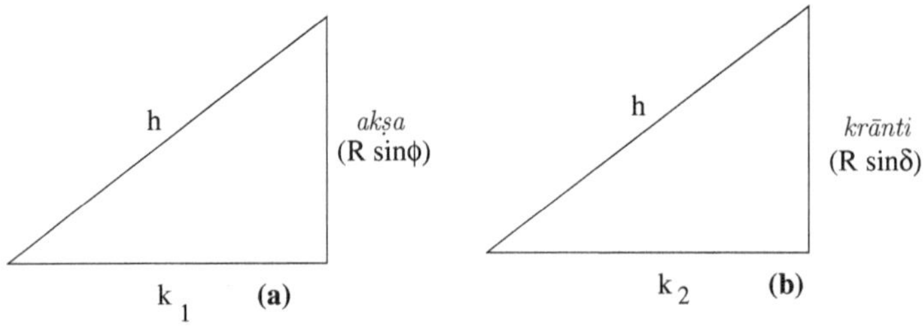

Fig. 3.31 Two triangles having the same hypotenuse defined while giving the expression for the *prabhā/chāyā* corresponding to the *koṇaśaṅku*.

The hypotenuse (h) defined above is used as the divisor for both the *bhujā* and the *koṭi* in later computations. Now, the sum or difference of the cross-products of the *bhujā* and the *koṭi* of the two triangles (with each term divided by the hypotenuse h) is stated to be the required *prabhā* (shadow). That is

$$prabh\bar{a} = \frac{ak\d{s}a \times k_2 \stackrel{+}{\sim} kr\bar{a}nti \times k_1}{h^2}. \tag{3.276}$$

Substituting for k_1, k_2 and h from (3.274) and (3.275), we have

$$R\sin z = \frac{R(\sin\phi\sqrt{\sin^2\phi + \frac{\cos^2\phi}{2} - \sin^2\delta} \stackrel{+}{\sim} |\sin\delta|\frac{\cos\phi}{\sqrt{2}})}{(\sin^2\phi + \frac{\cos^2\phi}{2})}. \tag{3.277}$$

It may be noted that we obtain the above equation right away by substituting $a = 45°$ in (3.269).

When the Sun's declination is north, it is possible to have two *koṇacchāyās*. This is discussed in verse 94. We explain this with the help of Fig. 3.32. Here S_1 and S_2 refer to the positions of the Sun before and after it crosses the prime vertical and when the *āśāgrā* is 45°. The angle A is measured from the prime meridian eastwards. But the *āśāgrā* (a) is measured from the prime vertical either to the north or to the south.

$$\sin z = \frac{(\sin\delta\cos\phi\cos A \stackrel{+}{\sim} \sin\phi\sqrt{\sin^2\phi - \sin^2\delta + \cos^2\phi\cos^2 A})}{(\sin^2\phi + \cos^2\phi\cos^2 A)}. \tag{3.278}$$

It may be noted that the above equation is the same as (3.277), when

$$|\cos a| = \frac{1}{\sqrt{2}}. \tag{3.279}$$

Incidentally, the above equation also clearly brings out the signs as set forth in the previous set of verses. Now we consider two different cases depending upon whether the Sun is in the northern ($\delta > 0$) or the southern ($\delta < 0$) hemisphere.

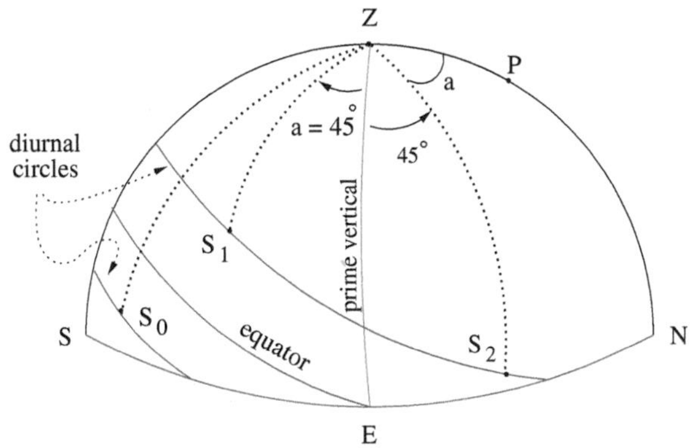

Fig. 3.32 Two different positions of the Sun (when $\delta > 0$), corresponding to which there is a possibility of a second *koṇaśaṅku* and *koṇacchāyā*.

Case 1: $\delta < 0$

When $\delta < 0$, $\sin\delta$ is negative. And when the Sun is above the horizon, corresponding to the position S_0 in Fig. 3.32, $A > 90°$, and therefore $\cos A < 0$. So, the first term (3.278) is always positive. We show shortly that the second term has to be positive. Hence, $\sin z$ is the sum of two positive terms.

Case 2: $\delta > 0$

When $\delta > 0$, corresponding to the position S_1 or S_2 in Fig. 3.32, $\sin\delta$ is positive. But, depending upon whether Sun is to the north or to the south of the prime vertical, two distinct cases are possible:

1. When the Sun is to the south $A > 90°$. Hence, $\cos A < 0$. So the first term in the above equation is negative. So, the second term has to be positive, as $\sin z$ should be positive when the Sun is above the horizon. In this case, $\sin z$ is clearly a difference of two positive quantities.
2. When the Sun is to the north, $A < 90°$. Hence $\cos A > 0$. So the first term in the above equation is positive. Hence $\sin z$ is the sum of two positive terms in this case.

Condition for the possibility of a *koṇacchāyā*: (when the Sun is to the north)

Due to the diurnal motion of the Sun, the *aśāgrā*, a, keeps on decreasing from the sunrise and becomes zero when the Sun is on the prime vertical. It can be shown that at sunrise

$$\sin a|_{\text{sunrise}} = \frac{\sin\delta}{\cos\phi}. \qquad (3.280a)$$

So, a *koṇacchāyā* is possible only when

$$\sin a|_{\text{sunrise}} \geq \frac{1}{\sqrt{2}}. \tag{3.280b}$$

Therefore, the condition for a *koṇacchāyā* reduces to

$$\sin\delta \geq \frac{\cos\phi}{\sqrt{2}}. \tag{3.281}$$

In this case, corresponding to the Sun's position S_2 in Fig. 3.32,

$$\cos A = \cos(90 - a) = \frac{1}{\sqrt{2}}, \tag{3.282}$$

and $\sin\delta\cos\phi\cos A$ is positive, as δ is positive. Hence the word *yogo'pi* (sum also), for northern declinations is used in verse 93b when the *koṇaśaṅku* is to the north.

For the other *koṇacchāyā*, corresponding to the Sun's position S_1 in Fig. (3.32), the first term in (3.278) is negative as $\cos A$ is negative, and hence we have to find the difference of the terms and not the sum.

In verse 94, the expression for the *nara* ($R\cos z$) is given. The text states:

$$nara = \frac{ak\d{s}a \times kr\bar{a}nti \pm k_1 \times k_2}{h^2}. \tag{3.283}$$

Substituting for k_1, k_2 and h from (3.274) and (3.275), we have

$$R\cos z = \frac{R(\sin\phi\sin\delta \mathrel{\substack{+\\ \sim}} \frac{\cos\phi}{\sqrt{2}}\sqrt{\sin^2\phi + \frac{\cos^2\phi}{2} - \sin^2\delta})}{(\sin^2\phi + \frac{\cos^2\phi}{2})}. \tag{3.284}$$

Using straightforward algebraic manipulations it can be shown that the above expression follows from (3.278) for the *chāyā* ($R\sin z$).

Condition for the occurrence of a second *koṇaśaṅku*/(*nara*):

Here again, as in the case of a *koṇacchāyā* ($R\sin z$), two *naras* are possible, when the product of the *akṣa* and the *krānti* is greater than that of their *koṭis*, and when the Sun is in the northern hemisphere. This is easily understood from the necessary and sufficient condition for the occurrence of a second *koṇacchāyā*, which is given in (3.281). In this case, the magnitude of the first term is greater than that of the second term in (3.284) and both '+' and '∼' can be taken in (3.284), and there are two solutions for $\cos z$ corresponding to two *koṇaśaṅkus*. However, when the product of the *akṣa* and the *krānti* is greater than that of their *koṭis*, and the declination is south, the first term is negative and its magnitude is greater than that of the second. In this case, $\cos z$ is negative. This implies that the *koṇaśaṅku* is not possible.

The shadow *prabhā/bhā/chāyā* and the hypotenuse *śruti* are obtained in *aṅgulas* by multiplying by 12 and dividing by the *śaṅku*. It was already shown [see (3.4)] that the shadow at any time is $12\tan z$, where z is the zenith distance of the Sun at that instant. Now,

$$prabh\bar{a} = 12\tan z = \frac{12\times R\sin z}{R\cos z} = \frac{12\times prabh\bar{a}}{\acute{s}a\dot{n}ku}. \tag{3.285}$$

Similarly,

$$\acute{s}ruti = \frac{12\times \acute{s}ruti}{\acute{s}a\dot{n}ku}. \tag{3.286}$$

That the *āśāgrā* must be equal to 45° for the *koṇaśaṅku* measurements is clearly stated by Śaṅkara Vāriyar in his *Yukti-dīpikā* as follows:

आशाग्राध्यर्धराशिज्या कोणशङ्कुप्रभाविधौ । [68]

In measurements related to the *koṇaśaṅku*, the *āśāgrā* is equal to 45° (the *adhyardharāśi*).[69]

Further, the procedure is clarified thus in *Yukti-dīpikā*:

लम्बवर्गार्धमूलं तु सा पुनः कोटिवृत्तगा ॥
कोटिः साक्षो भुजा, कर्णो वर्गयोगपदं द्वयोः ।
हारकः स पुनः क्रान्त्यक्षयोः कोटिघ्नयोर्मिथः[70] ॥
क्रान्त्यक्षकोट्यौ तत्कर्णवर्गभेदपदोद्भवे।
तत्र लब्धफलैक्यं यत् स्यात् छाया याम्यगोलगा॥ [71]

The square root of half the square or *lambaka* which is on the *koṭivṛtta* is the *koṭi*. The *bhujā* is the *akṣa*. The square root of the squares of the two is the *karṇa* hypotenuse. [That is also] the divisor for the mutual product of the *krānti*, *akṣa* and their *koṭis*.[72]

The *koṭis* of the *krānti* and *akṣa* are themselves obtained by subtracting their squares from the squares of the *karṇa* and taking the square root. The sum of the result obtained is the shadow in the southern hemisphere.

३.४० प्राग्लग्नानयनम्

3.40 Obtaining the orient ecliptic point

संस्कृतायनभानूत्थराशिगन्तव्यलिप्तिकाः ॥ १५ ॥
तद्राशिस्वोदयप्राणहता राशिकला हृताः ।

---

[68] {TS 1977}, p. 239.

[69] The term *adhyardharāśi* literally means: 'a *rāśi* increased by half [of it]', and it is equal to 45°, since a *rāśi* equals 30°.

[70] The prose order: द्वयोः वर्गयोगपदं (यः) स पुनः हारकः। कस्य? इत्याशङ्कायामाह - क्रान्त्यक्षयोः मिथः (= परस्परं) कोटिघ्नयोः (यत् फलं लभ्यते तस्येति)।

[71] {TS 1977}, p. 239.

[72] The *karṇa* is the divisor for each of the factors *krānti*, *akṣa* and their *koṭis* in the expression for $\cos z$ in (3.284).

असवो राशिशेषस्य गतासुभ्यस्त्यजेच्च तान् ॥ ९६ ॥
उत्तरोत्तरराशीनां प्राणाः शोध्याश्च शेषतः ।
पूरयित्वा रवे राशिं क्षिपेद्राशींश्च तावतः ॥ ९७ ॥
विशुद्धा यावतां प्राणाः, शेषास्त्रिंशद्गुणात् पुनः ।
तदूर्ध्वराशिमानाप्तान् भागान् क्षिप्त्वा रवौ तथा ॥ ९८ ॥
षष्टिघ्नाच्च पुनः शेषात् तन्मानाप्तकला अपि ।
एवं प्राग्लग्नमानेयम् अस्तलग्नं तु षड्भयुक् ॥ ९९ ॥
व्यत्ययेनायनं कार्यं मेषादित्वप्रसिद्धये ।

saṃskṛtāyanabhānūttharāśigantavyaliptikāḥ||95||
tadrāśisvodayaprāṇahatā rāśikalā hṛtāḥ|
asavo rāśiśeṣasya gatāsubhyastyajecca tān||96||
uttarottararāśīnāṃ prāṇāḥ śodhyāśca śeṣataḥ|
pūrayitvā rave rāśiṃ kṣipedrāśīṃśca tāvataḥ||97||
viśuddhā yāvatāṃ prāṇāḥ śeṣāstriṃśadguṇāt punaḥ|
tadūrdhvarāśimānāptān bhāgān kṣiptvā ravau tathā||98||
ṣaṣṭighnācca punaḥ śeṣāt tanmānāptakalā api|
evaṃ prāglagnamāneyam astalagnaṃ tu ṣaḍbhayuk||99||
vyatyayenāyanaṃ kāryaṃ meṣāditvaprasiddhaye|

From the longitude of the Sun corrected for *ayana*, the number of minutes to be elapsed in that *rāśi* [are calculated]. That is multiplied by the duration of the rising of that *rāśi* and is divided by the number of minutes in a *rāśi*. These are *prāṇas* corresponding to the remaining *rāśi* and they have to be subtracted from the *prāṇas* elapsed [since the sunrise].

From the remainder, the durations of the risings of the *rāśis* that follow have to be subtracted. Having added (*pūrayitvā*) the degrees remaining in that *rāśi* to the Sun, [the degrees corresponding to] that number of *rāśis*, whose rising times were subtracted, are to be added. The remaining *prāṇas* are multiplied by 30 and divided by the duration of rising of that *rāśi*. The result obtained is once again added to the Sun.

The remainder when multiplied by 60 gives the result in minutes. Thus the *prāglagna*, the orient ecliptic point, should be obtained. The *astalagna*, the setting ecliptic point, is obtained by adding six signs to that. To know the longitude of the ecliptic points from the *Meṣādi*, the *ayana* correction has to be applied reversely.

These verses give the procedure for finding the *prāglagna*, which is also referred to simply as the *lagna* at times. The term *lagna* (orient ecliptic point) means 'coinciding' or 'associated' with. In this context it refers to the longitude of the point of the ecliptic that is coinciding with the horizon at any point time during the day. The point on the eastern part of the horizon is called the *prāglagna* and the point on the western part is called the *astalagna*.

The procedure given here may be understood with the help of Fig. 3.33. Here S represents the Sun in the eastern part of the hemisphere, Γ the vernal equinox and R_1, R_2 etc. the ending points of the first *rāśi*, second *rāśi* and so on. h refers to the time elapsed after sunrise and H the hour angle of the Sun.

Let λ_s be the *sāyana* longitude of the Sun. Suppose the Sun is in the i-th *rāśi* (in Fig. 3.33, it is in the first *rāśi*), whose rising time at the observer's location is given by T_i. If θ_{R_i} is the angle remaining to be covered by the Sun in that *rāśi* (in minutes), then the time required for that segment of the *rāśi* to come above the horizon is given by

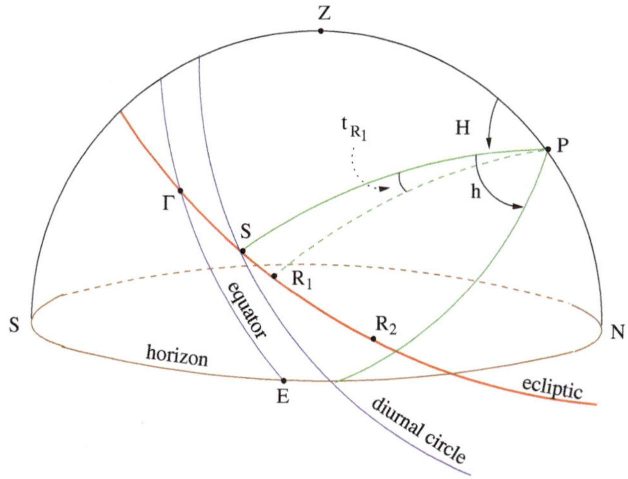

Fig. 3.33 Determination of the *prāglagna* (orient ecliptic point): approximate method.

$$t_{R_i} = \frac{\theta_{R_i} \times T_i}{30 \times 60}. \tag{3.287}$$

Subtracting this time t_{R_i} from the time elapsed since sunrise h, we have

$$h' = h - t_{R_i}. \tag{3.288}$$

From h' the times required for the subsequent *rāśis* to rise, T_{i+1}, T_{i+2}, etc. are subtracted till the remainder r stays positive. That is,

$$r = h' - T_{i+1} - T_{i+2} \ldots - T_{j-1} \qquad (r\ +ve). \tag{3.289}$$

Suppose we are in the j-th *rāśi* whose rising time is T_j. Then, the portion of R_j that would have come above the horizon in the remaining time r is given by

$$\theta_{E_j} = \frac{r \times 30}{T_j} \qquad \text{(in degrees)}. \tag{3.290}$$

Now, the longitude of the *prāglagna* (L) is

$$L = \lambda_s + \theta_{R_i} + 30 + 30 + \ldots + \theta_{E_j}. \tag{3.291}$$

By definition the *astalagna* is

$$astalagna = prāglagna + 180°. \tag{3.292}$$

This is because the horizon divides the ecliptic exactly into two parts. For the same reason, this may also be obtained by subtracting 180 degrees. The *lagnas* obtained by the above procedure are *sāyana-lagnas*. To obtain the *nirayaṇa* ones, one needs to subtract the *ayanāṃśa*. This is what is stated in the text as '*vyatyayenāyanaṃ kāryaṃ meṣāditvaprasiddhaye*'.

३.४१ प्राग्लग्नानयने स्थूलता

3.41 Inaccuracy in determining the *prāglagna*

एकस्मिन्नपि राशौ तु क्रमात् कालो हि भिद्यते ॥ १०० ॥
तेन त्रैराशिकं नात्र कर्तुं युक्तं यतस्ततः ।
एवमानीतलग्नस्य स्थूलतैव न सूक्ष्मता ॥ १०१ ॥

ekasminnapi rāśau tu kramāt kālo hi bhidyate || 100 ||
tena trairāśikaṃ nātra kartuṃ yuktaṃ yatastataḥ |
evamānītalagnasya sthūlataiva na sūkṣmatā || 101 ||

The [rising] time differs systematically even in the same *rāśi*. Hence, it is not appropriate to apply the rule of three here. Therefore the *lagna* obtained by the above procedure will be only approximate, and not exact.

In getting the longitude of the *prāglagna* by the procedure described in the previous section, we have made use of the rule of three at two stages: (i) in obtaining the time corresponding to a certain angle, and (ii) in obtaining the angle corresponding to a certain time, as noted from (3.287) and (3.290) respectively. It is known that the rising times of different *rāśis* are not the same. Here it is pointed out that the different segments of even the same *rāśi* take different times for rising above the horizon.[73]

In fact, an exact procedure for determining the value of the *prāglagna* is described a little later in the text. This is done by introducing two quantities, namely the *kālalagna* and the *madhyalagna*, which are defined in the following two sections.

३.४२ काललग्नानयनम्

3.42 Determination of the *kālalagna*

सायनार्कभुजाप्राणाः प्राग्वत् स्वचरसंस्कृताः ।
काललग्नं तदेवाद्ये, द्वितीये तु तदूनितम् ॥ १०२ ॥
राशिषट्कं, पदेऽन्यस्मिन् तद्युतं चरमे पुनः ।
तदूनं मण्डलं लग्नकालः स्यादुदये रवेः ॥ १०३ ॥
द्युगतप्राणसंयुक्तः कालो[74] विषुवदादिकः ।

sāyanārkabhujāprāṇāḥ prāgvat svacarasaṃskṛtāḥ |
kālalagnaṃ tadevādye, dvitīye tu tadūnitam || 102 ||

---

[73] This is due to the fact that the declination varies continuously as the longitude changes.

[74] अत्र कालशब्दः काललग्नवाचकः इति भाति । तथा च, द्युगतप्राणसंयुक्तः स्वचरसंस्कृतः सायनार्कभुजाप्राणाः, इष्टकालसमुत्पन्नः विषुवदादिकः काललग्नो भवतीत्यर्थः ॥

rāśiṣaṭkaṃ pade'nyasmin tadyutaṃ carame punaḥ |
tadūnaṃ maṇḍalaṃ lagnakālaḥ syādudaye raveḥ || 103 ||
dyugataprāṇasaṃyuktaḥ kālo viṣuvadādikaḥ |

The right ascension of the *sāyana* Sun corrected by the ascensional difference is the *kālalagna* in the first quadrant. In the second it is six signs minus that. In the other [third] quadrant, this added to six signs [is the *kālalagna*] and in the last [quadrant] the difference of it from one circle (360 degrees) is the *kālalagna* when the Sun rises.

[The above] added to the *prāṇas* elapsed gives the *kāla* (*kālalagna*) measured from the vernal equinox [at any desired time].

The time difference between the rising of the Sun and the vernal equinox is called the *kālalagna*, in the first instance. We denote it by the symbol L'. The procedure for computing the *kālalagna* when the Sun is in the first and the second quadrants can be understood with the help of Figs 3.34(a) and 3.34(b).

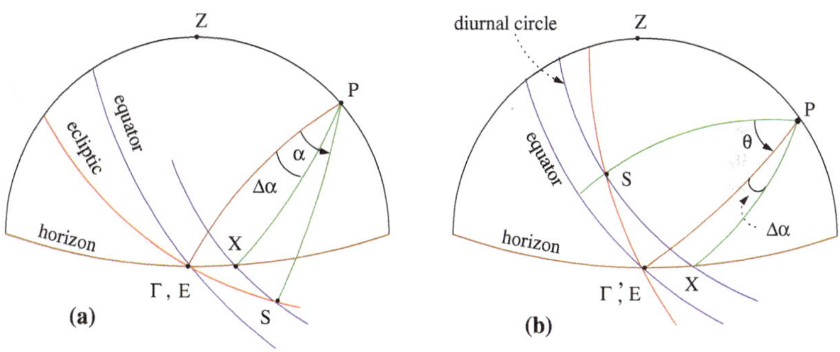

Fig. 3.34 Determination of the *kālalagna* when the Sun lies in the I and the II quadrants.

In these figures, Γ and Γ' represent the vernal and the autumnal equinoxes respectively which are rising at E. S, P and E denote the Sun, the celestial pole and the east point of the horizon. X is the point at which the Sun rises. Suppose the Sun is in the first quadrant. Then $E\hat{P}S = \alpha$ is the R.A. of the Sun, and $E\hat{P}X = \Delta\alpha$ is the ascensional difference. Then the time taken by the Sun to rise after the rise of Γ is given by

$$X\hat{P}S = L' = \alpha - \Delta\alpha. \tag{3.293}$$

This is the expression for the *kālalagna* when the Sun is in the first quadrant. When the Sun is in the other quadrants, let θ be the angle between the secondary to the equator through S and the 6 o'clock circle passing through E. The R.A.s of the Sun in the second, third and fourth quadrants are: $\alpha = (180-\theta)$, $(180+\theta)$ and $(360-\theta)$ respectively.

In Fig. 3.34(b), the Sun is in the second quadrant. The angle $\theta + \Delta\alpha$ clearly represents the time taken by Γ' to rise, after the sunrise. As Γ would be coinciding with the west point on the horizon, the *kālalagna* is given by

$$L' = 180 - (\theta + \Delta\alpha). \tag{3.294}$$

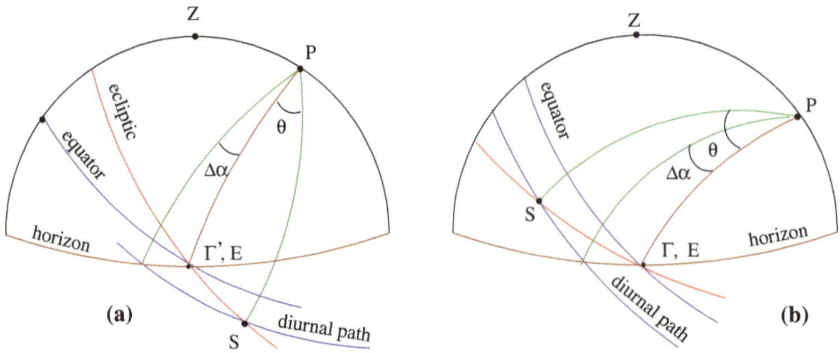

Fig. 3.35 Determination of the *kālalagna* when the Sun lies in the third and the fourth quadrants.

When the Sun is in the third quadrant, as shown in the Fig. 3.35(a), it is clear that the time taken by it to rise after Γ' is $\theta+\Delta\alpha$. Therefore the time taken by S to rise after Γ is

$$L' = 180 + (\theta + \Delta\alpha). \tag{3.295a}$$

When the Sun is in the fourth quadrant, the time taken by Γ to rise after the Sun is $= \theta - \Delta\alpha$ (refer Fig. 3.35(b)). Hence, the time taken by S to rise after Γ is

$$L' = 360 - (\theta - \Delta\alpha). \tag{3.295b}$$

In the above, the *kālalagna* is the time difference between the sunrise and the rise of Γ. From now onwards, the *kālalagna* or *kāla* is used to denote the time interval between the desired instant and the rise of Γ. Then, the *kāla* at any time

$$L'' = L' + \textit{prāṇas} \text{ elapsed after sunrise}, \tag{3.296}$$

which is the time after the rising of Γ, or the time measured from Γ.

३.४३ दृक्क्षेपानयनम्
3.43 Determination of the zenith distance of the *vitribhalagna*

अन्त्यद्युज्याहताक्षाद्यत् त्रिज्याप्तं यश्च लम्बकः ॥ १०४ ॥
काललग्नोत्थकोटिघ्नः करार्थाब्ध्युरगैर्हृतः ।
दृक्क्षेपस्तद्भिदैक्यं च काले कर्किमृगादिके ॥ १०५ ॥
विश्लेषे लम्बजाधिक्ये सौम्यो याम्योऽन्यदा सदा ।
तत्त्रिज्याकृतिविश्लेषात् मूलं दृक्क्षेपकोटिका ॥ १०६ ॥

antyadyujyāhatākṣādyat trijyāptaṃ yaśca lambakaḥ || 104 ||
kālalagnotthakoṭighnaḥ karārthābdhyuragairhṛtaḥ |
dṛkkṣepastadbhidaikyaṃ ca kāle karkimṛgādike || 105 ||
viśleṣe lambajādhikye saumyo yāmyo'nyadā sadā |
tattrijyākṛtiviśleṣāt mūlaṃ dṛkkṣepakoṭikā || 106 ||

> The *akṣa* multiplied by the *antyadyujyā* (the Rcosine of the obliquity of the ecliptic) and divided by the *trijyā*, and the *lambaka* multiplied by the *koṭi* of the *kālalagna* and divided by 8452, [are kept separately]. The *dṛkkṣepa* is the difference between or the sum of the two, depending upon whether the *kālalagna* is within the 6 signs beginning from *Karkaṭaka* or from *Mṛga*.
>
> Where the difference is greater than the one which is obtained from the *lambaka* [or *lambaja*],[75] the *dṛkkṣepa* is north. Otherwise it is always in the south. The square root of the square of the difference between it and the square of the *trijyā* is the *koṭi* of the *dṛkkṣepa*.

In the above verses, the expression for the *dṛkkṣepa* or the Rsine of the zenith distance of the *vitribhalagna* is given. The term *vitribhalagna*[76] (nonagesimal) refers to the point on the ecliptic which is exactly 90 degrees from the *lagna*, in the visible part of the hemisphere. The need for the expression for the zenith distance of the *vitribhalagna* arises from the fact that it is used in the computation of the exact expression for the *lagna*, which will be presented in a later section. This also plays an important role in the calculation of eclipses.

Here, two intermediate quantities x and y are defined whose sum or difference gives the expression for the Rsine of the zenith distance of the *vitribhalagna*. These quantities are defined as:

$$x = \frac{antyadyujy\bar{a} \times ak\d{s}a}{trijy\bar{a}} = \frac{R\cos\varepsilon \times R\sin\phi}{R}, \tag{3.297}$$

and
$$y = \frac{lambaka \times k\bar{a}lalagnako\d{t}i}{8452} = \frac{R\cos\phi \times R|\cos L''| \times \sin\varepsilon}{R} \tag{3.298}$$

Here $\frac{R}{\sin\varepsilon} = \frac{R^2}{R\sin 24}$ is taken to be 8452.

The expression for the *dṛkkṣepa* ($= R\sin ZV$ in Fig. 3.36) is given to be

$$d\d{r}kk\d{s}epa = x \pm y. \tag{3.299}$$

Substituting for x and y and using the notation $ZV = z_v$, we have

$$R\sin z_v = R\cos\varepsilon\sin\phi \pm R\cos\phi|\cos L''|\sin\varepsilon, \tag{3.300}$$

where the sign chosen is '+' when L'' is in the first or the fourth quadrant, and '−' when L'' is in the second or the third quadrant.

Proof:

In Fig. 3.36, Z, K and P represent the zenith, the pole of the ecliptic and the pole of the equator respectively. Γ is the vernal equinox and V the *vitribhalagna*. When Γ is on the horizon, both K and V are on the prime meridian. This situation is depicted in Fig. 3.36(a). At a later time during the day, as Γ keeps rising above the

---

75 लम्बात् अथवा लम्बकात् जायते इति लम्बजः ।

76 विगतं त्रिभं यस्मात् लग्नात्, तत् वित्रिभलग्नम् ।

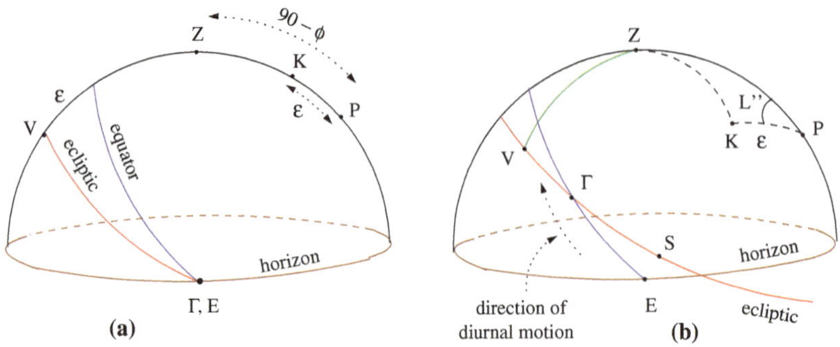

Fig. 3.36 Determination of the zenith distance of the *vitribhalagna*.

horizon and keeps moving towards the prime meridian along the equator as shown in Fig. 3.36(b), the point K also traces a small circle, of radius ε, around the pole of the equator P.

Consider an instant of time L'' units after the rise of Γ. At this time, the hour angle of K would also be L''. That is, $K\hat{P}Z = L''$. Considering the spherical triangle KPZ, and applying the cosine formula,

$$\cos KZ = \cos\varepsilon \sin\phi + \cos\phi \sin\varepsilon \cos L''. \tag{3.301}$$

Since $KV = 90$, $KZ = 90 \pm z_v$.[77] Therefore $|\cos KZ| = \sin z_v$, and we have

$$\sin z_v = |\cos\varepsilon \sin\phi + \cos\phi \sin\varepsilon \cos L''| \tag{3.302}$$

where $L'' = L' +$ *prāṇas* elapsed after sunrise as defined in the previous section.

It may be seen that the above equation is the same as (3.300) prescribed in the verses. Comparing (3.302) and (3.300), it is clear that the '+' sign should be taken when $\cos L''$ is positive, i.e. when L'' is in the first or the fourth quadrants (within 6 signs beginning from *Mṛga*, as stated in the text). Similarly the '−' sign should be taken when $\cos L''$ is negative, or when L'' is within 6 signs beginning from *Karka* or Capricorn.

When we take the '+' sign, L'' is within the first or the fourth quadrant and $KZ < 90°$. Then V is south of the prime vertical, that is, the *dṛkkṣepa* is south. Even when we take the '−' sign, when the *lambajyā* $|\cos\phi \sin\varepsilon \cos L''| < \cos\varepsilon \sin\phi$, $\cos KZ$ is positive and $KZ < 90°$. In this case also, the *dṛkkṣepa* is south.

However when we take the '−' sign and $|\cos\phi \sin\varepsilon \cos L''| > \cos\varepsilon \sin\phi$, then $\cos KZ$ is negative and $KZ > 90°$. In this case if we draw a figure as in Fig. 3.36, K would be below the horizon and V would be north of the prime vertical, or the *dṛkṣepa* is north. Finally, it is mentioned that

[77] In Fig. 3.36, the pole of the ecliptic K is indicated above the horizon and hence $KZ = 90 - z_v$. However, it is possible that K is below the horizon, in which case the '+' sign should be taken.

$$\textit{dṛkkṣepakoṭikā} = R\cos z_v = \sqrt{R^2 - (R\sin z_v)^2}. \quad (3.303)$$

This is elaborated in *Laghu-vivṛti* as follows:

तत्र तयोर्विश्लेषे तु कृते सति यदा अक्षज्ययोर्लग्नफलतः लम्बज्योत्पन्नस्य आधिक्यं तदा सौम्यदिग्गता दृक्क्षेपज्या, अन्यदा सदा याम्यैव।

There, when we find the difference between the two [x and y] if it is noted that the product involving the Rcosine of the latitude (the *lambajyotpannasya*) is greater than the product involving the Rsine of the latitude (the *akṣajyayorlagnaphalataḥ*), then the *dṛkkṣepa* is towards the north (*saumyadiggatā*), otherwise it is always towards the south.

३.४४ स्फुटोदयास्तलग्नानयनम्

3.44 Exact determination of the ecliptic point that is rising or setting

मध्याह्नाद्वा नतप्राणाः निशीथाद्वोन्नतासवः ।
एतद्बाणोनिता त्रिज्या चरज्याढ्या नता यदि ॥ १०७ ॥
उन्नताश्चेच्चरज्योना गोले याम्ये विपर्ययात् ।
द्युज्या लम्बकघातघ्ना त्रिज्याप्ता च पुनर्हृता ॥ १०८ ॥
कोट्या दृक्क्षेपजीवाया लब्धचापं रवौ क्षिपेत् ।
तल्लग्नं प्राक्कपाले स्यात् निशि चेत् तद्विवर्जितम् ॥ १०९ ॥
प्रत्यग्गतेऽस्तलग्नं स्यात् व्यस्तमेव दिवानिशोः ।
प्राक्पश्चाल्लग्नयोर्मध्यं लग्नं दृक्क्षेपसंज्ञितम् ॥ ११० ॥

madhyāhnādvā nataprāṇāḥ niśīthādvonnatāsavaḥ |
etadbāṇonitā trijyā carajyāḍhyā natā yadi || 107 ||
unnatāśceccarajyonā gole yāmye viparyayāt |
dyujyā lambakaghātaghnā trijyāptā ca punarhṛtā || 108 ||
koṭyā dṛkkṣepajīvāyā labdhacāpaṃ ravau kṣipet |
tallagnaṃ prākkapāle syāt niśi cet tadvivarjitam || 109 ||
pratyaggate'stalagnaṃ syāt vyastameva divāniśoḥ |
prākpaścāllagnayormadhyaṃ lagnaṃ dṛkkṣepasaṃjñitam ||110||

The *prāṇas* corresponding to the *nata* (hour angle) or those corresponding to the *unnata* may be obtained from either midday or midnight. If the *nata* is obtained, then the *bāṇa* (versed sine, *utkramajyā*) of it is to be subtracted from the *trijyā*, and the result is added to the *carajyā* [to obtain x]. If the *unnata* [is obtained], then the *carajyā* is to be subtracted from the result. [This is the procedure for the northern hemisphere]. For the southern hemisphere the set of operations are to be reversed.

[This x] is multiplied by the product of the *dyujyā* and the *lambaka* and is divided by the *trijyā* and the *koṭi* of the *dṛkkṣepa*. The arc of the result is to be applied positively to the Sun. This gives the *lagna* in the eastern part. For [the computations with mid-] night, the arc has to be subtracted from the Sun.

Since the processes have to be reversed for the day and night, if it (the Sun) is in the western part, the result gives the setting point of the ecliptic. The midpoint of the rising and setting *lagnas* is the *lagna* called the *dṛkkṣepa*.

A procedure for obtaining the *prāglagna* at any point in time during the day was described earlier in verses 95–99 of this chapter. In the subsequent verses however it was stated that the procedure described then was only approximate. In verses 102–6, the method to find two new quantities, namely the *kālalagna* and the *dṛkkṣepa* were given. The concept of the *kālalagna* is introduced as a prerequisite to arrive at the expression for the *dṛkkṣepa*, which in turn is introduced as a prerequisite to arrive at the exact expression for the *lagna*. The procedure for arriving at the exact *lagna* value is now described in verses 107–10. The procedure is as follows.

Initially, the hour angle of the Sun (east or west) is found from the half-duration of the day. If t_d is the half-duration of day, then the *nata* (hour angle) of the Sun is given by

$$\begin{aligned} H &= t_d - \text{time elapsed since sunrise} \\ \text{or} \qquad &= t_d - \text{time yet to elapse till sunset.} \end{aligned} \tag{3.304}$$

An intermediate quantity x is defined as

$$\begin{aligned} x &= trijy\bar{a} - b\bar{a}\d{n}a \text{ of } nata + carajy\bar{a} \\ &= R - (R - R\cos H) + R\sin\Delta\alpha \\ &= R\cos H + R\sin\Delta\alpha. \end{aligned} \tag{3.305}$$

Then, the *jyā* of $\Delta\theta$ whose arc has to be applied to the Sun to get the *lagna* is defined to be

$$jy\bar{a}\,\Delta\theta = \frac{x \times dyujy\bar{a} \times lambaka}{trijy\bar{a} \times d\d{r}kk\d{s}epako\d{t}i}. \tag{3.306}$$

Substituting for x in the above expression we have

$$R\sin\Delta\theta = \frac{(R\cos H + R\sin\Delta\alpha) \times R\cos\delta\; R\cos\phi}{R \times R\cos ZV}, \tag{3.307}$$

where $R\cos ZV$ is the *dṛkkṣepakoṭi* at the desired instant. From Chapter 2 ([see (2.84) in Section 2.11) we know that $\sin\Delta\alpha = \tan\phi\tan\delta$. Substituting for $\sin\Delta\alpha$ in the above equation and simplifying, we have

$$R\sin\Delta\theta = \frac{R(\cos\phi\cos\delta\cos H + \sin\phi\sin\delta)}{\cos ZV}. \tag{3.308}$$

From the above equation the arc $\Delta\theta$ is to be obtained and applied to the Sun to get the *prāglagna*. If λ_s is the longitude of the Sun, then the *prāglagna*, which is generally referred to as the *lagna*, is given by

$$lagna = \lambda_s + \Delta\theta \qquad (\text{at } udaya) \tag{3.309}$$

$$= \lambda_s - \Delta\theta \qquad (\text{at } asta). \tag{3.310}$$

On the other hand, if we are interested in arriving at the ecliptic points, from *unnata* (the hour angle of the Sun with reference to the midnight), then we need to do the reverse process. That is

$$\begin{aligned} lagna &= \lambda_s - \Delta\theta \qquad \text{(at } udaya\text{)} \\ &= \lambda_s + \Delta\theta \qquad \text{(at } asta\text{)}. \end{aligned} \tag{3.311}$$

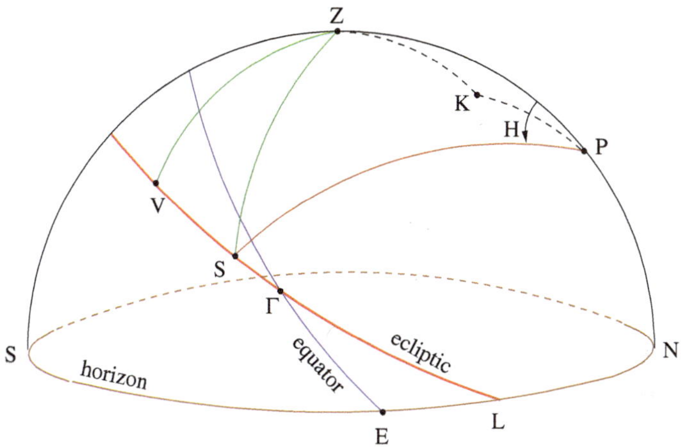

Fig. 3.37 Determination of the *prāglagna* (orient ecliptic point)–exact method.

Proof:

In Fig. 3.37, K and V represent the pole of the ecliptic and the *vitribhalagna* respectively. Let ϕ be the latitude of the observer and δ the declination of the Sun. In the triangle PZS, $PS = 90 - \delta$ (with δ −ve) and $ZP = 90 - \phi$. Now, applying the cosine formula to this triangle, we get

$$\cos ZS = \sin\phi \sin\delta + \cos\phi \cos\delta \cos H. \tag{3.312}$$

Now, in the spherical triangle VZS, VZ is the part of the vertical drawn form the pole of the ecliptic. Hence $Z\hat{V}S = 90$. Applying the cosine formula to this triangle, we have

$$\cos ZS = \cos VS \cos ZV. \tag{3.313}$$

Therefore

$$\cos VS = \frac{\cos ZS}{\cos ZV}. \tag{3.314}$$

By definition, the point V is at 90° from the *prāglagna* L. Hence

$$\cos VS = \sin SL = \sin(\lambda_l - \lambda_s), \tag{3.315}$$

where λ_l and λ_s are the longitudes of the *prāglagna* and the Sun respectively. Using (3.312) and (3.315) in (3.314), we have

$$\sin(\lambda_l - \lambda_s) = \frac{\cos\phi\cos\delta\cos H + \sin\phi\sin\delta}{\cos ZV}, \qquad (3.316)$$

which is the same as (3.308) given in the text, when we identify $\lambda_l - \lambda_s$ with $\Delta\theta$. Moreover, to get the longitude of the *prāglagna*, we see that we need to add $\Delta\theta$ to the longitude of the Sun. That is,

$$\lambda_l = \lambda_s + \Delta\theta. \qquad (3.317)$$

This is exactly the prescription given for finding the *prāglagna* from the hour angle (*nata*), determined for the eastern part of the hemisphere. The *astalagna* point would be to the west of the Sun in the western part of the hemisphere. Clearly, the corresponding arc $\Delta\theta$ has to be subtracted from Sun's longitude to obtain the *astalagna*. Thus we see that the procedure described in the text for determining *prāglagna* or *astalagna* is exact.

When the Sun is in the eastern part of the celestial sphere and below the horizon, it would be east of the *udaya-lagna* and H would be measured with respect to midnight. Then

$$\begin{aligned} lagna &= \lambda_s - \Delta\theta \qquad \text{(at } udaya\text{)} \\ lagna &= \lambda_s + \Delta\theta \qquad \text{(at } asta\text{)}. \end{aligned} \qquad (3.318)$$

The point V is the midpoint of the portion of the ecliptic above the horizon. Its longitude will be the average of the *prāglagna* and *astalagna*. This is also equal to the *prāglagna* $-90°$, or the *astalagna* $+90°$. The longitude of V, which is usually referred to as the *vitribhalagna*, is also called *dṛkkṣepalagna*.

३.४५ मध्यलग्नानयनम्

3.45 Determination of the *madhyalagna*

काललग्नं त्रिराश्यूनं मध्यकालस्ततः पुनः ।
लिप्ताप्राणान्तरं नीत्वा तद् दोश्चापे तु योजयेत् ॥ १११ ॥
ततश्चासून् नयेत् प्राग्वत् तल्लिप्तान्तरमुद्धरेत् ।
कालदोर्धनुषि क्षेप्यं ततः प्राणकलान्तरम् ॥ ११२ ॥
कालदोर्धनुषि क्षिप्त्वा तच्चापमविशेषयेत् ।
मध्यलग्नं तदेव स्यात् तत्काले प्रथमे पदे ॥ ११३ ॥
द्वितीयादिषु च प्राग्वत् मध्यलग्नमिहानयेत् ।

kālalagnaṃ trirāśyūnaṃ madhyakālastataḥ punaḥ |
liptāprāṇāntaraṃ nītvā tad doścāpe tu yojayet || 111 ||
tataścāsūn nayet prāgvat talliptāntaramuddharet |
kāladordhanuṣi kṣepyaṃ tataḥ prāṇakalāntaram || 112 ||

kāladordhanuṣi kṣiptvā taccāpamaviśeṣayet |
madhyalagnaṃ tadeva syāt tatkāle prathame pade || 113 ||
dvitīyādiṣu ca prāgvat madhyalagnamihānayet |

The *kālalagna* deficient by three *rāśis* (90 degrees) is the *madhyakāla*. Having obtained the *prāṇakalāntara* in minutes (*liptās*) from this, it may be added to the arc of the sine of that (the *madhyakāla*).

Once again obtain the *asus* as before and find the difference in minutes (*liptās*). This has to be added to the arc of the *madhyakāla*. From that the *prāṇakalāntara* [has to be determined].

Having applied this to the arc of the *madhyakāla*, the arc may be found iteratively. This is the *madhyalagna* at that instant, in the first quadrant. For the second and other quadrants, the *madhyalagna* may be obtained as earlier.

The term *madhyakāla* refers to the right ascension (R.A.) of the point on the equator which is situated on the prime meridian. In Fig. 3.38 this point is denoted by T. M represents the meridian ecliptic point and Γ the vernal equinox. The *madhyalagna* is the longitude of the meridian ecliptic point. Here the vernal equinox is shown to lie in the western part of the hemisphere.

Let α_T be the R.A. of the meridian equatorial point T, and H be the hour angle (H.A.) of Γ . By convention, the R.A. is measured along the equator eastward from Γ and the H.A. westwards from the prime meridian. From the figure, it is obvious that the R.A. of T is equal to the H.A. of Γ. That is, $\alpha_T = H$. By definition, the term

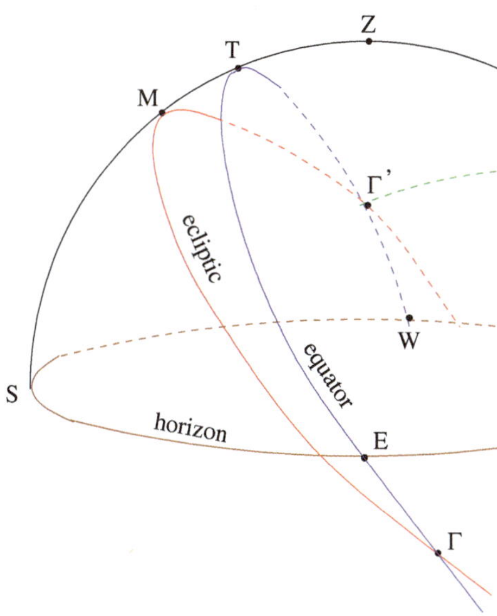

Fig. 3.38 Determination of the *madhyalagna* (meridian ecliptic point)—iterative method.

kālalagna (L'') refers to the difference in the time interval between the rise of Γ and the desired time. This is equal to $90 + \alpha_T$. Thus, we have the prescription given in the text to subtract 90° from the L'' (in angular measure), in order to obtain the R.A.

of the meridian equatorial point T. That is,

$$\alpha_T = L'' - 90. \tag{3.319}$$

Now, the problem is to find the *madhyalagna*, the longitude λ of the meridian ecliptic point M with the knowledge of α_T. Since the longitude is measured along the ecliptic and the R.A. along the equator, obviously $\lambda \neq \alpha_T$. Here an iterative procedure is described by which λ can be obtained from α_T. Earlier in the chapter (see (3.38)), a relation between the R.A. and the longitude was given in the form

$$\sin\alpha = \frac{\cos\varepsilon \sin\lambda}{\cos\delta}. \tag{3.320}$$

Taking the inverse, we have

$$\alpha = \sin^{-1}\left(\frac{\cos\varepsilon \sin\lambda}{\cos\delta}\right) = f(\lambda). \tag{3.321}$$

The iteration procedure given in the text may be described as follows: Let λ be the correct value of the longitude to be found by successive approximations. As a first approximation we take the value of λ to be the R.A. of T itself. That is,

$$\lambda_1 = \alpha_T. \tag{3.322}$$

The corresponding R.A. is $\alpha_1 = f(\lambda_1)$. Next, we find the *prāṇakalāntara* ($\delta\alpha_1$), which is to be added to α_T to get the second approximation of λ.

$$\delta\alpha_1 = \lambda_1 - \alpha_1 = \lambda_1 - f(\lambda_1) = \alpha_T - f(\alpha_T). \tag{3.323}$$

Now, in the second step,

$$\begin{aligned}\lambda_2 &= \alpha_T + \delta\alpha_1 \\ &= \alpha_T + (\alpha_T - f(\alpha_T)) \\ &= \alpha_T + (\lambda_1 - f(\lambda_1)).\end{aligned} \tag{3.324}$$

With the second approximate value of longitude (λ_2), we again calculate the *prāṇakalāntara* ($\delta\alpha_2$) which is to be added to the original value α_T to get the third approximation λ_3. We find

$$\alpha_2 = f(\lambda_2). \tag{3.325}$$

Therefore

$$\begin{aligned}\delta\alpha_2 &= \lambda_2 - \alpha_2 \\ &= \lambda_2 - f(\lambda_2).\end{aligned} \tag{3.326}$$

Now, in the third step,

$$\lambda_3 = \alpha_T + \delta\alpha_2$$

$$= \alpha_T + (\lambda_2 - f(\lambda_2)). \qquad (3.327)$$

The process is carried out until $\lambda_{n-1} \approx \lambda_n = \lambda$.

Justification of the procedure

From (3.320), we note that α is essentially a function of the longitude λ, since δ itself is function of λ (see (3.37)). Hence, the R.A. of the meridian ecliptic point α_T may be expressed as

$$\alpha_T = f(\lambda). \qquad (3.328)$$

In the first approximation, $\lambda = \lambda_1 = \alpha_T$. In the next approximation, let $\lambda = \lambda_2 = \lambda_1 + \delta\alpha_1 = \alpha_T + \delta\alpha_1$. Hence

$$\alpha_T = f(\alpha_T + \delta\alpha_1) = f(\alpha_T) + \delta\alpha_1 \times f', \qquad (3.329)$$

where $f' = \frac{df}{d\alpha_T}$. Therefore

$$\delta\alpha_1 = \frac{\alpha_T - f(\alpha_T)}{f'}. \qquad (3.330)$$

But $f' = \frac{df(\alpha_T)}{d\alpha_T} \approx 1$, as $f(\alpha_T) \approx \alpha_T$. This is all right as $\delta\alpha_1$ is the first-order correction and it is natural that f' is taken to the zeroth order.

Hence $\delta\alpha_1 = \alpha_T - f(\alpha_T)$, which leads to

$$\lambda_2 \approx \alpha_T + (\lambda_1 - f(\lambda_1)) \approx \alpha_m + (\alpha_T - f(\alpha_T)). \qquad (3.331)$$

This coincides with the expression in (3.324). In the next approximation, let $\lambda = \lambda_3 = \lambda_2 + \delta\alpha_2$. Hence

$$\begin{aligned} \alpha_T &= f(\lambda_3) \\ &= f(\lambda_2 + \delta\alpha_2) \\ &= f(\lambda_2) + \delta\alpha_2 \times f'. \end{aligned} \qquad (3.332)$$

Therefore

$$\delta\alpha_2 = \frac{[\alpha_T - f(\lambda_2)]}{f'}. \qquad (3.333)$$

If we assume again that $f' = 1$, then

$$\delta\alpha_2 = \alpha_T - f(\lambda_2) \qquad (3.334)$$

$$\text{and} \quad \lambda_3 = \lambda_2 + (\alpha_T - f(\lambda_2))$$

$$\text{or} \quad \lambda_3 = \alpha_T + (\lambda_2 - f(\lambda_2)), \qquad (3.335)$$

as stated. The next stages of iteration can also be understood in this manner. However, we are unable to fully justify the procedure as the approximation $f' = 1$ is not justified in the higher orders.

३.४६ अविशेषं विना मध्यलग्नानयनम्

3.46 Determining the *madhyalagna* without iteration

अविशेषं विना मध्यलग्नमानीयते यथा ॥ ११४ ॥
मध्यकालस्य कोटिज्या परमापक्रमाहता ।
त्रिज्यालब्धकृतिं त्यक्त्वा कालकोटित्रिजीवयोः ॥ ११५ ॥
वर्गाभ्यां शिष्टमूले द्वे कोटिज्या स्यात् द्विमौर्व्यपि ।
कोटिज्या त्रिज्ययोर्घातात्[78] द्युज्यावाप्तं तु चापितम् ॥ ११६ ॥
कालासवो मध्यलग्नभुजा तद्धीनभत्रयम् ।
पदव्यवस्था सुगमैवाद्यमध्यविलग्नवत् ॥ ११७ ॥

aviśeṣaṃ vinā madhyalagnamānīyate yathā || 114 ||
madhyakālasya koṭijyā paramāpakramāhatā |
trijyālabdhakṛtiṃ tyaktvā kālakoṭitrijīvayoḥ || 115 ||
vargābhyāṃ śiṣṭamūle dve koṭijyā syāt dvimaurvyapi |
koṭijyā dvijyayorghātāt dyujyāvāptaṃ tu cāpitam || 116 ||
kālāsavo madhyalagnabhujā taddhīnabhatrayam |
padavyavasthā sugamaivādyamadhyavilagnavat || 117 ||

[Here we describe a procedure] by which the *madhyalagna* can be arrived at without doing iteration. The *koṭi* of the *madhyakāla* is multiplied by the Rsine of the maximum declination and divided by the *trijyā*. The square of this is subtracted from the squares of the *kālakoṭi* and the *trijyā*. The square roots of the two resulting quantities are the *koṭijyā* and the *dvimaurvī* [respectively].

The arc of the product of the *koṭijyā* and the *trijyā* divided by the *dyujyā* is the *kālāsavaḥ*. This subtracted from three signs (90 degrees) is the *madhyalagna*. [The value of this] in different quadrants can be easily found as in the case of the *madhyalagna* [explained earlier].

The term *aviśeṣakarma*[79] refers to the process of obtaining a stable result by employing an iterative procedure. Hence the phrase '*aviśeṣaṃ vinā*' means 'without using an iterative process'. What is described in verses 114–7 is a procedure by which the *madhyalagna* can be obtained without doing any iteration. For this, an intermediate quantity, x, is defined by the relation

---

[78] The reading in both the printed editions is: द्युज्ययोर्घातात्। From the commentary *Laghuvivṛti*, it is clear that the multiplier for the *koṭijyā* is the *trijyā* and not the *dyujyā*.

[79] The word *viśeṣa* means 'distinction'. Hence *aviśeṣa* is 'without distinction'. Though the meanings of the words *viśeṣa* and *aviśeṣa* are opposed to each other, the latter should not be taken to mean *tulya* or 'completely identical'. In the context of mathematical calculations, it only means '*without distinction to a desired degree of accuracy*'. In other words, in the *aviśeṣakarma*, the iterative process needs to be carried out only up to a point wherein the two successive values of the results are 'without distinction' for a desired degree of accuracy. Once this accuracy is reached, the process may be terminated.

$$x = \frac{R\cos\alpha_T \times R\sin\varepsilon}{R}. \tag{3.336}$$

With x, two more quantities, namely the *koṭijyā* (p) and the *dvimaurvī* (q) are defined.

$$p = \sqrt{(R\cos\alpha_T)^2 - x^2}$$

$$\text{and} \qquad q = \sqrt{R^2 - x^2}. \tag{3.337}$$

Now, the *kālāsava* is defined to be the arc of the product of the *koṭijyā* and the *trijyā* divided by the *dvimaurvī* or the *dyujyā*. For convenience we write

$$R\sin\theta = \frac{R \times p}{q}, \tag{3.338}$$

where θ is *kālāsava*. The *madhyalagna* is then given by $90 - \theta$. Substituting for p, q and x in the above expression and simplifying, we have

$$R\cos M = \frac{R\cos\alpha_T \cos\varepsilon}{\sqrt{1 - \cos^2\alpha_T \sin^2\varepsilon}}. \tag{3.339}$$

Proof:

The relation between the R.A. α, the longitude λ and the declination δ of the Sun given by (3.39) may be written as

$$\sin\lambda = \frac{\sin\alpha\cos\delta}{\cos\varepsilon}. \tag{3.340}$$

Using the relation $\sin\delta = \sin\varepsilon\sin\lambda$ to replace δ in terms of λ, the above equation reduces to

$$\sin\lambda = \frac{\sin\alpha}{\cos\varepsilon}\sqrt{1 - \sin^2\varepsilon\sin^2\lambda}\,. \tag{3.341}$$

Squaring both sides and simplifying, we have

$$\sin\lambda = \frac{\sin\alpha}{\sqrt{1 - \cos^2\alpha\sin^2\varepsilon}}. \tag{3.342}$$

With some algebraic manipulation it can be shown that

$$\cos\lambda = \frac{\cos\alpha\cos\varepsilon}{\sqrt{1 - \cos^2\alpha\sin^2\varepsilon}}, \tag{3.343}$$

which is the same as (3.339), once we identify $\lambda \to M$ and $\alpha \to \alpha_T$. Also, using the 4-parts formula, the relation between α and λ can be shown to be

$$\tan\lambda = \tan\alpha \sec\varepsilon. \tag{3.344}$$

It is straightforward to see that (3.344) is the ratio of (3.342) and (3.343) and hence (3.339) can be obtained by using the above relation also.

Chapter 4
चन्द्रग्रहणप्रकरणम्
Lunar eclipse

४.१ चन्द्रभूच्छाययोः योगकालः

4.1 Time of conjunction of the Moon and the Earth's shadow

अर्कस्फुटं सचक्रार्धं भूच्छायास्फुटमुच्यते ।
सूर्यास्तमयकालोत्थौ छायाचन्द्रौ समीपगौ ॥ १ ॥
उदये वाथ विन्यस्य तद्योगोऽत्रनिरूप्यताम् ।
चन्द्रेऽधिके गतो योगः न्यूने चैष्य इति स्थितिः ॥ २ ॥
तदन्तरं तु षष्टिघ्नं गत्यन्तरहृतं तयोः ।
योगकालो[1] घटीपूर्वो गतो गम्योऽपि वा क्रमात् ॥ ३ ॥

arkasphuṭaṃ sacakrārdhaṃ bhūcchāyāsphuṭamucyate |
sūryāstamayakālotthau chāyācandrau samīpagau || 1 ||
udaye vātha vinyasya tadyogo'tranirūpyatām |
candre'dhike gato yogaḥ nyūne caiṣya iti sthitiḥ || 2 ||
tadantaraṃ tu ṣaṣṭighnaṃ gatyantarahṛtaṃ tayoḥ |
yogakālo ghaṭīpūrvo gato gamyo'pi vā kramāt || 3 ||

The true position of the Earth's shadow is said to be the sum of the true position of the Sun and a half-circle (180 degrees). Having determined the position of the Moon and the [Earth's] shadow either at sunrise or at sunset, whichever is closer [to the conjunction], the time of their conjunction may be determined. If [the longitude of] the Moon is greater then the conjunction is over, and if it is less then it is yet to occur. Their difference [in longitude] multiplied by 60 and divided by their difference in daily motion [gives] the time for conjunction (the *yogakāla*) expressed in *ghaṭīs* etc. that has already elapsed or is yet to elapse respectively.

---

[1] The compound word योगकाल can be derived in two ways: (i) (छायाचन्द्रमसोः) योगस्य कालः (the time of conjunction of the Moon and the shadow) or (ii) (छायाचन्द्रमसोः) योगार्थं कालः (the time for conjunction, either in the forward direction or reverse direction). In other words, योगार्थं कालः can connote योगात्पूर्वं गम्यः कालः (the time that is yet to elapse till conjunction) or योगानन्तरं गतः कालः (the time that has elapsed after conjunction). From a careful analysis of the content of verse 3 and again of verse 7 (in the next section)—where the word *yogakāla* has been employed once more—it becomes evident that the author has employed the word in the latter sense and not in the former. That is, by *yogakāla* he means the quantity Δt given by equation (4.3).

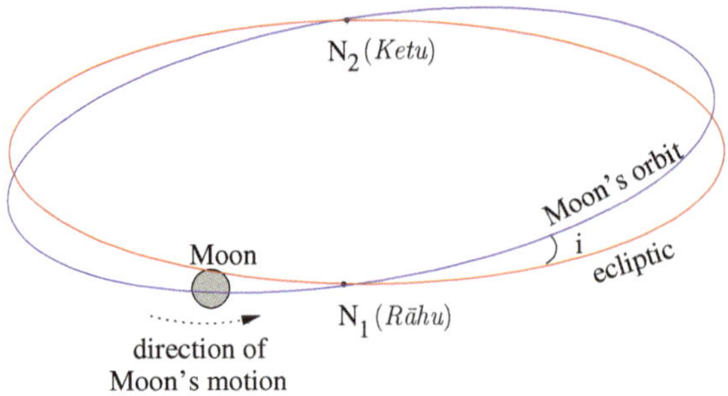

Fig. 4.1*a* Schematic representation of the situation of the Moon's orbit with respect to the ecliptic.

The Moon's orbit is inclined to the ecliptic as shown in Fig. 4.1*a*. The angle of inclination, denoted by *i*, is taken to be $270'(4\frac{1}{2}^{\circ})$ in most of the Indian astronomical texts including *Tantrasaṅgraha*.[2]

Moon's nodes and their retrograde motion

The points of intersection of the ecliptic and the Moon's orbit, N_1 and N_2 (see Fig. 4.1*a*), are the nodes of the orbit. As the Moon crosses node N_1 along the direction indicated in figure, it is ascending towards the north celestial pole, and hence node N_1 is called the ascending node. As it crosses N_2, it is descending towards the south pole and hence N_2 is called the descending node.

In fact, it is these two nodes that are called *Rāhu* and *Ketu* in Indian astronomy. The nodes themselves are in motion[3] and their motion is *retrograde*. That is, the direction of motion of the nodes is the opposite of that of the motion of the Moon, Sun and other planets. The time taken by the nodes to complete one full revolution is about 6793 days, or 18.6 years.

Possibility of a lunar eclipse

The Earth's shadow always moves along the ecliptic and its longitude will be exactly 180° plus that of the longitude of the Sun. When the Moon is close to the shadow and both of them are near a node, then there is a possibility of a lunar eclipse. This situation is depicted in Fig. 4.1*b*, where *C* represents the *chāyā* (shadow), and *A* and *B* are the positions of the Moon before and after the lunar eclipse.

---

[2] It is known today that the inclination of the Moon's orbit varies slightly with time, and that its average value is around 5.1°.

[3] This motion is mainly due to the variation in the gravitational force on the Moon exerted by the Earth, due to its equatorial bulge.

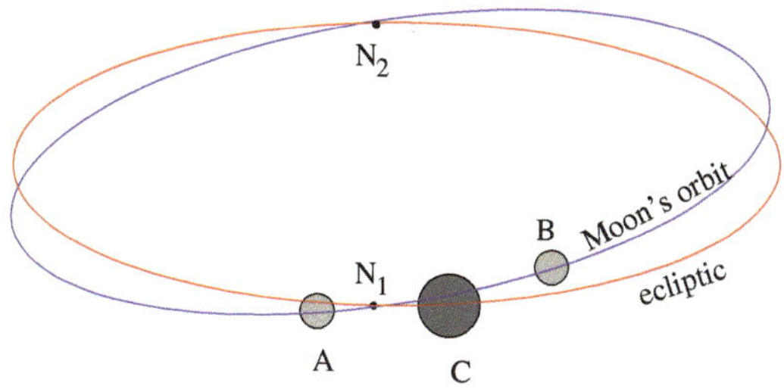

Fig. 4.1*b* Determination of the instant of conjunction.

Computation of the instant of conjunction

In verses 1–3, the procedure for the determination of the instant of conjunction of the shadow and the Moon is given. Usually, the longitudes of the planets are calculated at sunrise on a particular day. Let λ_s, λ_m and λ_c be the longitudes of the Sun, the Moon and the *chāyā* respectively. Then, obviously,

$$\lambda_c = \lambda_s + 180. \tag{4.1}$$

When the longitudes of the Moon and the Earth's shadow are the same, the Sun will be exactly at 180 degrees from the Moon. Since the Sun and the Moon are diametrically opposite each other at this instant, they are said to be in opposition. In order to determine this instant, the true longitudes of the Sun (λ_s) and the Moon (λ_m), are first calculated at sunrise on a full Moon day. Then, the difference in longitudes of the Moon and the *chāyā*, given by

$$\Delta\lambda = \lambda_m - \lambda_c, \tag{4.2}$$

is computed. The sign of $\Delta\lambda$ indicates if the instant of opposition is over or is yet to occur.

1. If $\Delta\lambda < 0$, it means that the instant of opposition is yet to occur as the Moon moves eastward with respect to the Sun.[4]
2. If $\Delta\lambda > 0$, it means that the instant of opposition is already over.

The positions of the Moon corresponding to these two situations are indicated by *A* and *B* in Fig. 4.1*b*. Δt, the time interval between sunrise and the instant of opposition, is computed using the relation

$$\Delta t = \frac{|\Delta\lambda|}{d_m - d_s} \times 60, \tag{4.3}$$

where d_m and d_s are the daily motions of the Sun and the Moon respectively. The above expression for Δt (in *ghaṭikās*) is obviously based upon the rule of three given

[4] It may be recalled that the shadow also moves eastward owing to the motion of the Sun, but at a rate much slower than that of the Moon.

by

$$d_m - d_s : 60 \ (ghaṭikās)$$
$$|\Delta\lambda| = |\lambda_m - \lambda_c| : \Delta t \ \text{(in } ghaṭikās\text{)}.$$

Having determined Δt, the time of opposition of the Sun and the Moon or equivalently the conjunction of the Moon and the Earth's shadow—the end of the full Moon day, which is the same as the middle of the eclipse denoted by t_m—is obtained using the relation

$$t_m = \text{sunrise time} \pm \Delta t. \tag{4.4}$$

We have to use '+' if the instant of opposition is yet to occur and '−' otherwise. The time given by (4.4) is only approximate, and the reason for the same has been stated in *Yukti-dīpikā* to be the continuous variation in the speed with which the Sun and the Moon move.[5] In order to obtain the exact instant of opposition, the text prescribes an iterative procedure which is explained in the following verses.

४.२ अविशेषेण स्फुटयोगकालानयनम्

4.2 Determination of the exact moment of conjunction by iteration

तात्कालिकौ पुनर्नीत्वा मध्यार्केन्दुस्फुटावपि ।
उदयाद्योगकालेन नीयते चेत्तदुक्तवत् ॥ ४ ॥
तात्कालिकार्कनिष्पन्नचरसंस्कार इष्यते ।
अस्तकालोक्तवत् तस्मात् चाल्यते चेत् चरोद्भवः ॥ ५ ॥
इष्टकालार्कतश्चैवं कार्यं प्राणकलान्तरात् ।
दोर्भेदाच्चापि संस्कारः रवीन्दू तौ स्फुटौ तदा ॥ ६ ॥
योगकालस्ततो नेयः तन्निघ्ना स्वा स्फुटा गतिः ।
षष्ट्याप्ता स्वस्फुटे योज्या गम्ये योगेऽन्यथान्यथा ॥ ७ ॥
समलिप्तौ भवेतां तौ पर्वान्तसमयोद्भवौ ।

tātkālikau punarnītvā madhyārkendusphuṭāvapi |
udayādyogakālena nīyate cettaduktavat || 4 ||
tātkālikārkaniṣpannacarasaṃskāra iṣyate |
astakāloktavat tasmāt cālyate cet carodbhavaḥ || 5 ||
iṣṭakālārkataścaivaṃ kāryaṃ prāṇakalāntarāt |
dorbhedāccāpi saṃskāraḥ ravīndū tau sphuṭau tadā || 6 ||
yogakālastato neyaḥ tannighnā svā sphuṭā gatiḥ |
ṣaṣṭyāptā svasphuṭe yojyā gamye yoge'nyathānyathā || 7 ||
samaliptau bhavetāṃ tau parvāntasamayodbhavau |

The mean and the true [longitudes of the] Sun and the Moon are once again obtained at the instant of conjunction. If the instant of conjunction is determined from [the position of the Sun and the Moon at] sunrise, then as described [earlier] it is desired that the value of the

---

[5] प्रतिक्षणं प्रभिन्नैव स्फुटभुक्तिर्द्युचारिणाम्।
कालभुक्त्योः मिथस्तस्मात् नानुपातः प्रवर्तते॥ ({TS 1977}, p. 252)

cara be determined at that instant [of conjunction], and applied [to obtain the true sunrise time]. If otherwise [the instant of conjunction is determined from the position of the Sun and the Moon at sunset], and the value of the *cara* obtained at sunset is applied as per the procedure described for [application at] sunset. From the *prāṇakalāntara* and the desired sine (*dorbheda*[6]), determined at the instant of conjunction, the corrections have to be done. Then the true positions of the Sun and the Moon are obtained. The time of conjunction is once again calculated from that. This (*yogakāla*) is multiplied by the true daily motion and divided by 60. The results are added or subtracted, according to whether the conjunction is yet to occur or has already occurred. Thus [by using an iterative procedure] the true longitudes of them (the Earth's shadow and the Moon[7]) will be rendered equal [even] in minutes at the end of the full Moon day (*parva*[8]).

The instant of conjunction calculated using (4.4) is only approximate, as Δt used in the expression is found using a simple rule of three that presumes a uniform motion of the Sun and the Moon, which is not true. In order to consider this non-uniform motion into account, an iterative procedure to determine the true instant of conjunction is described here.

As per the computational scheme followed by Indian astronomers, the instant of sunrise or sunset is the reference point for finding the time of any event. Hence, the instant of true sunrise is first to be determined accurately. It was noted in Chapter 2 that this involves the application of the *cara* (ascensional difference), and the equation of time, where the latter has two parts, namely the correction due to the equation of centre and the correction due to the *prāṇakalāntara*. Here it is prescribed that the *cara* and the equation of time are to be determined at the instant of conjunction, in order to find the instant of true sunrise or sunset as the case may be.

Next, the approximate value of the instant of conjunction is found and also the true longitudes of the Sun and Moon, while their true daily motions are also determined at this instant. The second approximate value of the instant of conjunction is determined using (4.3). The true longitudes and the daily motions are again computed at this instant, to obtain the third approximate value. The iteration process is carried till two successive values of the instant of conjunction are the same to the desired accuracy.

That the difference in the motion of the Sun and the Moon is a continuously varying quantity has been explicitly mentioned in *Laghu-vivṛti*, while giving an *avatārikā*[9] to these verses:

अथ दिनगत्यन्तरकलानां षष्टिघटिकाभिः सह यः सम्बन्धनियमः सः तदवयवेषु अन्यादृशः स्यात् तस्य प्रतिक्षणं नानारूपत्वात्। इत्येवमानीतस्य योगकालस्य अस्फुटत्वमाशङ्क्य विशेषेणैव स्फुटीकर्तुमाह।

However, the relation that exists between the difference in the daily motion [of the Sun and the Moon] with 60 *ghaṭikās* will be quite different from the one that holds for its

[6] The term *bheda* is sometimes used as an equivalent to *viśeṣa* (a particular). In the present context, the particular *doḥ* = *bhujā* that is referred to is the equation of centre, which is generally referred to as the *doḥphala* or *bhujāphala*.

[7] The Sun and the Moon in the case of a solar eclipse.

[8] New Moon day in the case of a solar eclipse.

[9] This refers to the succinct note or observation made by the commentator before introducing a chapter or section or successive set of verses dealing with a topic.

parts, since that (velocity of the Sun and the Moon) varies at every instant (*pratikṣaṇam nānārūpatvāt*). So much is this so, that there are grounds to doubt the accuracy of the time of conjunction thus determined, and hence a special [iterative] procedure is prescribed.

४.३ रवीन्द्वोः कक्ष्याव्यासार्धयोजनम्

4.3 Radii of the orbits of the Sun and the Moon in *yojanas*

दशाभ्यस्ता त्रिजीवेन्दोः कक्ष्याव्यासार्धयोजनम् ॥ ८ ॥
तच्चन्द्रभगणाभ्यस्तं भानोः स्वभगणोद्धृतम् ।

daśābhyastā trijīvendoḥ kakṣyāvyāsārdhayojanam || 8 ||
taccandrabhagaṇābhyastaṃ bhānoḥ svabhagaṇoddhṛtam |

The *trijyā* (the radius measured in minutes) multiplied by 10 is the [mean] radius of the orbit of the Moon in *yojanas*. This multiplied by the number of revolutions of the Moon and divided by the revolutions of the Sun is the Sun's [mean] orbital radius.

Let r_m and r_s be the (mean) radii of the orbits of the Moon and the Sun and n_m and n_s represent the number of revolutions made by them in a *Mahāyuga*. Then r_m, r_s in *yojanas* are given by

$$r_m = \textit{trijyā} \times 10 = 3438 \times 10 = 34380, \tag{4.5}$$

$$r_s = 34380 \times \frac{n_m}{n_s} = 34380 \times \frac{57753320}{4320000} \approx 459620. \tag{4.6}$$

The mean radius of the Sun's orbit has been obtained above by assuming that the linear velocities of the Sun and the Moon are the same.[10]

४.४ रवीन्दुभुवां बिम्बव्यासार्धयोजनम्

4.4 Radii of the Sun, Moon and the Earth in *yojanas*

भूवृत्तादुदितात् प्राग्वत् व्यासस्तस्यापि नीयताम् ॥ ९ ॥
दिग्वेदाब्धिमितो भानोः व्यासस्तिथ्यग्नयो विधोः ।

bhūvṛttāduditāt prāgvat vyāsastasyāpi nīyatām || 9 ||
digvedābdhimito bhānoḥ vyāsastithyagnayo vidhoḥ |

Let the diameter of the Earth be obtained from the value of its circumference as [stated] earlier. The diameter of the Sun is 4410 [*yojanas*] and that of the Moon is 315.

In Chapter 1, verse 29, the value of the circumference of the Earth was stated to be 3300 *yojanas*. Hence the diameter of the Earth D_e is

$$D_e = \frac{3300}{\pi} \simeq 1050.42 \textit{ yojanas}. \tag{4.7}$$

[10] It is indeed one of the fundamental assumptions of Indian astronomy that the mean linear velocities of all the planets are the same.

In the above equation we have taken the value of π to be the ratio of 355 and 113 as suggested by Śaṅkara Vāriyar in his *Laghu-vivṛti*.[11] The diameters of the Sun and Moon are specified to be 4410 and 315 *yojanas* respectively.

४.५ रवीन्द्वोः स्फुटयोजनकर्णः
4.5 Actual distances of the Sun and the Moon in *yojanas*

कालकर्णहतं स्वस्वकक्ष्याव्यासार्धयोजनम् ॥ १० ॥
त्रिज्याप्तं स्यात् भगोलस्य मण्डलस्य च मध्ययोः ।
योजनैरन्तरालं स्यात् तत्तत्कालोद्भवं स्फुटम् ॥ ११ ॥

kālakarṇahataṃ svasvakakṣyāvyāsārdhayojanam || 10 ||
trijyāptaṃ syāt bhagolasya maṇḍalasya ca madhyayoḥ |
yojanairantarālaṃ syāt tattatkālodbhavaṃ sphuṭam || 11 ||

The radii of their own orbits multiplied by the *kālakarṇa* and divided by the *trijyā* are the actual distances of separation between the centres of their discs and the centre of the *bhagola*, in *yojanas*, [as] the correct values (*sphuṭa*) are the ones determined from time to time (*tattatkālodbhavaṃ*).

The radii of the orbits of the Sun and Moon given earlier are only their mean values. The actual distances keep varying owing to the eccentricities of their orbits, and are actually proportional to the *manda-karṇa*, or the *kālakarṇa* as it is called here. The term *kālakarṇa* is defined in the commentary *Laghu-vivṛti* to be the *aviśeṣakarṇa*, the hypotenuse determined by iterative procedure.[12]

अर्केन्द्वोः कक्ष्याव्यासार्धयोजनं समनन्तरोक्तं कलारूपेण अविशिष्टमन्दकर्णेन निहत्य त्रिज्यया विभजेत्। तत्र लब्धं भगोलघनमध्यस्य स्वबिम्बघनमध्यस्य च अन्तरालं योजनात्मकं स्फुटं भवति; स्फुटयोजनकर्णः इत्यर्थः। तत्तत्कालोद्भवमित्यनेन तस्य प्रतिक्षणं नानारूपत्वमुक्तम् ।

The radii of the orbits of the Sun and Moon in *yojanas*, just mentioned, is multiplied by the *aviśiṣṭa-manda-karṇa* (iterated *manda*-hypotenuse) and divided by the *trijyā*. The result is the true distance in *yojanas* between the centres of the celestial sphere and the centre of the body (Sun or Moon); that is, it is the *sphuṭa-yojana-karṇa*. By stating that it has to be obtained for the particular time, it is indicated that it varies from instant to instant.

The word *aviśiṣṭa-manda-karṇa*, used in the commentary, refers to the word *kāla-karṇa* K in the text and is given by

---

[11] Śaṅkara Vāriyar observes:

यतो वृत्तपरिधेः त्रीशैर्निहतात् अर्थेष्वग्निभिर्विभज्य लब्धः तद्व्यास इति प्रागेवोक्तं ··· ।

The diameter is what is obtained by multiplying the circumference of the circle by 113 and dividing by 355 as stated earlier.

[12] This procedure is described in verses 41 and 42 of Chapter 2. A shortcut to the iterative method—due to Mādhava—is also presented there in verses 43 and 44.

$$\textit{aviśiṣṭa-manda-karṇa} = \frac{\textit{trijyā}^2}{\textit{viparīta-karṇa}}. \tag{4.8}$$

Using the expression for the *viparītakarṇa* (inverse hypotenuse) given earlier (Chapter 2, verse 44) we have

$$K = \frac{R^2}{\sqrt{R^2 - r_0^2 \sin^2\theta_{mk}} - r_0\cos\theta_{mk}}, \tag{4.9}$$

where θ_{mk} is the *manda-kendra*.[13] If r_m and r_s represent the mean radii of the orbits of the Moon and the Sun (*kakṣyāvyāsārdha*), then their actual distances of separation d_{1m} and d_{1s} from the centre of the *bhagola* (celestial sphere) in *yojanas* are given by

$$d_{1m} = \frac{r_m \times K}{R} \quad \text{and} \quad d_{1s} = \frac{r_s \times K}{R}. \tag{4.10}$$

The suffix '1' employed in the above expressions indicates that these values correspond to what are known as the 'first' *sphuṭa-yojana-karṇas*, which have to be distinguished from the 'second' *sphuṭa-yojana-karṇas* defined in the following section. In fact, d_{1m} and d_{1s} represent the distances of the Moon and the Sun in their eccentric orbits, from which the 'second' *sphuṭa-yojana-karṇas* are obtained.

४.६ रवीन्द्वोः द्वितीयस्फुटयोजनकर्णः

4.6 Second approximation to the radii of the orbits of the Sun and the Moon in *yojanas*

उच्चोनशशिकोटिज्यादलं पर्वान्तजं स्फुटम् ।
स्फुटयोजनकर्णे स्वं जह्यात् कर्क्यादिजं ततः ॥ १२ ॥
स भूम्यन्तरकर्णः स्यात् तेन बिम्बकलां नयेत् ।
स्फुटयोजनकर्णे स्वे मासान्ते शशिवद्रवेः ॥ १३ ॥
व्यस्तं पक्षान्तजं कार्यं रविभूम्यन्तराप्तये ।

ucconaśaśikoṭijyādalaṃ parvāntajaṃ sphuṭam |
sphuṭayojanakarṇe svaṃ jahyāt karkyādijaṃ tataḥ || 12 ||
sa bhūmyantarakarṇaḥ syāt tena bimbakalāṃ nayet |
sphuṭayojanakarṇe sve māsānte śaśivadraveḥ || 13 ||
vyastaṃ pakṣāntajaṃ kāryaṃ ravibhūmyantarāptaye |

Half of the *koṭijyā* of the difference of the longitude of the Moon and its apogee, calculated at the moment of opposition, has to be added to or subtracted from the value of the (first) *sphuṭa-yojana-karṇa* depending upon whether the *manda-kendra* is *Mṛgādi* or *Karkyādi*. This is the actual distance of separation between the Earth and the Moon. From this the diameter of the Moon's disc must be obtained.

13 θ_{mk} is the same as $\theta_0 - \theta_m$ used in Sections 2.17 and 2.18 of Chapter 2.

> In the case of the Sun, to obtain its distance of separation from the Earth, the same process as was adopted for the Moon may be followed at the end of the [lunar] month (that is, at new Moon) and the reverse process may be followed at the end of the bright fortnight.

The first correction to be applied for obtaining the actual distance of separation between the centres of the Sun and the Moon—from the centre of the Earth during an eclipse—was described in the previous section. Here the second correction is given as

$$d_{2m} = d_{1m} + \frac{R\cos\theta_{mk}}{2}, \tag{4.11}$$

where θ_{mk} is the *manda-kendra*, determined at the instant of conjunction or opposition. The above expression is used for determining the actual distance of separation between the Earth and the Moon, for both solar and lunar eclipses.

In the case of the Sun, the actual distance of separation between the centres of the Earth and the Sun, d_{2s}, is given as:

$$\begin{aligned} d_{2s} &= d_{1s} + \frac{R\cos\theta_{mk}}{2} \qquad \text{(solar eclipse)} \\ d_{2s} &= d_{1s} - \frac{R\cos\theta_{mk}}{2} \qquad \text{(lunar eclipse).} \end{aligned} \tag{4.12}$$

The *yojana-karṇa* including the second correction is the *dvitīya-sphuṭa-yojana-karṇa* or simply the *dvitīya-sphuṭa-karṇa*. This arises because of the so-called *dvitīya-sphuṭīkaraṇa* or the second correction for the longitude of the Moon (similar to the 'evection' correction), which is discussed further in Chapter 8. In the case of the Moon, this corresponds to a new correction to the longitude (besides the equation of centre) which also affects the distance. In the case of the Sun, it alters the distance by a relatively smaller factor, without affecting the longitude.

As explained in *Yuktibhāṣā* (chapter 15), this correction arises from the fact that the centre of the celestial sphere (the *bhagola-madhya*) does not coincide with the centre of the Earth. It is at a distance of $\frac{R\cos(\lambda_s - U)}{2}$ (in *yojanas*) from the centre of the Earth in the direction of the Sun, where λ_s represents the longitude of the Sun and U the longitude of the apogee (*ucca*) of the Moon.[14] The true longitudes of the Sun and the Moon hitherto considered are actually with reference to the *bhagola-madhya*. The *dvitīya-sphuṭīkaraṇa* transforms them into longitudes (see section 8.1 below) with respect to the centre of the Earth. Here, we confine our attention only to the change in distance due to the above factor as is relevant in the discussion of eclipses.

In the case of a lunar eclipse, $\lambda_s = \lambda_m + 180^\circ$ at the instant of opposition. Hence,

$$R\cos(\lambda_m - U) = R\cos\theta_{mk} = -R\cos(\lambda_s - U). \tag{4.13a}$$

But for a solar eclipse, at the instant of conjunction, $\lambda_s = \lambda_m$.

---

[14] This is actually true only when $\cos(\lambda_s - U)$ is positive. When it is negative, the centre of the celestial sphere is at a distance of $\frac{|R\cos(\lambda_s - U)|}{2}$, in a direction opposite to the direction of the Sun.

$$R\cos(\lambda_m - U) = R\cos\theta_{mk} = R\cos(\lambda_s - U). \tag{4.13b}$$

First, let us consider the case when $\cos(\lambda_s - U)$ is positive.

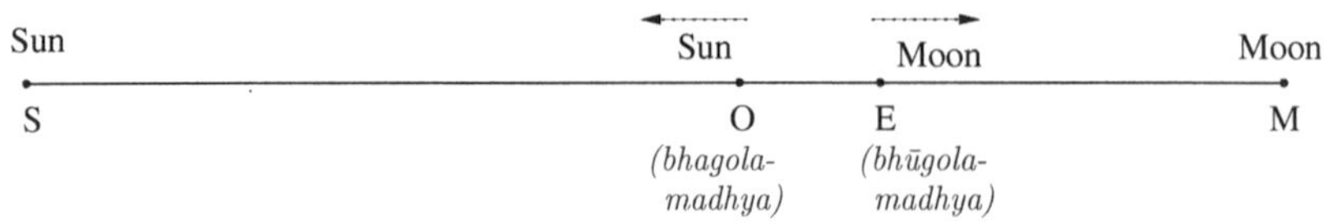

Fig. 4.2a Computation of the *dvitīya-sphuṭa-karṇa* in a lunar eclipse.

In Fig. 4.2*a*, *E* is the centre of the Earth (the *bhūgola-madhya*) and *O* the centre of the celestial sphere (the *bhagola-madhya*). *M* represents the Moon, and *S* the Sun. $OM = d_{1m}$ and $OS = d_{1s}$. Now, the distance of separation between the *bhagola-madhya* and the *bhūgola-madhya* is given by

$$OE = \frac{R\cos(\lambda_s - U)}{2} = -\frac{R\cos\theta_{mk}}{2}. \tag{4.14}$$

Then, the *dvitīya-sphuṭa-karṇa* d_{2m} of the Moon, which is the true distance of the Moon from the centre of the Earth, is

$$\begin{aligned} d_{2m} = EM &= OM - OE \\ &= d_{1m} + \frac{R\cos\theta_{mk}}{2}. \end{aligned} \tag{4.15}$$

Similarly, the *dvitīya-sphuṭa-karṇa* d_{2s} of the Sun, which represents the true distance of the Sun from the centre of the Earth, is given by

$$\begin{aligned} d_{2s} = ES &= OS + OE \\ &= d_{1s} - \frac{R\cos\theta_{mk}}{2}. \end{aligned} \tag{4.16}$$

As mentioned earlier, during a solar eclipse the Sun and the Moon are in the same direction ($\lambda_m = \lambda_s$). Therefore (4.14) becomes

$$OE = \frac{R\cos(\lambda_s - U)}{2} = +\frac{R\cos\theta_{mk}}{2}. \tag{4.17}$$

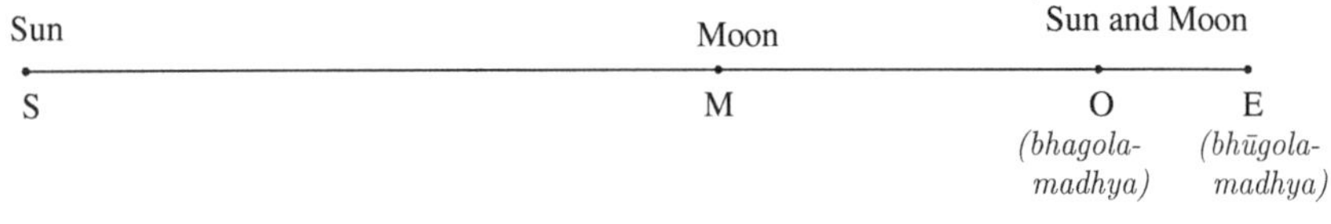

Fig. 4.2b Computation of the *dvitīya-sphuṭa-karṇa* in a solar eclipse.

In the case of solar eclipse, equations (4.15) and (4.16) take the form

$$d_{2m} = EM = OM + OE$$
$$= d_{1m} + \frac{R\cos\theta_{mk}}{2}, \quad (4.18)$$

and

$$d_{2s} = ES = OS + OE$$
$$= d_{1s} + \frac{R\cos\theta_{mk}}{2}. \quad (4.19)$$

It is straightforward to show that these relations are valid even when $\cos(\lambda_s - U)$ is negative, when O is at a distance $\frac{|R\cos(\lambda_s-U)|}{2}$ from E, in a direction opposite to the direction of the Sun.

The nomenclature *dvitīya-sphuta-yojana-karṇa* is introduced in *Laghu-vivṛti* thus:

एवं कृतो रवेः बिम्बघनमध्यस्य घनभूमध्यस्य च अन्तरालं योजनात्मकं भवति;
द्वितीयस्फुटयोजनकर्ण इति यावत् ।

Doing so gives the distance of separation between the centres of the solar disc and the Earth in *yojanas*; in fact, [this is] the *dvitīya-sphuṭa-yojana-karṇa*.

४.७ अर्केन्द्वोः बिम्बकलाव्यासः

4.7 Angular diameters of the orbs of the Sun and the Moon in minutes

बिम्बस्य योजनव्यासं विष्कम्भार्धहतं हरेत् ॥ १४ ॥
स्वभूम्यन्तरकर्णेन लिप्ताव्यासः शशीनयोः ।

bimbasya yojanavyāsaṃ viṣkambhārdhahataṃ haret || 14 ||
svabhūmyantarakarṇena liptāvyāsaḥ śaśīnayoḥ |

Let the [linear] diameters of the discs of the Sun and the Moon in *yojanas* multiplied by the *trijyā* be divided by their own distances of separation [to obtain] their angular diameters, in minutes.

Let D_s and D_m be the linear diameters of the Sun and the Moon. The formulae for obtaining the angular diameters α_s and α_m of the Sun and the Moon, from their linear diameters, are given to be

$$\alpha_s = \frac{D_s \times R}{d_{2s}} \quad \text{and} \quad \alpha_m = \frac{D_m \times R}{d_{2m}}, \quad (4.20)$$

where the denominators refer to the *dvitīya-sphuṭa-yojana-karṇa* of the Sun and the Moon whose computation has been described in the previous section.

Since the angular diameters $\alpha_s(\alpha_m)$ in minutes correspond to a distance R from the centre of the Earth, where R is *trijyā*, whereas the linear diameters $D_s(D_m)$

correspond to a distance of $d_{2s}(d_{2m})$, the above relations follow straightaway from the rule of three.

The commentary *Laghu-vivṛti* reminds us that the values of the linear diameters are mentioned in verse 10 of this chapter. This fact is recalled in the commentary *Laghu-vivṛti* as follows:

'दिग्वेदाब्धिमितो भानोः व्यासस्तिथ्यग्नयो विधोः' इत्युक्तरूपौ ··· ।

As stated earlier, the diameter of the Sun is 4410 *yojanas* and that of the Moon is 315 ....

४.८ भूच्छायायाः दैर्घ्यम्

4.8 Length of the Earth's shadow

रविभूम्यन्तरं खेषु पङ्क्तिघ्नं खर्तुनिर्जरैः ॥ १५ ॥
हृतं भगोलविष्कम्भात् भूच्छायादैर्घ्ययोजनम् ।

ravibhūmyantaraṃ kheṣu paṅtighnaṃ khartunirjaraiḥ || 15 ||
hṛtaṃ bhagolaviṣkambhāt bhūcchāyādairghyayojanam |

The distance of separation between the Earth and the Sun multiplied by 1050 and divided by 3360 is the length of the *chāyā*, the Earth's shadow, in *yojanas*.

In Fig. 4.3, S and E refer to the centres of the Sun and the Earth respectively. C represents the tip of the Earth's shadow in the shape of a cone. The length of the shadow from the centre of the Earth is nothing but the height of the cone denoted by l_c in the figure. It is given to be

$$l_c = \frac{d_{2s} \times 1050}{3360}. \tag{4.21}$$

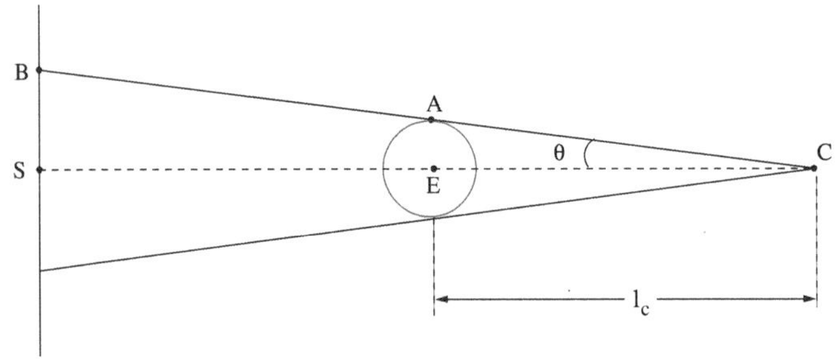

Fig. 4.3 Determination of the length of the Earth's shadow.

This may be understood as follows. Considering the similar triangles AEC and BSC, we have

$$\tan\theta = \frac{AE}{EC} = \frac{AE}{l_c} \tag{4.22}$$

$$= \frac{BS}{SC} = \frac{BS}{SE + l_c}. \tag{4.23}$$

Hence

$$\frac{AE}{l_c} = \frac{BS}{SE + l_c}. \tag{4.24}$$

Solving for l_c, we have

$$l_c = \frac{SE \times AE}{BS - AE}, \tag{4.25}$$

where SE refers to the actual distance of the Sun from the Earth. This is taken to be the *dvitīya-sphuṭa-karṇa* d_{2s}. From Fig. 4.3 it is obvious that BS and AE are the semi-diameters of the Sun and the Earth, whose diameters in *yojanas* are given as 4410 and 1050. Substituting these values in (4.25), we obtain (4.21).

४.९ भूच्छायायाः बिम्बकलाव्यासः

4.9 Angular diameter of the Earth's shadow in minutes

चन्द्रभूम्यन्तरं त्यक्त्वा शेषे भूव्यासताडिते ॥ १६ ॥
छायादैर्घ्यहृते व्यासः चन्द्रवत् तमसः कलाः ।

candrabhūmyantaraṃ tyaktvā śeṣe bhūvyāsatāḍite || 16 ||
chāyādairghyahṛte vyāsaḥ candravat tamasaḥ kalāḥ |

Subtracting the distance of separation between the Earth and the Moon (*dvitīya-sphuṭa-karṇa*) [from the length of the *chāyā*], and multiplying the reminder by the diameter of the Earth and dividing it by the length of the shadow, gives the [diameter of the] shadow as in the case of the Moon in minutes [at the distance of the Moon's orbit].

If l_c be the length of the *chāyā* (the Earth's shadow), and D_c and D_e are the linear diameters of the shadow and the Earth respectively, then the formula given for the diameter of the shadow at the distance of the Moon's orbit may be written as

$$D_c = \frac{(l_c - d_{2m}) \times D_e}{l_c}. \tag{4.26}$$

The above result may be understood with the help of Fig. 4.4. Here E represents the centre of the Earth, M the centre of the Moon and C the tip of the shadow. EC represents the length of the shadow, l_c, and EM the *sphuṭa-yojana-karṇa*, d_{2m}. From the triangle AEC,

$$\tan\theta = \frac{AE}{EC} = \frac{AE}{l_c}. \tag{4.27}$$

Similarly from the triangle BMC,

$$\tan\theta = \frac{BM}{MC} = \frac{BM}{l_c - d_{2m}}. \tag{4.28}$$

Therefore

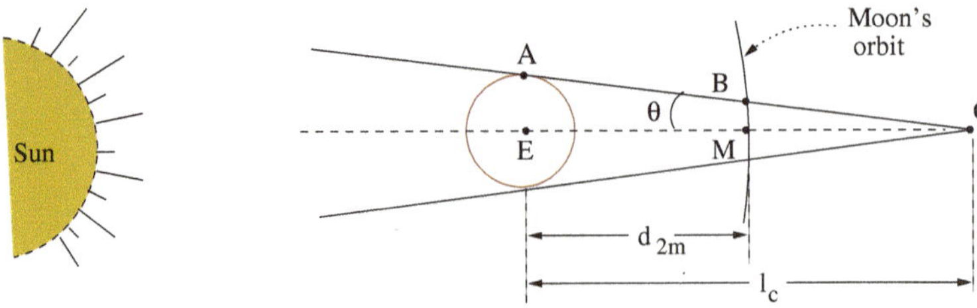

Fig. 4.4 Determination of the angular diameter of the Earth's shadow.

$$\frac{BM}{l_c - d_{2m}} = \frac{AE}{l_c}$$

$$\text{or} \qquad 2 \times BM = \frac{(l_c - d_{2m})\, 2 \times AE}{l_c}. \tag{4.29}$$

This is the same as (4.26), since BM represents the radius of the *chāyā* (shadow) at the distance of the Moon's orbit and AE is the radius of the Earth.

In this context, a graphic description of the shadow is given in *Yukti-dīpikā* as follows:

ज्योतिर्गोलो महाव्यासः भासयत्यखिलं रविः ।
आत्माभिमुखम् अन्यार्धेच्छायां कुर्वन् क्रमात् कृशाम् ॥
अतोऽर्काभिमुखं दीप्तं भूम्यर्धं मलिनं परम् ।
छाया च तस्या भूव्याससमादौ क्रमशः कृशा ॥ [15]

The Sun, being a ball of effulgence, with a large diameter (*mahā-vyāsa*), illuminates one half of all the objects facing towards him, [thereby] generating a shadow on the other half which thins out gradually.

Hence that half of the Earth facing the Sun is bright and the other half is dark. The shadow of that (the Earth's disc) has a diameter equal to the diameter of the Earth at the beginning and gradually becomes thin.

४.१० चन्द्रविक्षेपः स्फुटभुक्तिश्च

4.10 Moon's latitude and true daily motion

पातोनेन्दोर्भुजाजीवा व्योमताराहता हृता ॥ १७ ॥
त्रिज्यया सौम्ययाम्येन्दोः क्षिप्तिः सा[16] च स्फुटा गतिः ।
भगोलचन्द्रकर्णघ्ने भूचन्द्रान्तरयोजनैः ॥ १८ ॥
हृते स्फुटे इह ग्राह्ये क्षितिभुक्ती स्थितेर्दले ।

[15] {TS 1977}, p. 258.

[16] Here there is a possibility of confusion as the word **'सा'** could be associated either with *kṣiptiḥ* (occurring before), or with *gati* (occurring later). However, according to the context, it is to be associated with *kṣiptiḥ*, the latitude of the Moon, and not *gati*, the true daily motion.

pātonendorbhujājīvā vyomatārāhatā hṛtā || 17 ||
trijyayā saumyayāmyendoḥ kṣiptiḥ sā ca sphuṭā gatiḥ |
bhagolacandrakarṇaghne bhūcandrāntarayojanaiḥ || 18 ||
hṛte sphuṭe iha grāhye kṣiptibhuktī sthiterdale |

The Rsine of the longitude of the node subtracted from the Moon is multiplied by 270 and divided by the *trijyā*. This gives the latitude of the Moon lying to the north or south [of the ecliptic]. This and the true daily motion, multiplied by the *bhagola-candra-karṇa*, and divided by the actual distance of separation between the Earth and the Moon in *yojanas* (d_{2m}), are the true values of the latitude and the daily motion at the middle of the eclipse, which are to be considered [for computational purposes].

The formula given for the latitude β of the Moon is,

$$\beta = \frac{270 \times R\sin(\lambda_m - \lambda_n)}{R}, \tag{4.30}$$

where λ_m and λ_n are the longitudes of the Moon and its node respectively. R is the *trijyā*, whose value is taken to be 3438 minutes. 270 is the inclination of the Moon's orbit in minutes.

It is mentioned here that the values of the latitude and the true daily motion obtained at the middle of the eclipse must be corrected to get more accurate values, which are to be used for the computations of half durations etc. If β' and $\dot{\lambda}'_m$ are the corrected values of the latitude and true daily motion, then they are given by

$$\beta' = \beta \times \frac{d_{1m}}{d_{2m}} \qquad \text{and} \qquad \dot{\lambda}'_m = \dot{\lambda}_m \times \frac{d_{1m}}{d_{2m}}. \tag{4.31}$$

In the above relation d_{1m} represents the *prathama-sphuṭa-karṇa* or *bhagola-candra-karṇa*, which is the distance of the Moon from the centre of the *bhagola*, and d_{2m} the *dvitīya-sphuṭa-karṇa* or the *bhūgola-candra-karṇa*, which is the distance of the Moon from the centre of the Earth.

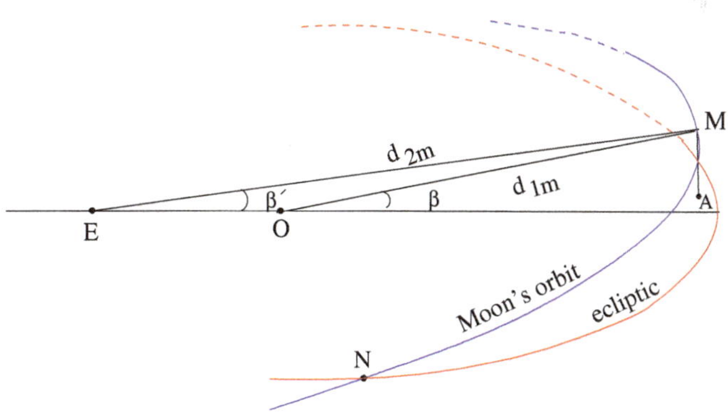

Fig. 4.5 Correction to the latitude and daily motion of the Moon.

The rationale behind (4.31) may be understood with the help of Fig. 4.5. Here O and E refer to the centres of the *bhagola* (the celestial sphere) and the *bhūgola* (the Earth) respectively. β' and β are the latitudes of the Moon as measured from these points. The expression for β given above in (4.30) is with reference to the centre of the *bhagola*, O. It is easily seen from the figure that

$$AM = EM \sin\beta' = OM \sin\beta$$

$$\text{or} \qquad \sin\beta' = \sin\beta \frac{OM}{EM}. \tag{4.32}$$

As β and β' are small, the sines may be replaced by the arcs and the above equation reduces to

$$\beta' = \beta \frac{OM}{EM} = \beta \frac{d_{1m}}{d_{2m}}. \tag{4.33}$$

Similarly, if $\dot{\lambda}_m$ and $\dot{\lambda}'_m$ are the (angular) daily motions with reference to O and E respectively, then

$$\dot{\lambda}'_m = \dot{\lambda}_m \frac{d_{1m}}{d_{2m}}, \tag{4.34}$$

which is based on the assumption that the planets have a common linear velocity irrespective of their distances.

४.११ ग्रहणस्य सदसद्भावः
4.11 The occurrence and non-occurrence of an eclipse

क्षिप्तिः सा चन्द्रभूच्छायाबिम्बैक्यार्धाधिका यदि ॥ १९ ॥
ग्रहणं नैव चन्द्रस्य हीना चेदस्य सम्भवः ।

kṣiptiḥ sā candrabhūcchāyābimbaikyārdhādhikā yadi || 19 ||
grahaṇaṃ naiva candrasya hīnā cedasya sambhavaḥ |

If that latitude [i.e., the latitude of the Moon as determined earlier] is greater than the sum of the semi-diameters of the Moon and the Earth's shadow, then there is no lunar eclipse; if it is less, then there is a possibility [of an eclipse].

In Fig. 4.6(a), A and X refer to the centres of the shadow and the Moon's disc respectively. AM' is the semi-diameter of the shadow and MX that of the Moon. AX is the latitude of the Moon at the instant of opposition. If the latitude of the Moon at this instant is exactly equal to or greater than the sum of the semi-diameters of the shadow and the Moon then there will be no eclipse. That is, if

$$AX \geq (AM' + MX) \tag{4.35}$$

at the instant of opposition, then there will be no eclipse, as no portion of the Moon ever enters the shadow.

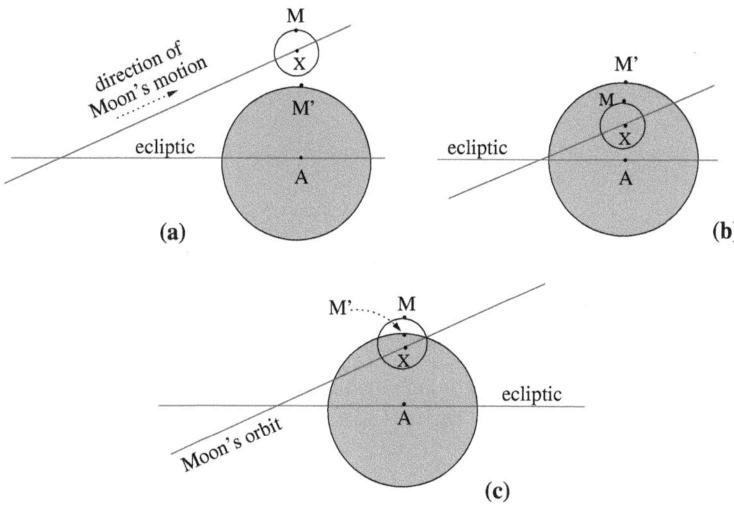

Fig. 4.6 The Earth's shadow and the Moon when there is (a) no lunar eclipse, (b) a total lunar eclipse and (c) a partial lunar eclipse.

४.१२ पूर्णग्रहणनियमः
4.12 The condition for the occurrence of a total eclipse

चन्द्रबिम्बोनभूच्छाया बिम्बार्धादधिका यदि ॥ २० ॥
सर्वग्रासो न चैव स्यात् हीना चेद् ग्रस्यतेऽखिलम् ।

candrabimbonabhūcchāyā bimbārdhādadhikā yadi || 20 ||
sarvagrāso na caiva syāt hīnā ced grasyate'khilam |

(If the latitude) [at *parvānta*] is greater than half the angular diameter of the shadow diminished by the Moon's disc, then total eclipse will not occur. If it is less, then the Moon will be eclipsed totally.

The condition on the latitude of the Moon for the occurrence of a total lunar eclipse may be explained with the help of Fig. 4.6(b). Here, AM' and MX represent the angular semi-diameters of the shadow and the Moon respectively. If the latitude is less than or equal to the difference between the respective semi-diameters of the shadow and the Moon then the eclipse will be total. That is, if

$$AX \leq (AM' - MX) \tag{4.36}$$

at the instant of opposition then it will be a total lunar eclipse. Obviously, if

$$(AM' - MX) < AX < (AM' + MX) \tag{4.37}$$

then there will be a partial eclipse. This situation is depicted in Fig. 4.6(c). The different cases discussed are summarized in Table 4.1.

Case	Condition to be satisfied at instant of opposition (*parvānta*)	Possibility/nature of eclipse
(a)	Latitude $\geq$ Sum of the semi-diameters, or $AX \geq (AM' + MX)$	(no eclipse)
(b)	Latitude $\leq$ Diff. between the semi-diameters, or $AX \leq (AM' - MX)$	(total eclipse)
(c)	Latitude in between the sum and diff., or $(AM' - MX) < AX < (AM' + MX)$	(partial eclipse)

Table 4.1 Three distinct possibilities that arise at the instant of opposition.

It is pointed out in *Laghu-vivṛti* that the string '*kṣiptiḥ sā*', referring to latitude, occurring at the beginning of the earlier verse, is to be related or carried on into this verse too.[17]

'क्षिप्तिः सा' इत्यत्रापि सम्बध्यते ।

Here (in this verse) also [the words] '*kṣiptiḥ sā*' are related or connected.

४.१३ स्थित्यर्धस्पर्शमोक्षयोः कालः

4.13 The time of half-duration, the first and the last contact

सम्पर्कार्धकृतेस्त्यक्त्वा क्षेपवर्गं पदीकृतम् ॥ २१ ॥
षष्टिघ्नं भानुशीतांश्वोः हृतं गत्यन्तरेण यत् ।
स्थित्यर्धनाडिकाद्यं तत्, पर्वान्ते तद्युतोनिते ॥ २२ ॥
मोक्षः स्पर्शश्च चन्द्रस्याप्यविशिष्टौ स्फुटौ तु तौ ।

samparkārdhakṛtestyaktvā kṣepavargaṃ padīkṛtam || 21 ||
ṣaṣṭighnaṃ bhānuśītāṃśvoḥ hṛtaṃ gatyantareṇa yat |
sthiyardhanāḍikādyaṃ tat, parvānte tadyutonite || 22 ||
mokṣaḥ sparśaśca candrasyāpyaviśiṣṭau sphuṭau tu tau |

Having subtracted the square of the latitude from the square of the sum of the semi-diameters [of the shadow and the Moon], the square root is found. This is multiplied by 60 and divided by the difference in the daily motion of the Sun and the Moon. The result is the half duration [of the eclipse] in *nāḍikās*, etc. This added to the instant of opposition and subtracted from it gives the time of end of the eclipse and the time of commencement [respectively]. These values of the half-durations, when iterated by the *aviśeṣa-karma*, lead to more accurate values.

The expression for the half-duration of the eclipse and the procedure to determine the instants of the beginning and the end of the eclipse are explained in the above verses. These may be understood with the help of Fig. 4.7. Here O represents the centre of the shadow, and X the centre of the Moon's disc as it is about to enter into the shadow.

[17] This process of carrying a word or a string of words on to the later verses is common in Sanskrit literature and is technically called "*anuvṛtti*."

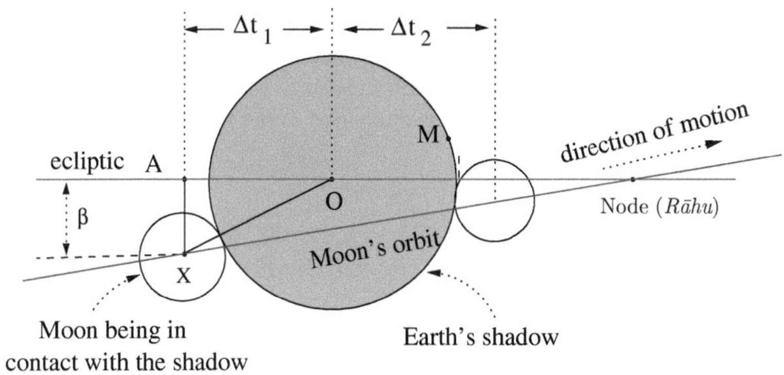

Fig. 4.7 First and the second half-durations of a lunar eclipse.

The total duration of the eclipse may be conceived as made up of two parts:

1. the time interval between the instant at which the Moon enters the shadow and the instant of opposition (Δt_1) and,
2. the time interval between the instants of opposition and complete release (Δt_2).

The suffixes 1 and 2 refer to the first and the second half-durations of the eclipse respectively. Though one may think naively that these two durations must be equal, this is not so because of the continuous change in the angular velocities of the Sun and the Moon and its nodes.

In Fig. 4.7, AX and OX represent the latitude (β) of the Moon and the sum of the semi-diameters (S) of the shadow and the Moon respectively. If $\dot{\lambda}_s$ and $\dot{\lambda}_m$ refer to the angular velocities of the Sun and the Moon per day, then the difference in their daily motion called the *gatyantara* or the *bhuktyantara* is given by

$$gatyantara = \dot{\lambda}_m - \dot{\lambda}_s. \tag{4.38}$$

The approximate value of the first half-duration of the eclipse in *nāḍikās* is found using the relation

$$\Delta t_0 = \frac{OA \times 60}{\text{Diff. in daily motion}} = \frac{\sqrt{OX^2 - AX^2}}{\dot{\lambda}_m - \dot{\lambda}_s} \times 60 = \frac{\sqrt{S^2 - \beta^2}}{\dot{\lambda}_m - \dot{\lambda}_s} \times 60. \tag{4.39}$$

Here the factor 60 represents the number of *nāḍikās* in a day. In the above expression, β is the latitude of the Moon at the middle of the eclipse. As the instant of opposition is known, the latitude of the Moon at the instant of opposition can be easily calculated. However, the instant of the beginning of the eclipse is yet to be determined, and hence the latitude of the Moon at the beginning is not known. Moreover, the latitude of the Moon is a continuously varying quantity. This being the case, it is quite clear that the result given by (4.39) is only approximate and clearly there is a necessity for an improvised technique. What is prescribed in the text is an iterative procedure for finding the half-duration. As a first approximation, the latitude known at the instant of opposition is taken to be β and Δt_0 is determined. The iterative procedure to be adopted is described in the following verses.

४.१४ स्थित्यर्ध-स्पर्शमोक्षाणामविशेषः

4.14 Iteration for obtaining the half-duration, and the time of the first and the last contact

स्थित्यर्धघ्ने स्फुटे भुक्ती षष्ट्याप्ते भानुचन्द्रयोः ॥ २३ ॥
शोधयेत् समलिप्तेन्दौ सूर्याच्च स्पार्शिकावुभौ ।
स्वभुक्तिमन्यथा पाते तदा तात्कालिकीकृताः[18] ॥ २४ ॥
स्पर्शेन्दोश्च पुनः क्षेपस्थित्यर्धं तद्गतिः शशी ।
प्राग्वदेवासकृत् कार्या निश्चलत्वदिदृक्षुणा ॥ २५ ॥
स्थित्यर्धगतिमर्केन्द्वोः क्षिप्त्वा तौ मौक्षिकौ नयेत् ।
अन्यत् सर्वं समं मोक्षे स्थित्यर्धस्याविशेषणे ॥ २६ ॥

sthityardhaghne sphuṭe bhuktī ṣaṣṭyāpte bhānucandrayoḥ || 23 ||
śodhayet samaliptendau sūryācca spārśikāvubhau |
svabhuktimanyathā pāte tadā tātkālikīkṛtāḥ || 24 ||
sparśendośca punaḥ kṣepasthityardhaṃ tadgatiḥ śaśī |
prāgvadevāsakṛt kāryā niścalatvadidṛkṣuṇā || 25 ||
sthityardhagatimarkendvoḥ kṣiptvā tau maukṣikau nayet |
anyat sarvaṃ samaṃ mokṣe sthityardhasyāviśeṣaṇe || 26 ||

The true daily motions of the Sun and the Moon (the *bhuktis*) multiplied by the half-duration and divided by 60 and subtracted from the positions of the Sun and the Moon [at the *parvānta*] give their positions at the time of contact (the *spārśika*). In the case of the node, its daily motion [has to be applied] in the reverse manner. Then [are obtained] the values at that time (of contact).

From the position of the Moon [and the node and shadow] obtained at the time of contact, again the latitude, the half-duration and the [daily] motions [are computed]. The Moon [and others] are to be obtained iteratively by one who is interested in their steady values (*niścalatva-didṛkṣuṇā*). By adding the distance traversed by the Sun and the Moon in half-duration [to their longitudes at the *parvānta*], their longitudes at the time of release are to be obtained. For finding the half-duration of release iteratively, everything else is the same.

The positions of the Sun and the Moon at the time of contact are obtained by subtracting their motions during the first-half duration from their values at the instant of opposition. The motion of the Sun/Moon is obtained by multiplying their true daily motion ($\dot{\lambda}_s, \dot{\lambda}_m$) by the half duration—the first approximation of which has been found as given by (4.39)—and dividing by 60 (the number of *nāḍikās* in

---

[18] The reading in the published text ({TS 1958}, p. 105) is: सदा तात्कालिकी कृतौ। This reading seems to be incorrect, and hence it has been corrected. The commentary *Laghu-vivṛti*—which runs as: एवं कृताः त्रयोऽपि ते तात्कालिका भवन्ति—is also in consonance with the correction implemented above.

a day). This is done for the node also (but applied in reverse, as its motion is retrograde), whose longitude is required for the computation of Moon's latitude. From the relative positions of O, X and A freshly determined, the first half-duration is again calculated. This is the second approximation to it. The iteration procedure is carried on till the successive approximations to the half-durations are not different from each other to a desired level of accuracy.

The procedure is the same for computing the second half-duration (*mokṣakāla*), except that the positions of the Sun and Moon at the time of the *mokṣa* (release) are obtained by adding their motions during the second half-duration to their values at the instant of opposition.

४.१५ स्थित्यर्धाभ्यां स्पर्शमोक्षौ

4.15 The time of the first and the last contact from the half-duration

स्पर्शस्थितिदलं शोध्यं पर्वान्तादितरत् क्षिपेत् ।
स्पर्शमोक्षौ तु तौ स्यातां मध्ये स्यात् परमग्रहः ॥ २७ ॥

sparśasthitidalaṃ śodhyaṃ parvāntāditarat kṣipet |
sparśamokṣau tu tau syātāṃ madhye syāt paramagrahaḥ || 27 ||

The half-duration of contact (*sparśasthitidala*) has to be subtracted from the instant of opposition and the other one [the half-duration of release] must be added. The two [values obtained] are the times of contact and release of the eclipse. The maximum obscuration (*paramagraha*[19]) is at the middle [of the eclipse].

If t_m be the instant of opposition, then the beginning and ending moments of the eclipse, which are referred to as the *sparśakāla* (instant of first contact) (t_b) and the *mokṣakāla* (instant of release) (t_e) respectively, are given by

$$t_b = t_m - \Delta t_1$$
$$\text{and} \qquad t_e = t_m + \Delta t_2, \tag{4.40}$$

where Δt_1 and Δt_2 are the first and the second half-durations of the eclipse determined by iteration. It was mentioned earlier (section 4.13) that in general these two durations will not be equal.

It is further stated here that the instant of maximum obscuration the *paramagrāsakāla*, t_p, is exactly in between the *sparśakāla* and the *mokṣakāla*. That is,

$$\begin{aligned} t_p &= t_b + \frac{\Delta t_1 + \Delta t_2}{2} \\ &= t_e - \frac{\Delta t_1 + \Delta t_2}{2}. \end{aligned} \tag{4.41}$$

[19] More commonly referred to as the *paramagrāsa*.

Since $\Delta t_1 \neq \Delta t_2$ in general, $t_p \neq t_m$. Let the difference between these two instants be given by

$$\delta t_p = t_p - t_m.$$

An alternate procedure for determining δt_p is discussed in both the commentaries, *Laghu-vivṛti* and *Yukti-dīpikā*. Here we explain the procedure given in *Laghu-vivṛti*.

···अत्र परमग्रासकाल एव स्फुटपर्वान्त इति। तदानयनार्थं पातोनार्कस्य दोर्ज्यां परमक्षेपकोट्या निहत्य इष्टविक्षेपकोट्या विभज्य लब्धं यत् फलं तस्य इष्टविक्षेपस्य च वर्गयोगतो यत् पदं तच्चापं[20] पातोनार्कभुजाचापेन सह विश्लिष्य शिष्टात् षष्टिघ्नात् अर्कचन्द्रयोः गत्यन्तरकलाभिर्विभज्य लब्धं कलादिकं युग्मपदगते सति इष्टविक्षेपे पूर्वानीते स्फुटपर्वान्ते प्रक्षिपेत्। ओजपदगते सति ततो विशोधयेत्। एवं कृत एव पर्वान्तः सूर्येन्द्वोर्ग्रहणे स्फुटो भवतीति प्रसिद्धात् पर्वान्तात् प्रदेशान्तरस्थित एवासौ स्फुटः पर्वान्तः।अत एव वक्ष्यति 'अल्पश्चेत् परमग्रासः चलेत् स्थितिदलेऽधिके' - इति।

Here [it is to be understood] that the instant of maximum obscuration (*paramagrāsakāla*) is the true instant of opposition (*sphuṭa-parvānta*). To obtain that, the Rsine of the longitude of the Sun minus the node must be multiplied by the Rcosine of the maximum latitude of the Moon and divided by Rcosine of the instantaneous latitude of the Moon. The result and the instantaneous latitude are squared and the square root [of the sum] is found. The arc corresponding to this is found and it is subtracted from the arc corresponding to the Rsine of the longitude of the Sun minus the node. The remainder is multiplied by 60 and divided by the difference in the daily motions [of the Sun and the Moon]. This quantity is added to the instant of opposition determined earlier, if the latitude of the Moon is in the even quadrant. It has to be subtracted if the latitude is in the odd quadrant. The instant of opposition thus determined is the exact instant of opposition in both solar and lunar eclipses. This will be different from the instant of opposition usually determined. It is therefore mentioned [later in the same chapter, verse 43]: 'If the maximum obscuration *paramagrāsa* is small, then it (*paramagrāsa*) will be shifted to [i.e, will happen in] the greater half-duration.'

The same procedure is described in *Yukti-dīpikā* and in fact Śaṅkara Vāriyar attributes this to Nīlakaṇṭha. The following verses are cited by Śaṅkara Vāriyar in his *Laghu-vivṛti* and are attributed by him to *Grahaṇanirṇaya* of Nīlakaṇṭha:[21]

परमक्षेपकोटिघ्नः पातोनार्कभुजागुणः ।
स्वेष्टविक्षेपकोट्याप्तः तत्क्षेपकृतियोगतः[22] ॥
पदं यच्चापितं यच्च पातोनार्कभुजाधनुः ।
तद्विश्लेषं हतं षष्ट्या गत्यन्तरहृतं क्षिपेत् ॥
पर्वान्ते युक्पदे क्षेपे शोधयेद्विषमे पदे ।
एवं कृतो हि पर्वान्तः सूर्येन्द्वोर्ग्रहणे स्फुटः ॥

---

[20] The reading in the printed text is: तच्चापपातोनार्कभुजाचापेन।

[21] Though this text *Grahaṇanirṇaya* has been listed among the works of Nīlakaṇṭha by Prof. K. V. Sarma, no manuscript of it has been traced to date.

[22] The term '*tat*' here refers to the quantity obtained earlier. Hence the *vigraha* of this compound has to be done as follows: तच्च क्षेपश्च = तत्क्षेपौ, तयोः कृत्योः योगः = तत्कृतियोगः, तस्मात्।

The content of the above verses can be explained as follows. First, an intermediate angle θ is defined thus:

$$\sin\theta = \sqrt{\beta_t^2 + \frac{\cos^2\beta_{max} \times \sin^2(\lambda_s - \lambda_n)}{\cos^2\beta_t}}, \qquad (4.42)$$

where β_t and β_{max} are the instantaneous and maximum latitudes of the Moon, and λ_s, λ_m and λ_n are the longitudes of the Sun, the Moon and the node respectively. Having found the arc θ corresponding to this, the quantity δt_p, which is to be applied to the instant of opposition (t_m), is defined to be

$$\delta t_p = \frac{((\lambda_m - \lambda_n) - \theta)}{\dot{\lambda}_m - \dot{\lambda}_s} \times 60 \qquad \text{(in } ghaṭikās\text{)}. \qquad (4.43)$$

As $\lambda_m = \lambda_s + 180°$, $|R\sin(\lambda_m - \lambda_n)| = |R\sin(\lambda_s - \lambda_n)|$. δt_p is to be added to the instant of opposition if the latitude is in the even quadrant, and subtracted from it if the latitude is in the odd quadrant.

Here we give a geometrical representation of the expression for θ appearing in (4.42). In this equation, we can replace λ_s by λ_m, since $\lambda_m = \lambda_s$ or $\lambda_s + 180°$ at the instant of opposition for an eclipse.

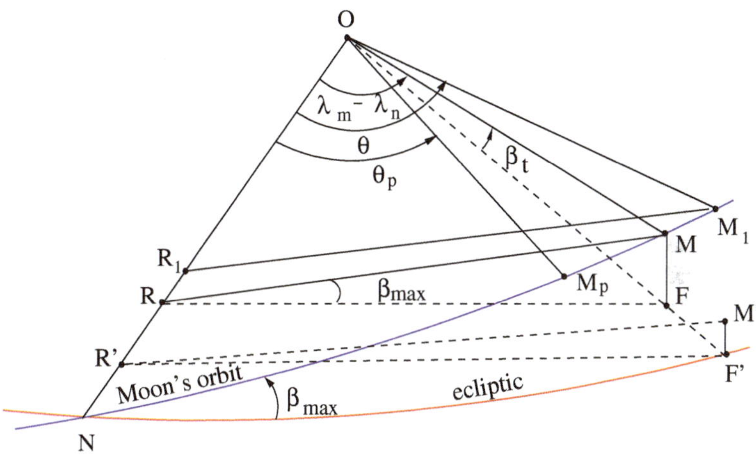

Fig. 4.8a Determination of the *paramagrāsakāla* from the instant of opposition obtained through an iterative process.

In Fig. 4.8*a*, O is the centre of the celestial sphere and N the ascending node of the Moon's orbit, whose inclination to the ecliptic is indicated as β_{max}. M is the position of the Moon at conjunction or opposition and its latitude at that instant is denoted by β_t. MR and MF are perpendiculars to the line of nodes and the plane of the ecliptic respectively. It is easily seen from the figure that

$$\begin{aligned} MR &= R\sin(\lambda_m - \lambda_n) \\ MF &= R\sin\beta_t \\ \text{and} \qquad OF &= R\cos\beta_t. \end{aligned} \tag{4.44}$$

The angle between MR and RF is the same as the angle between the planes of the Moon's orbit and the ecliptic, β_{max}. Now we extend OF to meet the ecliptic at F', and draw $F'R'$ parallel to FR. Then

$$\begin{aligned} RF &= MR\cos\beta_{max} \\ &= R\cos\beta_{max}\sin(\lambda_m - \lambda_n). \end{aligned} \tag{4.45}$$

Also,

$$\frac{R'F'}{RF} = \frac{OF'}{OF} = \frac{1}{\cos\beta_t}. \tag{4.46}$$

Using (4.45) in (4.46), we have

$$R'F' = \frac{R\cos\beta_{max}\sin(\lambda_m - \lambda_n)}{\cos\beta_t}. \tag{4.47}$$

We now draw $F'M' = R\sin\beta_t$ perpendicular to the plane of the ecliptic at F'. $R'M'$ is the hypotenuse in the right-angled triangle $M'R'F'$ and is given by

$$\begin{aligned} M'R' &= [M'F'^2 + R'F'^2]^{\frac{1}{2}} \\ &= \left[R^2\sin^2\beta_t + \frac{R^2\cos^2\beta_{max}\sin^2(\lambda_m - \lambda_n)}{\cos^2\beta_t}\right]^{\frac{1}{2}}. \end{aligned} \tag{4.48}$$

Again, we draw R_1M_1 perpendicular to ON from a point M_1 in the Moon's orbit, such that $R_1M_1 = R\sin\theta = M'R'$, as shown in the figure. Thus we have

$$R\sin\theta = \left[R^2\sin^2\beta_t + \frac{R^2\cos^2\beta_{max}\sin^2(\lambda_m - \lambda_n)}{\cos^2\beta_t}\right]^{\frac{1}{2}}. \tag{4.49}$$

This is the significance of the angle θ described in the text and appearing in (4.42). Also $\sin\beta_t = \sin\beta_{max}\sin(\lambda_m - \lambda_n)$. Therefore

$$\sin\theta = \sqrt{\sin^2\beta_{max} + \frac{\cos^2\beta_{max}}{\cos^2\beta_t}} \times \sin(\lambda_m - \lambda_n). \tag{4.50}$$

Clearly $|\theta| > |\lambda_m - \lambda_n|$.[23] Now, as per the prescription given in the text, the maximum obscuration occurs before the instant of opposition when the latitude is in the odd quadrant. Let us assume that the maximum obscuration occurs at M_p, corresponding to $N\hat{O}M_p = \theta_p$, such that

[23] This is because $\frac{\cos^2\beta_{max}}{\cos^2\beta_t} \geq \cos^2\beta_{max}$.

$$(\lambda_m - \lambda_n) - \theta_p = \theta - (\lambda_m - \lambda_n). \tag{4.51}$$

Though it is not clear how the position M_p of the Moon corresponds to its maximum obscuration during the eclipse, still we can get another rough estimate of the instant of maximum obscuration as follows:

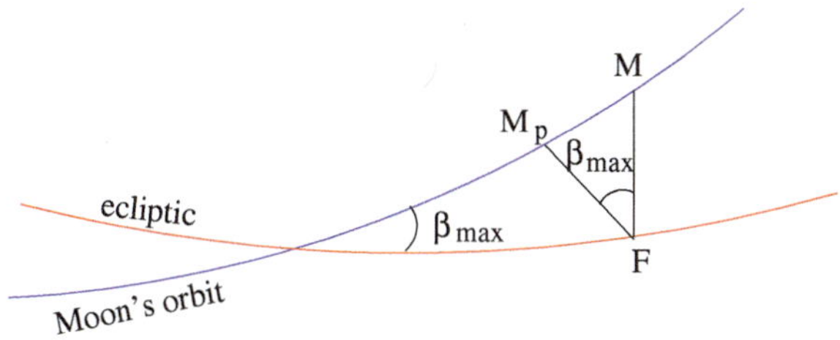

Fig. 4.8*b* Expression for the time difference between the *paramagrāsakāla* and the instant of opposition.

In Fig. 4.8*b*, at the instant of opposition, the Moon is at M. This position does not necessarily correspond to the minimum distance between the centres of the Moon and the shadow. The Moon is at the minimum distance from the ecliptic when it is at M_p such that FM_p is perpendicular to Moon's orbit. This can be taken to be the instant of maximum obscuration.

$$MF = R\sin\beta_t \simeq R\beta_t. \tag{4.52a}$$

Now $M\hat{F}M_p = \beta_{max}$, as MF and M_pF are perpendicular to the ecliptic and the Moon's orbit respectively. Hence

$$\begin{aligned} MM_p &= MF\sin\beta_{max} \\ &= R\sin\beta_t\sin\beta_{max}. \end{aligned} \tag{4.52b}$$

This amounts to a difference in longitude equal to $\sin\beta_t\sin\beta_{max}$, corresponding to the instants of conjunction and maximum obscuration. This corresponds to a time difference δt_p given by

$$\delta t_p = \frac{\sin\beta_t\sin\beta_{max}}{\dot{\lambda}_m - \dot{\lambda}_s} \times 60. \tag{4.53}$$

The expression for δt_p, deduced from $\sin\theta$ given by (4.42), as prescribed in the text, does not seem to be related to the above in any reasonable approximation.

Consider the expression for the *bimbāntara*, which is the separation between the centres of the Moon and the shadow used later in this chapter and also in Chapter 8. By minimizing that, we can obtain the value of FM_p, and consequently the following value of MM_p from that:

$$MM_p = FM_p\frac{\sin\beta_{max}}{\cos\beta_{max}}. \tag{4.54}$$

४.१६ रव्युदयास्तमये स्पर्शमोक्षयोः दृश्यादृश्यत्वम्

4.16 The visibility or otherwise of the the first and the last contact at sunrise and sunset

उदयास्तमयासन्ने स्पर्शे मोक्षे च संशयः ।
स्पर्शो दृश्य उतादृश्यः चन्द्रस्यास्तमयेत्विति ॥ २८ ॥
प्राग्वा पश्चाच्च मोक्षः स्यात् तत्प्रदेशोदयादिति ।

udayāstamayāsanne sparśe mokṣe ca saṃśayaḥ |
sparśo dṛśya utādṛśyaḥ candrasyāstamayetviti || 28 ||
prāgvā paścācca mokṣaḥ syāt tatpradeśodayāditi |

When the time of contact or release happens to be close to the time of sunrise/sunset, then there is bound to be a doubt [about the visibility of the contact or release]. The contact may or may not be visible when the Moon is setting. [Similarly,] regarding the visibility of release, [there will be doubt] as to whether it would occur earlier or later than the moonrise at that location.

When the lunar eclipse is close to the instant of sunrise (or moonset) or sunset (moonrise), the shadow disc and the lunar disc would be close to the horizon. Then, it is possible that the release or the first contact would be below the horizon and hence invisible. The criteria for the visibility of the first and last contact are given in the verses which follow.

४.१७ स्पर्शमोक्षयोः दृश्यादृश्यत्वनिर्णयः

4.17 The visibility or otherwise of *sparśa* and *mokṣa*

स्पर्शे रव्युदये कार्यो दृक्क्षेपः क्षितिरैन्दवी ॥ २९ ॥
व्यासार्धघ्नः स्फुटः क्षेपः सम्पर्कार्धहृतस्तु यः ।
तद्दृक्क्षेपधनुर्भेदः दिशोः साम्येऽन्यथा युतिः ॥ ३० ॥
तदूनभत्रयाज्जीवा तमोव्यासदलाहता ।
त्रिज्याप्ता लम्बने योज्या भूच्छायाशङ्कुलब्धये ॥ ३१ ॥
द्वितीयस्फुटभागस्य तिथ्यंशो लम्बनं त्विह ।
राशित्रयाधिके क्षेपदृक्क्षेपानीतकार्मुके ॥ ३२ ॥
अधिकस्य गुणात् प्राग्वत् लब्धं शोध्यं तु लम्बनात् ।
शेषस्तस्य तमःशङ्कुस्त्रिज्याकृत्या हतः स च ॥ ३३ ॥
हृतो घातेन सूर्यस्य द्युज्याया लम्बकस्य च ।
लब्धाः प्राणाः क्षपाशेषे यदि रव्युदयात् ततः ॥ ३४ ॥
स्पर्शः प्रागेव तर्ह्येव दृश्यः स्यान्न ततः परम् ।
एवं रव्यस्तकालोत्थक्षेपाद्याप्तेन शङ्कुना ॥ ३५ ॥

सिद्धैः प्राणैर्यदा मोक्षो रव्यस्तमयतः परम् ।
तदैव दृश्यतामेति ततः प्राङ्मोक्षणे न च [24] ॥ ३६ ॥

sparśe ravyudaye kāryo dṛkkṣepaḥ kṣiptiraindavī || 29 ||
vyāsārdhaghnaḥ sphuṭaḥ kṣepaḥ samparkārdhahṛtastu yaḥ |
taddṛkkṣepadhanurbhedaḥ diśoḥ sāmye'nyathā yutiḥ || 30 ||
tadūnabhatrayājjīvā tamovyāsadalāhatā |
trijyāptā lambane yojyā bhūcchāyāśaṅkulabdhaye || 31 ||
dvitīyasphuṭabhāgasya tithyaṃśo lambanaṃ tviha |
rāśitrayādhike kṣepadṛkkṣepānītakārmuke || 32 ||
adhikasya guṇāt prāgvat labdhaṃ śodhyaṃ tu lambanāt |
śeṣastasya tamaḥśaṅkustrijyākṛtyā hataḥ sa ca || 33 ||
hṛto ghātena sūryasya dyujyāyā lambakasya ca |
labdhāḥ prāṇāḥ kṣapāśeṣe yadi ravyudayāt tataḥ || 34 ||
sparśaḥ prāgeva tarhyeva dṛśyaḥ syānna tataḥ param |
evaṃ ravyastakālotthakṣepādyāptena śaṅkunā || 35 ||
siddhaiḥ prāṇairyadā mokṣo ravyastamayataḥ param |
tadaiva dṛśyatāmeti tataḥ prāṅmokṣaṇena ca || 36 ||

If the first contact is around the time of sunrise, then the *dṛkkṣepa* (Rsine of the zenith distance of the *vitribhalagna*) and the latitude of the Moon must be calculated.

The true latitude [of the Moon] is multiplied by the *trijyā* and divided by the sum of the semi-diameters of the Sun and the Moon [the result is called the *sphuṭa-kṣepa*]. If the directions of this and the *dṛkkṣepa* are the same, we find the difference, otherwise the two are added [and the value is noted]. The Rsine of this subtracted from 90 is multiplied by the semi-diameter of the shadow and divided by the *trijyā*. The result has to be added to the parallax in longitude for obtaining the *śaṅku* of Earth's shadow.

The parallax in longitude [to be] used here is one-fifteenth of the actual daily motion [of the Moon], the *dvitīya-sphuṭa-bhukti*. If the arc (ζ), obtained from the arcs corresponding to the *dṛkkṣepa* (z_v) and the *kṣepa* (θ),[25] is greater than 90 degrees [that is $\zeta > 90$], then the quantity obtained as earlier from the sine of the excess has to be subtracted from the parallax in longitude.

The remainder is the *śaṅku* corresponding to the shadow. This is multiplied by the square of the *trijyā* and divided by the product of the *dyujyā* and the *lambaka* of the Sun. If the result Δt_s, obtained in *prāṇas*, is less than the time remaining in the night to be elapsed Δt_u[26] before sunrise, then the first contact will be earlier [than moonset] and visible. Otherwise (if $\Delta t_s > \Delta t_u$), the first contact is not visible.

In the same way, from the *śaṅku* [which in turn is] obtained using the *dṛkkṣepa* and other quantities calculated at the time of sunset, the *prāṇas* [related to the visibility of the *mokṣa*] Δt_m[27] are calculated. If the last contact occurs later than sunset by an amount [Δt_m] thus obtained in *prāṇas*, then it will be [after the moonrise's and] visible. If it happens earlier, then the last contact will not be visible.

In the above verses, quantitative criteria for the visibility of the first contact and the last contact of the lunar eclipse are discussed. If the lunar eclipse were to com-

[24] The reading in both the printed editions is प्राङ्मोक्षणेन च ; whereas it must be प्राङ्मोक्षणे न च because the idea to be conveyed is: ततः प्राक् मोक्षणे मोक्षः दृश्यतां न एति।

[25] Though the term *kṣepa* literally means deflection—and generally it is taken to refer to the deflection from the ecliptic (β)—in the present context it refers to the arc θ. Perhaps θ is being referred to as the arc corresponding to the *kṣepa*, at it is obtained from the *kṣepa* (β).

[26] The suffix '*u*' refers to *udaya*.

[27] The suffix '*m*' refers to the last contact.

mence close to the sunrise time, then it might not be visible at all. In case it were to be visible, even then, only the first contact might be visible and not the last contact. Similarly, if the lunar eclipse were to end close to the sunset time, then only the last contact might be visible and not the first contact. Here the criteria for the visibility of the first contact around sunrise time and the last contact around the sunset time are clearly stated.

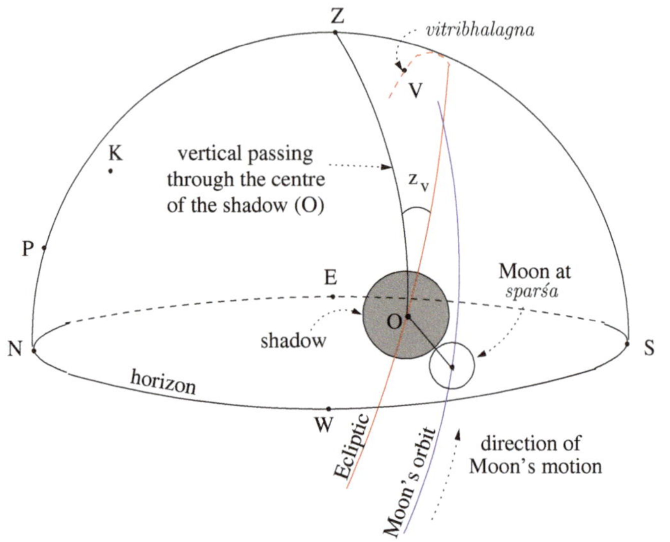

Fig. 4.9 Schematic sketch of the celestial sphere, when the first contact in a lunar eclipse is close to the sunrise time.

We explain the criteria for visibility with the help of Fig. 4.9. Here, P is the celestial pole and K the pole of the ecliptic. $NESW$ represents the horizon. The centre of the shadow O lies close to the west point on the horizon and the Moon is about to enter into the shadow. In other words, we have depicted the *sparśakāla*. The secondary to the ecliptic passing through K and Z meets the ecliptic at the point V, called the *vitribhalagna*, and its zenith distance is denoted by z_V. For the purposes of discussion, in the following O is taken to be on the horizon itself, as it is very close to it.

Vitribhalagna

When the centre of the Sun is on the horizon, then the centre of the shadow O will also be on the horizon. Now the zenith distance of the shadow $ZO = 90$. Since O is also a point on the ecliptic, it will be at 90 degrees from K, the pole of the ecliptic. Since $ZO = 90$ and $KO = 90$, the point O is the pole of the great circle passing through K and Z, which also passes through the *vitribhalagna* V. Thus, V is at 90 degrees from the orient ecliptic point—which is a point diametrically opposite to O—and is therefore called the *vitribhalagna*, the point which is at 90 degrees from the *lagna* (the orient ecliptic point). The zenith distance of the *vitribhalagna* (z_v) will be the same as $Z\hat{O}V$. That is, $ZV = Z\hat{O}V = z_v$.

Projection of the *samparkārdha* on the vertical

We need to find the projection of the sum of the semi-diameters of the Sun and the Moon called the *samparkārdha* (S_d) along the vertical to formulate the criteria for visibility. It is done in 3 steps: (i) finding the angle between S_d and the ecliptic; (ii) finding the angle between S_d and the vertical passing through the shadow; and (iii) finding the projection of S_d along the vertical. We explain these steps with the help of Fig. 4.10*a*. This figure is nothing but a section of the celestial sphere depicted in Fig. 4.9, redrawn with a different orientation. As in the previous figure, X represents the centre of the Moon. OX is the line joining the centres of the shadow and the Moon and is called the *samparkārdha*.

Step 1

Let OX make an angle θ with the ecliptic. Then from the triangle AOX we have

$$\sin\theta = \frac{AX}{OX} = \frac{\beta_t}{S_d}, \tag{4.55}$$

where β_t is the true latitude of the Moon and S_d the *samparkārdha*. In the text this relation is stated in the form:

$$R\sin\theta = \frac{\textit{sphuṭa-kṣepa} \times \textit{trijyā}}{\textit{samparkārdha}} = \frac{\beta_t \times R}{S_d}. \tag{4.56}$$

From (4.55), the arc θ is found, and it will be used in the succeeding steps.

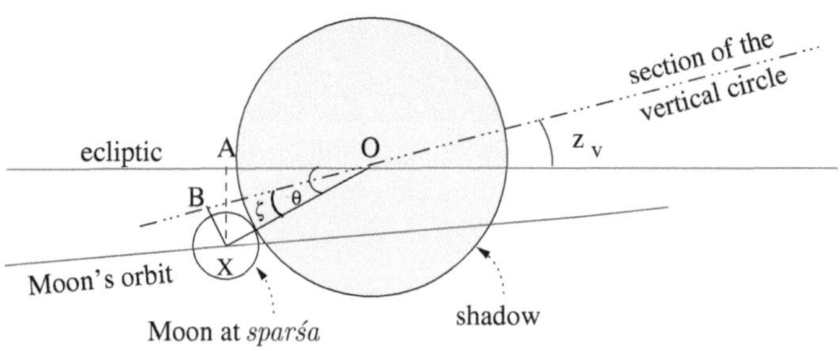

Fig. 4.10*a* Projection of the sum of the semi-diameters *samparkārdha* (OX) on the vertical circle during the first contact in a lunar eclipse when both the vertical and the *samparkārdha* are in the same direction.

Step 2

After determining θ, the angle made by S_d with the vertical passing through the centre of the shadow (ζ) is calculated. Since the angle made by the vertical with the ecliptic (z_v) is known, ζ is given by

$$\zeta = \theta \pm z_v. \tag{4.57}$$

In Fig. 4.10*a*, the relevant angle is $B\hat{O}X = \zeta$, and it is equal to $\theta - z_v$. In *Laghu-vivṛti* the choice of the sign to be employed is clearly stated.

तत्र लब्धस्य विक्षेपस्य केवलदृक्क्षेपस्य च चापयोः दिक्साम्ये विश्लेषं दिग्भेदे च योगं कुर्यात्।

If the directions are different, the sum of arcs of the *vikṣepa* ($R\sin\theta$) and the *dṛkkṣepa* ($R\sin z_v$) is to be found; if [the directions are] same, then their difference is to be calculated.

Here it is the directions of the vertical and the *samparkārdha* with the ecliptic which are referred to. In Fig. 4.10*a* both of them are shown to have the same direction, and hence $\zeta = \theta - z_v$. Sometimes it is possible that z_v is greater than θ. Hence, in general, ζ is given by

$$\zeta = |\theta - z_v|. \tag{4.58}$$

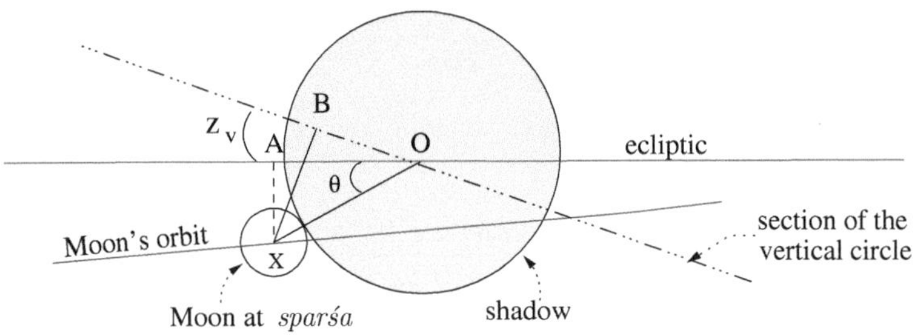

Fig. 4.10*b* Projection of the *samparkārdha* (OX) on the vertical circle during the first contact in a lunar eclipse when the vertical and the *samparkārdha* have different directions.

In Fig. 4.10*b* we have depicted the situation in which the vertical and the *samparkārdha* lie in opposite directions with respect to the ecliptic. Clearly, in this case, the arcs have to be added in order to find the projection of the *samparkārdha* along the vertical. That is,

$$\zeta = \theta + z_v. \tag{4.59}$$

Step 3

Having determined ζ, it now remains to find the projection of the *samparkārdha* along the vertical. In Figs. 4.10*a* and 4.10*b*, the projection is *BO*. From the triangle *BOX*,

$$\cos\zeta = \frac{BO}{OX} = \frac{BO}{S_d}$$

$$\text{or} \qquad BO = \cos\zeta \times S_d$$

$$= \frac{R\sin(90-\zeta)}{R} \times S_d. \qquad (4.60)$$

This is exactly the expression for the projection of the sum of semi-diameters of the shadow and the Moon along the vertical as described in the text.

Time measure of a segment along the vertical circle

For the time being, we ignore the Moon's parallax and find out the time measure of this segment *OB* along the vertical circle. This is the time taken by this segment to come down below the horizon.

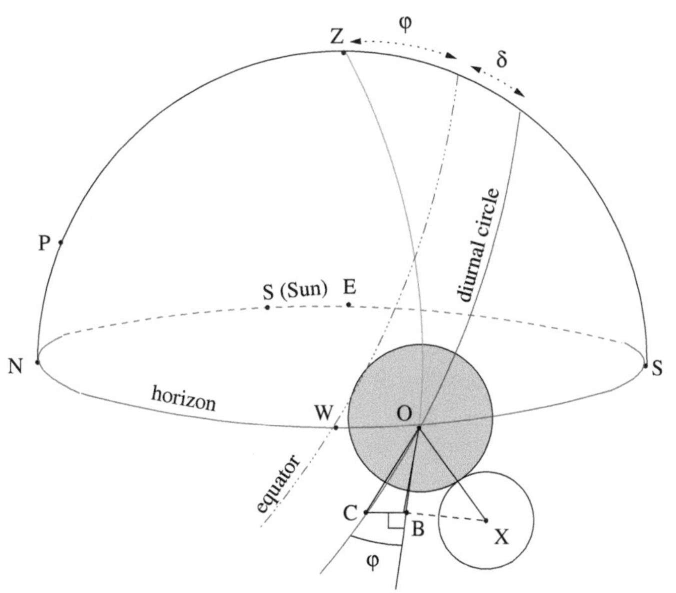

Fig. 4.11 Correspondence between the segment (*BO*) along the vertical circle and the segment (*CO*) along the diurnal circle.

In Fig. 4.11, OC is the segment of the diurnal circle, traced by the shadow, corresponding to the segment OB on the vertical circle. The angle between the diurnal circle and the horizon is the same as the angle between the equator and the horizon and it is equal to $90 - \phi$. Hence the angle $B\hat{O}C = \phi$.

Since the segment OB is small,[28] the triangle BOC may be considered as a planar triangle. Hence

$$\cos\phi = \frac{OB}{OC}$$
$$\text{or} \qquad OC = \frac{OB}{\cos\phi}. \tag{4.61}$$

The length of the arc in the equatorial circle, corresponding to the length of the arc OC in the diurnal circle, is given by

$$\frac{OC}{\cos\delta}, \tag{4.62}$$

where δ is the declination of the Sun. The time measure δt is precisely the segment on the equatorial circle corresponding to the segment OB on the vertical circle and is given by

$$\begin{aligned}\delta t &= \frac{OB}{\cos\phi\cos\delta} \\ &= \frac{OB \times R^2}{R\cos\phi \times R\cos\delta} \\ &= \frac{OB \times (trijy\bar{a})^2}{lambaka \times dyujy\bar{a}}.\end{aligned} \tag{4.63}$$

Ignoring the effect of parallax, δt is the time that must be available before sunrise for the first contact to be visible. This is what is stated in the text. The situation is schematically sketched in Fig. 4.12(a).

Effect of parallax on the visibility of the *sparśa*

The effect of parallax is to increase the zenith distance of an object. As a result the Moon and the shadow will suffer a downward shift along the vertical circles passing through them. Since the Moon is almost on the horizon, the shift suffered by the Moon may be taken to be its horizontal parallax. The same shift will be suffered by the shadow also. In Indian astronomy, the horizontal parallax (P) is taken to be *one-fifteenth* of the daily motion. Here it is specifically mentioned that this value should be taken for determining the criterion for the visibility of the first contact.

---

[28] At the most, OB can be *one* degree. This is so because OB is the projection of OX and $OX_{max} \approx 42' + 16' = 58'$ (less than 1 degree).

Thus we take

$$P = \frac{1}{15} \times \text{daily motion of the Moon.} \tag{4.64}$$

Converting P into time measure as we did earlier (4.63), and denoting this by δt_p, we have

$$\delta t_p = \frac{P \times R^2}{R\cos\phi \times R\cos\delta}. \tag{4.65}$$

This is the time that must be available before sunrise for the visibility of the first contact due to the effect of Moon's parallax alone.

Criterion for the visibility of *sparśa*

The criterion for the visibility of the first contact can be easily derived from (4.63) and (4.65). Denoting the sum of these two time measures by Δt_s we have

$$\Delta t_s = \frac{P' \times R^2}{R\cos\phi \times R\cos\delta}, \tag{4.66}$$

where $P' = P + OB$. We have already determined[29] the *sparśakāla* (t_b), the instant of the beginning of the eclipse, by iteration. From this instant, we calculate the time that is remaining in the night till the sunrise of the next day. We denote this time by Δt_u, which is given by

$$\Delta t_u = \text{sunrise time} - t_b. \tag{4.67}$$

Now, the criteria for the visibility of the first contact as shown in Fig. 4.12(b) is clearly

$$\Delta t_u \geq \Delta t_s,$$

$$\text{or} \qquad \Delta t_s \leq \Delta t_u. \tag{4.68}$$

If $\Delta t_s = \Delta t_u$, the apparent position of the Moon will be on the horizon at the *sparśakalā* and the first contact is visible. If $\Delta t_s > \Delta t_u$, the apparent centre of the lunar disc C will have already descended below the horizon, and the first contact will not be visible. The duration Δt_u is referred as the *kṣapāśeṣa*. The term *kṣapā* means night and *śeṣa* means the remainder. Hence the term *kṣapāśeṣa* means the time remaining in the night before sunrise. Hence the condition for the visibility of first contact is

$$\textit{kṣapāśeṣa} \geq \frac{P' \times R^2}{R\cos\phi \times R\cos\delta}. \tag{4.69}$$

Note:

It may so happen that $\zeta = \theta + z_v$ may be greater than 90 degrees. In such a case, the projection of the *samparkārdha* along the vertical (OB) will be upward along the vertical towards OD and above the horizon, as shown in Fig. 4.13. But the

---

[29] See Chapter 4, verse 27.

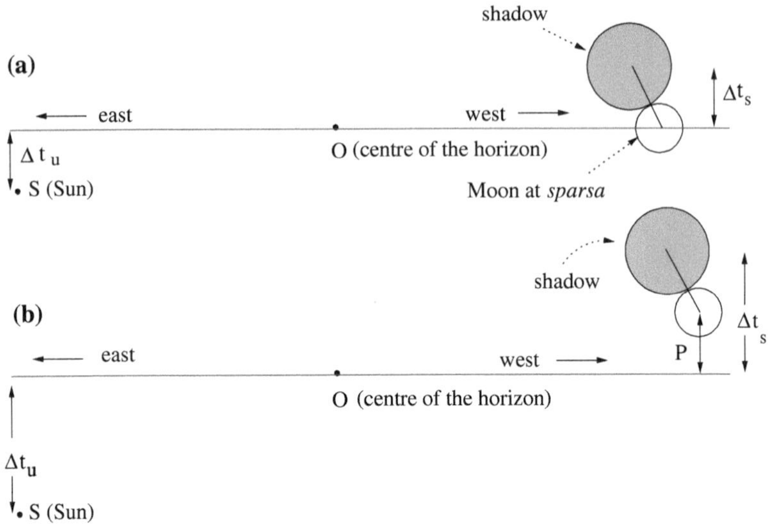

Fig. 4.12 Criterion for the visibility of first contact: (a) ignoring the effect of parallax; (b) including the effect of parallax.

parallactic shift is downwards along the vertical and towards OC. Since these two are in opposite directions, we need to take $P' = P - OB$ in (4.66).

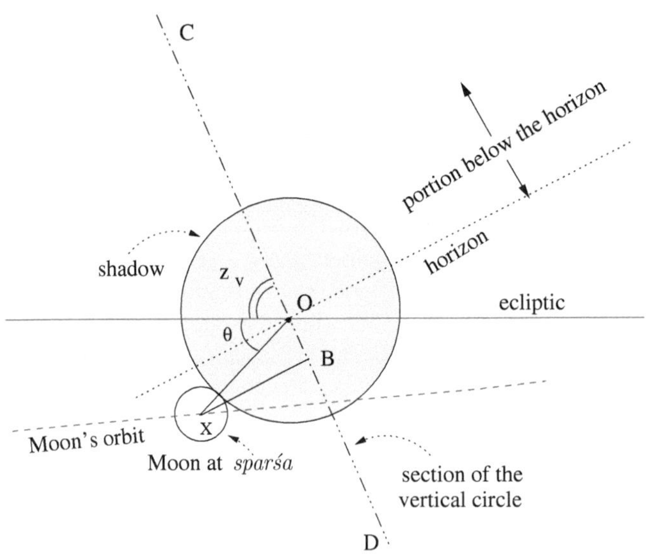

Fig. 4.13 Projection of the *samparkārdha* along the vertical when $\zeta = \theta + z_v$ is greater than 90 degrees.

Criterion for the visibility of the *mokṣa*

The criterion for the visibility of last contact can be arrived at in a similar manner. The only difference is that instead of considering the instant of first contact and the sunrise time, we need to work out the details with the instant of last contact and the sunset time (*astamanakāla*). The instant of last contact (t_e) has already been determined by iteration. Let Δt_a be the time interval between t_e and the sunset time. That is,

$$\Delta t_a = t_e - \text{sunset time}. \tag{4.70}$$

Now, the criterion for the visibility of the last contact, as shown in Fig. 4.14, clearly turns out to be

$$\Delta t_a \geq \Delta t_m, \tag{4.71}$$

where the expression for Δt_m is the same as that for Δ_s given by (4.66). Here also if

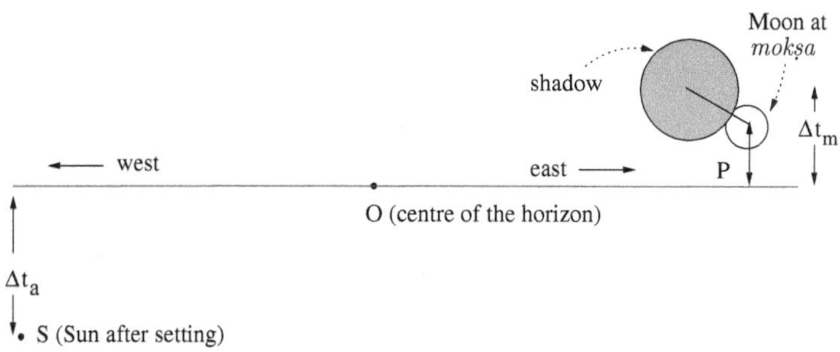

Fig. 4.14 Criterion for the visibility of last contact considering the effect of parallax.

$\Delta t_m = \Delta t_a$, the apparent centre of the Moon will be on the horizon at the last contact and it will be visible. If $\Delta t_m > \Delta t_a$, this point will have already descended below the horizon, and the last contact would not be visible.

४.१८ स्फुटबिम्बान्तरम्

4.18 Accurate distance of separation between the orbs

यद्वा तत्कालसूर्येन्दुस्फुटपातादिकं नयेत् ।
तत्रेन्दुधरणिच्छायास्फुटान्तरकलाहताः ॥ ३७ ॥
द्वितीयगतिभेदेन पूर्वभुक्त्यन्तरोद्धृता ।
स्फुटान्तरमिह ग्राह्यं तस्य क्षिप्तेश्च वर्गयोः ॥ ३८ ॥
योगे तच्छरभेदस्य वर्गं युक्त्वा पदीकृते ।
बिम्बान्तरं सुसूक्ष्मं स्यात् इष्टग्रासे तथैव च ॥ ३९ ॥
स्फुटान्तरस्य वर्गात् तु कृत्स्नव्यासहृतः शरः ।
क्षेपस्याप्येवमेव स्यात् स्वेषु वर्गयुतान् मुहुः ॥ ४० ॥

इष्टवर्गत्रिभागोनचापवर्गात् तथापि वा [30] ।
सम्पर्कार्धात् विशुद्धेऽस्मिन् ग्रासो बिम्बान्तरे सदा ॥ ४१ ॥

yadvā tatkālasūryendusphuṭapātādikaṃ nayet |
tatrendudharaṇicchāyāsphuṭāntarakalāhatāḥ || 37 ||
dvitīyagatibhedena pūrvabhuktyantaroddhṛtā |
sphuṭāntaramiha grāhyaṃ tasya kṣipteśca vargayoḥ || 38 ||
yoge taccharabhedasya vargaṃ yuktvā padīkṛte |
bimbāntaraṃ susūkṣmaṃ syāt iṣṭagrāse tathaiva ca || 39 ||
sphuṭāntarasya vargāt tu kṛtsnavyāsahṛtaḥ śaraḥ |
kṣepasyāpyevameva syāt sveṣu vargayutān muhuḥ || 40 ||
iṣṭavargatribhāgonacāpavargāt tathāpi vā |
samparkārdhāt viśuddhe'smin grāso bimbāntare sadā || 41 ||

Or alternately, find the longitudes of the Sun, the Moon and the node at a desired time. At that instant the difference between the longitudes of the Moon and the shadow is multiplied by the second difference in the daily motion (*dvitīya-sphuṭa-bhukti*) and divided by the first difference in the daily motion (*prathama-sphuṭa-bhukti*). The result should be taken as the true difference between the longitudes. To the sum of the squares of this (difference betwen longitudes) and the latitude, the square of the *śarabedha* is added. The square root of the result gives a more accurate value of the distance between the two objects [i.e. the centres of the Moon and the shadow]. This is the distance of separation at any desired obscuration.

The *śara* is obtained by dividing the square of the distance of separation by the full diameter. Even for the latitude, the *śara* is obtained in a similar manner. This is added to the square of the arc and the process is repeated.

Alternatively, by subtracting one-third of the square of the *iṣṭa* (*śara*)[31] from the square of the arc (*cāpa*) [and dividing by the diameter, a better approximation to the] *śara* may be obtained. The distance of separation thus obtained, when subtracted from the sum of the semi-diameters, always gives the portion obscured.

The traditional method of finding the distance of separation between the Moon and the shadow is approximate. This is due to the fact that the spherical triangle involved in the computation is taken to be a planar triangle, which, of course, is reasonable as the spherical triangle is small. In the above verses, an exact formula for the distance of separation is given. In this process a formula for the versed sine ($R(1-\cos\theta)$) is used, which is correct to $O(\theta^3)$. This is fairly accurate, as θ is small.

Let λ_s, λ_m and λ_n be the longitudes of the Sun, the Moon and the node at any desired instant during the eclipse. The longitude of the shadow (*chāyā*) λ_c is

$$\lambda_c = \lambda_s + 180. \tag{4.72}$$

Let $\Delta\lambda$ be the difference in longitude between the *chāyā* (shadow) and the Moon at any instant. Obviously

$$\Delta\lambda = \lambda_c - \lambda_m. \tag{4.73}$$

Here a procedure for obtaining a more accurate value of the difference in longitude is suggested which is also used later for computing the distance of separation between

[30] The prose order is a little difficult as it needs *anuvṛtti* from previous lines. It may be written as: अपि (च) इष्टवर्गत्रिभागोनचापवर्गात् तथा (पूर्वोक्तप्रकारेण) कृत्स्नव्यासहृतो वा शरः (भवति)।

[31] Here the word इष्ट refers to the शर, as that is the desired quantity.

the *chāyā* and the Moon. The new value $\Delta\lambda'$ is obtained from the old value $\Delta\lambda$ using the relation

$$\Delta\lambda' = \Delta\lambda \times \frac{\textit{dvitīya-sphuṭa-bhukti}}{\textit{prathama-sphuṭa-bhukti}}. \tag{4.74}$$

It is presumed that $\Delta\lambda'$ represents the true difference in longitudes more accurately as it involves the true difference in daily motion or the *dvitīya-sphuṭa-bhukti*, which incorporates the second correction to the Moon, namely the correction due to 'evection'. In Fig. 4.15(a), OX represents the distance of separation (D), the *bimbāntara*, between the *chāyā* and the Moon. As per the traditional method, the triangle AOX is considered to be a planar triangle while finding OX:

$$OX = \sqrt{OA^2 + AX^2}. \tag{4.75}$$

In the above relation, OA is the difference in the longitude of the *chāyā* and the Moon ($\Delta\lambda'$), and AX is the true latitude of the Moon (β_t). Though the values of both are known, the above relation is an approximate one because OA and AX are segments of great circles and not parts of a planar triangle.

In the above set of verses the text presents a different approach which does not involve such an approximation. Here we consider a different triangle OQX to determine OX. From X, $XQ = R\sin\beta_t$, is drawn perpendicular to the plane of the ecliptic, where β_t is the latitude. As OQ is in the plane of ecliptic, OQX forms a right-angled triangle with OX as the hypotenuse. Hence

$$OX = \sqrt{OQ^2 + QX^2}. \tag{4.76}$$

But OQ is not yet known and is found using the triangle OPQ (see Fig. 4.15(b)) which is right-angled at P. The point P is obtained by drawing a perpendicular from O to CA. Considering the triangle OPQ, we have

$$OQ^2 = OP^2 + PQ^2. \tag{4.77}$$

In the RHS of the above equation, $OP = R\sin\Delta\lambda'$ is known, where $\Delta\lambda'$ is the *sphuṭāntara* or the difference in longitudes. PQ is termed the '*śarabheda*' (difference in versines) in verse 39. The *śara* of an angle θ is $R(1-\cos\theta)$. In Fig. 4.15(b) PA is the *śara*[32] corresponding to the angle $O\hat{C}A = \Delta\lambda'$. Similarly AQ is the *śara* corresponding to the angle $A\hat{C}X = \beta_t$ (see Fig. 4.15(c)). Obviously $PQ = PA - AQ$. Since the angles $\Delta\lambda'$ and β_t are known,

$$\begin{aligned} PA &= R(1-\cos\Delta\lambda') \\ \text{and} \qquad AQ &= R(1-\cos\beta_t). \end{aligned} \tag{4.78}$$

Therefore

[32] Literally, the term *śara* means the 'arrow'. In the figure the section $OPBAO$ looks like a bow, with OB as the string. Since PA looks like an arrow, it is called the *śara*.

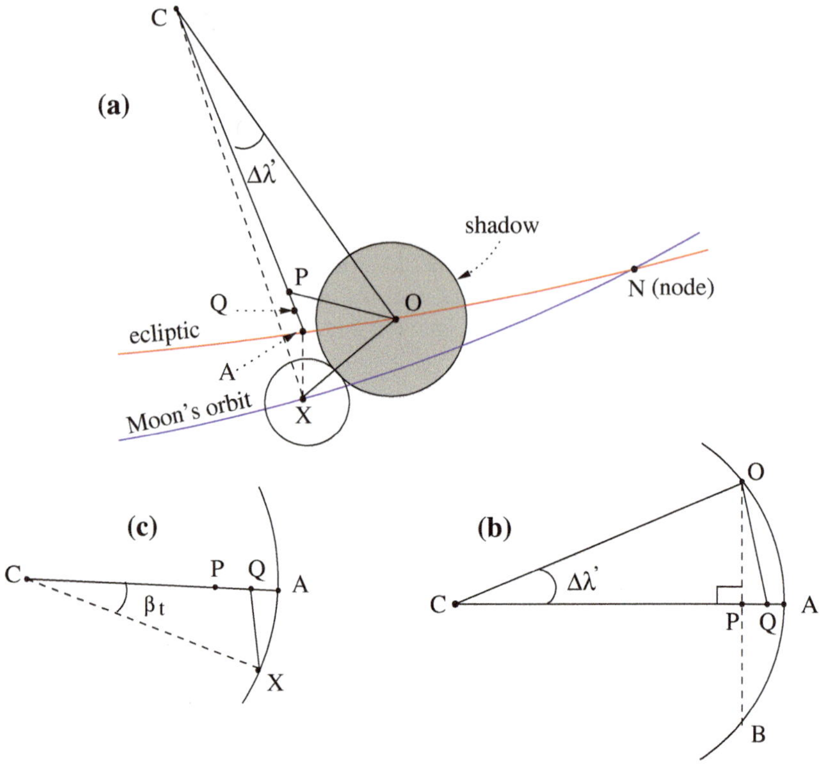

Fig. 4.15 Formula for finding the distance of separation between the centres of the *chāyā* and the Moon.

$$\begin{aligned} PQ &= R(1-\cos\Delta\lambda') - R(1-\cos\beta_t) \\ &= R\cos\beta_t - R\cos\Delta\lambda', \end{aligned} \tag{4.79}$$

is the *śarabheda*. Using (4.77) in (4.76) we have

$$\begin{aligned} OX &= \sqrt{OP^2+QX^2+PQ^2} \\ \text{or} \quad D &= \sqrt{(R\sin\Delta\lambda')^2+(R\sin\beta_t)^2+(R\cos\beta_t-R\cos\Delta\lambda')^2}. \end{aligned} \tag{4.80}$$

This is exact and is the same as the expression given in the text:

$$bimb\bar{a}ntara = \sqrt{sphu\d{t}\bar{a}ntara^2 + vik\d{s}epa^2 + \acute{s}arabheda^2}. \tag{4.81}$$

Formula for finding the *śara*

Now for small θ,

$$\acute{s}ara = R(1-\cos\theta) \approx \frac{R\theta^2}{2}. \tag{4.82}$$

This is correct to $O(\theta^3)$. Here the *śara* is taken to be $\frac{(R\theta)^2}{2R}$ in the first approximation. To calculate the separation between the discs in (4.81), we need to calculate two *śaras*: the *śara* related to the *sphuṭāntara* and the *śara* related to the *vikṣepa*. These two are obtained by replacing θ by $\Delta\lambda'$ and β_t in the above equation. In verse 40 the *śara* related to the *sphuṭāntara* is stated to be

$$\acute{s}ara = \frac{sphu\d{t}\bar{a}ntara^2}{vy\bar{a}sa}, \tag{4.83}$$

which is the first approximation. It is then mentioned that the same rule may be applied to the *vikṣepa*:

क्षेपस्याप्येवमेव स्यात्।
Even for the *kṣepa* it is the same way.

This approximate expression given for the *śara* in (4.83) is sought to be improved upon by employing any of the two following methods.

Method 1:

This is an iterative process. The first approximation to the *śara* is given by

$$\acute{s}ara_0 = \frac{(R\theta)^2}{2R}. \tag{4.84}$$

The second approximation is given to be

$$\acute{s}ara_1 = \frac{(R\theta)^2 + (\acute{s}ara_0)^2}{2R}. \tag{4.85}$$

In general,

$$\acute{s}ara_{n+1} = \frac{(R\theta)^2 + (\acute{s}ara_n)^2}{2R}. \tag{4.86}$$

The iterations have to be carried out till we get consecutive concordant values. This is what has been indicated by the use of the word '*muhuḥ*' in verse 40. The whole iterative procedure has been cryptically coded in one-quarter of the verse:

स्वेषु वर्गयुतां मुहुः ।[33]

The rationale behind this iterative process is not clear to us.

[33] Perhaps this has to be understood with *anuṣaṅgas* as

स्वेषु = शरेषु, पूर्वपूर्वलब्धशरानिति यावत्। वर्गयुतां = चापवर्गयुतां कृत्वा, कृत्स्नव्यासेन हृतः शरो भवतीत्यनुषज्यते। एवं यावदविशेषं कर्तव्यम् इति मुहुः इति पदेन सूचितं भवति।

Method 2:

Here again we start with the first approximation to the *śara*:

$$\acute{s}ara_0 = \frac{(R\theta)^2}{2R}. \tag{4.87}$$

A better approximation of the *śara* is given to be

$$\begin{aligned}
\acute{s}ara &= \frac{c\bar{a}pa^2 - \frac{1}{3}(\acute{s}ara_0)^2}{vy\bar{a}sa} \\
&= \frac{1}{2R}\left[(R\theta)^2 - \frac{1}{3}\left(\frac{(R\theta)^2}{2R}\right)^2\right] \\
&= R\left(\frac{1}{2!}\theta^2 - \frac{1}{4!}\theta^4\right).
\end{aligned} \tag{4.88}$$

Since

$$(1-\cos\theta) = \frac{\theta^2}{2!} - \frac{\theta^4}{4!} + O(\theta^6), \tag{4.89}$$

it is obvious that this approximation is correct up to fifth-order.

There is an interesting discussion on the computation of the *śara* in the form of question and answer in *Yukti-dīpikā*. The student asks:

सूक्ष्मं बिम्बान्तरं कर्तुं गृह्यते ज्याशरौ यतः ॥
नन्वत्र नीयते बाणः कथङ्कारं स्फुटान्तरात् ।
समस्तज्याकृतेर्बाणः कृत्स्नव्यासोद्धृतो यतः ॥ [34]

Since, in order to find a more accurate value of the separation between the discs, the *śaras* corresponding to the two angles were calculated; now, how is it that the *śara* is calculated from the *sphuṭāntara*? Is it not true that the *śara* has to be obtained by dividing the square of the Rsine (*samastajyākṛti*) by the diameter?

The teacher replies:

सत्यं, किन्त्विह कृत्स्नज्या प्रायः स्वधनुषा समा ।
ग्राह्यग्राहकयोर्भेदलिप्तानां स्वल्पतावशात् ॥[35]

True. But since the arc is small we take the Rsine of the arc to be the arc itself. The arc is small because the distance of separation in minutes, between the eclipsed and the eclipser, is small.

We know that (see Fig. 4.16*a*)

$$\begin{aligned}
(1-\cos\theta) &= 2\sin^2\frac{\theta}{2} \\
R(1-\cos\theta) &= 2R\sin^2\frac{\theta}{2}
\end{aligned}$$

[34] {TS 1977}, p. 273.

[35] {TS 1977}, p. 273.

$$= \frac{2\left(R\sin\frac{\theta}{2}\right)^2}{R}$$
$$= \frac{\left(2R\sin\frac{\theta}{2}\right)^2}{2R}$$
$$\simeq \frac{(R\theta)^2}{2R} \quad \text{(since } \theta \text{ is small).} \tag{4.90}$$

In his reply to the student's query the teacher essentially states that $2R\sin\frac{\theta}{2} \simeq R\theta$ for a small θ.

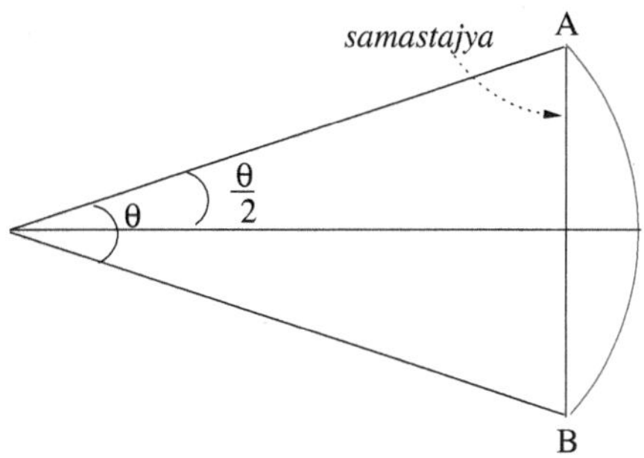

Fig. 4.16*a* The Rsine of an arc when the arc itself is small.

Portion of the Moon obscured

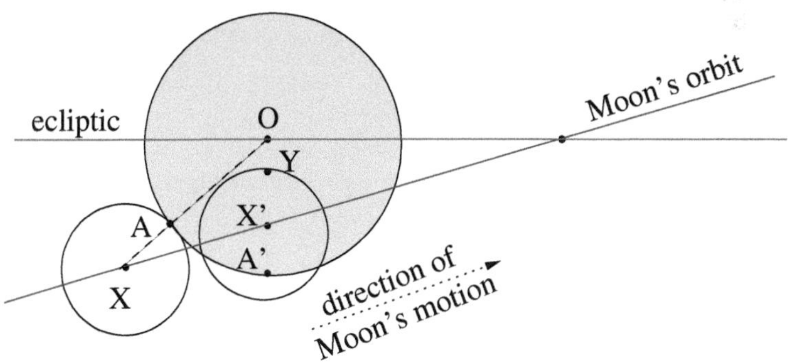

Fig. 4.16*b* Finding the portion of the Moon obscured at any instant of time during the eclipse.

The portion of the Moon obscured at any instant of time during the eclipse is called *grāsa*. In the second half of verse 41, it is given to be

$$grāsa = samparkārdha - bimbāntara. \tag{4.91}$$

The rationale behind this may be understood with the help of Fig. 4.16*b*. Here O is the centre of the shadow. X and X' are the centres of the Moon at the beginning of the eclipse and at a time t later. At the beginning of the eclipse,

$$samparkārdha = OA + AX. \tag{4.92}$$

This is the separation between the discs at the first contact. Hence the *grāsa* = 0. At a time t later, when the Moon's centre is at X',

$$samparkārdha = OA' + X'Y. \tag{4.93}$$

At this instant,

$$bimbāntara = OX' = OY + YX'. \tag{4.94}$$

Therefore

$$\begin{aligned} grāsa &= (OA' + X'Y) - (OY + YX') \\ &= OA' - OY \\ &= A'Y. \end{aligned} \tag{4.95}$$

From the figure it is clear that $A'Y$ is indeed the portion of the Moon which is obscured, thereby showing that the expression given by (4.91) is exact.

४.१९ ग्रहणस्य अदृश्यत्वावस्थितिः

4.19 State of the eclipse being invisible

स्वच्छत्वात् षोडशांशोऽपि ग्रस्तश्चन्द्रो न दृश्यते ।
लिप्तात्रयमपि ग्रस्तं तीक्ष्णत्वान्न विवस्वतः ॥ ४२ ॥

svacchatvāt ṣoḍaśāṃśo'pi grastascandro na dṛśyate |
liptātrayamapi grastaṃ tīkṣṇatvānna vivasvataḥ || 42 ||

Because of its brightness, even if *one-sixteenth* of the Moon is obscured it will not be noticed. In the case of the Sun, even if the measure of obscuration is up to 3′, it may not be noticeable because of its sharpness (brilliance).

Here it is stated that a lunar eclipse would be visible only if more than one-sixteenth of the Moon is obscured. Similarly it is said that a solar eclipse will be noticeable only if more than 3′ of the solar disc—which is about one-tenth of the solar disc—is obscured. These seem to be empirical criteria.

४.२० परमग्रासस्य स्फुटपर्वान्ततश्चलनम्

4.20 Shift of the instant of maximum obscuration from the instant of opposition

अल्पश्चेत् परमग्रासः चलेत् स्थितिदलेऽधिके ।
तस्माद्बिम्बान्तरेणैव ग्रासोऽन्वेष्यो महानिह ॥ ४३ ॥

alpaścet paramagrāsaḥ calet sthitidale'dhike |
tasmādbimbāntareṇaiva grāso'nveṣyo mahāniha || 43 ||

If the measure of obscuration is small, then the instant of maximum obscuration (*paramagrāsakāla*) will occur in the larger [part of the two] half-duration[s]. Therefore, the amount of maximum obscuration has to be calculated only from the separation between the discs [as determined earlier at the *paramagrāsakāla*].

A brief mention of maximum obscuration (*paramagrāsa*) was made in section 4.15 while discussing the instants of the first and the last contacts. An expression for the instant of maximum obscuration (*paramagrāsakāla*) is given in (4.41). Here it is implicitly stated that the *paramagrāsakāla* will be different from the instant of opposition at which the longitudes of the shadow and the Moon are equal. It is further mentioned that this instant occurs during the larger of the two half-durations.

In Fig. 4.17, Y and X' represent the centre of the Moon at the instant of opposition and a little later when the obscuration is maximum. It is evident from the figure that $OX' < OY$. Hence, the *paramagrāsakāla*, which depends only upon the distance between the centres of the shadow and the lunar disc, could be different from the instant of opposition.

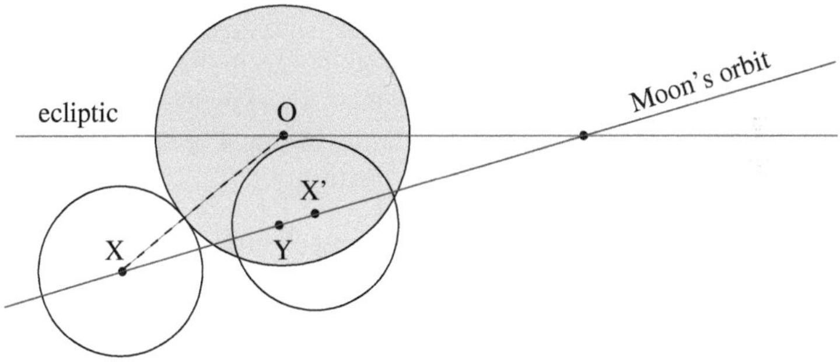

Fig. 4.17 Schematic sketch to illustrate that the instant of opposition could be different from the *paramagrāsakāla*.

४.२१ अक्षवलनं आयनवलनं च

4.21 Deflection due to latitude and that due to declination

नतज्याक्षज्ययोर्घातः त्रिज्याप्तस्तस्य कार्मुकम् ।
तदंशाः सौम्ययाम्यास्ते पूर्वापरकपालयोः ॥ ४४ ॥
राशित्रययुताद् ग्राह्यात् क्रान्त्यंशैर्दिक्समैर्युतात् ।
भेदेऽन्तराद्गुणस्तेन[36] हत्वा बिम्बान्तरं हरेत् ॥ ४५ ॥
त्रिज्यया वलनं स्पष्टं वृत्ते बिम्बान्तरोद्भवे ।

natajyākṣajyayorghātaḥ trijyāptastasyakārmukam |
tadaṃśāḥ saumyayāmyāste pūrvāparakapālayoḥ || 44 ||
rāśitrayayutād grāhyāt krāntyaṃśairdiksamairyutāt |
bhede'ntarādguṇastena hatvā bimbāntaraṃ haret || 45 ||
trijyayā valanaṃ spaṣṭaṃ vṛtte bimbāntarodbhave |

The arc of the product of the Rsines of the hour angle and latitude divided by the *trijyā* [is the *akṣavalana*]. Its value in degrees is taken to be south or north depending upon whether it lies in the eastern or western half of the celestial sphere.

From the longitude of the eclipsed object increased by 90 degrees, and from the maximum inclination of the ecliptic [the *āyana-valana* must be determined]. The Rsine is calculated from the sum when the directions are the same, and from the difference when they are different. Having multiplied the separation between the discs by this, divide by the *trijyā*. The result is the true *valana* in the circle whose radius is equal to the separation between the discs.

The term *valana* refers to the angle between the line joining the Moon and the centre of the shadow, and the local vertical (small) circle, which is parallel to the prime vertical. This is denoted by the angle ψ in Fig. 4.19. If the Moon's latitude is ignored, this is essentially the angle between the ecliptic and the prime vertical. This consists of two parts, namely the *akṣavalana* and the *āyanavalana*. The *akṣavalana* is the angle between the diurnal circle and the prime vertical which is also the angle between the secondaries to them, whereas the *āyanavalana* is the angle between the diurnal circle and the ecliptic. In Fig. 4.18, ξ is the *akṣavalana* and θ the *āyanavalana*. Then, according to the text,

$$R\sin\xi = R\sin\phi.\frac{R\sin H}{R} \tag{4.96}$$

$$\text{and} \qquad R\sin\theta = R\sin(90+\lambda)\sin\varepsilon. \tag{4.97}$$

We first consider the *akṣavalana*.

Expression for the *akṣavalana*

In Fig. 4.18, K is the pole of the ecliptic, P the north celestial pole, N the north point and M the Moon, which is taken to be situated on the ecliptic, as the Moon's latitude

[36] The prose order is: दिक्साम्ये (अक्षायनवलनयोः) युतात्, दिग्भेदे तयोः अन्तरात् (यल्लभ्यते), तस्य गुणं कृता = (ज्यामानीय), तेन बिम्बान्तरं हत्वा ··· ।

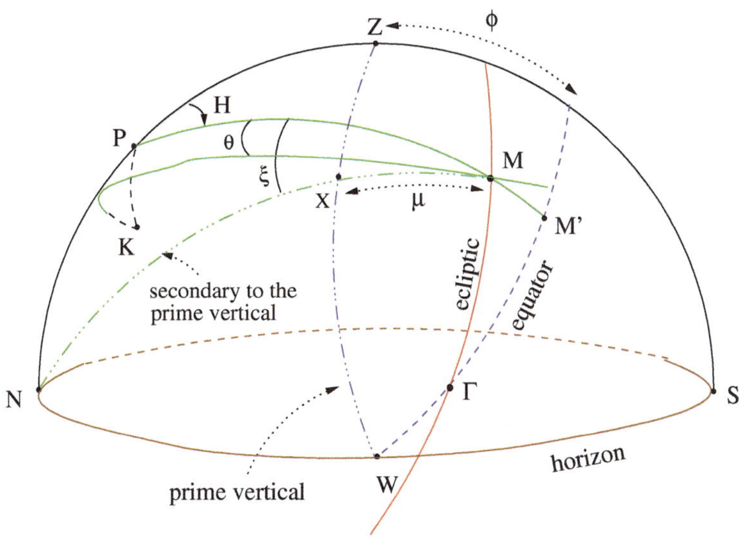

Fig. 4.18 Determination of the *akṣavalana* and the *āyanavalana*. The ecliptic and its secondary through the centre of the *chāyā* (shadow) are denoted by solid lines, whereas the equator and the meridian passing through the Moon are denoted by dashed lines.

is ignored. The arc $NP = \phi$ is the latitude of the observer and $N\hat{P}M = 180 - H$, where H is the hour angle of the Moon. $PM = 90 - \delta$ and $KM = 90$.

Consider a secondary to the prime vertical passing through the north point and the Moon. X is the point of intersection of this secondary with the prime vertical. The *akṣavalana*, which is the angle between the diurnal circle and the prime vertical, is also the angle between the secondaries to the prime vertical and the equator, which is denoted by ξ in the figure. Considering the spherical triangle NPM and applying the sine formula, we have

$$\frac{\sin\xi}{\sin NP} = \frac{\sin(180 - H)}{\sin NM} = \frac{\sin P\hat{N}M}{\sin PM}. \tag{4.98}$$

In the above equation, $NP = \phi$, $PM = 90 - \delta$ and $NM = 90 + \mu$, where μ is a part of the arc lying along the secondary to the prime vertical drawn from N to M. Since N is the pole of the vertical circle ZXW, $P\hat{N}M = ZX = z$ is the zenith distance of X. Hence the above equation becomes

$$\sin\xi = \frac{\sin\phi\sin H}{\cos\mu} = \frac{\sin\phi\sin z}{\cos\delta}. \tag{4.99}$$

This differs from the expression given in the text (4.96) by a factor of $\cos\mu$ in the denominator.

Expression for the *āyanavalana*

The *āyanavalana*, which is the angle between the ecliptic and the diurnal circle, is also the angle between the secondaries to the equator and the ecliptic passing through M. In Fig. 4.18, $K\hat{M}P = \theta$ is the *āyanavalana*. In the spherical triangle KPM,

$$K\hat{P}M = K\hat{P}\Gamma + \Gamma\hat{P}M' = 90 + \alpha, \tag{4.100}$$

where α is the R.A. of M. $K\hat{P}\Gamma = P\hat{K}\Gamma = 90$, because Γ is the pole of the great circle passing through K and P. Also,

$$P\hat{K}M = P\hat{K}\Gamma - \Gamma\hat{K}M = 90 - \lambda. \tag{4.101}$$

Applying sine formula to the spherical triangle KPM, we have,

$$\frac{\sin\theta}{\sin\varepsilon} = \frac{\sin K\hat{P}M}{\sin KM} = \frac{\sin P\hat{K}M}{\sin PM}$$

$$\text{or} \quad \frac{\sin\theta}{\sin\varepsilon} = \frac{\sin(90+\alpha)}{\sin(90-\beta)} = \frac{\sin(90-\lambda)}{\sin(90-\delta)}. \tag{4.102}$$

Hence

$$\sin\theta = \frac{\sin\varepsilon\sin(90+\alpha)}{\cos\beta} = \frac{\sin\varepsilon\sin(90-\lambda)}{\cos\delta} = \sin\varepsilon\,\frac{\sin(90+\lambda)}{\cos\delta}. \tag{4.103}$$

This differs from the formula given in the text (4.97) by the factor $\cos\delta$ in the denominator.

Application of the *valana*

Having determined the *akṣavalana* and the *āyanavalana*, we need to find the total *valana* (ψ). It is given to be

$$\psi = \xi - \theta \qquad (\text{if } K \text{ between } P \,\&\, N), \tag{4.104}$$

$$\psi = \xi + \theta \qquad (\text{if } P \text{ between } K \,\&\, N). \tag{4.105}$$

In Fig. 4.18, where K is between P and N, $\psi = \xi - \theta$ is the angle between KM and NM, which are the secondaries to the ecliptic and prime vertical passing through M. The dashed line AB in Fig. 4.19(a) represents the line of motion of the shadow. The line of motion of the Moon will be at a distance β from it. OF is the *bimbāntara* (S), or the distance of separation between the centres of the *chāyā* and the Moon, which varies during the eclipse. The total *valana* (ψ) obtained above is in the measure of circle whose radius is the *trijyā*. We shall now transform this into the measure of a circle whose radius is equal to the *bimbāntara* (the distance of separation between the *chāyā* and the Moon). The local east–west line passing through O is depicted in the figure.

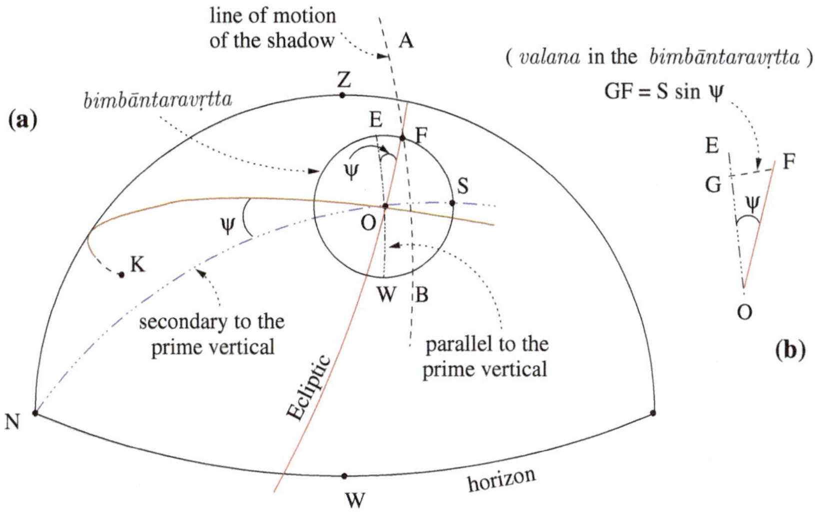

Fig. 4.19 Application of the sum of or the difference between the *akṣavalana* and the *āyanavalana*.

We draw FG perpendicular to the local east–west line OE as indicated in Fig. 4.19(b). FG, which is the distance between F and the local east–west line, is the true *valana* (inclination) in the circle whose radius is the separation between the discs;

$$\begin{aligned} FG &= \sin\psi \times OF \\ &= R\sin\psi \times \frac{OF}{R} \\ &= R\sin\psi \times \frac{bimb\bar{a}ntara}{trijy\bar{a}}. \end{aligned} \tag{4.106}$$

This is what is explained in verse 46a of the text.

Note:

In the discussion above, we have pointed out the errors in the expression for the *akṣavalana* and the *āyanavalana*. In fact, the same errors are to be found in *Yuktibhāṣā* also. Moreover, the latitude of Moon is neglected here. In *Yuktibhāṣā*, however, the effect of latitude is included through another term called the '*vikṣepavalana*'.

४.२२ ग्रहणपरिलेखनम्

4.22 Graphical representation of the eclipse

बिम्बान्तरसमेनादौ लिखेत् सूत्रेण मण्डलम् ॥ ४६ ॥
बिम्बान्तरं तु सम्पर्कदलं स्यात् स्पर्शमोक्षयोः ।
शलाकाङ्कितमध्या या वलनद्विगुणायता ॥ ४७ ॥
प्राग्रेखागतमध्याङ्कनेमिस्पृष्टोभयाग्रया ।
वलनं नीयतां स्पष्टं यथादिशमथानया ॥ ४८ ॥
प्रत्यग्रेखास्तमध्याङ्कन्यस्तया व्यस्तमेव च ।
रेखां पूर्वापरां कुर्यात् तद्वशादपि चेतराम् ॥ ४९ ॥
छायामार्गोऽनयोः पूर्वः विक्षेपान्तरिता विधोः ।
रेखाविक्षेपदिक्स्थातो मार्गौ तत्कालजौ तु तौ ॥ ५० ॥
लिप्ताव्यासदलेनेन्दोः स्वमार्गे परिधिं[37] लिखेत् ।
पौर्णमास्यां प्रतीच्यां तु प्रागेव प्रतिपद्यपि ॥ ५१ ॥
वृत्तमध्ये तमोबिम्बं स्वमार्गे सर्वदा लिखेत् ।
तमोबिम्बाद् बहिर्भूतं दृश्यमिन्दौ न तद्गतम् ।
हनुपार्श्वादिभेदादि दृश्यतामत्र सुस्फुटम् ॥ ५२ ॥

bimbāntarasamenādau likhet sūtreṇamaṇḍalam || 46 ||
bimbāntaraṃ tu samparkadalaṃ syāt sparśamokṣayoḥ |
śalākāṅkitamadhyā yā valanadviguṇāyatā || 47 ||
prāgrekhāgatamadhyāṅkanemispṛṣṭobhayāgrayā |
valanaṃ nīyatāṃ spaṣṭaṃ yathādiśamathānayā || 48 ||
pratyagrekhāstamadhyāṅkanyastayā vyastameva ca |
rekhāṃ pūrvāparāṃ kuryāt tadvaśādapi cetarām || 49 ||
chāyāmārgo'nayoḥ pūrvaḥ vikṣepāntaritā vidhoḥ |
rekhāvikṣepadiksthāto mārgau tatkālajau tu tau || 50 ||
liptāvyāsadalenendoḥ svamārge paridhiṃ likhet |
paurṇamāsyāṃ pratīcyāṃ tu prāgeva pratipadyapi || 51 ||
vṛttamadhye tamobimbaṃ svamārge sarvadā likhet |
tamobimbād bahirbhūtaṃ dṛśyamindau na tadgatam |
hanupārśvādibhedādi dṛśyatāmatra susphuṭam || 52 ||

In the first place draw a circle of radius equal to the separation between the discs with the help of a thread. At the time of the first contact and last contact, the separation between the discs would be equal to the sum of the semi-diameters [of the *chāyā* and the Moon].

With a marking at the middle take a stick (*śalākā*) whose length is equal to twice that of the *valana*. Place the stick eastwards such that the centre of the stick lies on the east portion of the line (*OE*), and its ends touch the circumference of the circle. By doing so let the *valana* be brought out clearly as per the direction. Then, with the same stick placed westwards, with its centre on the west line (*OW*), let the *valana* be brought out clearly in the reverse direction. Draw the east–west line and with the help of it, also the other ones parallel to it.

---

[37] The reading in both the printed editions is: परिधौ। This seems to be inappropriate in the present context. Perhaps, the right form of usage should be परिधिम्। Second case, because Nīlakaṇṭha here asks us to draw a circle (परिधि।) The word परिधौ is in the seventh case, which means on the circle. No operation to be performed on a circle is described here. Hence परिधिं, which is in the second case ending, is most likely to be the correct reading.

The first one (east–west line) is the trajectory of the shadow. The one along the direction of deflection [out of the other two] is the trajectory separated from it owing to deflection. Both these trajectories are only instantaneous (*tatkālajau*).

Draw a circle whose radius is the same as that of the Moon along its trajectory [drawn earlier]. This must be drawn in the west on the full Moon day and in the east on the *pratipad*. The shadow must always be drawn at the centre of the circle [on the north–south line], with its centre on its own trajectory. That portion of the Moon which lies outside the shadow is visible. Thus here the horns (*hanus*) of the Moon and the portion [east or west] in which they lie may be clearly visualized.

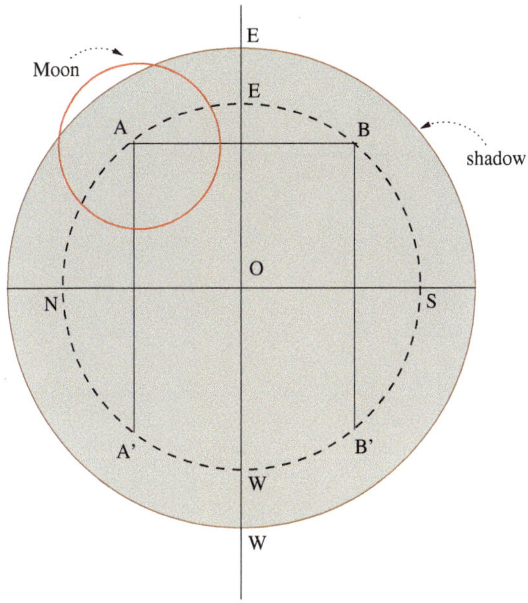

Fig. 4.20 Graphical representation of a lunar eclipse based on the calculation of the difference between the*akṣavalana* and the *āyanavalana*.

The procedure for the graphical representation of a lunar eclipse described above is explained with the help of Fig. 4.20. *ENWS* is a circle centered at *O* whose radius is equal to the 'separation between the discs'. Here *EW* should be the east–west line, and *NS* the north–south line. Take a stick of length *AB*, which is twice the *valana*, and place it along the north–south direction such that *A* and *B* are on the circumference. *AB* should be towards the east of *NS*, on the *pratipad* (just after the instant of conjunction), and towards the west of *NS* on full Moon day (just before the instant of conjunction). Draw *EW* through *E* and AA' and BB' parallel to this. Then *EW* represents the trajectory of the shadow. AA' or BB' represents the trajectory of the Moon depending upon the direction of the *valana* (northwards or southwards respectively). It is significant to note that these two trajectories are considered as instantaneous. This is because the separations between the discs, the *valana* etc. are all varying continuously.

The dark circle in the figure is the circle with the radius of the shadow whose centre is at O. In the figure, the centre of the Moon is shown to be at A (it could be A' or B or B', depending upon the *valana* and whether the desired instant is before or after the eclipse). Draw a circle with a radius equal to that of the lunar disc. Then the portion of the Moon outside the circle representing the shadow is visible, and the portion inside is eclipsed.

Chapter 5
रविग्रहणप्रकरणम्
Solar eclipse

५.१ लम्बनावनत्योः सदसद्भावः
5.1 The possibility or otherwise of *lambana* and *nati*

ग्रहे दृश्यार्धमध्यस्थे न किञ्चिदपि लम्बनम् ।
तस्माद् दृक्क्षेपलग्नं स्यात् प्राक् पश्चाद्वा स्थिते ग्रहे ॥ १ ॥
समोर्ध्वगापवृत्तस्थे नतेरपि न सम्भवः ।
तत्क्षेपकोटिवृत्ते क्वाप्यूर्ध्वसूत्रस्पृशि स्थिते ॥ २ ॥
सममण्डलमध्यात्तद्विप्रकर्षे नतिर्भवेत् [1]।

grahe dṛśyārdhamadhyasthe na kiñcidapi lambanam |
tasmād dṛkkṣepalagnaṃ syāt prāk paścādvā sthite grahe|| 1 ||
samordhvagāpavṛttasthe naterapi na sambhavaḥ |
tatkṣepakoṭivṛtte kvāpyūrdhvasūtraspṛśi sthite || 2 ||
samamaṇḍalamadhyāttadtviprakarṣe natirbhavet |

When the planet lies at the mid-point of the visible half [of the ecliptic] then there will be no parallax in longitude (*lambana*). [Then] it is also in the *dṛkkṣepa-lagna*. Parallax in longitude is possible only when the planet is to the east or west of this (*dṛkkṣepa-lagna*).

When the planet lies on the ecliptic that also happens to be a vertical circle (*samordhvaga-apavṛtta*), then there will be no parallax in latitude (*nati*) also. It is possible to have parallax in latitude only when the planet is displaced from the centre of the prime vertical (zenith) and lies in the *kṣepakoṭivṛtta* which is secondary to the ecliptic.

In Fig. 5.1, S represents the Sun on the ecliptic. K is the pole of the ecliptic and ZS is the vertical passing through S. The parallax of the Sun at S is given by

$$p = SS' = P \sin z, \tag{5.1a}$$

[1] Perhaps the prose order is: सममण्डलमध्यात् (सममण्डलमध्यं च खमध्यमेव) तद्विप्रकर्षे (तस्य ग्रहस्य विप्रकर्षे) तत्क्षेपकोटिवृत्ते (तस्य = अपमण्डलस्य, क्षेपकोटिवृत्ते = समतिरश्चीनवृत्ते) क्वापि ऊर्ध्वसूत्रस्पृशि स्थिते (ग्रहे) नतिर्भवेत्।

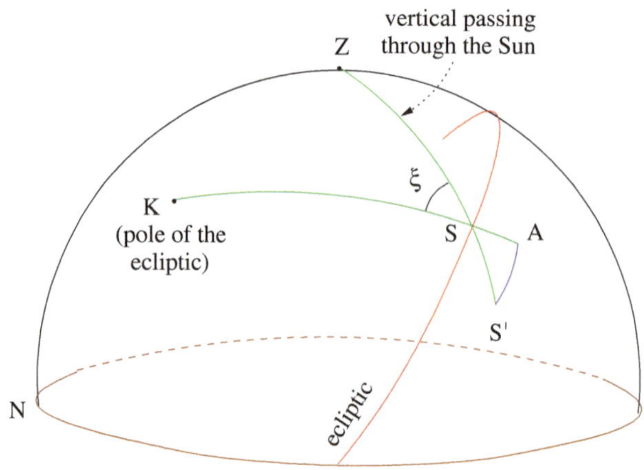

Fig. 5.1 Parallaxes in longitude and latitude.

where, P is the horizontal parallax and $z = ZS$ is the zenith distance of the Sun. KA is the secondary to the ecliptic passing through S. The *lambana* and *nati* are the components of the parallax along and perpendicular to the ecliptic. They are given by

$$lambana = S'A = SS' \sin \xi \tag{5.1b}$$

and

$$nati = SA = SS' \cos \xi, \tag{5.1c}$$

where ξ is the angle between the secondary KS to the ecliptic and the vertical ZS passing through S.

In Fig. 5.2a, P is the north celestial pole and K_0 and K_1 represent the positions of the north pole of the ecliptic at different instants of time. When the ecliptic pole is at K_0, the secondary to the ecliptic passing through Z, that is ZV_0SQNP, is the same as the prime meridian. W being the pole of this vertical circle, it coincides with the autumnal equinox Γ'. The point V_0 at the intersection of the ecliptic and the prime meridian is the *dṛkkṣepa-lagna* or *vitribhalagna* (nonagesimal), which is the point on the visible half of the ecliptic at 90° from the orient ecliptic point. As mentioned earlier, K_1 is the pole of the ecliptic at a later instant when the ecliptic intersects the horizon at B. Consider a secondary to the ecliptic passing through K_1 and Z. This intersects the ecliptic at V_1 and it can be easily seen that V_1 is the *dṛkkṣepa-lagna* at that instant.[2]

---

[2] Since the point B is at the intersection of the ecliptic and the horizon, $K_1B = 90°$ and also $ZB = 90°$. Since the two points K_1 and Z are at 90° from B, B is the pole of the great circle passing through K_1 and Z. Then, by definition, all the points in this circle must be at 90° from B, and hence $BV_1 = 90°$. Thus V_1 is the *dṛkkṣepa-lagna*.

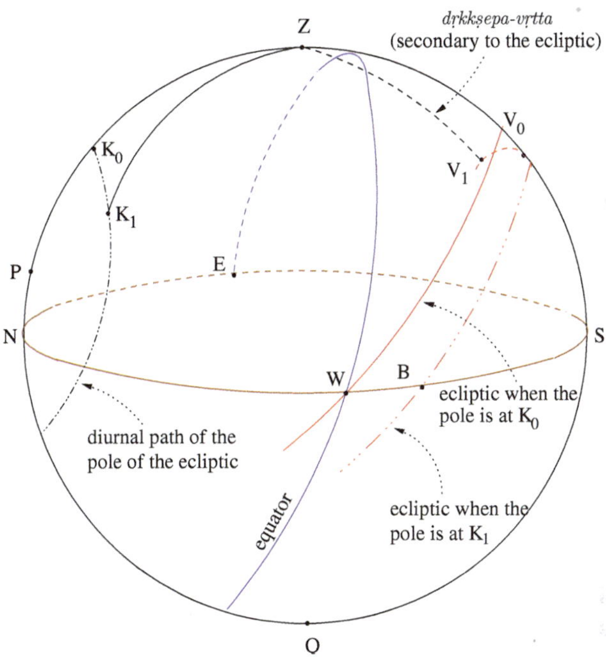

Fig. 5.2*a* Condition for the absence of *lambana* (parallax in longitude).

Condition for zero *lambana*

It was explained earlier (Section 3.8) that the effect of the parallax is to increase the zenith distance of the celestial object. If the object happens to be on the *dṛkkṣepa-vṛtta* (the vertical circle through the nonagesimal), the effect of the parallax is to deflect the object further down along the *dṛkkṣepa-vṛtta* (K_1ZV_1), as the *dṛkkṣepa-vṛtta* is a vertical circle. Since ZV_1 is a secondary to the ecliptic, this deflection does not result in any apparent change in the longitude of the celestial object. The celestial object must be to the east or west of the *dṛkkṣepa-lagna* for it to have non-zero parallax in longitude. Hence, the condition for the object to have zero parallax in longitude can be stated as when:

- *the celestial object is lying at the centre of the visible part of the ecliptic (dṛśyārdha-madhyastha)*; or equivalently
- *the object is lying on the dṛkkṣepa-vṛtta, which is a secondary to the ecliptic passing through Z.*

Condition for zero *nati*

Let us assume that the pole of the ecliptic K lies on the horizon at some instant of time during the day, as shown in Fig. 5.2*b*. Let the ecliptic intersect the horizon at

B. Then obviously $KZ = 90°$. Since ZB is also $90°$, the point B is the pole of the circle passing through K and Z. At this instant, Z itself is the *vitribhalagna* and ZB, which is a quarter of the ecliptic, is also a vertical to the horizon.

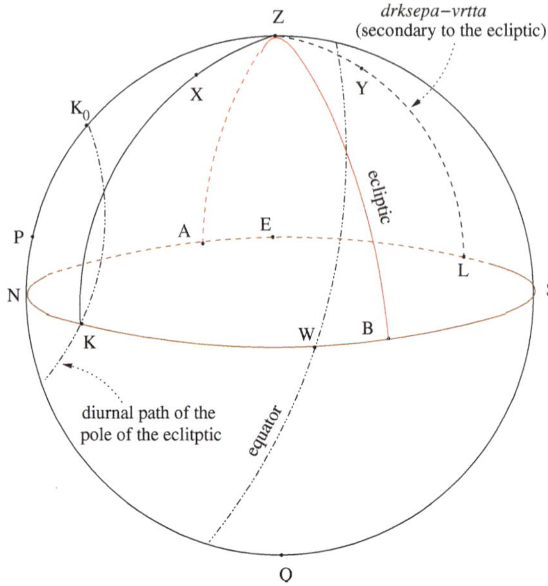

Fig. 5.2*b* Condition for the absence of parallax in latitude.

Now, if an object lies on the ecliptic which also happens to be a vertical, then—since the effect of parallax is only to deflect the object along the ecliptic further down towards the horizon—there will be a change only in the longitude of the object and not in the latitude. Hence the parallax in latitude will be zero if the object lies along the ecliptic when it is a vertical circle. Even when it (the ecliptic) is a vertical circle, the parallax in latitude need not be zero when the object does not lie on it, but is on a *kṣepakoṭi-vṛtta*. When the object is at the zenith, parallax in longitude and latitude are both zero. Taking all these into consideration, the condition for parallax in latitude to be zero is: *the celestial object must be along the ecliptic, which should also be a vertical circle.*

Condition for the absence of both *lambana* and *nati*

We saw that the parallax in longitude would be zero when the object is situated on the *dṛkkṣepa-vṛtta*. Later it was shown that the parallax in latitude will be zero when the object lies along the ecliptic and the ecliptic is a vertical circle. These two conditions will be simultaneously satisfied only at two points, namely the zenith (Z) and the nadir (Q). Since we are interested only in the visible part of the hemisphere, the observer's zenith, which is called the *khamadhya* in the Indian astronomical literature, is the only point where the parallax in longitude and latitude are both zero. Thus the condition for parallax in both longitude and latitude to be zero simultaneously is: *the celestial object must be at khamadhya* (zenith).

५.२ दृक्क्षेपदृग्गत्योरानयनम्

5.2 Finding the *dṛkkṣepa* and the *dṛggati*

राश्यष्टमांशलङ्कोत्थप्राणैश्चापि नतासुभिः ॥ ३ ॥
रवौ क्षयधने कृत्वा मध्यलग्नं स्फुटं नयेत् ।
तस्माद्विषुवदादेः स्वक्रान्तिकार्मुकमानयेत् ॥ ४ ॥
तदक्षचापसंयोगो दिक्साम्येऽन्तरमन्यथा ।
मध्यज्याख्या हि तज्ज्या स्यात् प्राग्लग्नस्याग्रमौर्विका ॥ ५ ॥
उदयज्या तयोर्घातात् त्रिज्याप्तं बाहुमौर्विका ।
मध्यज्यायास्त्रिजीवायाः कृतिभ्यां तत्कृतिं त्यजेत् ॥ ६ ॥
तन्मूलयोस्त्रिजीवाघ्नमाद्यमन्येन संहरेत् ।
लभ्यते तत्र दृक्क्षेपः तत्कोटिर्दृग्गतिर्मता ॥ ७ ॥

rāśyaṣṭamāṃśalaṅkotthaprāṇaiścāpi natāsubhiḥ || 3 ||
ravau kṣayaghane kṛtvā madhyalagnaṃ sphuṭaṃ nayet |
tasmādviṣuvadādeḥ svakrāntikārmukamānayet || 4 ||
tadakṣacāpasaṃyogo diksāmye'ntaramanyathā |
madhyajyākhyā hi tajjyā syāt prāglagnasyāgramaurvikā || 5 ||
udayajyā tayorghātā trijyāptaṃ bāhumaurvikā |
madhyajyāyāstrijīvāyāḥ kṛtibhyāṃ tatkṛtiṃ tyajet || 6 ||
tanmūlayostrijīvāghnamādyamanyena saṃharet |
labhyate tatra dṛkkṣepaḥ tatkoṭirdṛggatirmatā || 7 ||

By [making use of] the hour angle and also the ascensional difference of one-eighth of the *rāśi* for an observer at Laṅka, and applying it to the Sun positively or negatively, let the true *madhyalagna* be obtained. From that, may the declination of the *viṣuvadādi* [3] be calculated. The sum—[if the directions are same]—or difference, if the directions are different, of this and the latitude of the place is to be found. The Rsine of this is known as the *madhyajyā* [and] the *agrā* of the orient ecliptic point will be the *udayajyā*. The product of the two divided by the *trijyā* is the *bāhumaurvikā*. From the squares of the *madhyajyā* and *trijyā* subtract the square of this (the *bāhumaurvikā*). Finding the square roots of them, multiply the former by the *trijyā* and divide it by the latter. The resulting quantity will be the *dṛkkṣepa* and the *koṭi* of it is considered to be the *dṛggati*.

In the above verses, the *dṛkkṣepa* and the *dṛggati* are expressed in terms of the *madhyajyā* and the *udayajyā*. These are used in the computation of parallax in longitude and latitude. In Fig. 5.3, ZV represents the zenith distance of the *vitribhalagna*. $R\sin ZV$ is called the *dṛkkṣepa* and $R\cos ZV$ is called the *dṛggati*.

First, the *madhyalagna* or the longitude of the meridian ecliptic point (λ_{ml})—which refers to the longitude of the point of the ecliptic which intersects with the

[3] From the context, it seems that the word *viṣuvadādi* has been employed to refer to the meridian ecliptic point. However, the etymological derivation of this meaning is not clear.

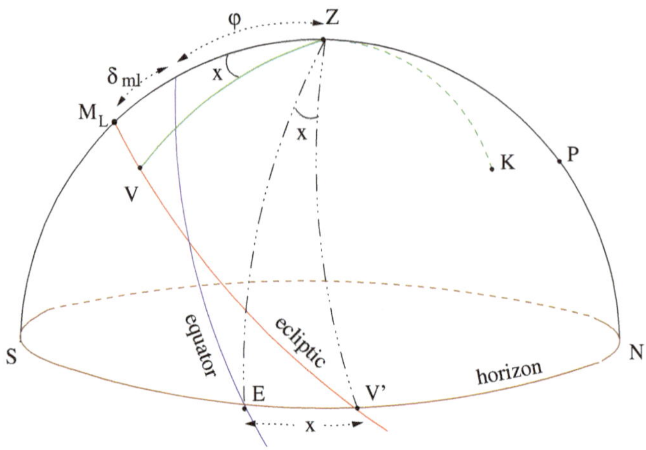

Fig. 5.3 Determination of the Rcosine of the zenith distance of the *vitribhalagna* called the *dṛkkṣepa*.

prime meridian—is determined. This point is denoted by M_L in Fig. 5.3. The longitude of this point has to be calculated from the hour angle of the Sun, and it is suggested that, for better accuracy in doing so, this should be computed from the rising time of one-eighth of a *rāśi* and not from that of the whole *rāśi*.

From the *madhyalagna*, the declination (δ_{ml}) corresponding to that point is obtained using the relation

$$\sin\delta_{ml} = \sin\varepsilon \sin\lambda_{ml}. \tag{5.2a}$$

Using this, the *madhyajyā*, which is the Rsine of the zenith distance of the meridian ecliptic point, is calculated:

$$madhyajy\bar{a} = R\sin z_{ml} = R\sin(\phi \pm |\delta_{ml}|), \tag{5.2b}$$

where ϕ is the latitude of the observer. Now the *udayajyā*, which is the *agrā* of the *udayalagna*, more often simply called the *lagna*, is given by

$$R\sin x = \frac{R\sin\delta_l}{\cos\phi}, \tag{5.3a}$$

where δ_l represents the declination of the *lagna*, the orient ecliptic point ($90 \pm x$ is the 'azimuth' of the *lagna*). Now the text defines an intermediate quantity called the *bāhumaurvikā* as follows:

$$b\bar{a}humaurvik\bar{a} = \frac{madhyajy\bar{a} \times udayajy\bar{a}}{trijy\bar{a}}. \tag{5.3b}$$

Making use of this, the *dṛkkṣepajyā* is defined as

$$dṛkkṣepajyā = trijyā \times \frac{\sqrt{madhyajyā^2 - (madhyajyā \times udayajyā)^2}}{\sqrt{trijyā^2 - (madhyajyā \times udayajyā)^2}}. \qquad (5.4a)$$

The *dṛkkṣepajyā* is also simply called the *dṛkkṣepa* or *dṛgjyā*. Using the notation $ZM_L = z_{ml}$, and $ZV = z_v$, the above equation reduces to:

$$\sin z_v = \frac{\sqrt{\sin^2 z_{ml} - (\sin z_{ml} \sin x)^2}}{\sqrt{1 - (\sin z_{ml} \sin x)^2}}, \qquad (5.4b)$$

where $\sin x$ is known from (5.3).

Proof:

The above result for the sine of the zenith distance of the *vitribhalagna* can be derived using spherical trigonometrical formulae. In Fig. 5.3, consider the spherical triangle ZVM_L. Using the sine formula we have

$$\frac{\sin VM_L}{\sin V\hat{Z}M_L} = \frac{\sin ZM_L}{\sin Z\hat{V}M_L}, \qquad (5.5a)$$

where $V\hat{Z}M_L = x$ and $Z\hat{V}M_L = 90$, as KZV is a secondary to the ecliptic. Further, using the notation $ZM_L = z_{ml}$, the above equation reduces to

$$\sin VM_L = \sin z_{ml} \sin x. \qquad (5.5b)$$

Applying the analogue to the cosine formula,

$$\sin a \cos B = \cos b \sin C - \sin b \cos c \cos A, \qquad (5.6)$$

with $a = ZM_L$, $b = VM_L$, $c = ZV, B = x$ and $A = 90$, we find

$$\sin ZM_L \cos x = \cos VM_L \sin ZV. \qquad (5.7)$$

Rewriting the above equation and using (5.5b) in it, we have

$$\begin{aligned} \sin z_v &= \frac{\sin z_{ml} \cos x}{\cos VM_L} \\ &= \frac{\sin z_{ml} \sqrt{1 - \sin^2 x}}{\sqrt{1 - \sin^2 VM_L}} \\ &= \frac{\sqrt{\sin^2 z_{ml} - (\sin z_{ml} \sin x)^2}}{\sqrt{1 - (\sin z_{ml} \sin x)^2}}, \end{aligned} \qquad (5.8)$$

which is the same as (5.4). The quantity *dṛggati* is stated to be

$$\begin{aligned} dr\underset{.}{g}gati &= \sqrt{trijy\bar{a}^2 - dr\underset{.}{k}k\underset{.}{s}epa^2} \\ &= \sqrt{R^2 - (R\sin z_v)^2} \\ &= R\cos z_v. \end{aligned} \tag{5.9}$$

It may be mentioned here that the *dṛkkṣepa* was found in order to calculate the *dṛggati*, which appears in the expression for the parallax in longitude, as will be shown in the following section.

५.३ लम्बनं पर्वान्तानयने तत्संस्कारश्च

5.3 Parallax in longitude and its application for finding the instant of conjunction

एकज्यावर्गतश्छेदः लब्धं दृग्गतिजीवया ।
छेदाप्तान्तरबाहुज्या भानुदृक्क्षेपलग्नयोः ॥ ८ ॥
लम्बनं नाडिकादिः स्यात् स्वर्णमूनाधिके रवौ ।
पर्वान्ते मुहुरप्येवं पर्वान्तमविशेषयेत् ॥ ९ ॥

ekajyāvargataśchedaḥ labdhaṃ dṛggatijīvayā |
chedāptāntarabāhujyā bhānudṛkkṣepalagnayoḥ || 8 ||
lambanaṃ nāḍikādiḥ syāt svarṇamūnādhike ravau |
parvānte muhurapyevaṃ parvāntamaviśeṣayet || 9 ||

The quantity obtained by dividing the square of Rsine of one [*rāśi*] by the Rsine of the *dṛggati* is called the *cheda*. The Rsine of the difference between the [longitudes of] the Sun and the *dṛkkṣepalagna* divided by the*cheda* is the parallax in longitude in *nāḍikās* etc. Depending on whether [the longitude of] the Sun is greater or smaller [than that of the *vitribhalagna*] the result must be added to or subtracted from the instant of conjunction. By repeating the above procedure, the instant of opposition has to be obtained iteratively.

An intermediate quantity called the *cheda* is defined by

$$cheda = \frac{(R\sin 30)^2}{dr\underset{.}{g}gati} = \frac{R}{4\times\cos z_v}. \tag{5.10}$$

The expression for the parallax in longitude ($\Delta\lambda$) in *nāḍikās* is given by

$$lambana = \frac{R|\sin(\lambda_s - \lambda_v)|}{cheda},$$

$$\text{or} \qquad \Delta\lambda = 4\times\cos z_v \times |\sin(\lambda_s - \lambda_v)|. \tag{5.11}$$

In the following we arrive at the expression for the parallax in longitude using spherical trigonometrical formulae. In Fig. 5.4, S represents the Sun and S' its deflected position due to parallax, which is along the vertical circle ZS. ξ is the angle between the secondary to the ecliptic and the vertical circle passing through the Sun. V is the *vitribhalagna*, $S'A$ is the parallax in longitude or the *lambana*, and SA is the parallax in latitude or the *nati*. They are given by

$$\begin{aligned} lambana &= S'A = SS' \sin\xi, \\ \text{and} \qquad nati &= SA = SS' \cos\xi. \end{aligned} \qquad (5.12)$$

The arc SV on the ecliptic is $\lambda_v - \lambda_s$. This is also the angle $Z\hat{K}S$. To find $\sin\xi$ we use the spherical triangle KZS and apply the sine formula. We have

$$\frac{\sin\xi}{\sin KZ} = \frac{\sin(\lambda_v - \lambda_s)}{\sin ZS}. \qquad (5.13)$$

In the above expression $KZ = 90 - z_v$, where z_v is the zenith distance of the *vitribhalagna*, and $ZS = z$, the zenith distance of the Sun. Hence

$$\sin\xi = \frac{\cos z_v \times \sin(\lambda_v - \lambda_s)}{\sin z}. \qquad (5.14)$$

It was mentioned earlier (Chapter 3, verses 10–11) that the parallax of any celestial

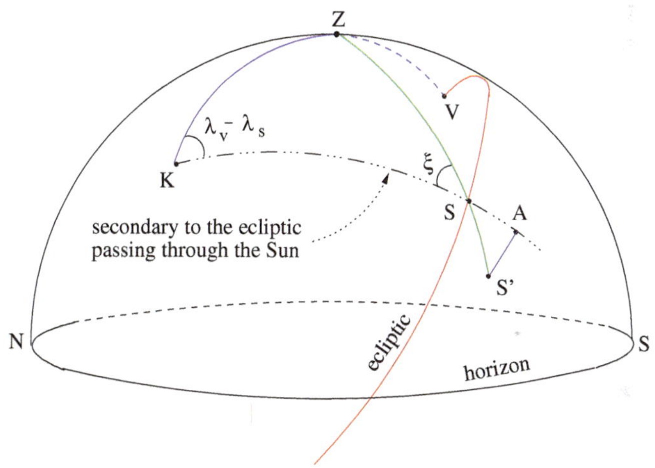

Fig. 5.4 Effect of parallax in longitude.

object is given by

$$p = P \times \sin z, \qquad (5.15)$$

where P is the horizontal parallax and z is the zenith distance of the object. It was also noted there that the horizontal parallax P is taken to be one-fifteenth of the daily motion of the object. Hence the parallax SS', in Fig. 5.4, is given by

$$SS' = p = \frac{\dot{\lambda}_s \times \sin z}{15}. \qquad (5.16)$$

Here we are actually interested in finding the difference between the lunar and solar parallaxes. For this, we first consider the parallax of the Sun. The factor $\dot{\lambda}_s$ in (5.16) represents the daily angular motion of the Sun. Hence the expression for the parallax in longitude of the Sun (l_s) becomes

$$l_s = \frac{\dot{\lambda}_s \times \sin z}{15} \times \frac{\cos z_v \times \sin(\lambda_s - \lambda_v)}{\sin z}. \tag{5.17}$$

Here the longitude of the Sun decreases due to the parallax when $\lambda_v > \lambda_s$, as is evident from Fig. 5.4 (S' is west of S). Hence we have $\lambda_s - \lambda_v$ as the argument of the sine.

Similarly, the parallax in longitude of the Moon (l_m) is given by

$$l_m = \frac{\dot{\lambda}_m \times \sin z}{15} \times \frac{\cos z_v \times \sin(\lambda_s - \lambda_v)}{\sin z}, \tag{5.18}$$

where $\dot{\lambda}_m$ is the daily angular motion of the Moon. Here we have taken the Moon also to be at S as it is a solar eclipse, and $\lambda_m = \lambda_s$. For finding the correction to the instant of conjunction, the relevant quantity is actually the difference in the parallaxes in longitude of the Sun and the Moon, which is given by

$$l_m - l_s = \frac{(\dot{\lambda}_m - \dot{\lambda}_s)}{15} \times \cos z_v \sin(\lambda_s - \lambda_v). \tag{5.19}$$

When this is divided by $\dot{\lambda}_m - \dot{\lambda}_s$, it gives the result in time units of days. The result when further multiplied by 60 gives the difference in *lambana* in *nāḍikās*. Thus the difference in *lambana* in *nāḍikās* is

$$l_m - l_s = 4 \times \cos z_v \sin(\lambda_s - \lambda_v). \tag{5.20}$$

Actually this incorporates the sign also, as

$$l_m - l_s = +ve \qquad (\lambda_s > \lambda_v), \tag{5.21}$$

and should be subtracted from the instant of opposition to obtain the true *parvānta*; and

$$l_m - l_s = -ve \qquad (\lambda_s < \lambda_v), \tag{5.22}$$

so that the magnitude of $l_m - l_s$ should be added to the instant of opposition to obtain the true instant of conjunction. In Fig. 5.4, $(l_m - l_s)$ is deflected westwards in longitude owing to parallax. Hence, the Moon is lagging behind the Sun at the calculated instant of conjunction and the true instant of conjunction is later. From now onwards, l stands for $l_m - l_s$, which is the difference in *lambanas* of the Moon and the Sun. In fact an iterative procedure is prescribed for obtaining t_m. This is indicated by the statement

मुहुरप्येवं पर्वान्तमविशेषयेत् ।

In *Laghu-vivṛti* the iterative procedure is described thus:

अत्र लम्बनसंस्कृतस्फुटपर्वान्ततः एव तात्कालिकार्कदृक्क्षेपलम्बनादिकं कर्तव्यम्। तत एव च लम्बनमित्यविशेषकरणेनैव स्फुटत्वादुक्तं 'मुहुरप्येवं पर्वान्तमविशेषयेत्' इति।

Here, the longitude of the Sun, the *dṛkkṣepa*, the parallax in longitude etc. should be calculated only at the instant of conjunction corrected by parallax in longitude. Again from that the parallax in longitude [should be calculated]. Thus, since the accuracy is obtained only through an iterative process (*aviśeṣakaraṇa*), it is said: the instant of conjunction should be repeatedly obtained in the same way.

Let t_{m0} be the mean instant of conjunction obtained by an iterative process before the application of the correction due to parallax in longitude. If l_1, l_2, l_3 ... be the successive *lambanas* (the differences in parallaxes in longitude of the Moon and the Sun in *nāḍikās*), l_1 being the value of the parallax in longitude at t_{m0}, then the successive iterated values of the instant of conjunction are given by

$$t_{mi} = t_{m0} + l_i \qquad (i = 1, 2, 3, \ldots). \tag{5.23}$$

The iteration must be continued till $t_{mi} - t_{mi-1} \approx 0$.

५.४ रवेर्नतिकलाः

5.4 Parallax in latitude of the Sun in minutes

अविशिष्टात्तु दृक्क्षेपात् स्फुटभुक्त्याहताद्रवेः ।
खस्वरेष्वेकभूताप्ता नतिर्दृक्क्षेपवद्दिशा ॥ १० ॥

aviśiṣṭāttu dṛkkṣepāt sphuṭabhuktyāhatādraveḥ |
khasvareṣvekabhūtāptā natirdṛkkṣepavaddiśā || 10 ||

From the product of the iterated value of the *dṛkkṣepa* and the true daily motion of the Sun divided by 51570, the parallax in latitude is obtained whose direction is the same as that of the *dṛkkṣepa*.

It was mentioned in the previous section that the value of the parallax in longitude is to be found iteratively. Since the expression for the parallax in longitude implicitly involves the value of the *dṛkkṣepa*, it is also to be found iteratively. Denoting the successive values of *dṛkkṣepa* by d_i, it is given by

$$d_i = R\sin z_{vi} \qquad (i = 1, 2, 3, \ldots), \tag{5.24}$$

where z_{vi} represents the zenith distance of the *vitribhalagna* at the i-th iteration. If z_v be the final iterated value of *vitribhalagna*, then the expression for the Sun's parallax in latitude given here is

$$nati = \frac{\dot{\lambda}_s \times R\sin z_v}{51570}, \tag{5.25}$$

where $\dot{\lambda}_s$ is the true daily motion of the Sun.

The rationale behind the above expression can be understood from Fig. 5.4. Considering the triangle KZS, and applying the cosine formula, we have

$$\begin{aligned} \cos KZ &= \cos KS \cos ZS + \sin KS \sin ZS \cos \xi \\ &= \sin ZS \cos \xi, \end{aligned} \tag{5.26}$$

since $KS = 90$. Further using the fact that $KZ = 90 - ZV = 90 - z_v$ and writing $ZS = z$, we get

$$\cos \xi = \frac{\sin z_v}{\sin z}. \tag{5.27}$$

Using the expression for SS' in (5.16) and substituting for $\cos \xi$ and SS' in (5.12), we get

$$\begin{aligned} nati &= \frac{\dot{\lambda}_s \sin z}{15} \times \frac{\sin z_v}{\sin z} \\ &= \frac{\dot{\lambda}_s \times R \sin z_v}{15R} = \frac{\dot{\lambda}_s \times R \sin z_v}{51570}. \end{aligned} \tag{5.28}$$

The direction of the parallax in latitude is mentioned to be the same as that of the *dṛkkṣepa*. In other words, if the *vitribhalagna* lies in the northern hemisphere, then the parallax in latitude will also be northwards, and if the *vitribhalagna* lies in the southern hemisphere, then the parallax in latitude will also be southwards. This can be easily understood from Fig. 5.4.

५.५ चन्द्रस्य नतिकला:

5.5 Parallax in latitude of the Moon in minutes

तत्कालेन्दुस्फुटात् क्षेपं प्राग्वत् कृत्वा स्फुटं ततः ।
क्षेपदृक्क्षेपचापैक्यात् साम्ये भेदेऽन्तराद्गुणः ॥ ११ ॥
द्वितीयस्फुटभुक्तिघ्नः हृतः खाद्रीषुभूशरैः ।
नतिः तत्क्षेपयोरैक्यं दिक्साम्येऽन्तरमन्यथा ॥ १२ ॥
तदर्कनतिलिप्तानां दिक्साम्येऽन्तरमेव च ।
दिग्भेदे चैक्यमेव स्यात् स्फुटा सूर्यग्रहे नतिः ॥ १३ ॥
अर्कस्य चेन्नतिः शिष्टा विश्लेषे व्यत्ययेन दिक् ।
चन्द्रस्यैव नतेर्ग्राह्या दिक् तस्याश्चान्यदा सदा ॥ १४ ॥

tatkālendusphuṭāt kṣepaṃ prāgvat kṛtvā sphuṭaṃ tataḥ |
kṣepadṛkkṣepacāpaikyāt sāmye bhede'ntarādguṇaḥ || 11 ||
dvitīyasphuṭabhuktighnaḥ hṛtaḥ khādrīṣubhūśaraiḥ |
natiḥ tatkṣepayoraikyaṃ diksāmye'ntaramanyathā || 12 ||
tadarkanatiliptānāṃ diksāmye'ntarameva ca |
digbhede caikyameva syāt sphuṭā sūryagrahe natiḥ || 13 ||
arkasya cennatiḥ śiṣṭā viśleṣe vyatyayena dik |
candrasyaiva natergrāhyā dik tasyāścānyadā sadā || 14 ||

> From the true longitude of the Moon at that instant (*sphuṭa-parvānta*), let the *vikṣepa* ([Moon's] latitude) be determined as before and then the true [latitude] be obtained. From the sum of or the difference between the *dṛkkṣepa* and the *vikṣepa*, depending upon whether their directions are the same or different, the Rsine is found. The result, multiplied by the second true daily motion [in minutes] and divided by 51570, is the actual parallax in latitude [of the Moon]. Find the sum of or the difference between this (parallax in latitude) and the *vikṣepa* depending on whether their directions are the same or different.
>
> The difference of this and the parallax in latitude of the Sun is found if they have the same directions. If they have the different directions, their sum is calculated. The result is the effective *nati* in the solar eclipse. When finding the difference, if the *nati* of the Sun remains, then the direction of the *nati* is to be taken reversely [i.e., opposite to that of the Moon's deflection]. Otherwise, the direction of the *nati* is the same as that of the *nati* of the Moon.

Till now, the word *nati* has been used for the parallax in latitude of a given object. Before proceeding further, it must be clarified here that the word *nati* is used in different senses from now onwards. However, all these connotations are associated with the deflection perpendicular to the ecliptic.

In the above verses, the procedure to obtain the effective deflection of the Moon from the Sun in the direction perpendicular to the ecliptic is given. This is used in the determination of the half-duration of the solar eclipse. It should be noted that the role played by the effective *nati* n_e in a solar eclipse is the same as the role played by the true latitude β_t in a lunar eclipse. We explain the given prescription for n_e with the help of Fig. 5.5. Here M represents the Moon and M' its position as seen by an observer on the surface of the Earth owing to the effect of parallax. V is the *vitribhalagna* (nonagesimal) and V' the point where the vertical circle through it meets the parallel to the ecliptic passing through the Moon. $M'A$ is the parallax in longitude of the Moon called the *lambana*, and MA is the *nati* which is the parallax in latitude; they are given by

$$\begin{aligned} lambana &= M'A = MM' \sin\xi \\ nati &= MA = MM' \cos\xi, \end{aligned} \tag{5.29}$$

where ξ is the angle between the vertical and the secondary to the ecliptic through M, and $MM' = P \sin z$, P being the horizontal parallax of the Moon and z its zenith distance. Since the horizontal parallax is taken to be one-fifteenth of the daily motion,

$$MM' = p = \frac{\dot{\lambda}_m \times \sin z}{15}, \tag{5.30}$$

where $\dot{\lambda}_m$ is the daily angular motion of the Moon. Let n_m be the parallax in latitude of the Moon. Then the expression for it is given as

$$n_m = \frac{\dot{\lambda}_m \times R\sin z'_v}{51570}, \tag{5.31}$$

where $z'_v = z_v \pm |\beta_t|$. The above expression is essentially the same as (5.25) except for two replacements: (i) $\dot{\lambda}_s$, the daily motion of the Sun, is replaced by $\dot{\lambda}_m$, the daily motion of the Moon; and (ii) z_v, the zenith distance of the *vitribhalagna* is replaced

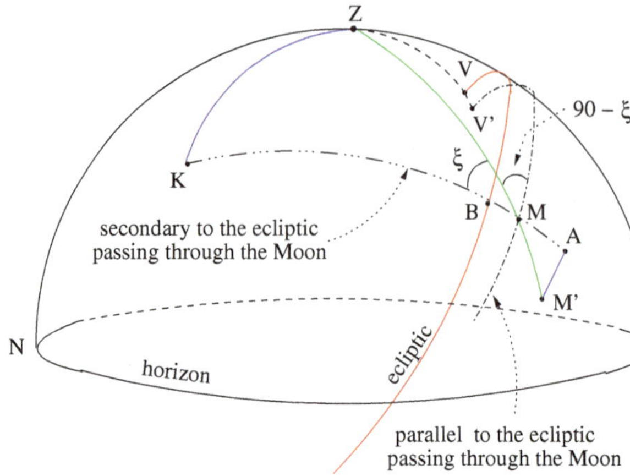

Fig. 5.5 The effect of parallax in the measurement of the Moon's latitude in a solar eclipse.

by z'_v, the zenith distance of the point V'. In arriving at the above expression an approximation is involved which can be shown to be very reasonable. Considering the triangle KZM in Fig. 5.5 and applying the cosine formula, we have

$$\cos KZ = \cos KM \cos ZM + \sin KM \sin ZM \cos \xi. \tag{5.32a}$$

This is approximated as

$$\cos KZ \approx \sin ZM \cos \xi, \tag{5.32b}$$

where we have used the approximation $KM \approx 90$. Actually $KM = 90 + \beta_t$, where β_t is the true latitude of the Moon. But since β_t is very small during an eclipse—in fact, it must be less than around $15'$ for an eclipse to occur—it is justified to take $KM = 90$. The same approximation is used in finding KZ. We have

$$\begin{aligned} KV' &= KV + VV' \\ &= 90 + \beta_t \\ &= KZ + ZV' \end{aligned} \tag{5.33}$$

Therefore $KZ = (90 + \beta_t) - ZV'$. We make the approximation $KZ \approx 90 - ZV'$. As mentioned earlier, since $VV' = \beta_t$ is very small, the approximation is quite good. Further, using the notation $ZM = z$ and $ZV' = z'_v = z_v \pm |\beta_t|$, we find

$$\begin{aligned} \cos KZ = \sin ZV' = \sin z'_v &= \sin z \cos \xi \\ \text{or} \qquad \cos \xi &= \frac{\sin z'_v}{\sin z}. \end{aligned} \tag{5.34}$$

Substituting for $\cos \xi$ and MM' in (5.29), we obtain the expression for the Moon's parallax in latitude given by (5.31).

The total *nati* of the Moon

The total *nati*, n_t, or the deflection of the Moon from the ecliptic, is

$$\begin{aligned} n_t &= |n_m + \beta_t| \qquad \text{(same direction)} \\ &= |n_m - \beta_t| \qquad \text{(opposite direction).} \end{aligned} \tag{5.35}$$

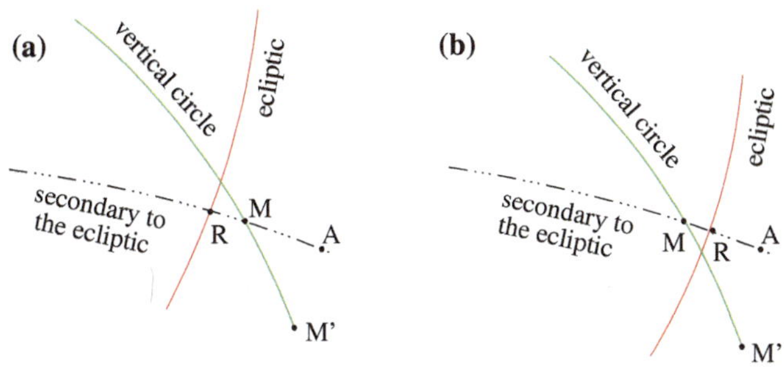

Fig. 5.6 The total deflection of the Moon from the ecliptic: (a) when the *vikṣepa* and the parallax in latitude have the same direction; (b) when the *vikṣepa* and the parallax in latitude have opposite directions.

This may be understood with the help of Fig. 5.6. Here RM represents the *vikṣepa* of the Moon. In (a) it is to the south of the ecliptic and in (b) it is to the north. But the deflection from the ecliptic in both the cases is to the south of the ecliptic. Hence in (a) the total deflection from the ecliptic n_t is the sum, $RA = MA + MR$, and in (b) the total deflection from the ecliptic is the difference $RA = MA - MR$.

The effective *nati*

We are actually interested in n_e, the effective deflection of the Moon from the Sun, because that is what determines the duration of a solar eclipse. In all cases, $n_e = n_t - n_s$. In the determination of n_e, one has to be very careful regarding the directions. All the possible cases are discussed in the verses of the text.

Case (i): n_s and n_t have opposite directions

Here n_s has a direction opposite to n_t. The magnitude of the effective deflection from the ecliptic is obtained by finding the sum of the magnitudes of n_s and n_t as shown in Fig. 5.7(a). It is given by $RA + SB$. That is

$$|n_e| = |n_t| + |n_s|. \tag{5.36}$$

The direction of the effective deflection from the ecliptic is the same as that of the Moon, that is RA.

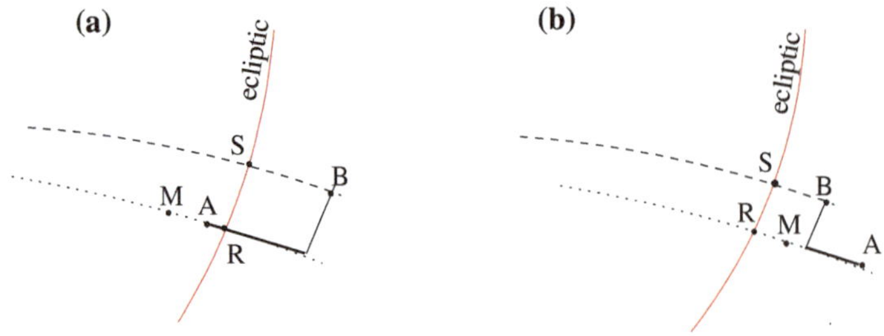

Fig. 5.7 The effective deflection from the ecliptic: (a) n_s to the south and n_t to the north; (b) n_s and n_t having the same direction. The dashed and the dotted lines represent the secondary to the ecliptic through the Sun and the Moon respectively.

Case (ii): n_s and n_t have the same direction with $n_t > n_s$

Here the effective deflection from the ecliptic is obtained by finding the difference between the two as shown in Fig. 5.7(b). It is represented by a thick line and is seen to be $RA - SB$. Here $n_e = n_t - n_s$ is in the 'same direction' as that of the Moon's deflection from the ecliptic, that is RA.

Case (iii): n_s and n_t have the same direction, but $n_s > n_t$

As before, the effective deflection from the ecliptic is obtained by finding the difference between the two as shown in Fig. 5.8. It is represented by a thick line and is seen to be $SB - RA$. Here $n_e = n_t - n_s$ is negative. Hence the direction of the resultant deflection from the ecliptic is opposite to that of the Moon (and the Sun, too). This is precisely what is meant by the statement:

अर्कस्य चेन्नतिः शिष्टा विश्लेषे व्यत्ययेन दिक् ।

This means that the effective deflection from the ecliptic is to be taken reversely, if there is a positive remainder after n_t is subtracted from n_s.

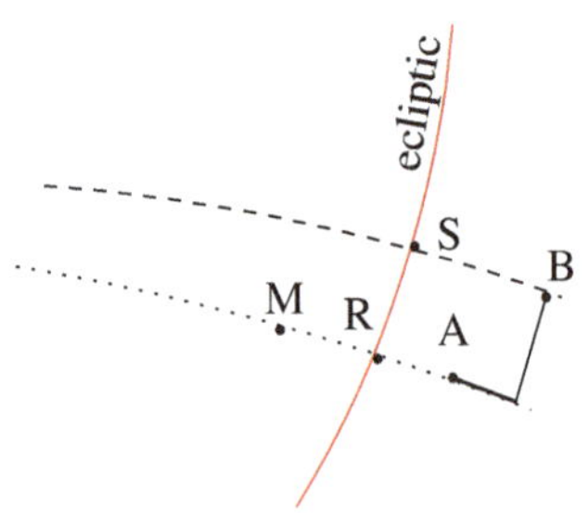

Fig. 5.8 The direction of the effective deflection from the ecliptic when n_s and n_t have the same direction, but $n_s > n_t$.

५.६ रविग्रहणस्य सदसद्भावः

5.6 The possibility of a solar eclipse

सम्पर्कार्धाधिका सा चेत् ग्रहणं नैव भास्वतः ।
सम्पर्कार्धात् त्यजेदूनां परमग्राससिद्धये ॥ १५ ॥

samparkārdhādhikā sā cet grahaṇaṃ naiva bhāsvataḥ |
samparkārdhāt tyajedūnāṃ paramagrāsasiddhaye || 15 ||

If that (effective deflection from the ecliptic) is greater than the sum of the semi-diameters, then there will be no solar eclipse. For obtaining the maximum obscuration, that must be subtracted from the sum of the semi-diameters.

In Fig. 5.9, A and X refer to the centres of the Sun and the Moon. AM and BX are their semi-diameters. AX is the effective deflection from the ecliptic n_e at the instant of conjunction. If n_e at this instant is exactly equal to or greater than the sum of the semi-diameters of the Sun and the Moon, then there will be no eclipse. This can be understood from Fig. 5.9(a). Here,

$$\begin{aligned} AX &= (AM+BX)+MB \\ n_e &= \text{Sum of the semi-diameters} + MB. \end{aligned} \tag{5.37}$$

Thus we see that the condition for the absence of solar eclipse is:

$$n_e \geq \text{Sum of the semi-diameters} \qquad \text{(at the middle of the eclipse)}. \tag{5.38}$$

Similarly, if n_e at the instant of conjunction is less than the sum of the semi-diameters, then there will be at least a partial eclipse. Such a situation is depicted in Fig. 5.9(b). Partial eclipse is possible when

$$\text{the obscured portion } MB > 0. \tag{5.39}$$

Now

$$\begin{aligned} MB &= BX - MX \\ &= BX - (AX - AM) \\ &= AM + BX - AX. \end{aligned} \tag{5.40}$$

Hence the condition for (at least) a partial eclipse can be written as

$$AX < AM + BX. \tag{5.41}$$

In other words,

$$n_e < \text{Sum of the semi-diameters} \qquad \text{(at the middle of the eclipse)}. \tag{5.42}$$

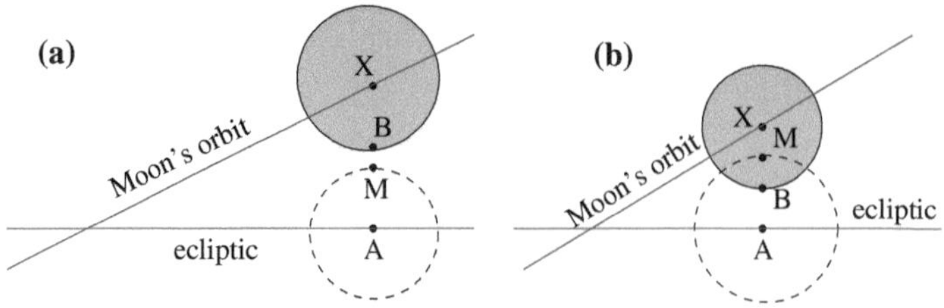

Fig. 5.9 The Sun and the Moon in the case of (a) no eclipse (b) a partial solar eclipse. While the dotted circle represents the Sun's disc, the shaded one represents that of the Moon.

५.७ स्थित्यर्धानयने लम्बनसंस्कारः

5.7 Application of *lambana* in finding the half-duration

रवीन्दुबिम्बसम्पर्कदलात् स्फुटनतेरपि ।
स्थित्यर्धं प्राग्वदानेयं तत्र तत्र च लम्बनम् ॥ १६ ॥
कपालैक्ये तु संयोज्यं स्थित्यर्धे लम्बनान्तरम् ।
लम्बनैक्यं[4] तु तद्भेदे तत् स्थित्यर्धे तु योजयेत् ॥ १७ ॥
प्राक्कपाले क्रमान्न्यूनं स्पर्शकालादिलम्बनम् ।
क्रमेण चाधिकं प्रत्यक्कपाले तेषु लम्बनम् ॥ १८ ॥
यत्रैतद्विपरीतं स्यात् शोध्यं तल्लम्बनान्तरम् ।
आसन्न उदयान्मोक्षे त्वधिकं मोक्षलम्बनम् ॥ १९ ॥
निशि चेन्मध्यकालोऽत्र विप्रकृष्टश्च मौक्षिकात् ।
मौक्षिकाल्लम्बनात् त्याज्यं तदा स्यान्मध्यलम्बनम् ॥ २० ॥
मोक्षस्थितिदलात् तत्र शोध्यं तल्लम्बनान्तरम् ।
एवमस्तमयासन्ने स्पर्शे तल्लम्बनेऽधिके ॥ २१ ॥
मध्यलम्बनकालात् तु त्याज्यं तत्र तदन्तरम् ।

ravīndubimbasamparkadalāt sphuṭanaterapi |
sthityardhaṃ prāgvadāneyaṃ tatra tatra ca lambanam|| 16 ||
kapālaikye tu saṃyojyaṃ sthityardhe lambanāntaram |
lambanaikyaṃ tu tadbhede tat sthityardhe tu yojayet || 17 ||
prākkapāle kramānnyūnaṃ sparśakālādilambanam |
krameṇa cādhikaṃ pratyakkapāle teṣu lambanam || 18 ||
yatraitadviparītaṃ syāt śodhyaṃ tallambanāntaram |
āsanna udayānmokṣe tvadhikaṃ mokṣalambanam || 19 ||
niśi cenmadhyakālo'tra viprakṛṣṭaśca maukṣikāt |
maukṣikāllambanāt tyājyaṃ tadā syānmadhyalambanam|| 20 ||
mokṣasthitidalāt tatra śodhyaṃ tallambanāntaram |
evamastamayāsanne sparśe tallambane'dhike || 21 ||
madhyalambanakālāt tu tyājyaṃ tatra tadantaram |

From the sum of the semi-diameters and the effective deflection from the ecliptic, the half-duration must be obtained as before. The parallax in longitude must also be obtained at those instants (the first contact, the middle contact and the last contact). If the hemisphere

---

[4] The reading in the printed text is लम्बनैक्ये , whereas लम्बनैक्यं seems to be the appropriate term. Hence we have given the same here.

is same, then the difference in parallaxes in longitude must be added to the half-duration. If they (the hemispheres) are different, then the sum of the parallaxes in longitude must be added to the half-duration. The parallaxes in longitude gradually decrease in the eastern hemisphere, whereas they gradually increase in the western hemisphere.

When the reverse is the case [that is, the successive parallaxes in longitude are increasing in the western hemisphere and decreasing in the eastern], then the difference in the parallaxes in longitude must be subtracted. When the last contact is close to sunrise, then the parallax in longitude at release is greater. Then the *madhyakāla* is in the night and the parallax in longitude in the middle has to be subtracted from that at release.

The difference must be subtracted from the instant of last contact. Similarly when the first contact is close to the sunset, because the parallax in longitude at first contact is greater than that (in the middle), the difference has to be subtracted from the *madhyakāla*.

Here the corrections to the half-durations of the eclipse due to parallax in longitude are discussed. Let l_1, l_2 and l_3 be the parallaxes in longitude (see Section 5.3) calculated at the first contact, the middle and the *mokṣakālas*. The expressions for the half-durations (see (5.47) and (5.48)) can be understood with the help of Fig. 5.10. Here S_1, S_2 and S_3 are the positions of the Sun on the ecliptic without

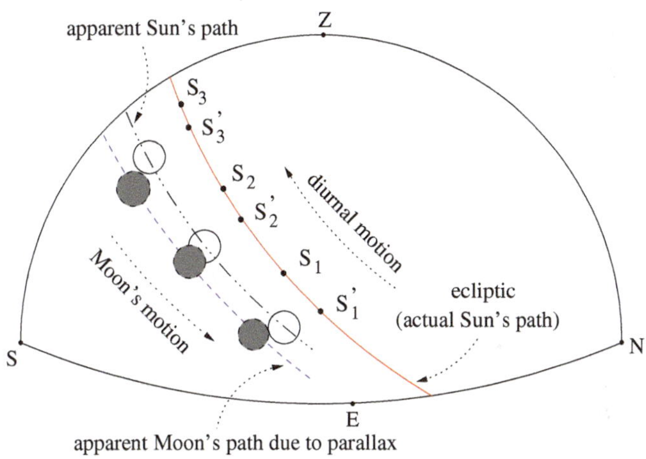

Fig. 5.10 The first contact, the middle and the last contact of a solar eclipse occurring in the eastern hemisphere.

parallax at the first contact, the middle and the last contact. S'_1, S'_2 and S'_3 are the projections of the apparent positions of the Sun along the ecliptic including the effect of parallax. We will clarify the meanings of S_i and S'_i shortly.

$$S_i S'_i = l_i \qquad i = 1, 2 \;\&\; 3, \tag{5.43}$$

are the parallaxes in longitude at the first contact, the middle and the last contact. In the absence of parallax, let Δt_1 and Δt_2 be the first and second half-durations. They are given by

$$\Delta t_1 = S_1 S_2 \qquad \text{and} \qquad \Delta t_2 = S_2 S_3. \tag{5.44}$$

Here we are using the symbols S_i and S'_i to denote the time-instants of the first contact, the middle and the last contact, and the corresponding ones corrected by the parallax in longitude. Hence, S_1S_2 corresponds to the true time difference between the positions of the Sun at S_1 and S_2, and not the arc S_1S_2 along the ecliptic. Similarly $S_iS'_i$ stands for the parallax in longitude l_i, which is actually the difference in *lambanas* between the Moon and the Sun $(l_{m_i} - l_{s_i})$ in *nāḍikās* and not the arc $S_iS'_i$. Let $\Delta t'_1$ and $\Delta t'_2$ be the first and second half-durations including the effect of parallax. We now proceed to discuss the different cases that can arise.

Case (i): *The first contact, the middle and the last contact in the eastern hemisphere*

The first and second half-durations are given by

$$\begin{aligned} \Delta t'_1 &= S'_1S'_2 \\ &= S'_1S_2 - S'_2S_2 \\ &= S'_1S_1 + S_1S_2 - S'_2S_2 \\ &= S_1S_2 + (S'_1S_1 - S'_2S_2) \\ &= \Delta t_1 + (l_1 - l_2). \end{aligned} \tag{5.45}$$

Similarly,

$$\begin{aligned} \Delta t'_2 &= S'_2S'_3 \\ &= S'_2S_3 - S'_3S_3 \\ &= S'_2S_2 + S_2S_3 - S'_3S_3 \\ &= S_2S_3 + (S'_2S_2 - S'_3S_3) \\ &= \Delta t_2 + (l_2 - l_3). \end{aligned} \tag{5.46}$$

The above equations (5.45) and (5.46) are valid if the first contact, the middle and the last contact take place in the eastern hemisphere as shown in Fig. 5.10.

Case (ii): *The first contact, the middle and the last contact in the western hemisphere*

This is depicted in Fig. 5.11. The expressions for the half-durations in this case are similar to the previous case and are given by

$$\begin{aligned} \Delta t'_1 &= S'_2S'_1 \\ &= S'_2S_1 - S'_1S_1 \\ &= S'_2S_2 + S_2S_1 - S'_1S_1 \\ &= S_2S_1 + (S'_2S_2 - S'_1S_1) \\ &= \Delta t_1 + (l_2 - l_1). \end{aligned} \tag{5.47}$$

Similarly,

$$\begin{aligned} \Delta t'_2 &= S'_3S'_2 \\ &= S'_3S_2 - S'_2S_2 \end{aligned}$$

$$
\begin{aligned}
&= S_3'S_3 + S_3S_2 - S_2'S_2 \\
&= S_3S_2 + (S_3'S_3 - S_2'S_2) \\
&= \Delta t_2 + (l_3 - l_2). \qquad (5.48)
\end{aligned}
$$

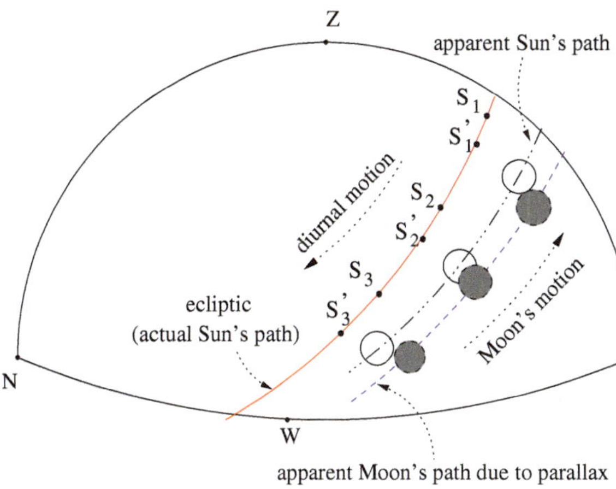

Fig. 5.11 The first contact, the middle and the last contact of a solar eclipse occurring in the western hemisphere.

Case (iii): *The first contact, the middle and the last contact in different hemispheres*
A typical case of the first contact and the middle happening in different hemispheres is shown in Fig. 5.12. Here the expression for the half-duration is given by

$$
\begin{aligned}
\Delta t_1' &= S_1'S_2' \\
&= S_1'S_2 + S_2'S_2 \\
&= S_1'S_1 + S_1S_2 + S_2'S_2 \\
&= S_1S_2 + (S_1'S_1 + S_2'S_2) \\
&= \Delta t_1 + (l_1 + l_2). \qquad (5.49)
\end{aligned}
$$

Similarly it can be shown that if the middle and the last contact happen in different hemispheres, then the expression for the half-duration will be

$$
\begin{aligned}
\Delta t_2' &= S_2'S_3' \\
&= S_2S_3 + (S_2'S_2 + S_3'S_3) \\
&= \Delta t_2 + (l_2 + l_3). \qquad (5.50)
\end{aligned}
$$

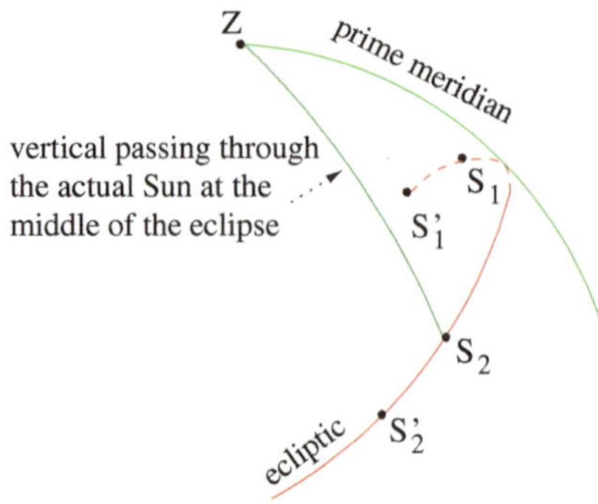

Fig. 5.12 The first contact and the middle of a solar eclipse occurring in different hemispheres.

Magnitudes of different *lambanas*

Now $S_iS_i' = l_i$, $i = 1,2,3$. As the effect of parallax increases with the zenith distance of the object, it is easily seen that

$$\begin{aligned} l_i &> l_{i+1} \qquad \text{(in eastern hemisphere)} \\ l_i &< l_{i+1} \qquad \text{(in western hemisphere).} \end{aligned} \tag{5.51}$$

Hence the parallaxes in longitude keep decreasing in the eastern hemisphere and keep increasing in the western hemisphere, over time. These inequalities are satisfied in most situations. However, care has to be exercised when either the first contact or the last contact happens to be near the time of sunset or sunrise. Such 'border-line' cases are discussed in the verses 19–22a.

When the instant of last contact is near the sunrise time, then the *madhyakāla* (instant of conjunction) will definitely be towards the end of the night, and the Sun/Moon will be below the horizon ($z > 90°$). Then $R\sin z$ would have a greater value at the last contact than in the middle. Hence the parallax in longitude at release will be larger than that in the middle. Therefore the second half-duration is given by

$$\begin{aligned} \Delta t_2' &= S_2'S_3' \\ &= S_2S_3 - (S_3'S_3 - S_2'S_2) \\ &= \Delta t_2 - (l_3 - l_2). \end{aligned} \tag{5.52}$$

This is the same as (5.46), except that $l_3 > l_2$, so that $l_3 - l_2$ must be subtracted from Δt_2.

Similarly, when the instant of first contact is near the sunset time, then the *madhyakāla* will definitely be towards the beginning of the night. Here again the parallax in longitude at first contact will be larger than that in the middle. Hence the first half-duration is given by

$$\Delta t_1' = S_1'S_2'$$

$$= S_1S_2 - (S_1'S_1 - S_2'S_2)$$
$$= \Delta t_1 - (l_1 - l_2). \tag{5.53}$$

५.८ स्पर्शकालाविशेषः

5.8 Time of *sparśa* by an iterative process

स्थित्यर्धमेवमानीतं स्पार्शिकं पर्वणस्त्यजेत् ॥ २२ ॥
अविशिष्टात् पुनस्तस्मिन् स्पर्शेऽपि नतिलम्बने ।
स्थित्यर्धमपि चानीय तन्नत्यानीतया सकृत् ॥ २३ ॥
स्थित्यर्धं स्पर्शमध्योत्थलम्बनान्तरसंस्कृतम् ।
लम्बनैक्ययुतं वा स्यात् स्पर्शमध्यकपालयोः ॥ २४ ॥
भेदे तेन विहीनो यो मध्यकालोऽविशेषितः ।
तत्रापि सकृदानीय नतिस्थितिदलादिकम् ॥ २५ ॥
लम्बनान्तरमैक्यं वा कुर्यात् स्थितिदलेऽत्र च ।
तेनापि स्पर्शकालादीन् प्राग्वदेवानयेन्मुहुः ॥ २६ ॥
कालं तं निश्चलीकृत्य मोक्षकालं तथानयेत् ।

sthityardhamevamānītaṃ spārśikaṃ parvaṇastyajet || 22 ||
aviśiṣṭāt punastasmin sparśe'pi natilambane |
sthityardhamapi cānīya tannatyānītayā sakṛt || 23 ||
sthityardhaṃ sparśamadhyotthalambanāntarasaṃskṛtam |
lambanaikyayutaṃ vā syāt sparśamadhyakapālayoḥ || 24 ||
bhede tena vihīno yo madhyakālo'viśeṣitaḥ |
tatrāpi sakṛdānīya natisthitidalādikam || 25 ||
lambanāntaramaikyaṃ vā kuryāt sthitidale'tra ca |
tenāpi sparśakālādīn prāgvadevānayenmuhuḥ || 26 ||
kālaṃ taṃ niścalīkṛtya mokṣakālaṃ tathānayet |

The first half-duration (*spārśikaṃ sthityardhaṃ*) of the eclipse may be subtracted from the instant of conjunction obtained by iteration. At the instant of first contact which is obtained, once again deflection from the ecliptic and parallax in longitude are obtained. The half-duration (Δt_{11}) is also calculated only once (*sakṛt*[5]) with the deflection from the ecliptic thus obtained.

To this half-duration, the sum of or the difference between the parallaxes in longitude is applied. The difference must be added [if the first contact and the middle of the eclipse occur in the same hemisphere] and the sum if they occur in different hemispheres. The result must be subtracted from the iterated value of the instant of conjunction. The deflection from the ecliptic and half-duration etc. may be obtained without iteration once again at this instant.

To this half-duration (Δt_{12}) the sum of or the difference beween the parallaxes in longitude must be applied. At the resulting time once again [the *nati*, half-duration etc. may be obtained]. Thus the beginning moment of the eclipse may be determined iteratively till a stable result is obtained (*niścalīkṛtya*). The ending moment (instant of last contact) may also be determined in the same manner.

[5] The word सकृत् is used deliberately to emphasize that, here, one need not find the half-duration by iteration.

In the above verses, an iterative procedure for determining the beginning and the ending moment of the eclipse (the first contact and the instant of last contact) is described. Let t_b, t_m and t_e be the actual beginning, middle and ending moments of the eclipse. Of the three, t_m has already been obtained by iteration. Here the objective is to find t_b and t_e by an iterative process from t_m which is supposed to be known accurately.

Let l_1, l_2 and l_3 be the parallaxes in longitude at the beginning, the middle and the ending moment of the eclipse. Of the three l_2 has already been found iteratively and it is only l_1 and l_3 that need to be calculated after each iteration. We shall denote by Δt_1 and Δt_2 the final iterated values of the first and the second half-durations of the eclipse. The intermediate values of the half-durations and the parallaxes in longitude are denoted with two suffixes. The first is used to keep track of which half-duration is being calculated (the first or second), and the second to denote the iteration count. Similarly in the case of parallax in longitude; for instance, Δt_{13} represents the third iterated value of the first half-duration. Similarly again, l_{14} refers to the parallax in longitude calculated at the beginning moment of the eclipse after four iterations. With this background we now explain the iterative process in detail.

Iterative process

We explain this process by considering case (ii) of Section 5.7, wherein the first contact, the middle and the last contact all occur in the western hemisphere, where $\Delta t_1'$ and $\Delta t_2'$ are given by (5.47) and (5.48). The other cases can be considered similarly. Let us denote the value of the half-duration of the eclipse determined with the deflection from the ecliptic at t_m as Δt_0. This value is approximate only because deflection from the ecliptic is a continuously varying quantity. Nevertheless, it serves as a starting point for the calculation. In finding the half-duration, the value of deflection from the ecliptic at the first contact or the last contact was taken to be the value at the instant of conjunction, that is t_m. This is obviously approximate. Hence as the first step for beginning the iteration, we take

$$\Delta t_0 = \Delta t_{10} = \Delta t_{20}. \tag{5.54}$$

Now, the first approximation to the instant of first contact is given by

$$t_{b1} = t_m - \Delta t_{10}. \tag{5.55}$$

At t_{b1}, the deflection from the ecliptic and parallax in longitude are calculated. We denote them by n_{e_1} and l_{11} respectively. With n_{e_1}, the half-duration, *sparśa-sthityardha*, without parallax in longitude correction is calculated using the formula

$$\delta t_1 = \frac{\sqrt{S^2 - n_{e_1}^2}}{\dot{\lambda}_m - \dot{\lambda}_s}, \tag{5.56}$$

where S in the numerator represents the sum of the semi-diameters of the Sun and the Moon, and $\dot{\lambda}_m - \dot{\lambda}_s$ in the denominator is the difference in their daily motions determined at that instant. This half-duration has to be corrected for the parallax in longitude. Thus, the first approximation to the first half-duration is given by

$$\Delta t_{11} = \delta t_1 + (l_2 - l_{11}). \tag{5.57}$$

Hence, the second approximation to the beginning moment of the eclipse is

$$t_{b2} = t_m - \Delta t_{11}. \tag{5.58}$$

At t_{b2}, once again the deflection from the ecliptic and the parallax in longitude are calculated. Denoting them by n_{e_2} and l_{12}, the half-duration (without parallax correction) is calculated using the formula

$$\delta t_2 = \frac{\sqrt{S^2 - n_{e_2}^2}}{\dot{\lambda}_m - \dot{\lambda}_s}. \tag{5.59}$$

With this, the second approximation to the first half-duration is given by

$$\Delta t_{12} = \delta t_2 + (l_2 - l_{12}). \tag{5.60}$$

And the third approximation to the beginning moment of the eclipse is given by

$$t_{b3} = t_m - \Delta t_{12}. \tag{5.61}$$

The above iterative process must be continued till we get stable values of Δt_{1i} to the desired accuracy. That is,

$$\Delta t_{1i} \approx \Delta t_{1\,i-1}. \tag{5.62}$$

When this happens, $t_{bi} \approx t_{b\,i-1} = t_b$. Now, t_b is the beginning moment of the eclipse, called the instant of first contact.

Rationale behind the iterative process

To determine the instant of first contact, the first half-duration is calculated including the effect of parallax and subtracted from the instant of conjunction. However, the formula for the half-duration involves the deflection from the ecliptic and the parallax in longitude at the instant of first contact, which are not known and are yet to be determined. This explains why an iterative process is used. In the first step, the deflection from the ecliptic and parallax in longitude are assumed to be the same as at t_m. From this, the instant of first contact is obtained. In the second step, the deflection from the ecliptic and the parallax in longitude are calculated at this approximate value of instant of first contact, and from these the instant of first contact in the next approximation is determined, and so on.

५.९ मोक्षकालाविशेषः
5.9 Time of *mokṣa* by an iterative process

अविशिष्टे तु पर्वान्ते स्थित्यर्धं तत् क्षिपेन्मुहुः ॥ २७ ॥
लम्बनान्तरमैक्यं वा मोक्षस्थितिदलेऽपि च ।
तत्तन्नतिकृते कार्यं प्राग्वत् तच्चाविशेषयेत्[6] ॥ २८ ॥
तद्युते मध्यकालेऽस्य मोक्षो वाच्यो विवस्वतः ।

aviśiṣṭe tu parvānte sthityardhaṃ tat kṣipenmuhuḥ || 27 ||
lambanāntaramaikyaṃ vā mokṣasthitidale'pi ca |
tattannatikṛte kāryaṃ prāgvat taccāviśeṣayet || 28 ||
tadyute madhyakāle'sya mokṣo vācyo vivasvataḥ |

That half-duration must be added to the iterated value of the instant of conjunction [and deflection from the ecliptic etc. must be calculated at that instant]. To the half-duration that is obtained from the deflection from the ecliptic calculated at that instant, here again the difference of the parallaxes in longitude or their sum must be applied. That [second half-duration] must be found iteratively. That added to the middle of the eclipse must be declared as the ending moment of the solar eclipse.

The iterative procedure for the determination of the instant of last contact is very similar to that of the instant of first contact. Since the procedure as well as the rationale has been described in detail in the previous section, here we just outline the iterative scheme in the form of equations for the sake of completeness. The first approximation to the instant of last contact is given by

$$t_{e_1} = t_m + \Delta t_{20}. \tag{5.63}$$

At t_{e_1}, the deflection from the ecliptic and the parallax in longitude are calculated. We denote them by n_{e_1} and l_{31} respectively. With n_{e_1}, the second half-duration the *mokṣasthityardha*, without parallax in longitude correction is calculated using the formula

$$\delta t_2 = \frac{\sqrt{S^2 - n_{e_1}^2}}{\dot{\lambda}_m - \dot{\lambda}_s}. \tag{5.64}$$

This is corrected for the parallax in longitude, and the first approximation to the second half-duration is given by

$$\Delta t_{21} = \delta t_2 + (l_{31} - l_2). \tag{5.65}$$

Thus, the second approximation to the instant of last contact is

$$t_{e_2} = t_m + \Delta t_{21}. \tag{5.66}$$

Again at t_{e_2}, deflections from the ecliptic and parallax in longitude are calculated. Denoting them by n_{e_2} and l_{32}, the half-duration is calculated using the formula

[6] The prose order is: अपि च, तत्तन्नतिकृते मोक्षस्थितिदले लम्बनान्तरमैक्यं वा मुहुः क्षिपेत्। तच्च प्राग्वत् अविशेषयेत् ।

$$\delta t_2 = \frac{\sqrt{S^2 - n_{e_2}^2}}{\dot{\lambda}_m - \dot{\lambda}_s}. \tag{5.67}$$

With this, the second approximation to the instant of last contact is given by

$$\Delta t_{22} = \delta t_2 + (l_{32} - l_2). \tag{5.68}$$

And the third approximation to the instant of last contact is

$$t_{e_3} = t_m + \Delta t_{22}. \tag{5.69}$$

The iteration is continued till the successive iterates converge to a stable value

$$\Delta t_{2i} \approx \Delta t_{2\,i-1} = \Delta t_2. \tag{5.70}$$

Then the instant of last contact is given by

$$t_e = t_m + \Delta t_2. \tag{5.71}$$

५.१० विमर्दार्धं निमीलोन्मीलनं च

5.10 Half-duration of obscuration and the time of submergence and emergence

चन्द्रबिम्बात् रवेर्बिम्बे त्यक्ते शिष्टस्य यद्दलम् ॥ २९ ॥
ततः स्फुटनतिर्हीना यदि स्यात् सकलग्रहः ।
चन्द्रबिम्बे रवेर्बिम्बात् त्यक्ते शिष्टस्य यद्दलम् ॥ ३० ॥
ततो यदि नतिर्हीना दृश्या स्यात् परिधिस्तदा ।
बिम्बभेदार्धवर्गात्तु[7] नतिर्वर्गोनितात् पदम् ॥ ३१ ॥
षष्टिघ्नं गतिभेदाप्तं विमर्दार्धं रवेरपि ।
चन्द्रेऽल्पेऽन्तर्ग्रहार्धं स्यात् प्राग्वत्ते चाविशेषयेत् ॥ ३२ ॥
मध्यकालाद् विमर्दार्धे शुद्धेऽत्रापि निमीलनम् ।
क्षिप्ते चोन्मीलनं तद्वत् पूर्तिश्छेदश्च नेमिगः ॥ ३३ ॥

candrabimbāt raverbimbe tyakte śiṣṭasya yaddalam || 29 ||
tataḥ sphuṭanatirhīnā yadi syāt sakalagrahaḥ |
candrabimbe raverbimbāt tyakte śiṣṭasya yaddalam || 30 ||
tato yadi natirhīnā dṛśyā syāt paridhistadā |
bimbabhedārdhavargāttu natirvargonitāt padam || 31 ||
ṣaṣṭighnaṃ gatibhedāptaṃ vimardārdhaṃ raverapi |
candre'lpe'ntargrahārdhaṃ syāt prāgvatte cāviśeṣayet|| 32 ||
madhyakālād vimardārdhe śuddhe'trāpi nimīlanam |
kṣipte conmīlanaṃ tadvat pūrtiśchedaśca nemigaḥ || 33 ||

---

[7] In both the printed editions, the reading is: बिम्बे भेदार्धवर्गात्तु, which is incorrect. It is most likely that the correct reading is: बिम्बभेदार्धवर्गात्तु।

If the Sun's disc is subtracted from the Moon's disc, and the *sphuṭa-nati* (true parallax in latitude) is less than half of the remainder thereof, then it is a total eclipse. If the Moon's disc is subtracted from the Sun's disc, and the *sphuṭa-nati* is less than half of the remainder thereof, then the periphery (of the Sun) will be seen.

The square root of the difference of the squares of the difference in the semi-diameters of the Sun and the Moon and the effective *nati*, multiplied by 60 and divided by the difference between their daily motions [that is, of the Sun and the Moon], is the half-duration of totality of the solar eclipse. If the Moon's disc is small, then the above measure is equal to the half-duration of annularity. These half-durations have to be found iteratively as described earlier.

From the middle of the eclipse, by subtracting and adding the half duration of totality, the instants of the beginning and the end of totality are obtained. Similarly the instants of *pūrti* (the instant of the beginning of annularity) and *cheda* (the instant of the end of annularity) are obtained in the case of an annular eclipse.

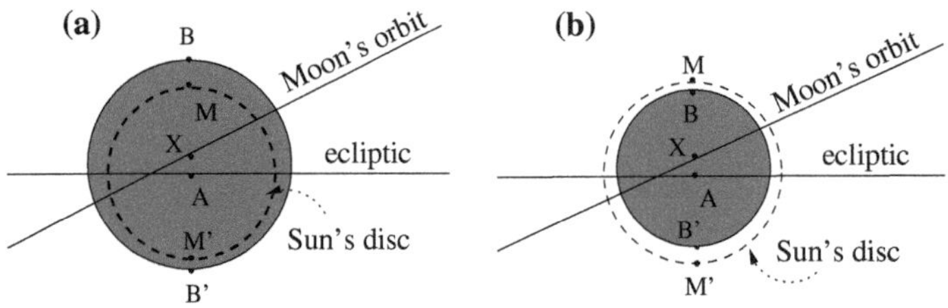

Fig. 5.13 The Sun and the Moon in the case of (a) a total solar eclipse and (b) an annular eclipse.

We depict the total and annular solar eclipses in Figs 5.13(a) and 5.13(b) respectively. Here A and X represent the centres of the discs of the Sun and the Moon. Now the condition for a total solar eclipse is

$$\begin{aligned} AM' &< AB' \\ \text{or} \quad AM' &< XB' - AX \\ \text{or} \quad AX &< XB' - AM', \end{aligned} \tag{5.72}$$

that is,

$$\textit{nati} < \text{radius of lunar disc} - \text{radius of solar disc.}$$

Similarly in Fig. 5.13(b), the condition for an annular eclipse is

$$\begin{aligned} AB &< AM \\ \text{or} \quad AX + XB &< AM \\ \text{or} \quad AX &< AM - XB \end{aligned} \tag{5.73}$$

that is,

$$\textit{nati} < \text{radius of solar disc} - \text{radius of lunar disc.}$$

The procedures for calculating the half-duration of the total solar eclipse and the annular solar eclipse are similar to those given in the case of the total lunar eclipse.

They are also similar to those for a partial solar eclipse except that the sum of the semi-diameters of the solar and lunar discs should be replaced by their difference. The half-durations are given by

$$\delta t = \frac{\sqrt{D^2 - n_e^2}}{\dot{\lambda}_m - \dot{\lambda}_s}, \tag{5.74}$$

where D is the difference in the semi-diameters of the lunar and solar discs, and n_e the effective deflection from the ecliptic at the beginning or end of totality.

५.११ सूर्यदृक्कर्णः

5.11 The *dṛkkarṇa* of the Sun

द्रष्टुर्भूपृष्ठगस्येन्दुबिम्बोऽर्काच्च महान् भवेत् ।
नानात्वात् प्रतिदेशं तत् नेयं बिम्बं स्वदेशजम् ॥ ३४ ॥
मध्यकालभुजाज्यायाः प्राग्वद्दृक्क्षेपमानयेत् ।
भानोर्दृक्क्षेपलग्ना ज्या[8] हता दृक्क्षेपशङ्कुना ॥ ३५ ॥
त्रिज्याप्ता दृग्गतिर्भानोः यत् तद्दृक्क्षेपवर्गयोः ।
योगात् पदं तदैक्योनं त्रिज्यावर्गाच्च यत्पदम् ॥ ३६ ॥
छायाशङ्कू रवेस्ताभ्यां भूव्यासार्धस्य योजनैः ।
हताभ्यां त्रिज्यया लब्धे दोःकोटी योजनात्मिके ॥ ३७ ॥
रविभूम्यन्तरात् कोटिं त्यक्त्वा तद्बाहुवर्गयोः ।
योगात् पदं भवेद्भानोः दृक्कर्णो योजनात्मकः ॥ ३८ ॥

draṣṭurbhūpṛṣṭhagasyendubimbo'rkācca mahān bhavet |
nānātvāt pratideśaṃ tat neyaṃ bimbaṃ svadeśajam || 34 ||
madhyakālabhujājyāyāḥ prāgvaddṛkkṣepamānayet |
bhānordṛkkṣepalagnā jyā hatā dṛkkṣepaśaṅkunā || 35 ||
trijyāptā dṛggatirbhānoḥ yat taddṛkkṣepavargayoḥ |
yogāt padaṃ tadaikyonaṃ trijyāvargācca yatpadam || 36 ||
chāyāśaṅkū ravestābhyāṃ bhūvyāsārdhasya yojanaiḥ |
hatābhyāṃ trijyayā labdhe doḥkoṭī yojanātmike || 37 ||
ravibhūmyantarāt koṭiṃ tyaktvā tadbāhuvargayoḥ |
yogāt padaṃ bhavedbhānoḥ dṛkkarṇo yojanātmakaḥ || 38 ||

[At the time of a solar eclipse] the dimension of the Moon's disc will be larger than that of the Sun for an observer on the surface of the Earth. Since this differs from place to place, the dimension at one's own location must be determined.

Find the *dṛkkṣepa* from the *madhya-kālabhujājyā* as mentioned earlier. The product of the *dṛkkṣepa-śaṅku* and the Rsine of the difference of the Sun and the *vitribhalagna* divided by the *trijyā* is the *dṛggati* of the Sun. The sum of the squares of this and the *dṛkkṣepa* is found. The square root of this and that of the *trijyā* squared minus this square are the *chāyā* (shadow) and the *śaṅku* (gnomon) of the Sun. These [the *chāyā* and *śaṅku*]

---

[8] This reading, found in both the printed editions, seems to be faulty. The meaning intended to be conveyed is: भानोः दृक्क्षेपलग्नस्य च यदन्तरं तस्य ज्या।

multiplied by the radius of the Earth in *yojanas* and divided by the *trijyā* are the *doḥ* and *koṭi* in *yojanas*.

The *koṭi* is subtracted from the distance of separation between the Sun and the Earth. The square root of the sum of the squares of this and the *bāhu* is the *dṛkkarṇa* of the Sun in *yojanas*.

Here the procedure for finding the *dṛkkarṇa* of the Sun is given. The term *dṛk* in this context refers to the observer. As the term *karṇa* is used to refer to the hypotenuse, the term *dṛkkarṇa* refers to the hypotenuse joining the Sun with the observer (OS in Fig. 5.14(a)). In short, the problem posed in the text is to obtain OS from ES. In order to determine OS, a few intermediate quantities are introduced. The quantities *dṛggati* and *dṛkkṣepa* are defined as follows:

$$\begin{aligned} \textit{dṛggati} &= \frac{R\sin(\lambda_s-\lambda_v)R\cos z_v}{R} \\ \textit{dṛkkṣepa} &= R\sin z_v. \end{aligned} \tag{5.75}$$

The *dṛggati* given by (5.75) is different from the *dṛggati* defined earlier in verse 7 of this chapter. It is then stated that

$$\begin{aligned} \textit{chāyā} &= \sqrt{\textit{dṛggati}^2+\textit{dṛkkṣepa}^2} \\ \text{i.e.} \qquad R\sin z_s &= \sqrt{(R\sin(\lambda_s-\lambda_v)\cos z_v)^2+(R\sin z_v)^2} \end{aligned} \tag{5.76}$$

$$\begin{aligned} \text{and} \qquad \textit{śaṅku} &= \sqrt{\textit{trijyā}^2-\textit{chāyā}^2} \\ \text{i.e.} \qquad R\cos z_s &= \sqrt{R^2-R\sin z_s^2}, \end{aligned} \tag{5.77}$$

where z_s is the zenith distance of the Sun.

To understand the rationale behind the above expressions let us consider the spherical triangle ZVS shown in Fig. 5.14(b). Here K is the pole of the ecliptic, Z the zenith, V the *vitribhalagna* and S the Sun. Since $Z\hat{V}S=90$, the cosine formula applied to this triangle gives

$$\cos(ZS)=\cos(ZV)\cos(SV). \tag{5.78}$$

Using the notation $ZS=z_s$, $ZV=z_v$ and $SV=\lambda_s-\lambda_v$, the above equation becomes

$$\cos z_s=\cos z_v\cos(\lambda_s-\lambda_v). \tag{5.79}$$

Squaring both the sides, and writing the cosines in terms of sines, we have

$$\begin{aligned} 1-\sin^2 z_s &= \cos^2 z_v(1-\sin^2(\lambda_s-\lambda_v)) \\ &= \cos^2 z_v-\cos^2 z_v\sin^2(\lambda_s-\lambda_v) \\ \text{or} \qquad \sin^2 z_s &= (1-\cos^2 z_v)+\cos^2 z_v\sin^2(\lambda_s-\lambda_v) \\ &= \cos^2 z_v\sin^2(\lambda_s-\lambda_v)+\sin^2 z_v, \end{aligned}$$

$$\text{or} \qquad \sin z_s = \sqrt{\cos^2 z_v \sin^2(\lambda_s - \lambda_v) + \sin^2 z_v}. \tag{5.80}$$

The above equation is the same as the expression for the *chāyā* in (5.76). The aim of the whole exercise of finding the *chāyā* and the *śaṅku* is to find ON and EN in Fig. 5.14(a), which in turn is used in finding the *dṛkkarṇa* OS. In verse 37, ON and EN are referred to as the *doḥ* (sine) and the *koṭi* (cosine) respectively. Let R_e be the radius of the Earth (OE).

Now, in Fig. 5.14(a), draw ZN' perpendicular to ES. As $EZ = R$ and $Z\hat{E}S = z_s$, we have $ZN' = R\sin z_s$ and $EN' = R\cos z_s$. Now the triangles EON and EZN' are similar. Hence

$$\begin{aligned} doh = ON &= \frac{EO}{EZ} \times ZN' \\ &= \frac{R_e}{R} \times R\sin z_s \\ &= \frac{bh\bar{u}vy\bar{a}s\bar{a}rdha}{trijy\bar{a}} \times ch\bar{a}y\bar{a} \end{aligned} \tag{5.81a}$$

$$\begin{aligned} \text{and} \qquad koti = EN &= \frac{EO}{EZ} \times EN' \\ &= \frac{R_e}{R} \times R\cos z_s \\ &= \frac{bh\bar{u}vy\bar{a}s\bar{a}rdha}{trijy\bar{a}} \times \acute{s}a\dot{n}ku. \end{aligned} \tag{5.81b}$$

These are the relations that have been stated in verse 37. Since R_e is in *yojanas*, the dimensions of the *doḥ* and the *koṭi* will be in *yojanas*. The expression of the *dṛkkarṇa*, which is the distance between the observer and the Sun, is given to be

$$dr kkarna = \sqrt{(ravibh\bar{u}myantara - koti)^2 + (doh)^2}. \tag{5.82}$$

The rationale behind (5.82) can again easily be understood from Fig. 5.14(a). Here ES (the *ravibhūmyantara*) and OS (the *dṛkkarṇa*) are the distances of the Sun from the centre of the Earth and the observer. Denoting them by D_s and d_s, and considering the triangle NOS, we have

$$\begin{aligned} OS &= \sqrt{NS^2 + ON^2} \\ d_s &= \sqrt{(ES - EN)^2 + ON^2} \\ &= \sqrt{(D_s - R_e\cos z_s)^2 + (R_e\sin z_s)^2}\,. \end{aligned} \tag{5.83}$$

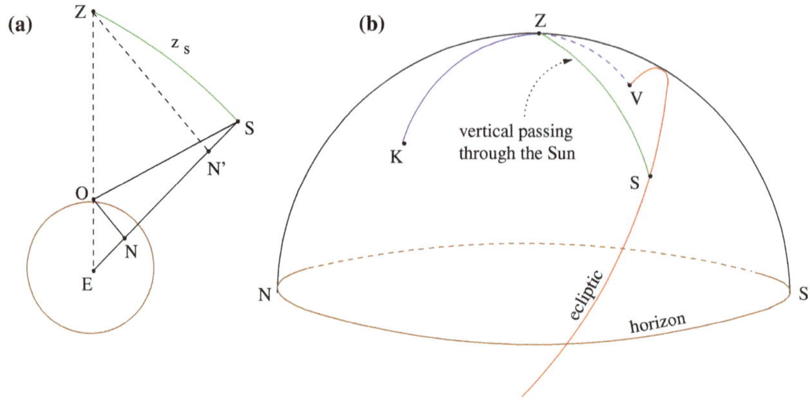

Fig. 5.14 Spherical triangle considered to find the distance of the Sun from the observer.

५.१२ चन्द्रदृक्कर्णः
5.12 The *dṛkkarṇa* of the Moon

क्षेपदृक्क्षेपचापैक्यं नतिः साम्येऽन्यथान्तरम् ।
राशित्रयं यदूनं[9] यत् तज्ज्याशङ्कुः पराभिधः ॥ ३९ ॥
चन्द्रोनलग्नबाहूना त्रिज्या जूकादिजान्विता ।
क्षेपकोट्या हता भक्ता त्रिज्यया बाण उच्यते ॥ ४० ॥
ततो दृक्क्षेपकोटिघ्नं त्रिज्याप्तं परतस्त्यजेत्[10] ।
शेषः शङ्कुः शशाङ्कस्य ततो दृग्ज्या च पूर्ववत् ॥ ४१ ॥
भूव्यासार्धहते ते च त्रिज्याप्ते कोटिदोःफले ।
दृक्कर्णोऽर्कवदिन्दोस्तत् स्वभूम्यन्तरयोजनैः ॥ ४२ ॥

kṣepadṛkkṣepacāpaikyaṃ natiḥ sāmye'nyathāntaram |
rāśitrayaṃ yadūnaṃ yat tajjyāśaṅkuḥ parābhidhaḥ || 39 ||
candronalagnabāhūnā trijyā jūkādijānvitā |
kṣepakoṭyā hatā bhaktā trijyayā bāṇa ucyate || 40 ||
tato dṛkkṣepakoṭighnaṃ trijyāptaṃ paratastyajet |
śeṣaḥ śaṅkuḥ śaśāṅkasya tato dṛgjyā ca pūrvavat || 41 ||
bhūvyāsārdhahate te ca trijyāpte koṭidoḥphale |
dṛkkarṇo'rkavadindostat svabhūmyantarayojanaiḥ || 42 ||

The *nati* [of the Moon] is the sum of the *vikṣepa* and the *dṛkkṣepa* if both have the same direction, and their difference if they have opposite directions. The Rsine of that (*nati*) subtracted from three signs is called the *paraśaṅku*.

The Rsine of the difference of the *lagna* and the Moon's longitude is added to the *trijyā* if the difference is the *tulādi* (greater than 180 degrees), and subtracted from the *trijyā* otherwise. This quantity multiplied by the Rcosine of the latitude of the Moon and divided by the *trijyā* is called the *bāṇa* ('arrow').

[9] The *vigraha* is: येन ऊनं = यदूनम् ।
[10] परशङ्कुः इति पूर्वं यदुक्तं तस्मात् त्यजेत्।

This (the *bāṇa*) multiplied by the *koṭi* of the *dṛkkṣepa* and divided by the *trijyā* is to be subtracted from the *para* (the *paraśaṅku*). The remainder is the *śaṅku* of the Moon. From that the *dṛgjyā* may be obtained as earlier. These quantities (the *śaṅku* and *dṛgjyā*) multiplied by the radius of the Earth and divided by the *trijyā* are the *koṭi* and *doḥphala*. From this, the distance of separation between the Moon and the Earth or the *dṛkkarṇa* of the Moon may be determined in the same way as in the case of the Sun.

The determination of the *dṛkkarṇa* of the Moon is a little more complicated than that of the Sun. We explain this with the help of Figs 5.15 and 5.16. In Fig. 5.15, the point V' is the point of intersection of the vertical through the *vitribhalagna* and the circle parallel to the ecliptic passing through the Moon. Hence

$$VV' = AM = -\beta_t, \tag{5.84}$$

is the magnitude of the true latitude of the Moon, called the *vikṣepa*. The '−' sign in the above equation indicates that the *vikṣepa* is southwards (as shown in the figure). Here an intermediate quantity called the *nati* (n_m) is introduced, which is not to be confused with the parallax in latitude of the Moon. It is defined to be the sum of the *dṛkkṣepa* and the *vikṣepa*:

$$n_m = ZV' = ZV - \beta_t = z_v \pm |\beta_t|, \tag{5.85}$$

where the sign should be '+' if both have the same direction and '−' if they have opposite directions. The term *paraśaṅku* is defined in verse 39*a* as

$$paraśaṅku = R\sin(90 - n_m) = R\cos n_m. \tag{5.86}$$

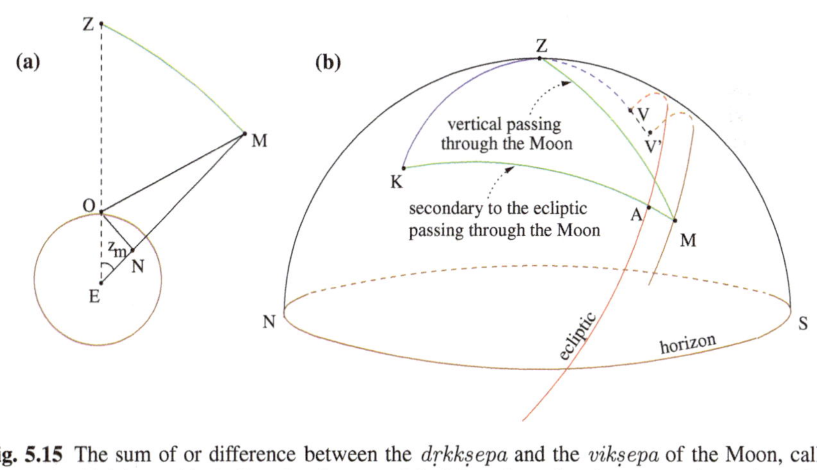

Fig. 5.15 The sum of or difference between the *dṛkkṣepa* and the *vikṣepa* of the Moon, called the *nati*, which is used in finding the distance of the Moon from the observer on the surface of the Earth.

Let λ_m be the *sāyana* (tropical) longitude of the Moon and λ_l that of the *lagna*. Using them, an intermediate quantity (x) is defined, which in turn is used to define another quantity, the *bāṇa*:

$$x = R \pm |R\sin(\lambda_l - \lambda_m)|, \tag{5.87}$$

where the sign should be '+' when $\lambda_l - \lambda_m > 180$ and '−' otherwise. In other words, $x = R - R\sin(\lambda_l - \lambda_m)$. In the measure of radians, x may be written as $x = 1 - \sin(\lambda_l - \lambda_m)$. Now, the *bāṇa* is defined as

$$bāṇa = \frac{x \times R\cos\beta_t}{R}. \tag{5.88}$$

Using this *bāṇa*, the *śaṅku* of the Moon ($R\cos z_m$) is given by

$$śaṅku = paraśaṅku - \frac{bāṇa \times dṛkkṣepakoṭi}{trijyā} \tag{5.89}$$

$$\begin{aligned} R\cos z_m &= R\cos n_m - \frac{bāṇa \times R\cos z_v}{\mathrm{R}} \\ &= R\cos n_m - x \times \cos\beta_t \times \cos z_v \\ &= R\cos n_m - R[1 - \sin(\lambda_l - \lambda_m)]\cos\beta_t \times \cos z_v. \end{aligned} \tag{5.90}$$

We now prove the result from spherical trigonometry.

Proof:

From (5.85),

$$\begin{aligned} \cos n_m &= \cos(z_v - \beta_t) \\ &= \cos z_v \cos\beta_t + \sin z_v \sin\beta_t \\ \text{or} \qquad \cos n_m - \cos z_v \cos\beta_t &= \sin z_v \sin\beta_t. \end{aligned} \tag{5.91}$$

Considering the spherical triangle ZAM in Fig. 5.16(a), and applying the cosine formula we have,

$$\cos ZM = \cos ZA \cos\beta_t - \sin ZA \sin\beta_t \cos\theta, \tag{5.92}$$

where θ is the angle between the vertical and the secondary to the ecliptic at A. The same cosine formula applied to the triangle ZAV yields

$$\cos ZA = \cos AV \cos ZV \qquad (\text{since } Z\hat{V}A = 90). \tag{5.93}$$

Applying the sine formula to the same triangle,

$$\frac{\sin ZV}{\sin Z\hat{A}V} = \frac{\sin ZA}{\sin 90}. \tag{5.94}$$

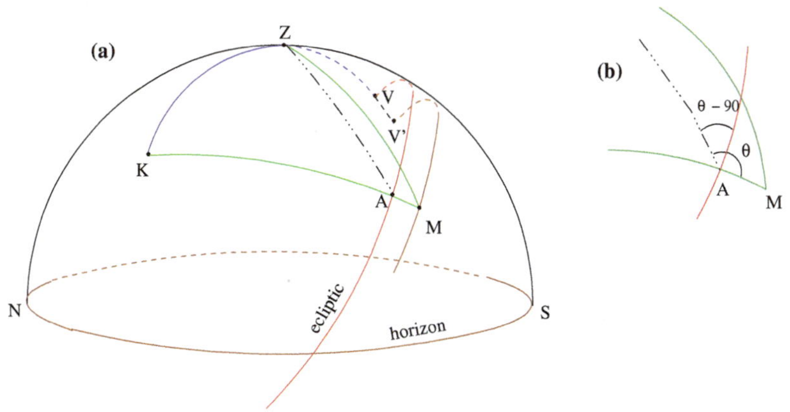

Fig. 5.16 Spherical triangle considered in order to find the *śaṅku* of the Moon, that is used in finding the distance of the Moon from the observer on the surface of the Earth.

But $\sin Z\hat{A}V = \sin(\theta - 90) = -\cos\theta$. Therefore

$$\sin ZV = -\sin ZA \cos\theta. \tag{5.95}$$

Using (5.91), (5.93) and (5.95) in (5.92) with $ZV = z_v$, we have

$$\begin{aligned} \cos ZM &= \cos AV \cos z_v \cos\beta_t + \sin z_v \sin\beta_t \\ &= \cos AV \cos z_v \cos\beta_t + \cos n_m - \cos z_v \cos\beta_t. \end{aligned} \tag{5.96}$$

In the above equation, $\cos AV = \cos(\lambda_m - \lambda_v)$. But $\lambda_v = \lambda_l - 90$. Therefore,

$$\cos AV = \cos(90 - (\lambda_l - \lambda_m)) = \sin(\lambda_l - \lambda_m). \tag{5.97}$$

Substituting for $\cos AV$ in (5.96)

$$\begin{aligned} \cos ZM &= \cos n_m - \cos z_v \cos\beta_t + \cos z_v \cos\beta_t \sin(\lambda_l - \lambda_m), \\ \text{or} \qquad \cos z_m &= \cos n_m - [1 - \sin(\lambda_l - \lambda_m)] \cos z_v \cos\beta_t, \end{aligned} \tag{5.98}$$

which is the same as (5.90).

After giving the expression for *śaṅku*, it is stated that the *dṛgjyā* may be obtained as earlier (*tato dṛgjyā ca pūrvavat*). The commentator explains this as follows.

ततो दृग्ज्या च पूर्ववत् तत्त्रिज्यावर्गविश्लेषमूलेनैव कर्तव्या ।

As done earlier, the *dṛgjyā* has to be obtained by subtracting its square from the square of the *trijyā* and finding the square root.

In other words,

$$dṛgjyā = \sqrt{trijyā^2 - śaṅku^2}$$

$$= \sqrt{R^2 - (R\cos z_m)^2} = R\sin z_m. \tag{5.99}$$

The purpose of finding the *śaṅku* and *dṛgjyā* is to find EN and ON in Fig. 5.15(a), which are required to find the *dṛkkarṇa* of the Moon OM. In verse 41, EN and ON are referred to as the *doḥphala* and the *koṭiphala* respectively. If R_e is the radius of the Earth (OE), then they are to be obtained in the same way, as in the case of the Sun, from the relations

$$doḥphala = \frac{bhūvyāsārdha \times dṛgjyā}{trijyā} = \frac{R_e \times R\sin z_m}{R} = R_e \sin z_m \tag{5.100}$$

and

$$koṭiphala = \frac{bhūvyāsārdha \times śaṅku}{trijyā} = \frac{R_e \times R\cos z_s}{R} = R_e \cos z_m. \tag{5.101}$$

Since R_e is in *yojanas* the dimensions of the *doḥ* and *koṭi* are also in *yojanas.* Then it is noted that the *dṛkkarṇa* of the Moon may be calculated as in the case of the Sun. This is clarified in *Laghu-vivṛti* thus:

> **अतः तदेव कोटिफलं चन्द्रस्य द्वितीयस्फुटयोजनकर्णतो विशोध्य शिष्टस्य तद्दोःफलस्य च वर्गयोगमूलं चन्द्रस्य दृक्कर्णो योजनात्मको भवति।**
>
> Therefore, the square root of the sum of the squares of the same *koṭiphala* of the Moon subtracted from the *dvitīya-sphuṭa-yojana-karṇa* and the *doḥphala* becomes the *dṛkkarṇa* of the Moon in *yojanas.*

In other words,

$$dṛkkarṇa = \sqrt{(dvitīya\text{-}sphuṭa\text{-}yojana\text{-}karṇa - koṭiphala)^2 + doḥphala^2}. \tag{5.102}$$

This is evident from Fig. 5.15(a). Here EM is referred to as the *dvitīya-sphuṭa-yojana-karṇa* (discussed in Chap. 4), and OM is the *dṛkkarṇa* to be found. We denote them by D_{2m} and d_m respectively. Considering the triangle NOM, we have

$$\begin{aligned} OM &= \sqrt{MN^2 + ON^2} \\ \text{or} \qquad d_m &= \sqrt{(EM - EN)^2 + ON^2} \\ &= \sqrt{(D_{2m} - R_e\cos z_m)^2 + (R_e \sin z_m)^2}. \end{aligned} \tag{5.103}$$

५.१३ दृग्गोलगतत्वसम्पादनम्

5.13 Transformation to the observer-centred celestial sphere

> त्रिज्याघ्नाद्योजनव्यासात् तेनाप्ता बिम्बलिप्तिकाः ।
> इष्टेन्दुः समलिप्तेन्दोः द्वितीयस्फुटभोगतः ॥ ४३ ॥

इष्टकेवलपर्वान्तद्युगतान्तरकालजात् ।
विक्षेपः केवलाच्चन्द्रात् प्राग्वत् त्रिज्याहृतो हतः ॥ ४४ ॥
योजनैर्विवरे चन्द्रभगोलघनमध्ययोः ।
दृक्कर्णयोजनैर्भक्तो दृग्गोले क्षेप इष्यताम् ॥ ४५ ॥
केवलादेव दृक्क्षेपात् भूव्यासार्धेन ताडितात् ।
विधोर्योजनदृक्कर्णभक्तात्र नतिलिप्तिकाः ॥ ४६ ॥
तद्वदेव च दृक्क्षेपात् स्वदृक्कर्णेन भास्वतः ।
रवीन्द्वोर्नतिभेदः स्यात् सर्वदैव नतिर्विधोः ॥ ४७ ॥
तद्भूपृष्ठोदितक्षेपयुतिः साम्येऽन्यथान्तरम् ।
एवं भूपृष्ठगान्नीत्वा नतिं बिम्बद्वयं तथा ॥ ४८ ॥
सर्वग्रासो विनिर्णेयो नाम्ना मध्यतमस्तथा ।
ग्रहणं वाप्यभावो वा वाच्यो मानैः स्फुटैरिह ॥ ४९ ॥

trijyāghnādyojanavyāsāt tenāptā bimbaliptikāḥ |
iṣṭenduḥ samaliptendoḥ dvitīyasphuṭabhogataḥ || 43 ||
iṣṭakevalaparvāntadyugatāntarakālajāt |
vikṣepaḥ kevalāccandrāt prāgvat trijyāhṛto hataḥ || 44 ||
yojanairvivare candrabhagolaghanamadhyayoḥ |
dṛkkarṇayojanairbhakto dṛggole kṣepa iṣyatām || 45 ||
kevalādeva dṛkkṣepāt bhūvyāsārdhena tāḍitāt |
vidhoryojanadṛkkarṇabhaktātra natiliptikāḥ || 46 ||
tadvadeva ca dṛkkṣepāt svadṛkkarṇena bhāsvataḥ |
ravīndvornatibhedaḥ syāt sarvadaiva natirvidhoḥ || 47 ||
tadbhūpṛṣṭhoditakṣepayutiḥ sāmye'nyathāntaram |
evaṃ bhūpṛṣṭhagānnītvā natiṃ bimbadvayaṃ tathā || 48 ||
sarvagrāso vinirṇeyo nāmnā madhyatamastathā |
grahaṇaṃ vāpyabhāvo vā vācyo mānaiḥ sphuṭairiha || 49 ||

The diameter in *yojanas* multiplied by the *trijyā* and divided by that (the *dṛkkarṇa*) is the diameter of the disc in minutes [with respect to the *dṛggola*]. From the *dvitīya-sphuṭa-bhukti* (second corrected rate of motion) of the Moon determined at the instant of conjunction (*samaliptendoḥ*), and the time difference between the desired instant and uncorrected instant of conjunction [i.e. the instant of conjunction not corrected for *lambana*], the longitude of the Moon at the desired instant is obtained.

The latitude of the Moon is obtained from its uncorrected longitude (not corrected for parallax in longitude) as earlier. It is divided by the *trijyā* and multiplied by the distance of separation between the centre of the Earth and the Moon. The above divided by the *dṛkkarṇa* in *yojanas* is the latitude in the *dṛggola*.

The *dṛkkṣepa* multiplied by the radius of the Earth and divided by the *dṛkkarṇa* of the Moon in *yojanas* will be the parallax in latitude [of the Moon] in minutes. Similarly from the *dṛkkṣepa* of the Sun and its own *dṛkkarṇa* [its parallax in latitude in minutes has to be obtained]. The difference between the *natis* of the Sun and the Moon will always be the effective parallax in latitude of the Moon.

The sum of or the difference between it (the effective parallax in latitude of the Moon) and the latitude as seen by an observer on the surface must be found depending upon whether they have the same direction or opposite directions.

Thus, having obtained the deflection from the ecliptic and the diameter of the two discs for an observer on the surface of the Earth, the [instant of] totality of the eclipse has to be determined, which is also called the middle of the eclipse. Similarly the occurrence or non-occurrence of the eclipse should be pronounced only by considering these actual values obtained thus.

The angular diameter of an object for an observer at the centre of the Earth can be calculated from its linear diameter, specified in the texts in *yojanas*, and its geocentric distance. In order to get the values for an observer on the surface of the Earth, a correction has to be applied to this value, since the angular diameter of a celestial object measured by an observer depends upon the distance of the observer from the object. This is true not only for the distances of the Sun and the Moon, but also for the values of the latitude and the deflection from the ecliptic. In the following we explain the corrections prescribed here to obtain the observer-centric values from the geocentric values.

Correction to the diameter

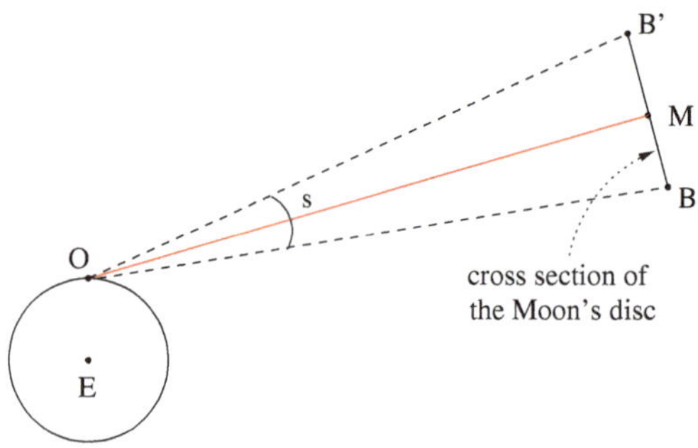

Fig. 5.17 The angular diameter of the Moon as seen by an observer on the surface of the Earth.

In Fig. 5.17, M is the centre of the Moon's disc. B' and B are its top and bottom edges. E is the centre of the Earth and O the observer on the surface of the Earth. BB' is the diameter of the Moon's disc in *yojanas* (specified in the text), OM is the *dṛkkarṇa* of the Moon in *yojanas*, and s is the angular diameter in radians. Then the angular diameter of the Moon in minutes as seen by an observer on the surface of the Earth is given to be

$$R \times s = \frac{R \times BB'}{OM}. \tag{5.104}$$

Similarly, the angular diameter of the Sun is obtained using the Sun's linear diameter in *yojanas* whose *dṛkkarṇa* was found earlier.

Correction to the latitude

The latitude of the Moon at a desired instant depends upon its longitude at that instant. Hence the longitude of the Moon at the desired instant is determined accurately first. If λ_{m0} is the longitude of the Moon at the instant of conjunction, and t_m and t_d denote the instant of conjunction and the desired instant, then the longitude of the Moon at the desired instant is given by

$$\lambda_{md} = \lambda_{m0} + \frac{(t_d - t_m) \times \dot{\lambda}_m}{60}. \tag{5.105}$$

In the above expression $\dot{\lambda}_m$ represents the Moon's *dvitīya-sphuṭa-bhukti* (second corrected rate of motion). The latitude of the Moon at the desired instant t_d is obtained using the formula

$$\beta = i \, \sin(\lambda_{md} - \lambda_n), \tag{5.106}$$

where λ_n is the longitude of the Moon's node at the desired instant. This latitude corresponds to an observer at the centre of the Earth. The latitude as seen by an observer on the surface of the Earth is given by

$$\beta_t = \frac{\beta \times \textit{dvitīya-sphuṭa-karṇa}}{\textit{dṛkkarṇa}}. \tag{5.107}$$

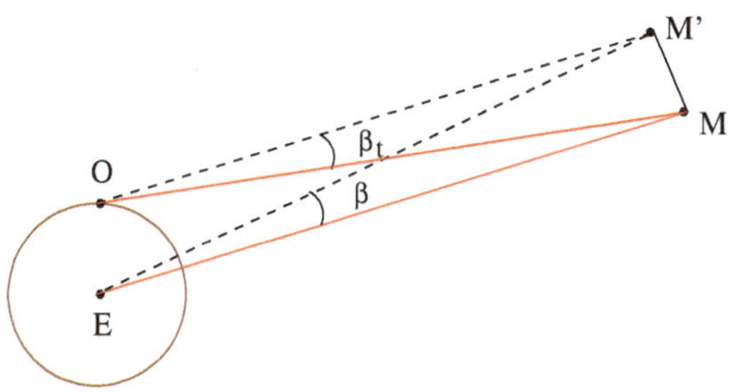

Fig. 5.18 The latitude of the Moon as seen by an observer on the surface of the Earth.

The above expression for the true latitude may be understood with the help of Fig. 5.18. Here M represents the actual position of the Moon, and M' is the point on the ecliptic whose longitude is the same as that of M. β and β_t are indicated in the figure. Considering the triangles MEM' and MOM', we have

$$MM' = \beta \times EM = \beta_t \times OM. \tag{5.108}$$

Therefore,

$$\beta_t = \frac{\beta \times EM}{OM}, \tag{5.109}$$

which is the same as (5.107), once it is recognized that EM is the *dvitīya-sphuṭa-karṇa* (the distance of separation between the centre of the Moon and the centre of the Earth) in *yojanas*, and OM is the *dṛkkarṇa* (the distance of separation between the centre of the Moon and the observer on the surface of the Earth), again in *yojanas*.

Correction to the *nati*

In Fig. 5.19(a), M is the Moon and z_m and z'_m are the actual and the apparent zenith distances of the Moon. If $R_e = OE$ is the radius of the Earth, then from the triangle MOE we have

$$\frac{\sin z_m}{OM} = \frac{\sin p_m}{R_e}. \tag{5.110}$$

Therefore, the parallax of the Moon is given by

$$\begin{aligned} z'_m - z_m = p_m &\approx \sin p_m \\ &= \frac{\sin z_m \times R_e}{OM}, \end{aligned} \tag{5.111}$$

where OM is the *dṛkkarṇa* of the Moon. This is shown in Fig. 5.19(a). The parallax

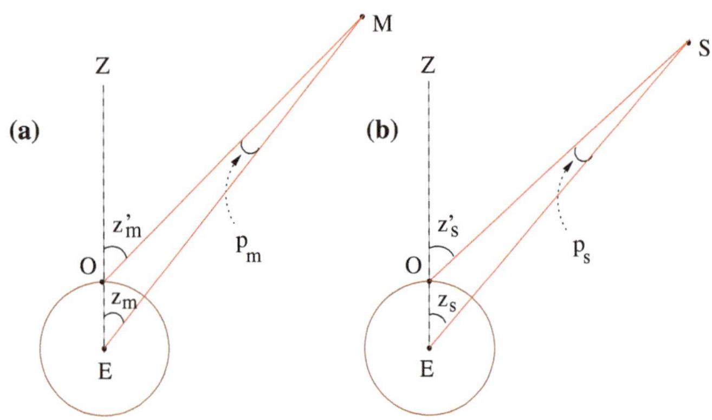

Fig. 5.19 The parallax of the Moon and the Sun.

is along the vertical through the Moon. For finding the Moon's *nati* (n_m), which is the component of the lunar parallax p_m perpendicular to the ecliptic, this has to be multiplied by

$$\cos\xi = \frac{\sin z_v}{\sin z_m}, \tag{5.112}$$

where $R \sin z_v$ is the *dṛkkṣepa* (refer to Sections 5.3 and 5.4 and Fig. 5.4 for details). Hence

$$\begin{aligned} n_m &= p_m \cos\xi \\ &= p_m \times \frac{\sin z_v}{\sin z_m} \\ &= \frac{R_e}{OM} \times \sin z_v, \end{aligned} \tag{5.113}$$

where we have used (5.111). Similarly the parallax of the Sun (Fig. 5.19(b)) is given by

$$z_s' - z_s = p_s \approx \sin p_s \tag{5.114}$$

$$= \frac{\sin z_s \times R_e}{OS}, \tag{5.115}$$

and the solar parallax in latitude n_s is given by

$$n_s = \frac{R_e}{OS} \times \sin z_v. \tag{5.116}$$

The net parallax in latitude is

$$n_n = (n_m \sim n_s). \tag{5.117}$$

The effective deflection from the ecliptic (n_e), which has to be considered for finding the half-duration of the eclipse or the duration of its totality etc. is obtained by finding the sum of or the difference between this and the true latitude of the Moon β_t, as calculated earlier. This is the same as the effective deflection from the ecliptic discussed in Section 5.5; that is,

$$n_e = \beta_t \pm n_n, \tag{5.118}$$

where the choice of sign is '+' if β_t and n_n have the same direction and '−' otherwise. This may be understood with the help of Fig. 5.20. Here S and M represent the geocentric positions of the Sun and the Moon respectively. $n_m = MM'$ and $n_s = SS'$ are the *natis* of the Moon and the Sun. $S'M'$ is the effective deflection from the ecliptic to be calculated. In Fig. 5.20(a),

$$\begin{aligned} SM' &= SM + MM' \\ &= SS' + S'M + MM' \\ &= SS' + S'M'. \\ \text{Therefore} \qquad S'M' &= SM' - SS' \\ &= SM + (MM' - SS') \\ &= \beta_t + n_n. \end{aligned} \tag{5.119}$$

The situation in which the two have opposite directions is shown in Fig. 5.20(b). In this figure,

$$\begin{aligned} S'M &= S'S + SM' + MM' \\ &= SS' + SM \\ &= S'M' + MM'. \\ \text{Therefore} \qquad S'M' &= S'M - MM' \\ &= SM + (SS' - MM') \end{aligned}$$

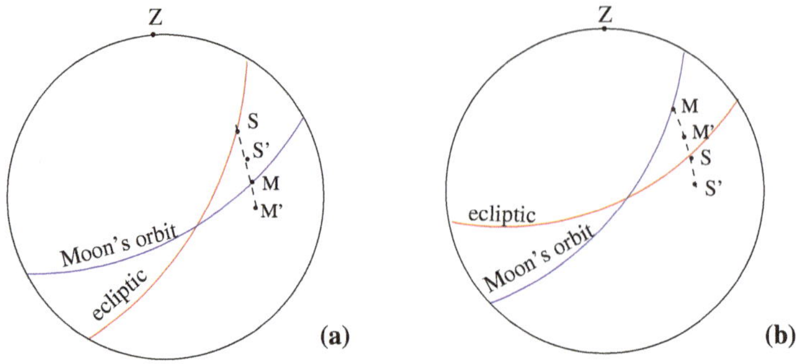

Fig. 5.20 The effective deflection from the ecliptic in a solar eclipse.

$$
\begin{aligned}
&= SM - (MM' - SS') \\
&= \beta_t - n_n. \qquad (5.120)
\end{aligned}
$$

५.१४ मध्यकालनिर्णयः

5.14 Determination of the middle of the eclipse

तत्कालचन्द्रदृक्क्षेपलग्नान्तरभुजागुणात् ।
अर्कवद् दृग्गतिः साध्या भूव्यासार्धहते तयोः ॥ ५० ॥
दृग्गती स्वस्वदृक्कर्णयोजनैर्विहृते कलाः ।
धनं दृक्क्षेपलग्नात् प्राक् सूर्येन्द्वोः ऋणमन्यथा ॥ ५१ ॥
एवं कृतार्कशीतांश्वोः साम्ये स्यात् सन्निकृष्टता ।

tatkālacandradṛkkṣepalagnāntarabhujāguṇāt |
arkavad dṛggatiḥ sādhyā bhūvyāsārdhahate tayoḥ || 50 ||
dṛggatī svasvadṛkkarṇayojanairvihṛte kalāḥ |
dhanaṃ dṛkkṣepalagnāt prāk sūryendvoḥ ṛṇamanyathā || 51 ||
evaṃ kṛtārkaśītāṃśvoḥ sāmye syāt sannikṛṣṭatā |

As in the case of Sun, the *dṛggati* of the Moon has to be obtained from the product of the *koṭi* of the *dṛkkṣepa* with the Rsine of the difference between its longitude and the longitude of the *dṛkkṣepa*. Their *dṛggati* must be multiplied by the radius of the Earth and divided by their own *dṛkkarṇa* in *yojanas*. The results [which are nothing but *lambana*] in minutes must be added to the longitudes of the Sun and the Moon if they lie to the east of the *dṛkkṣepa-lagna* and subtracted otherwise [if they lie to the west of the *dṛkkṣepa-lagna*]. Only when the [longitudes of the] Sun and the Moon thus obtained are equal will they be in close proximity.

What is described here is the procedure for determining the instant at which the longitudes of the Sun and the Moon are equal to each other as seen by an observer on the surface of the Earth. That is the instant of conjunction for the observer. Though an iterative method was described earlier for determining the instant of conjunction (verse 9), there the focus was on determining the parallax in longitude and it was implicitly assumed that the horizontal parallax is one-fifteenth of the daily motion of the Sun.

Here, a condition is given which must be satisfied at the middle of the eclipse with respect to an observer on the surface of the Earth (the *dṛggola*). The horizontal parallax is taken to be $\frac{R_e}{d_m}$ or $\frac{R_e}{d_s}$, in terms of the distances from the *dṛggola* observer. As stated earlier, the middle of the eclipse is when the corrected longitudes of the Sun and Moon are equal.

The *dṛggati* of the Sun and the Moon are defined to be:

$$\frac{R\cos z_v \times R\sin(\lambda_s - \lambda_v)}{R} \tag{5.121}$$

$$\text{and} \quad \frac{R\cos z_v \times R\sin(\lambda_m - \lambda_v)}{R}, \tag{5.122}$$

where z_v and λ_v are the zenith distance and longitude of the *vitribhalagna*, and λ_s and λ_m are the longitudes of the Sun and the Moon respectively.

It is mentioned that these quantities have to be multiplied by the radius of the Earth (R_e) and divided by their own *dṛkkarṇas* given in equations (5.83). Denoting them by $\Delta\lambda_m$ and $\Delta\lambda_s$, we have

$$\Delta\lambda_m = \frac{R_e \cos z_v \times R\sin(\lambda_m - \lambda_v)}{d_m} \tag{5.123}$$

$$\Delta\lambda_s = \frac{R_e \cos z_v \times R\sin(\lambda_s - \lambda_v)}{d_s}. \tag{5.124}$$

It can be shown that $\Delta\lambda_s$ and $\Delta\lambda_m$ are nothing but the effect of parallax in longitudes of the Sun and the Moon, expressed in minutes.

Proof:

In Fig. 5.21(a), M is the Moon and d_m its *dṛkkarṇa*. z_m is the zenith distance of the Moon and p_m its parallax. From the planar triangle OEM we have

$$\frac{\sin p_m}{R_e} = \frac{\sin z_m}{d_m}. \tag{5.125}$$

Therefore,

$$MM' = p_m \approx \sin p_m = \frac{R_e}{d_m} \sin z_m. \tag{5.126}$$

The parallax in longitude is given by

$$\begin{aligned} M'A &= MM' \sin\xi \\ \Delta\lambda_m &\approx \sin p_m \sin\xi \\ &= \frac{R_e}{d_m} \sin z_m \sin\xi. \end{aligned} \tag{5.127}$$

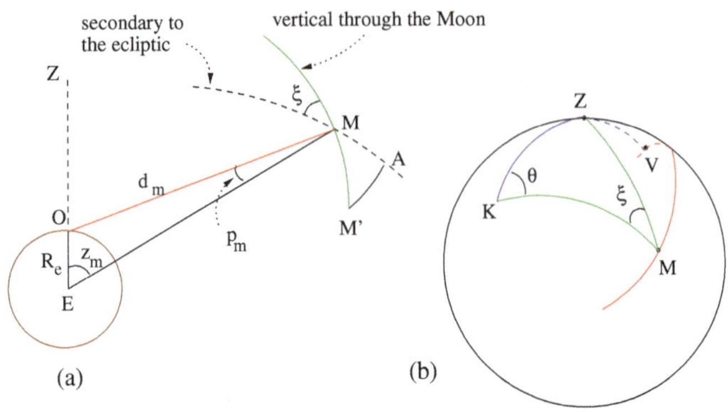

Fig. 5.21 (a) Parallax in the longitude of the Moon as seen by the observer at the centre of the *dṛggola*. (b) Spherical triangle formed by the pole of the ecliptic, the zenith and the Moon.

Considering the triangle *KZM* in Fig. 5.21(b) and applying the sine formula,

$$\frac{\sin\xi}{\sin(90-z_v)} = \frac{\sin\theta}{\sin z_m}. \tag{5.128}$$

Therefore,

$$\sin\xi\sin z_m = \sin\theta\cos z_v. \tag{5.129}$$

Since $\theta = (\lambda_m - \lambda_v)$, the above equation becomes

$$\sin\xi\sin z_m = \sin(\lambda_m - \lambda_v)\cos z_v. \tag{5.130}$$

Using the above equation in (5.127), we have

$$\Delta\lambda_m = \frac{R_e}{d_m}\sin(\lambda_m - \lambda_v)\cos z_v, \tag{5.131}$$

Similarly for the Sun it can be shown that

$$\Delta\lambda_s = \frac{R_e}{d_s}\sin(\lambda_s - \lambda_v)\cos z_v. \tag{5.132}$$

It can be easily seen that (5.131) and (5.132) are the same as (5.123) and (5.124) given in the text, but for the fact that the former are in radians while the latter are in minutes. The corrections (in minutes) have to be applied to the longitudes of the Sun and the Moon to obtain their longitudes as seen by the observer on the surface of the Earth. That is,

$$\begin{aligned}\lambda_s' &= \lambda_s + \Delta\lambda_s\\ \lambda_m' &= \lambda_m + \Delta\lambda_m.\end{aligned} \tag{5.133}$$

Here $\Delta\lambda_s(\Delta\lambda_m)$ is positive when $\lambda_s(\lambda_m) > \lambda_v$. Hence the magnitude of $\Delta\lambda_s(\Delta\lambda_m)$ is added to $\lambda_s(\lambda_m)$ to obtain the longitude corrected for parallax. Similarly, $\Delta\lambda_s(\Delta\lambda_m)$ is negative when $\lambda_s(\lambda_m) < \lambda_v$. Hence the magnitude of $\Delta\lambda_s(\Delta\lambda_m)$ is subtracted from $\lambda_s(\lambda_m)$ to obtain the corrected longitude, $\lambda'_s(\lambda'_m)$. Only when $\lambda'_s = \lambda'_m$ will it be the middle of the eclipse.

While commenting on the first half of the 52nd verse, *Laghu-vivṛti* notes:

एवं कृतलम्बनयोः अर्कचन्द्रयोः साम्ये सत्येव तद्बिम्बघनमध्ययोः परः सन्निकर्षः इति । तत्रैव ग्रहणमध्येनापि भाव्यम् ।

Only if the longitudes of the Sun and the Moon thus corrected for parallax in longitude are equal, will the two be in closest proximity. The middle of the eclipse must also be [understood to be] at that instant.

५.१५ अर्कचन्द्रबिम्बान्तरम्

5.15 Distance of separation between the Sun and the Moon

कृतलम्बनलिप्तार्कचन्द्रयोर्विवरस्य च ॥ ५२ ॥
कृत्योः स्फुटनतेश्चैक्यात् मूलं बिम्बान्तरं ग्रहे ।

kṛtalambanaliptārkacandrayorvivarasya ca || 52 ||
kṛtyoḥ sphuṭanateścaikyāt mūlaṃ bimbāntaraṃ grahe |

The square root of the sum of the squares of the difference in the longitudes of the Sun and the Moon, thus corrected for parallax in longitude and the effective *nati*, is the distance of separation between the two discs in an eclipse.

In Fig. 5.22, *S* and *M* are the positions of the Sun and the Moon corrected for parallax in longitude as described in the last section. n_e refers to the effective deflection of the Moon from the ecliptic. *A* is the point of intersection of the ecliptic and the secondary to the ecliptic passing through the Moon. *B* is a point on *AM* such that *AB* is the parallax in latitude of the Sun. Then *BM* is the effective *nati*, n_e. If λ'_s and λ'_m are the longitudes of the Sun and the Moon as seen by the *dṛggola* observer, then the distance of separation between the Sun and the Moon is given by

$$d^2 = \sqrt{(\lambda'_m - \lambda'_s)^2 + n_e^2}. \tag{5.134}$$

It must be noted that the above formula is applicable at any instant of time during the eclipse.

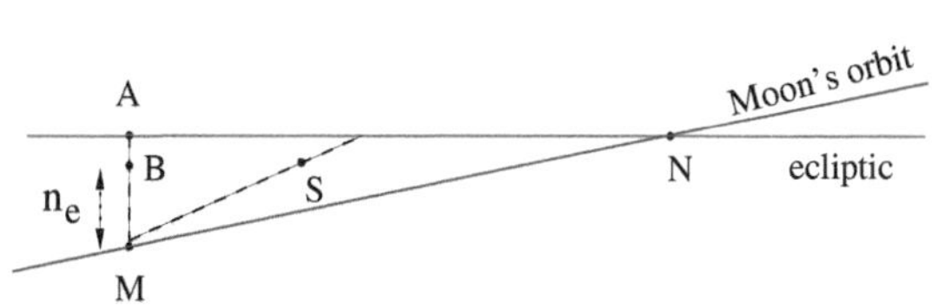

Fig. 5.22 Distance of separation between the centres of the Sun and the Moon's discs.

५.१६ ग्रहणदृश्यत्वस्य आदेश्यत्वम्
5.16 Announcement of the visibility of the eclipse

एवमस्तमये स्पर्शे मोक्षे वाप्युदये रवेः ॥ ५३ ॥
अल्पश्चेद् ग्रास आनीतः न वाच्यो द्वादशांशतः ।
अल्पेऽर्कगतितिथ्यंशयुताल्पव्यासखण्डतः ॥ ५४ ॥
शङ्कौ मध्यतमस्सर्वग्रहणं वा न लक्ष्यते ।
अस्तोदयार्कचन्द्रौ तत्प्राणभोगोनसंयुतौ ॥ ५५ ॥
कृत्वा तत्काललग्नं तैः नयेद् बिम्बान्तरं द्वयोः ।
अधिके बिम्बभेदार्धात् प्राक्पश्चाद् ग्रासमध्यतः ॥ ५६ ॥
नैव ते ग्रहणे दृश्ये सर्वं मध्यतमोपि वा ।

evamastamaye sparśe mokṣe vāpyudaye raveḥ || 53 ||
alpaśced grāsa ānītaḥ na vācyo dvādaśāṃśataḥ |
alpe'rkagatitithyaṃśayutālpavyāsakhaṇḍataḥ || 54 ||
śaṅkau madhyatamassarvagrahaṇaṃ vā na lakṣyate |
astodayārkacandrau tatprāṇabhogonasaṃyutau || 55 ||
kṛtvā tatkālalagnaṃ taiḥ nayed bimbāntaraṃ dvayoḥ |
adhike bimbabhedārdhāt prākpaścād grāsamadhyataḥ || 56 ||
naiva te grahaṇe dṛśye sarvaṃ madhyatamopi vā |

Thus, when the first contact occurs close to the sunset, or the last contact occurs close to the sunrise, and if the *grāsa* obtained by the method described earlier is found to be less than one-twelfth the diameter of the sun, then the occurrence of a solar eclipse should not be announced.

If the *śaṅku* is less than the sum of one-fifteenth of the daily motion of the Sun and its semi-diameter, then the annular or total eclipse will not be visible.

From the longitudes of the Sun and the Moon corrected by their own *prāṇakalāntaras*, the *kālalagna* has to be obtained. From them [the *vitribhalagna*, *dṛkkṣepajyā*, parallax in longitude and the *nati*], the distance of separation between the two discs may be obtained. If this is greater than the difference in the semi-diameters of the discs, then neither of the two eclipses, total or annular, will be visible.

Here in the text a few conditions regarding the visibility of the solar eclipse are given. Regarding the announcement of the occurrence of the eclipse, it is stated that if

$$grāsa < \frac{1}{12}\text{Sun's disc}, \tag{5.135}$$

then the eclipse is not visible, though it actually occurs. This result appears to be an empirical criterion, as no justification is provided.

Regarding the total or annular eclipses, two conditions are stated. One is the condition on the *śaṅku* (the Rcosine of the zenith distance of the Sun/Moon) and the other, the condition on the difference between their semi-diameters, which are explained below.

Condition on *śaṅku* for total or annular eclipse

Let x denote the sum of one-fifteenth of the daily motion of the Sun and its semi-diameter. That is,

$$x = \frac{1}{15}\dot{\lambda}_s + \text{semi-diameter of smaller disc.} \tag{5.136}$$

In the RHS of the above equation, the first term represents the horizontal parallax of the Sun, which is the parallax at sunrise or sunset. The condition for the visibility of the totality or annularity is given to be

$$\acute{s}a\dot{n}ku > x. \tag{5.137}$$

Obviously the condition is applicable for those eclipses whose *madhyakāla* (middle of the eclipse) is very close to sunrise/sunset. Consider Fig. 5.23(*a*) and (*b*) in

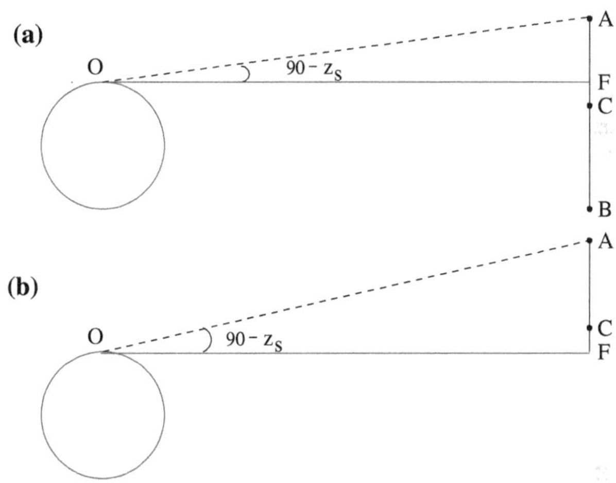

Fig. 5.23 Criteria for the visibility of the totality/annularity of a solar eclipse.

which $x = AC$ is the semi-diameter of the solar disc for a terrestrial observer plus the horizontal parallax. $A\hat{O}F = 90° - z_s$, where z_s is the zenith distance. Now, as $A\hat{O}F$ is small,

$$\begin{aligned} AF &= A\hat{O}F \qquad \text{(in minutes)} \\ &= R\sin(90 - z_s) \\ &= R\cos z_s \\ &= \acute{s}a\dot{n}ku. \end{aligned} \tag{5.138}$$

In Fig. 5.23(a), as C is below the horizon,

$$AF < AC = x \qquad \text{or} \qquad \acute{s}a\dot{n}ku < x. \tag{5.139}$$

In this case, the totality/annularity is not visible. However, in Fig. 5.23(b) where C is above the horizon the totality/annularity will be visible. For this to happen,

$$AF > AC \qquad \text{or} \qquad \acute{s}a\dot{n}ku > x. \tag{5.140}$$

Condition on the difference of semi-diameters

The condition for the totality/annularity that will be presented in this section is not very different from the one discussed earlier. In fact, the difference is only in the details. Here, the difference in the semi-diameters is determined more accurately and this is due to the fact that the longitude of the *vitribhalagna* is found more precisely by finding the *kālalagna*—the procedure for which is discussed in detail in Chapter 3. From the *kālalagna*, the *prāglagna* (the orient ecliptic point) may be determined accurately. With this the *vitribhalagna*, the *dṛkkṣepa-jyā* and hence the parallax in longitude and the deflection from the ecliptic are obtained. After obtaining the parallaxes in longitude (l_s and l_m) they are applied to the longitudes of the Sun and the Moon in order to determine their values more precisely at the true sunrise/sunset. That is,

$$\lambda_s' = \lambda_s \pm l_s \tag{5.141}$$
$$\lambda_m' = \lambda_m \pm l_m, \tag{5.142}$$

where the choice of signs in '+' at sunrise and '−' at sunset. From them, the separation between the discs, d, is calculated:

$$d = \sqrt{(\lambda_m' - \lambda_s')^2 + n_e^2}, \tag{5.143}$$

where n_e is the effective deflection from the ecliptic. If d_s and d_m are the diameters of the solar and the lunar discs then the condition given for the visibility of totality/annularity may be mathematically represented by

$$d < \frac{d_m \sim d_s}{2}. \tag{5.144}$$

The rationale behind the above expression can be understood with the help of Fig. 5.24. Here

$$\begin{aligned} AX = d &= \sqrt{AT^2 + XT^2} \\ &= \sqrt{(\lambda_m' - \lambda_s')^2 + n_e^2}, \end{aligned} \tag{5.145}$$

since $AT = \lambda_m' - \lambda_s'$ and $XT = n_e$. Just as in Section 5.10, the condition for visibility of totality is

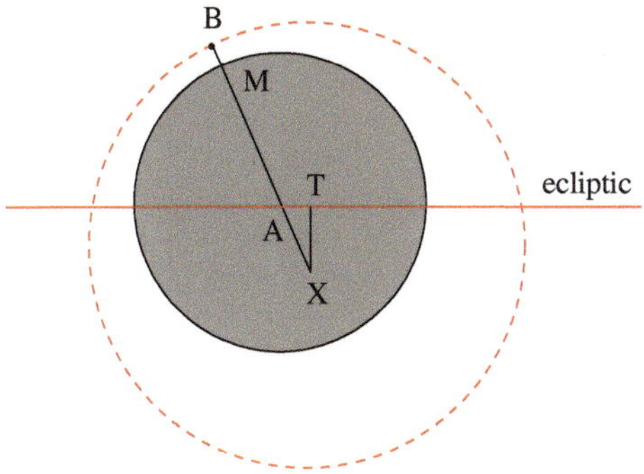

Fig. 5.24 Condition for the visibility of the totality/annularity.

$$MX = AX + AM < XB \tag{5.146}$$

$$\text{or} \qquad AX < XB - AM$$

$$\text{that is,} \qquad d < \frac{d_m - d_s}{2}. \tag{5.147}$$

Similarly for annularity, the condition is

$$d < \frac{d_s - d_m}{2}. \tag{5.148}$$

५.१७ ग्रहणपरिलेखनम्

5.17 Graphical representation of the eclipse

स्पर्शे मध्ये च मोक्षे चाप्यन्यत्रेष्टेऽपि वा पृथक् ॥ ५७ ॥
वलनद्वयमानीय प्राग्वत् तद्योगभेदजात् ।
गुणादेकाङ्कभूभक्तं वलनं स्यात् स्फुटन्त्विह ॥ ५८ ॥
वृत्तं धृतिमितास्येन कर्कटेनालिखेत् क्षितौ ।
दिशौ पूर्वापरे व्यस्तं लेखनेऽफलके यदि ॥ ५९ ॥
वलनं पूर्ववन्नीत्वा रवेः पन्थाश्च तद्वयात् ।
नतेर्दिशि विधोस्तस्मात् स्फुटनत्यन्तरेऽपरः [11] ॥ ६० ॥
कार्यस्तद्वृत्तमध्येऽथ रविबिम्बं स्फुटं लिखेत् ।
मात्वा तत्केन्द्रगैकाग्रबिम्बान्तरशलाकया ॥ ६१ ॥
बिन्दुं कृत्वा विधोर्मार्गे तद्बिम्बं तत्र संलिखेत् ।
स्पर्शे प्रत्यङ्मुखीं मोक्षे शलाकां प्राङ्मुखीं नयेत् ॥ ६२ ॥
एवमेवेष्टकालेऽपि प्राक् पश्चाद् ग्रासमध्यतः ।

[11] The prose order is: तस्मात् = पथः स्फुटनत्यन्तरे, अपरः (पन्थाः) कार्यः । (सः) विधोः (पन्थाः) ।

चन्द्रबिम्बाद्बहिर्भूतो भागो दृश्योऽर्कमण्डले ॥ ६३ ॥
तदन्तर्गतभागस्तु ग्रस्तस्तेनासितः सदा ।

sparśe madhye ca mokṣe cāpyanyatreṣṭe'pi vā pṛthak || 57 ||
valanadvayamānīya prāgvat tadyogabhedajāt |
guṇādekāṅkabhūbhaktaṃ valanaṃ syāt sphuṭantviha || 58 ||
vṛttaṃ dhṛtimitāsyena karkaṭenālikhet kṣitau |
diśau pūrvāpare vyastaṃ lekhane'phalake yadi || 59 ||
valanaṃ pūrvavannītvā raveḥ panthāśca taddvayāt |
naterdiśi vidhostasmāt sphuṭanatyantare'paraḥ || 60 ||
kāryastadvṛttamadhye'tha ravibimbaṃ sphuṭaṃ likhet |
mātvā tatkendragaikāgrabimbāntaraśalākayā || 61 ||
binduṃ kṛtvā vidhormārge tadbimbaṃ tatra saṃlikhet |
sparśe pratyaṅmukhīṃ mokṣe śalākāṃ prāṅmukhīṃ nayet ||62||
evameveṣṭakāle'pi prāk paścād grāsamadhyataḥ |
candrabimbādbahirbhūto bhāgo dṛśyo'rkamaṇḍale || 63 ||
tadantargatabhāgastu grastastenāsitaḥ sadā |

After determining the two *valanas* separately at the [time of] the first contact, the middle and the *mokṣa*, or at any desired instant, their sum or difference has to be found as described earlier.[12] The Rsine of this divided by 191 is the true *valana* here [in the case of a solar eclipse]. A circle with a radius equal to 18 units should be drawn on the Earth [a flat surface]. The east and west directions have be marked in the opposite sense when the sketch is made on a plank. The [direction of the] *valana* has to be obtained as earlier [by making marks on the circumference of the circle on either side of the east-west line etc., as described for a lunar eclipse]. Then the path traced by the Sun has to be drawn in the direction of the parallax in latitude. At a distance of *sphuṭa-nati* from that (path of the Sun), another path for the Moon has to be drawn.

Then, at the centre of that circle, the disc of the Sun may be drawn clearly. Then, with a piece of thin pointed stick (the *śalākā*) whose measure is equal to the distance of separation between the two discs, mark a point in the path of the Moon. Then draw the Moon's disc there. The *śalākā* has to be pointed towards the west during the beginning of the eclipse (the first contact) and towards the east during the end of the eclipse (the last contact). The same is true of any instant which is prior to or later than the instant of the mid-eclipse. The portion [of the Sun] which lies outside the Moon's disc is visible. The portion which lies inside is the portion eclipsed, and hence is always dark.

As in the previous chapter on lunar eclipse, consider the angle between the ecliptic and vertical through the Sun/Moon. Suppose the sum or difference of the two *valanas* is ψ. Then $R\sin\psi$ is the *valana* in a circle of radius equal to the *trijyā* (R). Then the *valana* corresponding to a circle of radius 18 is

$$sphuṭa\text{-}valana = 18\sin\psi = \frac{R\sin\psi}{191}, \tag{5.149}$$

as the value of R is taken to be $3438 = 191 \times 18$. Here the value of the radius of the circle has been chosen to be 18 only for the sake of convenience. In the last quarter of verse 58, it is mentioned:

वलनं स्यात् स्फुटन्त्विह॥

This will be the true *valana* here (*iha*).

---

[12] Chapter 4, verses 44, 45.

While commenting upon the word '*iha*', the following observation is made in *Laghu-vivṛti*:

> इहेत्यनेन अस्य चन्द्रग्रहणतो भेदो दर्शितः। यतः तत्र बिम्बान्तरेण निहत्य त्रिज्यया विभक्तं स्फुटं वलनम्।
>
> By using the word *iha*, distinction from the lunar eclipse has been shown. Because there the *valana* is [obtained by] multiplying it ($R\sin\psi$) by the separation between the discs and dividing by the *trijyā* [to obtain the *valana* corresponding to a circle of radius equal to the separation between the discs].

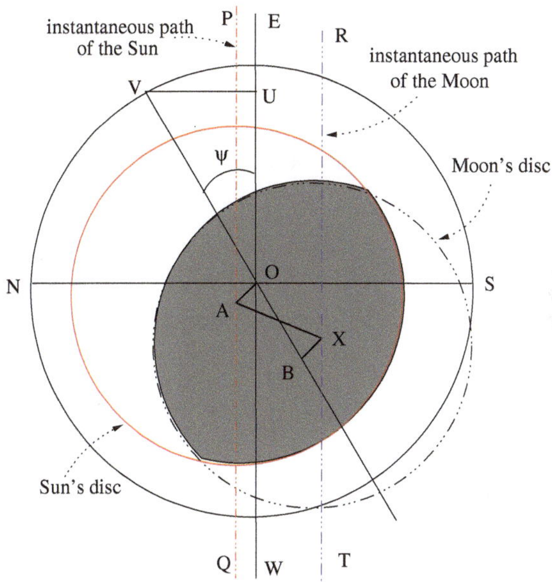

Fig. 5.25 Graphical representation of a solar eclipse. The solar and lunar discs are drawn with A and X as centres. The shaded portion is the eclipsed part of the Sun.

In Fig. 5.25, $ENWS$ is a circle of radius 18 units with O as the centre. EW is along the local east-west direction, and NS is along the north-south direction. Draw a line UV perpendicular to EW such that

$$UV = 18\sin\psi, \tag{5.150}$$

where $\psi = U\hat{O}V$ is the angle corresponding to the *valana*. Then VO represents the direction of the ecliptic as ψ is the angle between the ecliptic and the local east-west direction. A is the centre of the Sun's disc and OA, perpendicular to VO, is the parallax in latitude of the Sun, which is the distance of the Sun from the ecliptic due to parallax. X is the centre of the Moon's disc. It is located such that

(i) AX is the distance between the solar and lunar discs and

(ii) $AO+XB$ (where XB is perpendicular to EO) is the *sphuṭa-nati*, which is the effective deflection from the ecliptic n_e discussed earlier.

Now, draw *PQ* and *RT* passing through *A* and *X* and parallel to the east-west line. Then *PQ* and *RT* are the instantaneous paths of the Sun and the Moon. Regarding the depiction of the path traced by the Sun and the Moon, the commentator in his *Laghu-vivṛti* observes:

···सर्वं पूर्वोक्तं स्मारितम्। तत्र तथाकृतायाः पूर्वापररेखायाः पार्श्वतः यथादिशं स्वावनत्यन्तरितां अपरामपि प्रागपरां रेखां कुर्यात्। स इह रवेर्मार्गः। ततः चन्द्रार्कयोः अवनत्यन्तरयोः योगान्तरोत्पन्नस्फुटावनतितुल्येऽन्तरे तद्दिश्यपरामपि प्रागपरां रेखां कुर्यात्। स विधोर्मार्गः।

...all that was described earlier is recalled. There, to the line thus drawn parallel to the east-west line, as per the direction [determined], another line is to be drawn east-west at a distance of its own parallax in latitude from the ecliptic. That is the [instantaneous] path of the Sun. Then again draw an east-west line along that direction at a distance equal to the difference in the deflection from the ecliptic of the Sun and the Moon. That is the path of the Moon.

The term *tadvṛttamadhye* in verse 61 needs to be clarified. This literally means '*at the centre of that circle*'. Here the commentary states that:

एवं कृतयोः तयोः मार्गयोः तस्य धृतिमितव्यासार्धस्य मध्ये रविबिम्बं स्ववृत्तव्यासार्धेन स्फुटतरं लिखेत् ।

Thus it is clear that here the term 'that circle' refers to the circle drawn with a radius equal to 18 units. However, the term 'centre' in the verse should not be taken literally to mean the centre of that circle. In fact, it refers to the centre of the path of the Sun (*A*) as drawn which actually intersects the north–south line (*NS*) passing through the centre of the circle.

Chapter 6
व्यतीपातप्रकरणम्
Vyatīpāta

६.१ व्यतीपातसम्भवः
6.1 The possibility of *vyatīpāta*

अर्केन्द्वोर्हीयते चैका यदान्या वर्धते क्रमात् ।
क्रान्तिज्ययोस्तदा साम्ये व्यतीपातो न चान्यथा [1] ॥ १ ॥
वैधृतोऽयनसाम्ये स्यात् लाटः स्यादेकगोलयोः ।

arkendvorhīyate caikā yadānyā vardhate kramāt |
krāntijyayostadā sāmye vyatīpāto na cānyathā || 1 ||
vaidhṛto'yanasāmye syāt lāṭaḥ syādekagolayoḥ |

Of [the two objects] the Sun and the Moon, when [the magnitude of the declination of] one is decreasing and the other is increasing steadily, and when the [magnitudes of] the Rsines of their declinations become equal, then it is *vyatīpāta* and not otherwise; [The same is called] *vaidhṛta* if the *ayanas* are the same and *lāṭa* when the hemispheres are the same.

Condition for the occurrence of *vyatīpāta*

Let δ_s and δ_m be the declinations of the Sun and the Moon at any given time. Then the condition to be satisfied for the occurrence of *vyatīpāta* is given to be

$$|\delta_s| = |\delta_m|, \tag{6.1}$$

with the constraint that the variation in the two declinations should be having opposite gradients. That is, if $|\delta_s|$ is increasing, $|\delta_m|$ should be decreasing and vice versa. Such a situation is schematically depicted in Fig. 6.1.

[1] The prose order of this verse is: यदा अर्केन्द्वोः (मध्ये) एका क्रान्तिः क्रमात् हीयते अन्या च वर्धते तदा (क्रान्त्योः) साम्ये व्यतीपातः अन्यथा न च।

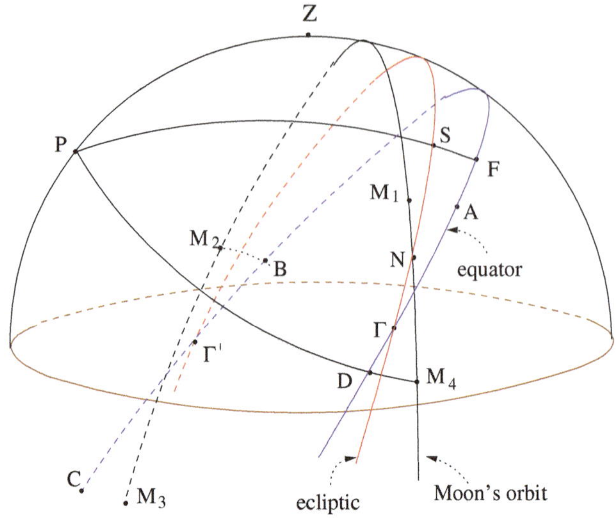

Fig. 6.1 Positions of the Sun and the Moon during *vyatīpāta*.

Occurrence of *lāṭa* and *vaidhṛta*

In the Fig. 6.1, Z represents the zenith, P the north celestial pole, Γ the vernal equinox and N the ascending node of the Moon's orbit. The Sun is at S whose declination $|\delta_s| = SF$. M_1, M_2, M_3 and M_4 correspond to different positions of the Moon, which lie in four different quadrants, at which

$$|\delta_m| = M_1A = M_2B = M_3C = M_4D = |\delta_s|. \tag{6.2}$$

Out of these four positions of the Moon—since $|\delta_s|$ is increasing at S—only M_2 and M_4 correspond to *vyatīpāta*, as $|\delta_M|$ is decreasing only at these two positions. Moreover, it may be noted that when the Moon is at M_2, the *ayanas* of the Sun and the Moon are different but lie in the same hemisphere. Hence it is an instance of *lāṭa-vyatīpāta*. On the other hand, when the Moon is at M_4, the *ayanas* of the Sun and the Moon are the same (both are northerly). Hence this is an example of *vaidhṛta-vyatīpāta*.

The commentary begins with the following *avatārikā*: [2]

एवं रवीन्द्वोः ग्रहणद्वयं दृग्गोलविषयं स्पष्टतरं प्रदर्शितम्। इदानीं भगोलविषयं तयोरेव क्रान्तिसाम्यजनितं व्यतीपातं प्रदर्शयितुमाह।

Thus the two eclipses of the Sun and the Moon, related to the observer-centred celestial sphere (*dṛggola*), were clearly demonstrated. Now in order to explain the concept of *vyatīpāta*—that arises owing to the equality of declinations of them [the Sun and the Moon]—related to the geocentric celestial sphere *bhagola*, [the following] is stated.

[2] The word *avatārikā* refers to succinct introductory remarks.

६.२ अर्केन्द्वोः क्रान्त्यानयनम्
6.2 Finding the declination of the Sun and the Moon

संस्कृतायनसूर्येन्द्वोः क्रान्तिज्ये पूर्ववन्नयेत् ॥ २ ॥

saṃskṛtāyanasūryendvoḥ krāntijye pūrvavannayet || 2 ||

From the *ayana*-corrected longitudes (*sāyana* longitudes) of the Sun and the Moon, let the Rsines of their declinations be determined as earlier.

In Fig. 6.2*a*, Γ is the vernal equinox, S the Sun, M the Moon and N_1 its ascending node. The meridian passing through the Sun meets the equator at G. If λ_s is the longitude of the Sun at S, then its declination is given by

$$R\sin\delta_s = R\sin\varepsilon\sin\lambda_s. \tag{6.3}$$

The secondary to the ecliptic passing through the Moon intersects the ecliptic at I. If λ_m and δ' are the longitude and declination of this point, then considering the triangle ΓIJ and applying the sine formula we obtain

$$R\sin\delta' = R\sin\varepsilon\sin\lambda_m. \tag{6.4}$$

It is the LHS of (6.3) and (6.4) that are referred to as Rsines of the declination (*krānti-jyā*) of the Sun and the Moon in the above verse. Though (6.4) does not give the actual declination of the Moon, which will be derived in the subsequent sections, it can be taken as a reasonable approximation when the latitude of the Moon is small. In fact, as we will see in the next section, the derivation of the actual expression for the Moon involves the declination of the point I given in (6.4) and that's precisely the reason for Nīlakaṇṭha's statement that it may be determined as earlier.

६.३ चन्द्रस्य इष्टक्रान्त्यानयने विशेषः
6.3 Speciality in the determination of the desired declination of the Moon

पातोनेन्दोर्भुजा जीवा परमक्षेपताडिता ।
त्रिज्याभक्ता विधोः क्षेपः तत्कोटिमपि चानयेत् ॥ ३ ॥
परमापक्रमकोट्या विक्षेपज्यां निहत्य तत्कोट्या ।
इष्टक्रान्तिं चोभे त्रिज्याप्ते योगविरहयोग्ये स्तः ॥ ४ ॥
सदिशोः संयुतिरनयोः वियुतिर्विदिशोरपक्रमः स्पष्टः ।
स्पष्टापक्रमकोटिर्द्युज्या विक्षेपमण्डले वसताम् ॥ ५ ॥
इत्युक्तात्र स्फुटा क्रान्तिः गृह्यतां गोलवित्तमैः ।

pātonendorbhujā jīvā paramakṣepatāḍitā |
trijyābhaktā vidhoḥ kṣepaḥ tatkoṭimapi cānayet || 3 ||
paramāpakramakoṭyā vikṣepajyāṃ nihatya tatkoṭyā |
iṣṭakrāntiṃ cobhe trijyāpte yogavirahayogye staḥ || 4 ||

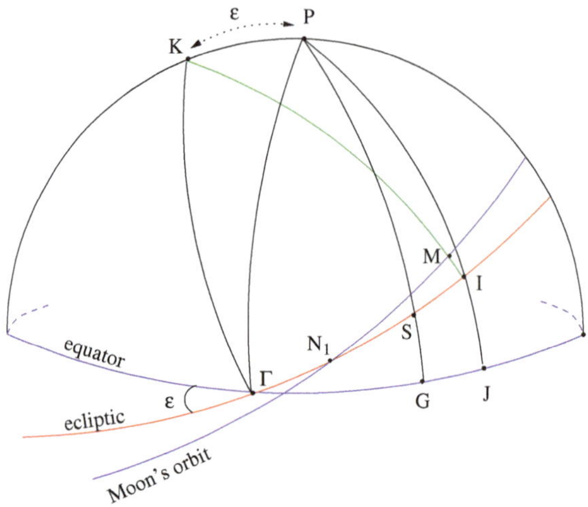

Fig. 6.2*a* Finding the declinations of the Sun and the Moon.

sadiśoḥ saṃyutiranayoḥ viyutirvidiśorapakramaḥ spaṣṭaḥ |
spaṣṭāpakramakoṭirdyujyā vikṣepamaṇḍale vasatām || 5 ||
ityuktātra sphuṭā krāntiḥ gṛhyatāṃ golavittamaiḥ |

The Rsine [of the longitude] of the node subtracted from [the longitude of] the Moon, multiplied by the maximum deflection [of the Moon's orbit] and divided by the *trijyā*, gives the latitude of the Moon (β). Let the Rcosine of it also be obtained.

Having multiplied the Rsine of the latitude of the Moon (the *vikṣepajyā*) by the cosine of the maximum deflection [of the ecliptic from the equator], and having multiplied the Rcosine of that (the latitude of the Moon) by the [Rsine of the] desired declination [of the Moon determined earlier], the two [products] divided by the *trijyā* are readily suited for addition or subtraction.

If these two are in the same direction then they must be added, and if they are in different directions then their difference must be found. [Now] the true declination [of the Moon is obtained]. The Rcosine of the true declination will be the day-radius (*dyujyā*) for those residing in the *vikṣepamaṇḍala*. Let the process of the [determination of] true declination [of the Moon] thus explained be understood by the experts in the spherics.

Considering the triangle PKM in Fig. 6.2*b* and applying the cosine formula, we have

$$\cos PM = \cos PK \cos KM + \sin PK \sin KM \cos PKM. \tag{6.5}$$

Let λ_m, β and δ_m be the longitude, the latitude and the declination of the Moon. Then

$$KM = 90 - \beta \qquad \text{and} \qquad PM = 90 - \delta_m. \tag{6.6}$$

P and K being the poles of the equator and the ecliptic, the arc $PK = \varepsilon$. Γ is the pole of the great circle passing through K and P. Therefore

$$\Gamma\hat{K}P = 90 \qquad \text{and} \qquad P\hat{K}M = 90 - \lambda_m. \tag{6.7}$$

Using (6.6) and (6.7) in (6.5) we obtain

$$\sin\delta_m = \cos\varepsilon\sin\beta + \sin\varepsilon\cos\beta\sin\lambda_m. \tag{6.8}$$

This is the true declination of the Moon with latitude β. The *krāntijyā* of the Moon denoted by δ', and given by (6.4), is the declination of a point on the ecliptic which has the same longitude as the Moon (point *I* in Fig. 6.2*a*). It can be easily seen that the RHS of (6.8) reduces to the RHS of (6.4) when $\beta = 0$. Now, using (6.4) in (6.8),

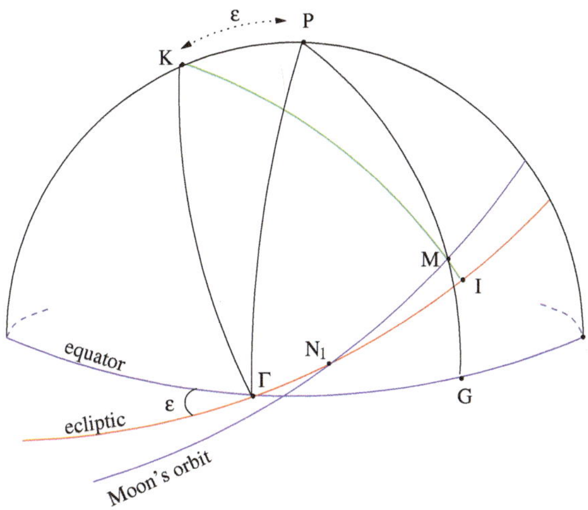

Fig. 6.2*b* Determination of the true declination of the Moon.

we have

$$\sin\delta_m = \cos\varepsilon\sin\beta + \cos\beta\sin\delta'. \tag{6.9}$$

This is the formula for the true declination (*spaṣṭāpakrama*) of the Moon given in the verses, where it is stated in the form

$$|R\sin\delta_m| = \frac{R\cos\varepsilon\,|R\sin\beta|}{R} \overset{+}{\sim} \frac{R\cos\beta\,|R\sin\delta'|}{R}. \tag{6.10}$$

The value of the latitude of the Moon (β) used in the above equation is found by using the formula

$$\sin\beta = 270\times\sin\lambda_m, \tag{6.11}$$

where the inclination of the Moon's orbit with the ecliptic is taken to be 270 minutes. Further, it may be noted from Fig. 6.2 that, when β and δ' are in the same direction, then the two terms are to be added, and if they are in different directions, that is, if $\sin\beta$ and $\sin\delta'$ have opposite signs, then $|\sin\delta_m|$ would be the difference of two positive terms. This is what is mentioned in the first half of verse 5 in the text, and is indicated by $\overset{+}{\sim}$ in (6.10).

In a passing remark the text also mentions that: 'The Rcosine of the true declination will be the day-radius (*dyujyā*) for those residing in the *vikṣepamaṇḍala*'. Here, the term *vikṣepamaṇḍala* refers to the declination circle of the Moon whose radius is given by

$$dyujy\bar{a} = R\cos\delta_m = \sqrt{R^2 - (R\sin\delta_m)^2}. \tag{6.12}$$

६.४ इन्दोः प्रकारान्तरेण क्रान्त्यानयनम्

6.4 Determination of the declination of the Moon by another method

अथवा क्रान्तिरानेया परक्रान्त्या[3] विधोरपि ॥ ६ ॥
परमक्षेपकोटिघ्नं जिनभागगुणं हरेत् ।
त्रिज्यया क्षेपवृत्तेऽस्य नाभ्युच्छ्रय इहाप्यते ॥ ७ ॥
पातस्य सायनस्याथ दोःकोटिज्ये उभे हते ।
क्षिप्त्या परमया त्रिज्याभक्ते स्यातां च तत्फले ॥ ८ ॥
अन्त्यद्युज्याहतं तत्र कोटिजं त्रिज्यया हरेत् ।
नाभ्युच्छ्रये च तत् स्वर्णं मृगकर्क्यादि पातजम् ॥ ९ ॥
तद्बाहुफलवर्गैक्यमूलं क्रान्तिः परा विधोः ।
त्रिज्याघ्नं दोःफलं भक्तं तया चलनमायनम् ॥ १० ॥
जूकक्रियादिगे पाते स्वर्णं तत् सायने विधौ ।
तद्बाहुज्या हता क्रान्त्या तदा परमया स्वया ॥ ११ ॥
त्रिज्याप्तापक्रमज्येन्दोः स्फुटा तात्कालिकी भवेत् ।

athavā krāntirāneyā parakrāntyā vidhorapi || 6 ||
paramakṣepakoṭighnaṃ jinabhāgaguṇaṃ haret |
trijyayā kṣepavṛtte'sya nābhyucchraya ihāpyate || 7 ||
pātasya sāyanasyātha doḥkoṭijye ubhe hate |
kṣiptyā paramayā trijyābhakte syātāṃ ca tatphale || 8 ||
antyadyujyāhataṃ tatra koṭijaṃ trijyayā haret |
nābhyucchraye ca tat svarṇaṃ mṛgakarkyādi pātajam || 9 ||
tadbāhuphalavargaikyamūlaṃ krāntiḥ parā vidhoḥ |
trijyāghnaṃ doḥphalaṃ bhaktaṃ tayā calanamāyanam || 10 ||
jūkakriyadige pāte svarṇaṃ tat sāyane vidhau |
tadbāhujyā hatā krāntyā tadā paramayā svayā || 11 ||
trijyāptāpakramajyendoḥ sphuṭā tātkālikī bhavet |

Otherwise the [true] declination of the Moon may be obtained from its maximum declination. The Rsine of 24 (degrees) multiplied by the Rcosine of maximum inclination is divided by the *trijyā*. The quantity obtained is called the *nābhyucchraya* of the *kṣepavṛtta*.

The Rsine and the Rcosine of the *sāyana* longitude of the node, multiplied by the maximum deflection [of the Moon's orbit] and divided by the *trijyā*, will be those *phalas* [i.e. the

[3] In another reading of the text, we find the term स्फुटक्रान्ति instead of परक्रान्ति। That the latter is correct gets confirmed from the procedure and formulae given in the text. The commentator Śaṇkara Vāriyar has also adopted the reading परक्रान्ति।

> *dohphala* and the *koṭiphala*]. Of them, the *koṭiphala* is multiplied by the Rcosine of the maximum declination of the Sun and divided by the *trijyā*. The result is added to or subtracted from the *nābhyucchraya* depending upon whether the [*sāyana*] longitude of the node lies within six *rāśis* beginning with *Mṛga* or *Karkaṭaka*. The square root of the sum of the squares of that and the *dohphala* is the maximum declination of the Moon.
>
> The *dohphala* multiplied by the *trijyā* and divided by that [i.e. the quantity obtained above] is defined as the *ayanacalana* [of the Moon]. This has to be added to or subtracted from the *sāyana* longitude of the Moon depending upon whether the node lies within six *rāśis* beginning with Libra (*Jūka*) or with Aries (*Kriyā*). The Rsine of that is multiplied by the maximum declination and divided by the *trijyā*. The result is the refined (*sphuṭā*) instantaneous [value of the] Rsine of the declination of the Moon.

An expression for the declination (δ_m) of the Moon which is similar to (6.3) is presented in the above verses. We may write such an expression as

$$\sin\delta_m = \sin I \sin\eta, \tag{6.13}$$

where $\eta = (\lambda_m - A)$; λ_m and A refer to the longitude and *ayanacalana* of the Moon. I represents the maximum declination of the Moon which keeps varying and depends upon the position of the Moon's ascending node along the ecliptic. It is also the inclination of the Moon's orbit with the equator. For instance, when the ascending node N_1 coincides with the vernal equinox, then the inclination of the Moon's orbit is

$$I = \delta_{max} = \varepsilon + i, \tag{6.14}$$

which is the same as the maximum declination attained by the Moon. On the other hand, when the ascending node coincides with the autumnal equinox then the inclination of the Moon's orbit is

$$I = \delta_{min} = \varepsilon - i. \tag{6.15}$$

Generally the value of the obliquity of the ecliptic, ε is taken to be 24° and the inclination of the Moon's orbit with the ecliptic, i, to be 4.5°.

From (6.13) it may be noted that the expression for the Moon's declination involves obtaining expressions for two intermediate quantities, namely

1. the maximum declination of the Moon in its orbit, which is called the *parā-krānti*, denoted by I, and
2. the right ascension of the point of intersection of the Moon's orbit and the equator. This is called the *ayanacalana* and is denoted by A.

The desired true declination of the Moon, denoted by δ_m, is expressed in terms of these quantities.

Expression for the *parā-krānti* and *ayanacalana*

The expression for the *parā-krānti*, in turn requires the defining of a few intermediate quantities. A term called the *nābhyucchraya* (x) is defined as

$$x = \frac{R\sin\varepsilon\, R\cos i}{R}, \tag{6.16}$$

Then the *doḥphala* (D) and the *koṭiphala* (K) are defined to be

$$D = \frac{R|\sin\lambda_n|\, R\sin i}{R},$$
$$\text{and} \qquad K = \frac{R|\cos\lambda_n|\, R\sin i}{R}. \tag{6.17}$$

We introduce yet another quantity (y), defined by

$$\begin{aligned} y &= R\cos\varepsilon \times K \\ &= \frac{R\cos\varepsilon\, |\cos\lambda_n|\, R\sin i}{R}. \end{aligned} \tag{6.18}$$

Using x and y, one more term (z) is defined to be

$$\begin{aligned} z &= x - y \qquad \text{when } 90 < \lambda_n \leq 270, \\ &= x + y \qquad \text{otherwise.} \\ \text{Essentially,} \qquad z &= x + R\cos\varepsilon\cos\lambda_n \sin i. \end{aligned} \tag{6.19}$$

Now the *parā-krānti*, the maximum declination I of the Moon, is given as

$$\begin{aligned} R\sin I &= \sqrt{z^2 + D^2} \\ &= \sqrt{(R\sin\varepsilon\cos i + R\cos\varepsilon\sin i\cos\lambda_n)^2 + (R\sin\lambda_n\sin i)^2}. \end{aligned} \tag{6.20}$$

The *ayanacalana* (A) of the Moon is defined in terms of the maximum declination through the relation

$$R\sin A = \frac{R \times D}{R\sin I}. \tag{6.21}$$

This is also referred to as the *vikṣepacalana*.

Expression for the *iṣṭakrānti*

Having obtained the *ayanacalana*, it is added to the true longitude of the Moon when $180^\circ \leq \lambda_n \leq 360^\circ$, and subtracted from it otherwise. The Rsine of the result is multiplied by the Rsine of the maximum declination and divided by the *trijyā* to get the Rsine of the desired declination. That is,

$$\begin{aligned} R\sin\delta_m &= \frac{R\sin I \times R\sin(\lambda_m \pm A)}{R} \\ &= R\sin I \sin\eta, \end{aligned} \tag{6.22}$$

where η is the angle of separation between the Moon and the point of intersection of its orbit with the equator, along the orbit of the Moon. In the following we provide the rationale behind (6.20), (6.21) and (6.22) with the help of Figs 6.3*a*, 6.3*b* and 6.3*c*.

Derivation of the expression for the *parākrānti*

While the *Yuktibhāṣā* derivation of the expression for the *parākrānti* is given in Appendix E, here we derive the same using modern spherical trigonometry.

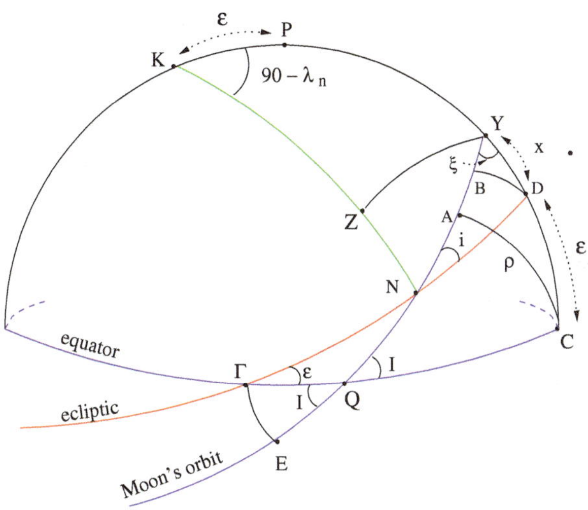

Fig. 6.3*a* Determination of the *parā-krānti*, the greatest declination that can be attained by the Moon at a given point in time.

In Fig. 6.3*a*, P is the celestial pole, K the pole of the ecliptic, Γ the vernal equinox and N the node of the Moon's orbit. Let I be the inclination of the Moon's orbit to the equator. Draw a great circle arc ΓE which is perpendicular to the Moon's orbit at E. Considering the triangle ΓEN and applying the sine formula, we have

$$\sin \Gamma E = \sin i \sin \lambda_n. \tag{6.23}$$

Here $\lambda_n = \Gamma \hat{K} N$ is the *sāyana* longitude of the node. Similarly, applying the sine formula to the triangle ΓEQ, we have

$$\sin \Gamma E = \sin I \sin \Gamma Q, \tag{6.24}$$

Hence

$$\sin I \sin \Gamma Q = \sin i \sin \lambda_n, \tag{6.25}$$

where I is the angle of inclination of the Moon's orbit with respect to the equator.

In the figure, C and D are points that are 90° away from Γ along the equator and ecliptic respectively. Let ρ be the arc from C, perpendicular to the Moon's orbit. Considering the triangle QAC, which is right-angled at A, and using the sine formula,

$$\begin{aligned} \sin\rho &= \sin I \sin QC \\ &= \cos\Gamma Q \sin I \qquad (\Gamma Q + QC = 90). \end{aligned} \tag{6.26}$$

Let $BD = \widetilde{K}$ be the arc from D, perpendicular to the Moon's orbit. Considering the triangle NBD, which is right-angled at B, and using the sine formula,

$$\begin{aligned} \sin\widetilde{K} &= \sin i \sin ND \\ &= \cos\lambda_n \sin i \qquad (\Gamma N + ND = 90). \end{aligned} \tag{6.27}$$

Let the Moon's orbit be inclined at an angle ξ to the prime meridian $KPYDC$. Let $YD = x$. Therefore, $YC = YD + DC = x + \varepsilon$. Now considering the triangles YBD and YAC and using the sine formula, we have

$$\sin\widetilde{K} = \sin x \sin\xi \qquad \text{and} \qquad \sin\rho = \sin(x+\varepsilon)\sin\xi. \tag{6.28}$$

Therefore,

$$\begin{aligned} \frac{\sin\rho}{\sin\widetilde{K}} &= \frac{\sin(x+\varepsilon)}{\sin x} \\ &= \frac{\sin x \cos\varepsilon + \cos x \sin\varepsilon}{\sin x} \\ &= \cos\varepsilon + \frac{\cos x}{\sin x}\sin\varepsilon. \end{aligned} \tag{6.29}$$

In the above equation, we would like to express $\frac{\cos x}{\sin x}$ in terms of other known quantities. From now on, all the intermediate steps till (6.35) are worked out for that purpose. Let $NY = \chi$ in the triangle NDY, which is right-angled at D. Using the sine formula, we have

$$\sin x = \sin\chi \sin i. \tag{6.30}$$

Let YZ be perpendicular to the secondary to the ecliptic passing through N. Considering the triangle NYZ which is right-angled at Z, we have

$$\sin YZ = \sin\chi \cos i. \tag{6.31}$$

Now $N\hat{K}Y = 90 - \lambda_n$. Further,

$$\begin{aligned} KY &= KP + PY \\ &= \varepsilon + (90 - (x+\varepsilon)) \\ &= 90 - x. \end{aligned} \tag{6.32}$$

Considering the triangle KYZ, which is right-angled at Z, we have

$$\begin{aligned}\sin YZ &= \sin KY \sin(90-\lambda_n) \\ &= \sin(90-x)\cos\lambda_n \\ &= \cos x \cos\lambda_n. \end{aligned} \tag{6.33}$$

From (6.31) and (6.33),

$$\sin\chi\cos i = \cos x\cos\lambda_n. \tag{6.34}$$

Replacing $\sin\chi$ in the above equation using (6.30), we have

$$\cos x\cos\lambda_n = \frac{\cos i}{\sin i}\sin x$$

or
$$\frac{\cos x}{\sin x} = \frac{\cos i}{\sin i\cos\lambda_n}. \tag{6.35}$$

Using the above in (6.29), we obtain

$$\frac{\sin\rho}{\sin\widetilde{K}} = \cos\varepsilon + \frac{\cos i}{\sin i\cos\lambda_n}\sin\varepsilon. \tag{6.36}$$

Further, eliminating $\sin\widetilde{K}$ using (6.27) in the above equation, we have

$$\sin\rho = \sin i\cos\lambda_n\cos\varepsilon + \cos i\sin\varepsilon. \tag{6.37}$$

From (6.26) and (6.37), we get

$$\sin I\cos\Gamma Q = \sin i\cos\lambda_n\cos\varepsilon + \cos i\sin\varepsilon. \tag{6.38}$$

Now squaring and adding (6.25) and (6.38), we obtain

$$\sin^2 I = (\sin\lambda_n\sin i)^2 + (\sin i\cos\lambda_n\cos\varepsilon + \cos i\sin\varepsilon)^2. \tag{6.39}$$

Therefore,

$$\sin I = \sqrt{(\sin\lambda_n\sin i)^2 + (\sin i\cos\lambda_n\cos\varepsilon + \cos i\sin\varepsilon)^2}. \tag{6.40}$$

This is the formula for the inclination of the Moon's orbit to the equator presented in the text as given in (6.20), which is also the maximum declination of the Moon (at any given time). It is known that the nodes of the Moon's orbit complete one revolution in about 18.6 years. During that period, it could happen that the Moon's orbit lies in between the ecliptic and the equator as indicated in Fig. 6.3*b*. In such a situation, the expression for $\sin\rho$ in (6.28) will have $\sin(\varepsilon-x)$ instead of $\sin(x+\varepsilon)$. The effect of this in the final expression for the *parā-krānti* (maximum declination) would be

$$\sin I = \sqrt{(\sin\lambda_n\sin i)^2 + (\cos i\sin\varepsilon - \sin i\cos\lambda_{n'}\cos\varepsilon)^2}, \tag{6.41}$$

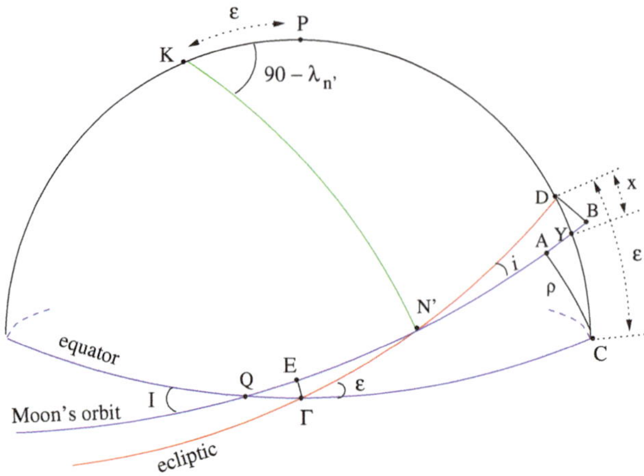

Fig. 6.3*b* Determination of the *parā-krānti*, when the Moon's orbit is situated between the equator and the ecliptic.

where N' is the descending node of the Moon's orbit and $\lambda_{n'} = \lambda_n + 180°$. Then it can easily be seen that the above equation is also the same as (6.20) given in the text.

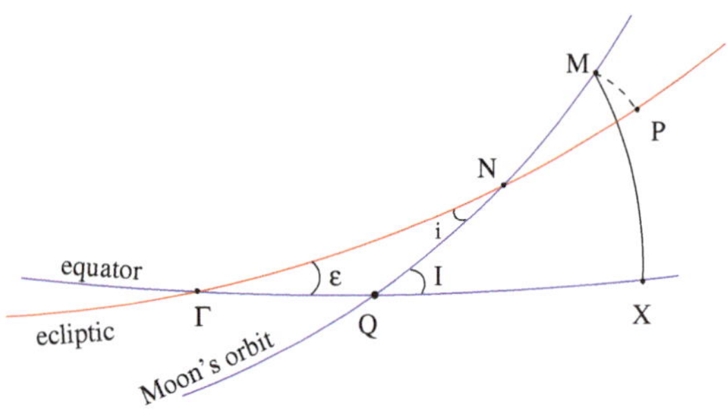

Fig. 6.3*c* Determination of the *iṣṭa-krānti*, the actual declination of the Moon at a given point in time.

In Fig. 6.3*c*, M represents the Moon and MX is its declination at a given instant. P is the point where the secondary to the ecliptic passing through M meets the ecliptic. Considering the triangle MQX, which is right-angled at X, and applying the sine formula,

$$\sin\delta_m = \sin MQ \, \sin I. \tag{6.42}$$

Now

$$\begin{aligned} MQ &= MN + NQ \\ &= MN + \Gamma N + NQ - \Gamma N \end{aligned}$$

$$\begin{aligned} &\approx NP + \Gamma N - (\Gamma N - NQ) \\ &= \Gamma P - (\Gamma N - NQ) \\ &\approx \lambda_m - \Gamma Q, \end{aligned} \tag{6.43}$$

where λ_m is the *sāyana* longitude of the Moon. In arriving at the above equation we have used two approximations:

1. $MN \approx NP$. This is a fairly good approximation since i, the inclination of the Moon's orbit is, very small.[4]
2. The other approximation is that $(\Gamma N - NQ) \approx \Gamma Q$. This again is reasonable as i is small.[5]

Applying the sine formula to the triangle ΓQN, we have

$$\sin \Gamma Q = \frac{\sin \lambda_n \sin i}{\sin I}. \tag{6.44}$$

It may be noted that the above equation is the same as (6.21) presented by Nīlakaṇṭha, once we identify ΓQ with the *ayanacalana* A. Obviously the term *ayanacalana* in this context refers to the right ascension of the point Q.

Again, because i is small, we may write

$$MQ \approx \lambda_m - \Gamma Q = \lambda_m - A. \tag{6.45}$$

Substituting for MQ in (6.42) we get

$$\sin \delta = \sin(\lambda_m - A) \sin I, \tag{6.46}$$

which is the same as the expression for the declination (6.22) given in the text.

६.५ व्यतीपातस्य सदसद्भावः

6.5 The occurrence or non-occurrence of *vyatīpāta*

संस्कृतक्षेपचलनसायनेन्दोः रवेः पदात् ॥ १२ ॥
ओजयुग्मतया भेदे व्यतीपातो न चान्यथा ।

saṃskṛtakṣepacalanasāyanendoḥ raveḥ padāt || 12 ||
ojayugmatayā bhede vyatīpāto na cānyathā |

Only if the longitude of the Moon, corrected for the change in *vikṣepa* and *ayana* [as described earlier], is such that the Sun and the Moon lie in the odd and the even quadrants [or vice versa] does *vyatīpāta* occur and not otherwise.

The condition for the possibility of the occurrence of *vyatīpāta* or otherwise, that was hinted at in—and hence to be inferred from—verses 1 and 2a of this chapter, is

[4] It may be recalled that the inclination is taken to be $270' = 4.5°$ in Indian astronomy.

[5] It needs to be verified numerically how good this approximation is.

being explicitly stated here. It is said that the Sun and the Moon must be in odd and even quadrants for the occurrence of *vyatīpāta*. In other words, the gradients with respect to the change in declination must have opposite signs during *vyatīpāta*.

The following verse in *Laghu-vivṛti* succinctly puts forth the criteria to be satisfied for *vyatīpāta* to occur:

क्रान्तिसाम्ये व्यतीपातो भवेद् भिन्नपदस्थयोः ।
नाभिन्नपदयोरर्कचन्द्रयोर्नाप्यतुल्ययोः ॥

> *Vyatīpāta* occurs only when the declinations [of the Sun and the Moon] are equal and they are in different quadrants. And not when they are in the same quadrant or when their declinations are not equal [in magnitude].

We explain this with the help of Fig. 6.4. Here S refers to the Sun and ST its declination. M_1 and M_4 represent the Moon when it lies in the I and the IV quadrant respectively. We have depicted their positions such that

$$AM_1 = ST = BM_4. \tag{6.47}$$

In other words, the magnitude of the declination of the Moon at M_1 is same as that at M_4, which is also equal to that of the Sun. When the Moon is at M_1 there is no *vyatīpāta*, because the declination gradients of the Sun and the Moon have the same sign. On the other hand, when the Moon is at M_4 there will be a *vyatīpāta* since the gradients have opposite signs, and it is *vaidhṛta* since the Sun and the Moon lie in different hemispheres.

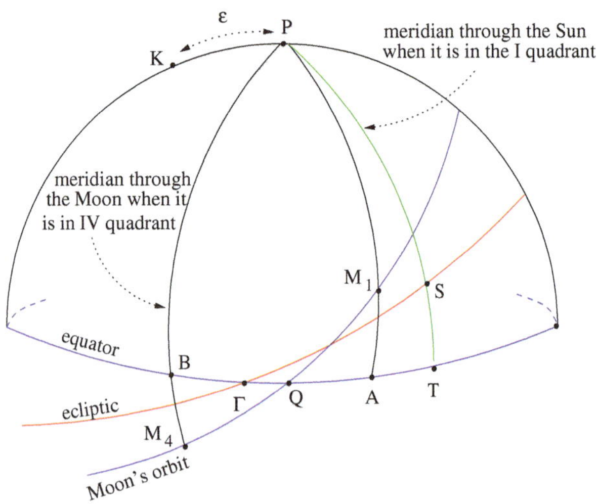

Fig. 6.4 Criterion for the occurrence of *vyatīpāta*.

For the sake of clarity and completeness we present in Table 6.1 all the different possible cases that could give rise to a *vyatīpāta*. The term 'quadrant', occurring as the heading of the first column of the table, has been given a special connotation that suits the present context. From this, the origins of the quadrants for the Sun and the Moon are taken to be the (ascending) points of intersection of their own orbits with the equator. They are referred to as *gola-sandhis.* In the case of the Sun it is the same as the vernal equinox, marked as Γ. The *gola-sandhi* of the Moon is marked by the point Q (see Fig. 6.4). This point moves at a much faster rate than Γ. It completes a cycle in about 18.6 years which amounts to about 20° per year.

Quadrant		Declination		*Ayana*		Nature of
Sun	Moon	Sun	Moon	Sun	Moon	*Vyatīpāta*
I	I	↑	↑	*uttara*	*uttara*	—
I	II	↑	↓	*uttara*	*dakṣiṇa*	*lāṭa*
I	III	↑	↑	*uttara*	*dakṣiṇa*	—
I	IV	↑	↓	*uttara*	*uttara*	*vaidhṛta*
II	I	↓	↑	*dakṣiṇa*	*uttara*	*lāṭa*
II	II	↓	↓	*dakṣiṇa*	*dakṣiṇa*	—
II	III	↓	↑	*dakṣiṇa*	*dakṣiṇa*	*vaidhṛta*
II	IV	↓	↓	*dakṣiṇa*	*uttara*	—
III	I	↑	↑	*dakṣiṇa*	*uttara*	—
III	II	↑	↓	*dakṣiṇa*	*dakṣiṇa*	*vaidhṛta*
III	III	↑	↑	*dakṣiṇa*	*dakṣiṇa*	—
III	IV	↑	↓	*dakṣiṇa*	*uttara*	*lāṭa*
IV	I	↓	↑	*uttara*	*uttara*	*vaidhṛta*
IV	II	↓	↓	*uttara*	*dakṣiṇa*	—
IV	III	↓	↑	*uttara*	*dakṣiṇa*	*lāṭa*
IV	IV	↓	↓	*uttara*	*uttara*	—

Table 6.1 The different possible cases for the occurrence of *vyatīpāta.*

६.६ व्यतीपाताभावनियमः

6.6 The criterion for the non-occurrence of *vyatīpāta*

अर्केन्द्वोः परमक्रान्त्योः अल्पा त्रिज्याहतान्यया ॥ १३ ॥
भक्ता ततोऽधिके बाहौ महाक्रान्तेर्न तुल्यता ।
तच्चापं भत्रयाच्छोध्यं तदाढ्योनायनान्तयोः ॥ १४ ॥
अन्तरालं गते तस्मिन् क्रान्त्योः साम्यं न जायते ।

arkendvoḥ paramakrāntyoḥ alpā trijyāhatānyayā || 13 ||
bhaktā tato'dhike bāhau mahākrānterna tulyatā |
taccāpaṃ bhatrayācchodhyaṃ tadāḍhyonāyanāntayoḥ || 14 ||
antarālaṃ gate tasmin krāntyoḥ sāmyaṃ na jāyate |

The lesser of the the maximum declinations of the Sun and the Moon is multiplied by the *trijyā* and divided by the other. If the Rsine of the greater is larger than the result, then there will be no equality.

The arc of that has to be subtracted from 90°. The result has to be added and subtracted from the *ayanāntas.* If 'that' lies in between, then the equality of the declinations does not take place.

Like many other verses in *Tantrasaṅgraha*, these have been written in a somewhat terse form and require a detailed explanation. The condition given here for the non-occurrence of *vyatīpāta* may be represented in the form

$$R\sin\lambda_+ > \frac{R\sin\delta_- \times R}{R\sin\delta_+}, \tag{6.48}$$

where R represents the *trijyā*, δ_+/δ_- is the larger/smaller of the *paramakrāntis* of the Sun and the Moon, and λ_+ the longitude of the Sun/Moon corresponding to δ_+ (measured from the point of intersection of its orbit with the equator). If the above condition is satisfied then there will be no *vyatīpāta*. The maximum declination of the Moon depends upon the situation of the lunar orbit, which in turn is determined by the location of the Moon's nodes. It is worth while discussing the variation of the maximum declination quantitatively before we take up (6.48).

Variation in the maximum declination of the Moon

Let δ_s^* and δ_m^* be the maximum declinations of the Sun and the Moon. While the maximum value of the Sun's declination is fixed—and is equal to the obliquity of the ecliptic, $\varepsilon = 24°$—the maximum declination of the Moon δ_m^* is a variable quantity. Its value depends upon the position of the ascending node (*Rāhu*, denoted by N_1) of the Moon's orbit with respect to the equinox. The range of its variation is given by

$$(\varepsilon - i) < \delta_m^* < (\varepsilon + i), \tag{6.49}$$

where i is the declination of the Moon's orbit, which is taken to be 4.5° in the text. When *Rāhu* coincides with the vernal equinox, then $\delta_m^* = (\varepsilon + i)$. On the other hand, when it coincides with the autumnal equinox, then $\delta_m^* = (\varepsilon - i)$. The two limiting cases are depicted in Figs 6.5*a* and 6.5*b* respectively.

As the Moon's orbit itself has a retrograde motion, around the ecliptic, the node of the Moon's orbit completes one revolution in about 18.6 years. Hence the interval between these two limiting cases depicted in Fig. 6.5 is nearly 9.3 years. We now analyse the two cases from the viewpoint of the occurence of *vyatīpāta* or otherwise.

Case i: $\delta_m^* \geq \delta_s^*$

When the maximum declination of the Moon is greater than the obliquity of the ecliptic, then invariably the magnitude of the declination of the Moon becomes equal to that of the Sun four times during the course of its sidereal period (section 6.1). Of the four instants at which the declinations are equal, only two correspond to *vyatīpāta*. These two *vyatīpātas*, namely *lāṭa* and *vaidhṛta*, necessarily occur during the course of a sidereal revolution of the Moon.

In Fig. 6.5*a*, we have depicted the limiting case in which the Moon's orbit has the maximum inclination ($I = \varepsilon + i$) to the ecliptic. U and D represent the *ayana*-

sandhis[6] in the northern and the southern hemispheres respectively. M_1 and M_2 represent the positions of the Moon, in the *uttarāyaṇa* and *dakṣiṇāyaṇa* (northern and southern courses of the Sun), when its declination is equal to that of the obliquity of the ecliptic. Let t_m be the time taken by the Moon to travel from M_1 to M_2. During this interval, the declinations of the Sun and the Moon will never become equal and hence there can be no *vyatīpāta*. This is because the declination of the Moon during this period will be greater than that of the Sun. As has been stated in the text:

अन्तरालं गते तस्मिन् क्रान्त्योः साम्यं न जायते।

When it (the Moon) is in that interval, the declinations do not become equal.

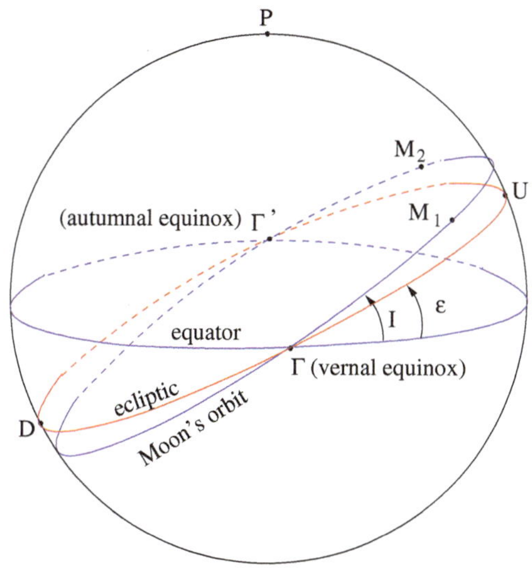

Fig. 6.5*a* Moon's orbit having the maximum inclination, $I = \varepsilon + i$, with the ecliptic.

Case ii: $\delta_m^* < \delta_s^*$

When the Moon's orbit lies completely in between the equator and the ecliptic, then, depending upon the longitude of the Sun, its declination could remain greater than the maximum declination of the Moon—which is the same as the inclination of the Moon's orbit with respect to the equator, denoted by I in Fig. 6.5*b*—for fairly long intervals of time. The said interval may extend even up to two to three months of time when the inclination I has the minimum value. During this period, the declination of the Moon doesn't become equal to that of the Sun and hence *vyatīpāta* does not occur.

In Fig. 6.5*b*, S_1 corresponds the position of the Sun when its declination is just equal to δ_m^*. As the Sun S_1 has northern motion, and is approaching the *ayanasandhi*

[6] The point of intersection of the two *ayanas*, namely the *uttarāyaṇa* and the *dakṣiṇāyana*.

U, its declination will be increasing during the next few days till it reaches the maximum ε. Having crossed the *ayanasandhi*, the Sun starts receding away from it and its declination starts decreasing. When the Sun is at S_2, again its declination will be equal to δ_m^*. Between S_1 and S_2, the declination of the Sun remains greater than δ_m^* and hence there will be no *vyatīpāta*.

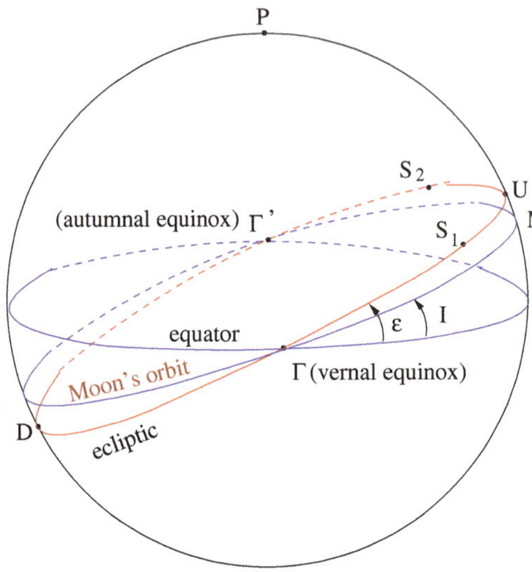

Fig. 6.5*b* Moon's orbit having the minimum inclination, $I = \varepsilon - i$, with the ecliptic.

However, the inclination of the Moon's orbit, which is the same as the maximum declination attained by the Moon, does not change significantly. The change in the maximum declination from $(\varepsilon + i)$ to the minimum value $(\varepsilon - i)$, a difference of $2 \times 4.5^\circ = 9^\circ$, takes place in about 9.3 years. This amounts to hardly one degree per year or around $5'$ per month, whereas the change in the declination of the Sun is around $480'$ per month. Hence, the change in the inclination of the Moon's orbit over a few weeks is negligible compared with that of the Sun. In the following, we make a rough estimate of the duration during which there will be no *vyatīpāta*.

Minimum period for which *vyatīpāta* does not occur

In the latter half of the 14th verse and the first half of the 15th verse, the criterion for the non-occurrence of *vyatīpāta* is given. From this, the minimum period during which *vyatīpāta* does not occur can be estimated. For numerical illustration, we choose the limiting case where the maximum declination of the Moon attains its minimum value as shown in Fig. 6.5*b*. In this case $\delta_m^* = 24.0 - 4.5 = 19.5$. The longitude of the Sun corresponding to this declination is

$$\lambda_s = \sin^{-1}\left(\frac{\sin 19.5}{\sin 24}\right) \approx 55^\circ. \qquad (6.50)$$

As the longitude of the Sun increases in the odd quadrants, the magnitude of its declination also increases. Hence, when the longitude[7] of the Sun is approximately in the range

$$55^\circ < \lambda_s < 125^\circ,$$

or when it is in the range

$$235^\circ < \lambda_s < 305^\circ, \qquad (6.51)$$

the magnitude of its declination will always be greater than the maximum declination the Moon can attain. Therefore, there will be no *vyatīpāta* during this period.

Since the rate of motion of the Sun is approximately 1° per day, under the limiting cases the minimum period for which a *vyatīpāta* does not occur is about 70 days. As the longitude of the Sun is 0° around March 21, this period approximately extends from the later half of the second week of May to the last week of July, when the Sun is in the northern hemisphere. When the Sun is in the southern hemisphere, this period would be from from the later half of November to the end of January approximately. With this background, we now proceed to explain the criterion given in the text.

Rationale behind Nīlakaṇṭha's criterion for the non-occurrence of *vyatīpāta*

The declination of the Sun and its longitude are related through the formula

$$\sin\delta_s = \sin\varepsilon\,\sin\lambda_s, \qquad (6.52)$$

where ε is the obliquity of the ecliptic, which is the same as the maximum declination of the Sun. In otherwords $\delta_s^* = \varepsilon$. The longitude of the Sun λ_s is measured from Γ along the ecliptic and is given by ΓS in Fig. 6.6.

The declination of the Moon is given by

$$\sin\delta_m = \sin I\,\sin QM = \sin I \sin\eta. \qquad (6.53)$$

Here, I is the inclination of the Moon's orbit with respect to the equator. As in the case of the Sun, I is the maximum declination of the Moon. That is, $\delta_m^* = I$. $QM = \eta$ is measured along the Moon's orbit from the point of intersection of the equator and the Moon's orbit. We have seen that $\eta \approx \lambda_m - A$, where λ_m is Moon's longitude, and A is its '*ayanacalana*'. Dividing (6.52) by (6.53) and rearranging, we have

$$\sin\eta \times \frac{\sin\delta_s}{\sin\delta_m} = \frac{\sin\varepsilon}{\sin I} \times \sin\lambda_s. \qquad (6.54)$$

[7] Since declination is involved, the longitudes that we talk about here are all *sāyana* longitudes.

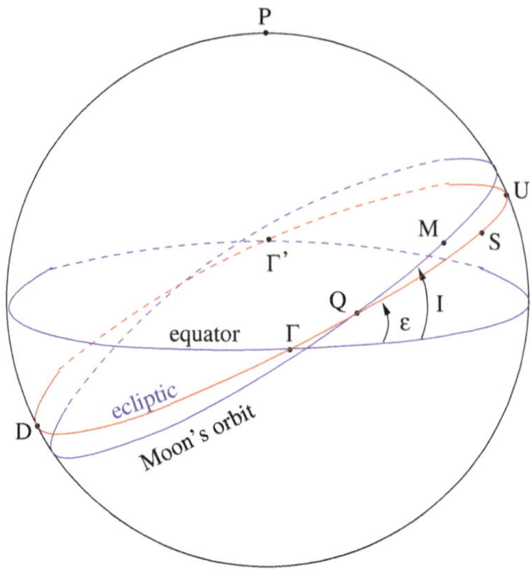

Fig. 6.6 Schematic sketch of the Moon's orbit and the ecliptic, when the maximum inclination of the Moon's orbit, I, is greater than ε.

Depending upon the position of the Moon's ascending node represented by Q in Fig. 6.6, either $\varepsilon > I$ or $\varepsilon < I$. The case $\varepsilon = I$ is true only at one instant, and is a very special case. The other two cases do prevail for an extended period of time. Now let us consider the case $\varepsilon < I$. *Vyatīpāta* occurs under these circumstances when

$$\sin\eta = \frac{\sin\varepsilon}{\sin I} \times \sin\lambda_s. \tag{6.55}$$

As $\sin\lambda_s \leq 1$, this implies that the condition for the occurrence of *vyatīpāta* is

$$\sin\eta \leq \frac{\sin\varepsilon}{\sin I}. \tag{6.56}$$

Hence there is no *vyatīpāta* if

$$\sin\eta > \frac{\sin\varepsilon}{\sin I}. \tag{6.57}$$

The above condition is the same as the one given in (6.48) once we identify that $\varepsilon = \delta_-$, $I = \delta_+$ and $\eta = \lambda_+$, as the maximum declination of the Sun is less than that of the Moon. Similarly, when $I < \varepsilon$, there is no *vyatīpāta* if

$$\sin\lambda_s > \frac{\sin I}{\sin\varepsilon}. \tag{6.58}$$

The equivalence of this condition with (6.48) is also clear once we identify that in this case $I = \delta_-$, $\varepsilon = \delta_+$ and $\lambda_s = \lambda_+$.

६.७ व्यतीपातास्य गतैष्यत्वनिर्णयः

6.7 Determining whether *vyatīpāta* has occurred or is yet to occur

दोर्ज्यां रवेः परक्रान्त्या हत्वा चान्द्र्या तया हरेत् ॥ १५ ॥
लब्धचापसमे चन्द्रबाहौ क्रान्तिगुणौ समौ ।
चन्द्रस्यौजपदस्थस्य दोर्धनुष्यधिके ततः ॥ १६ ॥
व्यतीपातो गतो न्यूने भावी युग्मपदेऽन्यथा ।
तदिष्टचन्द्रधनुषः स्वस्वभुक्तिघ्नमन्तरम् ॥ १७ ॥
गतियोगहृतं स्वर्णं दोषे गम्ये गतेऽपि च ।
सूर्येन्द्वोरन्यथा पाते तावत्कुर्यादिदं मुहुः ॥ १८ ॥
यावदर्कोत्थधनुषा तत्कालेन्दुधनुः समम् ।

dorjyāṃ raveḥ parakrāntyā hatvā cāndryā tayā haret || 15 ||
labdhacāpasame candrabāhau krāntiguṇau samau |
candrasyaujapadasthasya dordhanuṣyadhike tataḥ || 16 ||
vyatīpāto gato nyūne bhāvī yugmapade'nyathā |
tadiṣṭacandradhanuṣaḥ svasvabhuktighnamantaram || 17 ||
gatiyogahṛtaṃ svarṇaṃ doṣe gamye gate'pi ca |
sūryendvoranyathā pāte tāvatkuryādidaṃ muhuḥ || 18 ||
yāvadarkotthadhanuṣā tatkālendudhanuḥ samam |

Having multiplied the sine of the longitude of the Sun by its *parakrānti*, divide [that] by the *parakrānti* of the Moon. If the resulting arc [say x] is equal to the arc corresponding to the Moon (η), then the Rsine of the declination of the Sun and the moon will be equal.

If the arc corresponding to the Moon, in the odd quadrant, is greater than that (x), then *vyatīpāta* has already occurred; if less, then it is yet to occur. It is exactly the reverse [when the Moon is] in the even quadrant.

The difference of the arc corresponding to that ($R\sin x$) and that of the [*ayana*-corrected] Moon must be multiplied separately by their own [i.e. of the Sun and Moon] daily motions and divided by the sum of their daily motions. The results must be added to or subtracted from the Sun and the Moon depending upon whether the *doṣa* (a *vyatīpāta*) is yet to occur or has already occurred. In the case of the node it has to be applied inversely. The process has to be repeated till the arc obtained from the Sun becomes equal to the Moon's arc [found] at that time.

We saw earlier in (6.54) that the ratio of the declinations of the Sun and the Moon satisfy the relation

$$\frac{\sin\delta_s}{\sin\delta_m} = \frac{\sin\varepsilon\,\sin\lambda_s}{\sin I\,\sin\eta}. \tag{6.59}$$

Using the notation

$$\sin x = \frac{\sin\varepsilon}{\sin I}\times\sin\lambda_s \qquad \text{and} \qquad y = \frac{\sin\delta_s}{\sin\delta_m}, \tag{6.60}$$

(6.59) reduces to,

$$\sin\eta\times y = \sin x. \tag{6.61}$$

If $x = \eta$, the above equation implies that $y = 1$, that is, $\delta_s = \delta_m$. This is precisely the condition given here for the declinations of the Sun and Moon to be equal and is stated in the following words:

लब्धचापसमे चन्द्रबाहौ क्रान्तिगुणौ समौ ।

Though the term *cāpa* in general refers to arc, in the present verse it seems to have been used to refer to the sine of the arc. In other words, the term *labdhacāpa* in the above verse refers to $\sin x$. The term *candrabāhu* refers to $\sin \eta$. As mentioned earlier,

$$\text{if } x = \eta, \qquad \text{it is the middle of } vyat\bar{\imath}p\bar{a}ta.$$

The criteria as to whether a *vyatīpāta* has already occurred, or it is yet to occur are given by

$$\begin{array}{lll} & \text{if } x < \eta, & \text{already occured,} \\ \text{and} & \text{if } x > \eta, & \text{yet to occur,} \end{array}$$

in the odd quadrant. It is the other way round in the even quadrant, as $|\delta_m|$ decreases with time. Here η is the angular separation of the Moon from the point of intersection of the Moon's orbit and the equator. In Fig. 6.7 it is given by $QM_i = \alpha_i$ ($i = 0, 1$ and 2).

Rationale behind the given criteria

(a) Criterion for *vyatīpāta* to have occured

Suppose $x < \eta$ at some time $t = t_1$, then we should have $y < 1$ in order that (6.61) is satisfied. Now

$$y < 1 \qquad \Rightarrow \qquad \sin \delta_s < \sin \delta_m. \tag{6.62}$$

This situation is represented by the positions of the Sun and the Moon at S_1 and M_1 in Fig. 6.7. Since the Moon is in the odd quadrant and the Sun is in the even quadrant, the magnitude of the declination of the Moon keeps increasing and that of the Sun keeps decreasing. Since $|\delta_s| < |\delta_m|$ at $t = t_1$, there must be an earlier instant, $t = t_0$, at which $|\delta_s| = |\delta_m|$. This is precisely the condition for the occurrence of *vyatīpāta*. Thus we see that if $x < \eta$ and the Moon is in the odd quadrant, then *vyatīpāta* has already occurred.

(b) Criterion for *vyatīpāta* to occur later

If $x > \eta$, then from (6.61), $y > 1$. Now,

$$y > 1 \qquad \Rightarrow \qquad \sin \delta_s > \sin \delta_m. \tag{6.63}$$

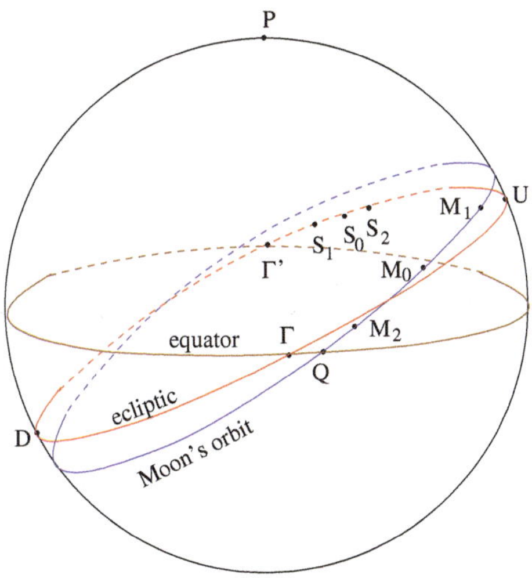

Fig. 6.7 Positions of the Sun and the Moon before *vyatīpāta*, at the instant of *vyatīpāta* and after *vyatīpāta*.

This situation is represented by the positions of the Sun and the Moon at S_2 and M_2 in Fig. 6.7. Again, since the Moon is in the odd quadrant and the Sun is in the even quadrant, the magnitudes of their declinations are increasing and decreasing respectively. Since at $t = t_2$, $|\delta_s| > |\delta_m|$, *vyatīpāta* is yet to occur at $t = t_0 > t_2$.

The situation in the even quadrants can be understood similarly. The above criteria are precisely those given in verses 16b and 17a to find out whether *vyatīpāta* has already occurred or it is yet to occur. In the succeeding verses 17b and 18, a procedure is given for finding the time interval (Δt) between the desired instant and the instant of *vyatīpāta*. Having determined this time interval, an iterative procedure for finding the longitudes of the Sun and the Moon at the instant of *vyatīpāta* is outlined.

The time interval between the desired instant and the middle of *vyatīpāta*

Let λ_s and λ_m be the longitudes of the Sun and the Moon a given instant t, and let the angular velocities (*gati*) of them at that instant be $\dot{\lambda}_s$ and $\dot{\lambda}_m$. It is seen from (6.60) that the quantity x is related to the Sun's longitude and η is related to the Moon's longitude. We denote the difference in arcs between x and η by $\Delta\theta$. That is,

$$x - \eta = \Delta\theta. \tag{6.64}$$

The significance of $\Delta\theta$ is that it refers to the angle by which the sum of the longitudes of the Sun and the Moon must increase for *vyatīpāta* to occur. It is mentioned that this has to be divided by the sum of the angular velocities of the Sun and the Moon. We denote the result in time units by Δt, which is given by

$$\Delta t = \frac{\Delta\theta}{\dot{\lambda}_m + \dot{\lambda}_s} \qquad \text{(in days)}$$
$$= \frac{\Delta\theta}{\dot{\lambda}_m + \dot{\lambda}_s} \times 60 \qquad \text{(in } ghaṭikas\text{)}. \tag{6.65}$$

The longitudes of the Sun and the Moon at the middle of *vyatīpāta*

The changes in the longitudes of the Sun and the Moon during the above time interval Δt are obtained by multiplying their daily motions with it. That is

$$\Delta\lambda_s = \dot{\lambda}_s \times \Delta t \tag{6.66}$$
$$\text{and} \qquad \Delta\lambda_m = \dot{\lambda}_m \times \Delta t. \tag{6.67}$$

If λ_s and λ_{s_0} are the longitudes of the Sun at the desired instant t and the middle of the *vyatīpāta*, then

$$\lambda_{s_0} = \lambda_s \mp \Delta\lambda_s. \tag{6.68}$$

Similarly, if λ_m and λ_{m_0} are the longitudes of the Moon at the desired instant and the middle of the *vyatīpāta*, then

$$\lambda_{m_0} = \lambda_m \mp \Delta\lambda_m. \tag{6.69}$$

Here we take the sign '$-$' if the *vyatīpāta* has already occurred and the sign '+' if it is yet to occur.

Iterative method

In the procedures described in the previous sections, it has been implicitly assumed that the rates of motion of the Sun and the Moon ($\dot{\lambda}_m$ and $\dot{\lambda}_s$) are constant, which is not true. Hence both Δt and the longitudes λ_{s0} and λ_{m0} obtained are only approximate. As a corrective measure to this, an iterative procedure for determining the longitudes of the Sun and the Moon at *vyatīpāta* is prescribed. The iterative method to be used here is indicated in verses 18b and 19a.

तावत् कुर्यादिदं मुहुः यावदर्कोत्थधनुषा तत्कालेन्दुधनुः समम्।

This [process] has to be repeated till the arc of the Moon at that time will be equal to that of the Sun.

The method indicated above, and further explained in the commentary, may be explained as follows. As Δt given by (6.65) is not exact, we denote it by Δt_1 to indicate that it is the first approximation to the actual value. Having determined Δt_1 we evaluate x and η at time

$$t_1 = t + \Delta t_1, \tag{6.70}$$

and denote their values as x_1 and η_1. The rates of motion of the Sun and the Moon are also evaluated at t_1 and are denoted by $\dot{\lambda}_{s1}$ and $\dot{\lambda}_{m1}$. With them, we find Δt_2

given by

$$\Delta t_2 = \frac{\Delta\theta_1}{\dot{\lambda}_{m1} + \dot{\lambda}_{s1}}, \tag{6.71}$$

where $\Delta\theta_1 = x_1 - \eta_1$. The second approximation to the actual instant of *vyatīpāta* is t_2 and is given by

$$t_2 = t_1 + \Delta t_2. \tag{6.72}$$

Again at t_2 the values of x and η denoted by x_2 and η_2 are to be determined. From their difference $\Delta\theta_2$, and the rates of motion of the Sun and the Moon, Δt_3 is found. The process is repeated and, in general,

$$\begin{aligned} \Delta\theta_i &= x_i - \eta_i \\ \Delta t_{i+1} &= \frac{\Delta\theta_i}{\dot{\lambda}_{mi} + \dot{\lambda}_{si}} \\ \text{and} \qquad t_{i+1} &= t_i + \Delta t_{i+1}. \end{aligned} \tag{6.73}$$

The iteration is continued till $\Delta t_r \approx 0$. At this instant (t'), $x = \eta$ to the desired accuracy. Hence, the longitude of the Moon that is determined from η in this process, which in turn is determined by finding x, would be the same as the longitude determined at t' directly. This is what is stated in verse 19a, quoted above. The instant of *vyatīpāta* is then given by

$$t' = t + \Delta t_1 + \Delta t_2 + \cdots + \Delta t_r. \tag{6.74}$$

Here, it should be noted that Δt_r can be positive or negative.

६.८ व्यतीपातमध्यः

6.8 The middle of *vyatīpāta*

क्रान्तिसाम्ये व्यतीपातमध्यकालः सुदारुणः ॥ १९ ॥

krāntisāmye vyatīpātamadhyakālaḥ sudāruṇaḥ || 19 ||

When the declinations [of the Sun and the Moon] are equal, that instant corresponds to the middle of *vyatīpāta*, which is quite dreadful.

६.९ व्यतीपातप्रारम्भः पर्यवसानञ्च

6.9 The beginning and the end of *vyatīpāta*

नवांशपञ्चकं तत्त्वभागौ बिम्बौ स्वभुक्तितः ।
सूर्येन्द्वोर्बिम्बसम्पर्कदलं षष्ट्या निहत्य यत् ॥ २० ॥
गतियोगोद्धृतं तद्धि व्यतीपातदलं विदुः ।
व्यतीपातदले तस्मिन् नाडिकादौ विशोधिते ॥ २१ ॥

मध्यकालाद् भवेत् तस्य प्रारंभसमयः स्फुटः ।
तद्युते मध्यकालेऽस्य मोक्षो वाच्यो हि धीमता ॥ २२ ॥

navāṃśapañcakaṃ tattvabhāgau bimbau svabhuktitaḥ |
sūryendvorbimbasamparkadalaṃ ṣaṣṭyā nihatya yat || 20 ||
gatiyogoddhṛtaṃ taddhi vyatīpātadalaṃ viduḥ |
vyatīpātadale tasmin nāḍikādau viśodhite || 21 ||
madhyakālād bhavet tasya prārambhasamayaḥ sphuṭaḥ |
tadyute madhyakāle'sya mokṣo vācyo hi dhīmatā || 22 ||

The daily motion of the Sun multiplied by 5 and divided by 9, and that of the Moon divided by 25, are the diameters of the discs (*bimbas*) of the Sun and the Moon. Half the sum of the discs multiplied by 60 and divided by the sum of their daily motions is considered to be the half-duration of the *vyatīpāta*.

By subtracting the half-duration of the *vyatīpāta*, in *nāḍikās* etc., from the middle of the *vyatīpāta*, the actual beginning moment is obtained. By adding the same to the the middle of the *vyatīpāta*, the ending moment has to be stated by the wise ones.

If $\dot{\lambda}_s$ and $\dot{\lambda}_m$ are the daily motions of the Sun and the Moon, expressed in minutes, then the angular diameters of their discs α_s and α_m are given as

$$\alpha_s = \frac{\dot{\lambda}_s \times 5}{9}, \qquad \alpha_m = \frac{\dot{\lambda}_m}{25}. \tag{6.75}$$

Now the angular diameter of the Sun

$$\alpha_s = \frac{D_s}{d_s}, \tag{6.76}$$

where D_s and d_s are the Sun's diameter and its distance from the Earth in *yojanas* respectively. The horizontal parallax of the Sun (P), whose value is taken to be one-fifteenth of daily motion of the Sun, is given by

$$P = \frac{R_e}{d_s} = \frac{1}{15}\dot{\lambda}_s. \tag{6.77}$$

Using this in (6.76),

$$\alpha_s = \frac{D_s}{R_e}\frac{\dot{\lambda}_s}{15} = \frac{2D_s}{D_e}\frac{\dot{\lambda}_s}{15}, \tag{6.78}$$

where $D_e = 2R_e$ is the diameter of the Earth. In Chapter 4, the values of D_s and D_e are given to be 4410 and 1050.42 *yojanas* respectively. Therefore

$$\alpha_s = \frac{2 \times 4410}{1050.42 \times 15}\,\dot{\theta}_s = 0.5598\,\dot{\lambda}_s. \tag{6.79}$$

It is this 0.5598 that is approximated by $\frac{5}{9} = 0.5556$ in the text. Similarly, the angular diameter of the Moon is given by

$$\alpha_m = \frac{2D_m}{D_e} \times \frac{\dot{\lambda}_m}{15}, \tag{6.80}$$

where D_m is the Moon's diameter in *yojanas*. As D_m is given to be 315 *yojanas*,

$$\begin{aligned}\alpha_m &= \frac{2\times 315}{1050.42\times 15}\,\dot{\lambda}_m \\ &= 0.04\,\dot{\lambda}_m \\ &= \frac{\dot{\lambda}_m}{25}.\end{aligned} \tag{6.81}$$

Using the angular diameters, the half-duration of the *vyatīpāta* is found using the formula

$$\Delta t = \frac{S\times 60}{\dot{\lambda}_m + \dot{\lambda}_s}, \tag{6.82}$$

where S is the sum of the semi-diameters of the Sun and the Moon and is given by

$$S = \frac{d_s + d_m}{2}. \tag{6.83}$$

Let t_b, t_m and t_e be the actual beginning, the middle and the ending moment of the *vyatīpāta*. Here t_m refers to the instant at which (6.1) is satisfied. Then the beginning and the ending moments are given by

$$t_b = t_m - \Delta t \qquad \text{and} \qquad t_e = t_m + \Delta t. \tag{6.84}$$

६.१० विष्कम्भादियोगान्त्यार्धानां त्याज्यत्वम्

6.10 Inauspiciousness of the later half of *viṣkambhayoga* and others

विष्कम्भादिषु योगेषु व्यतीपाताह्वयोऽपि यः ।
तस्य सप्तदशस्यान्त्यमर्धं चाप्यतिदारुणम् ॥ २३ ॥

viṣkambhādiṣu yogeṣu vyatīpātāhvayo'pi yaḥ |
tasya saptadaśasyāntyamardhaṃ cāpyatidāruṇam || 23 ||

The later half of the seventeenth *yoga* commencing with *viṣkambha*, also known as *vyatīpāta*, is extremely inauspicious.

Analogous to the 27 *nakṣatras*, 27 *yogas* (see Table 6.2) are defined in Indian astronomy. They correspond to intervals of time during which the sum of the longitudes of the Sun and the Moon increases by $13°20'$. It may be noted from Table 6.2 that the 17th *yoga* is called *vyatīpāta*. Perhaps, hereby due to the similarity in name, this is also considered inauspicious (particularly its later half).

In this context the following verse is quoted in the commentary *Laghu-vivṛti*:

सूर्येन्दुयोगे मैत्रस्य परार्धं सम्भवेद्यदि ।
सार्पमस्तकसंज्ञः स्यात् तदा दोषोऽतिनिन्दितः ॥

Among the *yogas* of the Sun and the Moon, the later half of *Maitra* is called *Sārpamastaka* and that period is considered to be highly inauspicious.

1. *viṣkambha*	10. *gaṇḍa*	19. *parigha*
2. *prīti*	11. *vṛddhi*	20. *śiva*
3. *āyuṣmān*	12. *dhruva*	21. *siddha*
4. *saubhāgya*	13. *vyāghāta*	22. *sādhya*
5. *śobhana*	14. *harṣaṇa*	23. *śubha*
6. *atigaṇḍa*	15. *vajra*	24. *śukla*
7. *sukarma*	16. *siddhi*	25. *brāhma*
8. *dhṛti*	17. *vyatīpāta*	26. *aindra*
9. *śūla*	18. *varīyān*	27. *vaidhṛti*

Table 6.2 The names of the 27 *yogas*.

Note:

In the above verse, the term *maitrasya* literally means ' belonging to *Maitra*'. According to the tradition, each *nakṣatra* is associated with a deity. The deity for the 17th *nakṣatra*, namely *Anūrādha*, is *Maitra*. Hence the 17th *nakṣatra* is called *Maitra*.

While discussing *vyatīpāta*, Bhāskara I states:

सूर्येन्दुयोगे चक्रार्धे व्यतीपातोऽथ वैधृतः ।
चक्रे च मैत्रपर्यन्ते विज्ञेयः सार्पमस्तकः ॥ [8]

When the sum of the [*nirayaṇa*] longitudes of the Sun and the Moon is half a circle (i.e. 180°) it is *vyatīpāta*; when the sum is a full circle (360°) it is *vaidhṛta*. If [the sum] extends to the end of *Maitra* (*Anūrādha nakṣatra*) then it is to be known as *sārpamastaka* [*vyatīpāta*].

६.११ व्यतीपातत्रयाणां त्याज्यत्वम्

6.11 Inauspiciousness of the three *vyatīpātas*

व्यतीपातत्रयं घोरं सर्वकर्मसु गर्हितम् ।
स्नानदानजपश्राद्धव्रतहोमादिकर्मसु ।
प्राप्यते सुमहच्छ्रेयः तत्कालज्ञानतस्ततः ॥ २४ ॥

vyatīpātatrayaṃ ghoraṃ sarvakarmasu garhitam |
snānadānajapaśrāddhavratahomādikarmasu |
prāpyate sumahacchreyaḥ tatkālajñānatastataḥ || 24 ||

The [period of the] three *vyatīpātas* (*lāṭa*, *vaidhṛta* and *sārpa-mastaka*) is [considered to be] dreadful and is inauspicious for performing all religious rites. But by acquiring the correct knowledge of these periods and performing certain deeds such as having a holy dip, performing charitable deeds or sacrificial deeds, doing penance, oath-taking, performing *homa* etc. one reaps great benefits.

[8] {LB 1974}, (II. 29), p. 39.

Chapter 7
दृक्कर्मप्रकरणम्
Reduction to observation

७.१ दृक्कर्मद्वयम् - आक्षम् आयनञ्च

7.1 The two visibility corrections–due to the latitude of the observer and due to the position on the ecliptic

विषुवद्भाघ्नविक्षेपात् द्वादशाप्तं विधोः स्फुटात् ।
उदये सौम्यविक्षेपे शोध्यमस्तमये धनम् ॥ १ ॥
व्यस्तं तद् याम्यविक्षेपे न मध्यस्थे विधाविदम् ।
सत्रिभग्रहजक्रान्तिभागघ्नाः क्षेपलिप्तिकाः ॥ २ ॥
विकलाः स्वमृणं क्रान्तिक्षेपयोः भिन्नतुल्ययोः ।
एवं कृतो ग्रहो लग्नं स्वोदये भवति स्फुटम् ॥ ३ ॥
स्वास्तेऽस्तलग्नमेवं स्यात् मध्यलग्नं खमध्यगे ।

viṣuvadbhāghnavikṣepāt dvādaśāptaṃ vidhoḥ sphuṭāt |
udaye saumyavikṣepe śodhyamastamaye dhanam || 1 ||
vyastaṃ tad yāmyavikṣepe na madhyasthe vidhāvidam |
satribhagrahajakrāntibhāgaghnāḥ kṣepaliptikāḥ || 2 ||
vikalāḥ svamṛṇaṃ krāntikṣepayoḥ bhinnatulyayoḥ |
evaṃ kṛto graho lagnaṃ svodaye bhavati sphuṭam || 3 ||
svāste'stalagnamevaṃ syāt madhyalagnaṃ khamadhyage |

The latitude of the Moon, multiplied by the equinoctial shadow and divided by 12, has to be subtracted from the true longitude of the Moon if the latitude is north at sunrise. At sunset it has to be added.

If the declination is south, then the application is reversed and there is no correction when the Moon is at the centre [zenith]. The product of latitude in minutes and the declination of the point whose longitude is 90 degrees plus that of the planet in degrees, [which is] in seconds, is added to or subtracted from the longitude of the planet depending upon whether the declination and the latitude are in different hemispheres or in the same hemisphere.

After this correction, the longitude of the planet thus obtained at the time of its own rise will be the *udayalagna* and the one that is obtained at its setting will be the *astalagna*. The longitude obtained when it is on the meridian will be the *madhyalagna*.

The aim is to find out the *lagna* (ascendant) corresponding to the instant when the planet is rising or setting. This would have been the same as the longitude of the planet, if the planet had no latitudinal deflection. Because of the deflection due to latitude, the *lagna* corresponding to the rising of the planet called the *udayalagna*, and the one corresponding to its setting called the *astalagna*, are different from the longitude of the planet at its rising or setting. To obtain the *lagna*, two corrections have to be applied, namely (i) the *ākṣa-dṛkkarma* (latitude correction) and (ii) the *āyana-dṛkkarma* (*āyana* correction).

Ākṣa correction

Let ϕ be the latitude of the place and β the latitude of the Moon. Then, the *ākṣa* correction a is given to be

$$\begin{aligned} a &= \frac{vik\d{s}epa \times vi\d{s}uvacch\bar{a}y\bar{a}}{12} \\ &= \frac{\beta \times 12 \tan\phi}{12} \\ &= \beta \times \tan\phi. \end{aligned} \tag{7.1}$$

This is applied to the longitude of the Moon as follows. If the latitude of the Moon is north,

$$\lambda' = \lambda - a \qquad \text{(Moon rising)} \tag{7.2}$$
$$= \lambda + a \qquad \text{(Moon setting).} \tag{7.3}$$

If the latitude of the Moon is south, then the application of the correction must be reversed.

Note: Though it is stated here that the above correction is to be applied to the Moon, it may be noted that the correction is applicable to all the planets in general. In fact, this is mentioned later in verse 7 of this chapter.

Rationale behind the *ākṣa* correction

In Fig. 7.1*a*, G is the planet, which is rising, whose longitude is λ and declination is δ. D is the point on the ecliptic where the secondary to the ecliptic passing through the planet intersects the ecliptic. Here $\Gamma D = \lambda$ represents the *sāyana* longitude of the planet. D' is the point on the ecliptic which is rising along with the planet. $\Gamma D'$ is to be found from ΓD. Let the secondary to the equator through G intersect the ecliptic at C. Then the difference

$$DD' = DC + CD'. \tag{7.4}$$

CD' is the *ākṣa-dṛkkarma* and CD is the *āyana-dṛkkarma*. The *ākṣa-dṛkkarma* is obtained as follows. Consider the triangles PGN and CGD'. Let

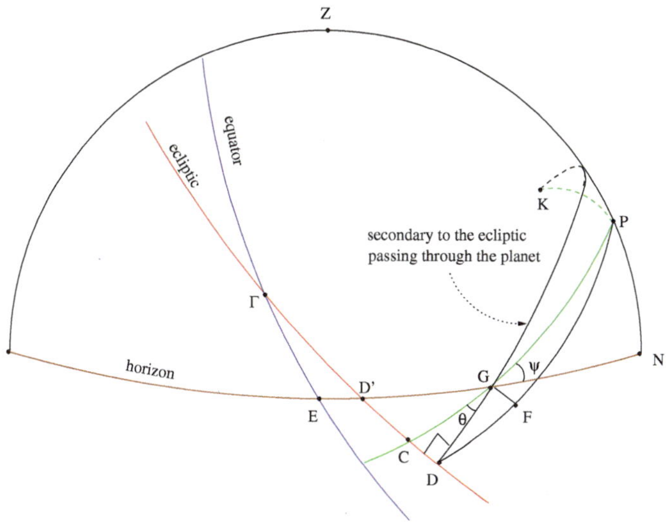

Fig. 7.1a *Ākṣa-* and *āyana-dṛkkarmas* when the latitude and the declination both have the same direction (both are positive).

$$C\hat{G}D' = P\hat{G}N = \psi. \tag{7.5}$$

In the spherical triangle *PGN*, $PG = 90 - \delta$, $PN = \phi$ and $P\hat{G}N = \psi$. Applying the sine formula, we have

$$\frac{\sin\psi}{\sin\phi} = \frac{\sin 90}{\sin(90-\delta)}\,. \tag{7.6}$$

Therefore

$$\sin\psi = \frac{\sin\phi}{\cos\delta}\,. \tag{7.7}$$

Since the latitude of the planet $GD = \beta$ is always small, the triangles CGD' and DGD' can be considered to be planar triangles. In the triangle *CGD*, let the angle $C\hat{G}D$ be θ. Here

$$\begin{aligned} C\hat{D'}G = D\hat{D'}G &= 90 - D\hat{G}D' \\ &= 90 - (C\hat{G}D + C\hat{G}D') \\ &= 90 - (\theta + \psi). \end{aligned} \tag{7.8}$$

Now

$$\frac{CD'}{\sin\psi} = \frac{CG}{\sin[90-(\theta+\psi)]} = \frac{CG}{\cos(\theta+\psi)}\,. \tag{7.9}$$

From the triangle *CGD*,

$$CG = \frac{GD}{\cos\theta} = \frac{\beta}{\cos\theta}\,. \tag{7.10}$$

Using (7.10) in (7.9), we get the expression for the *ākṣa-dṛkkarma* as

$$a = CD' = \frac{\beta \sin \psi}{\cos \theta \cos(\theta + \psi)}. \tag{7.11}$$

When δ is small, $\cos\delta \approx 1$ and $\sin\psi = \sin\phi$ or $\psi = \phi$. If we also take $\theta \approx 0$,

$$a = \beta \tan \phi, \tag{7.12}$$

which is the same as (7.1), the formula prescribed in the text.

Āyana correction

In order to find the expression for the *āyana-dṛkkarma*, δ', which is the declination of the point on the ecliptic corresponding to $\lambda + 90$, i.e. the *sāyana* longitude of the planet increased by 90 degrees, is to be determined. Now, the *āyana* correction x is given to be

$$x(\text{sec}) = \beta(\text{min}) \times \delta'(\text{degrees}). \tag{7.13}$$

If λ is the *sāyana* longitude of the planet, then it is stated that the *āyana* corrected longitude is given by

$$\lambda' = \lambda \pm x, \tag{7.14}$$

where the sign '$-$' is to be chosen if δ' and β are in the same hemisphere, and '$+$' is to be chosen if they are in different hemispheres.

Rationale behind the *āyana* correction

In the spherical triangle KDP (see Fig. 7.1a), $KP = \varepsilon$, $KD = 90$. Let $K\hat{D}P = \theta'$. Also,

$$PD = 90 - \delta,^{1} \qquad D\hat{K}P = G\hat{K}P = 90 - \lambda. \tag{7.15}$$

Using the sine formula we have

$$\sin\theta' = \frac{\sin\varepsilon \cos\lambda}{\cos\delta} = \frac{\sin\delta'}{\cos\delta}, \tag{7.16}$$

where δ' is the declination corresponding to the longitude $90 + \lambda$. When δ is small, $\sin\theta' = \sin\delta'$ or $\theta' = \delta'$.

We now consider the triangle GDF. Since it is small, we consider it to be planar and hence we have

$$\begin{aligned} GF &= GD \sin\theta' \\ &= \beta \sin\theta' \\ &\approx \beta\delta'. \end{aligned} \tag{7.17}$$

[1] We have taken the declination of the point, where the secondary to the ecliptic passing through the planet meets the ecliptic, to be δ. This seems to have been the practice.

As β is small, the *āyana-dṛkkarma* will become

$$x = CD \approx GF \approx \beta\delta'. \tag{7.18}$$

Here x, β and δ' are all in radians. In the text it is mentioned that x is in seconds, β in minutes and δ' in degrees. Here the commentator Śaṅkara Vāriyar observes:

> '···तत्कालसायनेन्दोः क्रान्तिचापमानीय ततः षष्ट्या विभज्य लब्धाः क्रान्तिभागाः तद्गुणिता इष्टविक्षेपकला एव दर्शनसंस्कारविकलाः ॥'.
>
> The visibility correction in seconds is obtained by first finding the arc corresponding to the declination from the *sāyana* longitude of the Moon and then converting the same to degrees (*krāntibhāga*) by dividing by 60, and multiplying it by the desired latitude in minutes.

If the relation (7.18), which is in radians, is to be expressed in seconds, it has to be multiplied by R (which will give it in minutes) and then by 60 (to get it in seconds):

$$\begin{aligned} x \times (R \times 60) &= (\beta \times \delta') \times (R \times 60) \\ x^*(\text{sec}) &= (\beta \times R) \times (\delta' \times 60). \end{aligned} \tag{7.19}$$

Now we rewrite[2] the above as

$$x^*(\text{sec}) \approx (\beta \times R) \times \left(\frac{\delta' \times R}{60}\right) \tag{7.20}$$

$$= \beta^*(\text{min}) \times \delta^*(\text{deg}). \tag{7.21}$$

Since β and δ' are in radians, $\beta^* = \beta \times R$ is in minutes, and $\delta^* = \frac{R\delta'}{60}$ is in degrees. This explains the different units specified for the different quantities appearing in (7.13).

Application of the *āyana-dṛkkarma*

When the declination and the latitude of the planet have the same direction—say both are positive as shown in Fig. 7.1*a*—then the *āyana-dṛkkarma* has to be applied to the longitude of the planet negatively. That is, if λ is the (*sāyana*) longitude before correction, then the longitude with the *āyana-dṛkkarma* (λ') will be

$$\begin{aligned} \lambda' = \Gamma C &= \Gamma D - CD \\ &= \lambda - x. \end{aligned} \tag{7.22}$$

---

[2] In rewriting, the following approximation seems to have been used:

$$\frac{R}{60} = \frac{3438}{60} = 57.3 \approx 60.$$

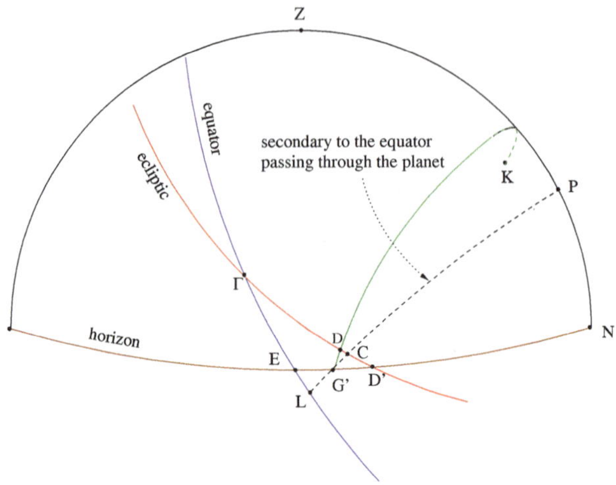

Fig. 7.1*b* *Ākṣa-* and *āyana-dṛkkarmas* when the latitude and the declination both have different directions (one positive and the other negative).

On the other hand, if the declination and latitude have opposite directions (Fig. 7.1*b*), then the correction has to be applied positively. In the figure, CD' represents the *ākṣa* correction and CD represents the *āyana* correction. In this case, it may be noted that $\beta = G'D$ is negative, whereas the declination is positive.

$$\begin{aligned} \lambda' = \Gamma C &= \Gamma D + DC \\ &= \lambda + |x| \\ &= \lambda + |\beta|\,|\delta'|. \end{aligned} \tag{7.23}$$

Absence of the *dṛkkarma*

In Figs. 7.1*a* and 7.1*b*, we have depicted the *dṛkkarma* corrections when the planet is rising. Similar corrections have to be applied at its setting. By applying these corrections we essentially get the *lagna* at the time of rising or setting of the planet. Since the rising and setting times of the different *lagnas* have already been discussed extensively in Chapter 3, here the whole exercise is meant for finding the rising and setting time of the planets.

When the planet is on the meridian, the *ākṣa-dṛkkarma* is zero. The formula for finding the *ākṣa-dṛkkarma* is given by

$$a = \frac{\beta \sin\psi}{\cos\theta \cos(\theta + \psi)}. \tag{7.24}$$

When the *graha* G is on the meridian, $\psi = P\hat{G}N = 0°$ or $180°$. Therefore, $\sin\psi = 0$ and hence $a = 0$. Thus, in the *madhyalagna* calculation, only the *āyana-dṛkkarma* contributes.

७.२ ग्रहाणां इष्टविक्षेपः
7.2 The desired latitude of the planets

मन्दस्फुटात् स्वपातोनात् भौमादीनां भुजागुणात् ॥ ४ ॥
परमक्षेपनिघ्ना स्यात् क्षेपोऽन्त्यश्रवणोद्धृतः ।
कृतनेत्रभुजङ्गाङ्गदिशो दशगुणाः क्रमात् ॥ ५ ॥
पातभागाः कुजादीनां अथ विक्षेपलिप्तिकाः ।
नवार्कर्त्वर्करवयो दशघ्नाः परमाः क्रमात् ॥ ६ ॥

mandasphuṭāt svapātonāt bhaumādīnāṃ bhujāguṇāt || 4 ||
paramakṣepanighnā syāt kṣepo'ntyaśravaṇoddhṛtaḥ |
kṛtanetrabhujaṅgāṅga diśo daśaguṇāḥ kramāt || 5 ||
pātabhāgāḥ kujādīnāṃ atha vikṣepaliptikāḥ |
navārkartvarkaravayo daśaghnāḥ paramāḥ kramāt || 6 ||

The product of the Rsine of the *manda-sphuṭa* of the planet Mars etc. from which the longitude of the node is subtracted, and the maximum deflection [of the planetary orbit], when divided by the last hypotenuse (*antya-śravaṇa*) gives the latitude of the planet. 4, 2, 8, 6 and 10, [each of them] multiplied by 10, are respectively the longitudes of the nodes in degrees of the five planets beginning with Mars. The maximum deflection [of Mars etc. from the ecliptic] in minutes are 9, 12, 6, 12 and 12, respectively, multiplied by 10.

The product of the maximum deflection of the planet and the Rsine of the *manda-sphuṭa* minus the node of the planet, i.e. $R\sin(\lambda_{ms} - \lambda_n)$, divided by the *trijyā* is the heliocentric latitude of the planet. When that is multiplied by the *trijyā* and divided by the last hypotenuse called the *antya-śravaṇa*, it gives the geocentric latitude of the planet known as the *iṣṭa-vikṣepa* (desired latitude). In other words, the relation given is

$$\textit{iṣṭa-vikṣepa} = \frac{\textit{parama-kṣepa} \times R\sin\,(\textit{mandasphuṭa} - \textit{pāta})}{\textit{antya-śravaṇa}}. \qquad (7.25)$$

The commentator mentions that the *manda-sphuṭa* referred to here is the longitude of the planet obtained after applying the full-*manda* correction. That is, the longitude obtained after the third stage of the correction in the case of exterior planets.[3] That the term *antya-śravaṇa* is used to refer to the *śīghra-karṇa*, which arises in the last operation of the four-step process, is clarified by the commentator as follows:

...अन्त्यश्रवणेन शीघ्रकर्णेन विभजेत्। तत्र लब्धः इष्टविक्षेपः। अत्र मन्दस्फुटशब्देन कृत्स्नमन्दफलसंस्कृतस्फुटमध्यम उच्यते। शीघ्रकर्णश्च अन्त्यस्फुटकर्मण्यानीतः। अत एवोक्तम् 'अन्त्यश्रवणोद्धृत' इति।

[3] For details regarding the four stages of the correction that is done to obtain the true longitude of the planet, the reader is referred to Chapter 2, verses 61–8.

...divide by the last hypotenuse which is *śīghra-karṇa*. The result thus obtained is the desired deflection. Here, by the word *manda-sphuṭa*, the mean planet corrected by the application of the whole *manda-phala* is referred to. The *śīghra-karṇa* is the one that is obtained in the final-correction process (*antya-sphuṭa-karma*). Hence it is said: 'divided by the *antya-śravaṇa*'.

Earlier in Chapter 2, we have discussed the revision of the planetary model by Nīlakaṇṭha, and his geometrical model of planetary motion. In this unified model for both exterior and interior planets, each planet moves in an eccentric circle around the *śīghrocca*, which is the mean Sun. The longitude of the planet on the eccentric orbit with respect to the mean Sun corresponds to the *manda-sphuṭa-graha*, which is the true heliocentric longitude. The mean Sun itself moves around the Earth uniformly in the plane of the ecliptic. Taking this into account, we obtain a longitude of the planet with respect to the Earth, which is the geocentric longitude. The same considerations apply for a unified model of latitudes, presented in the above verses.

The geocentric latitude at any desired instant called the *iṣṭa-vikṣepa*, given in (7.25), may be expressed as

$$\beta_E = \beta_{max} \times \frac{R\sin(\lambda_{ms} - \lambda_n)}{\text{śīghra-karṇa}}, \tag{7.26}$$

where λ_{ms} is the *manda-sphuṭa*, λ_n is the longitude of the node, β_{max} is the maximum deflection, and the *sīghrakarṇa* is the Earth–planet distance. The maximum deflections (in minutes) are 19, 120, 60, 120 and 120 for Mars, Mercury, Jupiter, Venus and Saturn respectively. The rationale behind the above expression may be understood as follows.

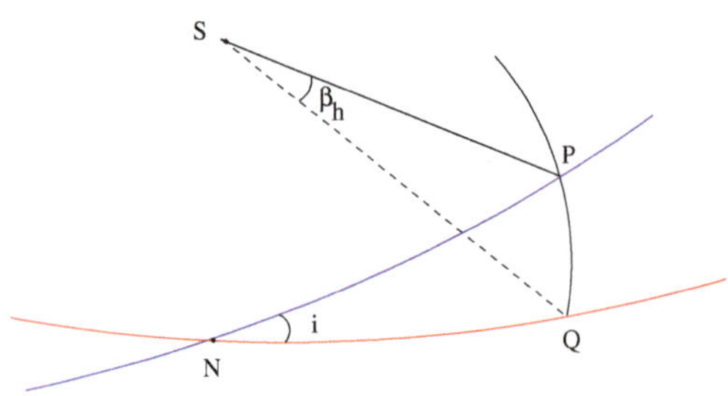

Fig. 7.2*a* Heliocentric latitude of the planet.

Consider the latitude of the planet with respect to the Sun, as shown in Fig. 7.2*a*. Here *P* and *N* refer to the planet and the node, and *S* is the mean Sun. The orbit of the planet is inclined at an angle *i* with respect to the mean Sun. Then the heliocentric latitude β_h is given by

$$\beta_h \approx i\sin(\lambda_{ms} - \lambda_n), \tag{7.27}$$

where the latitude and the inclination are assumed to be small. The relation between the geocentric latitude, β_E, which is measured with respect to the Earth, and β_h is depicted in Fig. 7.2*b*. Now the arc *PQ* may be written as

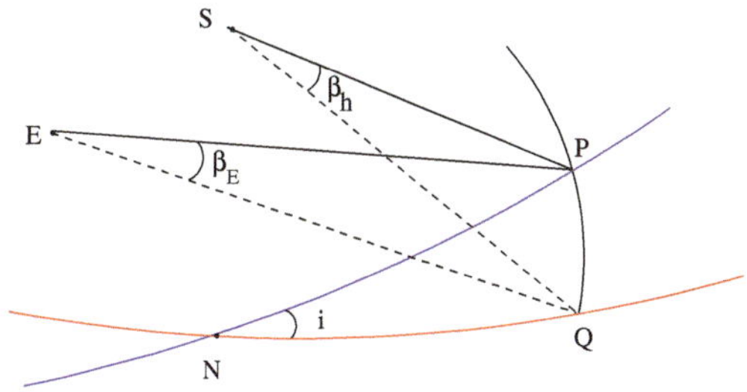

Fig. 7.2*b* Obtaining the geocentric latitude of a planet from its heliocentric latitude.

$$PQ = \beta_E \times EP. \tag{7.28}$$

$$\text{Also} \qquad PQ = \beta_h \times SP. \tag{7.29}$$

Hence

$$\beta_E = \beta_h \frac{SP}{EP}, \tag{7.30}$$

$$\text{or} \qquad \beta_E = i \sin(\lambda_{ms} - \lambda_n)\, \frac{SP}{EP}. \tag{7.31}$$

This is the model described in *Yuktibhāṣā* also. For exterior planets, $SP = R$ (the *trijyā*), and $EP =$ the *śīghra-karṇa*. Then

$$\beta_E = \frac{iR \sin(\lambda_{ms} - \lambda_n)}{\textit{śīghra-karṇa}}. \tag{7.32}$$

Comparing this expression with (7.25), the inclination i can be identified with the maximum deflection, β_{max}.

For the interior planets, $SP = r_s$, the radius of the *śīghra* epicycle. Then

$$\beta_E = \frac{i\, r_s \sin(\lambda_{ms} - \lambda_n)}{\textit{śīghra-karṇa}} = \frac{i\left(\frac{r_s}{R}\right) R \sin(\lambda_{ms} - \lambda_n)}{\textit{śīghra-karṇa}}.$$

Again comparing this with (7.25), β_{max} should be identified with $i\left(\frac{r_s}{R}\right)$. In other words i, which is the inclination of the orbit of an interior planet with respect to the ecliptic, should be identified with $\beta_{max}\left(\frac{R}{r_s}\right)$.[4] β_{max} is given as 2° for both, Mercury and Venus. Now, the mean values of $\frac{r_s}{R}$ for Mercury and Venus are $\frac{31}{80}$ and $\frac{59}{80}$. Then we find that the inclinations for Mercury and Venus are found to be 5° 10′ and 2° 43′ respectively. In fact, these values are essentially the same as the ones in *Āryabhaṭīya*.

[4] This point has been noted by D. A. Somayaji in his explanatory notes to *Siddhāntaśiromaṇi*; {SSR 2000}, p. 476.

Planet	Maximum deflection, β_{max} (textual)	Corresponding inclination, i (textual)	Inclination, i (modern)
Mercury	2°	$5^\circ\ 10'$	7°
Venus	2°	$2^\circ\ 46'$	$3^\circ\ 24'$
Mars	$1^\circ\ 30'$	$1^\circ\ 30'$	$1^\circ\ 51'$
Jupiter	1°	1°	$1^\circ\ 18'$
Saturn	2°	2°	$2^\circ\ 29'$

Table 7.1 Comparison of the textual values of the inclinations of the planetary orbits with the modern ones.

We compare the values of these inclinations with the modern values in Table 7.1. It can be seen from table that for exterior planets there is reasonable agreement between the stated values and the modern values. The interior planets fare worse, even after taking the factor of $\frac{R}{r_s}$ into account. This is understandable, particularly for Mercury, as their latitudes would have been difficult to observe.

७.३ ग्रहाणां दृक्कर्म
7.3 Reduction to observation of the true planets

विक्षेपाच्छशिवत् कार्ये तेषां दृक्कर्मणी उभे ॥ ७ ॥

vikṣepācchaśivat kārye teṣāṃ dṛkkarmaṇī ubhe || 7 ||

The two *dṛkkarmas* have to be carried out (for the planets also) as in the case of the Moon, using their own latitudes.

Since the *dṛkkarma* for the five star planets have been discussed earlier along with the Moon, while explaining verses 1–4 of this chapter, this is not elaborated here.

७.४ दृक्कर्मणि प्रकारान्तरम्
7.4 Alternate method for reduction to observation

विक्षेपदृक्क्षेपवधे[5] त्रिमौर्व्या निहत्य तत् कोटिवधेन भक्ते ।
धनुर्धनर्णं हरिदैक्यभेदात् तयोः शशाङ्कादुदयेऽन्यथास्ते ॥ ८ ॥
एवं वा युगपत् कार्यं दृक्कर्मयुगलं स्फुटम् ॥ ९ ॥

vikṣepadṛkkṣepavadhe trimaurvyā nihatya tat koṭivadhena bhakte |
dhanurdhanarṇaṃ haridaikyabhedāt tayoḥ
śaśāṅkādyudaye'nyathāste || 8 ||
evaṃ vā yugapat kāryaṃ dṛkkarmayugalaṃ sphuṭam || 9 ||

The product of the *vikṣepa* [of the planet] and the *dṛkkṣepa* is to be multiplied by the *trijyā* and divided by the product of the cosines of the *vikṣepa* and the *dṛkkṣepa*. The arc

[5] We feel that the text here should be विक्षेपदृक्क्षेपवधम् instead of विक्षेपदृक्क्षेपवधे, because the former seems to be grammatically appropriate.

of the result has to be added or subtracted [to the longitude of the planet] depending upon whether the *vikṣepa* and the *dṛkkṣepa* have the same direction or otherwise, when it is rising as in the case of the Moon. During setting the application has to be reversed.

In this way, the two *dṛkkarmas* may be carried out in a single step for getting the position.

The aim is to arrive at a single formula for the *dṛkkarma* (DD' of Fig. 7.1 or 7.3). It turns out that the formula is an exact result without involving the kind of approximations used in getting the *ākṣa* and *āyana-dṛkkarmas* (in verses 1–4). It is stated that the *dṛkkarma* is the arc corresponding to the following expression:

$$DD' = \text{Arc of}\left(\frac{\textit{vikṣepa} \times \textit{dṛkkṣepa} \times \textit{trijyā}}{\cos\,(\textit{vikṣepa}) \times \cos\,(\textit{dṛkkṣepa})}\right). \tag{7.33}$$

In modern notation, we have the following relation

$$DD' = \sin^{-1}(\tan\beta \tan z_v). \tag{7.34}$$

This expression is the same as the expression for the '*cara*' (ascensional difference), with the declination δ replaced by the celestial latitude β and the terrestrial latitude ϕ replaced by the arc of the *dṛkkṣepa* z_v.

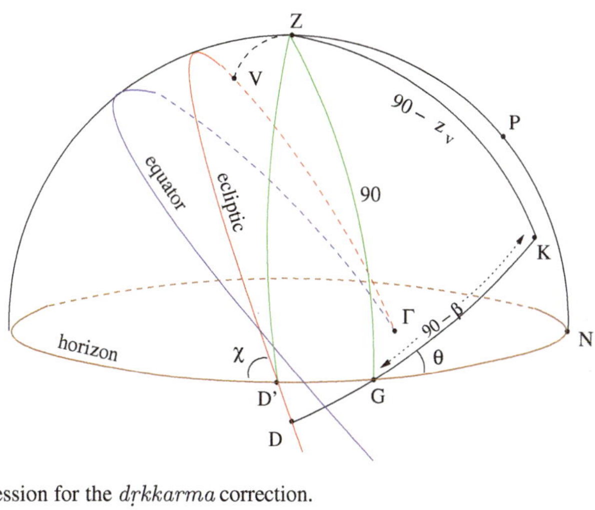

Fig. 7.3 Expression for the *dṛkkarma* correction.

Proof:

In what follows, we give the derivation of the *dṛkkarma* correction using spherical trigonometry. For a derivation of this result using the traditional approach, the reader is referred to Chapter 14 of *Yuktibhāṣā*.[6] In Fig. 7.3, V refers to the *vitribhalagna*, which is a point on the ecliptic which is 90° from the *lagna* D'. The sine of the zenith distance of the *vitribhalagna*, denoted by $\sin z_v$, is called the *dṛkkṣepa*. In the triangle VZD',

$$D'V = 90, \qquad ZD' = 90, \qquad \text{and} \qquad V\hat{D'}Z = ZV = z_v. \tag{7.35}$$

In the spherical triangle $DD'G$, $GD = \beta$ and $D\hat{D'}G = 90 - V\hat{D'}Z = 90 - z_v$. Let θ be the instantaneous angle between the secondary to the ecliptic through G and the horizon. Using the sine formula, we have

$$\frac{\sin DD'}{\sin\theta} = \frac{\sin\beta}{\sin(90 - z_v)}. \tag{7.36}$$

Therefore

$$\sin DD' = \frac{\sin\theta \sin\beta}{\cos z_v}. \tag{7.37}$$

Considering the spherical triangle ZGK and applying the cosine formula, we have

$$\cos ZK = \cos\beta \sin\theta. \tag{7.38}$$

Since $ZK = 90 - z_v$, we have

$$\sin\theta = \frac{\sin z_v}{\cos\beta}. \tag{7.39}$$

Using (7.39) in (7.36), we get

$$\sin DD' = \tan\beta \tan z_v, \tag{7.40}$$

which is the same as the formula (7.34) given in the text.

७.५ काललग्नं कालभागश्च

7.5 *Kālalagna* and the divisions of time

एवं कृतस्य चन्द्रादेः स्वोदयेऽस्तमयेऽपि वा ।
सायनस्य रवेश्चापि काललग्नं नयेद् द्वयोः ॥ १० ॥
तदन्तरभवैर्भागैः दृश्यः स्याद् द्वादशादिभिः ।
न्यूनैः खेटो न दृश्यः स्यात् भानुरश्मिहतप्रभः ॥ ११ ॥
द्वादशात्यष्टयो विश्वे रुद्राङ्कतिथयः क्रमात् ।
चन्द्रादिकालभागास्तैः दृश्या स्वार्कान्तरोद्भवैः ॥ १२ ॥

evaṃ kṛtasya candrādeḥ svodaye'stamaye'pi vā |
sāyanasya raveścāpi kālalagnaṃ nayed dvayoḥ || 10 ||
tadantarabhavairbhāgaiḥ dṛśyaḥ syād dvādaśādibhiḥ |

[6] {GYB 2008}, pp. 822–6.

nyūnaiḥ kheṭo na dṛśyaḥ syāt bhānuraśmihataprabhaḥ || 11 ||
dvādaśātyaṣṭayo viśve rudrāṅkatithayaḥ kramāt |
candrādikālabhāgāstaiḥ dṛśyā svārkāntarodbhavaiḥ || 12 ||

Let the *kālalagna* of the *sāyana* (tropical) Sun and that of the Moon and the other planets, with the above corrections incorporated, be found at their rising or setting.

If the difference in degrees [stated below] is greater than twelve etc. then the planet is visible. If less, it will not be visible [because of] its glow being suppressed by the brilliance of the Sun.

If the differences between the *kālalagna* of the Sun and those of the other planets starting with the Moon are 12, 17, 13, 11, 9 and 15 respectively, then they are visible.

Here the minimum angular separations that are required for the visibility of the Moon and the five star planets in terms of the difference in the *kālalagnas* are specified. The minimum angular separations for the different planets are given in Table 7.2.

Name of planet	Min. angular separation, ψ_m (in deg)
Moon	12
Mars	17
Mercury	13
Jupiter	11
Venus	9
Saturn	15

Table 7.2 Angular separation required for the visibility of the planets.

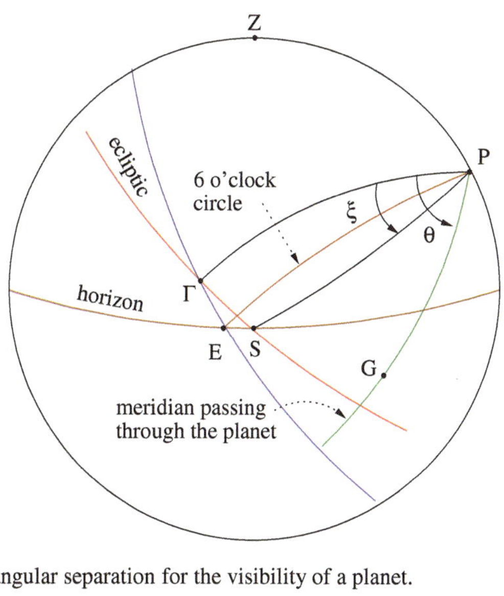

Fig. 7.4 Minimum angular separation for the visibility of a planet.

In Fig. 7.4, S is the Sun and G the planet. If ξ and θ are the *kālalagnas* of the Sun and the planet respectively, then the difference between their *kālalagnas* is given by

$$\psi = |\theta - \xi|. \tag{7.41}$$

The criterion for the visibility of a planet given by Nīlakaṇṭha is that $\psi \geq \psi_m$, where ψ_m is the minimum angular separation, whose value is listed for each planet in Table 7.2. If $\psi < \psi_m$, then the planet is invisible. This seems to be an empirical prescription.

७.६ ग्रहाणां उदयास्तमयदृश्यादृश्यता

7.6 Visibility or otherwise of the planets during their rising and setting

अस्तं यान्तीड्यशन्यारा: पश्चात् प्रागुदयन्ति च ।
ज्ञशुक्रौ वक्रिणावेवं अन्यथा शीघ्रगा रवेः ॥ १३ ॥
पश्चाच्चेत् षड्भयुक्तार्कग्रहयोरन्तरोद्भवैः ।
कालभागैरिह ज्ञेयं दृश्यत्वं वाप्यदर्शनम् ॥ १४ ॥
स्वारूढभांशलिप्ताद्यैः स्वस्वविक्षेपतोऽपि च ।
ज्योतिषामितरेषां च कालांशैर्नीयते स्वकैः ।
दर्शनादर्शने वाच्ये सम्यगेव परीक्षकैः ॥ १५ ॥

astaṃ yāntīḍyaśanyārāḥ paścāt prāgudayanti ca |
jñaśukrau vakriṇāvevaṃ anyathā śīghragā raveḥ || 13 ||
paścāccet ṣaḍbhayuktārkagrahayorantarodbhavaiḥ |
kālabhāgairiha jñeyaṃ dṛśyatvaṃ vāpyadarśanam || 14 ||
svārūḍhabhāṃśaliptādyaiḥ svasvavikṣepato'pi ca |
jyotiṣāmitareṣāṃ ca kālāṃśairnīyate svakaiḥ |
darśanādarśane vācye samyageva parīkṣakaiḥ || 15 ||

Mars, Jupiter and Saturn set in the east and rise in the west. Even Mercury and Venus do so when they are in retrograde motion. It is otherwise in the case of those which move faster than the Sun (namely the Moon, Mercury and Venus in their direct motion).

If (the rising or setting) is to the west, then the visibility or otherwise is to be decided by finding the difference between the *kālalagna* of the planet and that of the Sun increased by 180 degrees.

Even in the case of other celestial objects, depending upon their latitude and the *rāśi* etc. in which they are located, and their own *kālalagna*, their visibility or non-visibility should be declared by the observers correctly.

At the first glance, the prose order of verse 13 would seem to be:

ईड्यशन्यारा: पश्चात् अस्तं यान्ति; प्रागुदयन्ति च। वक्रिणौ ज्ञशुक्रौ (अपि) एवम्। रवेः शीघ्रगा अन्यथा।

However, closely examining the phenomenon explained by these verses with diagrams, we concluded that the prose order must be:

ईड्यशन्याराः प्राक् अस्तं यान्ति; पश्चादुदयन्ति च। वक्रिणौ ज्ञशुक्रौ (अपि) एवम्। रवेः शीघ्रगा अन्यथा।

The translation of the verse given above is also in the light of the above understanding. This is at variance with the commentary given by Śaṅkara Vāriyar. It appears that the terms *udaya* and *astamaya* used in the above verses are not to be understood in the usual sense of the rising and setting of the planets at the observer's horizon. They are actually used to refer to the visibility and the non-visibility of the planets when their directions are close to that of the Sun.

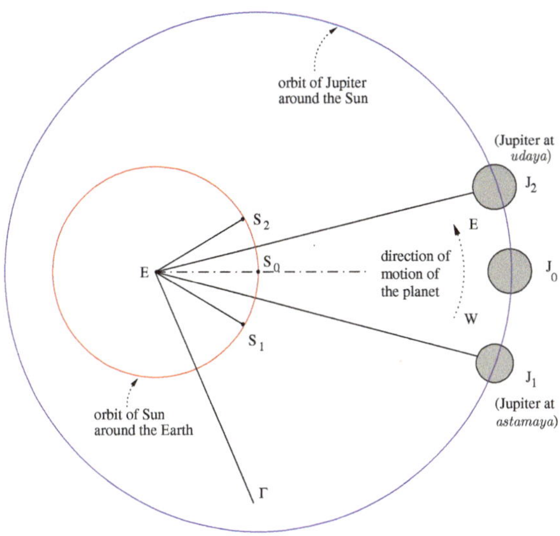

Fig. 7.5 Rising and setting of an exterior planet.

In Fig. 7.5, let J_1 and J_2 be the positions of the Jupiter when it becomes invisible (owing to the brilliance of the Sun) and once again becomes visible, respectively. Let S_1 and S_2 be the corresponding positions of the Sun. When the Sun is at S_1 the planet Jupiter is at J_1 and just becomes invisible. Here it must be noticed that the direction of Jupiter is to the east compared with that of the Sun. In other words, Jupiter sets when it is to the east of the Sun.

Similarly the rising (becoming visible again) of Jupiter happens when it is to the west of the Sun. This is indicated by the position of Jupiter at J_2 which is to the west of the Sun at S_2. The above picture is also true for inner planets when they are in retrograde motion (see Fig. 7.6). When the Sun is at S_1, Venus is at V_1, just setting, and it is to the east of the Sun as seen from the Earth. Similarly, when the Sun is at S_2, Venus is at V_2. It is just rising and lies to the west of the Sun as mentioned in the verse: *vakriṇau jñaśukrau (api) evaṃ.*

On the other hand, when the inner planets are executing direct motion, as shown in Fig. 7.7, they set in the west and rise in the east. The orbit of the Moon is also shown in the figure. Here M_1 and V_1 are the positions of the Moon and Venus when

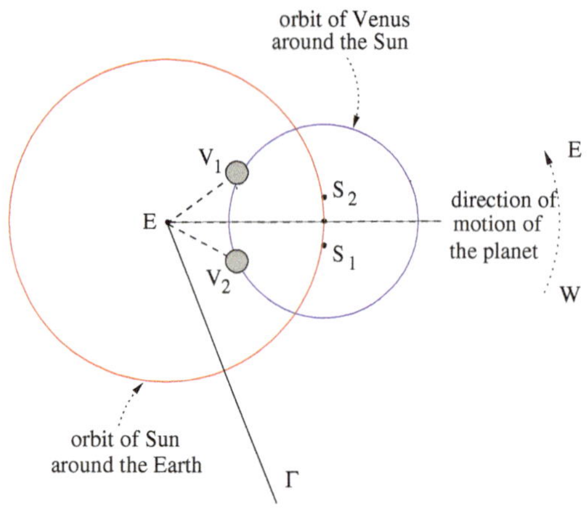

Fig. 7.6 Rising and setting of an interior planet in retrograde motion.

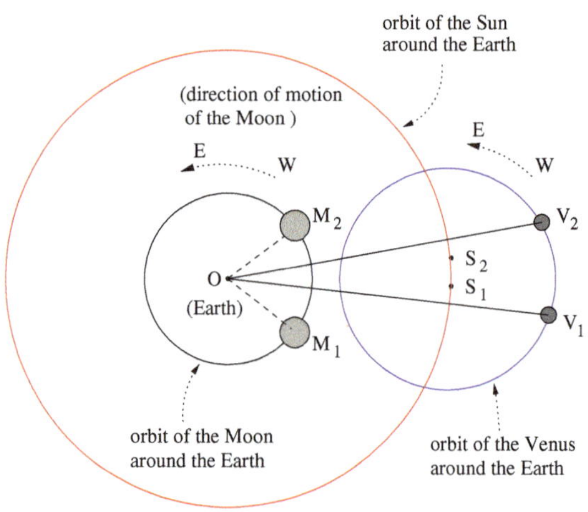

Fig. 7.7 Rising and setting of the Moon and the interior planet in its direct motion.

they lose their visibility (set). It may be noted that the Moon and Venus lie to the west of the Sun at S_1. By the time the Sun moves from S_1 to S_2, say, the Moon and the Venus would have moved to the positions M_2 and V_2, where they once again become visible (by rising). This is what is stated as:

> अन्यथा शीघ्रगा रवेः ।
>
> the directions (of rising and setting) are reversed for those (planets) which move faster than the Sun.

Chapter 8
शृङ्गोन्नतिप्रकरणम्
Elevation of lunar horns

८.१ भूचन्द्रान्तरस्फुटीकरणम्

8.1 Correcting the distance of separation between the Earth and the Moon

व्यर्केन्दुबाहुकोटिज्ये हतेऽपीन्दूच्चभास्वतः ।
कोट्यर्धेन त्रिजीवाप्तदशघ्नेन्दुकलाश्रुतौ ॥ १ ॥
अयनैक्ये च भेदे च स्वर्णं कोटिजमेतयोः ।
तद्बाहुफलवर्गैक्यमूलमिन्दुधरान्तरम् ॥ २ ॥
त्रिज्याघ्नं बाहुजं तेन भक्तं स्वर्णं विधोः स्फुटे ।
कर्क्येणादौ विधूच्चोनरवौ शुक्लेऽन्यथाऽसिते ॥ ३ ॥

vyarkendubāhukoṭijye hate'pīndūccabhāsvataḥ |
koṭyardhena trijīvāptadaśaghnendukalāśrutau || 1 ||
ayanaikye ca bhede ca svarṇaṃ koṭijametayoḥ |
tadbāhuphalavargaikyamūlamindudharāntaram || 2 ||
trijyāghnaṃ bāhujaṃ tena bhaktaṃ svarṇaṃ vidhoḥ sphuṭe |
karkyeṇādau vidhūcconaravau śukle'nyathā'site || 3 ||

The *bhujājyā* and *koṭijyā* of the difference between the longitudes of the Sun and the Moon is multiplied by half of the *koṭijyā* of the difference between the longitudes of the Sun and the *mandocca* of the Moon and divided by the *trijyā*. Of these two, the one obtained from the *koṭijyā* is applied to 10 times the hypotenuse of the Moon in minutes, positively or negatively depending on whether the *ayanas* are the same or different. The square root of the sum of the squares of that and the *bāhuphala* is the distance of separation between the Moon and the Earth (in *yojanas*). Whatever is obtained from the *bhujājyā* [or *bāhuphala*] has to be multiplied by *trijyā* and divided by that [*dvitīya-sphuṭa-karṇa*]. The result obtained must be applied positively or negatively to the true (*manda*-corrected) Moon, depending upon whether the longitude of Sun minus the *mandocca* of the Moon lies within the six signs beginning with *Karka* or *Mṛga* in a bright fortnight, and reversely in a dark fortnight.

The second major correction to the Moon's longitude is the so called 'evection' term, some form of which was first introduced by Ptolemy in his *Almagest*. Among the Indian astronomical works that are extant today, this correction first appears—in very much the form it is used in modern astronomy—in *Laghumānasa* of Mañjulācārya.[1] The set of verses given above present the evection correction which is to be applied to the Moon in a general situation, and not necessarily only in the computation of eclipses. In fact, the variation that arises in the Moon's distance owing to this term has already been considered in the earlier chapters on lunar and solar eclipses, in discussing the '*dvitīya-sphuṭa-yojana-karṇa*' (the second true distance in *yojanas*). In other words, the expression for the distance of separation between the Earth and the Moon given there takes this evection correction into account (though this is not explicitly stated). We now proceed to explain the formula presented in the text.

Let λ_s and λ_m be the longitudes of the Sun and the Moon. Then during an eclipse the two are related by

$$\begin{aligned} &\lambda_m = \lambda_s + 180^\circ \qquad \text{(lunar)} \\ \text{and} \quad &\lambda_m = \lambda_s \qquad\qquad \text{(solar).} \end{aligned} \tag{8.1}$$

Let λ_u be the true longitude of the *mandocca* (apogee) of the Moon. Now we define two quantities x and y as follows:

$$x = \frac{R\sin(\lambda_m - \lambda_s) \times \dfrac{R\cos(\lambda_s - \lambda_u)}{2}}{R} \tag{8.2}$$

$$\text{and} \quad y = \frac{R\cos(\lambda_m - \lambda_s) \times \dfrac{R\cos(\lambda_s - \lambda_u)}{2}}{R}. \tag{8.3}$$

Here x is called the *bāhuphala*, and y the *koṭiphala*. As discussed in Chapter 4, the mean value of the Earth–Moon distance is $10R$ or 34380 *yojanas*. When the equation of centre is included, the distance would be $10K$, where K is the *mandakarṇa*. The *koṭiphala* is to be applied to $10K$ to obtain an intermediate quantity K' that will be used in making a new estimate of the distance,

$$K' = 10K + y. \tag{8.4}$$

The sign of the correction is incorporated in the above expression. This is because y is positive when both $(\lambda_m - \lambda_s)$ and $(\lambda_s - \lambda_u)$ lie between -90° and $+90^\circ$ or between 90° and 270°, that is, both have the same *ayana*. It is only in this range that the product of the cosines is positive as both of them are positive or negative in this range. When they have different *ayanas*, y is negative. The distance of separation between the Earth and the Moon is given as

$$D_m = \sqrt{K'^2 + x^2}. \tag{8.5}$$

[1] It has been ascribed by later commentators to a work of Vaṭeśvara (904 CE), manuscripts of which have not been traced so far.

The geometrical picture corresponding to the evection correction—known as second correction or *dvitīya-sphuṭa*—is described in *Yuktibhāṣā*.[2] In this model, the centre of the *bhagola* is displaced from the centre of the Earth. The *manda-sphuṭa* (*manda*-corrected longitude) of the Moon obtained earlier was with respect to the centre of the *bhagola*. Now we have to determine the true Moon with respect to the centre of the Earth.

The procedure for the second correction is similar to the calculation of the *manda-sphuṭa* with the centre of the *bhagola* serving as the *ucca*, which is specified to be in the direction of the Sun. The distance between this and the centre of the Earth, which is the radius of the epicycle, is a continuously varying quantity and is given by

$$r = \frac{R}{2}\cos(\lambda_s - \lambda_u) \qquad \text{(in } yojanas\text{)}, \tag{8.6}$$

where, as stated earlier, λ_s and λ_u are the longitudes of the Sun and the apogee of Moon (*candrocca*). Here, the mean distance between the Moon and the centre of the *bhagola* is $10R = 34380$ *yojanas*. The actual distance between the same points is $10K$, where K is the *manda-karṇa* in minutes.

For the present, we ignore the Moon's latitude. In Fig. 8.1, C is the centre of the Earth, separated from the centre of the *bhagola* (C_z) by a distance given by (8.6). A is the *Meṣādi*, and $A\hat{C}C_Z = \lambda_s$. The *manda-sphuṭa* of the Moon is at M_1. In other words

$$\begin{aligned} A\hat{C}_Z M_1 &= \lambda_m \\ \text{and} \qquad C_Z M_1 &= 10K, \end{aligned} \tag{8.7}$$

where K is the *manda-karṇa* in minutes. It is clear that $C\hat{C}_Z N = \lambda_m - \lambda_s$. CM_1, the

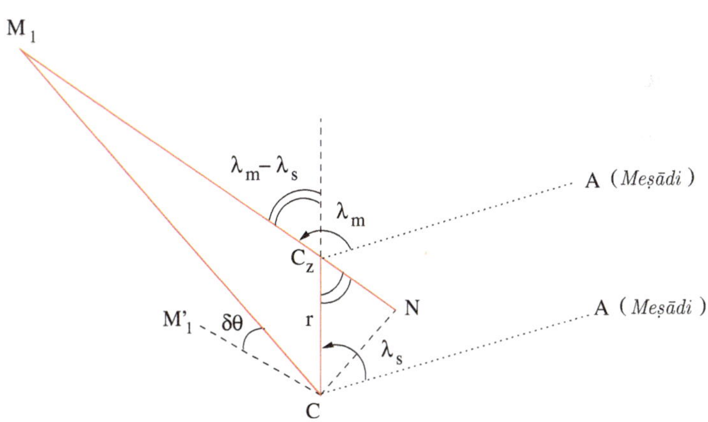

Fig. 8.1 The second correction for the Moon.

[2] {GYB 2008}, pp. 584–7, 786–8, 975–80.

dvitīya-sphuṭakarṇa in *yojanas*, is the distance between the *manda-sphuṭa* and the centre of the Earth. The *bhujāphala* and *koṭiphala* are given by

$$
\begin{aligned}
CN = x &= r\sin(\lambda_m - \lambda_s) \\
&= \frac{R}{2}\cos(\lambda_s - \lambda_u)\sin(\lambda_m - \lambda_s), \\
\text{and} \qquad C_Z N = y &= r\cos(\lambda_m - \lambda_s) \\
&= \frac{R}{2}\cos(\lambda_s - \lambda_u)\cos(\lambda_m - \lambda_s).
\end{aligned}
\tag{8.8}
$$

Then, the *dvitīya-sphuṭakarṇa* (the second true distance in *yojanas*) is given by

$$
\begin{aligned}
CM_1 = D_m &= \sqrt{(M_1N)^2 + CN^2} \\
&= \sqrt{(M_1C_Z + C_ZN)^2 + CN^2} \\
&= \sqrt{(\textit{manda-karṇa} + \textit{koṭiphala})^2 + \textit{bhujāphala}^2} \\
&= \left[\left(10K + \frac{R}{2}\cos(\lambda_s - \lambda_u)\cos(\lambda_m - \lambda_s)\right)^2 \right. \\
&\qquad \left. + \left(\frac{R}{2}\cos(\lambda_s - \lambda_u)\sin(\lambda_m - \lambda_s)\right)^2\right]^{\frac{1}{2}}.
\end{aligned}
\tag{8.9}
$$

As Śaṅkara notes in *Laghu-vivṛti*,

एवं तत्कोटिफलसंस्कृतस्य दशगुणितचन्द्रमन्दस्फुटकर्णस्य तद्बाहुफलस्य च वर्गयोग-मूलं इन्दुधरयोरन्तरं योजनात्मकं द्वितीयस्फुटकर्ण इत्यर्थः ।

Thus, it is to be understood that the square root of the sum of the squares of that *bāhuphala*, and the *karṇa* of the *manda-sphuṭa* of the Moon, which is multiplied by 10 and corrected by the *koṭiphala*, will be the *dvitīya-sphuṭakarṇa*, which corresponds to the distance of separation between the Earth and the Moon in *yojanas*.

Now the longitude of the Moon as seen from the centre of the Earth (*bhūgola*) is

$$
\begin{aligned}
\lambda'_m &= A\hat{C}M_1 \\
&= A\hat{C}M'_1 - M_1\hat{C}M'_1 \\
&= A\hat{C_z}M_1 - M_1\hat{C}M'_1 \\
&= \lambda_m - \delta\theta.
\end{aligned}
\tag{8.10}
$$

In the right-angled triangle CM_1N,

$$
\begin{aligned}
CN &= CM_1 \,\sin\,(C\hat{M_1}N) \\
&= CM_1 \sin\,(M_1CM'_1) \\
&= D_m \sin\delta\theta.
\end{aligned}
\tag{8.11}
$$

From (8.11) and (8.8) we have

$$R\sin\delta\theta = \frac{R\times x}{D_m} = \frac{R\dfrac{\frac{R}{2}\sin(\lambda_m-\lambda_s)\times\frac{R\cos(\lambda_s-\lambda_n)}{2}}{R}}{D_m}, \qquad (8.12)$$

where D_m is given in (8.9). This is what is stated in the text. Though the verse does not make it explicit whether the arc of this is to be taken before applying it to the Moon's *manda-sphuṭa*, from the discussion of this correction in *Yuktibhāṣā* it is clear that it is indeed the arc which has to be applied.

When it is a bright fortnight, $0 \leq (\lambda_m-\lambda_s) \leq 180^\circ$ and $\sin(\lambda_m-\lambda_s)$ is positive. Now x is negative when $(\lambda_s-\lambda_u)$ is between 90° and 270° and positive otherwise. Then the correction term in λ'_m, which is $-\delta\theta$, is positive or negative respectively in the above two ranges for $(\lambda_s-\lambda_u)$. In the dark fortnight, $180^\circ \leq (\lambda_m-\lambda_s) \leq 360^\circ$ and $\sin(\lambda_m-\lambda_s)$ is negative and the signs are interchanged.

८.२ स्फुटचन्द्रगतिः
8.2 The true motion of the Moon

मध्यभुक्तिर्दशघ्नेन्दोः त्रिज्याघ्ना योजनैर्हृता ।
भूचन्द्रान्तरगैर्भुक्तिः विधोरस्य स्फुटा मता ॥ ४ ॥

madhyabhuktirdaśaghnendoḥ trijyāghnā yojanairhṛtā |
bhūcandrāntaragairbhuktiḥ vidhorasya sphuṭā matā || 4 ||

The mean motion of the Moon multiplied by 10 and the *trijyā*, and [then] divided by the distance of separation between the Earth and the Moon in *yojanas*, is considered to be the true motion of the Moon.

If it is taken that the linear velocity of the Moon (and indeed of all the planets) is a constant, then the product of the true distance (D_m) and the true daily motion (d'_m) will be equal to the product of the mean distance ($10R$) and the mean daily motion (d_m). That is,

$$D_m \times d'_m = 10R \times d_m$$

$$\text{or} \qquad d'_m = \frac{d_m \times 10 \times R}{D_m}, \qquad (8.13)$$

which is what is given in the text.

८.३ पूर्वानीतस्फुटचन्द्रगत्यादीनां विनियोगः

8.3 Application of the true motion of the Moon etc. obtained earlier

ग्राह्योऽयमेव भूस्थानां द्रष्टॄणां चन्द्रमाः सदा[3]।
तिथिनक्षत्रयोगादौ लाटे चान्यो भगोलगः ॥ ५ ॥

grāhyo'yameva bhūsthānāṃ draṣṭṝṇāṃ candramāḥ sadā |
tithinakṣatrayogādau lāṭe cānyo bhagolagaḥ || 5 ||

This [what has been stated above] should be taken as the true Moon for the observers on the surface of the Earth. For (calculating) the *tithi, nakṣatra, yoga* etc. and the *lāṭa*, the other Moon obtained for the *bhagola* is to be used.

The positions of the Moon, its rate of motion, etc. as observed from the centre of the celestial sphere, called the *bhagola-madhya*, will be different from those observed from the centre of the Earth, called the *bhūgola-madhya*. Here Nīlakaṇṭha specifies the problems for which the *bhagola* values have to be used, and those for which the *bhūgola* values have to be used.

The calculation of the *tithi, nakṣatra,* etc. are all with respect to the centre of the celestial sphere (the *bhagola*). Hence the parameters obtained for an observer centred at the *bhagola* have to be used. In other words, in their computation the second correction for the Moon is not taken into account. However, for problems like the *grahaṇa, śṛṅgonnati* etc. the *bhūgola* values have to be used, that is, the second correction is taken into account. In these problems, which are very much dependent on the position of the observer on the surface, there is a further correction due to parallax, which has been discussed earlier.

This demarcation as to where to use the *dvitīya-sphuṭa-candra* (the second-corrected Moon) and where to use the *prathama-sphuṭa* has also been explained by Śaṅkara Vāriyar in his *Laghu-vivṛti*:

भूपृष्ठावस्थितद्रष्टृकेन्द्रेण ग्रहप्रापिणा प्रतिक्षणमन्यादृशसंस्थानेन दृङ्मण्डलेन यदा ग्रहः परिच्छिद्यते यथा ग्रहणद्वये चन्द्रार्कौ, यथा वा शृङ्गोन्नतौ चन्द्रः तत्र, उक्तवत् द्वितीयस्फुटसिद्ध एव चन्द्रो ग्राह्यः। अन्यत्र तिथिनक्षत्रयोगादौ लाटे च भगोलगत एव, प्रथमस्फुटसिद्ध एव इत्यर्थः।

When the planet is measured with reference to the *dṛṅmaṇḍala*, which is centred around the observer on the surface of the Earth and touches the planet, and which keeps changing every second, as in the case of the Sun and the Moon during an eclipse or the Moon in the case of the *śṛṅgonnati*, it is only the *dvitīya-sphuṭa-candra* that has to be considered, as mentioned earlier. Elsewhere, as in the case of [the computation of] the *tithi, nakṣatra, yoga* etc., as well as the *lāṭa-yoga*, it is only the first-corrected Moon, computed with reference to the *bhagola*, that needs to be considered.

---

[3] The prose order is: भूस्थानां द्रष्टॄणाम् अयमेव चन्द्रमाः सदा ग्राह्यः। Here the term *sadā* is to be understood to refer to the calculation of eclipse, *sṛṅggonnati* etc.

८.४ विक्षेपः नतिश्च
8.4 The latitude and the zenith distance

पातं विशोध्य चान्यस्मात् व्यतीपातोक्तवर्त्मना ।
आनीतमिष्टविक्षेपं योजनश्रुतिताडितम् ॥ ६ ॥
भूचन्द्रान्तरगैर्हृत्वा लब्धः क्षेपस्तथैव च ।
रवीन्द्वोः पृथगानीय नतिलम्बनलिप्तिकाः ॥ ७ ॥
प्राग्वद् भूपृष्ठविक्षेपः स चेन्दुनतिसंस्कृतः ।
बिम्बान्तरे नतिर्ग्राह्या विधोरर्कस्य चेत् स्वकाः ॥ ८ ॥

pātaṃ viśodhya cānyasmāt vyatīpātoktavartmanā |
ānītamiṣṭavikṣepaṃ yojanaśrutitāḍitam || 6 ||
bhūcandrāntaragairhṛtvā labdhaḥ kṣepastathaiva ca |
ravīndvoḥ pṛthagānīya natilambanaliptikāḥ || 7 ||
prāgvad bhūpṛṣṭhavikṣepaḥ sa cendunatisaṃskṛtaḥ |
bimbāntare natirgrāhyā vidhorarkasya cet svakāḥ || 8 || [4]

Having subtracted the node from the other [i.e. the longitude of the Moon obtained without applying the *dvitīya-sphuṭa*], the *vikṣepa* is to be obtained as per the procedure outlined in *vyatīpāta*. This should be multiplied by the hypotenuse in *yojanas* (the *manda-sphuṭa-karṇa*) and divided by the distance of separation between the Earth and the Moon (the *dvitīya-sphuṭa-karṇa*). The value that is obtained is the *vikṣepa* [as seen from the *bhūgola*].

Similarly, after obtaining the *nati, lambana* etc. in minutes, for the Sun and the Moon separately [corresponding to the *bhagola*], the latitude corresponding to the observer on the surface of the Earth (*bhūpṛṣṭha-vikṣepa*) has to be obtained as earlier. In finding the distance of separation between the discs, it is this *vikṣepa* corrected by the deflection from the ecliptic of the Moon that should be taken as the [true] deflection from the ecliptic of the Moon. In the case of the Sun, the parallax in latitude obtained as such should be taken as the [true] parallax in latitude.

The latitude of the Moon corresponding to the *bhagola* is multiplied by the *manda-karṇa* and divided by the *dvitīya-sphuṭa-karṇa* to obtain the latitude corresponding to the *bhūgola-madhya.* This *vikṣepa* is corrected by the *nati* of the Moon (deflection due to parallax) to obtain its true deflection from the ecliptic (the *bhūpṛṣtha-vikṣepa*). This is the deflection from the ecliptic which is to be used in the computation of the distance between the solar and lunar discs.

८.५ चन्द्रार्कबिम्बान्तरानयनम्
8.5 Finding the distance of separation between the solar and lunar discs

कृतलम्बनचन्द्रार्कविवरात् ज्याशरौ नयेत् ।
तज्ज्यामिन्दुनतीषुघ्नां त्रिज्याप्तां गुणतस्त्यजेत् ॥ ९ ॥
नतीषुफलकृत्योश्च भेदान्मूलमिषौ क्षिपेत् ।

---

[4] Here the term बिम्बान्तरे, should perhaps be interpreted to mean बिम्बान्तरानयने.

गुणबाणौ तथाभूतौ उच्येते विवरोद्भवौ ॥ १० ॥
विवरोत्थशरस्यार्कनतिबाणस्य चान्तरम् ।
अन्तरज्या च या यच्च नतिज्याविवरं तयोः ॥ ११ ॥
योग एव दिशोर्भेदे नत्योरत्र शशीनयोः ।
त्रयाणां वर्गयोगस्य मूलं बिम्बान्तरं स्फुटम् ॥ १२ ॥
कृतलम्बनचन्द्रार्कविवरे भत्रयाधिके ।
तदन्तरदलस्यैव ज्या ग्राह्या न शरस्तदा ॥ १३ ॥
रवीन्दुनतिबाणघ्ना द्विष्ठा त्रिज्याहृता पृथक् ।
योज्ये स्वस्वफलोने ते[5] जीवा सैवान्तरोद्भवा ॥ १४ ॥
नतीषुफलकृत्योर्ये भेदमूले तदन्तरम् ।
द्वितीयं चरमा प्राग्वत् नतियोगो भिदापि वा ॥ १५ ॥
त्रयाणामपि वर्गैक्यात् मूलं बिम्बान्तरं तदा ।
सदा बिम्बान्तरार्धस्य चापं द्विगुणमन्तरम् ॥ १६ ॥
रवीन्द्वोर्वलये तत्र द्रष्टृमध्योभयस्पृशि ।

kṛtalambanacandrārkavivarāt jyāśarau nayet |
tajjyāmindunatīṣughnāṃ trijyāptāṃ guṇatastyajet || 9 ||
natīṣuphalakṛtyośca bhedānmūlamiṣau kṣipet |
guṇabāṇau tathābhūtau ucyete vivarodbhavau || 10 ||
vivarotthaśarasyārkanatibāṇasya cāntaram |
antarajyā ca yā yacca natijyāvivaraṃ tayoḥ || 11 ||
yoga eva diśorbhede natyoratra śaśīnayoḥ |
trayāṇāṃ vargayogasya mūlaṃ bimbāntaraṃ sphuṭam || 12 ||
kṛtalambanacandrārkavivare bhatrayādhike |
tadantaradalasyaiva jyā grāhyā na śarastadā || 13 ||
ravīndunatibāṇaghnā dviṣṭhā trijyāhṛtā pṛthak |
yojye svasvaphalone te jīvā saivāntarodbhavā || 14 ||
natīṣuphalakṛtyorye bhedamūle tadantaram |
dvitīyaṃ caramā prāgvat natiyogo bhidāpi vā || 15 ||
trayāṇāmapi vargaikyāt mūlaṃ bimbāntaraṃ tadā |
sadā bimbāntarārdhasya cāpaṃ dviguṇamantaram || 16 ||
ravīndvorvalaye tatra draṣṭṛmadhyobhayaspṛśi |

From the difference in the longitudes of the Sun and the Moon, corrected for the parallax in longitude, the Rsine and Rversine have to be obtained. That Rsine multiplied by the Rversine of the deflection from the ecliptic of the Moon and divided by the *trijyā*, [referred to later as the *natīṣuphala*] should be subtracted from the [same] Rsine.

The square root of the difference between the squares of the Rversine of the deflection from the ecliptic and the *phala* (*natīṣuphala*) is to be added to the Rversine. [The quantities] thus obtained from the difference [of the Sun and the Moon] are called the *guṇa* and the *bāṇa* [respectively].

The difference between the Rversine of Sun's parallax in latitude and the *bāṇa* [is the first quantity]. The [one related to the] Rsine of the difference [, or the *guṇa*, is the second quantity]. The difference between the Rsines of the deflection from the ecliptic of the two [, i.e., the Sun and the Moon, is the third quantity]. Here if the directions of the deflection from the ecliptic of the Sun and the Moon are different, then it is only the sum that must be considered [and not the difference]. The square root of the sum of the squares of the three

[5] This has to be understood as follows: पृथक् (स्थापिते) स्वे (=ज्ये) स्वफलेन (=त्रिज्या-हृतरविनतिबाणघ्नज्यया, त्रिज्याहृतेन्दुनतिबाणघ्नज्यया च) ऊने, योज्ये।

[quantities defined above] is the true distance between the discs (the *bimbāntara*) [of the Sun and the Moon].

If the difference between the longitudes of the Sun and the Moon corrected for parallax in longitude is greater than 90°, then only the Rsine of half the difference of them (the Sun and the Moon) has to be considered and not the *śara*.

The Rsine [thus found], kept in two places separately and multiplied by the deflection from the ecliptic of the Sun and the Moon, should be divided by the *trijyā*. These [two] have to be subtracted individually from the Rsine [thus obtained] and the results added. That is the Rsine obtained from the difference [and this is the first *phala*]. The square roots of the difference of the squares of the *śaras* of the deflection from the ecliptic and the *phalas* [are also obtained]. The difference between them (the square roots) is the second [*phala*]. The last [or the third *phala*] is either the sum or the difference as earlier. The square root of the sum of the squares of the three [quantities obtained above] is then the true distance between the discs [of the Sun and the Moon].

The angular separation between the lines emanating from the observer and passing through the Sun and the Moon is always equal to twice the arc corresponding to half the separation between the discs.

Here the procedure for finding the exact angular separation between the Sun and the Moon is described. The angular separation is not the difference between the longitudes of the Sun and the Moon, as the two bodies do not lie on the ecliptic. The Moon has a latitudinal deflection, as its orbit is inclined to the ecliptic by about 5°. Moreover, both the Sun and the Moon have apparent latitudinal deflections due to parallax. In Fig. 8.2, the actual positions of the Sun and the Moon are represented by S' and M'. It may be noted that both are off the ecliptic plane. $S'\hat{O}M'$ is the actual angular separation which is to be determined.

In the figure, O represents the observer on the surface of the Earth and not the centre of the Earth. MQ is drawn perpendicular to OS. Let $\theta = S\hat{O}M$ be the difference between the longitudes of the Sun and the Moon, that is

$$\theta = \lambda_m - \lambda_s, \tag{8.14}$$

where λ_m and λ_s are the longitudes of the Moon and the Sun corrected for the parallax in longitude. Now we find out the *jyā* and *śara* of this angle:

$$jy\bar{a} = R\sin\theta = MQ \tag{8.15}$$
$$\acute{s}ara = R(1-\cos\theta) = SQ. \tag{8.16}$$

Depending upon the value of θ ($< 90°$ or $> 90°$), two apparently different formulae are given. The true angular separation between the Sun and the Moon is referred to as (the *sphuṭa-bimbāntara*). Let M'' and S'' be the projections of M' and S' on the plane of the ecliptic. If η_m is the net latitudinal deflection of the Moon (its latitude + parallax) denoted by MM', then the *śara* corresponding to it is given by

$$MM'' = R(1-\cos\eta_m). \tag{8.17}$$

Similarly the *śara* corresponding to the parallax in latitude of the Sun is

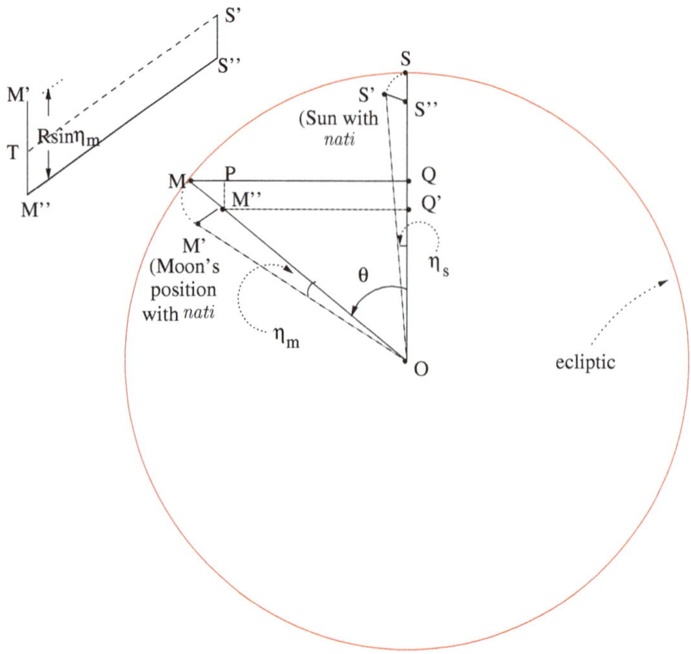

Fig. 8.2 The exact angular separation between the Sun and the Moon when their difference in longitude $\theta < 90°$.

$$SS'' = R(1 - \cos\eta_s). \tag{8.18}$$

With these parameters, a number of auxiliary quantities such as the *natīṣuphala*, *guṇa*, *bāṇa* etc. are defined in order to arrive at the required angular separation. We explain all of the them in the following by considering the two cases $\theta < 90°$ and $\theta > 90°$ separately.

Case 1: $0 < \theta \leq 90°$

From the *jyā*, (defined earlier in (8.15)) we have to find the the *natīṣuphala*, which is given by

$$\begin{aligned} natīṣuphala &= \frac{jyā \times indunatīṣu}{trijyā} \\ x &= \frac{R\sin\theta \times R(1-\cos\eta_m)}{R}. \end{aligned} \tag{8.19}$$

In Fig. 8.2, $M''P$ is the perpendicular from M'' on MQ. Therefore

$$PM = MM'' \times \cos(O\hat{M}Q)$$

$$= R(1-\cos\eta_m)\times\cos(90-\theta)$$
$$= R\sin\theta\times(1-\cos\eta_m), \qquad (8.20)$$

which is the same as the *natīṣuphala* given in (8.19). Thus we see that the *natīṣuphala* is nothing but the projection of MM'' along MQ. Now the difference between the *jyā* and the *natīṣuphala* is termed the *guṇa*. That is,

$$guṇa = g = R\sin\theta - x. \qquad (8.21)$$

It can be easily seen that the *guṇa* is nothing but PQ in Fig. 8.2. Having found PQ, SQ' is determined. The latter is termed the *bāṇa* (U).

$$\begin{aligned} U = SQ' &= SQ + QQ' \\ &= R(1-\cos\theta) + PM'' \qquad (\text{as } QQ' = PM'') \\ &= R(1-\cos\theta) + \sqrt{(MM'')^2 - PM^2} \\ &= R(1-\cos\theta) + \sqrt{(R(1-\cos\eta_m))^2 - x^2}. \end{aligned} \qquad (8.22)$$

Substituting for x and simplifying, we get

$$U = R(1-\cos\eta_m\cos\theta). \qquad (8.23)$$

We have to find the angle $S'\hat{O}M'$. In order to arrive at it, three quantities—referred to as *rāśis*—are defined. The square root of the sum of the squares of the three *rāśis* gives the exact separation (chord length) between the Sun and the Moon, or the *bimbāntara*. It can be shown that the three *rāśis* r_i, $i = 1,2,3$, defined are nothing but the sides of the different right-angled triangles considered to arrive at the required result. The first *rāśi* is defined as

$$\begin{aligned} r_1 &= U - R(1-\cos\eta_s) \\ &= R(\cos\eta_s - \cos\eta_m\cos\theta). \end{aligned} \qquad (8.24)$$

In Fig. 8.2,

$$\begin{aligned} S''Q' &= SQ' - SS'' \\ &= U - R(1-\cos\eta_s) \\ &= r_1. \end{aligned} \qquad (8.25)$$

Thus we see that the first *rāśi* is one side of the right-angled triangle $S''Q'M''$.

The second *rāśi* is defined in just one quarter of verse 11 as: '*antarajyā ca yā*'. However, the explanation necessary to understand this statement is provided in *Laghu-vivṛti*, where it is defined as:

नतीषुफलसंस्कृता चन्द्रार्कान्तरालज्या द्वितीयो राशिः ।

The Rsine of the difference of the Sun and the Moon corrected by the *natīṣuphala* is the second *rāśi*.

This implies that

$$\begin{aligned} r_2 &= R\sin\theta - x \\ &= R\cos\eta_m \sin\theta. \end{aligned} \tag{8.26}$$

Now $MQ = R\sin\theta$ and $PM = x$. Hence

$$PQ = PM - MQ = R\sin\theta - x. \tag{8.27}$$

Thus the second *rāśi* is PQ, which is also equal to $M''Q'$, the other side of the right-angled triangle $S''Q'M''$. Now

$$\sqrt{r_1^2 + r_2^2} = \sqrt{(S''Q')^2 + (M''Q')^2} = S''M'', \tag{8.28}$$

is the hypotenuse of the triangle $S''Q'M''$.

The third *rāśi* is given to be the difference in the *natijyās* of the Sun and the Moon. That is,

$$\begin{aligned} r_3 &= R\sin\eta_m \pm R\sin\eta_s \\ &= R(\sin\eta_m \pm \sin\eta_s). \end{aligned} \tag{8.29}$$

where the sign is chosen to be '+' when the deflections from the ecliptic have opposite directions, and '−' when they have the same direction. From Fig. 8.3, $M'M'' = R\sin\eta_m$ and $S'S'' = R\sin\eta_s$. Therefore

$$\begin{aligned} M'T &= M'M'' \pm S'S'' \\ &= r_3. \end{aligned} \tag{8.30}$$

It may be noted that r_3 forms one side of the right-angled triangle $S'TM'$, whose other side $S'T = S''M''$ is given by (8.28). The separation between the discs is defined to be

$$\mathscr{B} = \sqrt{r_1^2 + r_2^2 + r_3^2}. \tag{8.31}$$

Substituting the expressions for the three *rāśis* given by (8.24), (8.26) and (8.29) and simplifying, we have

$$\mathscr{B} = R\sqrt{2(1 - (\sin\eta_m \sin\eta_s + \cos\eta_m \cos\eta_s \cos\theta))}. \tag{8.32}$$

From Fig. 8.2,

$$\begin{aligned} \sqrt{(S''M'')^2 + (M'T)^2} &= \sqrt{(S''Q')^2 + (M''Q')^2 + (M'T)^2} \\ &= \sqrt{(S'T)^2 + (M'T)^2} \\ &= \sqrt{r_1^2 + r_2^2 + r_3^2} \\ &= S'M'. \end{aligned} \tag{8.33}$$

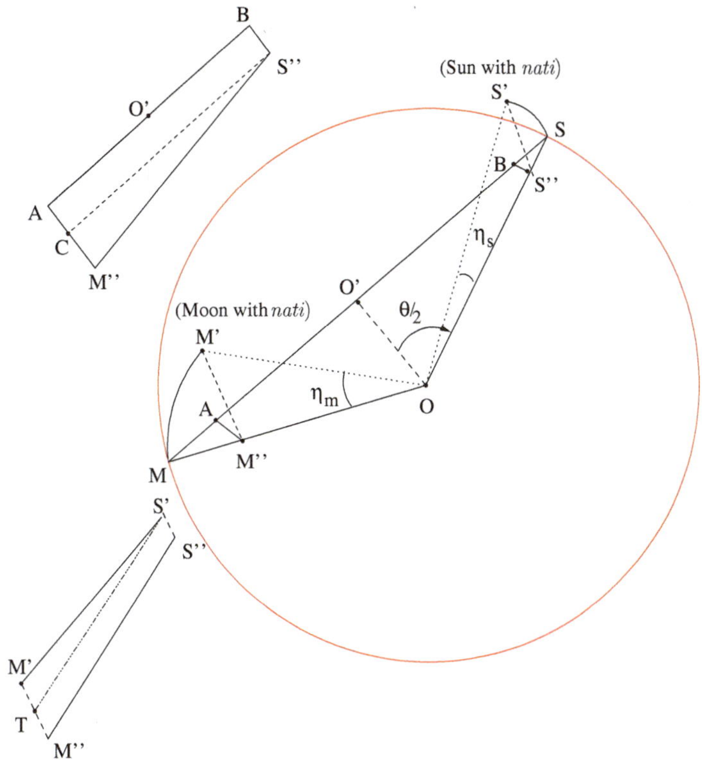

Fig. 8.3 The exact angular separation between the Sun and the Moon when their difference in longitude $\theta > 90°$.

Thus the expression given in the text for the separation between the discs $\mathscr{B}$ is for the chord joining the true positions of the Sun and the Moon as observed by an observer on the surface of the Earth. The angle corresponding to this chord length gives the required angular separation between the observed positions of the Sun and the Moon. This will be discussed after considering the second case, namely $\theta > 90°$.

Case 2: $90° < \theta \leq 180°$

In this case, illustrated in Fig. 8.3, it is mentioned that we need to consider $R\sin\frac{\theta}{2}$ instead of $R(1-\cos\theta)$ and that this has to be multiplied by the *bāṇas* corresponding to the *natis* of the Sun and the Moon. These results have to be stored. They are referred to as the *natīṣuphalas* by the commentator and are given by

$$x_m = R\sin\frac{\theta}{2} \times (1-\cos\eta_m)$$

$$\text{and} \qquad x_s = R\sin\frac{\theta}{2} \times (1-\cos\eta_s). \qquad (8.34)$$

In writing the above expressions the *trijyā* has been factored out from both the numerator and the denominator. In Fig. 8.3 we notice that

$$SS'' = R(1-\cos\eta_s) \tag{8.35}$$

$$\text{and} \qquad MM'' = R(1-\cos\eta_m). \tag{8.36}$$

$S''B$ and $M''A$ are perpendiculars to the line joining S and M.

$$\text{If} \qquad S\hat{O}M = \theta, \qquad \text{then} \qquad S''\hat{S}B = 90 - \frac{\theta}{2}. \tag{8.37}$$

Considering the triangle $S''SB$, by construction, SB is the projection of SS'' along MS. Thus

$$SB = R(1-\cos\eta_s)\sin\frac{\theta}{2}. \tag{8.38}$$

This is the same as the expression given for the *natīṣuphala* of the Sun. Similarly it can be shown that $MA = x_m$. Thus we see that the *natīṣuphalas* are nothing but the projections of the *śaras* of the *natis* of the Sun and the Moon along the chord joining S and M.

Now the *guṇas* are defined by

$$\begin{aligned} g_m &= R\sin\frac{\theta}{2} - x_m = R\sin\frac{\theta}{2}\cos\eta_m \\ \text{and} \qquad g_s &= R\sin\frac{\theta}{2} - x_s = R\sin\frac{\theta}{2}\cos\eta_s. \end{aligned} \tag{8.39}$$

In Fig. 8.3,

$$\begin{aligned} O'B &= O'S - SB \\ &= R\sin\frac{\theta}{2} - R\sin\frac{\theta}{2}\times(1-\cos\eta_s) \\ &= R\sin\frac{\theta}{2}\cos\eta_s. \end{aligned} \tag{8.40}$$

Similarly,

$$O'A = R\sin\frac{\theta}{2}\cos\eta_m. \tag{8.41}$$

Thus, from (8.39), (8.40) and (8.41) we see that the *guṇas* corresponding to the Sun and the Moon given earlier are nothing but $O'B$ and $O'A$. Their sum is defined to be the first *rāśi*:

$$\begin{aligned} AB &= AO' + O'B \\ r_1 &= R\sin\frac{\theta}{2}(\cos\eta_m + \cos\eta_s). \end{aligned} \tag{8.42}$$

The *utkramajyās* (versines) corresponding to the *natis* of the Sun and the Moon are

$$U_s = R(1-\cos\eta_s)$$
$$\text{and} \qquad U_m = R(1-\cos\eta_m). \tag{8.43}$$

With them the following quantities are defined:

$$h_s = \sqrt{U_s^2 - x_s^2},$$
$$h_m = \sqrt{U_m^2 - x_m^2}. \tag{8.44}$$

Their difference is taken to be the second *rāśi*:

$$r_2 = h_m - h_s. \tag{8.45}$$

Substituting the appropriate expressions for the *utkramajyās* and the *natīṣuphalas* ((8.43) and (8.34)), we get

$$r_2 = R\cos\frac{\theta}{2}(\cos\eta_m - \cos\eta_s). \tag{8.46}$$

Again from Fig. 8.3,

$$\begin{aligned} S''B &= \sqrt{(SS'')^2 - SB^2} \\ &= \sqrt{(R(1-\cos\eta_s))^2 - (R\sin\frac{\theta}{2}\ (1-\cos\eta_s))2} \\ &= \sqrt{U_s^2 - x_s^2} \\ &= h_s. \end{aligned} \tag{8.47}$$

Similarly, $M''A = h_m$. Thus

$$r_2 = M''A - S''B = M''C. \tag{8.48}$$

Clearly,

$$\begin{aligned} M''S'' &= \sqrt{(S''C)^2 + (CM'')^2} \\ &= \sqrt{(AB)^2 + (CM'')^2} \\ &= \sqrt{r_1^2 + r_2^2}\,. \end{aligned} \tag{8.49}$$

As in the previous case, the third *rāśi* is taken to be

$$r_3 = M'M'' \pm S'S'' = M'T. \tag{8.50}$$

The separation between the discs $M'S'$ is defined to be

$$
\begin{aligned}
M'S' &= \sqrt{(AB)^2 + (M''A - S''B)^2 + (M'M'' \pm S'S'')^2} \\
&= \sqrt{S'T^2 + M'T^2} \\
&= \sqrt{(S''M'')^2 + M'T^2} \\
\mathscr{B} &= \sqrt{(r_1^2 + r_2^2) + r_3^2} \\
&= R\sqrt{2(1 - (\sin\eta_m \sin\eta_s + \cos\eta_m \cos\eta_s \cos\theta))}. \qquad (8.51)
\end{aligned}
$$

Though the final expression for the separation between the discs is the same as in the previous case (8.32), it must be noted that the expressions for r_1 and r_2 are quite different in the two cases.

Now that the separation between the discs has been found for both cases, the only thing that remains is to convert this into angular measure. Let ϕ be the actual angular separation between S' and M' as indicated in Fig. 8.4*a*, that is, $S'\hat{O}M' = \phi$. Then it is easy to see that the separation between the discs $M'S'$ is given by

$$
\begin{aligned}
\mathscr{B} &= 2R\sin\left(\frac{\phi}{2}\right) \\
\text{or} \qquad \phi &= 2 \times R\sin^{-1}\left(\frac{\mathscr{B}}{2}\right). \qquad (8.52)
\end{aligned}
$$

This is the expression, stated in the text, for the angular separation. Note that ϕ can always be taken to be less than 180° (see Fig. 8.4*a*). As ϕ is very nearly equal to $|\lambda_m - \lambda_s|$, it will be easy to check whether $\phi \leq 90°$ or $90° < \phi < 180°$.

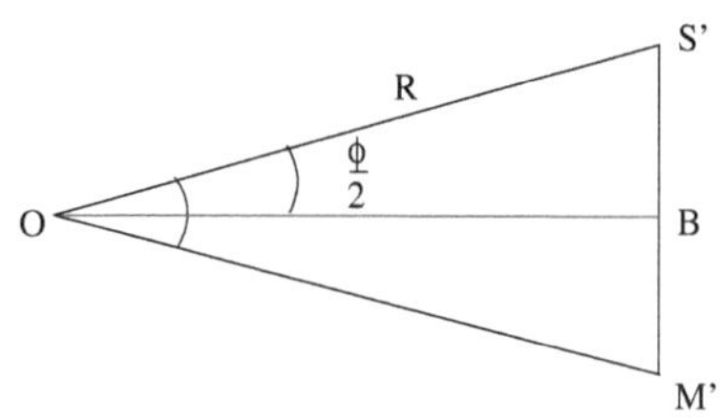

Fig. 8.4*a* Converting the separation between the discs into angular measure.

Bimbāntara through coordinate geometry

The three *rāśis*

$$r_1 = S''Q', \; r_2 = M''Q' \; \text{ and } \; r_3 = M'T$$

in Case 1, and

$$r_1 = AB, \; r_2 = M''A - S''B \; \text{ and } \; r_3 = M'M'' \pm S'S''$$

in Case 2, are the projections of the chord $S'M'$ along three mutually perpendicular directions. They are just the differences in coordinates of S' and M' along the three directions, and the whole exercise is very similar to what is done in three-dimensional coordinate geometry. For comparison, we derive the expression for the separation between the discs in (8.32) and (8.51) using modern coordinate geometry.

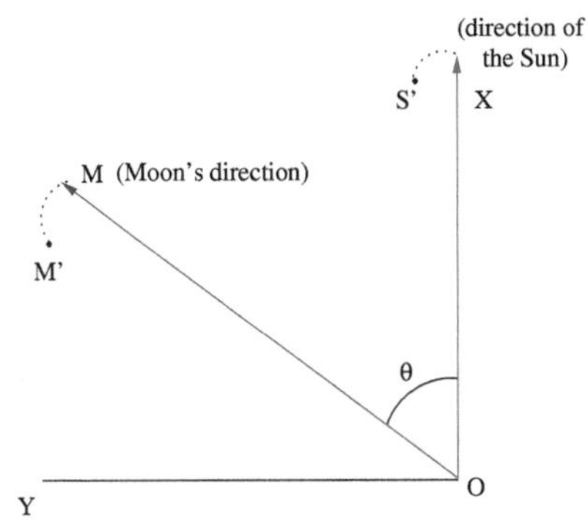

Fig. 8.4*b* The exact angular separation between the Sun and the Moon using coordinate geometry.

In Fig. 8.4*b*, the ecliptic is taken to be the x–y plane. Further, the x-axis is taken to be along the direction of the Sun. The line OM, drawn at an angle θ with respect to the x-axis, is taken to be the direction representing the longitude of the Moon. The actual positions of the Sun and the Moon as seen by an observer are off the plane owing to parallax. Their positions are represented by the points S' and M'. For convenience both are taken to be lying above the plane (+ve z-axis). Taking η_s and η_m to be the parallactic shifts in latitude of the Sun and the Moon, their coordinates are given by

$$\begin{aligned} S' &= (\cos\eta_s,\ 0,\ \sin\eta_s) \\ \text{and} \quad M' &= (\cos\eta_m\cos\theta,\ \cos\eta_m\sin\theta,\ \sin\eta_m). \end{aligned} \tag{8.53}$$

Therefore, the distance between them is given by

$$\begin{aligned} S'M' &= \sqrt{(\cos\eta_s - \cos\eta_m\cos\theta)^2 + (\cos\eta_m\sin\theta)^2 + (\sin\eta_s - \sin\eta_m)^2} \\ &= R\sqrt{2(1-(\sin\eta_m\sin\eta_s + \cos\eta_m\cos\eta_s\cos\theta))}, \end{aligned} \tag{8.54}$$

which is the same as the expression for the separation between the discs given by (8.32) and (8.51).

८.६ सितमानानयने बिम्बान्तरसंस्कारः

8.6 The correction to the angular separation for finding the Moon's phase

तद्बाहुज्या च कोटिज्या चन्द्रदृक्कर्णताडिते ॥ १७ ॥
भुजाकोटिफले भक्ते रविदृक्कर्णयोजनैः ।
ताभ्यां त्रिभज्यया कर्णाच्छीघ्रन्यायेन दोःफलम् ॥ १८ ॥
चापितं धनमेवात्र बिम्बान्तरधनुष्यदः ।
कर्क्येणादौ त्रिभज्यायां स्वर्णं कोटिफलं त्विह ॥ १९ ॥
सितमानार्थमेवैवं रवीन्द्वन्तरमिष्यते ।

tadbāhujyā ca koṭijyā candradṛkkarṇatāḍite || 17 ||
bhujākoṭiphale bhakte ravidṛkkarṇayojanaiḥ |
tābhyāṃ tribhajyayā karṇācchīghranyāyena doḥphalam || 18 ||
cāpitaṃ dhanamevātra bimbāntaradhanuṣyadaḥ |
karkyeṇādau tribhajyāyāṃ svarṇaṃ koṭiphalaṃ tviha || 19 ||
sitamānārthamevaivaṃ ravīndvantaramiṣyate |

By multiplying the Rsine and Rcosine of the [angular separation between the discs] by the *dṛkkarṇa* [in *yojanas*] of the Moon, and dividing by the *dṛkkarṇa* of the Sun in *yojanas*, the *bhujā* and the *koṭiphalas* are obtained.

From them and the *trijyā*, the *karṇa* may be obtained. From the *karṇa* and *bāhuphala*, by following the procedure discussed in *śīghra-saṃskāra*, the arc may be obtained and it must always be applied positively to the angular separation between the discs. [In determining the *karṇa*] the *koṭiphala* has to be applied positively or negatively, depending upon whether the argument lies within six signs beginning from *Karka* or from *Mṛga*. It is only for obtaining the phase of the Moon that [this correction to the] angular separation between the Sun and the Moon is required.

Let ϕ be the angular separation between the centres of the solar and the lunar discs[6] and d_s and d_m be the *dṛkkarṇas* (distances from the Earth)[7] of the Sun and the Moon. Then the *bāhuphala* (b) and the *koṭiphala* (k) are defined by

$$b = \frac{R|\sin\phi| \times d_m}{d_s}$$
$$k = \frac{R|\cos\phi| \times d_m}{d_s}. \tag{8.55}$$

From these two and the *trijyā*, the *karṇa* (K) is found to be

$$K = \sqrt{b^2 + (R \pm k)^2}, \tag{8.56}$$

where we choose the sign '$-$' when $270 \leq \phi \leq 90$, and '$+$' when $90 \leq \phi \leq 270$. Then the *cāpa* (arc) corresponding to the correction is to be determined from the *śīghra-karṇa-nyāya*, that is,

[6] ϕ should not be confused with the terrestrial latitude.

[7] The term *dṛkkarṇa* refers to the distance of separation between the observer and the celestial body in *yojanas*.

$$R\sin\theta = \frac{b\times R}{K}$$

$$\text{or} \qquad \theta = R\sin^{-1}\left(\frac{b\times R}{K}\right). \tag{8.57}$$

This is always applied positively to the angular separation ϕ obtained earlier. That is, the true angular separation is given by

$$\phi' = \phi + \theta. \tag{8.58}$$

Note: Here, it is mentioned that this procedure must certainly be adopted in the calculation of the phase of the Moon. It is implicit from the use of the word '*eva*', in verse 20*a*, that it need not be done elsewhere. The significance of ϕ' is explained in the following section.

८.७ सितमानम्

8.7 The measure of the phase

उत्क्रमज्या ततो ग्राह्या क्रमज्या च समे पदे ॥ २० ॥
बिम्बमानाहताद् बाणात् त्रिज्याढ्या च पदादिके ।
तद्गुणात् कृत्स्नविष्कम्भभक्तमेव सितं सदा ॥ २१ ॥

utkramajyā tato grāhyā kramajyā ca same pade || 20 ||
bimbamānāhatād bāṇāt trijyāḍhyā ca padādike |
tadguṇāt kṛtsnaviṣkambhabhaktameva sitaṃ sadā || 21 ||

From that (corrected angular separation between the discs) the Rversine is to be found [in the odd quadrants] and the Rsine in the even quadrants. If ϕ' is greater than a quadrant (90°), then the *trijyā* is added to the Rsine of the excess; or else the Rversine [is considered]. Either of them is multiplied by the the diameter of the disc of the Moon (*bimbamāna*). The result divided by twice the *trijyā* (*viṣkambha*) always gives the measure of the bright phase of the Moon.

The corrected angular separation between the discs ϕ', given by (8.58), can also be taken to be less than 180°. With ϕ', we have to compute

$$\mathit{utkramajy\bar{a}} = R(1-\cos\phi'), \tag{8.59}$$

when ϕ' is in the odd quadrant. On the other hand, the sum of the *kramajyā* (Rsine) and the *trijyā*, that is,

$$R(1+\sin\alpha), \qquad (\alpha = \phi' - 90), \tag{8.60}$$

has to be calculated when ϕ' is in the even quadrant.

Formula for the phase of the Moon

Now, if a_m is the angular diameter of the disc of the Moon in minutes, then the following is the expression for the phase, which is defined as the fraction of the illuminated portion of the lunar disc:

$$\frac{R(1-\cos\phi')\times a_m}{2R} \qquad 0 \le \phi' < 90$$

$$\text{and} \quad \frac{R(1+\sin\alpha)\times a_m}{2R} \qquad 90 \le \phi' < 180. \tag{8.61}$$

As $\alpha = \phi' - 90$ from (8.60), we see that the latter expression is the same as the previous one. As the argument of the trigonometric function is less than 90 degrees, in Indian astronomy, the two ranges are considered separately. Thus the formula for the illuminated portion of the lunar disc reduces to

$$\frac{(1-\cos\phi')a_m}{2} \tag{8.62}$$

in all cases.

Rationale behind the formula

In Fig. 8.5, E, S and M represent the Earth, Sun and the Moon respectively. Let $M\hat{S}E = \theta$ and $SM = x$. Now from the triangle MUS,

$$\begin{aligned} x &= \sqrt{MU^2+US^2} \\ &= \sqrt{(d_s - d_m\cos\phi)^2 + (d_m\sin\phi)^2} \\ &= \frac{d_s}{R}\sqrt{\left(R - \frac{d_m\cos\phi\, R}{d_s}\right)^2 + \left(\frac{d_m\sin\phi\, R}{d_s}\right)^2} \\ &= \frac{d_s}{R}\sqrt{(R\pm k)^2 + b^2} \\ &= \frac{d_s}{R} K. \end{aligned} \tag{8.63}$$

Now $x\sin\theta = MU = d_m\sin\phi$. Therefore,

$$\begin{aligned} R\sin\theta &= \frac{R\, d_m\sin\phi}{x} \\ &= R\frac{b}{K} \qquad \text{[using (8.57)]}. \end{aligned} \tag{8.64}$$

This is the angle θ that has to be added to ϕ to get ϕ'. It may be noted that the angular separation between the Sun and the Earth at the location of the Moon is

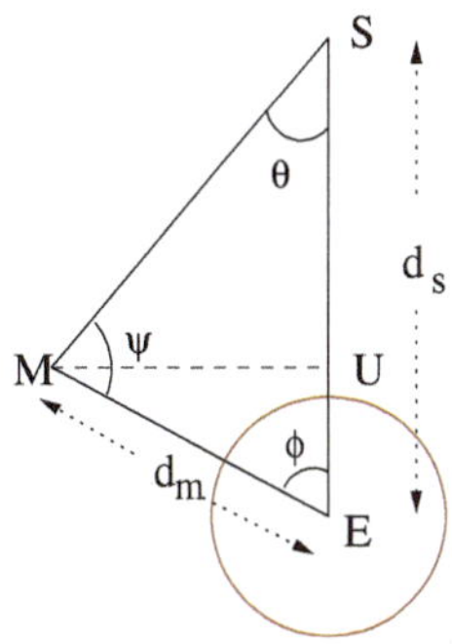

Fig. 8.5 The illuminated portion of the Moon as seen by the observer.

$\psi = 180 - (\theta + \phi)$. Hence

$$\cos\psi = -\cos(\theta + \phi) = -\cos\phi'. \tag{8.65}$$

This is used in the next subsection.

Phase of the Moon from spherical trigonometry

Consider the Earth-Sun-Moon system as shown in Fig. 8.6. By definition, the illuminated portion of the Moon's disc—also known as the phase of the Moon—is the arc CB. This corresponds to the segment CF across the diameter:

$$\begin{aligned} CF &= CM + MF \\ &= CM + MB\cos\psi \\ &= \frac{a_m}{2}(1 + \cos\psi) \qquad (\text{since } CM = MB = \frac{a_m}{2}), \end{aligned} \tag{8.66}$$

where a_m is the diameter of the Moon's disc. Using (8.65) in the above equation, the phase of the Moon is

$$\frac{a_m}{2}(1 - \cos\phi'), \tag{8.67}$$

which is precisely the formula given in the text.

८.८ शृङ्गोन्नतिः
8.8 Deflection of the horn (phase of the Moon)

प्राग्वच्छायाभुजां भानोः स्वाग्राभ्यां च विधोर्नयेत् ।
शङ्क्वग्रं सौम्यदिक्कं स्यात् अदृश्यार्धगते ग्रहे ॥ २२ ॥
दृक्कर्णानयने शङ्कोः फलमप्यत्र योजयेत् ।

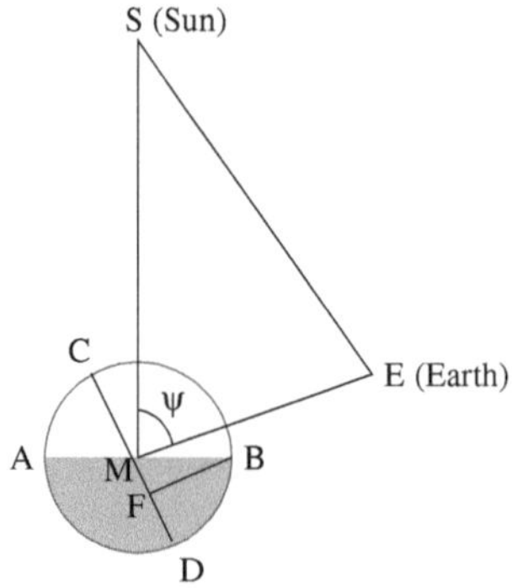

Fig. 8.6 The modern definition of the phase (illuminated portion) of the Moon.

व्यासार्धात् तद्धतात् छायाभक्ते क्षितिजगे उभे ॥ २३ ॥
योगस्तद्धनुषोः कार्यः यथायुक्त्यन्तरं तथा ।
छायाबाह्वोर्दिशोर्भेदे योगः साम्येऽन्तरं तथा ॥ २४ ॥
विश्लेषचन्द्रबाहुश्चेत् शिष्टः स्याद् व्यत्ययेन दिक् ।
सूर्यस्यैव ततोऽन्यत्र ग्राह्या दिग् योगभेदयोः ॥ २५ ॥
स्वभूम्यन्तरनिघ्नस्वदृग्ज्या दृक्कर्णभाजिता ।
अर्केन्द्वोः सुस्फुटा दृग्ज्या द्रष्टृभूपृष्ठगस्य हि ॥ २६ ॥
बाहुचापान्तरज्याघ्ना दृग्ज्या त्रिज्याहृता रवेः ।
चन्द्रबिम्बार्धनिघ्नाथ बिम्बान्तरभुजाहृता ॥ २७ ॥
उन्नतिश्चन्द्रशृङ्गस्य नतिर्वार्धगुणात्मिका ।
वर्गत्रयैक्यमूलस्य दलस्य द्विगुणं धनुः ॥ २८ ॥
यत्तस्य बाहुजीवात्र बिम्बान्तरभुजोदिता ।

prāgvacchāyābhujāṃ bhānoḥ svāgrābhyāṃ ca vidhornayet |
śaṅkvagraṃ saumyadikkaṃ syāt adṛśyārdhagate grahe || 22 ||
dṛkkarṇānayane śaṅkoḥ phalamapyatra yojayet |
vyāsārdhāt taddhatāt chāyābhakte kṣitijage ubhe || 23 ||
yogastaddhanuṣoḥ kāryaḥ yathāyuktyantaraṃ tathā |
chāyābāhvordiśorbhede yogaḥ sāmye'ntaraṃ tathā || 24 ||
viśleṣacandrabāhuścet śiṣṭaḥ syād vyatyayena dik |
sūryasyaiva tato'nyatra grāhyā dig yogabhedayoḥ || 25 ||
svabhūmyantaranighnasvadṛgjyā dṛkkarṇabhājitā |
arkendvoḥ susphuṭā dṛgjyā draṣṭṛbhūpṛṣṭhagasya hi || 26 ||
bāhucāpāntarajyāghnā dṛgjyā trijyāhṛtā raveḥ |
candrabimbārdhanighnātha bimbāntarabhujāhṛtā || 27 ||
unnatiścandraśṛṅgasya natirvārdhaguṇātmikā |
vargatrayaikyamūlasya dalasya dviguṇaṃ dhanuḥ || 28 ||
yattasya bāhujīvātra bimbāntarabhujoditā |

Like before, find the *chāyābhujā* (projection of the shadow perpendicular to the east-west line) of the Sun and that of the Moon from its own *agrās* (the *śaṇkvagrā* and *arkāgrā*). The *śaṇkvagrā* will be in the north direction if the planet is in the invisible hemisphere.

Here [when the planet is in the invisible hemisphere], while obtaining the *dṛkkarṇa*, the *śaṇkuphala* must be added [to the *dvitīya-sphuṭa-karṇa* (K_d)]. By multiplying them [the *chāyā-bhujas* of the Sun and the Moon] by the *trijyā* and dividing by the *chāyā*, the two quantities would have been converted to the ones corresponding to the *kṣitija*.

The arcs of the two have to be added or subtracted as is appropriate (*yathāyukti*). [That is,] if the two have different directions, then they have to be added and if they have the same direction then their difference has to be found.

While finding the difference, if the Rsine [of the zenith distance] of the Moon is remaining [that is, if $z_m > z_s$], then the directions have to be reversed. Otherwise [that is, if $z_s > z_m$, both in finding the sum and difference], the direction of the azimuth of the Sun ($R\sin A_s$) is taken to be the direction [of the resulting quantity, say x].

The *dṛgjyās* of the Sun and the Moon multiplied by the distance between them and the centre of the Earth, and the two products, each divided by its own *dṛkkarṇa*, are indeed the true values of the *dṛgjyā* of the Sun and the Moon for an observer on the surface of the Earth.

The Rsine of the sum or difference of the arcs (x) is multiplied by the [true value of the] *dṛgjyā* of the Sun and divided by the *trijyā*. This is multiplied by the radius of the lunar disc and divided by the Rsine of the distance between the discs (*bimbāntara-bhujā*). The result is the Rsine of the elevation of the cusp (*śṛṅgonnati*) or depression of the cusp (*śṛṅga-avanati*) of the Moon.

The square root of the sum of the squares of the three [quantities] is halved, converted into arc and doubled. The Rsine of that is called the *bimbāntara-bhujā*.

Obtaining an expression for the measure of the elevation of the Moon's horn (*śṛṅgonnati*) or its depression (*śṛṅgāvanati*) is quite an involved process and hence it is described in several steps. We discuss them in order below.

Chāyābhujā, *arkāgrā* and *śaṅkvagrā* and the relation among them

The terms *chāyā-bhujā*, also known as *mahābāhu*, *śaṅkvagrā* and *arkāgrā* have all been defined in Section 3.20. The relation between them was also discussed there. For the sake of convenience, we recapitulate some of them here. In Fig. 8.7, when the planet is at G_1, $ZG_1 = z$ is the zenith distance and $P\hat{Z}G_1 = A$ is the azimuth. Then *chāyābhujā* is $|R\sin z\cos A|$. I_1 is the foot of perpendicular from G_1 on the plane of the horizon. From I_1 draw I_1J_1 perpendicular to the east-west line. It is easily seen that

$$
\begin{aligned}
I_1\hat{O}J_1 &= a = 90 - A \\
\text{and} \qquad I_1J_1 &= \textit{chāyābhujā} \\
&= R\sin z\cos A \\
&= R\sin z\sin a. \qquad (8.68)
\end{aligned}
$$

The *udaya-sūtra* is the line joining the rising and setting points. It is parallel to the east-west line. The perpendicular distance between the *udaya-sūtra* and the east-west line is the *arkāgrā* ($R\cos A$), which has been shown (see (3.88)) to be $\frac{R\sin\delta}{\cos\phi}$.

The perpendicular distance of I_1 from the *udaya-sūtra*, represented by I_1K_1 in the figure, is called the *śaṅkvagrā*. Since the inclination of the diurnal circle to the vertical is same as that of the equator, which is equal to ϕ, $I_1\hat{G}_1K_1 = \phi$. Therefore

$$\tan\phi = \frac{I_1K_1}{I_1G_1} = \frac{I_1K_1}{\cos z}. \tag{8.69}$$

Hence,

$$I_1K_1 = \acute{s}a\dot{n}kvagr\bar{a} = \cos z \tan\phi. \tag{8.70}$$

When the planet is at G_1, $ark\bar{a}gr\bar{a} = J_1K_1$. We now summarize the relation between the *chāyābhujā*, *arkāgrā* and *śaṅkvagrā*.

Case i: *Chāyābhujā* north, δ north (planet at G_1)

In this case,

$$\begin{aligned} I_1J_1 &= J_1K_1 - I_1K_1 \\ ch\bar{a}y\bar{a}bhuj\bar{a} &= ark\bar{a}gr\bar{a} - \acute{s}a\dot{n}kvagr\bar{a}. \end{aligned} \tag{8.71}$$

Case ii: *Chāyābhujā* south, δ north (planet at G_2)

In this case,

$$\begin{aligned} I_2J_2 &= I_2K_2 - J_2K_2 \\ ch\bar{a}y\bar{a}bhuj\bar{a} &= \acute{s}a\dot{n}kvagr\bar{a} - ark\bar{a}gr\bar{a}. \end{aligned} \tag{8.72}$$

Further, $I_2\hat{O}J_2 = a = A - 90^\circ$.

Case iii: *Chāyābhujā* south, δ south (planet at G_3)

$$\begin{aligned} I_3J_3 &= I_3K_3 + J_3K_3 \\ ch\bar{a}y\bar{a}bhuj\bar{a} &= \acute{s}a\dot{n}kvagr\bar{a} + ark\bar{a}gr\bar{a}. \end{aligned} \tag{8.73}$$

In this case also, $I_3\hat{O}J_3 = a = A - 90^\circ$. It is clear that the point I will be north of the *udaya-sūtra* (that is, the *śaṅkvagrā* is north) only when the planet is below the horizon. This is because, the equator and the diurnal circle tilt away from N above the horizon, and towards N below the horizon.

Calculation of the *dṛkkarṇa*

The term *dṛkkarṇa* here refers to the distance of the Sun from the observer. This can be found from the *dvitīya-sphuṭa-karṇa*, K_d, the distance of the Sun from the centre of the Earth, and the zenith distance, z_s. In Fig. 8.8, O is the observer and C the centre of the Earth, whose radius is R_e. S is the Sun below the horizon whose zenith distance is $z_s > 90^\circ$. Let d_s be the *dṛkkarṇa* of the Sun. The *dvitīya-sphuṭa-karṇa* $K_d = CS$. Hence $MC = |R_e \cos z_s|$. Considering the triangle OSM,

$$d_s^2 = OS^2$$

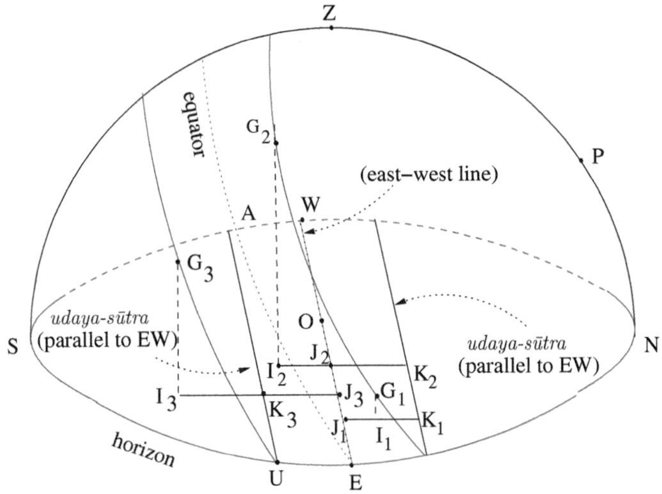

Fig. 8.7 Relation between the *arkāgrā*, *āśāgrā* and *śaṅkvagrā* when planet has northern declination.

$$= SM^2 + OM^2$$
$$= (K_d + |R_e \cos z_s|)^2 + (R_e \sin z_s|)^2. \tag{8.74}$$

Therefore

$$d_s = \sqrt{(K_d + |R_e \cos z_s|)^2 + (R_e \sin z_s|)^2}. \tag{8.75}$$

Here the *śaṅkuphala* ($|R_e \cos z_s|$) is added to K_d in the first term. When the Sun is

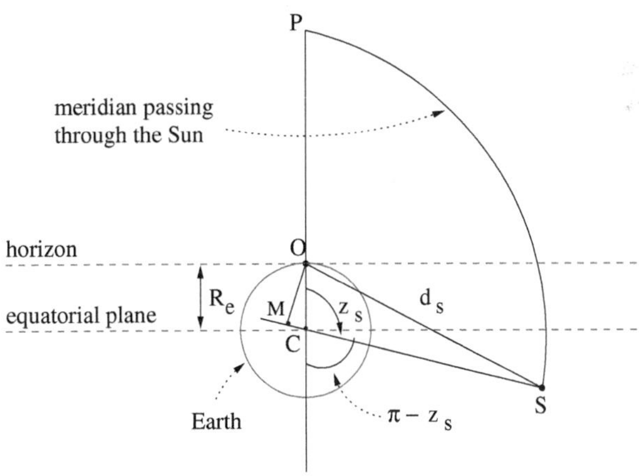

Fig. 8.8 Calculation of *dṛkkarṇa*, of the Sun.

above the horizon ($z_s < 90°$), it can easily be seen that

$$d_s = \sqrt{(K_d - |R_e \cos z_s|)^2 + (R_e \sin z_s)^2}. \tag{8.76}$$

That is, the *śaṅkuphala* is subtracted from K_d.

Difference between the azimuths of the Sun and the Moon

In Fig. 8.9, S and M represent the feet of the perpendiculars from the Sun and the Moon on the observer's horizon. SA and MB are the perpendiculars from the Sun and the Moon respectively on the EW line. If z_s is the zenith distance of the Sun, then

$$OS = ch\bar{a}y\bar{a} = R \sin z_s. \tag{8.77}$$

If A_s and A_m are the azimuths of the Sun and the Moon, then

$$SA = ch\bar{a}y\bar{a}b\bar{a}hu = R \sin z_s \sin a_s, \tag{8.78}$$

where $a_s = A_s \pm 90°$. The ratio of the *chāyābāhu* to the *chāyā* multiplied by the *trijyā* gives $R \sin a_s$. Similarly we find $R \sin a_m$. Then the sum of or difference between the arcs of the two is calculated . That is,

$$\alpha = a_m \overset{+}{\sim} a_s. \tag{8.79}$$

We take the '+' sign, when the Sun and the Moon are in different hemispheres, i.e. projections of the Sun and the Moon fall on either side of the EW line and the '$\sim$' when both lie in the north or south. In Fig. 8.9, both are shown to be lying to the north. If SA and MB are in different directions, or $SA > MB$, the projected point S

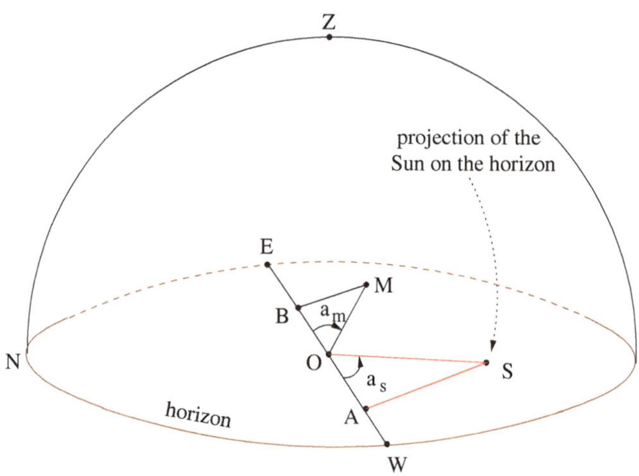

Fig. 8.9 The sum of or difference between the azimuths of the Sun and the Moon.

will be north/south of M if S is north/south. If SA and MB are in the same direction and if $MB > SA$, S will be south/north of M if S is north/south.

Zenith distances of the Sun and the Moon as seen by the observer

In Fig. 8.10(a), C is the centre of the Earth, O is the observer and S is the Sun. z_s is the zenith distance of the Sun with respect to the centre of the Earth and z_s' is that seen by the observer. Now

$$d_s \sin z_s' = K_{ds} \sin z_s. \tag{8.80}$$

Since the zenith distance z_s, d_s the *dṛkkarṇa* of the Sun, and $K_{ds} = CS$ the *dvitīya-sphuṭa-karṇa* are known, z_s' can be calculated. A similar relation holds for the Moon also (see Fig. 8.10(b)). Thus, for an observer on the surface of the Earth, the Rsines

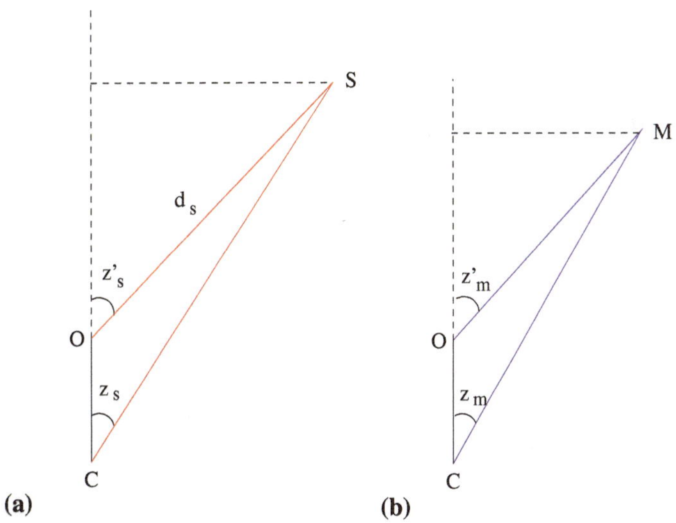

Fig. 8.10 Calculation of the observer's zenith distance (a) of the Sun, (b) of the Moon.

of the zenith distances of the Sun and the Moon are given by

$$R\sin z_s' = \frac{R\sin z_s \times K_{ds}}{d_s}$$

$$\text{and} \qquad R\sin z_m' = \frac{R\sin z_m \times K_{dm}}{d_m} \tag{8.81}$$

respectively.

Expression for the *śṛṅgonnati*

The *śṛṅgonnati*, $R_m \sin\beta$, is given by

$$R_m \sin\beta = \frac{R\sin\alpha \times R\sin z_s'}{R \times R\sin\phi} \times R_m, \tag{8.82}$$

where α is the sum or difference of the azimuths of the Sun and the Moon given by (8.79), ϕ is the angular separation between the Sun and the Moon, and R_m is the radius of the lunar disc. Essentially β is the angle of elevation/depression of the line of cusps of the Moon. $R\sin\phi$ is termed the *bimbāntarabhujā*. In verse 29a it is stated that the *bimbāntara*, the distance of separation between the Sun and the Moon (denoted by d), is the square root of sum of the squares of three *rāśis*. It is further mentioned that d and ϕ are related through the relation

$$d = 2R\sin\left(\frac{\phi}{2}\right). \tag{8.83}$$

Rationale behind the formula for the *śṛṅgonnati*

In Fig. 8.11(a), O represents the observer, M the centre of the Moon and S the Sun. The figure schematically depicts the situation where the Sun has already set and the Moon is about to set. This can be taken to roughly represent the scenario that prevails during the last quarter of the dark fortnight.

In Fig. 8.11(b), we have depicted the cross-sectional view of the Moon. M is the centre of the Moon. C_1C_2 is the line of cusps which is perpendicular to both MS and ME, which are the lines joining the Moon to the Sun and Earth respectively. C_1 and C_2 are the poles of the circle $XYQBZPX$, lying in the plane of the paper.

The illuminated portion of the Moon is the hemisphere facing the Sun, with $C_1YC_2C_1$ as the boundary. In this, the portion above the great circle $C_1XC_2C_1$ will be invisible to the observer, and the illuminated portion of the Moon as seen by the observer is the union of the two spherical triangles C_1XY and C_2XY. In other words, the cross-sectional view of the Moon as seen by the observer will be the interstice between the two arcs C_1XC_2 and C_1HGFC_2 (shown shaded in the figure). For an observer on the Earth, this portion looks as if two similar horns have been cemented together at the bottom. The tips of the horns are C_1 and C_2. Since the term *śṛṅga* is used for horns in Sanskrit, the phenomenon is called the *śṛṅgonnati*.

Normally the elevation of one of the *śṛṅgas* will be higher than that of the other. The *śṛṅgonnati* is the angle between the line of cusps and the horizontal plane. In the following we derive the expression for the *śṛṅgonnati* using modern spherical trigonometry and compare that with the expression given in the text. In Fig. 8.12, S and M represent the Sun and the Moon in the same way as in the previous figures, but on the surface of the celestial sphere. The point C is the pole of the great circle passing through M and S. The dotted lines OS and OM represent the directions of

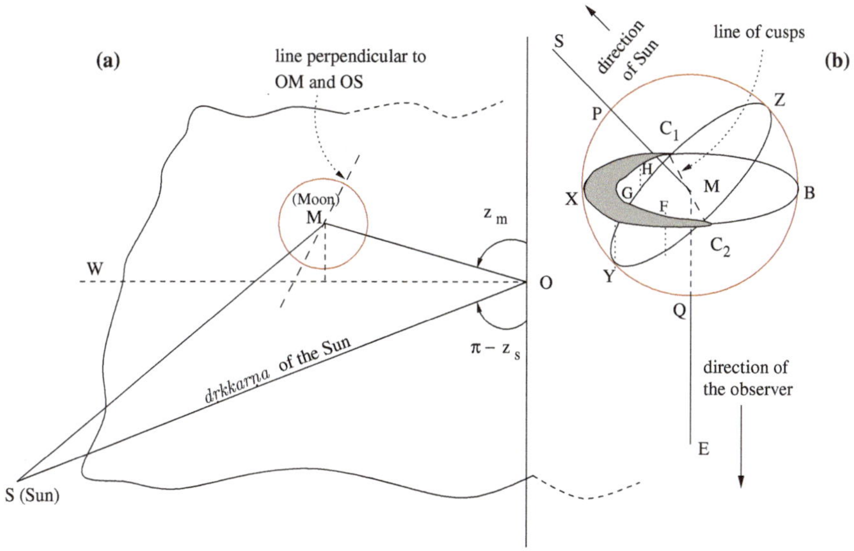

Fig. 8.11 (a) Schematic sketch of the Sun and the Moon with respect to the observer's horizon. (b) The phase of the Moon as seen by the observer for the situation depicted in (a).

the Sun and the Moon as seen by the observer. ϕ is the angular separation between the Sun and the Moon. The difference in their azimuths is given by α.

As may be seen from Fig. 8.11(b), the line of cusps C_1C_2 is perpendicular to OM and MS. In other words, it is perpendicular to the plane containing the observer, the Sun and the Moon. The direction of the Sun as seen from the Moon, and the direction as seen from the Earth, will be almost the same because the Moon is very close to the Earth as compared with the Sun. Hence, the lines MS and OS can be taken to be parallel. By construction, the line OC is parallel to the line of cusps C_1C_2. Hence (refer to Fig. 8.12(b) and (a)),

$$C\hat{O}X = C_1\hat{M}D = \beta. \tag{8.84}$$

Therefore

$$C\hat{O}Z = 90 - \beta = ZC. \tag{8.85}$$

Considering the triangle ZCM and using the cosine formula,

$$\sin\beta = \sin z_m \sin i_m, \tag{8.86}$$

where i_m is the spherical angle $Z\hat{M}S$. Here we need to know the angle i_m in terms of other known angles. For this we consider the triangle ZMS. Applying the sine formula, we get

$$\sin i_m = \frac{\sin z_s' \sin\alpha}{\sin\phi}. \tag{8.87}$$

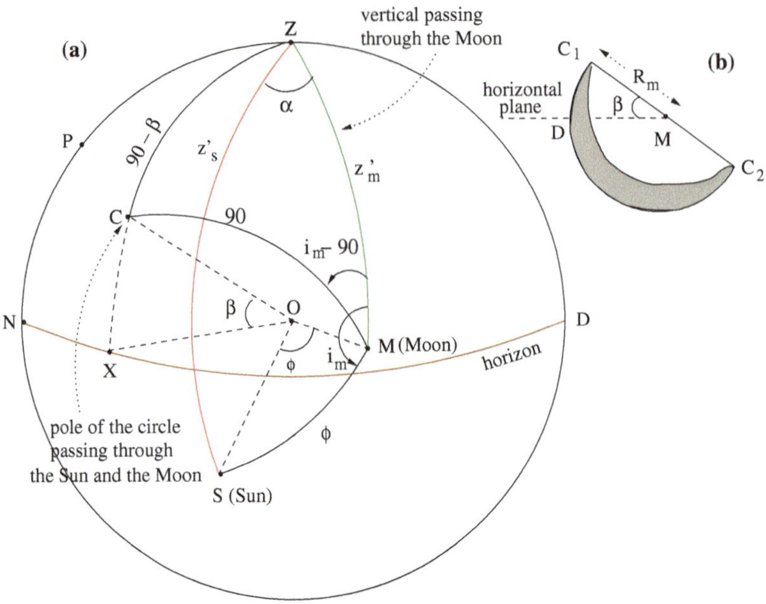

Fig. 8.12 (a) Schematic sketch for finding the expression of the *śṛṅgonnati* using modern spherical trigonometric formulae. (b) The inclination of the line of cusps of the Moon with respect to the plane of the horizon.

Substituting this in (8.86), we have

$$\sin\beta = \frac{\sin z'_s \sin\alpha}{\sin\phi}\sin z_m. \tag{8.88}$$

When the Moon is on the horizon, $z_m = 90$, and the above equation reduces to

$$\sin\beta = \frac{\sin z'_s \sin\alpha}{\sin\phi}, \tag{8.89}$$

which is same as the formula given in the text (8.82). Hence, it appears that the expression for the *śṛṅgonnati* (the angle between the line of cusps and the horizontal plane) in the verses is valid only when the Moon is on the horizon, that is $z_m = 90°$.

८.९ शृङ्गोन्नतेः परिलेखनम्

8.9 Graphical representation of the *śṛṅgonnati*

चन्द्रबिम्बार्धमानेन लिखेद्वृत्तं तु तद्गते ॥ २९ ॥
रेखे द्वे दिग्विभागार्थं, प्रत्यग्रेखाग्रतः पुनः ।
नीत्वा शृङ्गोन्नतेर्मानं प्राग्वदर्धगुणात्मकम् ॥ ३० ॥

चन्द्रादर्कदिशीन्दोस्तु परिधौ प्राग्विपर्ययात् ।
बिन्दुं कृत्वा लिखेद्रेखां तन्मार्गेण सितं नयेत् ॥ ३१ ॥
प्रत्यगग्रात् सिते पक्षे प्रागग्रादसितेऽपि च ।
सितान्ते बिन्दुमाधाय तिर्यग्रेखाग्रयोस्ततः ॥ ३२ ॥
बिन्दुं कृत्वा लिखेद्वृत्तं बिन्दवो नेमिगा यथा ।
वृत्तान्तराकृतिश्चन्द्रः शृङ्गोन्नत्या प्रदर्श्यताम् ॥ ३३ ॥
व्यस्तदिक्कोऽर्कबाहुः स्यात् तयोर्नानाकपालयोः ।
प्रत्यासन्नरवेर्भागात् इहाप्यन्तर्नयेत् सितम् ॥ ३४ ॥
अन्यस्मादसितं वापि सर्वमन्यद् यथोदितम् ।

candrabimbārdhamānena likhedvṛttaṃ tu tadgate || 29 ||
rekhe dve digvibhāgārthaṃ, pratyagrekhāgrataḥ punaḥ |
nītvā śṛṅgonnatermānaṃ prāgvadardhaguṇātmakam || 30 ||
candrādarkadiśīndostu paridhau prāgviparyayāt |
bindum̐ kṛtvā likhedrekhāṃ tanmārgeṇa sitaṃ nayet || 31 ||
pratyagagrāt site pakṣe prāgagrādasite'pi ca |
sitānte bindumādhāya tiryagrekhāgrayostataḥ || 32 ||
bindum̐ kṛtvā likhedvṛttaṃ bindavo nemigā yathā |
vṛttāntarākṛtiścandraḥ śṛṅgonnatyā pradarśyatām || 33 ||
vyastadikko'rkabāhuḥ syāt tayornānākapālayoḥ |
pratyāsannaraverbhāgāt ihāpyantarnayet sitam || 34 ||
anyasmādasitaṃ vāpi sarvamanyad yathoditam |

Draw a circle with a radius equal to that of the disc of the Moon. Draw two lines [from the centre perpendicular to each other] so as to mark the directions. From the tip of the west line, with a measure equal to the cusp of the Moon, which is nothing but half of the Rsine of elevation given earlier, mark a point on the circumference along the direction of the Sun from the Moon. [This has to be repeated] with the east point in the reverse order. Draw a line passing through these points, and [hence] find the bright phase.

From the west end during the waxing period, and from the east end during waning, mark points which represent the end of the bright portion. Also mark the end points of the *tirya-grekhā*. Now draw a circle such that the three points lie on its circumference. The [bright phase of the] Moon in the shape of the area inscribed/sandwiched between the two circles should thus be demonstrated through the elevation of the cusps.

If the two [i.e. the Sun and the Moon] are in different hemispheres, then their *chāyābāhus* will be in the opposite directions. Even then the bright phase has to be marked towards the direction of the Sun. The dark phase has to be shown from the other direction [that is the direction away from the Sun]. The rest of the process is as described earlier.

The graphical representation of the Moon's disc is done in two stages as explained below.

Marking the elevation on the Moon's disc and drawing the line of cusps

Having drawn a circle (see Fig. 8.13) whose radius is equal to that of the Moon's disc (in some scale), the north–south line *NS* parallel to the horizon, and the east-west line *EW* perpendicular to that, are drawn as shown in the figure. Then a point *B* in the direction of the Sun, on the circumference of the disc, is marked such that

$$GB = r_m \sin\beta. \tag{8.90}$$

This point B will be to the north or south of the EW line, depending upon the direction of the Sun with respect to the Moon at that instant. GB is the depression or elevation of the lunar cusps with respect to the Moon's centre. Similarly the point A is marked on the circumference in the direction opposite that of the Sun, such that $GB = HA$.

Having marked the points B and A on either side of the EW line, we draw the line BA, and a line CD perpendicular to it. CD is called the *tiryagrekhā* and it is the same as the line of cusps.

Representation of the illuminated portion

Then we locate a point F on AB such at BF represents the measure of the phase of the Moon. Through the points C, F and D we draw a circle. The portion inscribed between this circle CFD and the Moon's disc $CWBD$ is the illuminated portion of the Moon as seen by the observer (the shaded portion in the figure). Fig. 8.13(a) is the graphical representation of the Moon in the first quarter of the bright phase. During the last quarter of the bright phase, the disc will be as shown in (b).

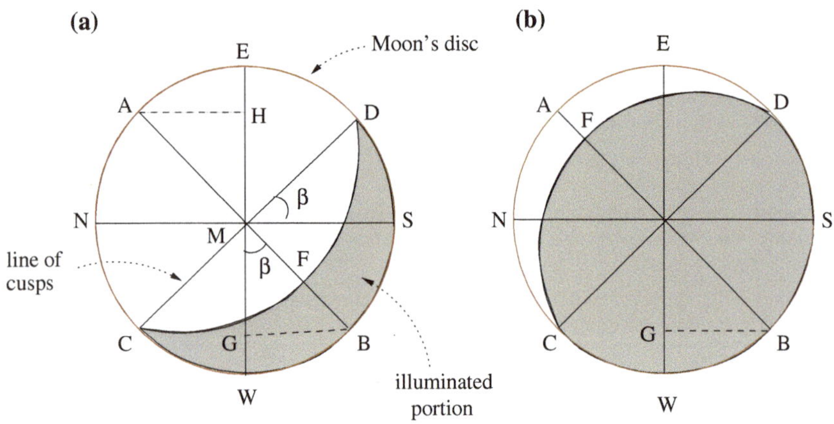

Fig. 8.13 Graphical representation of the Moon's disc (a) during the first quarter of the bright fortnight, (b) during the last quarter of the bright fortnight.

If the Sun and the Moon are in different hemispheres, the *chāyābāhus* will be in opposite directions. However, even in this case, the *sita* should be marked only with reference to the Sun's direction (to represent the bright phase). Even in the bright phase, if the dark portion of the Moon has to be represented, it has to be marked from the east (point A in the figure). The same process is repeated even for the dark fortnight, with the only difference being that, in graphical representation, the bright portion has to be marked from point A. In other words, it must be marked from the east point E as shown in Fig. 8.14.

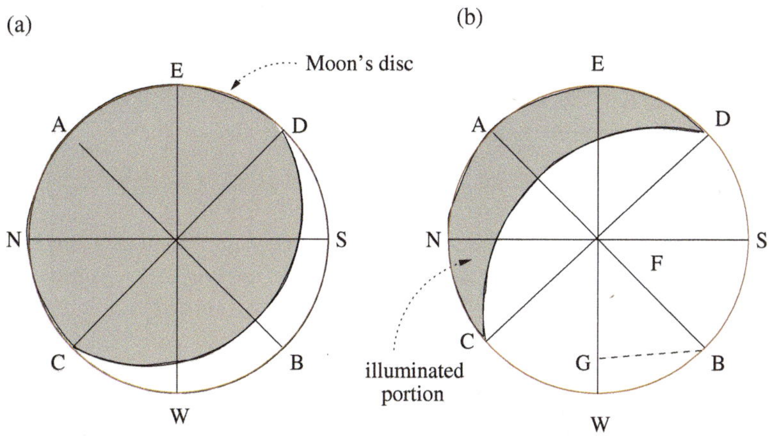

Fig. 8.14 Graphical representation of the Moon's disc (a) during the first quarter of the dark fortnight, (b) during the last quarter of the dark fortnight.

८.१० अर्कास्तमयात्परं चन्द्रस्योदयादिगणना
8.10 Time of moonrise etc. after the sunset

उदयास्तमयाविन्दोः अविशेषेण सिध्यतः ॥ ३५ ॥
मध्यप्राप्तिश्च कालश्च छाययैष्यो गतोऽपि वा ।

udayāstamayāvindoḥ aviśeṣeṇa sidhyataḥ || 35 ||
madhyaprāptiśca kālaśca chāyayaiṣyo gato'pi vā |

The rising and setting times of the Moon are obtained by the process of iteration. In the same way, the time at which it crosses the meridian and the time that has elapsed or is yet to elapse [since rising or till setting, are to be found by iteration].

Here it is stated that the time difference between sunset and moonrise (in the dark fortnight), or sunrise and moonrise (in the bright fortnight), is to be determined using an iterative procedure. To be specific, consider the bright fortnight.

Let λ_s be the longitude of the true Sun when it is rising, at the instant t_0. We assume that t_0 has been calculated taking the equation of time and the *carāsava* (ascensional difference) into account, from the instant at which the mean Sun transits the meridian at the desired place, as explained in Chapter 2. Let $\lambda_m(t_0)$ be the longitude of the true Moon at t_0. We have to find the exact instant at which the true Moon rises. This is done using an iterative procedure as follows:

1. Let $\theta_0 = \lambda_m(t_0) - \lambda_s$. This can be converted into time units from the rising times of the *rāśis*. Now find $\Delta t_0 = \theta_0$ (in time units). Then $t_1 = t_0 + \Delta t_0$ is the first approximation to moonrise.
2. Find the rate of motion of the Moon ($\dot{\lambda}_m$) at t_0. Now $\dot{\lambda}_m \times \Delta t_0 = \delta\lambda_m$ is the increase in the longitude of the Moon in the time interval Δt_0.
3. Let $\lambda_m + \delta\lambda_m = \lambda_m(t_1)$.
4. From $\theta_1 = \lambda_m(t_1) - \lambda_s$, we find $\Delta t_1 = \theta_1$ (in time units). Then $t_2 = t_0 + \Delta t_0 + \Delta t_1$ is the second approximation to the moonrise.

The whole process is repeated till Δt_n becomes negligible. Note that Δt_i can also be negative at any stage.

The time difference between the meridian transits of the Moon and Sun can also be determined by iteration in this manner, except that the *carāsus* are absent in this case. In the same way, the instant corresponding to the Moon on the meridian can be obtained, using the time elapsed since sunrise or the time to elapse till sunset from the desired time.

८.११ कुजादीनां कक्ष्याद्यानयनम्

8.11 Obtaining the dimensions of the orbits of Mars and other planets

रविवच्चन्द्रकक्ष्यायाः नेयान्येषां हि सा ततः ॥ ३६ ॥
भेदे समागमादौ च लम्बनाद्येवमेव हि ।
शीघ्रकर्णघ्नकक्ष्यायाः तद्वृत्तेन सितज्ञयोः ॥ ३७ ॥
आप्ता हि स्फुटकक्ष्या स्यात् तद्वशाल्लम्बनादि च ।

ravivaccandrakakṣyāyāḥ neyānyeṣāṃ hi sā tataḥ || 36 ||
bhede samāgamādau ca lambanādyevameva hi |
śīghrakarṇaghnakakṣyāyāḥ tadvṛttena sitajñayoḥ || 37 ||
āptā hi sphuṭakakṣyā syāt tadvaśāllambanādi ca |

As in the case of the Sun, the [dimension] of the orbits of the Moon and other planets have to be obtained. It is only from their orbits that the parallax in longitude at the instant of opposition, conjunction etc. have to be obtained in a similar manner [as described for eclipses]. [In the case of Mars, Jupiter and Saturn], the mean radii of their orbit (*kakṣyā-vyāsārdha-yojanas*) multiplied by the *śīghra-karṇa* [and divided by the *trijyā*] gives the true orbital radii (*sphuṭakakṣyās*). In the case of Mercury and Venus their mean orbital radii in *yojanas*, multiplied by the *śīghra-karṇa* and divided by the mean radii of their own orbits, give the true values of their orbital radii (*sphuṭakakṣyās*). And from that the parallax in longitude etc. [must be calculated].

Here it is stated that the average values of the radii of the planets in *yojanas* are to be determined from the orbit of the Moon, just as in the case of the Sun (chapter 4, verse 8).

According to the standard assumption in Indian astronomy, the mean linear velocities of all the planets have the same value. If R_p and R_m are the mean radii of the planet and the Moon in *yojanas*, and N_p and N_m are the number of revolutions made by them in a *yuga*, then the total distance covered by each in a *yuga* is the same, and hence

$$2\pi R_m \times N_m = 2\pi R_p \times N_p. \tag{8.91}$$

Thus, the mean radius of each planetary orbit (the *kakṣyā-vyāsārdha-yojana*) in *yojanas* is

$$R_p = \frac{R_m \times N_m}{N_p}. \tag{8.92}$$

This has to be used in the calculation of angular differences between the planets, in determining their conjunctions, in the calculation of parallax etc., for an observer on the surface of the Earth.

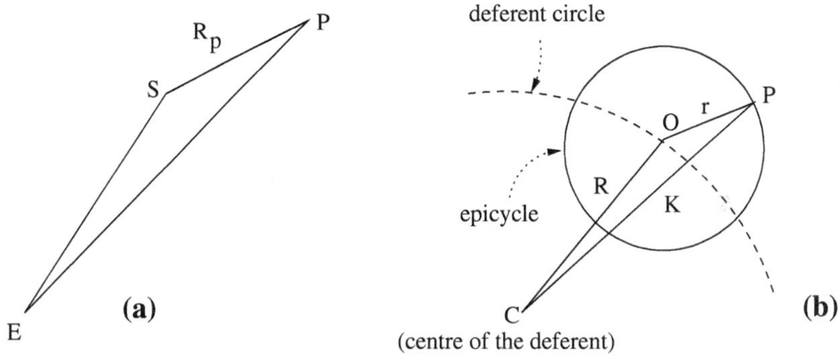

Fig. 8.15 Ratio of the Sun–planet distance to the Earth–planet distance of an inner planet (a) in *yojanas*, (b) in terms of the epicycle radius.

That the radii obtained from the procedure given in verses 36b and 37a are not the orbital radii around the Earth, is clear from the verses that follow it namely 37b and 38a. The commentary of this verse runs as follows:

सितज्ञयोरुक्तवदानीतं कक्ष्याव्यासार्धयोजनं स्वशीघ्रकर्णेन निहत्य शीघ्रवृत्तप्रमित-स्ववृत्तव्यासार्धेन विभजेत्। तत्र लब्धा स्फुटकक्ष्या भवति।

In the case of Mercury and Venus, the radius of the *kakṣyā-vṛtta* in *yojanas* has to be multiplied by the *śīghra-karṇa* and divided by the radius of its own orbit measured in units of the *śīghra-vrtta*. The result obtained will be the true radius of the orbit.

What is noteworthy here is the use of the phrase *'śīghravṛttapramita-svavṛtta-vyāsārdhena'*. Particularly, the word *sva-vṛtta* ('own' circle) clearly indicates Nīlakaṇṭha's geometrical picture of the motion of the inner planets. According to this picture, the epicycle which figures in the *śīghra-saṃskāra* is the same as the orbit of the inner planet.

In the procedure for finding the actual distance of the planet from the Earth, the *kakṣyāvyāsārdha-yojana* R_p, calculated using (8.92), is not to be taken as the mean Earth–planet distance, but instead as the mean *śīghrocca*–planet distance. If

E is the Earth and P the inner planet (see Fig. 8.15(a)), then the formula given by Nīlakaṇṭha for the Earth–planet distance may be written as

$$EP = \frac{K_s}{r} \times R_p, \tag{8.93}$$

where K_s is the *śīghra-karṇa* and r is the *svavṛtta-vyāsārdha* (radius of the 'own orbit'), which is the same as the radius of the epicycle in the *śīghra-saṃskāra* in the case of Mercury and Venus.

As regards the outer planets, the commentator states:

अन्येषां तु प्राग्वल्लब्धं कक्ष्याव्यासार्धं स्वशीघ्रकर्णगुणितं त्रिज्यया विभक्तम् - इति विशेषः ।

The above statement may be written down as

$$EP = \frac{K_s}{R} \times R_p, \tag{8.94}$$

where EP is the distance in *yojanas* of the outer planet from the Earth and K_s is the *śīghra-karṇa* as in the earlier formula (see Fig. 8.16). It may be noted that here the *trijyā*, R, appears in the denominator instead of the epicycle radius.

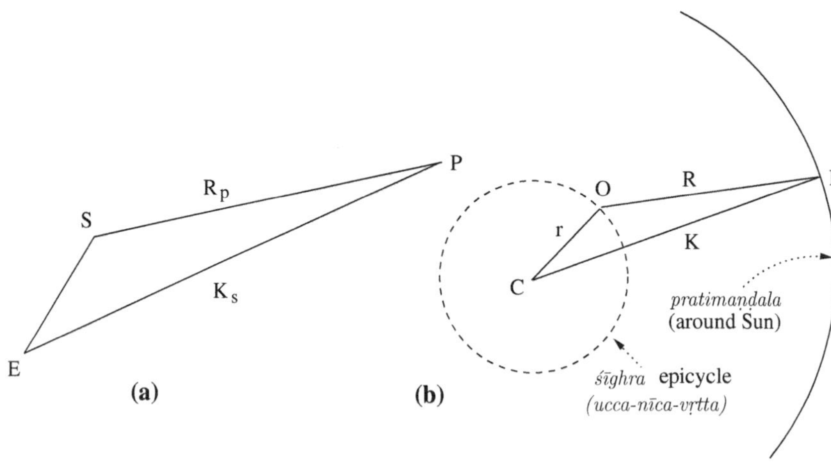

Fig. 8.16 Ratio of the Earth–planet distance to the planet–Sun distance of an outer planet (a) in *yojanas*, (b) in terms of the epicycle radius.

Note: In these verses Nīlakaṇṭha has introduced a modification of the traditional prescription for planetary distances. According to him the *kakṣyā-vyāsārdha-yojanas* computed from its revolution number according to (8.92), is actually the mean *śīghrocca*–planet distance and not the mean Earth–planet distance. The latter is to be found from (8.93) and (8.94).

Since these verses have come at the very end of the *Tantrasaṅgraha*, Nīlakaṇṭha has not worked out the implications of this revised prescription. Nīlakaṇṭha also briefly alludes to the above prescription for planetary distances in his *Golasāra* and *Āryabhaṭīya-bhāṣya*. However, even this revised prescription for planetary distances is not really consistent with the cosmological model that Nīlakaṇṭha has expounded definitively in his later work, *Grahasphuṭānayane vikṣepavāsanā*. This issue is discussed further in Appendix F.

८.१२ बिम्बमानादीनां दृक्साम्यपरीक्षणम्

8.12 Verifying the measures of the discs with the observed values

चक्रांशाद्यङ्किते वृत्ते तन्मध्यासक्तचक्षुषा ॥ ३८ ॥
ज्ञेयं ग्रहान्तरं बिम्बदृग्ज्या चापादिकं स्फुटम् ।

cakrāṃśādyaṅkite vṛtte tanmadhyāsaktacakṣuṣā || 38 ||
jñeyaṃ grahāntaraṃ bimbadṛgjyā cāpādikaṃ sphuṭam |

Considering a circle with 360 degrees, (seconds) etc. marked on it, and with the eye placed at its centre, the angular separation between the planets and the arc of the Rsines of their zenith distances (*dṛgjyās*) etc. are to be determined accurately.

Essentially, it is stated that with a flat circular ring of arbitrarily large radius and with degrees and seconds marked along the circumference and with a provision for viewing from the centre, suitably mounted, accurate angular measurements such as the separation between two planets, zenith distance etc. can be made.

The commentator explains how the set-up should be made in order to carry out observations, as follows:

कथमिति चेत्, तद्वृत्तमध्यं चक्षुर्गोलसन्निहितं कृत्वा अभीष्टग्रहद्वयस्फुटपरिधिकं यथा भवति तथा कृते तस्मिन् अभीष्टग्रहान्तरालपरिधिभागे यावन्तोऽम्शाः कला वा संभवन्ति तावदेव तयोर्बिम्बान्तरम्। एवं तत्तद्बिम्बमानं दृग्ज्याचापादिकमपि स्फुटमेव विज्ञातुं शक्यमिति।

If it is asked how it is, [then we say:] Keeping the centre of the circle close to the eyeball and focusing on the two desired planets in their orbits, whatever is measured to be the [angular] distance of separation in degrees or minutes is actually the separation between the discs. Thus the measure of their own discs as well as the arc corresponding to the *dṛgjyā* can all be correctly obtained.

८.१३ उपसंहारवचनम्

8.13 Concluding words

गोलः कालक्रिया चापि द्योत्यतेऽत्र मया स्फुटम् ॥ ३९ ॥
लक्ष्मीशनिहितध्यानैः इष्टं सर्वं हि लभ्यते ॥ ४० ॥

golaḥ kālakriyā cāpi dyotyate'tra mayā sphuṭam || 39 ||
lakṣmīśanihitadhyānaiḥ iṣṭaṃ sarvaṃ hi labhyate || 40 ||

The spherics and also the computations related to [the determination of] time [from the motion of planets] are clearly being expounded here. Those who meditate intensely on the Lord of *Lakṣmī* indeed obtain all that they desire.

It is mentioned in this concluding verse that the author has in this work, besides other things, also explained the topics that are generally covered under the sections on the *golapāda* and the *kālakriyāpāda* in other works such as *Āryabhaṭīya*.

Since only the calculational procedures for obtaining different physical quantities are being discussed in the text, the commentator Śaṅkara Vāriyar raises the objection as to what would have prompted Nīlakaṇṭha to make the above statement. Having raised this question he presents an answer to it in the latter half of the verse. The explanation runs as follows:

यद्यप्यस्मिन् गणितविशेष एव सम्यक्प्रदर्शितः, तथाप्यस्य समीचीनां युक्तिपदवीं अपह्नुवानेन मया गोलो गणितगम्यक्षेत्रविशेषः कालक्रिया च तत्परिच्छेदिका, स्पष्टतरमेव प्रकाशिते स्याताम्। नन्वन्यप्रदर्शनेन अन्यस्य प्रकाशनं कथम्? इत्यत आह - लक्ष्मीशनिहितध्यानैरिति।...प्रकरणसमाप्तिसमयाहर्गणश्च दर्शितः 'लक्ष्मीश-निहितध्यानैः' इत्यक्षरसङ्ख्यया - इति।

Though in this [work] it is only the calculational procedures that have been explained in detail, nevertheless in the process of explaining their rationale, the spherics (*gola*), which is the geometrical picture implied by the calculational procedures, and the computations related to time which delimit it (the *gola*) have also been explained as well. May be so [that the two were also explained]. But how is it that by explaining one thing something else also gets explained? Therefore he says – by meditating upon the Lord of Lakṣmī ...The author has also indicated the date of completion of the work through [the word] '*lakṣmīśanihitadhyānaiḥ*' by [using] the letter numeral (*akṣarasaṅkhyā*).

The material to be covered in the *Golapāda* includes the geometric models implied by the calculational procedures. The commentator, in clarifying the intent of the author, mentions that while explaining their rationale (by experienced teachers) the spherical astronomical descriptions and the computations of planetary motions which are related to them will get elaborated upon. He also mentions that if one has the grace of the Lord, then the geometrical pictures implied by the calculational procedures can be understood clearly even without a detailed description of them.

Further, the commentator points out that the chronogram of the date of completion of the work is also indicated by the word *lakṣmīśanihitadhyānaiḥ* in *akṣarasaṅkhyā* (letter numeral), which is the same as in the *kaṭapayādi* number system. According to this system this word corresponds to the number 1680553, which refers to the number of days elapsed since the commencement of *Kaliyuga*, namely *kalyahargaṇa*. This corresponds to *Meṣa* 1, 4601 *kalyabda* or March 27, 1500 CE. Since in the invocatory verse in Chapter 1 it was indicated that the composition of this work was started on *kalyahargaṇa* 1680548, which corresponds to March 22, 1500, it seems that Nīlakaṇṭha composed this great treatise on mathematical astronomy in just five days!

Appendix A
Representation of numbers

Long before the ten numerals: 1, 2, 3, ...0, were introduced and the notation for representing numbers became standardized, words like *ekam, dve, trīṇi, catvāri,* ... (one, two, three, four, ...) seem to have been employed by the Vedic people. Listings of a series of odd numbers and multiples of four are found in *Yajurveda,* [1] indicating the antiquity of numeration by words. In fact, in *Kṛṣṇa-yajurveda* one also finds a listing of powers of 10 in several places.[2]

Besides listing numbers, the Vedic corpus also presents evidences to show that additive and subtractive principles were employed while coining words to connote numbers. For instance, one finds words like *saptaviṃśati* ($7+20=27$), *dvātriṃśat* ($2+30=32$), *ekonaviṃśati* ($20-1=19$) and the like. It is not difficult to see that this can easily be extended to represent large numbers having three, four and more digits.

However, it does not take long to realize that this method of employing words to represent numbers becomes extremely cumbersome, particularly with the increase in the number of digits. Thus there is a need for developing more efficient ways of representing numbers. The need will be felt all the more when one deals frequently with large magnitudes, which is the case in subjects like astronomy. Besides this need mentioned above, Indian astronomers and mathematicians had to meet with one more constraint, namely the rules imposed by metrical compositions. These two requirements, besides other considerations, made them invent different schemes for representing numbers, among which (i) the *Kaṭapayādi* and (ii) the *Bhūtasaṅkhyā* systems are the most commonly employed ones.

Before we proceed to explain these systems in detail, it may be mentioned here that these systems, apart from being simply an alternate way of representing numbers, have several advantages over the word-numeral scheme described above. This will become amply evident from the description of these systems and the numerical examples provided in following sections.

---

[1] *Kṛṣṇa-yajurveda Taittirīya-saṃhitā* 4.6.11. A similar listing is found in *Yajurveda Vājasaneyī-saṃhitā* (18.24–25) also.

[2] See for instance, *Kṛṣṇa-yajurveda Taittirīya-saṃhitā* 4.4.10 and 7.2.20.

A.1 *Kaṭapayādi* system

The name '*Kaṭapayādi*' for this system of representing numbers stems from the fact that here the Sanskrit alphabets *ka, ṭa, pa, ya* etc. are used to denote the numbers. According to this system, the vowels standing alone, represent the number zero. However, the same vowels in conjunction with the consonants have no numerical significance. It is only the 33 consonants *k, kh, g, gh, …, ś, ṣ, s, h* that are associated with the numbers. The mapping of these consonants with different numbers is listed in Table A.1.

Number	1	2	3	4	5	6	7	8	9	0
Consonants	*k*	*kh*	*g*	*gh*	*ṅ*	*c*	*ch*	*j*	*jh*	*ñ*
used	*ṭ*	*ṭh*	*ḍ*	*ḍh*	*ṇ*	*t*	*th*	*d*	*dh*	*n*
to represent	*p*	*ph*	*b*	*bh*	*m*	–	–	–	–	–
numbers	*y*	*r*	*l*	*v*	*ś*	*ṣ*	*s*	*h*	*ḷ*[3]	–

Table A.1 The *Kaṭapayādi* system of numeration.

The following verse found in *Sadratnamālā* of Śaṅkaravarman (c. 1830 CE) succinctly summarizes the system:

नञावचश्च शून्यानि सङ्ख्याः कटपयादयः ।
मिश्रे तूपान्तहल्संख्या न च चिन्त्यो हलः स्वरः ॥

[The letters] *n, ñ* and the vowels [when standing alone] denote zeros. [The consonants] commencing from *ka, ṭa, pa* and *ya* denote the numbers [1, 2, 3, …] in order. In the case of conjunct consonants (*miśre tu*) only the last consonant represents the number. The vowel suffixed to a consonant should not be counted.

It is believed that this system of using alphabets to represent numbers is as old as 4th century CE. This is because the *cāndravākyā*, the chronograms associated with the Moon for reading its longitudes on different days simply from a look-up table—beginning with '*gīrṇaḥ śreyaḥ*', '*dhenavaḥ śrīḥ*' etc.—which are based on this system, were composed by the Kerala astronomer *Vararuci*, who is traditionally ascribed to the above period. In this system, as we read the chronogram and try to decipher the coded number, it is to be borne in mind that it is the least significant decimal place that is given first, and the highest the last. For example, take the word *āyurārogyasaukhyaṃ*, as indicated in the table here, represents the number

ā	*yu*	*rā*	*ro*	*gya*	*sau*	*khyam*
0	1	2	2	1	7	1

[3] This is a special character—very rarely employed in the classical Sanskrit literature—whose representation in *Devanāgari* script is ळ. However, the Kerala astronomers once in a while seem to employ this to represent the number nine. One such example using this character is presented in Table A.2.

1712210. In the *carnatic* (south Indian) system of music, the first two letters of the name of a *meḷakartā rāga* indicates the rank number of its *meḷakartā* in the *Kaṭapayādi* system.[4] Table A. 2 presents a few illustrative examples from the texts on Indian mathematics and astronomy.

Word/Words	Number represented
विद्वान्	44
तुन्नबलः	3306
कवीशनिचयः	160541
सर्वार्थशीलस्थिरः	2735747
निर्विद्धाङ्गनरेन्द्ररुक्	22203940
सुगन्धिनगनुत्	30937
भद्राङ्गभव्यासनः	714324
ऊनधनकृद्भूरेव	42410900
धीगापाङ्गजळाङ्गस्त्री	23983139
नानाज्ञानतपोधरः	29160000
हे विष्णो निहितं कृत्स्नम्[5]	1680548
लक्ष्मीशनिहितध्यानैः	1680553

Table A.2 A few examples of the *Kaṭapayādi* system of representing numbers.

A.2 *Bhūtasaṅkhyā* system

The word *Bhūtasaṅkhyā* is a compound word which has two constituents, namely *bhūta* and *saṅkhyā*—referring to a 'being' and a 'number' respectively. Thus the compound *Bhūtasaṅkhyā*, which can be derived as '*bhūtānāṃ saṅkhyā*', means 'the number associated with beings'. In fact this system uses words commonly employed in Sanskrit which are widely known to be associated with specific numbers such as:

1. The physical entities such as Earth, Sun, Moon, planets, stars, ocean, mountain, fire, sky, direction etc.

---

[4] For example, in the names of the *rāgas* **धीरशङ्कराभरणं** and **मेचकल्याणी**—popularly referred to as simply **शङ्कराभरणं** and **कल्याणी**—the first two syllables represent the numbers 29 and 65 respectively.

[5] This is the first quarter of the verse with which the text *Tantrasaṅgraha* commences. It has been pointed out by the commentator Śaṅkara Vāriyar that this also serves the purpose of being a chronogram—representing the *Kalyahargaṇa* corresponding to the date of commencement of the the work. The next example is the third quarter of the concluding verse of *Tantrasaṅgraha*, which again, as per the commentator, is the *Kalyahargaṇa* corresponding to the date of completion of the work. This indicates that the entire work, consisting of about 432 verses, has been composed in just five days!

2. The parts of a human body such as eyes, ears, jaws, knees, hands, fingers, teeth, nails etc.
3. The animals, such as serpent, horse, elephant etc.
4. The names of the gods, such as Śiva, Indra etc. and sometimes historical figures such as Manu, Rāma, Jina etc.
5. The season, fortnight, month, week, etc.

These are used to denote the numbers 1, 2, 3 etc. Since all the things listed above share the common property of 'being' (*bhūta*), this system of representation of numbers is called the *bhūtasaṅkhyā* system. Table A.3 presents a few examples of numbers given by Bhāskarācārya in his *Siddhāntaśiromaṇi* using this system.

Word	Number represented
खाद्रिरामाग्नयः	3370
वेदवेदाङ्कचन्द्राः	1944
वेदचन्द्रद्विवेदाब्धिनागाः	544214
भुजङ्गनन्दद्विनगाङ्गबाणषट्कृतेन्दवः	146567298
खाभ्रगगनामरेन्द्रियक्ष्माधराद्रिविषयाः	577533000

Table A.3 Numbers specified using the *Bhūtasaṅkhyā* system of representation.

This system, which is quite different from the alphabetical system of representation described earlier, has its own advantages and disadvantages. One of the distinct advantages, particularly from the viewpoint of an author of a text, would be that here it may be a lot easier to meet the metrical compulsions of verses used in the texts on astronomy and mathematics. As the language is extremely rich in synonyms, an author could choose any synonym that would suit the metre to represent a given number. However, from the viewpoint of the reader, this system may be considered disadvantageous for at least two reasons:

1. The lack of familiarity with the connotation of a specific *bhūta* representing a particular number would present difficulties in deciphering the number.
2. Even if one were somewhat familiar, the lack of knowledge of synonyms could pose serious problems—not to mention the difficulties that could arise owing to improper splitting of the words.

Of course, ignorance on the part of a reader is no reason to blame the system. Notwithstanding the 'disadvantages' mentioned above, this system has its own appeal, charm and beauty. The table below presents a list of *bhūtasaṅkhyās* that have been employed by astronomers in their texts. Note that the list should only be considered as representative and not exhaustive.

Number	*Bhūtas* used to refer to number
0	– *khaṃ, ākāśa, nabha, vyoma, antarikṣaṃ* … synonyms of sky/space

	– *bindu* (a dot)
1	– *indu, candra, himāṃśu, mṛgāṅka, śaśāṅka, śaśadhara, ...* all synonyms of the Moon – *pṛthvī, kṣiti, vasundharā, ku, dharaṇi, dharā ...* synomyms of the Earth – *nāyaka, mahīpāla, bhūpāla, ...* synonyms of a king, including *pitāmaha* (the creator *Brahmā*)
2	– *akṣi, cakṣu, nayana, netra ...* synonyms of eyes – *bāhu, bhuja, hasta, ...* synonyms of hands – words referring other parts of the body such as *karṇa* (ears), *jānu* (knees) and *kuca* (breasts) etc. – words like *aśvinau, ratīputrau* etc. which are known to be pairs from the *purāṇās* are also used
3	– *agni, anala, hutāśana, śikhin, vahni ...* synonyms of fire. – *bhuvana, jagat, loka ...* synonyms of 'world'. – *Rāma* (signifying the three *Rāmas*: *Paraśurāma, Ayodhyārāma* and *Balarāma*) – *hotṛ* (signifying the three important priests of the sacrifice *Adhvaryu, Hotā* and *Udgātā*)
4	– *abdhi, udadhi, jaladhi, vāridhi, payodhi, arṇava ...* synonyms of ocean – *śruti, veda, āmnāya ...* synonyms of *veda* – words like *yuga* (aeon), *āśrama* (stages of life), *varṇa* (broad classification of humans), *dik* (direction) etc. that are known to be four in number. – the word *kṛta*, being the name of the first of the group of four *yugās* is also used
5	– *iṣu, śara, bāṇa, ...* synonyms of arrow (supposed to be shot by Cupid to arouse desire) – *indriya, akṣa ...* synonyms of the sense organs – *viṣaya* (denoting the five sense objects) – *mahābhūta* (denoting the five basic elements: Earth, water, fire, air and space) – *prāṇa* (denoting the five types of wind in the body—*prāṇa, apāna, vyāna, udāna* and *samāna*
6	– *aṅga* (signifying the six subsidiary parts of *Veda*) – *ṛtu* (signifying the six seasons) – *kāraka* (signifying six relatants of an action)

– *rasa* (signifying six types of tastes: *madhura, āmla, lavaṇa, kaṭu, kaṣāya* and *tikta*)
– *ari* (signifying six enemies to be conquered: *kāma, krodha, lobha, moha, mada* and *mātsarya*)
– *darśana* (signifying six major philosophical systems)

7 – *aga, acala, adri, giri, bhūdhara, kṣmādhara* … synonyms of mountains
– *aśva, turaga, vājin, haya* … synonyms of horses
– words such as *ṛṣi, muni* (a particular group of seven sages), *svara* (fundamental notes in music), *dvīpa* (major islands) and *vāra* (weekdays)

8 – *hastin, gaja, diggaja, kuñjara, dantin, ibha* … synonyms of elephants
– *nāga, sarpa, takṣa, ahi* … synonyms of serpents
– words like *vasu* (types of wealth), *siddhi* (special powers), *maṅgala* (auspicious things) etc. that are known to represent eight things.

9 – *randhra, chidra* … synonyms of holes signifying the number of holes present in the human body (seven in the face and two used for excretion)
– *aṅka* (the digit), *graha* (planets), *durgā, go, nanda*

10 – *aṅguli* (fingers), *āśā, dik* (direction), *avatāra* (incarnations of Lord *Viṣṇu*), *rāvaṇaśira* (heads of the demon Rāvaṇa), *paṅkti* (rows) etc.

11 – *īśa, iśvara, rudra, śaṅkara, śiva, hara* … synonyms of Lord Siva
– *akṣauhiṇī* (a huge regiment of an army)

12 – *sūrya, arka, bhānu, āditya, divākara* … synonyms of Sun
– *māsa* (month), *rāśi* (zodical signs)

13 – *viśva, viśvedevāḥ, atijagatī, aghoṣa*

14 – *indra, śakra, manu, loka*

15 – *tithi, dina, pakṣa* (number of days in a fortnight)

16 – *aṣṭi, kalā* (one part of the lunar disc, which is conceived as made up of 16 parts)

17	– *atyaṣṭi*
23	– *vikṛti*
24	– *arhat, jina* (generic term for a Jaina saint), *gāyatrī* (metre having 24 syllables)
25	– *tattva* (the fundamental principles that the world is constituted of—taken to be 25 in *Sāṅkhya* philosophy)
27	– *nakṣatra, bhaṃ, tārā* ... synonyms of stars
32	– *rada, danta* ... synonyms of teeth
33	– *amara, deva, sura* ... synonyms of deities
48	– *jagatī* (metre having 48 syllables)

Appendix B
Spherical trigonometry

For an observer on the surface of the Earth, the sky appears to be the surface of a large sphere, with the celestial objects situated on it. For solving problems in positional astronomy, we need to know the properties of triangles drawn on spherical surfaces. This is the subject-matter of spherical trigonometry.

B.1 Great and small circles

A circle drawn on the surface of a sphere whose radius is equal to the radius of the sphere—or, equivalently, whose centre coincides with the centre of the sphere—is called a great circle. A great circle can also be conceived of as the intersection of a sphere with a plane passing through its centre. For instance, if the Earth is considered as a sphere, the equator on its surface is a great circle. All the meridian circles passing through the north and the south poles are also great circles.

If we consider any two points on the surface of a sphere that are not diametrically opposite, there is only one great circle that passes through them. If, however, the points happen to be diametrically opposite, then an infinite number of great circles can be drawn passing through them—just like the meridian or longitude circles on the surface of the Earth.

A small circle on the surface of a sphere is a circle whose centre does not coincide with the centre of the sphere. For instance, the Tropic of Cancer and Tropic of Capricorn, which are parallel to the equator, are small circles. In fact, all latitudinal circles are small circles, as their radii are smaller than the equator; their centres lie along the axis of the Earth and do not coincide with the centre of the Earth.

B.2 Spherical triangles

When two great circle arcs meet at a point, the 'spherical angle' between them is the angle between the tangents to them at that point. From now onwards, we refer to a spherical angle just as an angle.

A spherical triangle is a closed figure formed on the surface of a sphere by the pairwise intersection of three great circular arcs. As spherical astronomy is concerned primarily with the positions of the celestial objects on the surface of the celestial sphere, studying the properties of spherical triangles is of great importance in spherical astronomy. In Fig. B.1, ABC is a spherical triangle formed by the great circle arcs, AB, BC and CA. The spherical angles are denoted by A, B and C. Similarly, $A'B'C'$ is a spherical triangle.

The sides of a spherical triangle are the lengths of the great circle arcs forming it, divided by the radius of the sphere. Defined this way, the sides are the angles subtended by the great circle arcs at the centre, in radians. As the sides are angles, they are often expressed in degrees also. In the spherical triangle ABC, the sides BC, CA and AB are denoted by a, b and c.

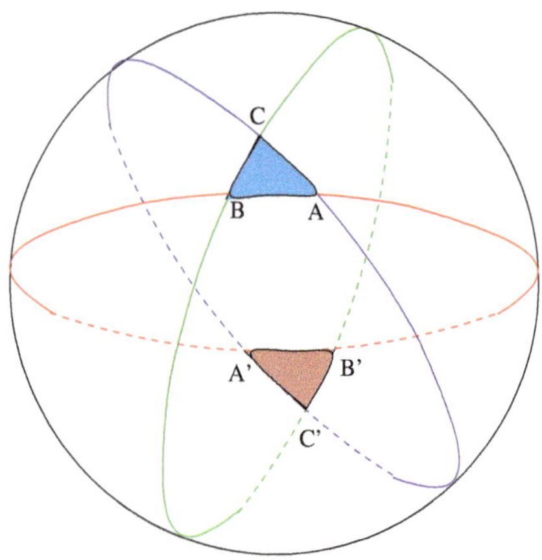

Fig. B.1 Spherical triangles formed by the intersection of three great circles on the surface of a sphere.

A spherical triangle can be conceived of as the spherical analogue of the planar triangle, for they have many properties in common. However, the geometry of a sphere is quite different from the geometry of a plane. Hence the properties of a spherical triangle are quite different from those of a planar triangle. However, as will be shown a little later, some of the fundamental formulae for a spherical triangle reduce to that of a plane triangle, when the sides of the former are 'small' compared with the radius of the sphere.

Properties of a spherical triangle

The following properties of a plane triangle are applicable to a spherical triangle as well:

1. The largest/smallest angle formed at the vertex is opposite the largest/smallest side of the triangle.
2. The sum of two sides of the triangle is always larger than the third side.

The important differences between the spherical and plane triangles that are of immediate utility in the study of spherical astronomy are, in summary:

1. While the sum of the three angles in a plane triangle is always equal to the sum of two right angles, in the case of a spherical triangle it is always greater than that. The sum is not constant and the upper bound happens to be the sum of six right angles. In other words, in a spherical triangle

$$180^\circ < A + B + C < 540^\circ. \tag{B.1}$$

2. While in a plane triangle the sides a, b and c are specified in units of length, in the case of a spherical triangle they are usually specified in terms of angles. The sum of the three sides in a spherical triangle satisfies the following inequality:

$$0^\circ < a + b + c < 360^\circ. \tag{B.2}$$

Fundamental formulae for spherical triangles

There are several formulae connecting the sides and angles of a spherical triangle. Out of them four are considered fundamental and they are frequently referred to, while providing explanations in the text. In the following we explain these formulae without providing any derivations.[1]

Cosine formula

If ABC is the spherical triangle, with sides a, b, c, then the law of cosines is given by

$$\cos a = \cos b \cos c + \sin b \sin c \cos A. \tag{B.3}$$

Clearly, there are two companions to the above formula. They are easily obtained by cyclically changing the sides and the angles, and are given by

$$\cos b = \cos c \cos a + \sin c \sin a \cos B \tag{B.4}$$

$$\cos c = \cos a \cos b + \sin a \sin b \cos C. \tag{B.5}$$

[1] The derivation of these formulae, for instance, may be found in W. M. Smart, *Textbook on Spherical Astronomy*, Cambridge University Press, 1965.

The law of cosines – more often referred to as the cosine formula – is analogous to the ordinary law of cosines used in plane trigonometry. It has two direct practical applications: (i) it straightaway gives the third side of a spherical triangle if the other two sides and the included angle are known, and (ii) it gives all the angles if all the three sides are known. Further, it may be noted that the above rule reduces to the planar law when the sides of the spherical triangle are small. It is well known that when θ is small,

$$\sin\theta \to \theta; \qquad \text{and} \qquad \cos\theta \to 1 - \frac{\theta^2}{2}. \tag{B.6}$$

Using the above approximation, (B.3) reduces to

$$a^2 = b^2 + c^2 - 2bc\cos A, \tag{B.7}$$

which is none other than the cosine formula for a plane triangle.

Sine formula

The relation between the ratio of the sides to that of the angles of a spherical triangle is given by

$$\frac{\sin a}{\sin A} = \frac{\sin b}{\sin B} = \frac{\sin c}{\sin C}. \tag{B.8}$$

When the sides a, b and c are small, it is quite evident that the above formula reduces to

$$\frac{a}{\sin A} = \frac{b}{\sin B} = \frac{c}{\sin C}, \tag{B.9}$$

which is the sine formula for a plane triangle.

Four-parts formula

Considering any four consecutive parts of a spherical triangle, which obviously includes two sides and two angles, they can be shown to satisfy the following relation:

$$\begin{aligned}\cos\,(\textit{inner side})\cos\,(\textit{inner angle}) = {} & \sin\,(\textit{inner side})\cot\,(\textit{other side}). \\ & - \sin\,(\textit{inner angle})\cot\,(\textit{other angle}).\end{aligned} \tag{B.10}$$

Of the two sides and two angles that we consider consecutively, the side which is flanked by two angles is called the '*inner side*' and the angle which is contained by two sides is called the '*inner angle*'. For instance, consider the four consecutive parts B, a, C and b in the spherical triangle ABC in Fig. B.2. Here a is the inner side and b is the other side. Similarly, C is the inner angle and B is the other angle. Then, for these four parts, the four-parts formula is

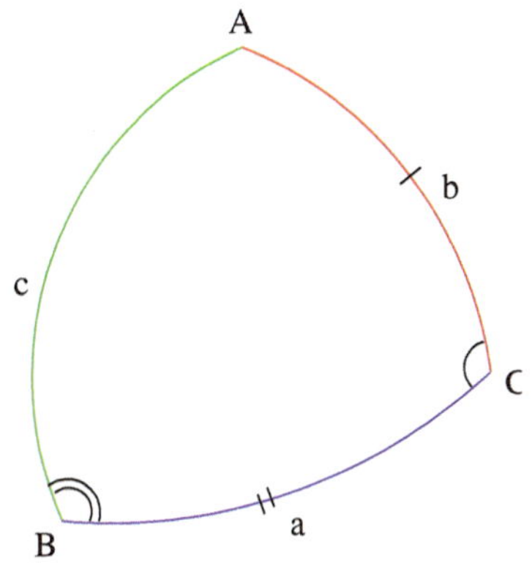

Fig. B.2 A spherical triangle with markings on two consecutive sides and consecutive angles.

$$\cos a \cos C = \sin a \cot b - \sin C \cot B. \tag{B.11}$$

Analogue of the cosine formula

One more formula, involving the three sides and two angles, that is often found to be useful in solving problems is

$$\sin a \cos B = \cos b \sin c - \sin b \cos c \cos A. \tag{B.12}$$

This formula generally goes by the name of the 'analogue of the cosine formula' as it is simply obtained by substituting the cosine formula in itself. For instance, the above formula is obtained by substituting (B.3) in (B.4).

Appendix C
Coordinate Systems

Anyone who observes the sky even for short periods of time will have the impression that the objects in it are in continuous motion. This motion consists of two parts. One of them is the apparent motion of all celestial objects, including stars, from east to west, which is actually due to the rotation of the Earth from west to east. This is the diurnal motion. The other is due to the relative motion of any particular celestial object like the Sun, Moon or a planet with respect to the seemingly fixed background of stars.

Just as one uses latitude and longitude (two numbers) to specify any location on the surface of the Earth, so also one employs different coordinate systems to specify the location of celestial objects on the celestial sphere at any instant. In this appendix, we will explain the three commonly employed coordinate systems—namely, the horizontal, the equatorial and the ecliptic.

C.1 Celestial sphere

All the celestial objects seem to be situated on the surface of a sphere of very large radius, with the observer at the centre. This is the celestial sphere. Though fictitious, the celestial sphere is the basic tool in discussing the motion (both diurnal and relative) of celestial objects.

In Fig. C.1, C is the centre of the Earth and O the observer on the surface of the Earth whose northerly latitude is ϕ. The tangential plane drawn at the location of the observer, represented by NOS, is the *horizon*. Only those celestial objects that are above the horizon can be seen by the observer. The point on the celestial sphere that is directly overhead, and in the direction of the plumb-line, is the *zenith*, denoted by Z. The plumb-line direction is the *nadir*.

As the Earth rotates about the axis PQ, it appears as if the entire celestial sphere rotates in the opposite direction about P_1, which is the point of interesection of the extension of QP with the celestial sphere. The line OP_2 is parallel to CP_1. Since the radius of the Earth is very small compared with the radius of the celestial sphere,

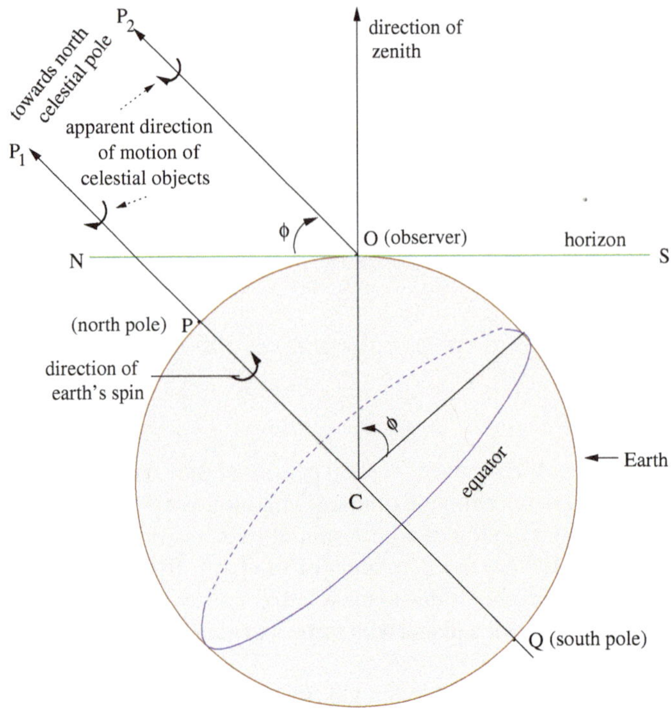

Fig. C.1 The horizon and the north celestial pole as seen by the observer on the surface of the Earth.

the points P_2 and P_1 would be indistinguishable on the celestial sphere. All the celestial bodies seem to be rotating around the axis OP_2 with a period equal to the period of rotation of the Earth (nearly 4 seconds less than 24 hours). The point P_2 is generally denoted by P and is called the *north celestial pole*. The *south celestial pole* is denoted by Q. The celestial sphere for the observer with latitude ϕ is shown in Fig. C.2.

C.2 Locating an object on the celestial sphere

An object situated at any point on the surface of the celestial sphere, which is a two dimensional surface, can be uniquely specified by two angles. Based on the choice of the fundamental great circle—the horizon, the celestial equator or the ecliptic—we have the following sytems listed in Table C.1.

Each of these systems has its own advantages and the choice depends upon the problem at hand, somewhat like the choice of coordinate system that is made in order to solve problems in physics. Table C.2 presents the Sanskrit equivalents of the

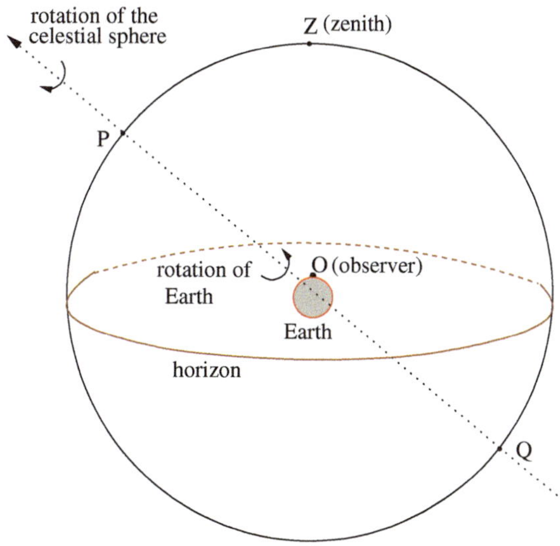

Fig. C.2 The celestial sphere for an observer in the northern hemisphere with latitude ϕ.

Coordinate system	Fundamental plane/circle	Poles of circle	Coordinates and notation used
Horizontal	Horizon	Zenith/nadir	Altitude and azimuth (a, A)
Equatorial	Celestial equator	Celestial poles	Declination and right ascension/hour angle (δ, α) or (δ, H)
Ecliptic	Ecliptic	Ecliptic poles	Celestial latitude and longitude (β, λ)

Table C.1 The different coordinate systems generally employed to specify the location of a celestial object.

different coordinates and the fundamental reference circles employed for specifying a celestial object.

The horizontal system

In this system, which is also known as the *alt-azimuth* system, the horizon is taken to be the fundamental reference place. In Fig. C.3, the great circle passing through the zenith and the north celestial pole P intersects the horizon at N and S, the north and the south points. E and W are the east and the west points, which are 90° away from the north and the south points. These four points together represent the four cardinal directions for an observer.

The circles passing through the zenith and perpendicular to the horizon are called *vertical* circles and the vertical passing through E and W is called the prime vertical.

Coordinates		Reference circles	
Modern name	Skt equivalent.	Modern name	Skt equivalent
Altitude	उत्क्रम	Horizon	क्षितिज
Azimuth	उदग्रा	Prime meridian	दक्षिणोत्तरवृत्त
Hour angle	नत	Prime meridian	दक्षिणोत्तरवृत्त
Declination	क्रान्ति	Celestial equator	विषुवद्वृत्त/घटिकावृत्त
Right Ascension	काल	Celestial equator	विषुवद्वृत्त/घटिकावृत्त
Declination	क्रान्ति	its secondary	तद्विपरीत
Longitude	भोग	Ecliptic	क्रान्तिवृत्त
Latitude	विक्षेप	its secondary	तद्विपरीत

Table C.2 Sanskrit equivalents for different coordinates and the reference circles.

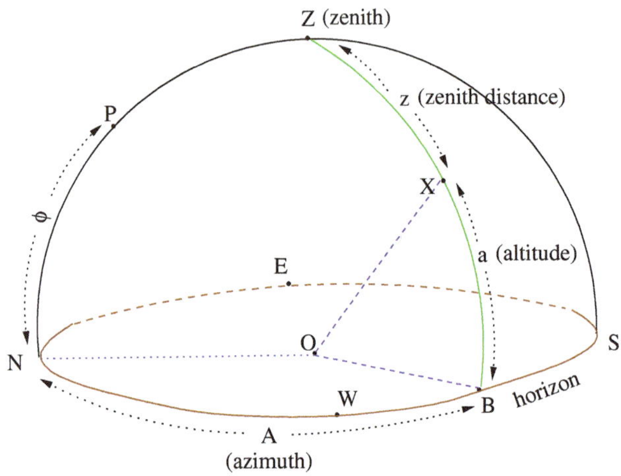

Fig. C.3 Altitude, azimuth and zenith distance in the horizontal system.

The point X in Fig. C.3 represents a star and ZX is the vertical passing through the star which intersects the horizon at B. Now we define two angles called the altitude and azimuth

$$\begin{aligned} \text{altitude } (a) &= X\hat{O}B \qquad (\text{range: } 0-90^\circ) \\ \text{azimuth}(A) &= N\hat{O}B \qquad (\text{range: } 0-360^\circ W). \end{aligned} \tag{C.1}$$

These two angles—one measured along the horizon and the other along the vertical circle perpendicular to the horizon—completely specify the location of the star. Sometimes, in place of altitude, *zenith distance*, given by

$$z = 90 - a, \tag{C.2}$$

could be specified. The main disadvantage of the horizontal system is that it is observer-dependent. Two observers situated at different locations on the Earth will come up with different coordinates.

The equatorial system

In this system, the celestial equator is taken to be the fundamental plane with reference to which the coordinates are specified. The celestial equator is a great circle whose plane is perpendicular to *OP*. Clearly, its plane is parallel to that of the Earth's equator. This would be inclined to the horizon by an angle equal to the co-latitude $(90-\phi)$ of the observer (see Fig. C.4).

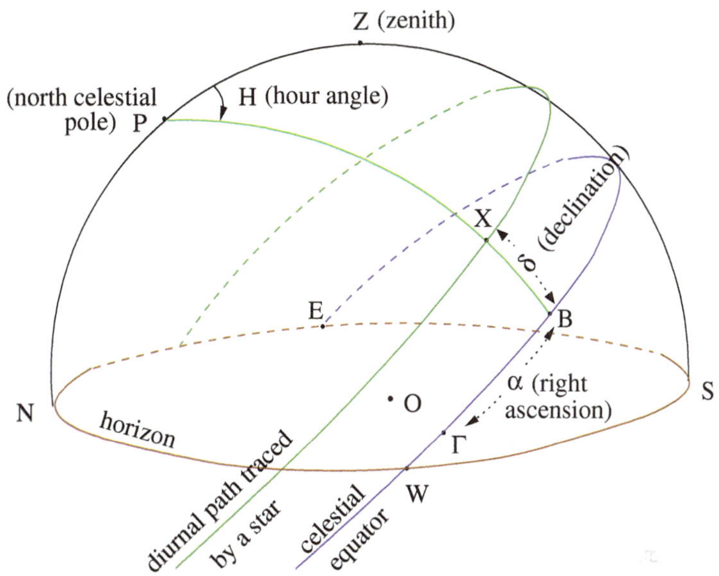

Fig. C.4 Declination, hour angle and right ascension in the equatorial system.

All circles passing through the pole *P* and perpendicular to the equator are known as *meridian* circles. Of them, the meridian circle passing through the zenith of the observer is of special significance and is known as the *prime meridian*. If *X* is the star whose coordinates are to be specified in this system, then we draw a meridian passing through the star *X* and the north celestial pole *P*. This intersects the equator at *B*. Now, the two quantities *declination* and *hour angle*[1] of the star are defined as

[1] While in modern astronomy the hour angle is specified in terms of hours 0–24, in Indian astronomical texts it is given in terms of *ghaṭikās*, whose measure is taken to be 24 minutes. Thus 60 *ghaṭikās* are equal to 24 hours. In texts such as *Laghujātaka*, the term *hora* is employed to specify the duration of an hour.

follows:

$$\begin{aligned} \text{declination } (\delta) &= X\hat{O}B \qquad (\text{range: } 0-90^\circ N/S)\\ \text{hour angle } (H) &= Z\hat{P}X \qquad (\text{range: } 0-360^\circ/24\,h\,W). \end{aligned} \tag{C.3}$$

To enable various formulae derived to be valid for both north and south declinations, it is convenient to treat δ as an algebraic quantity. δ is positive when it is north, and negative when it is south. We can obtain (δ, H) from (a, A) using spherical trigonometrical formulae. For instance, considering the triangle ZPX in Fig. C.4, where $PX = 90^\circ - \delta$, $PZ = 90^\circ - \phi$, $PZX = A$ and $ZPX = H$ and applying the cosine formula, we have

$$\begin{aligned} \cos(PX) &= \cos(PZ)\cos(ZX) + \sin(PZ)\sin(ZX)\cos(PZX)\\ \text{or} \qquad \sin\delta &= \sin\phi\sin a + \cos\phi\cos a\cos A, \end{aligned} \tag{C.4}$$

and

$$\begin{aligned} \cos(ZX) &= \cos(ZP)\cos(PX) + \sin(ZP)\sin(PX)\cos(ZPX)\\ \text{or} \qquad \sin a &= \sin\phi\sin\delta + \cos\phi\cos\delta\cos H. \end{aligned} \tag{C.5}$$

Now, if a and A are known then δ can be determined from (C.4). Then using (C.5) H can be determined. It may be noted that these formulae can be used to obtain (a, A) from (δ, H) also.

Of the two quantities (δ, H), though δ is independent of the observer—as the celestial equator is common to all the observers on the surface of the Earth—H is not so. This is because H is defined to be the angle between the prime meridian and the meridian passing through the star (measured westwards). Though the latter is observer-independent the former is not, as the prime meridian passes through the zenith of the observer.

To make the coordinates observer-independent, instead of hour angle, *right ascension* defined by

$$\text{right ascension } (\alpha) = \Gamma\hat{P}X \qquad (\text{range: } 0-360^\circ\, E), \tag{C.6}$$

is employed. In contrast to the hour angle H which is measured westwards, the right ascension is measured eastwards. Here, the point Γ shown in the figure represents the point of intersection of the equator and the ecliptic (to be discussed in the next subsection), and is observer-independent.

Further from Fig. C.4, it is clear that for the object X,

$$\text{H.A. } (X) + \text{R.A. } (X) = \text{H.A. } (\Gamma). \tag{C.7}$$

The above equation is valid for any choice of X. In other words, the sum of the hour angle and right ascension of any celestial object is always equal to the hour angle of the vernal equinox.

The ecliptic system

It has been known from ancient times that the Sun traces out a closed path on the celestial sphere each year. This apparent path of the Sun in the background of the stars is called the *ecliptic*. The system of coordinates which makes use of the ecliptic as the fundamental reference plane is known as the ecliptic system. In this system, two angles called the celestial longitude and the celestial latitude, or simply the longitude and the latitude, are used to specify the location of an object on the celestial sphere. These are defined as (see Fig. C.5).

$$\begin{aligned} \text{latitude } (\beta) &= X\hat{O}B \qquad \text{(range: } 0-90^\circ\text{N/S)} \\ \text{longitude } (\lambda) &= \Gamma\hat{K}X \qquad \text{(range: } 0-360^\circ/24\,\text{h East)}. \end{aligned} \tag{C.8}$$

Here K is the pole of the ecliptic. β is positive when it is north, and negative when it is south.

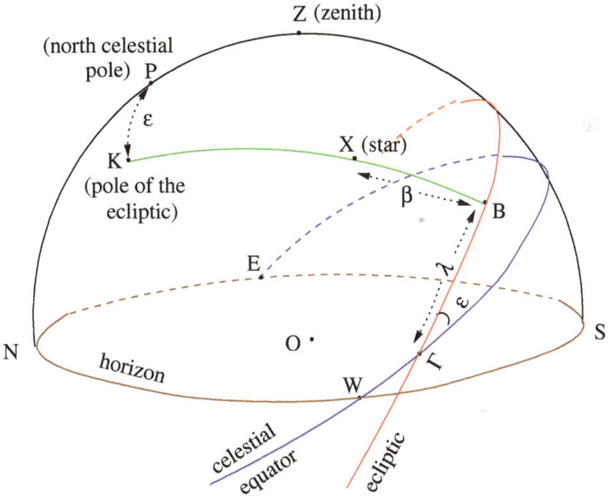

Fig. C.5 Celestial latitude and longitude in the ecliptic system.

Because the axis of rotation of the Earth is tilted (by roughly 23.5°) with respect to the plane of its orbital motion, the ecliptic, which is the path of the Sun on the celestial sphere, is a circle which is inclined with respect to the celestial equator. This inclination, denoted by ε, is known as the *obliquity of the ecliptic*. The ecliptic and the celestal equator inersect at two points known as the *vernal equinox* and *autumnal equinox*. The Sun's motion on the ecliptic is eastwards. At the vernal equinox Γ, it moves from south to north, or its declination changes sign from $-$ to $+$.

Among the various great circles represented on the celestial sphere, the ecliptic is very important. This is because the Sun moves along the ecliptic, and the inclinations of the orbits of all the planets and the Moon with the ecliptic are small.

Using the formulae of spherical trigonometry, it can be shown that the ecliptic coordinates (β,λ) and the equatorial coordinates (δ,α) are related through the following equations:

$$\sin\beta = \sin\delta\cos\varepsilon + \cos\delta\sin\varepsilon\sin\alpha \tag{C.9}$$

$$\sin\delta = \sin\beta\cos\varepsilon + \cos\beta\sin\varepsilon\sin\lambda. \tag{C.10}$$

C.3 Precession of equinoxes

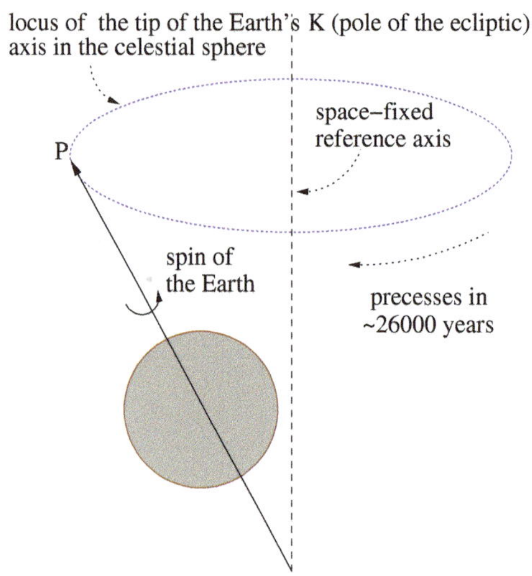

Fig. C.6 The locus of points traced by the equatorial axis due to precession.

Because of the gravitational forces of the Sun and the moon on the equatorial bulge of the rotating Earth, the rotational axis of the Earth moves with respect to a space-fixed reference frame as shown in Fig. C.6, like the axis of a precessing top. As a result of this, the relative orientations of the ecliptic and the celestial equator in space keep steadily varying, maintaining the angle of inclination around an average value of 23.5°. In otherwords, the equinoctial points Γ and Ω as indicated in Fig. C.7 move backwards (westwards) along the ecliptic. This phenomenon is known as the *precession of equinoxes*. In this figure, K is the pole of the ecliptic. The tip of the axis of rotation of the Earth moves around the circle $P \to P_1 \to P_2 \to P_3 \to P$.

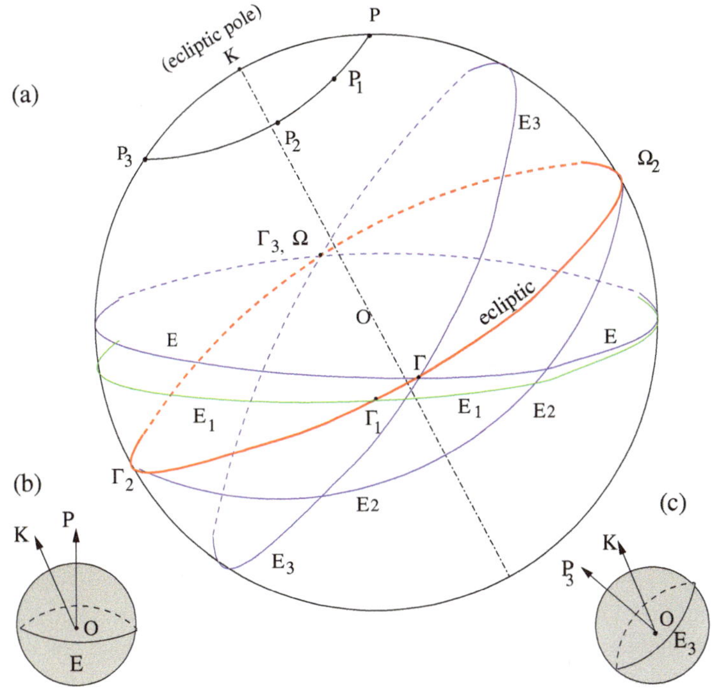

Fig. C.7 (a) Different orientations of the celestial equator with respect to the ecliptic due to precession; (b) and (c) orientation of the equator and its axis at different epochs about 13000 years apart.

As per the current estimates, the axis of rotation of the Earth completes one full revolution about the pole of the ecliptic in around 25800 years. This means that the vernal equinox Γ slides back along the ecliptic in the same period. The rate of motion is given by

$$\frac{360 \times 60 \times 60}{25800} \approx 50'27'' \qquad \text{per year,}$$

which amounts to 1° in about 72 years.

This phenomenon of precession seems to have been noted in the Indian tradition from early times. For instance, Varāhamihira (c. 505 CE) in his *Bṛhatsaṃhitā* observes:

आश्लेषार्धाद्दक्षिणमुत्तरमयनं रवेर्धनिष्ठाद्यम् ।
नूनं कदाचिदासीत् येनोक्तं पूर्वशास्त्रेषु ॥
साम्प्रतमयनं सवितुः कर्कटकाद्यं मृगादितश्चान्यत्।
उक्ताभावो विकृतिः प्रत्यक्षपरिक्षणैर्व्यक्तिः ॥

दूरस्थचिह्नवेधादुदयेऽस्तमयेऽपि वा सहस्रांशोः ।
छायाप्रवेशनिर्गमचिह्नैर्वा मण्डले महति ॥[2]

> There was indeed a time when the Sun's southerly course began from the middle of the star *Āśleṣa* and the northerly one from the commencement of the star *Dhaniṣṭhā*, as it has been so stated in ancient works.
>
> At present the southerly course of the Sun starts from the beginning of Cancer and the other from the initial point of sign Capricorn. The actual state of affairs which goes against the old statement can be verified by direct observation. The Sun's change of course can be detected by marking everyday the position of a distant object either at sunrise or sunset, or by watching and marking the entry and exit of the shadow of the gnomon fixed at the centre of a big circle drawn on the ground.

The above passage besides giving the positions of the solstices during Varāhamihira's time also clearly brings out Varāha's understanding of the phenomenon of precession. The simple techniques recommended by him for the determination of the current positions of the solstices also enable us to discern his ingenuity. This also highlights the fact that he was not blindly following the statements of the ancient writers, but as a true astronomer was making observations for himself to verify them, which is quite essential for the advancement of science in general and astronomy in particular.

Many of the Indian astronomical works talk about the phenomenon of precession. They give slightly differing values for the rate of precession, which are all centred around the value of $54'$ per year.

Sāyana and *nirayaṇa* longitudes

As noted in the previous section, the point of intersection of the ecliptic and the celestial equator, namely the vernal equinox, constantly drifts westwards at the rate of around $50''$ per year. There are two kinds of longitude specifications, namely *sāyana* and *nirayaṇa*, based on whether the vernal equinox or the beginning point of the *Meṣa rāśi* (which is the same as the beginning point of the *Aśvinī nakṣatra*) is taken as the reference point on the ecliptic for the measurement of longitudes.

The term *ayana* refers to motion in Sanskrit. Hence *sāyana* means 'with motion' and *nirayaṇa* means 'without motion'. In the *nirayaṇa* system, the longitude of any celestial object is measured from the beginning point of the *Meṣa rāśi* (which is a *fixed point* in space) in the eastward direction, whereas in the *sāyana* system it is measured with respect to the vernal equinox (which is a *moving point* with respect to fixed stars). This explains the origin of the terminology *sāyana* and *nirayaṇa* longitudes. While the longitudes differ, the celestial latitudes are the same in both the systems, as latitude is a measure of the distance of the celestial object along a the circle perpendicular to the ecliptic.

---

[2] {BS 1981}, (III.1–3), p. 23.

Ayanāṃśa

By doing a back computation, it is generally found that around the year 285 CE the vernal equinox coincided with the beginning point of the *Aśvinī nakṣatra*, so that the *sāyana* and the *nirayaṇa* longitudes were the same at that time. Due to the precession of equinox, at present[3] the position of vernal equinox is $24^\circ\ 00'\ 16''$ west of the *Meṣādi*. In other words, the *sāyana* longitude of the *Meṣādi* is $24^\circ\ 00'\ 16''$. The difference in longitude between the two reference points on the ecliptic is known as the *ayanāṃśa*:

$$\textit{ayanāṃśa} = \textit{sāyana}\ \text{longitude} - \textit{nirayaṇa}\ \text{longitude}.$$

This means that one has to add the *ayanāṃśa* to the *nirayaṇa* longitude to get the *sāyana* longitude (see Fig. C.8)

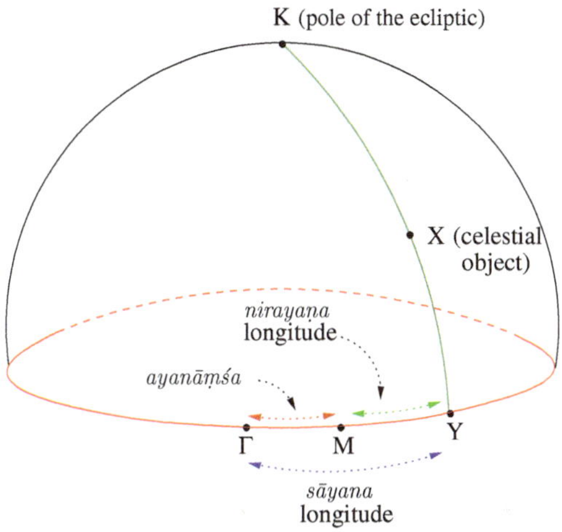

Fig. C.8 *Nirayaṇa* and *sāyana* longitudes. Γ is the vernal equinox and *M* is the *Meṣādi*.

In Fig. C.8, the arcs MY and ΓY are the *nirayaṇa* and *sāyana* longitudes of X. ΓM, which is the difference between them, is the *ayanāṃśa*. The *ayanāṃśa* for the beginning of each month is provided in some almanacs, like *Rāṣṭriya Pañcāṅga* published by the Government of India. So it is easy to convert from one system to another. Most of the *Pañcāṅgas* published in various parts India use the *nirayaṇa* system for fixing the dates and times of observance of almost all religious and social functions.

[3] As on March 22, 2010.

C.4 Equation of time

In ancient times day-to-day activities of human beings were much linked with the position of the Sun, and its diurnal motion across the sky was the primary means by which people kept track of time. Sundials were employed for this purpose. However, the time indicated by them is *apparent solar time*, based on the actual position of the Sun; this is to be distinguished from the *mean solar time*, based upon the motion of a fictitious body called the *mean Sun*, which we will define shortly.

The time interval between two successive passages of the Sun at the observer's meridian is defined as an *apparent solar day*. The length of an apparent solar day is not constant for two reasons:

1. The Sun does not move with uniform speed along its apparent orbit.
2. The orbit of the Sun is inclined to the equator—along which all the time measurements are done—by about 23.5°.

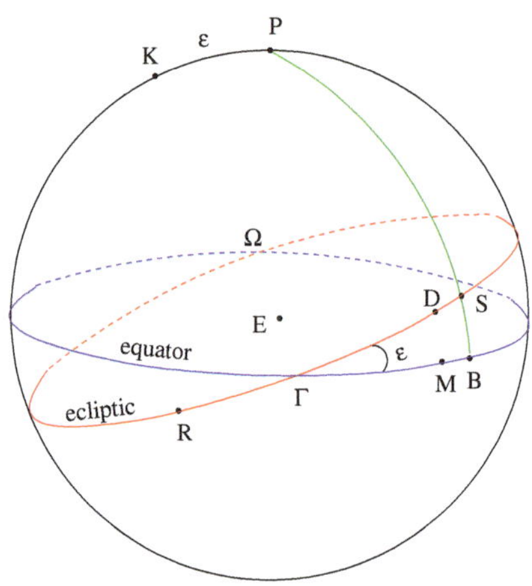

Fig. C.9 Positions of the 'true' Sun (S), the 'mean' Sun (M) and the 'dynamical mean' Sun (D).

In order to introduce a day which is of constant duration, ancient astronomers introduced a fictitious body called the 'mean Sun' which moves on the *equator* along the direction of increasing right ascension at a uniform speed. The time interval between two successive passages of this fictitious body across the observer's meridian is a constant quantity which is defined as a *mean solar day*. This is divided into 24 hours. All our clocks are set to measure this mean solar time.

The 'dynamical mean Sun' moves along the ecliptic at a constant angular speed, which is the same as the average angular speed of the true Sun and the mean Sun. In Fig. C.9, S, M and D represent the true/actual Sun, the mean Sun and the dynamical mean Sun respectively.

1. When the Sun is at the perigee, the dynamical mean Sun starts off its motion along the ecliptic with a uniform angular speed in such a way that it would meet the true Sun again at the perigee after one complete revolution.
2. The motion of the mean Sun is such that it starts off its motion along the equator—with uniform angular speed, when the dynamical mean Sun is at the vernal equinox—in such a way that it would meet the dynamical mean Sun again at the vernal equinox after one complete revolution.

The difference between the right ascension of the mean Sun (*RAMS*) and that of the actual Sun ($RA\odot$) is defined as the *equation of time* (ε) . That is,

$$\varepsilon = RAMS - RA\odot \tag{C.11}$$

$$= H.A.\odot - H.A.M.S. \tag{C.12}$$

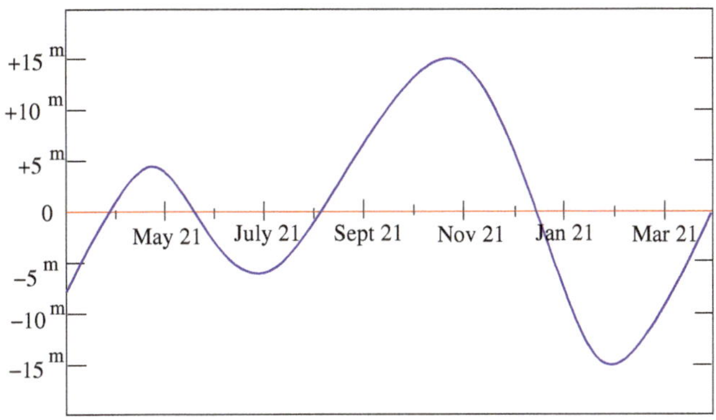

Fig. C.10 Variation in the equation of time over a year.

A graph depicting the variation in the equation of time over the period of an year is shown in Fig. C.10. The equation of time is the time difference between the meridian transits of the true Sun and the mean Sun.

Appendix D
Solution of the ten problems: a couple of examples from *Yuktibhāṣā*

A 'declination type' formula is the only essential ingredient in the solution of all the spherical trigonometry problems in *Yuktibhāṣā*. The problem can be posed thus: what is the distance of a point on a great circle from the plane of another great circle which intersects it?

In Fig. D.1, two circles with a common radius R and a common centre O intersect at points X and X'. Let i be the angle of inclination between the two circles. It may be noted that the maximum separation between the two circles given by $CD = Ri$ occurs when $CX = DX = 90°$.

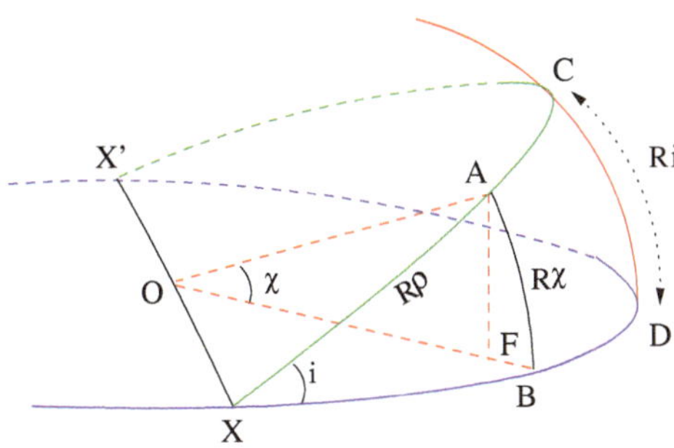

Fig. D.1 Measure of the arc connecting two intersecting circles.

Consider a point A on one of the circles such that arc $XA = R\rho$. Draw a great circle arc $AB = R\chi$ such that it is perpendicular to the second circle XDX' at B. Then $R\sin\chi$, denoted by AF in the figure, is the perpendicular distance between A and the second circle and is given by

$$R\sin\chi = R\sin i \, \sin\rho. \tag{D.1}$$

It can be easily seen that the above relation reduces to the familiar relation for the declination

$$\sin\delta = \sin\varepsilon \, \sin\lambda, \tag{D.2}$$

if, for example, we consider the two circles XCX' and XDX' to be the ecliptic and the celestial equator respectively, i to be the obliquity of the ecliptic ε, ρ to be the longitude λ and χ to be the declination δ.

Thus $R\sin\chi$ can be found if the arc $R\rho$ is given; conversely, the arc $R\rho$ can be found when the perpendicular distance $R\sin\chi$ is known. This is the '*trairāśika*' that is invariably used in the solution of all the ten problems discussed in *Yuktibhāṣā*.

Now there are five quantities: (i)*śaṅku* (gnomon) $R\cos z$, (ii) *nata-jyā* (Rsine hour angle) $R\sin H$, (iii) *apakrama* (declination) $R\sin\delta$, (iv) *āśāgrā* (amplitude) $R\sin a$, where $a = 90^\circ \sim A$, A being the azimuth, and (v) *akṣajyā* (Rsine latitude) $R\sin\phi$. When three of them are known, the other two are to be determined. This can happen in ten different ways, and so the topic is referred to as 'the ten problems'. We shall outline the *Yuktibhāṣā* derivation of the solution of the first two problems, where the *śaṅku* and the *nata*, and the *śaṅku* and the *apakrama*, are derived in terms of the other three quantities.

D.1 Problem one: to derive the *śaṅku* and *nata* from the other three quantities

We now discuss the method to derive the *śaṅku* and the *nata-jyā*, when the declination, *āśāgrā* and latitude are known.

In Fig. D.2, X is the planet. The great circle through Z and X is the *iṣṭa-digvṛtta*, cutting the horizon at A. If $WA = a$ is the arc between the west point and A, the *aśāgrā* is $R\sin a$. Let B be between N and W, at 90° from A. Then the great circle through Z and B is the *viparīta-digvṛtta*. Consider the great circle through B and the north celestial pole P. This is the *tiryagvṛtta*, which is perpendicular to both the *iṣṭa-digvṛtta* and the celestial equator. This is so because this circle passes through the poles of both the *digvṛtta* and the celestial equator (B and P respectively).

Let the *tiryagvṛtta* intersect the *iṣṭa-digvṛtta* and the celestial equator at C and D respectively. Let the arc $BP = x$. Then, as B is the pole of the *iṣṭa-digvṛtta*, $BC = 90$ or $PC = 90 - x$. As $PD = 90$, $CD = x$. This is indeed the angle between the *digvṛtta* and the celestial equator at Y ($X\hat{Y}U$). The distance between P on the meridian and the *viparīta-digvṛtta* ZB is given by

$$R\sin PF = R\sin a\,\cos\phi, \qquad \text{(D.3)}$$

as $PZ = 90 - \phi$, and $P\hat{Z}B$, the inclination of the *viparīta-digvṛtta* with the meridian, is a.

Let the angle between the *tiryagvṛtta* and the horizon be i. Then the angle between the *tiryagvṛtta* and the *viparīta-digvṛtta* is $90 - i$. It follows that $R\sin PF$ is also given by

$$R\sin PF = R\sin x\cos i. \qquad \text{(D.4)}$$

Equating the above two expressions,

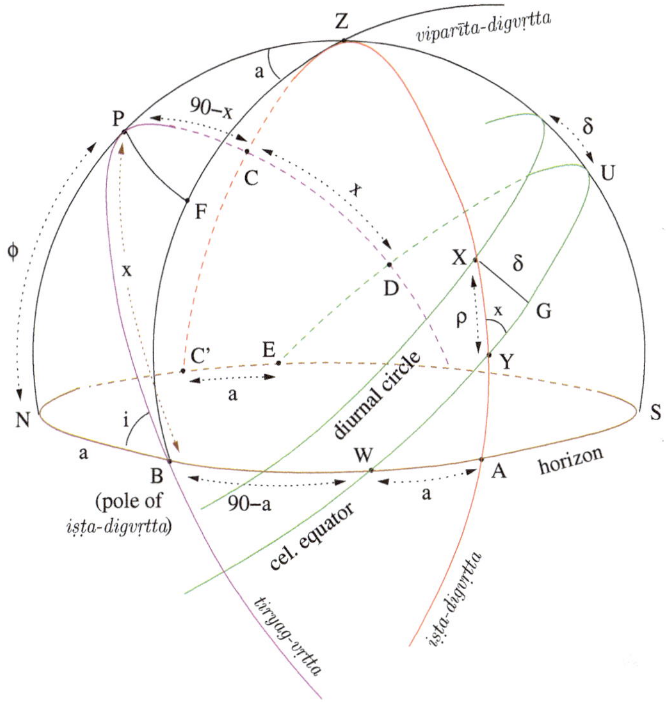

Fig. D.2 The important circles and their secondaries considered in the 'ten problems'.

$$R\sin x \cos i = R\sin a \cos\phi. \tag{D.5}$$

Now $PN = \phi$ is the perpendicular arc from P on the *tiryagvṛtta*, on the horizon, which is inclined to it at angle i. Therefore,

$$R\sin x \sin i = R\sin\phi. \tag{D.6}$$

From (D.5) and (D.6), we get

$$R\sin x = \sqrt{R^2\sin^2 a\cos^2\phi + R^2\sin^2\phi}, \tag{D.7}$$

which is what has been stated. This is the maximum separation between the *iṣṭa-digvṛtta* and the celestial equator, as the angle between them is x.

Now the arc BC on the *tiryagvṛtta* and the arc BC' on the horizon are both 90°. Hence arc $CC' = i$, the angle between the two *vṛttas*. Then $CZ = 90 - i$, and as C is at 90° from Y, the intersection between the celestial equator and the *iṣṭa-digvṛtta*, $ZY = i$. Hence the ascent of the *tiryagvṛtta* from the horizon on the *digvṛtta* $= i$ is the same as the descent of the equator from the zenith on the *digvṛtta*. Let the arc $XY = \rho$. XG is the perpendicular arc from X on the *digvṛtta* on the celestial equator.

$$\begin{aligned} R\sin(XG) = R\sin\delta &= R\sin(XY)\ \sin x \\ &= R\sin\rho \sin x. \end{aligned} \tag{D.8}$$

Now the perpendicular arc from Z on the *digvṛtta* on the celestial equator $= ZU = \phi$. Therefore

$$\begin{aligned} R\sin ZU = R\sin\phi &= R\sin(ZY)\sin x \\ &= R\sin i \sin x. \end{aligned} \tag{D.9}$$

$R\sin\rho$ and $R\sin i$ are called the *sthānīyas* or the 'representatives' of the *apakrama* and *akṣajyā* on the *digvṛtta*. Now the zenith distance

$$\begin{aligned} z = ZX &= ZY - XY \\ &= i - \rho. \end{aligned} \tag{D.10}$$

Therefore

$$\begin{aligned} R\sin z = R\sin(i-\rho) &= R\sin i\cos\rho - R\cos i\sin\rho \\ &= \frac{(R\sin\phi\cos\rho - R\sin\delta\cos i).R}{R\sin x}. \end{aligned} \tag{D.11}$$

Consider the *koṭis* of the $R\sin\phi$ and $R\sin\delta$ on a circle of radius $R\sin x$ (which are denoted as *koṭi*$'$):

$$\begin{aligned} \textit{koṭi}'(\phi) &= \sqrt{R^2\sin^2 x - R^2\sin^2\phi} \\ &= \sqrt{R^2\sin^2 x - R^2\sin^2 i\sin^2 x} \\ &= R\cos i\sin x. \end{aligned} \tag{D.12}$$

Similarly,

$$\begin{aligned} \textit{koṭi}'(\delta) &= \sqrt{R^2\sin^2 x - R^2\sin^2\delta} \\ &= \sqrt{R^2\sin^2 x - R^2\sin^2\rho\sin^2 x} \\ &= R\cos\rho\sin x. \end{aligned} \tag{D.13}$$

Hence we have

$$R\sin z = \frac{(R\sin\phi\ \textit{koṭi}'(\delta) - R\sin\delta\ \textit{koṭi}'(\phi))R}{R^2\sin^2 x}. \tag{D.14}$$

This is the shadow $R\sin z$ at the desired place, which is expressed in terms of the declination δ, the latitude ϕ and the *āśāgrā*, as x is given in terms of ϕ and a by

$$R\sin x = \sqrt{R^2\sin^2 a\cos^2\phi + R^2\sin^2\phi}. \tag{D.15}$$

The gnomon $R\cos z$ is given by

$$\begin{aligned} R\cos z &= R\cos(i-\rho) \\ &= R(\cos i \cos\rho + \sin i \sin\rho) \\ &= \frac{(ko\d{t}i'(\phi)ko\d{t}i'(\delta) + R\sin\phi\, R\sin\delta)R}{R^2 sin^2 x}. \end{aligned} \tag{D.16}$$

When the declination δ is south and $\delta > 90^\circ - \phi$, the diurnal circle is below the horizon and there is no gnomon. When the northern declination is greater than the latitude, midday is to the north of the zenith and the gnomon in the southern direction. However, in this case, the gnomon will occur only when the *āśāgrā* is north, i.e. A is north of W.

Now from (D.7) and (D.12), $ko\d{t}i'(\phi)$ reduces to

$$ko\d{t}i'(\phi) = R\cos\phi \sin a. \tag{D.17}$$

Similarly, from (D.7) and (D.13)

$$ko\d{t}i'(\delta) = R\sqrt{\sin^2\phi + \cos^2\phi \sin^2 a - \sin^2\delta}. \tag{D.18}$$

Hence,

$$R\cos z = \frac{R\,(\sin\phi \sin\delta + \cos\phi \sin a \sqrt{\sin^2\phi + \cos^2\phi \sin^2 a - \sin^2\delta}\,)}{(\sin^2\phi + \cos^2\phi \sin^2 a)}. \tag{D.19}$$

When the declination is north and the planet X is to the north of the prime vertical, one can show that $z = 180 - (i+\rho)$ and we would get

$$R\cos z = \frac{R\,(\sin\phi \sin\delta - \cos\phi \sin a \sqrt{\sin^2\phi + \cos^2\phi \sin^2 a - \sin^2\delta}\,)}{(\sin^2\phi + \cos^2\phi \sin^2 a)}. \tag{D.20}$$

When the declination is south, $z = i + \rho$ and we would get (D.19) again where it is understood that δ is negative.

Thus in all cases

$$R\cos z = \frac{R\,(\sin\phi \sin\delta \pm \cos\phi \sin a \sqrt{\sin^2\phi + \cos^2\phi \sin^2 a - \sin^2\delta}\,)}{(\sin^2\phi + \cos^2\phi \sin^2 a)}. \tag{D.21}$$

Koṇa-śaṅku (corner shadow)

The term *koṇa* means corner. In this context, it refers to the corner between any two cardinal directions, such as north-east, south-west etc. Technically, the *koṇa-śaṅku* or corner shadow occurs when the *āśāgrā* $= 45^\circ$. In this case, from (D.14) and (D.15) we have

$$R\sin x = \sqrt{\frac{1}{2}R^2\cos^2\phi + R^2\sin^2\phi} \tag{D.22}$$

$$R\sin z\sin x = \frac{R\sin\phi R\cos'\delta - R\sin\delta\, R\cos'\phi}{R\sin x}. \tag{D.23}$$

Derivation of *nata-jyā*

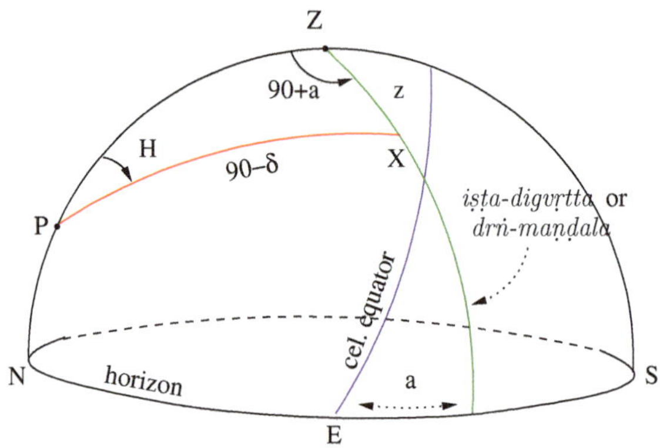

Fig. D.3 The *iṣṭa-digvṛtta* passing through a planet.

In Fig. D.3, X is the planet whose declination is δ. Let H be the hour angle. Since $PX = 90-\delta$, the distance between X and the north–south circle will be

$$\begin{aligned} &= R\sin H\sin(90-\delta) \\ &= R\sin H\cos\delta. \end{aligned} \tag{D.24}$$

But the maximum angle between the north–south circle and the *iṣṭa-digvṛtta* on which X is situated at a distance z from the zenith is $90-a$. Therefore the distance between X and the north–south circle is also

$$\begin{aligned} &= R\sin z\sin(90-a) \\ &= R\sin z\cos a = \textit{chāyā-koti}. \end{aligned} \tag{D.25}$$

Equating the two expressions, we get

$$R\sin H\cos\delta = R\sin z\cos a = \textit{chāyā-koti}.$$

Therefore the *nata-jyā* is given by

$$R\sin H = \frac{ch\bar{a}y\bar{a}ko\d{t}i}{\cos\delta} = \frac{ch\bar{a}y\bar{a}ko\d{t}i \times trijy\bar{a}}{dyujy\bar{a}}. \tag{D.26}$$

D.2 Problem two: the *śaṅku* and *apakrama*

Here, the *śaṅku* and *krānti* (*apakrama*) are to be derived in terms of the *nata-jyā*, *āśāgrā* and *akṣa*.

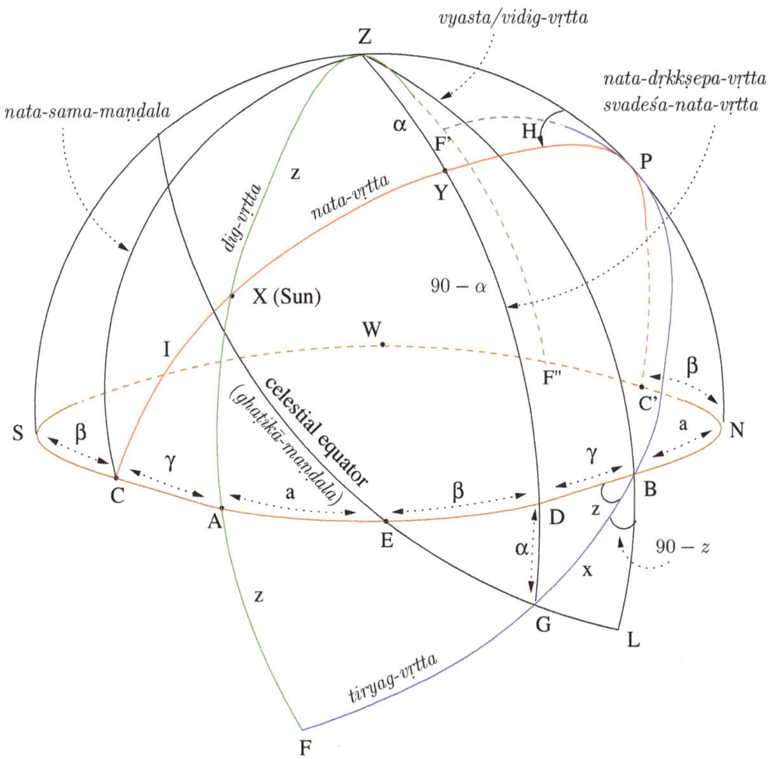

Fig. D.4 Some important great circles and their secondaries.

In Fig. D.4, the *nata-vṛtta* is the great circle passing through P and X (the Sun) which intersects the horizon at C. Now, draw the *nata-samamaṇḍala*, which is a vertical through Z and C. D is a point on the horizon at 90° from C. The *nata-dṛkkṣepa-vṛtta* or *svadeśa-nata* is the vertical through D and the *iṣṭa-digvṛtta* is the vertical through X intersecting the horizon at A. B is a point 90° from A and the vertical through B is the '*vyasta*' or *viparīta* or *vidig-vṛtta*. The point of intersection of the equator and the *nata-dṛkkṣepa-vṛtta* is denoted by G.

Consider the great circle (*tiryag-vṛtta*) through B and G. We show that BG is perpendicular to both the *natavṛtta* and the *digvṛtta*. The *tiryagvṛtta* and the *iṣṭa-digvṛtta* intersect at F. Y is the point of intersection of the *nata-dṛkkṣepa-vṛtta* and the *nata-vṛtta*. Let $ZY = \alpha$. $R\sin ZY = R\sin\alpha$ is the *svadeśa-nata-jyā*. $YD = 90-\alpha$, $R\sin(YD) = R\cos\alpha$ is the *svadeśa-nata-koṭi*.

Since B is at 90° from Z and A, it is the pole of the *iṣṭa-digvṛtta*. Therefore $BF = BX = 90°$. Similarly, C is the pole of the *nata-dṛkkṣepa-vṛtta*, since $CD = CZ = 90°$. Therefore G is at 90° from C. G, being on the celestial equator, is at 90° from P. Therefore G is the pole of the *nata-vṛtta*. Hence BG passes through the poles of *nata-vṛtta* and *digvṛtta*. Thus, BG is the perpendicular to both the *nata-vṛtta* and the *iṣṭa-digvṛtta*.

Now X is the pole of the *tiryagvṛtta*, as it is at 90° from B and G.[1] Therefore $XF = 90°$. But $XA = 90-z$. Hence, $AF = z$, where z is the maximum separation between the horizon and the *tiryagvṛtta* (as $BA = BF = 90°$). Therefore, $z = D\hat{B}G$. The *tiryagvṛtta* meets the *iṣṭa-digvṛtta* also at F'. Then,

$$\begin{aligned} 180° = FF' &= ZF' + ZF \\ &= ZF' + ZA + AF \\ &= ZF' + 90 + z. \end{aligned}$$

Therefore, $ZF' = 90-z$ or $F'F'' = z$. This is the elevation of the *tiryagvṛtta* from the horizon on the *iṣṭa-digvṛtta*. As this maximum separation occurs at 90°, $BF' = 90°$. It is clear from the figure that the angle between the *tiryagvṛtta* and the *vidig-vṛtta* is $90-z$.

Now C is the pole of ZD. Therefore $CY = 90°$, and the angle at Y is 90°. Since the angle between ZP and YP is H and $ZP = 90-\phi$, the sine of the zenith distance of the point Y, denoted by α, is

$$\begin{aligned} \sin\alpha &= \sin(90-\phi)\sin H \\ &= \cos\phi\sin H. \end{aligned} \tag{D.27}$$

Therefore

$$\cos\alpha = \sqrt{1-\cos^2\phi\sin^2 H}. \tag{D.28}$$

Let $CS = \beta$ be the distance between the north–south circle and the *nata-vṛtta* at the horizon. It is easy to see that $NC' = ED = \beta$, where C' is the point on the horizon diametrically opposite to C.

Note:

1. C being the pole of ZDG, $DY = 90-\alpha$ is the angle between the *nata-vṛtta* and the horizon. Therefore

$$\sin\phi = \sin PN = \sin(90-\alpha)\sin(PC).$$

---

[1] The point X is at 90° from G, since G is the pole of the *nata-vṛtta*.

Hence

$$\sin PC = \frac{\sin\phi}{\cos\alpha}. \tag{D.29}$$

2. Now H is the angle between the north-south circle and the *nata-vṛtta*. Therefore,

$$\sin\beta = \sin(SC) = \sin H \sin PC. \tag{D.30}$$

Using (D.29) in the above equation, we get

$$\begin{aligned}\sin\beta &= \frac{\sin H \sin\phi}{\cos\alpha} \\ &= \frac{\sin\phi \sin H}{\sqrt{1-\cos^2\phi\sin^2 H}},\end{aligned} \tag{D.31}$$

using (D.28). This result will be used later.

Again, in Fig. D.4, $AE = a$ is *iṣṭāgrā*. The angle between the *nata-sama-vṛtta* and the *digvṛtta* on the horizon is given by $CA = \gamma$. It may be noted that this is also equal to the angle between the *nata-dṛkkṣepa-vṛtta* and the *vyasta-dṛkkṣepa-vṛtta*. Since B is the pole of the *digvṛtta*, clearly $\gamma = 90 - \beta - a$. Therefore

$$\begin{aligned}\sin\gamma &= \sin(90-\beta-a) \\ &= \cos(\beta+a) \\ &= (\cos\beta\cos a - \sin\beta\sin a).\end{aligned} \tag{D.32}$$

When *āśāgrā* a is to the north of east, $\gamma = 90-\beta+a$ and $\sin\gamma = \cos\beta\cos a + \sin\beta\sin a$. Thus $\sin\gamma$ is determined in terms of known quantities, since $\sin a$ is given and $\sin\beta$ is known from (D.31).

Now, let $GB = x$ and GL be the perpendicular arc from G to *vidig-vṛtta*. Then $\sin DG$, which is the same as $\sin ZY$, is given by

$$\sin\alpha = \sin z \sin x. \tag{D.33}$$

Also

$$\sin GL = \sin x \cos z, \tag{D.34}$$

as z and $90 - z$ are the angles between the *tiryagvṛtta* and the horizon, and the *tiryagvṛtta* and the *vidig-vṛtta*, respectively. But the angle between ZG and ZL is γ and $ZG = 90° + \alpha$. (For $GY = 90°$, G being the pole of *nata-vṛtta*.) Therefore

$$\begin{aligned}\sin GL &= \sin(90+\alpha)\sin\gamma \\ &= \sin\gamma\cos\alpha.\end{aligned} \tag{D.35}$$

Equating the two expressions for $\sin GL$, we get

$$\sin x \cos z = \sin\gamma\cos\alpha. \tag{D.36}$$

We had

$$\sin x \sin z = \sin\alpha. \tag{D.37}$$

From (D.36) and (D.37), we get

$$\sin x = \sqrt{\sin^2\alpha + \sin^2\gamma\cos^2\alpha}. \tag{D.38}$$

Using the above in (D.36) and (D.37), we have

$$\cos z = \frac{\sin\gamma\cos\alpha}{\sqrt{\sin^2\alpha + \sin^2\gamma\cos^2\alpha}}, \tag{D.39}$$

$$\text{and} \qquad \sin z = \frac{\sin\alpha}{\sin x}. \tag{D.40}$$

Now

$$\sin\beta = \frac{\sin\phi\sin H}{\cos\alpha}. \tag{D.41}$$

Therefore

$$\begin{aligned}
\cos\beta &= \sqrt{1-\sin^2\beta} \\
&= \sqrt{1-\frac{\sin^2\phi\sin^2 H}{\cos^2\alpha}} \\
&= \frac{\sqrt{\cos^2\alpha - \sin^2\phi\sin^2 H}}{\cos\alpha} \\
&= \frac{\sqrt{1-\cos^2\phi\sin^2 H - \sin^2\phi\sin^2 H}}{\cos\alpha} \\
&= \frac{\cos H}{\cos\alpha},
\end{aligned} \tag{D.42}$$

where we have used (D.28).

Hence, from (D.32), (D.41) and (D.42), we have

$$\begin{aligned}
\sin\gamma\cos\alpha &= (\cos\beta\cos a - \sin\beta\sin a)\cos\alpha \\
&= \cos H\cos a - \sin\phi\sin H\sin a.
\end{aligned} \tag{D.43}$$

We have already shown that

$$\sin\alpha = \cos\phi\sin H. \tag{D.44}$$

Substituting these in (D.39), we obtain the following expression for *śaṅku* in terms of *natajyā*, *āśāgrā* and *akṣa*:

$$R\cos z = \frac{(R\cos H\cos a - R\sin\phi\sin H\sin a)R}{\sqrt{R^2\cos^2\phi\sin^2 H + (R\cos H\cos a - R\sin\phi\sin H\sin a)^2}}. \qquad \text{(D.45)}$$

We have actually considered the *āśāgrā* 'a' to be south in Fig. D.4. When the *āśāgrā* 'a' is north, the '$-$' sign in (D.45) has to be replaced by '$+$' sign. Similarly, substituting in (D.40) we have

$$R\sin z = \frac{(R\cos\phi\sin H)R}{\sqrt{R^2\cos^2\phi\sin^2 H + (R\cos H\cos a - R\sin\phi\sin H\sin a)^2}}. \qquad \text{(D.46)}$$

These are the gnomon and the shadow respectively.

Now X is at the intersection of the *nata-vṛtta* and the *digvṛtta*, which makes angles H and $90-a$, respectively, with the north-south circle. $PX = 90-\delta$ and $ZX = z$. Equating the two expressions for the distance between X and the north–south circle, we get

$$R\cos\delta\sin H = R\sin z\cos a. \qquad \text{(D.47)}$$

Hence

$$R\cos\delta = \frac{R\sin z\, R\cos a}{R\sin H}, \qquad \text{(D.48)}$$

or

$$Dyujy\bar{a} = \frac{ch\bar{a}y\bar{a} \times \bar{a}\acute{s}\bar{a}gr\bar{a}\text{-}ko\d{t}i}{natajy\bar{a}},$$

from which the *apakrama* can be obtained as

$$R\sin\delta = \sqrt{R^2 - R^2\cos^2\delta}. \qquad \text{(D.49)}$$

Appendix E
Derivation of the maximum declination of the Moon

Here we outline the derivation of the maximum declination of the Moon as given in Chapter 13 on *vyatīpāta* in *Yuktibhāṣā.*

E.1 Occurrence of *Vyatīpāta*

Vyatīpāta is said to occur when the (magnitudes of the) declinations of the Sun and the Moon are equal, and when one of them is increasing and the other decreasing. This can happen when one of these bodies is in an odd quadrant and the other is in an even quadrant.

E.2 Derivation of declination of the Moon

A method of computing the declination of the Moon (which has a latitude) has already been described. Here, a new method to compute it is described in Section 6.3. The declination of the Sun is determined with the knowledge of the intersection point (Γ in Fig. E.1) and the maximum divergence $R\sin\varepsilon$ of the ecliptic and the celestial equator. Similarly, the declination of the Moon can be determined if we know (i) the point where the celestial equator and the *vikṣepa-vṛtta* (the lunar orbit) intersect, (ii) the maximum divergence between them, and (iii) the position of the Moon on the *vikṣepa-vṛtta.*

E.3 *Vikṣepa*

The *vikṣepa-vṛtta* will intersect the ecliptic at *Rāhu* (the ascending node of the Moon) and *Ketu* (the descending node) and diverge northwards and southwards respectively, from those points. A method to determine the intersection point of the celestial equator and the *vikṣepa-vṛtta*, and their maximum divergence, is described first in qualitative terms. For this, four distinct cases are discussed.

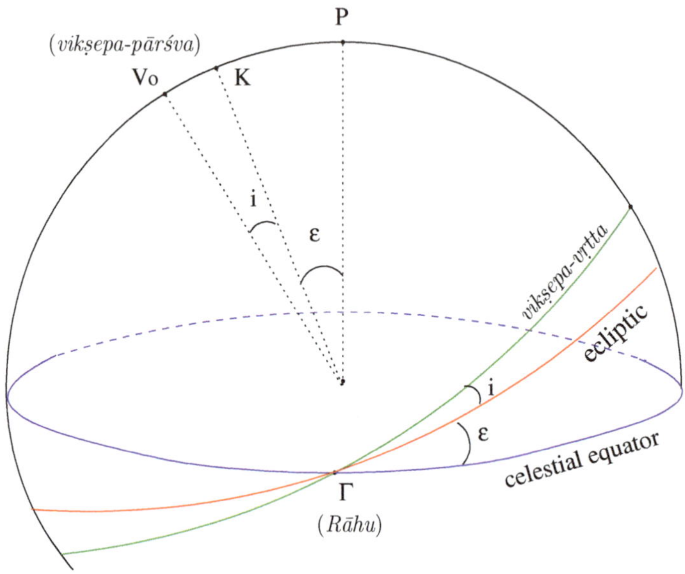

Fig. E.1 Moon's orbit when the node *Rāhu* coincides with the vernal equinox Γ.

Case 1: *Rāhu* at the vernal equinox:

Here the maximum declination (ε) on the ecliptic and maximum *vikṣepa* (i) on the *vikṣepa-vṛtta* are both on the north-south circle as shown in Fig. E.1. The maximum possible declination of the Moon on that day will be equal to the sum of these two ($\varepsilon + i$). Then, the declination of the Moon can be determined with the knowledge of its position on the *vikṣepa-vṛtta*, as the inclination of *vikṣepa-vṛtta* with the equator is ($\varepsilon + i$). The *vikṣepa-pārśva*[1] is the northern pole (V_0) of the *vikṣepa-vṛtta*. When *Rāhu* is at the vernal equinox, the distance between this and the north celestial pole is equal to ($\varepsilon + i$).

The *vikṣepa-pārśva* is the (north) pole of the *vikṣepa-vṛtta*, just as the north celestial pole is the pole of the celestial equator or the *rāśi-kūṭa* is the pole of the ecliptic. Whatever the position of *Rāhu*, the distance between the celestial pole and the *vikṣepa-pārśva* is equal to the maximum divergence between the equator and the *vikṣepa-vṛtta*.

Case 2: *Rāhu* at the winter (southern) solstice:

In this case, the *vikṣepa-vṛtta* would be deflected towards the north from the vernal equinox by the measure of maximum *vikṣepa* as shown in Fig. E.2. The

[1] Though generally the term *pārśva* refers to a side, in the present context it is used to refer to the pole.

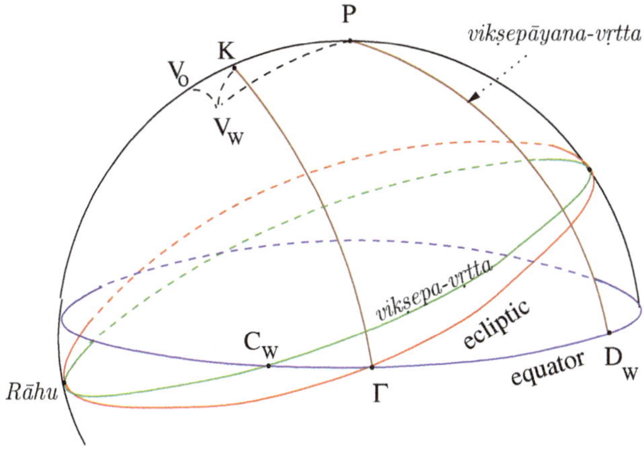

Fig. E.2 Moon's orbit when the node *Rāhu* coincides with the winter solstice.

vikṣepa-pārśva[2] would be deflected towards the west from V_0 and would be at V_W, with the arc length $KV_W = i$. The distance between the (celestial) pole P and V_W is the *vikṣepāyanānta* (I). The great circle passing through P and V_W is called the *vikṣepāyana-vṛtta*. Its intersection point (D_W) with the celestial equator would be deflected west from the north–south circle by the angle $K\hat{P}V_W$. The point of intersection of the *vikṣepa-vṛtta* and the *vikṣepāyana-vṛtta* corresponds to the maximum declination of the Moon in this set-up.

The *vikṣepa-viṣuvat* is the point of intersection of the *vikṣepa-vṛtta* and the celestial equator and is denoted by C_W. C_W is at 90° from D_W. $C_W\Gamma = K\hat{P}V_W$ is called *vikṣepa-calana*. C_W is situated west of the vernal equinox when *Rāhu* is at the winter solstice.

Case 3: *Rāhu* at the autumnal equinox:[3]

As depicted in Fig. E.3, the *vikṣepa-vṛtta* would intersect the north-south circle at a point north of the winter solstice by i, which is taken to be $4\frac{1}{2}^\circ$. The *vikṣepa-pārśva*, now at V', would also be deflected towards north from K, and the distance between V' and P would be $\varepsilon - i = 19\frac{1}{2}^\circ$. It is easy to see that the *vikṣepa-viṣuvat* would coincide now with the equinox and there will be no *vikṣepa-calana*.

Case 4: *Rāhu* at the summer (northern) solstice:

This situation is depicted in Fig. E.4. Here, the *vikṣepa-pārśva* V_E is deflected towards the east from V_0, with $KV_E = i$. The *vikṣepāyana-vṛtta* touches the equator at D_E, which is deflected east from the north–south circle. The *vikṣepa-viṣuvat* is at C_E and is east of the vernal equinox Γ.

Thus the location of the *vikṣepa-pārśva*, V, depends upon the position of *Rāhu*. However, it is always at a distance of maximum *vikṣepa* from the northern *rāśi-kūṭa*

---

[2] It may be noted that this point V_W lies on the other side of the celestial sphere.

[3] The autumnal equinox was approximately at the middle of the *Kanyā-rāśi* at the time of composition of *Yuktibhāṣā* (c. 1530 CE).

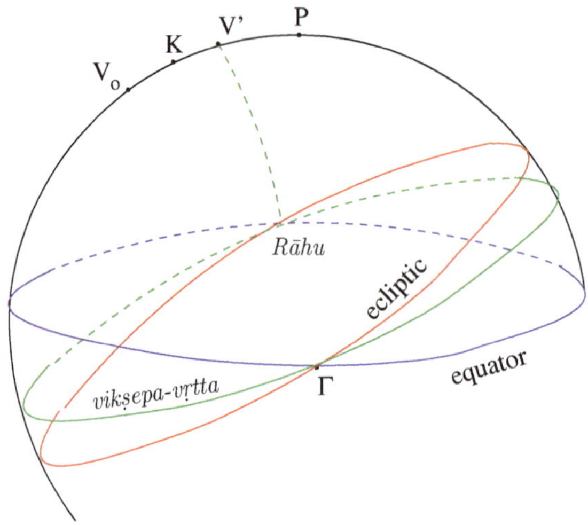

Fig. E.3 Moon's orbit when the node *Rāhu* coincides with the autumnal equinox.

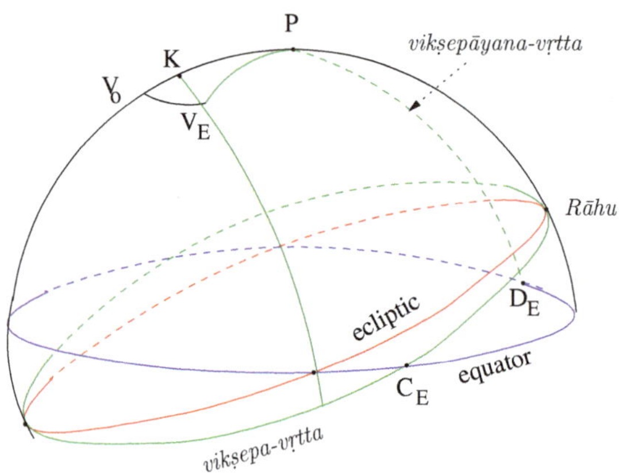

Fig. E.4 Moon's orbit when the node *Rāhu* coincides with the summer solstice.

($KV = i$). The location of the southern *vikṣepa-pārśva* with respect to the southern *rāśi-kūṭa* can be discussed along similar lines.

E.4 *Vikṣepa-calana*

Here the method to determine the distance between the (north) celestial pole and the *vikṣepa-pārśva* is described in broad terms first. Consider Fig. E.5. The *vikṣepa-pārśva* is at V_0 separated from K by the maximum *vikṣepa* i. Drop a perpendicular V_0T from V_0 to OK, where O is the centre of the sphere. As the arc $V_0K = i$,

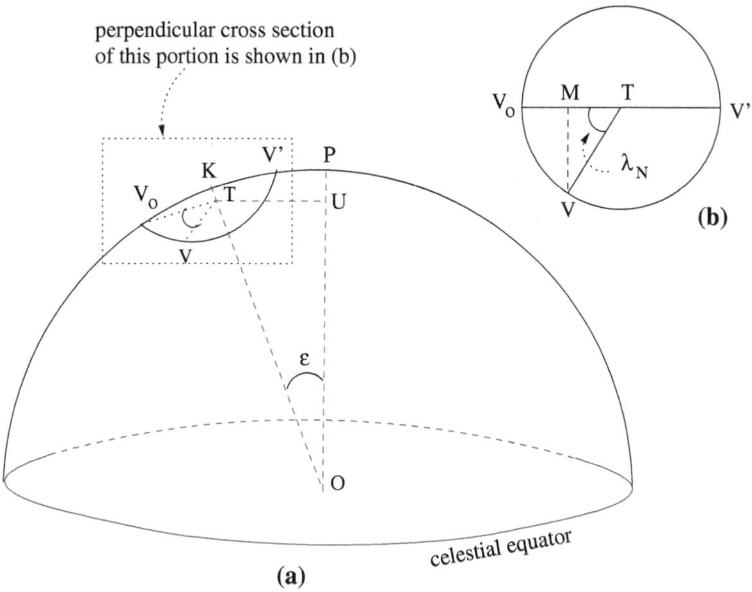

Fig. E.5 The distance between the *vikṣepa-pārśva* and the north celestial pole.

$V_0T = R\sin i$. Draw a circle with radius $R\sin i$ centred at T in the plane perpendicular to OT. This is the *vikṣepa-pārśva-vṛtta*. It may be noted that this circle (shown separately in Fig. E.5(b)), will be parallel to the plane of the ecliptic.[4] Mark a point V on this circle such that the angle corresponding to the arc V_0V is the longitude of *Rāhu*, λ_N. Drop a perpendicular TU from T to the *akṣa-daṇḍa* OP. Draw a circle with U as the centre and TU as the radius in the plane perpendicular to OP. The relationship between this circle and the *vikṣepa-pārśva-vṛtta* is the same as that of the *kakṣyāvṛtta* and the *ucca-nīcavṛtta*. Now

$$OT = R\cos i,$$
$$\text{and } TU = OT\sin\varepsilon = R\cos i \sin\varepsilon,$$

is the radius of the *kakṣyāvṛtta*. Draw VM perpendicular to V_0T. Then $VM = R\sin i \sin\lambda_N$ and $MT = R\sin i\cos\lambda_N$ play the role of the *bhujā-phala* and the *koṭiphala* respectively in the determination of VU, which is the *karṇa*. It must be noted that VM is along the east-west direction and perpendicular to the plane of the figure. It is the distance between V and the north–south circle. When the *Rāhu* is between *Makarādi* and *Karkyādi* (or equivalently λ_N is between 270° and 90°), the *koṭiphala* has to be added to the representative of the *trijyā*, which is TU. Similarly, when it is between *Karkyādi* and *Makarādi* (λ_N is between 90° and 270°), the *koṭiphala* is to be subtracted. (Actually the *koṭiphala* has to be projected along TU before this is done; this becomes clear in the next section.) When *Rāhu* is at

[4] In the figure VM is along the east-west line and is perpendicular to the plane of the figure.

the vernal equinox, *vikṣepa-pārśva* is at V_0 and VP would be maximum. Similarly, when *Rāhu* is at the autumnal equinox, *vikṣepa-pārśva* is at V' and VP is minimum.

The *vikṣepa-pārśva* is in the eastern part of the sphere (or to the east of the north–south circle) when *Rāhu* moves from the vernal equinox to the autumnal equinox (or λ_N is between 0° and 180°). Then the *vikṣepa-viṣuvat* is situated east of the equinox, and the *vikṣepa-calana* is to be subtracted (from the longitude of the Moon) while calculating the Moon's declination. Similarly, the *vikṣepa-viṣuvat* is situated west of the equinox, when *Rāhu* moves from the autumnal equinox to the vernal equinox (or λ_N is between 180° and 360°), and the *vikṣepa-calana* is to be added (to the longitude of the Moon) while calculating the Moon's declination.

E.5 *Karṇānayana*

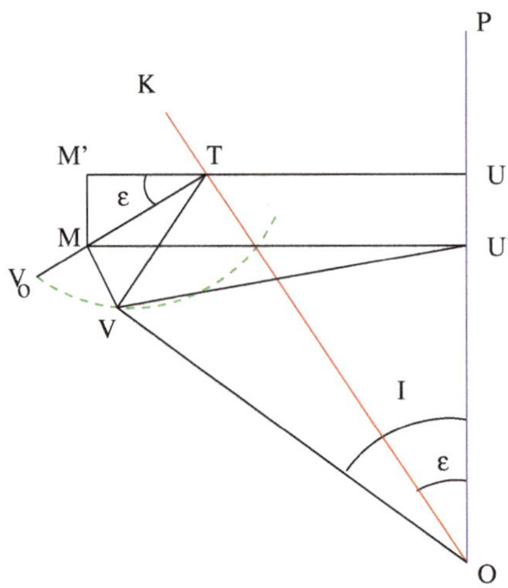

Fig. E.6 The inclination of the Moon's orbit with the equator.

In Fig. E.6, the points V_0, V, T (the centre of the *vikṣepa-pārśva-vṛtta*), M and U have the same significance as in Fig. E.5. MV is perpendicular to the plane of the figure. Draw MU' from M, perpendicular to the *akṣa-daṇḍa*, OP. VM is perpendicular to the plane of the figure and hence to OP, and MU' is also perpendicular to OP. Hence $VU'M$ is a triangle, right-angled at M, and in a plane perpendicular to OP. Therefore, VU' is perpendicular to OP and is the desired distance, $R\sin I$, between V and the *akṣa-daṇḍa*. Let MM' be perpendicular to UM', which is the extension of UT. The angle between TM' and TM is ε. It is clear that $MU' = M'U$. Therefore

$$\begin{aligned} M'U &= M'T + TU \\ &= MT\cos\varepsilon + R\cos i \sin\varepsilon \end{aligned}$$

$$= R\sin i \cos\lambda_N \cos\varepsilon + R\cos i \sin\varepsilon, \tag{E.1}$$

where MT is the *koṭiphala* discussed in the previous section. It may be seen that $MV = R\sin i \sin\lambda_N$, is the *bhujā-phala*. Then

$$\begin{aligned} VU' &= \sqrt{(MV)^2 + (MU')^2} \\ &= \sqrt{(R\sin i \sin\lambda_N)^2 + (R\sin i \cos\lambda_N \cos\varepsilon + R\cos i \sin\varepsilon)^2}. \end{aligned} \tag{E.2}$$

Clearly $VU' = R\sin I$, where I is the angle corresponding to the arc VP. Hence,

$$R\sin I = \sqrt{(R\sin i \sin\lambda_N)^2 + (R\sin i \cos\lambda_N \cos\varepsilon + R\cos i \sin\varepsilon)^2}. \tag{E.3}$$

This is the maximum declination, or the maximum divergence between the equator and the *vikṣepavṛtta* (the Moon's orbit).

Appendix F
The traditional Indian planetary model and its revision by Nīlakaṇṭha Somayājī [1]

It is now generally recognized that the Kerala school of Indian astronomers,[2] starting from Mādhava of Saṅgamagrāma (1340–1420 CE), made important contributions to mathematical analysis much before this subject developed in Europe. The Kerala astronomers derived infinite series for π, sine and cosine functions and also developed fast convergent approximations to them.[3]

Here we shall explain how the Kerala school also made equally significant discoveries in astronomy, and particularly in planetary theory. Mādhava's disciple Parameśvara of Vaṭaśśeri (c. 1380–1460) is reputed to have made continuous and careful observations for over 55 years. He is famous as the originator of the *Dṛg-gaṇita* system, which replaced the older *Parahita* system. He also discussed the geometrical picture of planetary motion as would follow from the traditional Indian planetary model.

---

[1] This appendix, prepared by K. Ramasubramanian, M. D. Srinivas and M. S. Sriram, is a revised and updated version of the following earlier studies on the subject: (i) K. Ramasubramanian, M. D. Srinivas and M. S. Sriram, Modification of the Earlier Indian Planetary Theory by the Kerala Astronomers (c. 1500) and the implied Heliocentric Picture of Planetary Motion, *Current Science* 66, 784–790, 1994. (ii) M. S. Sriram, K. Ramasubramanian and M. D. Srinivas (eds), *500 Years of Tantrasangraha: A Landmark in the History of Astronomy*, IIAS, Shimla 2002, pp. 29–102. (iii) Epilogue: Revision of Indian Planetary Model by Nīlakaṇṭha Somayājī, in *Gaṇita-yukti-bhāṣā* of Jyeṣṭhadeva, ed. and tr. K. V. Sarma with Explanatory Notes by K. Ramasubramanian, M. D. Srinivas and M. S. Sriram, 2 vols, Hindustan Book Agency, Delhi 2008; repr. Springer, 2009, vol II, pp. 837–856.

[2] For the Kerala school of astronomy, see for instance, K. V. Sarma, *A Bibliography of Kerala and Kerala-based Astronomy and Astrology*, Hoshiarpur 1972; K. V. Sarma, *A History of the Kerala School of Hindu Astronomy*, Hoshiarpur 1972.

[3] For overviews of the Kerala tradition of mathematics, see S. Parameswaran, *The Golden Age of Indian Mathematics*, Kochi 1998; G. G. Joseph, *The Crest of the Peacock: Non-European Roots of Mathematics*, 2nd edn. Princeton 2000; C. K. Raju, *Cultural Foundations of Mathematics: The Nature of Mathematical Proof and the Transmission of the Calculus from India to Europe in the 16th c. CE*, Pearson Education, Delhi 2007; Kim Plofker, *History of Mathematics in India: From 500 BCE to 1800 CE*, Princeton 2009; G. G. Joseph (ed.), *Kerala Mathematics: History and Possible Transmission to Europe*, B. R. Publishing, New Delhi 2009. See also the detailed mathematical notes in *Gaṇita-yukti-bhāṣā* cited above.

Nīlakaṇṭha Somayājī of Tṛkkaṇṭiyūr (c. 1444–1550), a disciple of Parameśvara's son Dāmodara, carried out a fundamental revision of the traditional planetary theory. In his treatise *Tantrasaṅgraha*, composed in 1500, Nīlakaṇṭha outlines the detailed computational scheme of his revised planetary model. For the first time in the history of Astronomy, Nīlakaṇṭha proposed that in the case of an interior planet (Mercury or Venus), the *manda*-correction or the equation of centre should be applied to what was traditionally identified as the *śighrocca* of the planet—which, in the case of interior planets, corresponds to what we currently refer to as the mean heliocentric planet. This was a radical departure from the traditional Indian planetary model where the *manda*-correction for an interior planet was applied to the mean Sun.[4]

In this way, Nīlakaṇṭha arrived at a much better formulation of the equation of centre and the latitudinal motion of the interior planets than was available either in the earlier Indian works or in the Islamic or the Greco-European traditions of astronomy till the work of Kepler, which was to come more than a hundred years later. In fact, in so far as the computation of the planetary longitudes and latitudes is concerned, Nīlakaṇṭha's revised planetary model closely approximates to the Keplerian model, except that Nīlakaṇṭha conceives of the planets as going in eccentric orbits around the mean Sun rather than the true Sun.

In his *Āryabhaṭīya-bhāṣya*, Nīlakaṇṭha explains the rationale behind his revision of the traditional planetary theory. This has to do with the fact (which was noticed by several Indian astronomers prior to Nīlakaṇṭha) that the traditional Indian planetary model employed entirely different schemes for computing the latitudes of the exterior and the interior planets. While the latitudes of the exterior planets were computed from their so-called *manda-sphuṭa* (which corresponds to what we currently refer to as the true heliocentric planet), the latitudes of the interior planets were computed from their so-called *śīghrocca*. Nīlakaṇṭha argued that since the latitude should be dependent on the deflection (from the ecliptic) of the planet itself and not of any other body, what was traditionally referred to as the *śīghrocca* of an interior planet should be identified with the planet itself. Nīlakaṇṭha also showed that this would lead to a unified treatment of the latitudinal motion of all the planets—interior as well as exterior.[5]

In *Āryabhaṭīya-bhāṣya*, Nīlakaṇṭha also discusses the geometrical picture of planetary motion implied by his revised model.[6] This geometrical picture, which is also stated by Nīlakaṇṭha succinctly in terms of a few verses in *Golasāra* and *Siddhānta-darpaṇa*, is essentially that the planets move in eccentric orbits (which

[4] It had also been a general feature of all ancient planetary theories in the Greco-European and the Islamic traditions of astronomy, till the work of Kepler, that the equation of centre for an interior planet was wrongly applied to the mean Sun.

[5] In fact, it has been noted in a later text, *Vikṣepagolavāsanā*, that Nīlakaṇṭha pioneered a revision of the traditional planetary theory in order to arrive at a unified formulation of the motion in latitude of both the interior and the exterior planets.

[6] The renowned Malayalam work *Gaṇita-yukti-bhāṣā* (c. 1530) of Jyeṣṭhadeva also gives a detailed exposition of the geometrical picture of planetary motion as per the planetary model of Nīlakaṇṭha outlined in *Tantrasaṅgraha*.

are inclined to the ecliptic) around the *śīghrocca*, which in turn goes around the Earth.

While discussing the geometrical picture of planetary motion, *Āryabhaṭīya-bhāṣya*, as well as *Golasāra* and *Siddhānta-darpaṇa*, consider the orbit of each of the planets individually and they are not put together in a single cosmological model of the planetary system. There is however an interesting passage in *Āryabhaṭīya-bhāṣya*, where Nīlakaṇṭha explains that the Earth is not circumscribed by the orbit of the interior planets, Mercury and Venus; and that the mean period of motion in longitude of these planets around the Earth is the same as that of the Sun, precisely because they are being carried around the Earth by the Sun. In fact, Nīlakaṇṭha seems to be the first savant in the history of astronomy to clearly deduce from his computational scheme—and not from any speculative or cosmological argument—that the interior planets go around the Sun and that the period of their motion around the Sun is also the period of their latitudinal motion.

In a remarkable short tract called *Grahasphuṭānayane vikṣepavāsanā*, which seems to have been written after *Āryabhaṭīya-bhāṣya* as it cites extensively from it, Nīlakaṇṭha succinctly describes his cosmological model, which is that the five planets, Mercury, Venus, Mars, Jupiter and Saturn, go around the mean Sun in eccentric orbits (inclined to the ecliptic), while the mean Sun itself goes around the Earth.[7] Following this, Nīlakaṇṭha also states that the dimensions of *śīghra* epicycles are specified by measuring the orbit of the mean Sun around the Earth in terms of the planetary orbit in the case of the exterior planets, and they are specified by measuring the planetary orbit (which is smaller) in terms of the orbit of the mean Sun in the case of the interior planets. This remarkable relation[8] follows clearly from the identification of the *śighrocca* of all the planets with physical mean Sun, a fact also stated by Nīlakaṇṭha in his *Āryabhaṭīya-bhāṣya.*

Towards the very end of the last chapter of *Tantrasaṅgraha*, Nīlakaṇṭha briefly considers the issue of planetary distances. Unlike the longitudes and latitudes of planets, the planetary distances were not amenable to observations in ancient astronomy and their discussion was invariably based upon some speculative hypothesis. In traditional Indian planetary theory, at least from the time of Āryabhaṭa, the mean planetary distances were obtained based on the hypothesis that all the planets go around the Earth with the same linear velocity—i.e. they all cover the same physical distance in any given period of time. In *Tantrasaṅgraha*, Nīlakaṇṭha proposes an alternative prescription for planetary distances which seems to be based on the principle that all the planets go around the *śighrocca* with the same linear velocity. He also briefly hints at this alternative hypothesis in his *Āryabhaṭīya-bhāṣya.* However, among the available works of Nīlakaṇṭha, there is no discussion of plan-

---

7 This cosmological model is the same as the one proposed by Tycho Brahe, albeit on entirely different considerations, towards the end of sixteenth century.

8 The *śīghra* epicycle is essentially the same as the epicycle associated with the so-called 'solar anomaly' in the Greco-European tradition of astronomy, and the above relation is the same as the one proposed by Nicholas Copernicus (perhaps around the same time as Nīlakaṇṭha) by identifying this epicycle as the orbit of the Earth around the Sun in the case of the exterior planets and as the orbit of the planet itself in the case of the interior planets.

etary distances as would follow from his revised cosmological model outlined in *Grahasphuṭānayane vikṣepavāsanā*.

Before taking up the various aspects of the revised planetary model of Nīlakaṇṭha it is essential to understand the traditional Indian planetary model, which had been in vogue at least from the time of Āryabhaṭa (c. 499). We shall therefore devote the initial sections of this appendix to a detailed exposition of the traditional Indian planetary theory and important developments in it prior to the work of Nīlakaṇṭha.

F.1 The traditional Indian planetary model: *Manda-saṃskāra*

In the Indian astronomical tradition, at least from the time of Āryabhaṭa (499 CE), the procedure for calculating the geocentric longitudes of the planets consists essentially of two steps:[9] first, the computation of the mean longitude of the planet known as the *madhyama-graha*, and second, the computation of the true or observed longitude of the planet known as the *sphuṭa-graha*.

The mean longitude is calculated for the desired day by computing the number of mean civil days elapsed since the epoch (this number is called the *ahargaṇa*) and multiplying it by the mean daily motion of the planet. Having obtained the mean longitude, a correction known as *manda-saṃskāra* is applied to it. In essence, this correction takes care of the eccentricity of the planetary orbit around the Sun. The equivalent of this correction is termed the 'equation of centre' in modern astronomy, and is a consequence of the elliptical nature of the orbit. The longitude of the planet obtained by applying the *manda*-correction is known as the *manda-sphuṭa-graha* or simply the *manda-sphuṭa*.

While *manda-saṃskāra* is the only correction that needs to be applied in case of the Sun and the Moon for obtaining their true longitudes (*sphuṭa-grahas*), in the case of the other five planets, two corrections, namely the *manda-saṃskāra* and *śīghra-saṃskāra*, are to be applied to the mean longitude in order to obtain their true longitudes. Here again, we divide the five planets into two groups: the interior, namely Mercury and Venus, and the exterior, namely Mars, Jupiter and Saturn—not necessarily for the purpose of convenience in discussion but also because they are treated differently while applying these corrections.

The *śīghra-saṃskāra* is applied to the *manda-sphuṭa-graha* to obtain the true geocentric longitude known as the *sphuṭa-graha*. As will be seen later, the *śīghra* correction essentially converts the heliocentric longitude into the geocentric longitude. We will now briefly discuss the details of the *manda-saṃskāra*, which will

[9] For a general review of Indian astronomy, see D. A. Somayaji, *A Critical Study of Ancient Hindu Astronomy*, Dharwar 1972; S. N. Sen and K. S. Shukla (eds), *A History of Indian Astronomy*, New Delhi 1985 (rev. edn 2000); B. V. Subbarayappa and K. V. Sarma (eds.), *Indian Astronomy: A Source Book*, Bombay 1985; S. Balachandra Rao, *Indian Astronomy: An Introduction*, Hyderabad 2000; B. V. Subbarayappa, *The Tradition of Astronomy in India: Jyotiḥśāstra*, PHISPC vol. IV, Part 4, Centre for Studies in Civilizations, New Delhi 2008.

be followed by a discussion on the *śīghra-saṃskāra* for the exterior and the interior planets respectively.

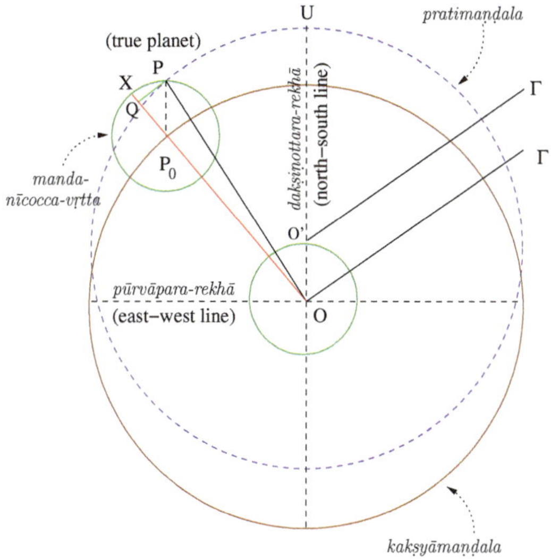

Fig. F.1 The epicyclic and eccentric models of planetary motion.

F.1.1 Epicyclic and eccentric models

As mentioned earlier, the *manda-saṃskāra* essentially accounts for the eccentricity of the planetary orbit. This may be explained with the help of Fig. F.1. Here, O is the centre of the *kakṣyāmaṇḍala*[10] on which the mean planet P_0 is assumed to be moving with mean uniform velocity. $O\Gamma$ is the reference line usually chosen to be the direction of *Meṣādi*. The *kakṣyā-maṇḍala* is taken to be of radius R, known as the *trijyā*.[11] The longitude of the mean planet P_0 moving on this circle is given by

$$\Gamma\hat{O}P_0 = madhyama\text{-}graha = \theta_0. \tag{F.1}$$

The longitude of the *manda-sphuṭa-graha* P given by $\Gamma\hat{O}P$ is to be obtained from θ_0, and this can be obtained by either by an eccentric or epicyclic model.

[10] The centre of the *kakṣyāmaṇḍala* is generally referred to as the *bhagola-madhya* (centre of the celestial sphere), and it coincides with the centre of the Earth in the case of the Sun and the Moon, when the 'second correction' which corresponds to the 'evection term' is ignored.

[11] The value of the *trijyā* is chosen such that one minute of arc in the circle corresponds to unit length. This implies that $2\pi R = 21600$ or $R \approx 3437.74$, which is taken to be 3438 in most of the Indian texts.

The procedure for obtaining the longitude of the *manda-sphuṭa-graha* by either of the two models involves the longitude of the *mandocca*. In Fig. F.1, OU represents the direction of the *mandocca* whose longitude is given by

$$\Gamma\hat{O}U = mandocca = \theta_m. \tag{F.2}$$

The modern equivalent of *mandocca* is *apoapsis*—apogee in the case of the Sun and the Moon and aphelion in the case of the five planets.

Around the mean planet P_0, a circle of radius r is to be drawn. This circle is known as the *manda-nīcocca-vṛtta*[12] or simply as *manda-vṛtta* (epicycle). The texts specify the value of the radius of this circle r ($r \ll R$), in appropriate measure, for each planet.

At any given instant of time, the *manda-sphuṭa-graha* P is to be located on this *manda-nīcocca-vṛtta* by drawing a line from P_0 along the direction of *mandocca* (parallel to OU). The point of intersection of this line with the *manda-nīcocca-vṛtta* gives the location of the planet P. Since this method of locating the *manda-sphuṭa-graha* involves the construction of an epicycle around the mean planet, it is known as the epicyclic model.

Alternatively, one could draw the *manda-nīcocca-vṛtta* of radius r centred around O, which intersects OU at O'. With O' as centre, a circle of radius R (shown by dashed lines in the figure) is drawn. This is known as *pratimaṇḍala* or the eccentric circle. Since P_0P and OO' are equal to r, and they are parallel to each other, $O'P = OP_0 = R$. Hence, P lies on the eccentric circle. Also,

$$\Gamma\hat{O}'P = \Gamma\hat{O}P_0 = madhyama\text{-}graha = \theta_0. \tag{F.3}$$

Thus, the *manda-sphuṭa-graha* P can be located on an eccentric circle of radius R centred at O' (which is located at a distance r from O in the direction of *mandocca*), simply by marking a point P on it such that $\Gamma\hat{O}'P$ corresponds to the the mean longitude of the planet. Since this process involves only an eccentric circle, without making a reference to the epicycle, it is known as the eccentric model. Clearly, the two models are equivalent to each other.

F.1.2 Calculation of *manda-sphuṭa*

The formula presented by the Indian astronomical texts for the calculation of the *manda-sphuṭa*—the longitude of the planet obtained by applying the *manda-saṃskāra* (equation of centre) to the mean longitude of the planet—and the underlying geometrical picture can be understood with the help of Fig. F.2.[13] Here,

[12] The adjective *nīcocca* is given to this *vṛtta* because, in this conception, it moves from *ucca* to *nīca* on the deferent circle along with the mean planet P_0. The other adjective *manda* is to suggest that this circle plays a crucial role in the explanation of the *manda-saṃskāra*.

[13] It may be noted that Fig. F.2 is the same as Fig. F.1, with certain circles and markings removed from the latter and certain others introduced in the former for the purposes of clarity.

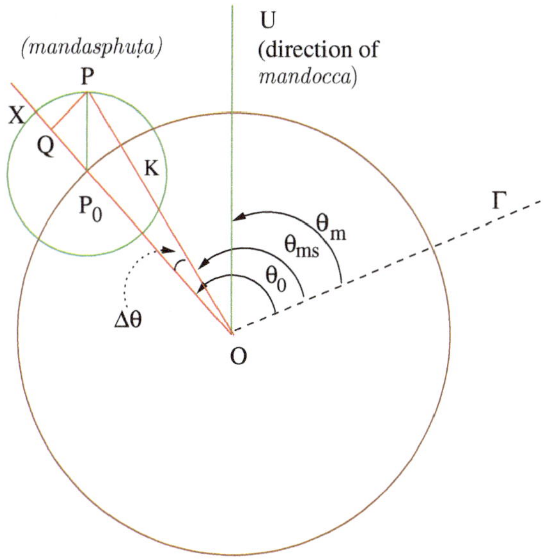

Fig. F.2 Geometrical construction underlying the rule for obtaining the *manda-sphuṭa* from the *madhyama* using the epicycle approach.

$\theta_{ms} = \Gamma \hat{O} P$ represents the *manda-sphuṭa* which is to be determined from the position of the mean planet (*madhyama-graha*) P_0. Clearly,

$$\begin{aligned} \theta_{ms} &= \Gamma \hat{O} P \\ &= \Gamma \hat{O} P_0 - P \hat{O} P_0 \\ &= \theta_0 - \Delta\theta. \end{aligned} \qquad \text{(F.4)}$$

Since the mean longitude of the planet θ_0 is known, the *manda-sphuṭa* θ_{ms} is obtained by simply subtracting $\Delta\theta$ from the *madhyama*. The expression for $\Delta\theta$ can be obtained by making the following geometrical construction. We extend the line OP_0, which is the line joining the centre of the *kakṣyāmaṇḍala* and the mean planet, to meet the epicycle at X. From P drop the perpendicular PQ onto OX. Then

$$\begin{aligned} U \hat{O} P_0 &= \Gamma \hat{O} P_0 - \Gamma \hat{O} U \\ &= \theta_0 - \theta_m \end{aligned} \qquad \text{(F.5)}$$

is the *manda-kendra* (*madhyama* – *mandocca*), whose magnitude determines the magnitude of $\Delta\theta$ (see (F.8)). Also, since P_0P is parallel to OU (by construction), $P\hat{P_0}Q = (\theta_0 - \theta_m)$. Hence, $PQ = r\sin(\theta_0 - \theta_m)$ and $P_0Q = r\cos(\theta_0 - \theta_m)$. Since the triangle OPQ is right-angled at Q, the hypotenuse $OP = K$ (known as the *manda-karṇa*) is given by

$$K = OP = \sqrt{OQ^2 + QP^2}$$

$$= \sqrt{(OP_0 + P_0Q)^2 + QP^2}$$
$$= \sqrt{\{R + r\cos(\theta_0 - \theta_m)\}^2 + r^2 \sin^2(\theta_0 - \theta_m)}. \qquad \text{(F.6)}$$

Again from the triangle POQ, we have

$$K \sin \Delta\theta = PQ$$
$$= r\sin(\theta_0 - \theta_m). \qquad \text{(F.7)}$$

Multiplying the above by R and dividing by K we have

$$R\sin\Delta\theta = \frac{r}{K} R\sin(\theta_0 - \theta_m). \qquad \text{(F.8)}$$

In the Āryabhaṭan school, the radius of the *manda* epicycle is assumed to vary in the same way as the *karṇa*, as explained for instance by Bhāskara I (c. 629) in his *Āryabhaṭīya-bhāṣya*, and also in his *Mahābhāskarīya*. Thus the relation (F.8) reduces to

$$R\sin\Delta\theta = \frac{r_0}{R} R\sin(\theta_0 - \theta_m), \qquad \text{(F.9)}$$

where r_0 is the mean or tabulated value of the radius of the *manda* epicycle.

F.1.3 *Aviśiṣṭa-manda-karṇa:* iterated hypotenuse

According to the geometrical picture of planetary motion given by Bhāskara I, the radius of the epicycle *manda-nīcocca-vṛtta* (r) employed in the the *manda* process is not a constant. It varies continuously in consonance with the hypotenuse, the *manda-karṇa* (K), in such a way that their ratio is always maintained constant and is equal to the ratio of the mean epicycle radius (r_0)—whose value is specified in the texts—to the radius of the deferent circle (R). Thus, according to Bhāskara, as far as the *manda* process is concerned, the motion of the planet on the epicycle is such that the following equation is always satisfied:

$$\frac{r}{K} = \frac{r_0}{R}. \qquad \text{(F.10)}$$

If this is the case, then the question arises as to how one can obtain the *manda-karṇa* as well as the the radius of the *manda-nīcocca-vṛtta* at any given instant. For this, Bhāskara provides an iterative procedure called *asakṛt-karma*, by which both r and K are simultaneously obtained. We explain this with the help of Fig. F.3*a*. Here P_0 represents the mean planet around which an epicycle of radius r_0 is drawn. The point P_1 on the epicycle is chosen such that PP_1 is parallel to the direction of the *mandocca*, OU.

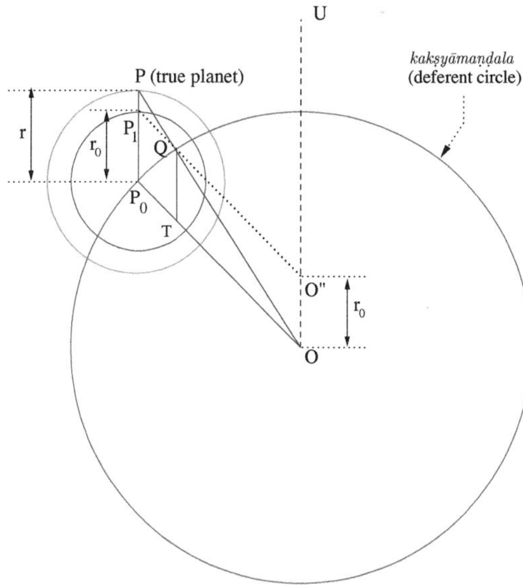

Fig. F.3*a* The variation of the radius of the *manda* epicycle with the *manda-karṇa*.

Now, the first hypotenuse (*sakṛt-karṇa*) is found from r_0 using the relation

$$OP_1 = K_1 = [(R\sin(\theta_0 - \theta_m))^2 + (R\cos(\theta_0 - \theta_m) + r_0)^2]^{\frac{1}{2}}. \qquad \text{(F.11)}$$

From K_1, using (F.10), we get the next approximation to the radius $r_1 = \frac{r_0}{R}K_1$, and the process is repeated. From r_1 we get the next approximation to the *karṇa*,

$$K_2 = [\{R\sin(\theta_0 - \theta_m)\}^2 + \{R\cos(\theta_0 - \theta_m) + r_1\}^2]^{\frac{1}{2}}, \qquad \text{(F.12)}$$

and from that we get $r_2 = \frac{r_0}{R}K_2$ and so on, till the radii and the *karṇas* do not change (*aviśeṣa*). The term *aviśeṣa* means 'not distinct'. In the present context it means that the successive *karṇas* are not distinct from each other. That is, $K_{i+1} \approx K_i = K$. If this is satisfied, then $r_{i+1} \approx r_i = r$. Consequently, the equation giving the *manda*-correction (F.8) becomes

$$R\sin\Delta\theta = \frac{r}{K}R\sin(\theta_0 - \theta_m) = \frac{r_0}{R}R\sin(\theta_0 - \theta_m). \qquad \text{(F.13)}$$

Thus the computation of the *manda-phala* involves only the mean epicycle radius and the value of the *trijyā*. It does not involve the value of the *manda-karṇa*. It can be shown that the iterated *manda-karṇa* is actually given (in the limit) by OP in Fig. F.3*a*, where the point P is obtained as follows.[14] Consider a point O'' at a distance of r_0 from O along the direction of *mandocca* OU and draw $O''P_1$ so that it meets the concentric at Q. Then produce OQ to meet the extension of P_0P_1 at P.

[14] See for instance, the discussion in {MB 1960}, pp. 111–9.

Mādhava of Saṅgamagrāma, the renowned mathematician and astronomer of the 14th century, by carefully analysing the geometry of the problem, came up with a brilliant method of finding the *aviśiṣṭa-manda-karṇa* without performing an iterative process, which is explained in the next section.

F.1.4 Mādhava's formula for the aviśiṣṭa-manda-karṇa

Mādhava's procedure for determining the *aviśiṣṭa-manda-karṇa* involves finding a new quantity called the *viparyaya-karṇa* or *viparīta-karṇa*. The term *viparīta-karṇa* literally means 'inverse hypotenuse', and is nothing but the radius of the *kakṣyāvṛtta* when the *manda-karṇa* is taken to be the *trijyā*, *R*. The following verses from *Tantrasaṅgraha* (II, 43–44) present the way of obtaining the *aviśiṣṭa-manda-karṇa* proposed by Mādhava that circumvents the iterative process.

विस्तृतिदलदोःफलकृतिवियुतिपदं कोटिफलविहीनयुतम्।
केन्द्रे मृगकर्किगते स खलु विपर्ययकृतो भवेत् कर्णः॥
तेन हृता त्रिज्याकृतिः अयत्नविहितोऽविशेषकर्णः स्यात् ।

The square of the *doḥphala* is subtracted from the square of the *trijyā* and its square root is taken. The *koṭiphala* is added to or subtracted from this depending upon whether the *kendra* (anomaly) is within six signs beginning from *Karki* (Cancer) or *Mṛga* (Capricorn). This gives the *viparyaya-karṇa*. The square of the *trijyā* divided by this *viparyaya-karṇa* is the *aviśeṣa-karṇa* (iterated hypotenuse) obtained without any effort [of iteration].

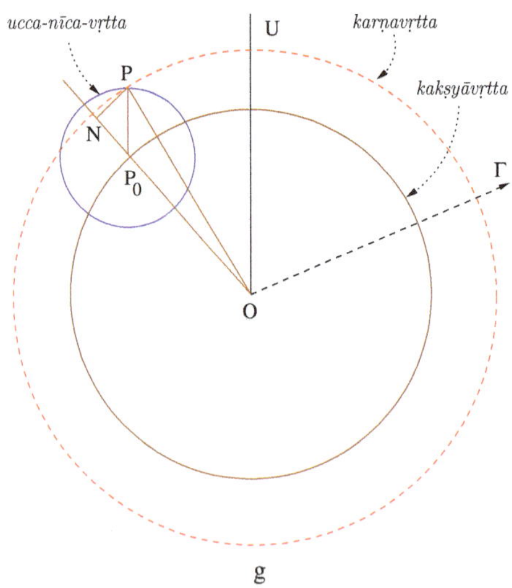

Fig. F.3*b* Determination of the *viparīta-karṇa* when the *kendra* is in the first quadrant.

The rationale behind the formula given for the *viparīta-karṇa* is outlined in the Malayalam text *Yuktibhāṣā*, and can be understood with the help of Figs F.3*a* and F.3*b*. In these figures P_0 and P represent the mean and the true planet respectively. N denotes the foot of a perpendicular drawn from the true planet P to the line joining the centre of the circle and the mean planet. NP is equal to the *doḥphala*. Let the radius of the *karṇavṛtta* OP be set equal to the *trijyā* R. Then the radius of the *uccanīca-vṛtta* P_0P is r_0, as it is in the measure of the *karṇavṛtta*. In this measure, the radius of the *kakṣyāvṛtta* $OP_0 = R_v$, the *viparīta-karṇa*, and is given by

$$\begin{aligned} R_v &= ON \pm P_0N \\ &= \sqrt{R^2 - (r_0 \sin(\theta_0 - \theta_m))^2} \pm |r_0 \cos(\theta_0 - \theta_m)|. \end{aligned} \qquad (F.14a)$$

Nīlakaṇṭha has also given another alternative expression for the *viparīta-karṇa* in terms of the longitude θ_{ms} of the *manda-sphuṭa*.

$$R_v = \sqrt{R^2 + r_0^2 - 2Rr_0 \cos(\theta_{ms} - \theta_m)}. \qquad (F.14b)$$

This is clear from the triangle OP_0P, where $OP_0 = R_v$, $OP = R$ and $P_0PO = \theta_{ms} - \theta_m$.

In Fig. F.3*a*, Q is a point where $O''P_1$ meets the concentric. OQ is produced to meet the extension of P_0P_1 at P. Let T be the point on OP_0 such that QT is parallel to P_0P_1. Then it can be shown that $OT = R_v$ is the *viparīta-karṇa*. Now, in triangle OQT, $OQ = R$, $QT = P_1P_0 = r_0$ and $O\hat{Q}T = P\hat{O}U = (\theta_{ms} - \theta_m)$ and we have

$$OT = \sqrt{R^2 + r_0^2 - 2Rr_0 \cos(\theta_{ms} - \theta_m)} = R_v. \qquad (F.14c)$$

Now, since triangles OQT and OPP_0 are similar, we have

$$\begin{aligned} \frac{OP}{OP_0} &= \frac{OQ}{OT} = \frac{R}{R_v} \\ \text{or,} \quad OP = K &= \frac{R^2}{R_v}. \end{aligned} \qquad (F.15)$$

Thus we have obtained an expression for the *aviśiṣṭa-manda-karṇa* in terms of the *trijyā* and the *viparīta-karṇa*. As the computation of the *viparīta-karṇa* as given by (F.14*a*) does not involve iteration, the *aviśiṣṭa-manda-karṇa* can be obtained in one stroke using (F.15) without having to go through the arduous iterative process.

F.1.5 *Manda-saṃskāra* for the exterior planets

We will now discuss the details of the *manda* correction for the case of the exterior planets, namely Mars, Jupiter and Saturn, as outlined in the traditional texts of Indian astronomy. The texts usually specify the the number of revolutions (*bhagaṇas*) made by the planets in a large period known as *Mahāyuga*. In Table F.1, we list the *bhagaṇas* as specified in the texts *Āryabhaṭīya* and *Tantrasaṅgraha*. In the same table, we have also given the corresponding sidereal period of the planet in civil days along with the modern values for the same.

Planet	Revolutions (in *Āryabhaṭīya*)	Sidereal period (in *Āryabhaṭīya*)	Revolutions (in *Tantrasaṅgraha*)	Sidereal period (in *Tantrasaṅgraha*)	Modern values of sidereal period
Sun	4320000	365.25868	4320000	365.25868	365.25636
Moon	57753336	27.32167	57753320	27.32168	27.32166
Moon's apogee	488219	3231.98708	488122	3232.62934	3232.37543
Moon's node	232226	6794.74951	232300	6792.58502	6793.39108
Mercury's *śīghrocca*	17937020	87.96988	17937048	87.96974	87.96930
Venus's *śīghrocca*	7022288	224.69814	7022268	224.70198	224.70080
Mars	2296824	686.99974	2296864	686.98778	686.97970
Jupiter	364224	4332.27217	364180	4332.79559	4332.58870
Saturn	146564	10766.06465	146612	10762.53990	10759.20100

Table F.1 The *bhagaṇas* and sidereal periods of the planets.

In the case of exterior planets, while the planets move around the Sun they also move around the Earth, and consequently, the mean heliocentric sidereal period of the planet is the same as the mean geocentric sidereal period. Therefore, the *madhyama-graha* or the mean longitude of the planet, as obtained from the above *bhagaṇas*, would be the same as the mean heliocentric longitude of the planet as understood today. Now the *manda-saṃskāra* is applied to the *madhyama-graha* to obtain the *manda-sphuṭa-graha*. As we will see below, this *manda* correction is essentially the same as the equation of centre in modern astronomy and thus the *manda-sphuṭa-graha* would essentially be the true heliocentric longitude of the planet.

It was shown above in (F.9) that the magnitude of the correction $\Delta\theta$ to be applied to the mean longitude is given by

$$R\sin\Delta\theta = \frac{r_0}{R}\, R\sin(\theta_0 - \theta_m), \tag{F.16}$$

If $\frac{r_0}{R}$ is small in the above expression, then $\sin\Delta\theta \ll 1$ and we can approximate $\sin\Delta\theta \approx \Delta\theta$. Hence (F.16) reduces to

$$\Delta\theta = \frac{r_0}{R}\sin(\theta_0 - \theta_m). \tag{F.17}$$

As $\Delta\theta = \theta_0 - \theta_{ms}$, in this approximation we have

$$\theta_{ms} \approx \theta_0 - \frac{r_0}{R}\sin(\theta_0 - \theta_m). \tag{F.18}$$

As outlined in Section F.8.1, in the Keplerean picture of planetary motion the equation of centre to be applied to the mean heliocentric longitude of the planet is given—to the first order in eccentricity—by the equation

$$\Delta\theta \approx (2e)\sin(\theta_0 - \theta_m). \tag{F.19}$$

Now, comparing (F.19) and (F.17), we see that the *manda* correction closely approximates the equation of centre as understood in modern astronomy if the values of $\frac{r_0}{R}$ are fairly close to $2e$.

The values of $\frac{r_0}{R}$ for different planets as specified in *Āryabhaṭīya* and *Tantrasaṅgraha* are listed in Table F.2. It may be noted here that the ratios specified in the texts are close to twice the value of the eccentricity ($2e$) associated with the planetary orbits. In Table F.2, the modern values of $2e$ are listed according to Smart.[15]

Name of the planet	*Āryabhaṭīya* $\frac{r_0}{R}$	Average	*Tantrasaṅgraha* $\frac{r_0}{R}$	Average	$2e$ Modern
Sun	$\frac{13.5}{360}$	0.0375	$\frac{3}{80}$	0.0375	0.034
Moon	$\frac{31.5}{360}$	0.0875	$\frac{7}{80}$	0.0875	0.110
Mercury	$\frac{31.5-9\lvert\sin(\theta_0-\theta_m)\rvert}{360}$	0.075	$\frac{1}{6}$	0.167	0.412
Venus	$\frac{18-9\lvert\sin(\theta_0-\theta_m)\rvert}{360}$	0.0375	$\frac{1}{14+\frac{R\lvert\sin(\theta_0-\theta_m)\rvert}{240}}$	0.053	0.014
Mars	$\frac{63+18\lvert\sin(\theta_0-\theta_m)\rvert}{360}$	0.200	$\frac{7+\lvert\sin(\theta_0-\theta_m)\rvert}{39}$	0.192	0.186
Jupiter	$\frac{31.5+4.5\lvert\sin(\theta_0-\theta_m)\rvert}{360}$	0.0938	$\frac{7+\lvert\sin(\theta_0-\theta_m)\rvert}{82}$	0.091	0.096
Saturn	$\frac{40.5+18\lvert\sin(\theta_0-\theta_m)\rvert}{360}$	0.1375	$\frac{39}{360}$	0.122	0.112

Table F.2 Comparison of *manda* epicycle radii and modern eccentricity values.

F.1.6 Manda-saṃskāra for interior planets

For the interior planets Mercury and Venus, since the mean geocentric sidereal period of the planet is the same as that of the Sun, the ancient Indian astronomers took the mean Sun as the *madhyama-graha* or the mean planet. Having taken the mean Sun as the mean planet, they also prescribed the application of the *manda* correction, or the equation of centre characteristic of the planet, to the mean Sun, instead of the mean heliocentric planet. Therefore, the *manda-sphuṭa-graha* in the case of

[15] W. M. Smart, *Textbook on Spherical Astronomy*, Cambridge University Press, 1965, pp. 422–3.

an interior planet, as computed from (F.17) in the traditional planetary model, is just the mean Sun, with a correction applied, and does not correspond to the true heliocentric planet.

However, the ancient Indian astronomers also introduced the notion of the *śīghrocca* for these planets whose period (see Table F.1) is the same as the mean heliocentric sidereal period of these planets. Thus, in the case of the interior planets, it is the longitude of the *śīghrocca* which will be the same as the mean heliocentric longitude of the planet as understood in the currently accepted model of the solar system. As we shall see below, the traditional planetary model made use of this *śīghrocca*, crucially, in the calculation of both the longitudes and latitudes of the interior planets.

F.2 *Śīghra-saṃskāra*

We will now show that the application of *śīghra-saṃskāra* is equivalent to the transformation of the *manda-sphuṭa* to the true geocentric longitude of the planet called the *sphuṭa-graha*. Just as the *mandocca* plays a major role in the application of *manda-saṃskāra*, so too the *śīghrocca* plays a key role in the application of *śīghra-saṃskāra*. As in the case of *manda-saṃskāra*, we shall consider the application of *śīghra-saṃskāra* for the exterior and interior planets separately.

F.2.1 Exterior planets

For the exterior planets, Mars, Jupiter and Saturn, we have already explained that the *manda-sphuṭa-graha* is the true heliocentric longitude of the planet. The *śīghra-saṃskāra* for them can be explained with reference to Fig. F.4*a*. Here *A* denotes the *nirayaṇa-meṣādi*, *E* the Earth and *P* the planet. The mean Sun *S* is referred to as the *śīghrocca* for exterior planets and thus we have

$$
\begin{aligned}
A\hat{S}P &= \theta_{ms} && (\textit{manda-sphuṭa})\\
A\hat{E}S &= \theta_s && (\text{longitude of } \textit{śīghrocca} \text{ (mean Sun)})\\
A\hat{E}P &= \theta && (\text{geocentric longitude of the planet}).
\end{aligned}
$$

The difference between the longitudes of the *śīghrocca* and the *manda-sphuṭa*, namely

$$\sigma = \theta_s - \theta_{ms}, \tag{F.20}$$

is called the *śīghra-kendra* (anomaly of conjunction) in Indian astronomy. From the triangle *EPS* we can easily obtain the result

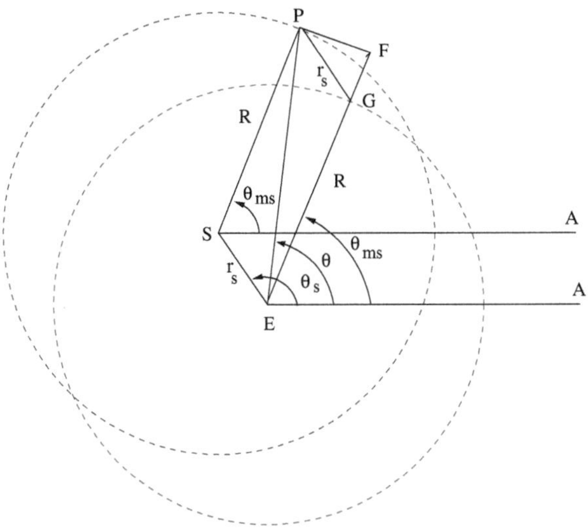

Fig. F.4*a* *Śīghra* correction for exterior planets.

$$\sin(\theta - \theta_{ms}) = \frac{r_s \sin\sigma}{[(R + r_s\cos\sigma)^2 + r_s^2 \sin^2\sigma]^{\frac{1}{2}}}, \qquad (F.21a)$$

which is the *śīghra* correction formula given by Indian astronomers to calculate the geocentric longitude of an exterior planet. It may be noted that the true or geocentric longitude of the planet known as the *śīghra-sphuṭa* is found in the same manner from the *manda-sphuṭa*, as the *manda-sphuṭa* is found from the mean planet, the *madhyama-graha*.

From Fig. F.4*a* it is clear that the *śīghra-saṃskāra* transforms the true heliocentric longitudes into geocentric longitudes only if the ratio of the radii of the epicycle and the deferent circle is equal to the ratio of the Earth–Sun and planet–Sun distances. That this is indeed very nearly so in the Indian texts, as may be seen from Table F.3. It my also be noted that (F.21*a*) has the same form as the formula for the difference between the geocentric and heliocentric longitudes for an exterior planet in the Keplerian model (see (F.46)) if $\frac{r_s}{R}$ is identified with the ratio of the Earth–Sun and planet–Sun distances. However, (F.21*a*) is still an approximation as it is based upon mean Sun and not the true Sun.

F.2.2 Interior planets

The *śīghra-saṃskāra* for the interior planets can be explained with reference to Fig. F.4*b*. Here E is the Earth and S (the *manda*-corrected mean Sun) is the *manda-sphuṭa-graha* and P, the so-called *śīghrocca*, actually corresponds to the (mean heliocentric) planet. We have

$$A\hat{E}S = \theta_{ms} \qquad (\textit{manda-sphuṭa})$$
$$A\hat{S}P = \theta_s \qquad (\text{longitude of } \textit{śīghrocca})$$
$$A\hat{E}P = \theta \qquad (\text{geocentric longitude of the planet}).$$

Again, the *śīghra-kendra* is defined as the difference between the *śīghrocca* and the *manda-sphuṭa-graha* as in (F.20). Thus, from the triangle EPS we get the same formula

$$\sin(\theta - \theta_{ms}) = \frac{r_s \sin\sigma}{[(R + r_s\cos\sigma)^2 + r_s^2 \sin^2\sigma]^{\frac{1}{2}}}, \qquad \text{(F.21}b\text{)}$$

which is the *śīghra* correction given in the earlier Indian texts to calculate the geocentric longitude of an interior planet. For the interior planets also, the value specified for $\frac{r_s}{R}$ is very nearly equal to the ratio of the planet–Sun and Earth–Sun distances, as may be seen from Table F.3.

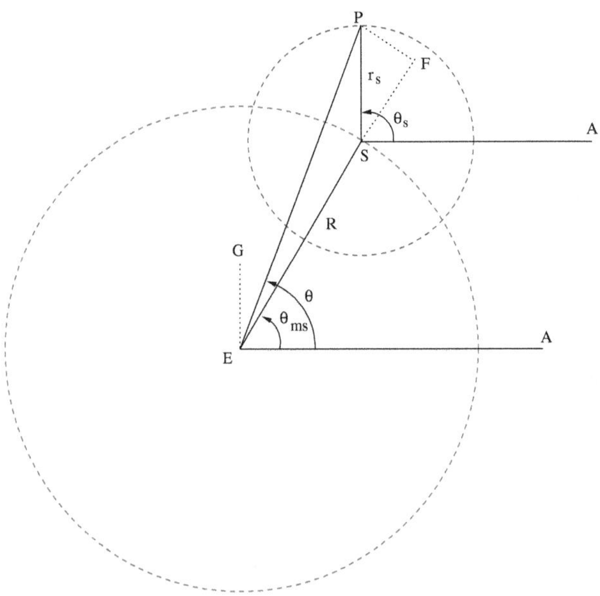

Fig. F.4*b* *Śīghra* correction for interior planets.

Since the *manda* correction or equation of centre for an interior planet was applied to the longitude of the mean Sun instead of the mean heliocentric longitude of the planet, the accuracy of the computed longitudes of the interior planets according to the ancient Indian planetary models would not have been as good as that achieved for the exterior planets. But for the wrong application of the equation of centre, equation (F.21*b*) has the same form as the formula for the difference between the geocentric longitude of an interior planet and the Sun in the Keplerian model (see (F.50)), if $\frac{r_s}{R}$ is identified with the ratio of the planet–Sun and Earth–Sun distances.

Name of the planet	*Āryabhaṭīya* $\frac{r_s}{R}$	Average	*Tantrasaṅgraha* $\frac{r_s}{R}$	Average	Modern value
Mercury	$\frac{139.5-9\lvert\sin(\theta_{ms}-\theta_s)\rvert}{360}$	0.375	$\frac{133-\lvert\sin(\theta_{ms}-\theta_s)\rvert}{360}$	0.368	0.387
Venus	$\frac{265.5-9\lvert\sin(\theta_{ms}-\theta_s)\rvert}{360}$	0.725	$\frac{59-2\lvert\sin(\theta_{ms}-\theta_s)\rvert}{80}$	0.725	0.723
Mars	$\frac{238.5-9\lvert\sin(\theta_{ms}-\theta_s)\rvert}{360}$	0.650	$\frac{7+\lvert\sin(\theta_{ms}-\theta_s)\rvert}{39}$	0.656	0.656
Jupiter	$\frac{72-4.5\lvert\sin(\theta_{ms}-\theta_s)\rvert}{360}$	0.194	$\frac{16-\lvert\sin(\theta_{ms}-\theta_s)\rvert}{80}$	0.194	0.192
Saturn	$\frac{40.5-4.5\lvert\sin(\theta_{ms}-\theta_s)\rvert}{80}$	0.106	$\frac{9-\lvert\sin(\theta_{ms}-\theta_s)\rvert}{80}$	0.106	0.105

Table F.3 Comparison of $\frac{r_s}{R}$, as given in *Āryabhaṭīya* and *Tantrasaṅgraha*, with the modern values of the ratio of the mean values of Earth–Sun and planet–Sun distances for the exterior planets and the inverse ratio for the interior planets.

F.2.3 Four-step process

In obtaining the expression (F.21) for the *śīghra* correction, we had taken SP, the Sun–planet distance, to be given by R. But actually SP is a variable and is given by the (iterated) *manda-karṇa* K. Hence the correct form of the *śīghra* correction should be

$$\sin(\theta_s-\theta_{ms}) = \frac{r_s \sin\sigma}{\{(K+r_s\cos\sigma)^2 + r_s^2\sin^2\sigma\}^{\frac{1}{2}}}, \qquad \text{(F.22)}$$

where K is the (iterated) *manda-karṇa*. Since K as given by (F.14) and (F.15) depends on the *manda* anomaly $\theta-\theta_m$, the *śīghra* correction as given by (F.22) cannot be tabulated as a function of the *śīghra* anomaly (σ) alone.

It is explained in *Yuktibhāṣā* (section 8.20) that, in order to simplify computation, the ancient texts on astronomy advocated that the computation of the planetary longitudes may be done using a four-step process—involving half-*manda* and half-*śīghra* corrections followed by the full *manda* and *śīghra* corrections. The *śīghra* corrections involved in the four-step process are based on the simpler formula (F.21) which can be read off from a table. According to *Yuktibhāṣā*, the results of the four-step process indeed approximate those obtained by the application of the *manda* correction followed by the *śīghra* correction where, in the latter correction, the effect of the *manda-karṇa* is properly taken into account as in (F.22).

F.2.4 Computation of planetary latitudes

Planetary latitudes (called *vikṣepa* in Indian astronomy) play an important role in the prediction of planetary conjunctions, the occultation of stars by planets etc. In Fig. F.5, P denotes the planet moving in an orbit inclined at an angle i to the ecliptic, intersecting the ecliptic at point N, the node (called the *pāta* in Indian astronomy). If β is the latitude of the planet, θ_h its heliocentric longitude and θ_n the heliocentric

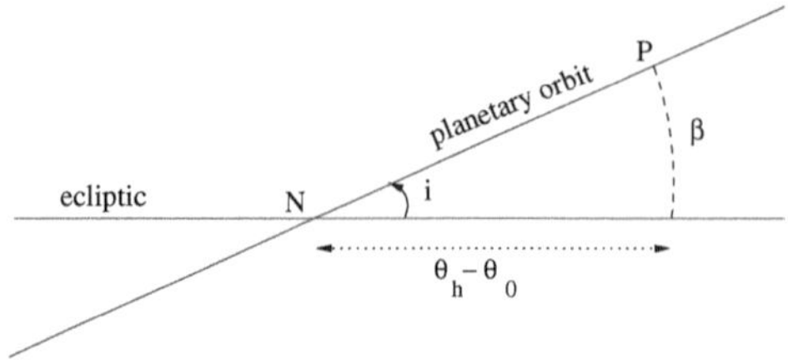

Fig. F.5 Heliocentric latitude of a planet.

longitude of the node, then it can be shown that

$$\sin\beta = \sin i \, \sin(\theta_h - \theta_n). \tag{F.23}$$

For small i we have

$$\beta = i \, \sin(\theta_h - \theta_n). \tag{F.24}$$

This is essentially the rule for calculating the latitude of a planet, as given in Indian texts, at least from the time of Āryabhaṭa.[16] For the exterior planets, it was stipulated that

$$\theta_h = \theta_{ms}, \tag{F.25}$$

the *manda-sphuṭa-graha*, which, as we saw earlier, coincides with the heliocentric longitude of the exterior planet. The same rule applied for interior planets would not have worked, because in the traditional Indian planetary model the *manda*-corrected mean longitude for the interior planet has nothing to do with its true heliocentric longitude. However, most of the Indian texts on astronomy stipulated that the latitude in the case of the interior planets is to be calculated from (F.24) with

$$\theta_h = \theta_s + manda \text{ correction}, \tag{F.26}$$

the *manda*-corrected longitude of the *śīghrocca*. Since the longitude of the *śīghrocca* for an interior planet, as we explained above, is equal to the mean heliocentric longitude of the planet, (F.26) leads to the correct relation that, even for an interior planet, θ_h in (F.24) becomes identical with the true heliocentric longitude. Thus we see that the earlier Indian astronomical texts did provide a fairly accurate theory for the planetary latitudes. But they had to live with two entirely different rules for calculating latitudes: one for the exterior planets given by (F.25), where the *manda-sphuṭa-graha* appears; and an entirely different one for the interior planets given by (F.26), which involves the *śīghrocca* of the planet, with the *manda* correction included.

This peculiarity of the rule for calculating the latitude of an interior planet was noticed repeatedly by various Indian astronomers, at least from the time of

---

[16] Equation (F.24) actually gives the heliocentric latitude and needs to be multiplied by the ratio of the geocentric and heliocentric distances of the planet to get the geocentric latitude. This feature was implicit in the traditional planetary models.

Bhāskara I (c. 629), who in his *Āryabhaṭīya-bhāṣya* drew attention to the fact that the procedure given in *Āryabhaṭīya* for calculating the latitude of an interior planet is indeed very different from that adopted for the exterior planets.[17] The celebrated astronomer Bhāskarācārya II (c. 1150) also draws attention to this peculiar procedure adopted for the interior planets, in his *Vāsanā-bhāṣya* on his own *Siddhāntaśiromaṇi*, and quotes the statement of Caturveda Pṛthūdakasvāmin (c. 860) that this peculiar procedure for the interior planets can be justified only on the ground that this is what has been found to lead to predictions that are in conformity with observations.[18]

F.3 Geometrical picture of planetary motion according to Parameśvara

The renowned Kerala astronomer Parameśvara of Vaṭasseri (1380–1460) has discussed in detail the geometrical model implied in the conventional planetary model of Indian astronomy in his super-commentary *Siddhāntadīpikā* (on Govindasvāmin's commentary on) *Mahābhāskarīya* of Bhāskara I. A shorter version is available in his commentary on *Āryabhaṭīya*, which is given below.

स्फुटविधियुक्तिस्सिध्येन्नैव विना छेद्यकेन विहगानाम्।
तस्मादिह संक्षेपाच्छेद्यककर्म प्रदर्श्यते तेषाम्॥
त्रिज्याकृतं कुमध्यं कक्ष्यावृत्तं भवेत्तु तच्छैघ्र्यम्।
शीघ्रदिशि तस्य केन्द्रं शीघ्रान्त्यफलान्तरे पुनः केन्द्रम्॥
कृत्वा विलिखेद्वृत्तं शीघ्रप्रतिमण्डलाख्यमुदितमिदम्।
इदमेव भवेन्मान्दे कक्ष्यावृत्तं पुनस्तु तत्केन्द्रात्॥
केन्द्रं कृत्वा मन्दान्त्यफलान्तरे वृत्तमपि च मन्ददिशि।
कुर्यात् प्रतिमण्डलमिदमुदितं मान्दं शनीड्यभूपुत्राः॥
मान्दप्रतिमण्डलगास्तत्कक्ष्यायां तु यत्र लक्ष्यन्ते।
तत्र हि तेषां मन्दस्फुटाः प्रदिष्टास्तथैव शैघ्रे ते॥
प्रतिमण्डले स्थितास्स्युस्ते लक्ष्यन्ते पुनस्तु शैघ्राख्ये।
कक्ष्यावृत्ते यस्मिन्भागे तत्र स्फुटग्रहास्ते स्युः।
एवं सिद्ध्यति तत्र स्फुटयुग्मं तत्र भवति दृग्भेदः।
यत्र खगा लक्ष्यन्ते तत्रस्था लक्षिता यतोऽन्यस्मिन्॥
क्रियतेऽत्र तन्निमित्तं मध्ये मान्दार्धमपि शैघ्रार्धम्।
शैघ्रं मान्दं मान्दं शैघ्रञ्चेति क्रमस्स्मृतोऽन्यत्र॥
मान्दं कक्ष्यावृत्तं प्रथमं बुधशुक्रयोः कुमध्यं स्यात्।
तत्केन्द्रान्मन्ददिशि मन्दान्त्यफलान्तरे तु मध्यं स्यात्।
मान्दप्रतिमण्डलस्य तस्मिन् यत्र स्थितो रविस्तत्र।
प्रतिमण्डलस्य मध्यं शैघ्रस्य तस्य मानमपि च गदितम्।

---

[17] {AB 1976}, p. 32, 247.

[18] {SSR 1981}, p. 402.

शीघ्रस्ववृत्ततुल्यं तस्मिंश्चरतस्सदा ज्ञशुक्रौ च।
स्फुटयुक्तिः प्राग्वत्स्यात् दृग्भेदः पूर्ववद्भवेदिह च॥
क्रियतेऽत्र तन्निमित्तं शैघ्रार्धं व्यत्ययेन मन्दोच्चे।
तत्सिद्धं मान्दं प्राक् पश्चाच्छैघ्रञ्च सूरिभिः पूर्वैः ॥[19]

> Since the rationale for the *sphuṭavidhi* (the scheme of computing the true planet) for the celestial bodies is not clear without the aid of *chedyaka* (diagrams), we present briefly the way of obtaining the diagrams.
>
> For Mars, Jupiter and Saturn, with the centre of the Earth as the centre, the *śīghra-kakṣyā-vṛtta* (concentric circle) is drawn with the *trijyā* ($R \sin 90$) as the radius. Then draw the *śīghra-pratimaṇḍala* (eccentric circle) with its centre located at a distance of the *śīghra-antyaphala* (maximum *śīghra*-correction) in the direction of the *śīghrocca*. The same will be the *manda*-concentric. From its centre go along in the direction of the *mandocca* a distance equal to the maximum *manda* correction, and with this as the centre draw a circle. This is referred as the *manda* eccentric circle. The planets Mars, Jupiter and Saturn move on this eccentric when reduced to the *manda*-concentric they are referred to as *manda-sphuṭa*, and when reduced to the *śīghra*-concentric they are *sphuṭa* (true planets). ...
>
> For Mercury and Venus, the *manda*-concentric is first drawn with the centre of the Earth as the centre. From that go along in the direction of *mandocca* a distance equal to the maximum *manda* correction and with that as the centre draw the *manda* eccentric circle. The point where the Sun is located on that eccentric is the centre of the *śīghra* epicycle and the radius of that circle is [not the *trijyā* but] as enunciated. In that *śīghra* epicycle, the Mercury and the Venus always move ...

The *chedyaka* procedure enunciated by Parameśvara is illustrated in Figs F.6 and F.7. In both these figures, O represents the observer, M the *mandocca* and P the planet whose longitude as measured from O is to be determined. In Fig. F.6, the circles C_1, C_2 and C_3 are all of radius R. The circle C_1, centred around the observer O, is the *śīghra-kakṣyā-maṇḍala* or the *śīghra*-concentric circle. The circle C_2 which is centred at the *śīghrocca* S is the *śīghra-pratimaṇḍala* (*śīghra*-eccentric circle). The distance of separation between these two circles denoted by OS is the *śīghrāntya-phala*, and corresponds to the radius of the *śīghra* epicycle. It has been clearly enunciated by Parameśvara that the *śīghra-pratimaṇḍala*, denoted by C_2 in the figure, itself serves as the *manda-kakṣyā-maṇḍala*, or the *manda*-concentric circle. The third circle C_3, which is centred around the *mandocca* M, is the *manda-pratimaṇḍala* or the *manda*-eccentric circle. The distance of separation between the centers of the *manda*-concentric and the *manda*-eccentric circles is equal to the radius of the *manda* epicycle and is also the *mandāntya-phala*, whose measure varies from planet to planet.

Parameśvara has depicted the geometrical picture of motion of the interior planets also by employing three circles, C_1, C_2 and C_3, as in the case of exterior planets, as shown in Fig. F.7. However, here these three circles have completely different connotations and, while C_1 and C_2 are of radius R, C_3 is of radius r_s, the radius of the *śīghra* epicycle. Here the circle C_1 centred around O, is the *manda-kakṣyā-maṇḍala*, or the *manda*-concentric circle. The circle C_2, which is centred around the *mandocca* M, is the *manda-pratimaṇḍala*, which serves as the locus for the

[19] {AB 1874}, pp. 60-1.

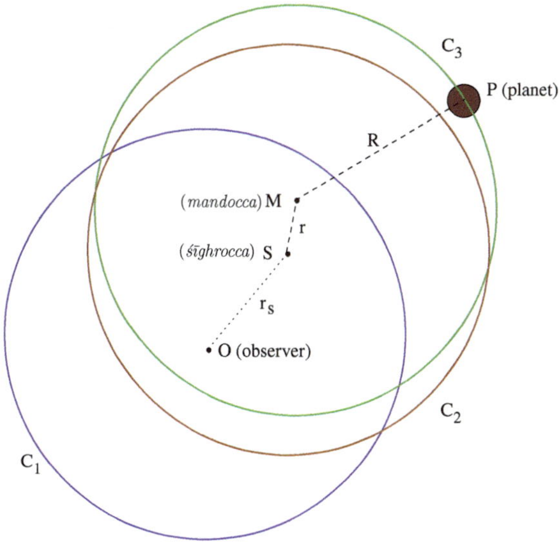

Fig. F.6 Geometrical picture of the motion of an exterior planet given by Parameśvara.

centre of the *śīghra-vṛtta* denoted by the circle C_3. The distance of separation between the centers of C_1 and C_2 is equal to the radius of the *manda* epicycle, and is also the *mandāntya-phala*. P represents the *śīghrocca* associated with the interior planet and S is the *manda*-corrected Sun on the *manda-pratimaṇḍala*.

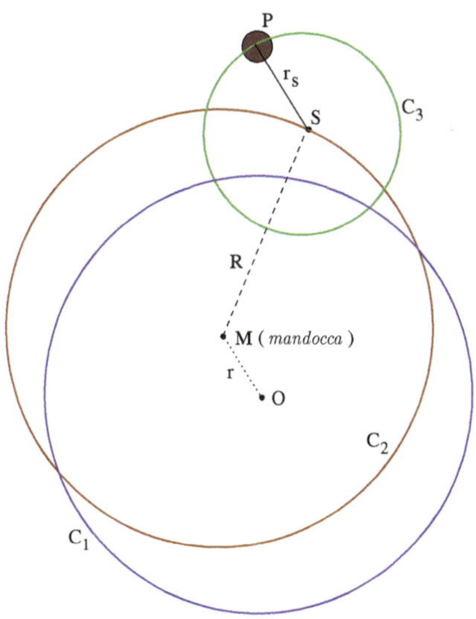

Fig. F.7 Geometrical picture of the motion of an interior planet given by Parameśvara.

It is important to note that, through his diagrammatic procedure, Parameśvara clearly illustrates the fact that, in the traditional planetary model, the final longitude that is calculated for an interior planet is actually the geocentric longitude of what is called the *śīghrocca* of the planet. From Figs F.6 and F.7 we can see easily that Parameśvara's geometrical picture of planetary motion is fairly accurate except for the fact that the equation of centre for the interior planets is wrongly applied to the mean Sun. Incidentally, it may also be noted that Parameśvara has given a succinct description of the same *chedyakavidhi* in his *Goladīpikā*.[20]

F.4 Nīlakaṇṭha's revised planetary model

Among the available works of Nīlakaṇṭha, his revised planetary motion is discussed in the works *Tantrasaṅgraha*, *Āryabhaṭīya-bhāṣya*, *Siddhānta-darpaṇa* and the *Vyākhyā* on it, *Golasāra* and the tract called *Grahasphuṭānayane vikṣepavāsanā*. Of these, *Golasāra* and *Siddhānta-darpaṇa* are presumed to have been written prior to the detailed work *Tantrasaṅgraha* composed in 1500. The *Āryabhaṭīya-bhāṣya* refers to *Golasāra* and *Tantrasaṅgraha*. The *Siddhānta-darpaṇa-vyākhyā* cites the *Āryabhaṭīya-bhāṣya*. In the same way, the small but important tract *Grahasphuṭānayane vikṣepavāsanā* includes long passages from the *Āryabhaṭīya-bhāṣya* and is clearly a later composition.

In *Tantrasaṅgraha*, Nīlakaṇṭha presents the revised planetary model and also gives the detailed scheme of computation of planetary latitudes and longitudes, but he does not discuss the geometrical picture of planetary motion. Towards the end of the last chapter of the work, Nīlakaṇṭha introduces a prescription for the *sphuṭakakṣyā* (the true distance of the planets). There seems to be just a brief (and incomplete) mention of this subject in *Golasāra* and *Āryabhaṭīya-bhāṣya*.

The geometrical picture of planetary motion is discussed in detail in the *Āryabhaṭīya-bhāṣya*. It is also succinctly presented in terms of a few verses in both *Golasāra* and *Siddhānta-darpaṇa*. Nīlakaṇṭha presents some aspects of his cosmological model while discussing the geometrical picture of the motion of the interior planets in his *Āryabhaṭīya-bhāṣya*. He presents a definitive but succinct account of his cosmological model in terms of a few verses in his later work *Grahasphuṭānayane vikṣepavāsanā*.

F.4.1 Identifying the mean Mercury and Venus

In the very first chapter of *Tantrasaṅgraha* (c. 1500), Nīlakaṇṭha introduces a major revision of the traditional Indian planetary model, according to which what were traditionally referred to as the *śīghrocca*s of the interior planets (Mercury and

[20] {GD 1916}, pp. 14–15.

Venus) are now identified with the planets themselves; and the mean Sun is taken as the *śīghrocca* of all the planets.

खाश्विदेवेषुसप्ताद्रिशराश्चेन्दोः, कुजस्य तु ।
वेदाङ्गाहिरसाङ्काश्विकराः, ज्ञस्य स्वपर्ययाः ॥
नागवेदनभस्सप्तरामाङ्कस्वरभूमयः ।
व्योमाष्टरूपवेदाङ्गपावकाश्च बृहस्पतेः ॥
अष्टाङ्गदस्रनेत्राश्विखाद्रयो भृगुपर्ययाः ।[21]

[The number of revolutions in a *mahāyuga*] of the Moon is 57753320. That of Mars is 2296864. The number of own revolutions of Mercury is 17937048. That of Jupiter is 364180. The number of revolutions of Venus is 7022268.

Here the commentator Śaṅkara Vāriyar observes:

अत्र स्वशब्देन पर्यायाणां भास्कराचार्याद्यभिमतं स्वशीघ्रोच्चसम्बन्धित्वं बुधस्य निरस्तम् ।[22]

Here, by the use of the word *sva* (own), the association of this number of revolutions with the *śīghrocca* of Mercury, as done by Bhāskara and others, is rejected.

It may be noted (see Table F.1) that, except for the above redefinition of the mean Mercury and Venus, the *bhagaṇas*, or the number of planetary revolutions in a *Mahāyuga*, are nearly same as those given in *Āryabhaṭīya*.

F.4.2 Computation of planetary longitudes

Nīlakaṇṭha presents the details of his planetary model in the second chapter of *Tantrasaṅgraha*. For the exterior planets, he essentially follows the traditional model. He also retains the four-step process, while noting that (the rationale for such a scheme seems to be essentially that) such has been the recommendation of the earlier masters:

मान्दं शैघ्रं पुनर्मान्दं शैघ्रं चत्वार्यनुक्रमात् ।
कुजगुर्वर्कजानां हि कर्माण्युक्तानि सूरिभिः ॥[23]

The earlier masters have stated that the *manda*, *śīghra* and again *manda* and *śīghra* are the four corrections that have to be applied in sequence in the case of Mars, Jupiter and Saturn (in order to obtain their geocentric longitude).

The actual procedure given by Nīlakaṇṭha is the following: If θ_0 is the mean longitude of the planet and θ_m that of its *mandocca*, then θ_1 (the longitude at the end of the first step of the four-step process) is found by applying the half-*manda* correction as follows:

---

[21] {TS 1958}, p. 8.
[22] {TS 1958}, p. 9.
[23] {TS 1958}, p. 41.

$$\theta_1 = M + \frac{1}{2} R\sin^{-1}\left[-\frac{r_0}{R} R\sin(\theta_0 - \theta_m)\right], \qquad \text{with}$$
$$\frac{r_0}{R} = \frac{[7 + |\sin(\theta_0 - \theta_m)|]}{39} \qquad \text{(for Mars)}$$
$$\frac{r_0}{R} = \frac{[7 + |\sin(\theta_0 - \theta_m)|]}{82} \qquad \text{(for Jupiter)}$$
$$\frac{r_0}{R} = \frac{39}{320} \qquad \text{(for Saturn).}$$

Then θ_2 is found by applying the half-*śīghra* correction with the mean Sun θ_s as the *śīghrocca* as follows:

$$\theta_2 = \theta_1 + \frac{1}{2} R\sin^{-1}\left[\frac{r_s}{K_{s1}} R\sin(\theta_s - \theta_1)\right], \qquad \text{with}$$
$$K_{s1} = [\{r_s \sin(\theta_1 - \theta_s)\}^2 + \{R + r_s \cos(\theta_1 - \theta_s)\}^2]^{\frac{1}{2}}$$
$$\left(\frac{r_s}{R}\right) = \frac{[53 - 2|\sin(\theta_1 - \theta_s)|]}{80} \qquad \text{(for Mars)}$$
$$\left(\frac{r_s}{R}\right) = \frac{[16 - |\sin(\theta_1 - \theta_s)|]}{80} \qquad \text{(for Jupiter)}$$
$$\left(\frac{r_s}{R}\right) = \frac{[9 - |\sin(\theta_1 - \theta_s)|]}{80} \qquad \text{(for Saturn).}$$

Then the *manda-sphuṭa* θ_{ms} is found by adding the whole *manda* correction obtained with θ_2 to θ_0:

$$R\sin(\theta_{ms} - \theta_0) = -\left(\frac{r_0}{R}\right) R\sin(\theta_2 - \theta_m).$$

Then the true planet *sphuṭa-graha* P is found by applying the whole of the *śīghra* correction to θ_{ms}.

$$R\sin(\theta - \theta_{ms}) = \left[\frac{r_s}{K_s} R\sin(\theta_s - \theta_{ms})\right]$$
$$\text{where} \qquad K_s = [\{r_s \sin(\theta_{ms} - \theta_s)\}^2 + \{R + r_s \cos(\theta_{ms} - \theta_s)\}^2]^{\frac{1}{2}}. \qquad \text{(F.27)}$$

Again, as we had noted earlier in connection with the traditional planetary model, in the above four-step process also the iterated *manda*-hypotenuse (*aviśiṣṭa-manda-karṇa*) does not appear and the *manda* and *śīghra* corrections can be read off from a table.

In the case of the interior planets, Nīlakaṇṭha presents just the two-step process: *manda-saṃskāra* followed by *śīghra-saṃskāra*. For the interior planets, if θ_0 is the longitude of the mean planet (as per his revised model), θ_m its *mandocca* and θ_s that of the mean Sun (*śīghrocca*), then the *manda* correction leading to the *mandasphuṭa* is given by

$$R\sin(\theta_{ms} - \theta_0) = -\frac{r_0}{R} R\sin(\theta_0 - \theta_m)$$

$$\frac{r_0}{R} = \frac{1}{6}, \frac{1}{\left[14 + \frac{|R\sin(\theta_0 - \theta_m)|}{240}\right]} \qquad \text{(for Mercury, Venus).}$$

It may be recalled that the *aviśiṣṭa-manda-karṇa* K is to be calculated using the Mādhava formula (F.15). The *śīghra* correction giving the true planet θ is given by

$$R\sin(\theta - \theta_s) = \left[\left(\frac{r_s}{R}\right)\left(\frac{K}{K_s}\right) R\sin(\theta_{ms} - \theta_s)\right]$$

$$\text{where} \qquad K_s = [R\sin(\theta_{ms} - \theta_s)^2 + \{R\cos(\theta_{ms} - \theta_s) + \left(\frac{r_s}{R}\right) K\}^2]^{\frac{1}{2}} \qquad \text{(F.28)}$$

$$\left(\frac{r_s}{R}\right) = \frac{[31 - 2|\sin(\theta_{ms} - \theta_s)|]}{80R} \qquad \text{(for Mercury)}$$

$$\left(\frac{r_s}{R}\right) = \frac{[59 - 2|\sin(\theta_{ms} - \theta_s)|]}{80R} \qquad \text{(for Venus).}$$

Note that in the above two-step process the *aviśiṣṭa-manda-karṇa* K shows up in the *śīghra* correction. In his discussion of the geometrical picture of planetary motion in the *Āryabhaṭīya-bhāṣya*, Nīlakaṇṭha presents the two-step process as the planetary model for all the planets. This has also been the approach of *Yuktibhāṣā*.

F.4.3 Planetary latitudes

In the seventh chapter of *Tantrasaṅgraha*, Nīlakaṇṭha gives the method for calculating the latitudes of planets, and prescribes that for all planets, both exterior and interior, the latitude is to be computed from the *manda-sphuṭa-graha*.

मन्दस्फुटात् स्वपातोनात् भौमादीनां भुजागुणात्।
परमक्षेपनिघ्ना स्यात् क्षेपोऽन्त्यश्रवणोद्धृतः ॥[24]

The Rsine of the *manda-sphuṭa* of the planet Mars etc., from which the longitude of its node is subtracted, is multiplied by the maximum latitude and divided by the last hypotenuse (the *śīghra* hypotenuse of the last step). The result is the latitude of the planet.

This is as it should be, for in Nīlakaṇṭha's model the *manda-sphuṭa-graha* (the *manda* corrected mean longitude) coincides with the true heliocentric longitude for both exterior and interior planets. In this way, Nīlakaṇṭha, by his modification of the traditional Indian planetary theory, solved the problem, long-standing in Indian astronomy, of there being two different rules for calculating the planetary latitudes.

In the above verse, Nīlakaṇṭha states that the last hypotenuse that arises in the process of computation of longitudes, namely the *śīghra-karṇa* K_s, is to be used as the divisor. In *Āryabhaṭīya-bhāṣya*, he identifies this as the Earth–planet distance (the *bhū-tārāgraha-vivara*). There, Nīlakaṇṭha has also explained how the computations of true longitude and latitude get modified when latitudinal effects are also

[24] {TS 1958}, p. 139.

taken into account. The true Earth-planet distance (the *bhū-tārāgraha-vivara*) is also calculated there in terms of the K_s and the latitude.[25]

From the above discussion it is clear that the central feature of Nīlakaṇṭha's revision of the traditional planetary model is that the *manda* correction, or the equation of centre for the interior planets, should be applied to the mean heliocentric planet (or what was referred to as the *śīghrocca* in the traditional Indian planetary model), and not the mean Sun. In this way Nīlakaṇṭha, by 1500 CE, had arrived at the correct formulation of the equation of centre for the interior planets, perhaps for the first time in the history of astronomy. Nīlakaṇṭha was also able to formulate a unified theory of planetary latitudes.

Just as was the case with the earlier Indian planetary model, the ancient Greek planetary model of Ptolemy and the planetary models developed in the Islamic tradition during the 8th–15th centuries postulated that the equation of centre for an interior planet should be applied to the mean Sun, rather than to the mean heliocentric longitude of the planet as we understand today.[26] Further, while the ancient Indian astronomers successfully used the notion of the *śīghrocca* to arrive at a satisfactory theory of the latitudes of the interior planets, the Ptolemaic model is totally off the mark when it comes to the question of latitudes of these planets.[27]

Even the celebrated Copernican revolution brought about no improvement in the planetary theory for the interior planets. As is widely known now, the Copernican model was only a reformulation of the Ptolemaic model—with some modifications borrowed from the Maragha school of astronomy of Nasir ad-Din at-Tusi (c. 1201–74), Ibn ash-Shatir (c. 1304–75) and others—for a heliocentric frame of reference, without altering his computational scheme in any substantial way for the interior planets. As an important study notes:

> 'Copernicus, ignorant of his own riches, took it upon himself for the most part to represent Ptolemy, not nature, to which he had nevertheless come the closest of all'. In this famous and just assessment of Copernicus, Kepler was referring to the latitude theory of Book V [of *De Revolutionibus*], specifically to the 'librations' of the inclinations of the planes of the eccentrics, not in accordance with the motion of the planet but by the unrelated motion of the Earth. This improbable connection between the inclinations of the orbital planes and the motion of the Earth was the result of Copernicus's attempt to duplicate the apparent latitudes of Ptolemy's models in which the inclinations of the epicycle planes were variable. In a way this is nothing new since Copernicus was also forced to make the equation of centre of the interior planets depend upon the motion of the Earth rather than the planet.[28]

Indeed, it appears that the correct rule for applying the equation of centre for an interior planet to the mean heliocentric planet (as opposed to the mean Sun), and a

[25] {ABB 1957}, pp. 6–7. This issue has also been discussed at great length in {GYB 2008}, pp. 495–500, 653–9, 883–9).

[26] See for example *The Almagest by Ptolemy*, translated by G. J. Toomer, London 1984.

[27] As a well-known historian of astronomy has remarked: 'In no other part of planetary theory did the fundamental error of the Ptolemaic system cause so much difficulty as in accounting for the latitudes, and these remained the chief stumbling block up to the time of Kepler' (J. L. E. Dreyer, *A History of Astronomy from Thales to Kepler*, New York 1953, p. 200).

[28] N. M. Swerdlow and O. Neugebauer, *Mathematical Astronomy in Copernicus' De Revolutionibus*, Part I, New York 1984, p. 483.

satisfactory theory of latitudes for the interior planets, were first formulated in the Greco-European astronomical tradition only in the early 17th century by Kepler.

We have already seen how the traditional Indian planetary model presented a fairly accurate computational scheme for calculating longitudes and latitudes for the exterior planets. With his revision of the traditional model, Nīlakaṇṭha arrived at a fairly accurate scheme for the interior planets also. In fact, as a computational scheme for calculating planetary longitudes and latitudes, Nīlakaṇṭha's model is indeed a good approximation to the Keplerian model of planetary motion.

F.4.4 Rationale for the revised planetary model

In his *Āryabhaṭīya-bhāṣya*, Nīlakaṇṭha explains the rationale behind his revision of the traditional planetary theory. This has to do with the fact (which, as we have mentioned above, was also noticed by several Indian astronomers prior to Nīlakaṇṭha) that the traditional planetary model employed entirely different schemes for computing the latitudes of the exterior and the interior planets. While the latitude of the exterior planets was computed from their so-called *manda-sphuṭa* (which corresponds to what we currently refer to as the true heliocentric planet), the latitudes of the interior planets was computed from their so-called *śīghrocca*. Nīlakaṇṭha argued that since the latitude should be dependent upon the deflection (from the ecliptic) of the planet itself and not of any other body, what was traditionally referred to as the *śīghrocca* of an interior planet should be identified with the planet itself. Nīlakaṇṭha also showed that this would lead to a unified treatment of the latitudinal motion of all the planets—interior as well as exterior.

In his commentary on verse 3 of *Golapāda* of Āryabhaṭa dealing with the calculation of latitudes, Nīlakaṇṭha discusses the special features that arise in the case of interior planets. It is here that he provides a detailed rationale for his revision of the traditional planetary model:

शीघ्रवशाच्च विक्षेप उक्तः। कथमेतद्युज्यते? ननु स्वबिम्बस्य विक्षेपः स्वभ्रमणवशादेव भवितुमर्हति, न पुनः अन्यभ्रमणवशादिति। सत्यम्। न पुनः अन्यस्य भ्रमणवशात् अन्यस्य विक्षेपः उपपद्यते। तस्मात् बुधः अष्टाशीत्यैव दिनैः स्वभ्रमणवृत्तं पूरयति।...
एतच्च नोपपद्यते यदेकेनैव संवत्सरेण तत्परिभ्रमणमुपलभ्यते नैवाष्टाशीत्या दिनैः। सत्यम्। भगोलपरिभ्रमणं तस्याप्येकेनैवाब्देन।...

एतदुक्तं भवति। तयोः भ्रमणवृत्तेन न भूः कबलीक्रियते। ततो बहिरेव सदा भूः। भगोलैकपार्श्वे एव तद्वृत्तस्य परिसमाप्तत्वात् तद्भगणेन न द्वादशराशिषु चारः स्यात्। तयोरपि वस्तुतः आदित्यमध्यम एव शीघ्रोच्चम्। शीघ्रोच्चभगणत्वेन पठिता एव स्वभगणाः। तथापि आदित्यभ्रमणवशादेव द्वादशराशिषु चारः स्यात्, शीघ्रवृत्तस्य कक्ष्यायाः महत्त्वात्। शीघ्रोच्चनीचवृत्तस्याप्येकभागगमेव स्वभ्रमणवृत्तम्। यथा कुजादीनामपि शीघ्रोच्चं स्वमन्दकक्ष्यामण्डलादिकमाकर्षति एवमेतयोरपि। अनयोः पुनः

तदाकर्षणवशादेव द्वादशराशिषु चार इति।[29]

> The latitudinal motion is said to be due to that of the *śīghrocca*. How is this appropriate? Isn't the latitudinal motion of a body dependent on the motion of that body only, and not on the motion of something else? The latitudinal motion of one body cannot be obtained as being due to the motion of another. Hence [we should conclude that] Mercury goes around its own orbit in 88 days ... However this also is not appropriate because we see it going around [the Earth] in one year and not in 88 days. True, the period in which Mercury completes one full revolution around the *bhagola* (the celestial sphere) is one year only [like the Sun] ...
>
> All this can be explained thus: Their [Mercury and Venus] orbits do not circumscribe the Earth. The Earth is always outside their orbit. Since their orbit is always confined to one side of the [geocentric] celestial sphere, in completing one revolution they do not go around the twelve signs (*rāśis*). Even for them in reality the mean Sun is the *śīghrocca*. It is only their own revolutions which are stated to be the revolutions of the *śīghrocca* [in *Āryabhaṭīya*]. It is only due to the revolution of the Sun [around the Earth] that they (i.e. the interior planets, Mercury and Venus) complete their movement around the twelve signs [and complete their revolution of the Earth]. Because the *śīghra* epicycle is larger than their orbit, their orbit is completed on one side of the *śīghra* epicycle. Just as in the case of Jupiter etc. [the exterior planets] the *śīghrocca* attracts [and drags around] the *manda*-orbits on which they move (the *manda-kakṣyā-maṇḍala*), in the same way it does for these [interior] planets also. And it is owing to this attraction that these [interior planets] move around the twelve signs.

There is also a later work of unknown authorship, *Vikṣepagolavāsanā*, which confirms that it was indeed Nīlakaṇṭha who proposed that the *manda* correction for the interior planets, Mercury and Venus, should be applied to the mean planets themselves and not to their *śīghrocca*, in order to arrive at a coherent and unified theory of planetary latitudes. The relevant verses of this work are the following:

पूर्वाचार्यैस्तु मान्दे अपि खलु परिधी भानुकक्ष्याकलाभिः
मात्वोक्ते तेन मान्देऽपि च दिनकरमध्यं स्वमध्यं प्रदिष्टम्।
मन्दोच्चोनार्कमध्यादुदितमृदुफलं क्षेपनीतौ चलोच्चे
कुर्वन्त्येतन्न युक्तं तदकरणमतो मानसे युक्तिमत् स्यात्॥

कुर्वन्त्यस्मिन् हि पक्षे तदिदमनुचितं भिन्नजातित्वहेतोः
तस्मात् गार्ग्येण मान्दे शशिसुतसितयोः मध्यमं स्वीयमध्यम्।
प्रोक्तं मान्दं च वृत्तं प्रमितमिह तयोः स्वीयकक्ष्याकलाभिः
शैघ्रे स्वान्मध्यवृत्तात् दिनकरवलयस्याधिकत्वेन युक्त्या।
मध्योच्चे तद्वृती चापि च विनिमयतः कल्पिते लाघवार्थम्॥[30]

> Indeed by the earlier *ācāryas*, even in the *manda* procedure, orbits [for Mercury and Venus] were stated by measuring them in terms of the orbit of the mean Sun, and hence for them their own mean position would be that of the mean Sun. For obtaining the latitudinal deflection (*kṣepanītau*) [of the planet] they were also applying the *manda*-correction (*mṛduphala*)—obtained by subtracting the *mandocca* [of the planet] from the mean

---

[29] {ABB 1957}, pp. 8-9.

[30] *Vikṣepagola-vāsanā* in {GVV 1979}, p. 52. As we shall see later, these verses closely follow the verses of Nīlakaṇṭha's *Grahassphuṭānayane vikṣepavāsanā*, in {GVV 1979}, p. 58.

Sun—to the *śīghrocca*. There is no rationale for this and that it was omitted in *Mānasa* (*Laghumānasa* of Mañjulācārya) seems quite reasonable. This approach followed by the earlier *ācāryas* is also inappropriate because the quantities [that which is used for finding the *mṛduphala* and that to which *mṛduphala* is applied] belong to different classes (*bhinnajāti*).

Therefore, it was proposed by Gārgya (Nīlakaṇṭha) that in the *manda* procedure it is their own mean position [and not the mean Sun] that should be considered as the mean position of Mercury and Venus. The dimension of *mandavṛtta* should also be taken to be given in terms of the measure of their own orbits (*svīyakakṣyā-kalābhiḥ*). In the *śīghra* process, since the orbit of the Sun is larger than their own mean orbit (*madhyavṛtta*), he also proposed that a simple way of formulating the correction would be by supposing that the mean and the *ucca* (*śīghrocca*) and their corresponding orbits (*kakṣyāvṛtta* and *śighravṛtta*) are indeed reversed.

F.5 Geometrical picture of planetary motion according to Nīlakaṇṭha

In his *Āryabhaṭīya-bhāṣya*, while commenting on verses 17–21 of the *Kālakriyāpāda*, Nīlakaṇṭha explains that the orbits of the planets, and the locations of various concentric and eccentric circles or epicycles associated with the *manda* and *śīghra* processes, are to be inferred from the computational scheme for calculating the true geocentric longitude (*sphuṭa-graha*) and the latitude of the planets (*vikṣepa*).

तत्र ताराग्रहाणां पुनरुच्चद्वयं परिधिद्वयं च प्रदर्शितम्। तत्र कः परिधिः कक्ष्यामण्डल-केन्द्रगः कस्मिन् प्रदेशे पुनरितरस्य स्थितिः इत्येतत् विक्षेपानयनकर्मणा स्फुटक्रम-वशाच्च निर्णेतुं शक्यम्।

We have explained that in the case of the *tārā-grahas* (the five planets) there are two *uccas* and two epicycles. There, issues such as which epicycle has a centre on the concentric and where the other epicycle is located, can be settled by (analysing) the procedure for finding out the true longitude and latitude of the planet.

F.5.1 Geometrical picture of the motion of the exterior planets

Nīlakaṇṭha first gives the following general outline of the geometrical picture of planetary motion:

अत्रायमभिसन्धिः। कक्ष्यामण्डलकेन्द्र एव शीघ्रपरिधेरपि केन्द्रम्। तत्परिधौ शीघ्रोच्चा-क्रान्तप्रदेशे मन्दपरिधिकेन्द्रं च। एवं परिधौ पुनर्मन्दोच्चप्रदेशे प्रतिमण्डलकेन्द्रं च। तच्च प्रतिमण्डलमाकाशकक्ष्यायाः स्वभगणावाप्तैर्योजनैस्तुल्यम्। तस्मिन्नेव ग्रहबिम्बमितरैः

समयोजनगतिर्भ्रमति। तत्तुल्यमेव तत् कक्ष्यामण्डलं शीघ्रपरिधौ उच्चप्रदेशे केन्द्रं कृत्वा परिलेखनीयम्। तत्रापि कर्णमण्डलं मन्दकर्णन्यायेन अविशेष्य परिलेखनीयम्।[31]

Here, what is intended to be conveyed is as follows: The centre of the *kakṣyā-maṇḍala* (concentric) is also the centre of the *śīghra* epicycle; on that epicycle, at the location of the *śīghrocca*, is the centre of the *manda* epicycle; in the same way, on that *manda* epicycle at the location of *mandocca* is the centre of the *pratimaṇḍala* (eccentric). (The circumference of) that *pratimaṇḍala* is equal to the circumference of the sky (*ākāśa-kakṣyā*) divided by the revolution number of the planet. The planetary orb moves with the same linear velocity, as that of the others, in that (*pratimaṇḍala*) only. The corresponding concentric (*kakṣyā-maṇḍala*) should be drawn with the same dimension with its centre on the *śīghra* epicycle at the location of *śīghrocca*. There also the circle of the hypotenuse is to be obtained by the process of iteration as per the rule for the *manda-karṇa*.

Later, while commenting on verse 3 of *Golapāda*, Nīlakaṇṭha explains how the above picture needs to be modified when the latitudinal motion is also taken into account. The main feature is that it is the *manda* epicycle together with the eccentric which is inclined to the ecliptic and not the *śīghra* epicycle (which represents the Earth–Sun relative motion):

भगोलमध्यनाभिकस्य कार्त्स्न्येन अपमण्डलमार्गगस्य शीघ्रवृत्तस्य परिधौ यः शीघ्रोच्चसमप्रदेशः तद्धि मन्दकर्मणि कक्ष्यामण्डलकेन्द्रमिति कालक्रियापादे एवे उक्तम्। तदेव मन्दोच्चनीचवृत्तस्य कर्णमण्डलस्य च केन्द्रम्। एवमेतनि त्रीणि मण्डलानि अपमण्डलमार्गमभितः अर्धशः उत्तरतो दक्षिणतश्च विक्षिप्तानि।[32]

It has already been stated in the *Kālakriyāpāda* that on the *śīghra-vṛtta*, which has its centre at the centre of the celestial sphere and is in the plane of the ecliptic, the point which corresponds to the *śīghrocca* is in fact the centre of the *kakṣyā-maṇḍala* (concentric) in the *manda* process. The same (*śīghrocca*) is also the centre of the *manda-nicocca-vṛtta* (the *manda* epicycle) and also of the (*manda*) *karṇa-maṇḍala* (the hypotenuse circle or the orbit). In this way these three circles (*manda* concentric, epicycle and hypotenuse circle) are inclined to the ecliptic towards both the north and the south.

Based on the description presented above, we arrive at the geometrical picture of motion—for an exterior planet—as shown in Fig. F.8*a*. In this figure, O represents the location of the observer and is considered to be the *bhagola-madhya* (the centre of the celestial sphere). The circle centred around O, with radius equal to the tabulated radius of the *śīghra* epicycle, r_s, is called the *śīghra-nīcocca-vṛtta*, on which the *śīghrocca* or the mean Sun S is located.

It is said that the *manda-nīcocca-vṛtta* (also called the *mandaparidhi*) is a circle with the *śīghrocca* as the centre. The *mandocca* U is located on this circle, whose radius is equal to the (variable) radius of the *manda* epicycle. The *pratimaṇḍala* on which the planet P moves is centred at the *mandocca*. SP is the *manda-karṇa* denoted by K and $\Gamma\hat{S}P$ is the *manda-sphuṭa*. $\Gamma\hat{O}P$ is the true geocentric planet known as the *śīghra-sphuṭa*. The distance of the planet from the centre of the *bhagola* is denoted by K_s and it is also the *śīghra-karṇa*.

Among the various circles depicted in Fig. F.8*a*, it is said that the circles centred around the *śīghrocca* S, namely the *manda-nīcocca-vṛtta*, the *manda-karṇa-vṛtta*

[31] {ABB 1931}, vol. II, p. 70.

[32] {ABB 1957}, p. 5.

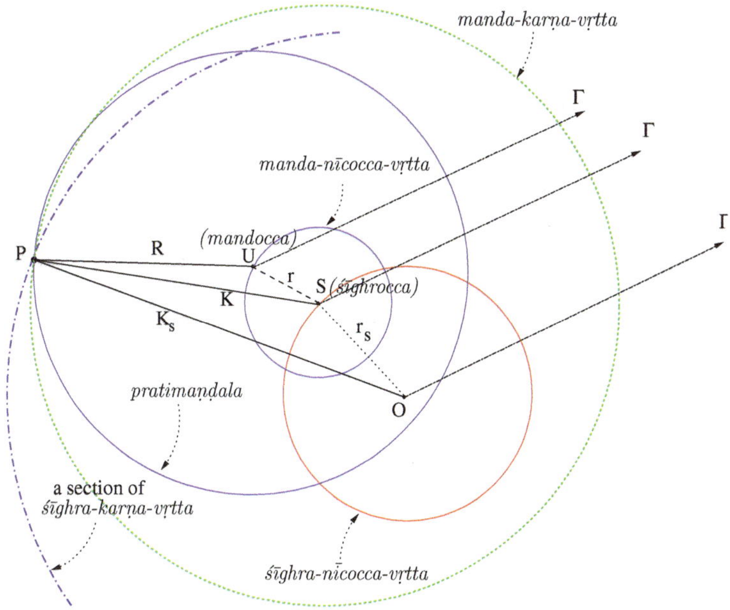

Fig. F.8*a* Geometrical picture of the motion of an exterior planet given by Nīlakaṇṭha.

and the *manda* concentric (which is not indicated in the figure), are inclined to the plane of the ecliptic towards the north and the south. The figure also depicts a section of the *śīghra-karṇa-vṛtta*—centred around *O*—which represents the instantaneous orbit (the orbit in which the planet moves at that instant) of the planet with respect to the Earth.

F.5.2 Geometrical picture of the motion of the interior planets

Nīlakaṇṭha explains in the commentary on verse 3 of *Golapāda* that the above geometrical picture of motion needs to be modified in the case of the interior planets. We have earlier (in Section F.4.4) cited a part of this discussion where Nīlakaṇṭha had noted that the interior planets go around the Sun in orbits that do not circumscribe the Earth, in a period that corresponds to the period of their latitudinal motion, and that they go around the zodiac in one year as they are dragged around the Earth by the Sun. Having identified the special feature of the orbits of the interior planets that they do not circumscribe the Earth, Nīlakaṇṭha explains that it is their own orbit, which is smaller than the *śīghra-nīcocca-vṛtta*, that is tabulated as the epicycle in a measure where the latter is 360 degrees.

तेनोभे अपि वृत्ते व्यत्यस्य कल्प्येते। तयोः शीघ्रोच्चनीचवृत्तस्य षष्टिशतत्रयांशेनैव परिमाय स्वकक्ष्यावृत्तमेव शीघ्रोच्चनीचवृत्तत्वेन कल्प्यते। मन्दोच्चनीचवृत्तञ्च तदंशेनैव परिमाय पठितम्।[33]

These two circles (the concentric and the *śīghra* epicycle) are now to be imagined in the contrary way. Of them, the concentric itself (being smaller than the epicycle) is given in units where the *śīghra* epicycle is taken to be 360°, and will now play the role of epicycle. The *manda* epicycle is also taken to be tabulated in terms of this (concentric).

Nīlakaṇṭha then goes on to explain the process of computation of the true longitude of these planets in the same manner as outlined in *Tantrasaṅgraha* and one that corresponds to the following geometrical picture of motion.

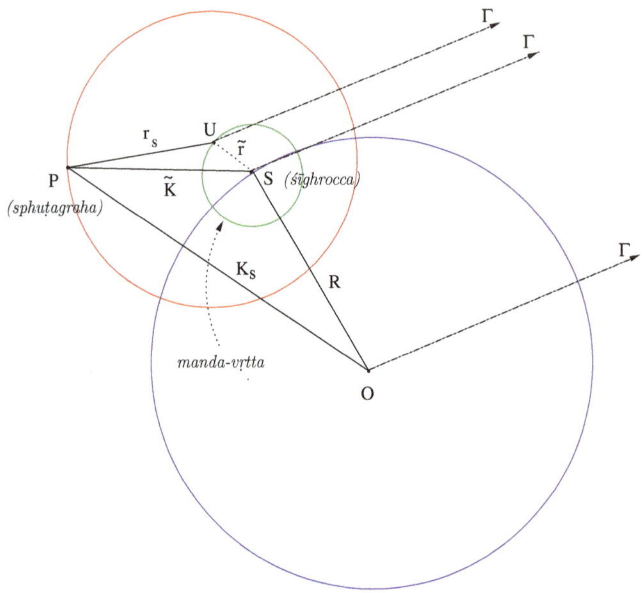

Fig. F.8*b* Geometrical picture of the motion of an interior planet given by Nīlakaṇṭha.

The geometrical picture of the motion of the interior planets as presented by Nīlakaṇṭha is shown in Fig. F.8*b*. Here, O is the observer, assumed to be at the centre of the celestial sphere (the *bhagola-madhya*). S is the *śīghrocca* which is taken to be the mean Sun for all the planets. P is the planet moving around the mean Sun in an eccentric orbit. This eccentric orbit is centred at U, the *mandocca*. The point U itself is conceived to be moving on the *manda-nīcocca-vṛtta* centred around S.

For interior planets the planet–Sun distance is smaller than the Earth–Sun distance. Hence, the radius of the planet's eccentric orbit (UP) is taken to be the radius of the *śīghrocca-nīca-vṛtta* r_s, and the radius of the mean Sun's orbit (OS) is taken

33 {ABB 1957}, p. 9.

to be the *trijyā*, R. Further, since the *mandapratimaṇḍala*, or the *manda* eccentric on which the planet moves, is of dimension r_s and not R, the (variable) *manda* epicycle r itself is to be scaled by a factor $\frac{r_s}{R}$ and will be $\tilde{r} = r\frac{r_s}{R}$. Correspondingly the (iterated) *manda-karṇa* K will also be scaled to $\tilde{K} = K\frac{r_s}{R}$.

Nīlakaṇṭha presents a clear and succinct statement of the geometrical picture of planetary motion for both interior and exterior planets in both *Golasāra* and *Siddhānta-darpaṇa*. The verses from the latter are cited below:

ग्रहभ्रमणवृत्तानि गच्छन्त्युच्चगतीन्यपि।
मन्दवृत्ते तदर्केन्द्वोः घनभूमध्यनाभिकम्॥
मध्यार्कगति चान्येषां तन्मध्यं शीघ्रवृत्तगम्।
तेषां शैघ्र्यं भचक्रान्न विक्षिप्तं गोलमध्यगम्॥
शैघ्र्यत्वेन तदंशैः स्वं प्रमायोक्तं ज्ञशुक्रयोः।
मन्दवृत्तस्य चैवात्र क्षयवृद्धी स्वकर्णवत्॥[34]

The [eccentric] orbits on which planets move (the *graha-bhramaṇa-vṛtta*) themselves move at the same rate as the apsides (the *ucca-gati*) on the *manda-vṛtta* [or the *manda* epicycle drawn with its centre coinciding with the centre of the *manda* concentric]. In the case of the Sun and the Moon, the centre of the Earth is the centre of this *manda-vṛtta*.

For the others [namely the planets Mercury, Venus, Mars, Jupiter and Saturn] the centre of the *manda-vṛtta* moves at the same rate as the mean Sun (*madhyārka-gati*) on the *śīghra-vṛtta* [or the *śīghra* epicycle drawn with its centre coinciding with the centre of the *śīghra concentric*. The *śīghra-vṛtta* for these planets is not inclined with respect to the ecliptic and has the centre of the celestial sphere as its centre.

In the case of Mercury and Venus, the dimension of the *śīghra-vṛtta* is taken to be that of the concentric and the dimensions [of the epicycles] mentioned are of their own orbits. The *manda-vṛtta* [and hence the *manda* epicycle of all the planets] undergoes increase and decrease in size in the same way as the *karṇa* [or the hypotenuse or the distance of the planet from the centre of the *manda* concentric].

As was noted earlier, the renowned Malayalam work *Gaṇita-yukti-bhāṣā* (c. 1530) of Jyeṣṭhadeva also gives a detailed exposition of the above geometrical picture planetary motion. The expressions for the longitudes for the exterior and interior planets obtained from the above pictures are essentially the same as the ones in the Keplerian model in (F.46) and (F.50).

F.6 Nīlakaṇṭha's cosmological model

While discussing the geometrical picture of planetary motion, *Āryabhaṭīya-bhāṣya* as well as *Golasāra* and *Siddhānta-darpaṇa* consider the orbit of each of the planets individually, and they are not put together in a single cosmological model of the planetary system.

There is of course a remarkable passage in *Āryabhaṭīya-bhāṣya* (which we have cited earlier (see Section F.4.4) while explaining Nīlakaṇṭha's rationale for the revision of the traditional planetary model) where Nīlakaṇṭha explains that the Earth

[34] {SDA 1978}, p. 18.

is not circumscribed by the orbit of the interior planets, Mercury and Venus; and that the mean period of motion in longitude of these planets around the Earth is the same as that of the Sun, precisely because they are being carried around the Earth by the Sun. In fact, Nīlakaṇṭha seems to be the first savant in the history of astronomy to clearly deduce from his computational scheme (and not from any speculative or cosmological argument) that the interior planets go around the Sun and that the period of their motion around the Sun is also the period of their latitudinal motion.

एतदुक्तं भवति। तयोः भ्रमणवृत्तेन न भूः कबलीक्रियते। ततो बहिरेव सदा भूः। भगोलैकपार्श्वे एव तद्वृत्तस्य परिसमाप्तत्त्वात् तद्भगणेन न द्वादशराशिषु चारः स्यात्। तयोरपि वस्तुतः आदित्यमध्यम एव शीघ्रोच्चम्। शीघ्रोच्चभगणत्वेन पठिता एव स्वभगणाः। तथापि आदित्यभ्रमणवशादेव द्वादशराशिषु चारः स्यात्, शीघ्रवृत्तस्य कक्ष्यायाः महत्त्वात्। शीघ्रोच्चनीचवृत्तस्याप्येकभागगमेव स्वभ्रमणवृत्तम्। यथा कुजादीनामपि शीघ्रोच्चं स्वमन्दकक्ष्यामण्डलादिकमाकर्षति एवमेतयोरपि। अनयोः पुनः तदाकर्षणवशादेव द्वादशराशिषु चार इति।

Nīlakaṇṭha presents his cosmological model very clearly in a remarkable short tract called *Grahasphuṭānayane vikṣepavāsanā*, which seems to have been written after *Āryabhaṭīya-bhāṣya* as it quotes extensively from it. Here he clearly integrates the geometrical picture of motion of different planets into a single model of the planetary system by identifying the *śīghrocca*, that each of the planets goes around, with the physical 'mean Sun moving on the orbit of the Sun'. Based on this identification, Nīlakaṇṭha also states that the ratio of the radius of the *śīghra* epicycle to that of the concentric is nothing but the ratio of the mean radius of the orbit of the Sun around the Earth to the mean radius of the orbit of the planet itself, in the case of the exterior planets, while it is the other way around in the case of the interior planets. He further explains that this difference between the exterior and interior planets is because, in the case of the interior planets, their orbit is smaller than the orbit of the Sun around the Earth and the dimensions of the epicycle and concentric have to be interchanged. In Nīlakaṇṭha's own words:

इन्द्वादेः स्वस्वपातद्वयत उदगवाग् अर्धशः क्रान्तिवृत्तात्
विक्षिप्ता मान्दकक्ष्या कथितनिजलवैः सर्वदा तुल्यसङ्ख्यैः।
तत्रेन्दोर्मान्दकक्ष्या ह्यपमवलयमध्यस्थकेन्द्रा कुजादे-
र्मान्दाः कक्ष्या भगोलस्थितदिनकरकक्ष्यास्थमध्यार्ककेन्द्राः॥

किञ्चारेड्यार्कजानां निजवृतिकलया मापयित्वार्ककक्ष्यां
शैघ्राण्युक्तानि वृत्तानि हि बुधसितयोस्त्वर्ककक्ष्याकलाभिः।
स्वां कक्ष्यां मापयित्वा पुनरिह कथिते शीघ्रवृत्ते यतोऽतो
भानोर्मध्यं स्वमध्यं भवति चलविधौ स्वीयमध्यं चलोच्चम्॥

पूर्वाचार्यैस्तु मान्दे अपि खलु परिधी भानुकक्ष्याकलाभिः
मात्वोक्ते तेन मान्देऽपि च दिनकरमध्यं स्वमध्यं तयोः स्यात्।
मन्दोच्चोनार्कमध्याद्दुदितमृदुफलं क्षेपनीतौ चलोच्चे

कुर्वन्त्यस्मिन् हि पक्षे तदिदमनुचितं भिन्नजातित्वहेतोः ॥

मान्दे स्वमध्यमेवात इह भृगुविधोर्मध्यमं कल्पनीयं
ग्राह्यं मान्दं च वृत्तं प्रमितमिह तयोः स्वीयकक्ष्याकलाभिः ।
शैघ्रे स्वान्मध्यवृत्तात् दिनकरवलयस्याधिकत्वादवश्यं
मध्योच्चे तद्वृती चापि च विनिमयतः कल्पनीये हि युक्त्या ॥[35]

The *manda-vṛttas* of the Moon and the others (the five planets) are deflected from the two nodes of their own orbits, half-way towards the north and the south of the ecliptic (*krānti-vṛtta*) by a measure that has been specified separately [for each planet] and which remains the same for all times. There [again] the *manda-vṛtta* of the Moon is centred at the centre of the ecliptic (*apamavalaya*), whereas the *manda-vṛttas* of Mars etc. (the five planets) are centred at the mean Sun which lies on the orbit of the Sun (*dinkara-kakṣyāstha-madhyārka*) situated in the celestial sphere (*bhagola*).

Moreover, in the case of Mars, Jupiter and Saturn, the [dimensions of their] *śīghra-vṛttas* have been stated by measuring the orbit of the [mean] Sun (*arka-kakṣyā*) in terms of minutes of (the dimensions of) their own orbits (*nija-vṛti-kalayā*). However in the case of Mercury and Venus, the [dimensions of their] *śīghra-vṛttas* have indeed (*punaḥ*) been stated by measuring their own orbits in terms of the minutes of (the dimension of) the orbit of the [mean] Sun (*arka-kakṣhyā-kalābhiḥ*). Since it is done this way (*yataḥ*), (*ataḥ*) the mean Sun becomes the mean planet in the *śīghra* procedure (*calavidhi*) and their own mean positions become the *śīghroccas* (*caloccas*).

Indeed, by the earlier *acāryas*, even in the *manda* procedure [their own] orbits [for Mercury and Venus] were stated by measuring them in terms of the orbit of the mean Sun, and hence for them their own mean position would be that of the mean Sun. Even in this school (*asmin hi pakṣe*) for obtaining the latitudinal deflection (*kṣepanītau*) [of the planet] they were applying the *manda* correction (*mṛduphala*) [which was] obtained by subtracting the *mandocca* [of the planet] from the mean Sun, to the *śīghrocca*. This is however inappropriate because these (the quantity used for finding the *mṛduphala* and the quantity to which the *mṛduphala* is applied) belong to different classes (*bhinnajāti*).

Therefore, even in the *manda* procedure it is their own mean position [and not the mean Sun] that should be considered as the mean position of Mercury and Venus. The dimension of the *manda-vṛtta* should also be taken to be given in terms of the measure of their own orbits (*svīya-kakṣyā-kalābhiḥ*). In the *śīghra* process, since the orbit of the Sun is larger than their own mean orbit (*madhyavṛtta*), one has to devise an intelligent scheme (*yuktyā*), in which the mean and the *ucca* (*śīghrocca*) and their corresponding orbits (*kakṣyā-vṛtta* and *śīghra-vṛtta*) are reversed.

The first verse clearly describes the cosmological model of Nīlakaṇṭha, which is that the five planets, Mercury, Venus, Mars, Jupiter and Saturn, go around the mean Sun in an eccentric orbit—inclined to the ecliptic (see Fig. F.9)—while the mean Sun itself goes around the Earth[36]. It is in the second verse that Nīlakaṇṭha makes the remarkable identification that

$$\frac{r_s}{R} = \frac{\text{mean Earth–Sun distance}}{\text{mean Sun–planet distance}} \quad \text{(for exterior planets)} \qquad \text{(F.29}a\text{)}$$

[35] {GVV 1979} 1979, p. 58. As we noted earlier, the initial verses of the anonymous tract *Vikṣepagolavāsanā* closely follow the above verses of Nīlakaṇṭha.

[36] As we noted earlier, this cosmological model is the same as the one proposed by Tycho Brahe, albeit on entirely different considerations, towards the end of sixteenth century.

$$\frac{r_s}{R} = \frac{\text{mean Sun–planet distance}}{\text{mean Earth–Sun distance}} \quad \text{(for interior planets).} \qquad \text{(F.29}b\text{)}$$

where r_s is the radius of the *śīghra* epicycle and R is the radius of the concentric. We had noted earlier in Section F.2 that the *śīghra*-process serves to transform the heliocentric longitudes to geocentric longitudes, precisely because the above relations (F.29*a*) and (F.29*b*) are indeed satisfied (see Table F.3), even though the traditional Indian astronomical texts did not conceive of any such relation between the radii of the *śīghra* epicycles and the mean ratios of Earth–Sun and Sun–planet distances. In fact, Nīlakaṇṭha seems to be the first Indian astronomer to explicitly state the relations (F.29*a* and F.29*b*), which seems to follow clearly from his identification of the *śīghrocca* of each planet with the physical 'mean Sun lying on the orbit of the Sun' (*dinakara-kakṣyāstha-madhyārka*).[37]

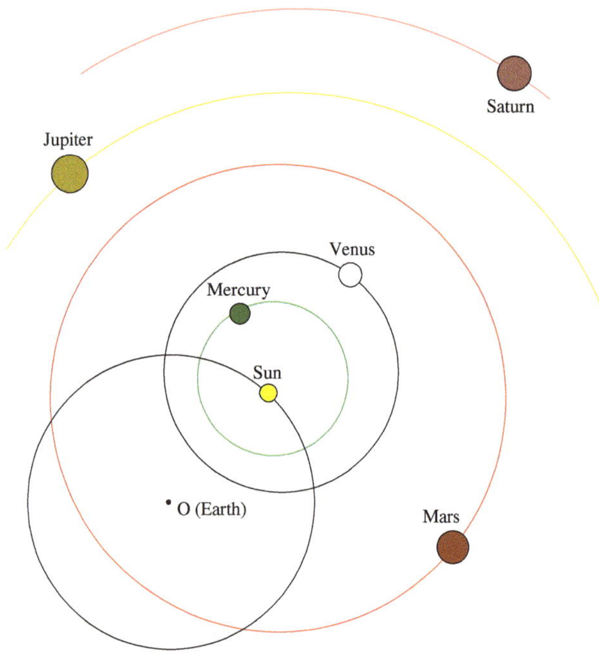

Fig. F.9 Nīlakaṇṭha's cosmological model showing the five planets moving in eccentric orbits around the mean Sun.

The last two verses above discuss the rationale behind the revised planetary model proposed by Nīlakaṇṭha and have been dealt with already in Section F.4.4. However, what is noteworthy in the context of the cosmological model of

[37] As we noted earlier, Nicholas Copernicus also seems to have arrived at the same relation (perhaps around the same time as Nīlakaṇṭha) by identifying the epicycle associated with the so-called 'solar anomaly' in the Ptolemaic model with the orbit of the Earth around the Sun in the case of the exterior planets and with the orbit of the planet itself in the case of the interior planets.

Nīlakaṇṭhais the clear statement that is found again in these verses that the orbits of the interior planets are indeed smaller than the orbit of the Sun (*dinakaravalaya*).

F.7 The problem of planetary distances

F.7.1 Planetary distances in traditional Indian astronomy

Unlike the longitudes and latitudes of planets, the planetary distances were not directly amenable to observation in ancient astronomy and their discussion was often based upon some speculative hypothesis. In traditional Indian planetary theory, at least from the time of Āryabhaṭa, the mean planetary distances were obtained based on the hypothesis that all the planets go around the Earth with the same linear velocity—i.e. they all cover the same physical distance in a given period of time.

Āryabhaṭa, indicates this principle in verse 6 of *Gītikāpāda* of *Āryabhaṭīya*, where he also mentions that one minute of arc in the orbit of the Moon measures 10 *yojanas* (which is a distance measure used in Indian Astronomy). In verse 7 of *Gītikāpāda* he gives the diameters of the Earth, Moon and the Sun in *yojanas*. The number of revolutions of the various planets (see Table F.1) are given in verses 3 and 4 of *Gītikāpāda*. Based on these, we can work out the *kakṣyā* (mean orbital circumference) and the *kakṣyāvyāsārdha* (orbital radii) of the Sun, Moon and the various planets as given in Table F. 4.

Planet	Diameter (*yojanas*)	Revolutions in a *Mahāyuga*	*Kakṣyā* (circumference) (in *yojanas*)	*Kakṣyāvyāsārdha* (radius)	Radius/Earth-diameter
Earth	1050				
Moon	315	57753336	216000	34380	65.5
Sun	4410	4320000	2887667	459620	875.5

Table F.4 *Kakṣyāvyāsārdhas* (orbital radii) of the Sun and the Moon given by Āryabhaṭa.

From Table F.4, we can see that the mean distance of the Moon has been estimated by the Indian astronomers fairly accurately (the modern value of the mean distance of Moon is about 60 Earth radii), but the estimate of the distance of Sun is short by a factor of around 30 (the modern value of the mean distance of Sun is around 23500 Earth radii).[38]

[38] The ancient astronomers' estimates of the Earth–Sun distance were all greatly off the mark. Ptolemy estimated the mean distance of the Sun to be 1210 Earth radii which is low by a factor of 20. The values given by Copernicus and Tycho were also of the same order. The value estimated by Kepler was short by a factor of 6. In 1672 the French astronomer Cassini arrived at a value which is within 10% of the actual mean distance.

Planet	Diameter (*yojanas*)	Revolutions in a *Mahāyuga*	*Kakṣyā* (circumference (in *yojanas*))
Moon	5,77,53,336	2,16,000	34,380
Sun	43,20,000	28,87,667	4,59,620
Mercury	1,79,37,020	6,95,473	1,10,696
Venus	70,22,388	17,76,421	2,82,747
Mars	22,96,824	54,31,195	8,64,481
Jupiter	3,64,224	3,42,50,133	54,51,480
Saturn	1,46,564	8,51,14,493	1,35,47,390

Table F.5 *Kakṣyāvyāsārdhas* (orbital radii) of the planets given by Āryabhaṭa.

The *kakṣyāvyāsārdhas* given in Table F.5 give the mean Earth–planet distance as per the planetary model of Āryabhaṭa. They essentially served the purpose of fixing the order of the various planets,[39] which is given by Āryabhaṭa in the *Kālakriyāpāda* of *Āryabhaṭīya*:

भानामधः शनैश्चरसुरगुरुभौमार्कशुक्रबुधचन्द्राः ।
एषामधश्च भूमिर्मेधीभूता खमध्यस्था॥[40]

Below the stars [are the orbits of] Saturn, Jupiter, Mars, Sun, Venus, Mercury and Moon. Below them is the solid Earth [suspended] in the middle of the space.

Āryabhaṭa gives a prescription for the true Earth–planet distance towards the end of *Kālakriyāpāda*:

भूताराग्रहविवरं व्यासार्धहृतः स्वकर्णसंवर्गः ।[41]

The Earth–planet distance is given by the product of the [*manda* and *śīghra*] *karṇas* of the planet divided by the radius (of the concentric).

Thus the prescription of Āryabhaṭa is that

$$\text{Earth–planet distance} = \frac{manda\text{-}karṇa \times śīghra\text{-}karṇa}{R}. \qquad \text{(F.30)}$$

Nīlakaṇṭha in his *Āryabhaṭīya-bhāṣya* explains that, since usually the *śīghra-karṇa* is evaluated with respect to a concentric of the standard radius, the above prescription of Āryabhaṭa implies that the Earth–planet distance is actually given by the *śīghra-karṇa* which is evaluated with respect to a concentric circle whose radius is given by the (iterated) *manda-karṇa*.[42] This is in accordance with his geometrical picture of planetary motion as given, say, in Fig. F.6.

[39] On the other hand, in the early Greco-European tradition, there was considerable ambiguity concerning the order of planets. Neither does Ptolemy discuss the issue of planetary distances in his *Almagest*. In his later work, *Planetary Hypothesis*, Ptolemy uses the principle that the orbit of each planet fills the entire space between those of the neighbouring planets to arrive at estimates of planetary distances.

[40] {AB 1976}, p. 102.

[41] {AB 1976}, p. 111.

[42] {ABB 1931}, pp. 53–4.

The above relation (F.30) gives the true Earth–planet distance in minutes, as usually the *manda-karṇa* and *śīghra-karṇa* are evaluated with respect to a concentric circle whose radius is given by the *trijyā*, $R \approx 3438'$. From this, the true Earth–planet distance (sometimes called the *sphuṭa-kakṣyā*) in *yojanas* is obtained by using the relation

$$\textit{Sphuṭa-kakṣyā}\ (\text{in } yjn) = \frac{\text{Earth–planet distance (in min)} \times \textit{kakṣyā-vyāsārdha}\ (\text{in } yjn)}{\text{Radius (in min)}}. \tag{F.31}$$

The above relation is based on the hypothesis employed in the traditional Indian planetary theory that the *kakṣyāvyāsārdha* given in Table F.5 represents the mean Earth–planet distance in *yojanas.*

F.7.2 Nīlakaṇṭha *on planetary distances*

In the fourth chapter of *Tantrasaṅgraha*, dealing with lunar eclipses, Nīlakaṇṭha gives the mean radius of the orbit of the Moon in *yojanas* to be the *trijyā* (radius) in minutes multiplied by 10, i.e. 34380 *yojanas.* He also states that the radii of the orbits of the Sun and the Moon are in inverse proportion to their *bhagaṇas*, or the number of revolutions in a *Mahāyuga.* He further gives the diameters of the Moon and Sun in *yojanas* to be 315 and 4410, respectively, and also states that the diameter of the Earth is to be found from the circumference of 3,300 *yojanas* given in verse 1.29. Table F. 6 gives diameters and mean distances in *yojanas.*

Planet	Diameter (*yojanas*)	Revolutions in a *Mahāyuga*	*Kakṣyā* (circumference) (in *yojanas*)	*Kakṣyā-vyā-sārdha* (radius) (in *yojanas*)	Radius/Earth-diameter
Earth	1050.4				
Moon	315	5,77,53,336	216,000	34,380	65.5
Sun	4410	43,20,000	28,87,667	4,59,620	875.5

Table F.6 *Kakṣyāvyāsārdhas* (orbital radii) of the Sun and the Moon given by Nīlakaṇṭha.

Nīlakaṇṭha then states that the *sphuṭa-yojana-karṇas*, the first approximations to the true distance of the centres of Sun and Moon from the centre of the Earth, are given by their mean distances multiplied by the iterated *manda-karṇa* divided by the radius. Finally he gives the *dvitīya-sphuṭa-yojana-karṇas*, the true distances taking into account the second correction, corresponding to the so-called evection term, for both Sun and Moon at times of conjunction and opposition. The general expression for *dvitīya-sphuṭa-yojana-karṇa* is given in the first two verses of Chapter

8. *Tantrasaṅgraha* does not discuss the corresponding geometrical picture of lunar motion, which is however dealt with in detail in *Yuktibhāṣā*[43].

Nīlakaṇṭha takes up the issue of planetary distances towards the very end of the last chapter (Chapter 8) of *Tantrasaṅgraha*. Here, he first notes that the mean radius of the orbit (*kakṣyāvyāsārdha*) of each planet is to be found in the same way as was prescribed in the case of the Sun in Chapter 4, namely by multiplying the *kakṣyāvyāsārdha* and the revolutions in a *Mahāyuga* of the Moon, and dividing the product by the revolutions of the planet in a *Mahāyuga*.

रविवद्यन्द्रकक्ष्याया नेयान्येषां हि सा ततः ।[44]

This is essentially the principle of traditional Indian astronomy that all the planets travel equal distances in their orbits in any given period of time, or that they all have the same linear velocity. Nīlakaṇṭha in fact states this principle explicitly in his *Siddhānta-darpaṇa* as follows:

ग्रहयोजनभुक्तिः स्याद् दशघ्नेन्दोः कलागतिः ।[45]

The velocity in minutes [per unit time] (*kalāgati*) of the Moon multiplied by 10 is the velocity of [each] planet in *yojanas* [per unit time] (*yojanabhukti*).

Based on the number of revolutions given in Chapter 1 of *Tantrasaṅgraha* we can calculate the mean orbital radii (*kakṣyāvyāsārdha*) of all the planets as given in Table F.7.

Planet	Revolutions in a *Mahāyuga*	*Kakṣyā* (circumference in *yojanas*)	*Kakṣyāvyāsārdha* (radius in *yojanas*)
Moon	57753320	216000	34380
Sun	4320000	2887666	459620
Mercury	17937048	695472	110696
Venus	7022268	1776451	282752
Mars	2296864	5431195	864465
Jupiter	364180	34254262	5452137
Saturn	146612	85086604	13542951

Table F.7 *Kakṣyāvyāsārdhas* (orbital radii) of the planets given by Nīlakaṇṭha.

While the values of the *kakṣyāvyāsārdha* given by Nīlakaṇṭha differ only marginally from those given in *Āryabhaṭīya* (see Table F.5), Nīlakaṇṭha's inter-

[43] {GYB 2008}, Section 11.36, pp. 584–7, 786–8, 975–80. It may be of interest to note that the maximum variation in the distance of Moon due to the second correction in Nīlakaṇṭha's model is only of the order of 10% and not the ridiculous figure of around 50% found in the Ptolemaic model of evection. Of course, the expression for the second correction given by Nīlakaṇṭha is essentially the same as the one given by Mañjulācārya (c. 932) and is more accurate and elegant than the Ptolemaic formulation of evection. See also M. S. Sriram, Planetary and Lunar Models in *Tantrasaṅgraha* and *Gaṇita-Yuktibhāṣā*, in *Studies in History of Indian Mathematics*, ed. by C. S. Seshadri, Hindustan Book Agency, New Delhi 2010, pp. 353–89.

[44] {TS 1958}, p. 154.

[45] {SDA 1976}, p. 13.

pretation of this *kakṣyāvyāsārdha* and his prescription for the true Earth-planet distance in *yojanas* (the *sphuṭa-kakṣyā*) are indeed very different from what we outlined earlier in connection with the traditional Indian planetary model. Nīlakaṇṭha presents his prescription for *sphuṭa-kakṣyā* rather tritely in just a single verse of *Tantrasaṅgraha*:

शीघ्रकर्णघ्नकक्ष्यायास्तद्वृत्तेन सितज्ञयोः ।
आप्ता हि स्फुटकक्ष्या स्यात् तद्वशाल्लम्बनादि च॥[46]

[In the case of Mars, Jupiter and Saturn], the mean radii of their orbits (*kakṣyāvyāsārdhas*) multiplied by the *śīghra-karṇa* [and divided by the *trijyā*] give the true orbital radii (*sphuṭa-kakṣyās*). In the case of Mercury and Venus their mean orbital radii (*kakṣyāvyāsārdha*) multiplied by the *śīghra-karṇa* and divided by the mean radii of their own orbits (*tad-vṛttas*), give the true values of their orbital radii (*sphuṭa-kakṣyās*). And from that the *lambana* etc. [must be calculated].

The above prescription has been clearly explained by Śaṅkara Vāriyar as follows:

सितज्ञयोरुक्तवदानीतं कक्ष्याव्यासार्धयोजनं स्वशीघ्रकर्णेन निहत्य शीघ्रवृत्तप्रमित-स्ववृत्तव्यासार्धेन विभजेत्। तत्र लब्धा स्फुटकक्ष्या भवति। अन्येषां तु प्राग्वल्लब्धं कक्ष्याव्यासार्धं स्वशीघ्रगुणितं त्रिज्यया विभक्तमिति विशेषः।[47]

In the case of Mercury and Venus, the mean radii of their orbit in *yojanas* (*kakṣyāvyāsārdha-yojana*) has to be multiplied by the *śīghra-karṇa* and divided by the radius of their own orbit which is the indeed the *śīghra-vṛtta*. The result is the true radius of the orbit (*sphuṭa-kakṣyā*) [in *yojanas*]. For the other planets (Mars, Jupiter and Saturn) the difference is that the mean radii (*kakṣyāvyāsārdhas*) [in *yojanas*] obtained as before and multiplied by their own *śīghra* [*karṇa*] should be divided by the radius of the concentric (the *trijyā*) [in order to obtain true radius of the orbit in *yojanas*].

There is a verse in *Golasāra* which seems to give a partially similar prescription for the case of interior planets:

स्वक्षितिविवरघ्नं तद्योजनमपि केवलान्त्यफलभाज्यम्॥[48]

[For Mercury and Venus] their distance from the Earth (their *śīghra-karṇa*) multiplied by their (mean orbit radius in) *yojanas* is to be divided only by their last *śīghra-phala* (or the radius of the *śīghra* epicycle).

Thus, Nīlakaṇṭha's prescription for the *sphuṭa-kakṣyā* or the true Earth–planet distance in *yojanas* can be expressed as follows:

$$\textit{Sphuṭa-kakṣyā} = \frac{\textit{kakṣyāvyāsārdha} \times \textit{śīghra-karṇa}}{\text{Radius}} \qquad \text{[exterior]} \quad \text{(F.32)}$$

$$\textit{Sphuṭa-kakṣyā} = \frac{\textit{kakṣyāvyāsārdha} \times \textit{śīghra-karṇa}}{\text{Radius of } \textit{śīghra} \text{ epicycle}} \qquad \text{[interior].} \quad \text{(F.33)}$$

---

[46] {TS 1958}, chapter 8, verses 37b–38a.

[47] {TS 1958}, p. 155.

[48] {GS 1970}, p. 23.

The expression for the *sphuṭa-kakṣyā* for the exterior planets seems to be the same as that given by (F.31) used in the traditional planetary models, while that for the interior planets (F.33) differs by the fact that the radius (of the concentric) in the denominator in (F.31) is replaced by the radius of the *śīghra* epicycle.[49] In other words, the *kakṣyāvyāsārdha* for Nīlakaṇṭha is a mean distance in *yojanas* which corresponds to the radius of the concentric in the case of the exterior planets; and it is a mean distance in *yojanas* corresponding to the radius of the *śīghra* epicycle in the case of interior planets. If we take a careful look at the geometrical picture of planetary motion given in Fig. F.8*a* and Fig. F.8*b*, we can easily see that, according to Nīlakaṇṭha, the *kakṣyāvyāsārdha* in *yojanas* (given in Table F.7), following the equal linear velocity principle, is not the mean Earth–planet distance, but is in fact the *śīghrocca*–planet distance.

This fact that the *kakṣyāvyāsārdha* in *yojanas*, obtained based on the principle that all the planets cover equal distances in equal times, should be understood as the mean *śīghrocca*–planet distance (and not the mean Earth–planet distance) has been clearly stated by Nīlakaṇṭha in the passage from *Āryabhaṭīya-bhāṣya* that we cited earlier while discussing the geometrical picture of planetary motion:

> कक्ष्यामण्डलकेन्द्र एव शीघ्रपरिधेरपि केन्द्रम्। तत्परिधौ शीघ्रोच्चाक्रान्तप्रदेशे मन्दपरिधिकेन्द्रं च। एवं परिधौ पुनर्मन्दोच्चप्रदेशे प्रतिमण्डलकेन्द्रं च। तच्च प्रतिमण्डलमाकाशकक्ष्यायाः स्वभगणावाप्तैर्योजनैस्तुल्यम्। तस्मिन्नेव ग्रहबिम्बमितरैः समयोजनगतिर्भ्रमति।
>
> The centre of the *kakṣyā-maṇḍala* (concentric) is also the centre of the *śīghra* epicycle; on that epicycle, at the location of the *śīghrocca*, is the centre of the *manda* epicycle; in the same way, on that *manda* epicycle at the location of *mandocca* is the centre of the *pratimaṇḍala* (eccentric). (The circumference of) that *pratimaṇḍala* is equal to the circumference of the sky (*ākāśa-kakṣyā*) divided by the revolution number of the planet. The planetary orb moves with the same linear velocity as that of the others in that (*pratimaṇḍala*) only.

In the above passage in *Āryabhaṭīya-bhāṣya*, Nīlakaṇṭha states that the planets are orbiting with equal linear velocity in eccentric orbits about the *śīghrocca*. In other words, the *kakṣyāvyāsārdhas* in *yojanas* given in Table F.7 refer to the mean *śīghrocca*–planet distances in Nīlakaṇṭha's model. This seems to be a major departure from the conventional identification of these *kakṣyāvyāsārdhas* (derived in inverse ratio with *bhagaṇas*) with mean Earth–planet distances.

Thus, both in his *Tantrasaṅgraha* (c. 1500 CE) and in the later work *Āryabhaṭīya-bhāṣya*, Nīlakaṇṭha seems to be clearly working towards an alternative cosmology, where the planets—Mercury, Venus, Mars, Jupiter and Saturn—all go around the *śīghrocca*. His attempt to modify the traditional prescription for the planetary distances is also a step in this direction. However, even this modified prescription for the planetary distances that Nīlakaṇṭha proposes in *Tantrasaṅgraha* and

[49] This important difference between the *sphuṭa-kakṣyās* for the exterior and interior planets, in Nīlakaṇṭha's theory, seems to have been overlooked by Pingree in his analysis of 'Nīlakaṇṭha's Planetary Models' (D. Pingree, Journal of Indian Philosophy 29, 187–95, 2001). Pingree uses the *Sphuṭa-kakṣyā* formula (F.32), as applicable to the exterior planets, to arrive at the upper and lower limits of the Earth–planet distance in the case of Venus.

Āryabhaṭīya-bhāṣya is not really consistent with the cosmological model that he clearly enunciates in his later tract *Grahasphuṭānayane vikṣepavāsanā.* It is herein that Nīlakaṇṭha identifies the *śīghrocca* with the physical mean Sun and also gives the relations (F.29*a*) and (F.29*b*) between the ratio of the radii of the *śīghra* epicycle and the concentric with the ratio of the Earth–planet and Earth–Sun distances. Since the size of *śīghra* epicycles have already been fixed (see the tabulated values of radii of *śīghra* epicycles both in traditional planetary theory and in Nīlakaṇṭha's model in Table F.3), there is no longer any freedom to introduce a separate new hypothesis for the determination of the *śīghrocca*–planet distances.

Therefore, Nīlakaṇṭha's relations (F.32) and (F.33) for the planetary distances (however revolutionary they may be in relation to the traditional planetary models) are not consistent with the cosmological model definitively stated by Nīlakaṇṭha in *Grahasphuṭānayane vikṣepavāsanā.* In fact, once the *śīghrocca* of all the planets is identified with the physical mean Sun, the planetary distances get completely determined by the dimensions of the *śīghra* epicycles which are related to the ratios of the mean Sun–planet and Earth–Sun distances. The true Earth-planet distances in *yojanas* would then be given by the following:

$$\textit{Sphuṭa-kakṣyā} = \frac{\textit{kakṣyāvyāsārdha} \text{ of the Sun} \times \textit{śīghra-karṇa}}{\text{Radius of } \textit{śīghra} \text{ epicycle}} \quad \text{[ext.]} \tag{F.34}$$

$$\textit{Sphuṭa-kakṣyā} = \frac{\textit{kakṣyāvyāsārdha} \text{ of the Sun} \times \textit{śīghra-karṇa}}{\text{Radius}} \quad \text{[int.].} \tag{F.35}$$

The above relations follow from the fact that the mean orbit of the Sun is the *śīghra* epicycle in the case of the exterior planet, while it would be the concentric in the case of the interior planet.

It would be interesting to see whether any of the later works of Nīlakaṇṭha (which are yet to be located) or any of the works of later Kerala astronomers deal with these implications of the cosmological model of Nīlakaṇṭha for the calculation of planetary distances.

F.8 Annexure: Keplerian model of planetary motion

The planetary models described above can be appreciated better if we understand how the geocentric coordinates of a planet are calculated in Kepler's model. The three laws of planetary motion discovered by Kepler in the early seventeenth century, which form the basis of our present understanding of planetary orbits, may be expressed as follows:

1. Each planet moves around the Sun in an ellipse, with the Sun at one of the foci.
2. The areal velocity of a planet in its orbit is a constant.
3. The square of the orbital period of a planet is proportional to the cube of the semi-major axis of the ellipse in which it moves.

Kepler's laws can be derived from Newton's second law of motion and the law of gravitation. It may be recalled that Kepler's laws are essentially kinematical laws, which do not make any reference to the concepts of 'acceleration' and 'force', as we understand them today. Even then, they capture the very essence of the nature of planetary orbits and can be used to calculate the planetary positions, once we know the parameters of the ellipse and the initial coordinates. Since the planetary models proposed in Indian astronomy are also kinematical in nature, it makes sense to compare the two. So in what follows we will attempt to summarize the computation of the geocentric longitude and latitude of a planet which follows from Kepler's laws. This will also help in understanding the similarity that exists between the Keplerian model and the computational scheme adopted by the Indian astronomers.

F.8.1 Elliptic orbits and the equation of centre

A schematic sketch of the elliptic orbit of a planet P, moving around the Sun S with the latter at one of its foci is shown in Fig. F.10. Here a and b represent the semi-major and semi-minor axes of the ellipse. Γ refers to the first point of Aries. $\theta_a = \Gamma\hat{S}A$ denotes the longitude of the aphelion (A) and $\theta_h = \Gamma\hat{S}P$ is the heliocentric longitude of the planet.

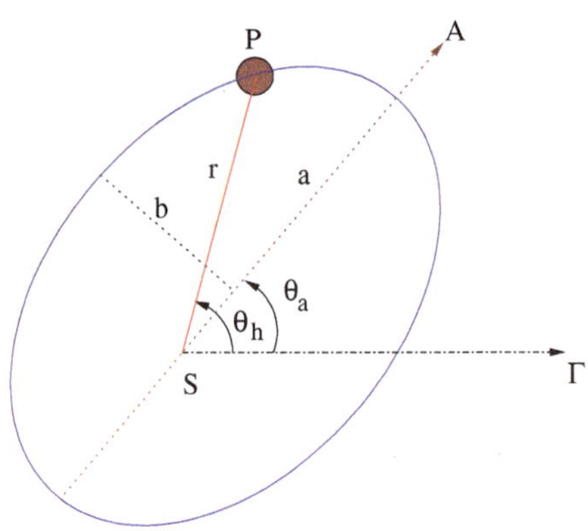

Fig. F.10 Elliptic orbit of a planet around the Sun.

The equation of the ellipse (in polar coordinates, with the origin at one of the foci), may be written as

$$\frac{l}{r} = 1 - e\cos(\theta_h - \theta_a), \qquad \text{(F.36)}$$

where e is the eccentricity of the ellipse and $l = a(1-e^2)$. Therefore

$$r = l[1 + e\cos(\theta_h - \theta_a)] + O(e^2),$$

$$r^2 = l^2[1 + 2e\cos(\theta_h - \theta_a)] + O(e^2). \tag{F.37}$$

As the area of an ellipse is πab, the areal velocity can also be written as $\frac{\pi ab}{T} = \frac{\omega ab}{2}$, where T is the time period and $\omega = \frac{2\pi}{T}$ is the mean angular velocity of the planet. Since the areal velocity of the planet at any instant is given by $\frac{1}{2}r^2\dot{\theta}_h$, and is a constant according to Kepler's second law, we have

$$r^2\dot{\theta}_h = \omega ab. \tag{F.38}$$

Using the above expression for r^2 in (F.37), we find

$$l^2\dot{\theta}_h[1 + 2e\cos(\theta_h - \theta_a)] = \omega ab + O(e^2). \tag{F.39}$$

Now $l = a\,(1 - e^2) = a + O(e^2)$ and $ab = a^2 + O(e^2)$. Hence

$$\dot{\theta}_h[1 + 2e\cos(\theta_h - \theta_a)] \approx \omega, \tag{F.40}$$

where the equation is correct to $O(e)$. Integrating with respect to time, we obtain

$$\begin{aligned} &\theta_h + 2e\sin(\theta_h - \theta_a)] \approx \omega t, \\ \text{or} \quad &\theta_h - \omega t = -2e\sin(\theta_h - \theta_a). \end{aligned} \tag{F.41}$$

The argument of the sine function in the above equation involves θ_h, the actual heliocentric longitude of the planet, which is to be determined from the mean longitude θ_0. However, θ_h may be expressed in terms of θ_0 to $O(e^2)$. On so doing, the above equation reduces to

$$\theta_h - \omega t = \theta_h - \theta_0 = -2e\sin(\theta_0 - \theta_a) + O(e^2). \tag{F.42}$$

It may be noted that in (F.42) we have written ωt as θ_0, as the mean longitude of the planet increases linearly with time, t. $\theta_0 - \theta_a$, the difference between the longitudes of the mean planet and the apogee/aphelion, is known as the 'anomaly'. It may be noted that this difference is termed the *manda-kendra* in Indian astronomy. Thus (F.42) gives the equation of centre which is the difference between the true heliocentric longitude θ_h and the mean longitude θ_0, correct to $O(e)$, in terms of the anomaly. It is straightforward to see that the equation of centre correction arises owing to the eccentricity of the orbit and that its magnitude depends upon the value of the anomaly.

F.8.2 Geocentric longitude of an exterior planet

The orbits of all the planets are inclined at small angles to the plane of the Earth's orbit around the Sun, known as the ecliptic. We will ignore these inclinations and assume that all the planetary orbits lie on the plane of the ecliptic while calculating the planetary longitudes, as the corrections introduced by these inclinations are known to be small. We will consider the longitude of an exterior planet, i.e. Mars, Jupiter or Saturn, first and then proceed to discuss separately the same for an interior planet, i.e. Mercury or Venus.

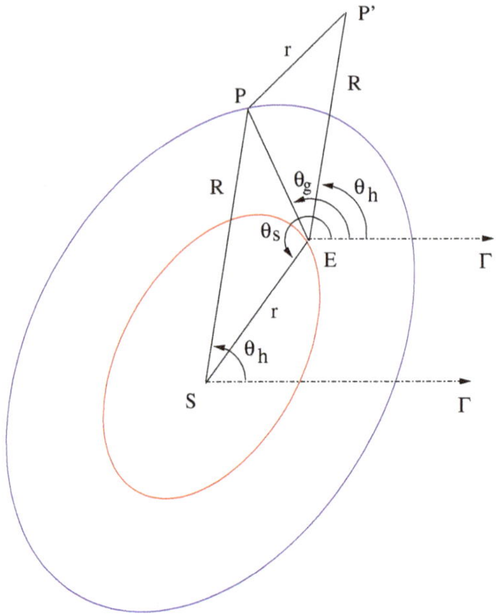

Fig. F.11 Heliocentric and geocentric longitudes of an exterior planet in Kepler's model.

The elliptic orbit of an exterior planet P and that of the Earth E around the Sun S are shown in Fig. F.11. Here, $\theta_h = \Gamma\hat{S}P$ is the true heliocentric longitude of the planet. $\theta_S = \Gamma\hat{E}S$ and $\theta_g = \Gamma\hat{E}P$ are the true geocentric longitudes of the Sun and the planet respectively, while r and R are the distances of the Earth and the planet from the Sun, which vary along their orbits.

We draw $EP' = R$ parallel to SP. Then, by construction, $P'P = r$ is parallel to ES. In the previous section (see (F.42)) it was described how θ_h is computed from the mean longitude θ_0, by applying the equation of centre. Now we need to obtain the true geocentric longitude θ_g from the heliolcentric longitude θ_h. It may be noted that

$$E\hat{P}S = P\hat{E}P' = \theta_g - \theta_h \qquad \text{and} \qquad E\hat{S}P = 180^\circ - (\theta_s - \theta_h). \tag{F.43}$$

In the triangle ESP,

$$\begin{aligned} EP^2 &= R^2 + r^2 - 2rR\cos[180^\circ - (\theta_s - \theta_h)], \\ \text{or} \qquad EP &= [(R + r\cos(\theta_s - \theta_h))^2 + r^2\sin^2(\theta_s - \theta_h)]^{\frac{1}{2}}. \end{aligned} \tag{F.44}$$

Also,

$$\frac{\sin(E\hat{P}S)}{ES} = \frac{\sin(E\hat{S}P)}{EP}. \qquad \text{(F.45)}$$

Using (F.43)–(F.44), we have

$$\sin(\theta_g - \theta_h) = \frac{r\sin(\theta_s - \theta_h)}{[(R + r\cos(\theta_s - \theta_h))^2 + r^2\sin^2(\theta_s - \theta_h)]^{\frac{1}{2}}}. \qquad \text{(F.46)}$$

Here $(\theta_s - \theta_h)$, the difference between the longitude of the Sun and that of the heliocentric planet, is known as the 'solar anomaly' or 'anomaly of conjunction'.[50] Thus (F.46) gives $(\theta_g - \theta_h)$ in terms of the solar anomaly. Adding this to θ_h, we get the true geocentric longitude θ_g of the planet.

F.8.3 Geocentric longitude of an interior planet

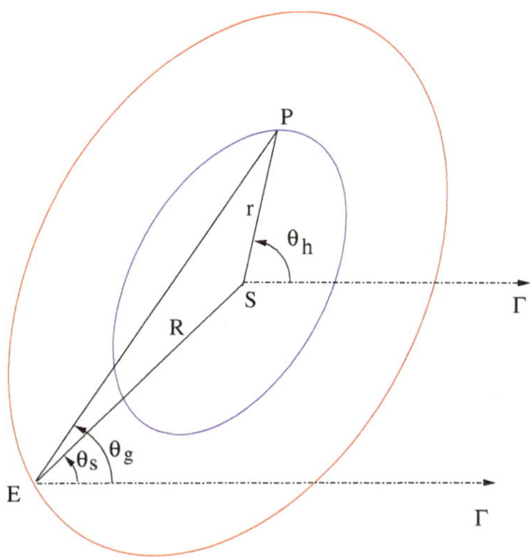

Fig. F.12 Heliocentric and geocentric longitudes of an interior planet in Kepler's model.

The elliptic orbit of an interior planet P and that of the Earth E around the Sun are shown in Fig. F.12. Here, $\theta_h = \Gamma\hat{S}P$ is the true heliocentric longitude of the planet, which can be computed from the mean heliocentric longitude and the equation of centre (see (F.42)). $\theta_s = \Gamma\hat{E}S$ and $\theta_g = \Gamma\hat{E}P$ are the true geocentric longitudes of the Sun and the planet respectively. As in the case of exterior planets, here too r

[50] The equivalent of this in Indian astronomy is the difference between the longitude of the *manda-sphuṭa* θ_{ms} and that of the *śīghrocca* θ_s, known as the *śīghra-kendra*.

and R represent the variable distances of the planet and the Earth from the Sun respectively.

It can easily be seen that

$$S\hat{E}P = \theta_g - \theta_s \qquad \text{and} \qquad E\hat{S}P = 180^\circ - (\theta_h - \theta_s). \tag{F.47}$$

Now considering the triangle ESP, we have

$$EP = [(R + r\cos(\theta_h - \theta_s))^2 + r^2 \sin^2(\theta_h - \theta_s)]^{\frac{1}{2}}. \tag{F.48}$$

Also,

$$\frac{\sin(S\hat{E}P)}{SP} = \frac{\sin(E\hat{S}P)}{EP}. \tag{F.49}$$

Using (F.47)–(F.49), we get

$$\sin(\theta_g - \theta_s) = \frac{r\sin(\theta_h - \theta_s)}{[(R + r\cos(\theta_h - \theta_s))^2 + r^2 \sin^2(\theta_h - \theta_s)]^{\frac{1}{2}}}. \tag{F.50}$$

Since all the parameters in the RHS of the above equation are known, the difference $(\theta_g - \theta_s)$ can be determined from this equation. Adding θ_s to this, we get the true geocentric longitude, θ_g of the planet. We now proceed to explain how the latitude of a planet is obtained in the Keplerian model.

F.8.4 Heliocentric and geocentric latitudes of a planet

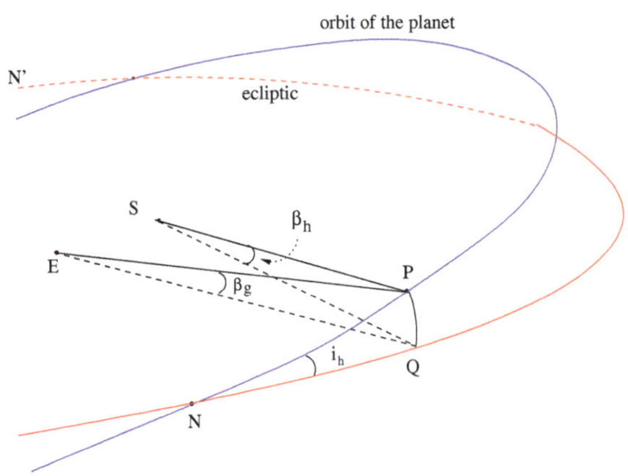

Fig. F.13 Heliocentric and geocentric latitudes of a planet in Kepler's model.

In Fig. F.13, the orbit of the planet P is shown to be inclined at an angle i_h to the ecliptic. N and N' are the nodes of the planetary orbit. PQ is the circular arc perpendicular to the ecliptic. Then the heliocentric latitude β_h is given by

$$\beta_h = \frac{PQ}{SP}. \tag{F.51}$$

If λ_P and λ_N are the heliocentric longitudes of the planet and the node, it can easily be seen that

$$\sin\beta_h = \sin i_h \, \sin(\lambda_P - \lambda_N) \qquad \text{or} \qquad \beta_h \approx i_h \, \sin(\lambda_P - \lambda_N), \tag{F.52}$$

as i_h and β_s are small. In the figure we have also shown the location of the Earth E. The latitude β_g (geocentric latitude) as measured from E would be different from the one measured from the Sun and is given by

$$\beta_g = \frac{PQ}{EP}. \tag{F.53}$$

From (F.51)–(F.53), we find that

$$\begin{aligned} \beta_g &= \beta_h \frac{SP}{EP} \\ &= \frac{i_h \, SP \, \sin(\lambda_P - \lambda_N)}{EP}, \end{aligned} \tag{F.54}$$

where EP, the true distance of the planet from the Earth, can be found from (F.44) or (F.48).

Glossary

adhika	Excess; additive.
adhikamāsa, adhi-māsa	Intercalary month: a lunar month in which no *saṅkrānti* (solar transit across zodiacal signs) occurs; considered to be excess and is not counted as a part of the lunar year.
āḍhya	Quantity that is to be added.
ādi	Beginning, starting point.
ādityamadhyama	(1) The mean Sun. (2) The mean longitude of the Sun.
agrā	Amplitude at rising, that is, the perpendicular distance of the rising point from the east–west line; the Rsine thereof.
agrāṅgula	*agrā* specified in *aṅgulas*.
Ahargaṇa	Count of days; number of civil days elapsed since the commencement of a chosen epoch.
āhatya	Having multiplied (same as *hatvā*).
ahorātra	Day (day + night); civil day.
ahorātravṛtta, dyuvṛtta	Diurnal circle: a small circle parallel to the celestial equator corresponding to a definite declination, along which a celestial body moves during the course of a day.
ākāśa	(1) Sky. (2) Number zero in the *Bhūtasaṅkhyā* system.
ākāśakakṣyā, ambarakakṣyā	Boundary circle of the sky, the circumference of which is the linear distance traversed by a planet in a *yuga*, equal to 12474720576000 *yojanas*.
akṣa	Terrestrial latitude (see also *vikṣepa*); Rsine of terrestrial latitude.
ākṣa	Relating to (terrestrial) latitude.
akṣacāpa	The arc corresponding to the terrestrial latitude.
akṣadṛkkarma	Correction to quantities due to the latitude of the observer.

akṣajīvā, akṣajyā	Rsine of the terrestrial latitude.
akṣakṣetra	Latitudinal triangle: right-angled triangle in which one of the angles is the latitude of the observer.
akṣamaurvikā	Same as *akṣajyā*.
akṣavalana	Deflection due to the latitude of the observer. Part of the inclination of the ecliptic to the local vertical, due to the observer's latitude.
amāvāsī, amāvāsyā	New Moon day, the end of which marks the commencement of a lunar month in the *amānta* system.
aṃhaspati	Name of the *adhimāsa* (lunar month without a solar transit) that is succeeded by a *kṣayamāsa* (lunar month with two solar transits), both of which are considered to be an integral part of the lunar year.
aṃśa	(1) Part. (2) Numerator. (3) Degree, $\frac{1}{360}$th of a circle. (4) Fraction.
aṅgula	A unit of measurement used to measure linear distances, taken to be approximately an inch.
antarāla	(1) Difference. (2) The perpendicular distance from a point to a straight line or plane. (3) Divergence. (4) Intervening.
antya	(1) 10^{15} (Place and number). (2) The digit of highest denomination. (3) The last term in a series.
antyakarṇa	The last hypotenuse in the iterative process for the computation of the *manda*-hypotenuse K, such that the relation $\frac{r_0}{R} = \frac{r}{K}$ is satisfied.
antyakrānti	Maximum declination, taken to be 24 degrees, which is the same as the inclination of the ecliptic to the celestial equator.
antyaśravaṇa	See *antyakarṇa*.
apakrama	Declination of a celestial body measured along the meridian circle from the equator towards the north/south pole; Rsine of the declination.
apakramajyā	Rsine of the declination.
apakramamaṇḍala, apakramavṛtta, apamaṇḍala	Ecliptic: the great circle in the celestial sphere along which the Sun moves in the background of stars, during the course of a year. This is the reference circle for the measurement of the celestial longitude.
aparapakṣa	The part of the lunar month from full moon to new moon, during which the Moon's phase wanes.
aparaviṣuvat	Autumnal equinox: the point at which the Sun coursing along the ecliptic crosses the celestial equator from the north to the south.

ardhajyā (*jyā*)	Rsine of an arc, which of half of the chord.
arkāgrā	(1) Measure of the amplitude in the arc of the celestial horizon lying between the east point and point where the heavenly body concerned rises. (2) The distance from the extremity of the gnomonic shadow and the equinoctial shadow.
arkāgrāṅgula	Measure of the *arkāgrā* in *aṅgulas*.
ārkṣa, *nākṣatra*	Related to a star, i.e. sidereal.
ārttavatsara	Tropical year, from *viṣuvat* (equinox) to *viṣuvat*; also referred to as *sāyanavatsara*.
āśāgrā	Amplitude: angle between the vertical passing through the celestial object and the prime vertical; Rsine thereof.
āśāgrākoṭi	Rcosine of amplitude.
asita	Not bright/white, generally used to refer to (1) the dark fortnight, (2) the non-illuminated portion of the moon during an eclipse.
aśra	(1) A side of a polygon. (2) An edge.
asta, *astamaya*	Setting, diurnal as well as heliacal.
astalagna	(1) *Lagna* (orient ecliptic point) at the time of a planet's setting. (2) Setting or occident ecliptic point.
asu, *prāṇa*	21600th part of a sidereal day, or 4 sidereal seconds, which is presumed to correspond to the time taken by a healthy human being to inhale and exhale once.
avalambaka	Plumb-line that marks the perpendicular to the horizon.
avama, *tithikṣaya*	Omitted/declined *tithi*: a *tithi* that commences after sunrise and ends before the next sunrise, during which special/auspicious events are not performed.
avāntarayuga (*yuga*)	Unit of time, viz. 576 years (210389 days) adopted by some Hindu astronomers (referred to simply as a *yuga*).
aviśeṣa, *aviśiṣṭa*	Literally, 'no distinction'; Generally employed to qualify a quantity obtained using an iterative process.
aviśeṣa-karṇa, *aviśiṣṭa-manda-karṇa*	Hypotenuse obtained by using an iterative process prescribed in connection with the *manda-saṃskāra* or the equation of centre correction.
ayana	(1) The solsticial points. (2) Northward and/or southward motion of the Sun or other planets towards these points.
ayanacalana	Motion of the equinoxes as well as solsticial points.
ayanadṛkkarma	Correction due to the obliquity of the ecliptic.
ayanāntonnati	Elevation of the solstices.
ayanasandhi, *ayanānta*	Solstices (summer/winter), where the northward and southward motions intersect.

ayanavalana	Part of the inclination of the ecliptic to the local vertical, due to its obliquity.
ayuta	10^4 (both number and place).
bāhu	(1) Rsine. (2) Number two in the *Bhūtasaṅkhyā* system. (3) Side of a geometrical figure (employed in the texts dealing with geometry).
bāhujyā	Rsine.
bāṇa	(1) Literally, arrow. (2) Rversed sine: $R(1-\cos\theta)$. (3) Number five in the *Bhūtasaṅkhyā*.
bha	Asterism: star.
bhacakra	Circle of asterisms; also called a *bhapañjara*.
bhāga	See *aṃśa*.
bhagaṇa	See *paryaya*.
bhagola	(1) Sphere of asterisms. (2) Zodiacal sphere, with its centre at the Earth's centre.
bhagola-madhya	Centre of the zodiacal sphere.
bhagola-śaṅku	Gnomon with reference to the centre of the *bhagola* (zodiacal sphere).
bhājaka	Divisor.
bhājya	Dividend.
bhakakṣyā	Path of the asterisms.
bhakūṭa	The poles of the ecliptic. Same as *rāśikūṭa*.
bhatraya	Three *rāśis*, that is, 90 degrees.
bhoga	See *bhukti*.
bhū, bhūmi	(1) Earth. (2) One side of a triangle or quadrilateral taken as reference/base that is placed on the Earth.
bhū-bhramaṇa	Earth's rotation.
bhūcchāyā	Earth's shadow.
bhūdina	(1) Terrestrial/civil day, the average time interval between two successive sunrises. (2) The number of civil days in a *yuga*/*kalpa*.
bhūgola	Earth-sphere.
bhūgola-madhya	Centre of the earth-sphere.
bhujā	(1) Opposite side of a right-angled triangle. (2) The *bhujā* of an angle is obtained from the degrees gone in the odd quadrants and to go in the even quadrants.
bhujājyā	Rsine of an angle, or the usual sine multiplied by the *trijyā* whose value is very close to 3438.

bhujāntaraphala	Correction for the equation of time due to the obliquity of the ecliptic.
bhujāphala	Equation of centre correction.
bhūjyā	See *kṣitijyā*.
bhukti	Motion; daily motion.
bhūmadhya	Centre of the Earth.
bhūmadhya-rekhā	Terrestrial equator.
bhūparidhi	Circumference of the Earth.
bhū-tārāgraha-vivara	Distance of separation between the Earth and a planet.
bhūvyāsārdha	Radius of the Earth.
bimba	Disc of a planet.
bimba-ghana-madhyāntara	Distance of separation between the centres of the discs of any two planets, especially the Sun and Moon.
bimbamāna	Measure of a planet's disc.
bimbāntara	See *bimbaghana-madhyāntara*; angular separation between the discs.
cakra	(1) Circle. (2) Cycle.
cakrakalā, *cakraliptā*	Minutes of arc contained in a circle which is equal to $360 \times 60 = 21600$.
cakrāṃśa	360 (number of degrees in a circle).
candragrahaṇa	Lunar eclipse.
cāndramāsa	(1) Lunar month. (2) The time interval between two successive new moons whose average value is ≈ 29.54 civil days.
candra-śṛṅgonnati	Elevation of Moon's cusps.
cāpa	(1) Arc of a circle. (2) Constellation *Dhanus*.
cāpabhujā	Rsine of an arc; as the argument of an Rsine is always less than 90 degrees in the Indian texts, the angle corresponding to the arc is measured from *Meṣādi* and *Tulādi* in the anti-clockwise direction in the first and third quadrants and in the clockwise direction in the second and fourth quadrants.
cāpakhaṇḍa	Segment of *cāpa*.
cāpakoṭi	Complementary arc of *bhujācāpa*.
cāpīkaraṇa	Calculating the arc of a circle from its Rsine or semichord.
cara	Ascensional difference: equal to the arc of the celestial equator lying between the 6 o'clock circle for a place with a specified latitude, and the horizon; usually expressed in *nāḍikās*.
cāra	Motion; same as *gati*.

cārabhoga	Direct daily motion.
caradala, carārdha	Half of a *cara*.
carajyā	Rsine of a *cara*.
carakalā	Minutes corresponding to a *cara*.
carakhaṇḍa	Segment of ascensional difference.
caraprāṇa	*cara* (ascensional difference) expressed in *prāṇas* (sidereal seconds).
carasaṃskāra	The correction due to the ascensional difference.
carāsava	See *caraprāṇa*.
caturyuga	Group of four *yugas* (see *yuga*).
chādaka	See *grāhaka*.
chādya	See *grāhya*.
chāyā	(1) Shadow. (2) Rsine of zenith distance.
chāyābāhu, chāyābhujā	Rsine of the gnomonic shadow; $R\sin z\sin a$, where z is the zenith distance and a is the *āśāgrā*.
chāyāgrā	Tip of the shadow cast by a gnomon.
chāyākarṇa	Hypotenuse of a right-angled triangle one of whose sides is the gnomon and the other is the shadow.
chāyākoṭi	Rcosine of the shadow of a gnomon.
chāyākoṭi-vṛtta	Circle described by the Rcosine of the shadow of a gnomon.
cheda	Denominator.
chedaka, chedya	(1) Figure. (2) Diagram. (3) Drawing.
dakṣiṇa	Southern.
dakṣiṇāyana	Southward motion (of the Sun) from the summer solstice to winter solstice.
dakṣiṇottaramaṇḍala, (-vṛtta)	Prime meridian – great circle passing through the poles of the equator and the zenith of the observer.
dakṣiṇottara-natavṛtta	See *ghaṭikā-natavṛtta*.
dakṣiṇottararekhā	North–south line; Meridian-circle.
dala	Half of any quantity (see for instance, *viṣkambhadala*).
darśana-saṃskāra	Visibility correction of planets.
daśa	10 (both number and place).
daśapraśna	Ten problems related to finding any two out of the five quantities: zenith distance, declination, hour angle, amplitude and the latitude, given the other three, from spherical trigonometry.

deśāntara	(1) Longitude. (2) Difference in terrestrial longitude. (3) Correction to the celestial longitude due to the observer's terrestrial longitude.
deśāntara-kāla	Time difference corresponding to the difference in terrestrial longitude.
deśāntara-saṃskāra	Correction related to the difference in longitude.
dhana	(1) Positive. (2) Additive.
dhanus	Arc of a circle.
dhruva	(1) Celestial pole (north or south). (2) Fixed initial positions or longitudes of planets at a chosen epoch.
dhruvanakṣatra	Pole star.
dhruvonnati	Elevation of the celestial pole.
digagrā	See *aśāgrā.*
digvaiparītya	Reversal of direction; perpendicularity.
dik	Direction, generally the four cardinal ones.
dikjñāna	Knowledge of the directions.
diksūtra	Straight lines indicating directions.
dinabhukti	The angle traversed (by any planet) per day.
divyābda	Divine year, equal to 360 ordinary years.
divyadina	Divine day, equal to one year.
doḥ	Literally, hand. See *bhujā/bāhu.*
doḥphala	Opposite side of a right–angled triangle conceived inside an epicycle of specified radius with one of the vertices coinciding with the centre of the epicycle, and the angle subtended at that vertex being the *manda-kendra* or *śīghra-kendra.*
dorjyā	Rsine.
dṛggati	Arc of the ecliptic measured from the central ecliptic point or its Rsine; Rsine altitude of the nonagesimal.
dṛggatijyā	Rsine *dṛkkṣepa.*
dṛggola	(1) Visible celestial sphere for an observer – the observer-centred celestial sphere. (2) The *khagola* and *bhagola* together.
dṛggolacchāyā	Shadow corresponding to the *dṛggola.*
dṛggolaśaṅku	Gnomon corresponding to the *dṛggola.*
dṛgjyā	Rsine of the apparent zenith distance ($R\sin z'$, where z' is the zenith distance corresponding to the observer).
dṛgvṛtta	Vertical circle passing through the zenith and the planet.
dṛkchāyā	Parallax.

dṛkkarma	Reduction of observations to the visible sphere.
dṛkkarṇa	Hypotenuse with the *dṛggolaśaṅku* and *dṛggolacchāyā* as sides.
dṛkkṣepa	(1) Ecliptic zenith distance. (2) Zenith distance of the nonagesimal (point on the ecliptic whose longitude is less than that of the *lagna* (ascendant) by 90 degrees) or its Rsine.
dṛkkṣepajyā	Rsine *dṛkkṣepa*.
dṛkkṣepakoṭi	Rcosine of *dṛkkṣepa*.
dṛkkṣepa-lagna	Central ecliptic point or nonagesimal—point on the ecliptic whose longitude is less than that of the *lagna* (ascendant or the ecliptic point on the eastern horizon) by 90 degrees.
dṛkkṣepa-vṛtta	(1) Vertical circle through the central ecliptic point. (2) Secondary to the ecliptic passing through the zenith.
dṛksiddha	That which is obtained by the observation.
dṛṅmaṇḍala	See *dṛgvṛtta*.
dvādaśāṅgula-śaṅku	A gnomon 12 *aṅgulas* in length used in the measurement of shadows.
dvādaśāṅgula-śaṅkucchāyā	Shadow of a 12-*aṅgula* gnomon.
dvitīya-sphuṭa	Second correction (generally associated with evection term for Moon).
dvitīya-sphuṭa-bhukti	True rate of motion (of the Moon) obtained by employing the second correction.
dyugaṇa	See *Ahargaṇa*.
dyujyā	Day-radius—radius of the diurnal circle, whose magnitude is $R\cos\delta$, δ being the declination of the celestial object
dyuvṛtta	See *Ahorātravṛtta*.
eka	(1) Unit. (2) Unit's position. (3) One.
ekadeśa	A portion of some quantity; for instance the segment of a straight line or an area and so on.
eṣya	That which is to be traversed.
gata	Elapsed quantity (days, time etc.).
gatacāpa	The arc already traversed.
gatagantavyaprāṇa	The *prāṇas* elapsed and yet to elapse.
gata-kali	Elapsed *Kali* years: number of years elapsed since the beginning of the *Kaliyuga* as the epoch.
gati	(1) Motion. (2) Rate of motion (of celestial bodies).
gatibheda	Difference in motions or rates of motion.
gatikalā	Motion expressed in minutes of arc.

ghana	(1) Cube of a number. (2) A solid object.
ghāta	Product of numbers.
ghaṭikā or *nāḍikā*	Unit of time which is equal to one-sixtieth of a sidereal day, approximately 24 minutes.
ghaṭikā-maṇḍala, ghaṭikā-vṛtta	Celestial equator, which is the same as the path traced by the star rising exactly in the east and setting exactly in the west.
ghaṭikā-natavṛtta	A great circle passing through the poles and perpendicular to the celestial equator. Also see *natavṛtta.*
golā	A sphere; generally used with prefixes such as '*bhū, bhā*' etc. For instance see *bhūgola, bhagola.*
golādi	Vernal equinox: the point of contact of the *ghaṭikāvṛtta* (equator) and the *apakramavṛtta* (ecliptic).
golakendra	Centre of *gola.*
golamadhya	Centre of the sphere.
graha	That which is in motion (*gacchatīti grahaḥ*), which includes the Sun, Moon, planets, the *uccas* (higher apsides) and the *pātas* (nodes).
graha-bhramaṇa-vṛtta	Literally, circle of motion of a planet. This is generally identified with the *pratimaṇḍala.*
grahabhukti	See *grahagati.*
grahagati	Daily motion of a planet.
grāhaka	Eclipsing body; also called *grāhakabimba.*
grahaṇa	Eclipse.
grahaṇa-kāla	Time or duration of an eclipse.
grahaṇa-madhya	Middle of an eclipse.
grahaṇa-pari-lekhana	Geometrical or graphical representation of the course of an eclipse.
graha-sphuṭa	True longitude of a planet.
grahāstodaya	Rising and setting of a planet.
grahayuti/yoga	Conjunction of planets.
grāhya	Eclipsed body; also called *grāhyabimba.*
grāsa	Obscuration—the maximum width of the overlap of two intersecting circles in an eclipse and the measure thereof.
grāsonavyāsa	The difference between the diameter and the eclipsed portion in an eclipse.
guṇa	(1) Multiplication. (2) Multiplier. (3) Rsine.
guṇaka, guṇakāra	Multiplier.

gurvakṣara	A time unit which is equal to one-sixtieth of a *vināḍī* or $\frac{24}{60}$ of a sidereal second.
hanana	Multiplication.
hāra, hāraka	Divisor.
haraṇa	Division.
haraṇaphala	Result of division, quotient.
hata	That which is multiplied.
hṛta	That which is divided.
icchā	Literally, desire; generally used to refer to the third of the three quantities, whose corresponding *phala* is to be determined by employing the rule of three.
icchāphala	The desired consequent; the fourth quantity, corresponding to *icchā* to be obtained by the rule of three.
indūcca	Higher apsis of the Moon.
indupāta	Node of the Moon.
iṣṭa	Desired quantity.
iṣṭabhujācāpa	Arc corresponding to the desired Rsine.
iṣṭadigvṛtta	Vertical circle passing through the zenith and the planet.
iṣṭadikchāyā	Shadow in the desired direction.
iṣṭadoḥkoṭidhanus	The complementary arc of any chosen arc.
iṣṭadyujyā	Desired *dyujyā* (Rcosine of declination).
iṣṭagrahaṇakāla	Desired moment during an eclipse/occultation.
iṣṭajyā	Rsine at the desired point on the circumference of a circle.
iṣṭāpakrama	Desired declination.
iṣṭāpakramakoṭi	Rcosine of the desired declination.
iṣṭasaṅkhyā	The desired number.
itarajyā	The other Rsine (ordinate).
itaretarakoṭi	The Rcosine (ordinate) of each other.
jaladhi	The number 4 in the *Bhūtasaṅkhyā* system; also 10^{14} (both number and place).
jhaṣa (matsya)	Figure in the form of a fish in geometrical construction such as intersecting circles.
jīvā	Rsine of an arc; $R\sin\theta$ where θ is the angle corresponding to the arc and R is the *trijyā*, which is the radius of a circle whose circumference is 21600 units; $R \approx 3438$.

jīve-paraspara-nyāya	Rule for obtaining the Rsine of the sum or difference of two angles, wherein the Rsine of one angle is multiplied by the Rcosine of the other and vice-versa. That is, $R\sin(A \pm B) = \frac{R\sin A R\cos B \pm R\cos A R\sin B}{R}$.
Jūka	The sign *Tulā* (Libra).
jyā	See *jīvā*; Perhaps earlier *jyā* referred to the chord corresponding to an arc, that is $2R\sin\frac{\theta}{2}$, where θ is the angle corresponding to the arc. But later, as in *Tantrasaṅgraha*, the *jyā* refers to $R\sin\theta$.
jyācāpāntara	Difference between an arc and its Rsine.
jyākhaṇḍa	Rsine difference.
jyāpiṇḍa	The Rsines of one, two etc. parts of a quadrant which is divided into a certain number of equal parts, generally 24.
jyārdha	Same as what came to be termed the *jyā*, that is, $R\sin\theta$; (see *jyā*).
jyāsaṅkalita	The summation of Rsines.
jyāvarga	Square of the Rsine.
jyotirgola	Sphere of celestial bodies.
jyotiścakra	Circle of asterisms.
kakṣyā	Orbit of a planet.
kakṣyāmaṇḍala, kakṣyāvṛtta	Deferent or concentric circle, on which the mean planet moves.
kakṣyā-vyāsārdha	Mean radius of the planetary orbit.
kalā	Minute of an arc (angular measure); also referred to as *liptā*, *liptikā*; $\frac{1}{21600}$th part of the circumference of a circle.
kalāgati	Motion expressed in minutes of arc.
kālalagna	(1) Time elapsed after the rise of the vernal equinox at any instant. (2) Time interval between the rise of the vernal equinox and the sunrise.
kalāvyāsa	Angular diameter (for instance, of Sun, Moon etc.) expressed in minutes.
Kaliyuga	The *yuga* (aeon) which commenced on February 18, 3102 BCE at sunrise at Laṅkā.
kalyādi	Beginning of the *Kali* epoch.
kalyādi-dhruva	Initial positions (longitudes) of planets at the beginning of the *Kali* epoch.
kalyahargaṇa	Number of civil days elapsed since the beginning of the *Kaliyuga*.

kapāla	Hemisphere, usually employed with an adjective like *prāk* (east), *paścima* (west), etc.
karaṇa	(1) Period corresponding to half a *tithi*. (2) Also used to refer to a class of astronomical texts that choose a recent epoch in contrast to the *siddhantic* works that choose the begining of the *kalpa* or the *kaliyuga* as the epoch.
Karka, Karki	Cancer.
Karkyādi	Six signs commencing from the sign Cancer.
kārmuka	Arc of a circle.
karṇa	(1) Hypotenuse of a right-angled triangle. (2) Radius vector.
karṇa-vṛtta	Hypotenuse circle: a circle drawn with hypotenuse as the radius in either the *manda* or the *śīghra* correction.
karṇa-vṛtta-jyā	Rsine in the hypotenuse circle.
kendra	(1) Centre of a circle (2) Anomaly–The angular separation of a planet from the *mandocca* or *śīghrocca*.
kendrabhramaṇa	Movement (rotational) of the *kendra*.
kendrabhukti	Daily motion of the anomaly.
khagola	Celestial sphere or globe.
khakakṣyā	See *ākāśakakṣyā*.
khamadhya	The centre of the sky; the zenith.
khaṇḍagrahaṇa	Partial eclipse.
khaṇḍajyā	The difference between two successive ordinates or Rsines; essentially the first differential of the *jyā*.
khaṇḍajyāntara	The difference of the differences, or the second differential of the *jyā*.
khaṇḍajyāyoga	Sum of Rsine differences.
kheṭa	That which wanders in space (planet).
koṇa	(1) Corner. (2) Direction in between any two cardinal directions (north-east, south-west etc.). (3) Angle.
koṇacchāyā	Corner shadow: shadow corresponding to the instant at which the planet intersects the *koṇavṛtta* (see *koṇavṛtta*).
koṇaśaṅku	*śaṅku* considered at the instant at which the planet passes through the *koṇavṛtta* (see *koṇavṛtta*).
koṇavṛtta	Vertical circle passing through the north-east and the south-west points, or south-east and north-west points of the horizon.
koṭi	(1) Adjacent side of a right-angled triangle. complement of *bhujā* that is Rcosine. (2) 10^7 (both number and place).

koṭicāpa	Arc corresponding to Rcosine of an arc; 90 degrees minus *cāpa*.
koṭijyā	Rcosine of an arc.
koṭikhaṇḍa	(1) Portion of *koṭi*. (2) The difference between two successive values: essentially, the first differential of the *koṭijyā*.
koṭimūla, koṭyagra	The points corresponding to the base, tip of the *koṭi*.
koṭiphala	Adjacent side of a right-angled triangle conceived inside an epicycle of specified radius with one of the vertices coinciding with the centre of the epicycle, and the angle subtended at that vertex being the *manda-kendra* or *śīghrakendra*.
koṭivṛtta	Rcosine circle – circle whose radius is the *koṭi*.
kramajyā	Rsine segments taken in order.
kramaśaṅku	Gnomon formed at the moment of passing the *koṇavṛtta*.
krānti	See *apakrama*.
krāntijyā	Rsine of the declination.
krāntikoṭi	Rcosine of the declination.
krāntimaṇḍala	See *apakramamaṇḍala*.
Kriyā	The sign *Meṣa* (Aries).
kṛṣṇapakṣa	Dark half of the lunar month; (see also *aparapakṣa*).
kṛti	(1) Square. (2) Composition.
kṣayatithi	Unreckoned *tithi*.
kṣepa	(1) Celestial latitude. (2) Additive quantity.
kṣetra	Planar geometrical figure.
kṣipti	See *kṣepa*.
kṣitija	Horizon–the tangential plane drawn at the location of the observer, passing through the four cardinal directions.
kṣitijyā, *kṣitimaurvikā*	Product of *carajyā* (ascensional difference) and *dyujyā* (Rcosine of declination)—which corresponds to the Rsine of *carāsus* (arc of the ascensional difference) on the diurnal circle whose separation from the equator is δ: $\frac{R\sin\phi\sin\delta}{\cos\phi}$.
Kulīra	See *Karki*.
lagna	Orient ecliptic point, that is, the longitude of the ecliptic point at the eastern horizon.
lagnasamamaṇḍala	Vertical circle passing through the orient ecliptic point.
lambaka, lambana	(1) Plumb-line. (2) Rsine of co-latitude, i.e., Rcosine of latitude (3) Parallax. (4) Parallax in longitude.
lambana-nāḍikā	Parallax in longitude in *nāḍikās* (24 sidereal minutes).
lambana-yojana	Parallax in terms of *yojanas*

Laṅkā	A fictitious place located on the earth's equator and on the reference meridian (passing through Ujjayinī) and defined to have zero terrestrial longitude.
laṅkākṣitija	Horizon at Laṅkā: equatorial horizon.
laṅkodaya	[Time of the] rising at Laṅkā.
laṅkodayajyā	Rsine corresponding to *Laṅkodaya*.
lāṭa	A type of *vyatīpāta*, which occurs when the longitudes of the Sun plus the Moon are equal to 180 degrees.
liptā	See *kalā*.
liptāvyāsa	Angular diameter in minutes.
madhya	(1) Literally, mean or middle portion. (2) 10^{16} (both number and place).
madhyabhukti, *madhyagati*	The mean rate of motion of planet obtained from the number of revolutions given for a *Mahāyuga*.
madhyagraha	Mean longitude of the planet.
madhyagrahaṇa	Mid-eclipse.
madhyāhna	Midday.
madhyāhnacchāyā	Midday-shadow.
madhyāhnāgrāṅgula	Measure of amplitude at noon in terms of *aṅgula*.
madhyajyā	Meridian sine, i.e. Rsine of the zenith distance when the planet crosses the prime meridian.
madhyakāla	Mean time, middle of an eclipse etc.
madhyalagna	Meridian ecliptic point—the point of the ecliptic on the prime meridian.
madhyalambana	Parallax in longitude in the middle.
madhyama	Mean longitude of a planet.
mahābāhu	Literally, great arm, which refers to $R\sin z$ where z is the zenith distance.
mahācchāyā	Literally, great shadow, which actually refers to the distance from the foot of the *mahāśaṅku* to the centre of the Earth; Rsine zenith distance.
mahājyā	The 24 Rsines used for computation.
mahāmeru	(1) The big mount *Meru*, taken to mark the terrestrial pole in the north, where the north polar star is right above; (2) Location situated 90 degrees north of Laṅkā.
mahāśaṅku	(1) Great gnomon. (2) The perpendicular dropped from the Sun to the horizon (when the radius of the celestial sphere is taken to be R), which is equal to Rsine altitude or Rcosine of zenith distance.

Makara	Capricorn.
Makarādi	The six signs commencing from *Makara* (Capricorn).
māna	(1) Measure (2) An arbitrary unit of measurement.
manda	(1) Slow (2) Associated with the equation of centre; (3) *mandocca*—apogee of slow motion. (4) Saturn.
manda-karma	*manda* correction in planetary computation; procedure for obtaining the equation of centre.
manda-karṇa	Hypotenuse associated with *manda* correction.
manda-karṇa-vṛtta	Circle with a radius equal to that of *manda-karṇa*.
manda-kendra	*manda* anomaly, that is the difference in the longitude between the *mandocca* (apogee or apsis) and the mean planet; mean anomaly.
maṇḍala	(1) Circle (2) Orb.
manda-paridhi	Circumference of the *manda-vṛtta* (epicycle associated with the equation of centre).
manda-phala	The equation of centre correction to be applied to the mean planet.
manda-saṃskāra	See *manda-karma*.
manda-sphuṭa, *manda-sphuṭa-graha*	The longitude of a planet obtained after applying the *manda* correction (equation of centre) to the mean longitude (known as *madhyagraha*).
manda-vṛtta	*manda* epicycle, that is, the epicycle associated with the equation of centre.
mandocca	Uppermost point in the *manda* epicycle; apogee; apsis.
mandoccanīca-vṛtta	See *manda-vṛtta* and *ucca-nīca-vṛtta*.
maṅgalācaraṇa	Invocation.
marut	See *pravahamāruta*.
māsa	Month.
matsya (jhaṣa)	Fish, fish-figure; see *jhaṣa*.
mauḍhya	Invisibility of a planet due to its direction/longitude being close to that of the Sun.
maurvikā	See *jyā*.
Meru	See *Mahāmeru*.
Meṣa	Aries.
Meṣādi	(1) First point of Aries. (2) Commencing point of the ecliptic. (3) Six signs beginning with *Meṣa*.
Mīna	Pisces.
Mithuna	Gemini.

mokṣa	Literally freedom, which actually refers to the process of emergence in an eclipse.
mokṣakāla	Instant of *mokṣa* (last contact).
mokṣalambana	Parallax in longitude at release.
Mṛga	The 10th sign: *Makara* (Capricorn).
Mṛgādi	The six signs beginning with Capricorn.
mūla	(1) The base or starting point of a line or arc. (2) Square root, cube root etc.
nabhomadhya	See *khamadhya*.
nābhyucchraya	Elevation of the *nābhi*.
nāḍikā	See *ghaṭikā*.
nāḍī-vṛtta	Celestial equator (see *ghatikāmaṇḍala*).
Nakra	Capricorn (generally refered to as *Makara*).
nakṣatra	Star; asterism; constellation.
nākṣatradina	Sidereal day, which is equal to the time interval between two successive transits of a particular star across the horizon or the meridian ($\approx 23^h56^m$ of a civil day).
nakṣatragola	The starry sphere.
nakṣatrakakṣyā, *bhakakṣyā*	Orbit of the asterisms, equal to 173260008 *yojanas*, denoted by the expression *janānunītiraṅgasarpa*, being 50 times the orbit of the Sun.
nākṣatravarṣa	Sidereal year, which is equal to the time interval between two successive transits of the Sun across the same star say *nirayaṇameṣādi*; also called a *nirayaṇa* year.
nata	(1) Hour angle, which gives the time interval between midday and current time. (2) Meridian zenith distance.
natabhāga	Zenith distance in degrees.
nata-dṛkkṣepavṛtta	Circle touching the zenith and prime vertical.
natajyā	Rsine of hour angle; occasionally, Rsine of zenith distance.
natakoṭijyā	Rcosine of the hour angle.
natamaṇḍala	See *natavṛtta*.
natapräṇa	Hour angle in *prāṇas*.
natasamamaṇḍala	Prime vertical.
natavṛtta	Great circle which intersects another great circle perpendicularly; for instance a *ghaṭikānatavṛtta* which is perpendicular to the *ghaṭikāvṛtta* (equator).
nati	Parallax in celestial latitude; deflection from (perpendicular to) the ecliptic.

natikalā	*nati* in minutes.
natilambanaliptā	Rcosine of parallax in celestial longitudes in terms of minutes of arc.
natiyoga	Sum of two parallaxes in celestial latitude.
natyantara	Difference between parallaxes in celestial latitude.
nemi	Circumference (of a circle).
nīca	The closest point in the *pratimaṇḍala* from the centre of the *kakṣyāmaṇḍala*.
nīcoccamaṇḍala	See *ucca-nīca-vṛtta*.
nimīlana	Immersion (in eclipse).
nirakṣa-deśa	(Place having) zero latitude; equatorial region.
nirakṣa-kṣitija	Equatorial horizon.
nirakṣa-rekhā	Terrestrial Equator.
nirakṣodaya	Rise of an object for an equatorial observer.
nirayaṇa	Without motion; sidereal or with respect to a fixed stars in '*nirayaṇa* longitude'.
nṛcchāyā	See *śaṅkucchāyā*.
ojapada	Odd quadrants (the first and the third).
pada	(1) Square root (2) Terms of a series (3) Quarter (4) Quadrant of a circle.
padīkṛta	When the square root is obtained.
pakṣa	Fortnight (bright or dark half of the lunar month consisting of 15 *tithis*).
palaprabhā/palabha	Equinoctial shadow.
paṅkti	Column; ten (number and place).
paramagrāsa	Maximum in an eclipse obscuration.
paramagrāsakāla	Instant of maximum obscuration in an eclipse.
para/parama-krānti	See *antyakrānti*.
paramāntarāla	Maximum distance of separation.
paramāpakrama	Greatest declination.
para/paramaśaṅku	Rsine of greatest altitude, that is, Rsine of meridian altitude.
paramasvāhorātra	Longest day in the year.
paribhramaṇa	A complete revolution of a planet along the zodiac with reference to a fixed star.
paridhi	See *nemi*.
pari-lekha/lekhana	Graphical or diagrammatic representation.
pārśva	Side; surface.

parvānta	Middle of the eclipse, that is, the instant when Moon is in conjunction with or in opposition to the Sun; ending moment of the new or full moon.
paryaya	(1) Count of a certain repeated process. (2) Number of revolutions of a planet in a *yuga*.
pāta	Node (generally ascending node).
paṭhitajyā	Tabulated Rsines (generally 24).
phala	(1) Fruit or Result. (2) The outcome of any calculation; most commonly employed in the rule of three.
piṇḍajyā	Whole Rsine.
pitṛdina	Day of the *pitṛs*, which is a lunar month.
prāglagna	Orient ecliptic point; longitude of the ecliptic point on the eastern horizon.
prākkapāla	The eastern hemisphere.
pramāṇa	(1) A measure. (2) Means of evidence. (3) Antecedent–the first term of a proportion (rule of three).
pramāṇaphala	The consequent: (see the second term in a proportion).
prāṇa	4 sidereal seconds; See *asu*.
prāṇakalāntara	Difference between the longitude and right ascension of the Sun in the *prāṇa* unit.
pratimaṇḍala	Eccentric circle, with a radius equal to the *trijyā*, *R*, but whose centre is shifted from the centre of the *kakṣyāmaṇḍala* along the direction of *mandocca*, by a certain measure that is specified for each planet.
pratipat	The first day of a lunar fortnight, also called *prathamā*.
pratyakkapāla	The (western) hemisphere other than the one (eastern) that is being considered.
pravahabhramaṇa	Revolution of the wind called *pravaha* or that of the planets due to *pravaha*.
pravahamāruta, *pravahavāyu*	Wind named *pravaha* (*prakarṣeṇa vahatīti pravahaḥ*), responsible for the diurnal motion of all the celestial bodies.
pṛṣṭha	Surface of some object; for instance, the surface of Earth is referred as *bhūpṛṣṭha*.
pūrṇimā	Full Moon day.
pūrvāpara-rekhā	East–west line.
pūrvāpara-vṛtta	Prime vertical (the circle passing through the zenith and the east and west points of the horizon).
pūrvaviṣuvat	Vernal equinox.
Rāhu	The ascending node of the Moon.

rāśi	Literally, a group. It refers to: (1) A number (which is a member of a group). (2) A zodiacal sign equal to 30 degrees in angular measure.
rāśicakra	Ecliptic.
rāśikūṭa	The place where all the *rāśis* meet (poles of the ecliptic).
rāśikūṭavṛtta	The circle passing through the *rāśikūṭas* and intersecting the ecliptic at intervals of one *rāśi*.
rāśipramāṇa	Measure of the *rāśi*.
rāśyudaya	Rising of the *rāśi*.
ṛkṣa, nakṣatra	Asterism, star-group.
ṛṇa	Negative or quantity to be subtracted.
rūpa	Unity or number one in the *Bhūtasaṅkhyā* system (literally, form, which is unique to every entity).
sadṛśa	(1) Of the same denomination or kind. (2) Similar.
sahasra	Thousand (both number and place).
sakṛtkarṇa	One-step hypotenuse.
śalākā	Thin, pointed stick.
samaghāta	Product of like terms.
samamaṇḍala	Prime vertical (circle passing through the zenith and the east and west points of the horizon).
samamaṇḍalachāyā	Rsine of zenith distance of a celestial body when it is on the prime vertical.
samasaṅkhyā	Even number.
sama-śaṅku, sama-maṇḍala-śaṅku	Rsine of altitude of a celestial body when it lies on the prime vertical.
samastajyā	Rsine of a full arc.
samparkārdha	Half the sum of the diameters of the eclipsed and eclipsing bodies; line of contact.
sampāta	Point of intersection
saṃsarpa	The lunar month preceding/succeeding a lunar month called *Aṃhaspati*.
saṃskāra	A correction to be applied (additive or subtractive) to get the desired/corrected value.
saṃvarga	Product.
saṃvatsara, saurasaṃvatsara	Tropical year, which is the time interval between two successive transits of the Sun across the vernal equinox.
saṅkramaṇa, saṅkrānti	Sun's transit from one *rāśi* to the next (refers to both the instant as well as the process).

śaṅku	(1) Gnomon (usually of 12 units). (2) Sometimes *mahāśaṅku* (great gnomon), the perpendicular dropped from the Sun to the horizon (= Rsine of altitude). (3) The number 10^{13}.
śaṅkucchāyā	Shadow of the gnomon.
śaṅkukoṭi	Compliment of altitude or zenith distance.
śaṅkvagrā	North-south distance of the rising or setting point from the tip of the shadow, i.e. the distance on the plane of the horizon from the rising–setting line.
śara	(1) Arrow. (2) Rversed sine, $R(1-\cos\theta)$.
śarabheda	Difference between two *śaras*.
śaronavyāsa	Diameter minus *śara*.
sārpamastaka	*vyatīpāta* when the Sun plus Moon is equal to $7°16'$.
saumya	Northern, literally, that which is related to *soma*, which also has the meaning of 'heaven' among others.
saumyagola	Northern hemisphere.
saura	Related to Sun; solar.
saurābda	Solar year.
sāvanadina	(1) Civil day. (2) Mean time interval between two successive sunrises.
sāyana	With motion; tropical or with respect to the vernal equinox, as in *sāyana* longitude.
śeṣa (*śiṣṭa*)	Remainder in an operation.
śīghra-bhujā-jyā	Rsine of the *śīghra* anomaly.
śīghra-karma	*śīghra* correction in planetary computation; procedure for obtaining the correction associated with the anomaly of conjunction.
śīghra-karṇa	(1) Hypotenuse associated with *śīghra* correction. (2) Geocentric radius vector.
śīghra-kendra	Anomaly of conjunction; ngular separation between *śīghrocca* and *manda-sphuṭa* (planet corrected for equation of centre) of a *tārāgraha* (actual planet) used to compute *śīghra-phala*.
śīghra-kendra-jyā	Rsine of the *śīghra* anomaly.
śīghra-paridhi	Circumference of the *śighra* epicycle.
śīghra-phala	The correction to be applied to the *manda-sphuṭa* (a planet corrected for the equation of centre) to obtain the geocentric longitude of the planet.
śīghra-saṃskāra	See *śīghra-karma*.

śīghra-sphuṭa	The longitude of a planet obtained by applying the *śīghra* correction.
śīghra-vṛtta	The *śīghra* epicycle, that is, the epicycle associated with the anomaly of conjunction.
śīghrocca	(1) Higher apsis (or the uppermost point) of the epicycle employed in the *śīghra* correction which represents the direction of the mean Sun for all planets (as per the geometrical picture of planetary motion described by Nīlakaṇṭha). (2) Apex of the planet moving faster.
śīghroccanīca-vṛtta	See *śīghra-vṛtta*.
śiñjinī	See *jyā*.
śiṣṭa	Remainder in an operation.
śiṣṭacāpa	The difference between the given *cāpa* and the nearest *mahājyācāpa* (arc whose Rsine is tabulated).
sita	(1) Bright. (2) Illuminated part of the Moon. (3) Venus.
sitapakṣa	Bright half of the lunar month.
śodhya	That which is to be subtracted.
sparśa	Literally, touch; first contact in an eclipse.
sparśakāla	Instant of first contact.
sparśalambana	Parallax in longitude at first contact.
sphuṭa-(graha)	True; actual/true position (of a planet).
sphuṭa-gati	True daily motion of a planet.
sphuṭa-graha	True longitude of a planet.
sphuṭa-kakṣyā	True value of the orbital radius.
sphuṭa-kriyā	Procedure for the computation of the true (geocentric) position/longitude of a planet.
sphuṭa-madhyāntarāla	Difference between the true and the mean longitudes of a planet.
sphuṭa-nati	True parallax in latitude; true deflection perpendicular to the ecliptic.
sphuṭāntara	Difference between the true longitudes.
sphuṭanyāya	Rationale behind the procedure employed in obtaining the true position of a planet.
sphuṭa-vikṣepa	Corrected celestial latitude.
śṛṅgonnati	Elevation of the lunar horns (cusps).
śruti	Hypotenuse; more commonly referred to as *karṇa*.
sthityardha	Half-duration of an eclipse.
śūnya	Zero (literally void/emptiness).

sūryagrahaṇa	Solar eclipse.
sūtra	(1) Line. (2) Direction. (3) Formula. (4) Aphorism.
sva(m)	(1) Addition. (2) Additive quantity.
svadeśakṣitija	Horizon at one's place.
svadeśanata	Meridian zenith distance at one's place.
svadeśanatakoṭi	Rcosine of *svadeśanata*.
svāhorātravṛtta	Diurnal circle.
svarṇa	(*sva* + *ṛṇa*) When added or subtracted.
svastika	Observer's zenith.
tāḍana	Multiplication.
tamas	(1) Shadow cone of the Earth at the Moon's distance. (2) Moon's nodes.
tārāgraha	Star planets, that is, the actual planets: Mercury, Venus, Mars, Jupiter and Saturn.
tatpara	Angular measure corresponding to one-sixtieth of a second (*vikalā*).
tiryagvṛtta	Oblique or transverse circle; for example, a great circle passing through the north and south celestial poles is a *tiryagvṛtta* of the *ghaṭikāmaṇḍala* (celestial equator).
tithi	Lunar day, a thirtieth part of a synodic lunar month, or the time interval during which the difference in the longitudes of the Moon and the Sun increases by 12 degrees.
tithikṣaya	See *avama*.
tithyanta	End of a *tithi*.
trairāśika	(1) Rule of three. (2) Direct proportion.
tribhajyā	Rsine of three *rāśis*, same as *trijyā*.
tribhuja	A three-sided figure; triangle.
trijyā, trirāśijyā	Rsine 90 degrees. The radius of the circle whose circumference is 21600 units, whose value is very nearly 3438 units (number of minutes in a radian).
trimaurvikā	See *trijyā*.
triśarādi	Set of odd numbers 3, 5, 7, etc.
Tulā	Libra.
Tulādi	The six signs commencing from *Tulā*.
tuṅga	Apogee or aphelion (literally, 'peak', *ucca*).

ucca	(1) Higher apsis pertaining to the epicycle (*manda* or *śīghra*). Equivalently, the farthest point in the *pratimaṇḍala* from the centre of the *kakṣyāmaṇḍala*. (2) The apogee of the Sun and the Moon, and the aphelion of the planets.
ucca-nīca-sūtra	The line joining the higher and lower apsides.
ucca-nīca-vṛtta	Epicycle: the circle moving up and down with its centre on the deferent circle (*kakṣyāmaṇḍala*) and which touches the *ucca* and the *nīca* points on the *pratimaṇḍala* during the course of its motion.
udaya	Rising; heliacal rising; rising point of a star or constellation at the horizon.
udayajyā	(1) Rsine of the amplitude of the rising point of the ecliptic. (2) Oriental sine. (3) Rsine of the amplitude of *lagna* in the east.
udayakāla	The moment of rising of a celestial body.
udayalagna	Rising sign; the orient ecliptic point.
udayasūtra	The line joining the rising and setting points.
Ujjayinī	City in central India, the meridian passing through which is taken to be the standard meridian (zero terrestrial longitude) in Indian texts.
ujjhitvā	Having subtracted.
unmaṇḍala	(1) Six o'clock circle; east-west hour circle; equinoctial colure; great circle passing through the north and south poles and the two east-west *svastika*. (2) *Laṅkākṣitija*: horizon at Laṅkā (equatorial horizon).
unmīlana	Opening, emersion in eclipse.
unnatajyā	Altitude of a planet: Rsine of 90 degrees minus zenith distance.
unnataprāṇa	The time in *prāṇas* yet to elapse for a planet to set.
upādhi	Assumption; limiting agent.
upāntya	Close to the end; penultimate (term).
upapatti (yukti)	Proof; rationale; demonstration; justification.
ūrdhva	The topmost, earlier or preceding.
ūrdhvādhorekhā	Line through the upper and lower points, the vertical.
utkramajyā	Rversed sine ($R(1-\cos\theta)$, where θ is the angle corresponding to the arc).
uttara	Northern.
uttarāyaṇa	Northward motion (of the Sun) from winter solstice to summer solstice.

vaidhṛta	A type of *vyatīpāta* that occurs when the sum of the longitudes of the Sun and the Moon equals 360 degrees.
vakragati/bhoga,	Retrograde motion of a planet.
valana	Deflection of a planet from the vertical due to *akṣa* or *ayana.*
varga	Square.
vihṛta	That which is divided.
vikalā	Second = $\frac{1}{60}$th of a minute of angular measure.
vikṣepa	(1) Latitudinal deflection (Rsine of celestial latitude). (2) Celestial latitude. (3) Polar latitude.
vikṣepacalana	Related to *ayanacalana.*
vikṣepakoṭivṛtta	Small circle corresponding to a specific celestial latitude parallel to the ecliptic.
vikṣepamaṇḍala	Orbit of a planet (inclined to the ecliptic).
vikṣipta	Deviated (from the ecliptic).
vikṣiptagrahakrānti	Declination of a planet with a latitudinal deflection.
viliptā	See *vikalā.*
vimardārdha	Half of total obscuration in an eclipse.
vināḍikā, vināḍī	$\frac{1}{60}$th of *nāḍikā* = 24 sidereal seconds.
vinimaya	Interchange.
viparītacchāyā	Reverse computation (of time) from gnomonic shadow.
viparītakarṇa	Reverse or inverse hypotenuse: $\frac{R^2}{K}$, where K is the *aviśiṣṭakarṇa* (iterated hypotenuse).
viparyaya	Inverse or reverse; also called *viparyāsa.*
viṣama	(1) Odd number or quadrant. (2) Difficult.
viśeṣa	Speciality, Difference.
viṣkambha	(1) Diameter. (2) The first of 27 daily *yogas.*
viṣkambhadala	Semi-diameter.
viśleṣa	Subtraction, difference.
vistarārdha	Semi-diameter or radius.
vistṛtidala	Semi-diameter (*vistṛti* is diameter).
viṣuvacchāyā	Equinoctial midday shadow, that is, the shadow of a gnomon measured at the meridian transit, when the Sun is at the equinox.
viṣuvadbhā	See *viṣuvacchāyā.*
viṣuvadbhāgra	Tip of the shadow on the equinoctial day.
viṣuvanmaṇḍala	See *ghaṭikāmaṇḍala.*
viṣuvat	Vernal or autumnal equinox.

viṣuvatkarṇa	Hypotenuse of equinoctial shadow.
vitribhalagna	See *dṛkkṣepa-lagna*.
vivara	Difference; gap, space in between.
viyoga	Subtraction.
vṛtta	Circle.
vṛttakendra	Centre of a circle.
vṛttanemi(-paridhi)	Circumference of a circle.
vṛttapāda	One-fourth of a circle, quadrant; 90 degrees.
vṛttapārśva	Pole: on of the ends of the axis around which a sphere is made to rotate.
vṛttapāta	The two points at which two great circles intersect.
vyāsa	Diameter of a circle.
vyāsa-dala/ardha	Semi-diameter, radius.
vyasta-karṇa	See *viparīta-karṇa*.
vyatīpāta	(1) The phenomenon when the magnitudes of the declinations ($\lvert\delta\rvert$) of the Sun and Moon are equal but the rates of change of $\lvert\delta\rvert$ are opposite in sign. (2) The time when the sum of the longitudes of the Sun and the Moon equals 180 degrees.
vyatīpāta-kāla	The time of occurrence of *vyatīpāta*.
yāmya	Southern (related to *Yama*).
yāmyagola	Southern half of celestial sphere.
yāmyottara-rekhā	See *dakṣiṇottara-rekhā*.
yoga	(1) Conjuction of two planets. (2) Sum. (3) Daily *yoga* (*nityayoga*): which are 27 in number and named *Viṣkambha, Prīti, Ayuṣmān*, etc. being the sum of the longitudes of the Sun and the Moon.
yogacāpa	Arc corresponding to the sum of two given semi-chords (Rsines).
yogakāla	(1) The time of conjunction of the Moon and the Sun/Ẹarth's shadow. (2) The time needed for/elapsed after conjunction.
yojana	Unit of linear measure, equal to a few miles, which has not been standardized and varies from text to text. In *Tantrasaṅgraha*, the circumference of the Earth is specified to be 3300 *yojanas*.
yojanagati	Daily motion in terms of *yojanas*.
yojanavyāsa	Diameter in *yojanas*.

yuga	Aeon; a large unit of time, for instance, *Kaliyuga* whose duration is 432000 years or *Mahāyuga* made of 4320000 years; could also refer to a short unit like 576 years as in *Tantrasaṅgraha*.
yugabhagaṇa	Number of revolutions made by a planet in the course of a *Mahāyuga* (4320000 years).
yugma	(1) Even. (2) The second and fourth quadrants in a circle.
yukti	Proof; rationale; reasoned justification.

Bibliography

Source Works

{AB 1874} *Āryabhaṭīyaṃ* of Āryabhaṭa with the commentary *Bhaṭadīpikā* of Parameśvara, ed. by B. Kern, Leiden 1874 (repr. 1906, 1973).

{AB 1976} *Āryabhaṭīya* of Āryabhaṭa, ed. with tr. and notes by K. S. Shukla and K. V. Sarma, INSA, New Delhi 1976.

{ABB 1930} *Āryabhaṭīya* of Āryabhaṭācārya with the *Mahābhāṣya* of Nīlakaṇṭha Somasutvan, Part I, *Gaṇitapāda*, ed. by Sāmbaśiva Śāstrī, Trivandrum Sanskrit Series 101, Trivandrum 1930.

{ABB 1931} *Āryabhaṭīya* of Āryabhaṭācārya with the *Mahābhāṣya* of Nīlakaṇṭha Somasutvan, Part II, *Kālakriyāpāda*, ed. by Sāmbaśiva Śāstrī, Trivandrum Sanskrit Series 110, Trivandrum 1931.

{ABB 1957} *Āryabhaṭīya* of Āryabhaṭācārya with the *Mahābhāṣya* of Nīlakaṇṭha Somasutvan, Part III, *Golapāda*, ed. by Śūranād Kuñjan Pillai, Trivandrum Sanskrit Series 185, Trivandrum 1957.

{ABB 1976} *Āryabhaṭīya* of Āryabhaṭa with the commentary of Bhāskara I and Someśvara, ed. by K. S. Shukla, INSA, New Delhi 1976.

{BG 1817} *Bījagaṇita* of Bhāskarācārya II, tr. by H. T. Colebrooke, in *Algebra with Arithmetic and Mensuration from the Sanskrit of Brahmagupta and Bhāskara*, London 1817.

{BG 1949} *Bījagaṇita* of Bhāskarācārya II, ed. with Bhāskara's *Vāsanā* by Achyutananda Jha, Varanasi 1949 (repr. 1994).

{BS 1981} *Bṛhatsaṃhitā* of Varāhamihira, with tr. and notes by M. Ramakrishna Bhat, Motilal Banarasidass, Delhi 1981 (repr. 1997).

{BSS 1902} *Brāhmasphuṭasiddhānta*, of Brahmagupta ed. with his own commentary by Sudhakara Dvivedi, Benares 1902.

{BSS 1966} *Brāhmasphuṭasiddhānta*, of Brahmagupta ed. with *Vāsanā* of Pṛthūdakasvāmin by Ram Swarup Sharma, 4 vols, New Delhi 1966.

{BSS 2003} *Brāhmasphuṭasiddhānta* of Brahmagupta, chapter XXI, ed. with English tr. by S. Ikeyama, *Ind. Jour. Hist. Sci.*, INSA, New Delhi 2003.

{CCG 1976} *Candracchāyāgaṇita* of Nīlakaṇṭha Somayājī, with auto-commentary, ed. and tr. by K. V. Sarma, Hoshiarpur 1976.

{DG 1963} *Dṛggaṇita* of Parameśvara, ed. by K. V. Sarma, Hoshiarpur 1963.

{GD 1916} *Goladīpikā* of Parameśvara, ed. by T. Ganapati Sastri, Trivandrum 1916.

{GD 1957} *Goladīpikā* of Parameśvara, ed. with auto-commentary *Vivṛti* by K. V. Sarma, Madras 1957.

{GS 1970} *Golasāra* of Nīlakaṇṭha Somayājī, ed. and tr. by K. V. Sarma, Hoshiarpur 1970.

{GVV 1979} *Grahasphuṭānayane vikṣepavāsana* of Nīlakaṇṭha Somayājī, ed. in *Gaṇitayuktayaḥ* by K. V. Sarma, Hoshiarpur 1979.

{GYB 1948} *Gaṇita-yukti-bhāṣā*, of Jyeṣṭhadeva, *Gaṇitādhyāya* ed. with notes in Malayalam by Ramvarma Maru Tampuran and A. R. Akhileswara Aiyar, Trichur 1948.

{GYB 1953} *Gaṇita-yukti-bhāṣā* of Jyeṣṭhadeva, ed. by K. Chandrasekharan, Government Oriental Manuscripts Library, Madras 1953.

{GYB 2008} *Gaṇita-yukti-bhāṣā* of Jyeṣṭhadeva, ed. and tr. by K. V. Sarma with Explanatory Notes by K. Ramasubramanian, M. D. Srinivas and M. S. Sriram, Hindustan Book Agency, New Delhi 2008 (repr. Springer 2009).

{JM 1977} *Jyotirmīmāṃsā* of Nīlakaṇṭha Somayājī, ed. by K. V. Sarma, Hoshiarpur 1977.

{KK 1934} *Khaṇḍakhādyaka* of Brahmagupta, ed. with tr. and notes by P. C. Sengupta, University of Calcutta 1934.

{KK 1970} ed. and tr. by Bina Chatterjee, 2 vols, INSA, New Delhi 1970.

{KA 1975} *Karaṇāmṛta* of Citrabhānu, ed. with commentary of Śaṅkara Vāriyar by V. Narayanan Nambudiri, Trivandrum 1975.

{KP 1953} *Karaṇapaddhati* of *Putumana Somayājī*, ed. by K. Sambasiva Sastri, Trivandrum 1937; ed. with explanatory notes in Malayalam by P. K. Koru, Cherp 1953.

{KU 1964} *Karaṇottama* of Acyuta Piṣāraṭi, ed. with auto-commentary *Vivaraṇa* by K. Raghavan Pillai, Trivandrum 1964.

{LB 1963} *Laghubhāskarīya*, of Bhāskara-I, ed. and tr. with notes K. S. Shukla, Lucknow 1963.

{LB 1974} *Laghubhāskarīya* of Bhāskara-I, with the *Vivaraṇa* of Śaṅkaranārāyaṇa, ed. by P. K. Narayana Pillai, University of Kerala, Trivandrum 1974.

{LM 1952} *Laghumānasa* of *Muñjāla* with *Vyākhyā* of Parameśvara, ed. by B. D. Apte, Pune 1944; 2nd edn 1952.

{LM 1990} *Laghumānasā* of Muñjāla, ed. with tr. and notes by K. S. Shukla, INSA, New Delhi 1990.

{LVT 1817} *Līlāvatī* of Bhāskarācārya II, tr. by H. T. Colebrooke, in *Algebra with Arithmetic and Mensuration from the Sanskrit of Brahmagupta and Bhāskara*, London 1817.

{LVT 1937} *Līlāvatī* of Bhāskarācārya II, ed., with Bhāskara's *Vāsanā* and *Buddhivilāsinī* of Gaṇeśa Daivajña, by V. G. Apte, 2 vols, Pune 1937.

{LVT 1975} *Līlāvatī* of Bhāskarācārya II, with a commentary *Kriyākramakarī* of Śaṅkara Vāriyar (only up to a portion of *Kṣetravyavahāra*), ed. by K. V. Sarma, Hosiarpur 1975.

{MB 1957} *Mahābhāskarīya* of Bhāskara I, ed., with commentary of Govindasvāmin and *Siddhāntadīpikā* of Parameśvara, by T. S. Kuppanna Sastri, Madras 1957.

{MB 1960} *Mahābhāskarīya* of Bhāskara I, ed. and tr. with notes K. S. Shukla, Lucknow 1960.

{PS 1971} *Pañcasiddhāntikā* of Varāhamihira, ed. and tr. by G. Thibaut and Sudhakara Dvivedi, Varanasi 1889; ed. and tr. by O. Neugebauer and D. Pingree, Copenhagen 1970–1.

{PS 1993} *Pañcasiddhāntikā* of Varāhamihira, ed. and tr. by T. S. Kuppanna Sastri and K. V. Sarma, Chennai 1993.

{RGSN 1977} *Rāśigolasphuṭanīti* of Acyuta Piśāraṭi, ed. and tr. by K. V. Sarma, Madras 1953; ed. and tr. by K. V. Sarma, Hoshiarpur 1977.

{SC 1973} *Sphuṭacandrāpti* of Mādhava, ed. and tr. by K. V. Sarma, Hoshiarpur 1973.

{SDA 1976} *Siddhantadarpaṇa* of Nīlakaṇṭha Somayājī, ed. by K. V. Sarma, Madras 1955; ed. with auto-commentary and tr. by K. V. Sarma, Hoshiarpur 1976.

{SDI 1957} *Siddhāntadīpikā* on *Mahābhāskarīyabhāṣya* of Bhāskara-I by Parameśvara, ed. by T. S. Kuppanna Sastri, Madras 1957.

{SDVT 1981} *Śiṣyadhīvṛddhidatantra* of Lalla, ed. by Sudhakara Dvivedi Benares 1886; ed. and tr. by Bina Chatterjee, 2 vols, INSA, New Delhi 1981.

{SNT 1974} *Sphuṭanirṇayatantra* of Acyuta Piśāraṭi, ed. with auto-commentary by K. V. Sarma, Hoshiarpur 1974.

{SSI 1957} *Sūryasiddhānta* with *Vivaraṇa* of Parameśvara, ed. by K. S. Shukla, Lucknow 1957.

{SSI 1995} *Sūryasiddhānta* with *Tattvāmṛta* (a lucid commentary with annotation and notes in Sanskrit) by Kapilesvara Sastry, Chaukhambha Sanskrit Bhavan, Varanasi 1995.

{SSR 1981} *Siddhāntaśiromaṇi* of Bhāskarācārya-II, ed. with Bhāskara's *Vāsanā* and Nṛsiṃha Daivajña's *Vāsanāvārtika* by Muralidhara Chaturvedi, Varanasi 1981.

{SSR 2000} *Siddhāntaśiromaṇi* of Bhāskarācārya-II, *Grahagaṇitādhyāya* tr. by D. Arkasomayaji, Rastriya Sanskrit Vidyapeetha, Tirupati 2000.

{TS 1958} *Tantrasaṅgraha* of Nīlakaṇṭha Somayājī, ed. with *Laghuvivṛti* by S. K. Pillai, Trivandrum 1958.

{TS 1977} *Tantrasaṅgraha* of Nīlakaṇṭha Somayājī, ed. with *Yukti-dīpikā* (for chapters I–IV) and *Laghu-vivṛti* (for chapters V–VIII) of Śaṅkara Vāriyar by K. V. Sarma, Hoshiarpur 1977.

{TS 1999} *Tantrasaṅgraha* of Nīlakaṇṭha Somayājī, tr. by V. S. Narasimhan, Ind. *Jour. Hist. Sci.*, INSA, New Delhi 1998–99.

{VS 1986} *Vaṭeśvarasiddhānta* of Vaṭeśvara ed. and tr. by K. S. Shukla, 2 vols, INSA, New Delhi 1985–86.

Secondary Works

1. S. Balachandra Rao, *Indian Mathematics and Astronomy: Some Landmarks*, Bangalore 1998 (3rd edn 2004).
2. S. Balachandra Rao, *Indian Astronomy: An Introduction*, Hyderabad 2000.
3. E. Burgess, *The Sûryasiddhânta: A Text-Book of Hindu Astronomy*, American Oriental Society, New Haven 1860 (repr. Motilal Banarsidass, Delhi 1989, 1997, 2000).
4. B. Datta and A. N. Singh, *History of Hindu Mathematics*, 2 vols, Lahore 1935, 1938 (repr. Asia Publishing House, Bombay 1962; repr. Bharatiya Kala Prakashan, Delhi 2004).
5. Dennis Duke, The Earth in India: the mathematical basis of ancient Indian planetary models, *Arch. Hist. Exact. Sci.,* **59**, 563–576, 2005.
6. Dennis Duke, The second lunar anomaly in ancient Indian Astronomy, *Arch. Hist. Exact. Sci.,* **61**, 147–157, 2007.
7. Dennis Duke, Mean motion and longitudes in Indian Astronomy, *Arch. Hist. Exact. Sci.,* **62**, 489–509, 2008.
8. G. Emch, M. D. Srinivas and R Sridharan (eds), *Contribution to the History of Mathematics in India*, Hindustan Publishing Agency, New Delhi 2005.
9. James Evans, *The History and Practice of Ancient Astronomy*, Oxford University Press, New York 1998.
10. R. C. Gupta, Second order interpolation in Indian mathematics up to the fifteenth century, *Ind. Jour. Hist. Sci.,* **4**, 86–98, 1969.
11. R. C. Gupta, Solution of the Astronomical Triangles as found in the *Tantrasaṅgraha*, *Ind. Jour. Hist. Sci.,* **9**, 86–99, 1974.
12. R. C. Gupta, Mādhava's Rule for finding angle between the ecliptic and the horizon and Āryabhaṭa's knowledge of it, in *History of Oriental Astronomy*, Cambridge, 1987, pp. 197–202.

13. S. Hariharan, Declination in Indian Astronomy and the Approach of Kerala Astronomers, *Jour. Asiatic Soc.*, **30**, 39–49, 1988.
14. T. Hayashi, T. Kusuba, and M. Yano, *Studies in Indian Mathematics: Series, Pi and Trigonometry* (in Japanese), Tokyo, 1997.
15. V. Nath, *Sawai Jai Singh and His Astronomy*, Motilal Banarsidass, New Delhi 1995.
16. S. Parameswaran, *The Golden Age of Indian Mathematics*, Swadeshi Science Movement, Kochi 1998.
17. D. Pingree, *Census of the Exact Sciences in Sanskrit*, series A, vols 1–5, American Philosophical Society, Philadelphia 1970–1994.
18. D. Pingree, Concentric with equant, *Arch. Int. Hist. Sci.*, **24**, 26–28, 1973.
19. D. Pingree, History of mathematical astronomy, in *Dictionary of Scientific Biography*, vol. XV, New York 1978, pp. 533–633.
20. D. Pingree, *Jyotiśśāstra: Astral and Mathematical Literature*, Wiesbaden 1981.
21. D. Pingree, Nīlakaṇṭhás Planetary Models, *Jour. Ind. Phil.*, **29**, 187-195, 2001.
22. Kim Plofker, *History of Mathematics in India: From 500 BCE to 1800 CE*, Princeton, 2009.
23. C. K. Raju, *Cultural Foundations of Mathematics: The Nature of Mathematical Proof and the Transmission of the Calculus from India to Europe in the 16th c. CE*, Pearson Education, Delhi 2007.
24. K. Ramasubramanian, M. D. Srinivas and M. S. Sriram, Modification of the earlier Indian planetary theory by the Kerala astronomers (c. 1500 AD) and the implied heliocentric picture of the planetary motion, *Current Science* **66**, 784–790, 1994.
25. K. Ramasubramanian and M. D. Srinivas, Development of calculus in India, in *Studies in the History of Indian Mathematics* ed. by C. S. Seshadri, Hindustan Book Agency, New Delhi 2010, pp. 201–286.
26. K. Ramasubramanian and M. S. Sriram, Correction to the Terrestrial Latitude in *Tantrasaṅgraha*, *Ind. Jour. Hist. Sci.*, **38**, 129–144, 2003.
27. T. A. Sarasvati Amma, *Geometry in Ancient and Medieval India*, Motilal Banarsidass, Delhi 1979.
28. K. V. Sarma, *A Bibliography of Kerala and Kerala-based Astronomy and Astrology*, Hoshiarpur 1972.
29. K. V. Sarma, *A History of the Kerala School of Hindu Astronomy*, Hoshiarpur 1972.
30. K. V. Sarma and S. Hariharan, *Yuktibhāṣā* of Jyeṣṭhadeva: A book of rationale in Indian mathematics and astronomy in analytical appraisal, *Ind. Jour. Hist. Sci.*, **26**, 185–207, 1991.
31. S. R. Sarma, *The Archaic and the Exotic: Studies in the History of Indian Astronomical Instruments*, Manohar, New Delhi 2008.
32. H. Selin, (ed.) *Astronomy Across Cultures: The History of Non-Western Astronomy*, Kluwer, Dordrecht 2000.

33. S. N. Sen and K. S. Shukla (eds), *A History of Indian Astronomy*, New Delhi 1985 (rev. edn 2000).
34. K. S. Shukla, Use of hypotenuse in the computation of the equation of center under the epicyclic theory in the school of Āryabhaṭa, *Ind. Jour. Hist. Sci.*, **8**, 43–57, 1973.
35. D. A. Somayaji, *A Critical Study of Ancient Hindu Astronomy*, Karnatak University, Dharwar 1972.
36. M. D. Srinivas, 'Indian approach to science: the case of *Jyotiśśāstra*', *PPST Bulletin*, 19–20, Madras 1990.
37. M. S. Sriram, K. Ramasubramanian and M. D. Srinivas (eds), *500 Years of Tantrasangraha: A Landmark in the History of Astronomy*, Shimla 2002.
38. M. S. Sriram, Planetary and lunar models in *Tantrasaṅgraha* and *Gaṇitayuktibhāṣā*, in *Studies in the History of Indian Mathematics*, ed. by C. S. Seshadri, Hindustan Book Agency, New Delhi 2010, pp. 353–389.
39. B. V. Subbarayappa and K. V. Sarma (eds), *Indian Astronomy: A Source Book*, Bombay 1985.
40. B. V. Subbarayappa, *The Tradition of Astronomy in India: Jyotiśśāstra*, PHISPC vol. IV, part 4, Center for Studies in Civilizations, New Delhi 2008.
41. N. M. Swerdlow, Ptolemy's theories of the latitude of the planets in the Almagest, Handy Tables and Planetary Hypotheses' in J. Z. Buchwald and A. Franklin (eds), *Wrong for the Right Reasons*, Springer, New York 2005, pp. 41–71.
42. N. M. Swerdlow and O. Neugebauer, *Mathematical Astronomy in Copernicus' De Revolutionibus*, Springer, New York 1984.
43. G. J. Toomer, *Ptolemy's Almagest*, Duckworth, London 1984.
44. C. M. Whish, On the Hindu quadrature of the circle and the infinite series of the proportion of the circumference to the diameter exhibited in the four Shastras, the Tantrasangraham, Yucti Bhasa, Carana Paddhati and Sadratnamala, *Trans. R. Asiatic. Soc.* (GB), **3**, 509–523, 1834.
45. M. Yano, Distance of planets in Indian astronomy in Il-Seong Nha and F. Richard Stephenson (eds), *Oriental Astronomy from Guo Shoujing to King Sejong* (Proceedings of the First International Conference on Oriental Astronomy), Yonsei University Press, Seoul 1997, pp. 113–120.

Index

[1] Here and in what follows, in the case of commonly occuring terms such as *Ahargaṇa, jyā*, etc. only their first few occurences are indicated.

Index of Half-verses

FSC
www.fsc.org
MIX
Papier | Fördert gute Waldnutzung
FSC® C083411

Kolibri 360 GmbH
Ferdinand-Jühlke-Straße 7
99095 Erfurt, Deutschland
produktsicherheit@kolibri360.de